Environmental Geology
An Earth System Science Approach

Environmental Geology
An Earth System Science Approach

SECOND EDITION

DOROTHY MERRITTS
Franklin and Marshall College

KIRSTEN MENKING
Vassar College

ANDREW DE WET
Franklin and Marshall College

W.H. Freeman and Company
A Macmillan Higher Education Company

Publisher: Kate Parker
Senior Acquisitions Editor: Bill Minick
Developmental Editor: Brittany Murphy
Senior Marketing Manager: John Britch
Marketing Assistant: Samantha Zimbler
Senior Media and Supplements Editor: Amy Thorne
Project Editor: Kerry O'Shaughnessy
Production Manager: Julia DeRosa
Text and Cover Designer: Blake Logan
Illustration Coordinator: Janice Donnola
Illustrations: Precision Graphics
Photo Editor: Elyse Rieder
Composition: Sheridan Sellers
Printing and Binding: LSC Communications

Credit is given to the following sources for the cover photo and title page: Andrew de Wet, part opener photos: **Part 1:** Katarina Stefanovic/ Getty Images; **Part 2:** Christian Beier/age fotostock/SuperStock; **Part 3:** Kelly Redinger/Design Pics/Getty Images; **Part 4:** Ron Niebrugge/ wildnatureimages.com; **Part 5:** AP Photo/LM Otero.

Library of Congress Control Number: 2014931281

ISBN-13: 978-1-4292-3743-7
ISBN-10: 1-4292-3743-0

Printed in the United States of America

W. H. Freeman and Company
41 Madison Avenue, New York, NY 10010
Houndmills, Basingstoke RG21 6XS, England
www.whfreeman.com

Brief Contents

Contents

When the first edition of this book was released in 1998, Earth's human population numbered just under 6 billion, the average Chinese citizen bicycled to work, the United States appeared to be running out of oil and natural gas, and carbon dioxide levels in the atmosphere were around 365 parts per million volume (ppmv). Sixteen years later, another billion people live on the globe, a growing Chinese middle class is buying automobiles, major advances in extractive technology have led to a new oil and gas boom in the United States, and atmospheric carbon dioxide has risen to 400 ppmv. The combination of the larger human population and our increasingly affluent lifestyle is putting enormous strains on Earth's systems. As a result of our activities, the polar ice caps are melting at an increasing rate, habitat loss, and exotic species invasions have damaged ecosystems, and an erratic hydrologic cycle is generating ever more extreme weather phenomena. At the same time that human impacts on the globe have mounted, the pace of environmental disasters appears to have quickened. The earthquakes, tsunamis, hurricanes, tornados, floods, wildfires, and droughts that have killed hundreds of thousands of people and displaced tens of millions more at the beginning of the new millennium are unspeakable tragedies and point toward an increasing vulnerability of the human population to geological hazards.

The complexity of the environmental crises that confront us in the 21st century necessitate an Earth system science approach if we are to find ways to understand and address them. The flooding of the city of New Orleans by Hurricane Katrina in 2005 is a perfect example. Fully comprehending that event requires understanding not only the atmospheric conditions that gave rise to the hurricane, but also the fluvial and biologic processes that built the Mississippi River delta wherein the city lies, and the impact of dam construction upstream, which has starved the delta of the sediment it needs to stay above the Gulf of Mexico's waves. Nor can we comprehend the disaster without understanding the subsidence that has occurred as a result of adjustment of Earth's crust to the delta sediment load and as a result of groundwater pumping, both of which have caused New Orleans to slip below sea level over the past century. We also can't forget oil and gas exploration, which has sliced canals through coastal wetlands, allowing salty ocean water to travel inland and impact brackish and freshwater marshes so that land loss brings hurricane storm surge ever closer to the city. As this one example shows, atmosphere, hydrosphere, pedosphere, biosphere, and lithosphere all were involved in conjunction with human activity to produce the costliest environmental disaster in United States history, and all must be considered in an integrated and interdisciplinary approach. It is just this sort of approach that underpins the text of this book and that we believe will guide humanity toward a better future.

We see truly hopeful outcomes of the past two decades in the growth of the sustainability movement, itself deeply rooted in Earth system science. The realization that seemingly small, mundane actions can lead to regional, continental, or even global consequences when multiplied by many millions or billions of actors is a direct outgrowth of advances in our understanding of the numerous processes that generate Earth's climate. This awareness led to deployment of solar panels and wind turbines around the world, to development of "green" building standards in the United States and other nations, to higher fuel and energy efficiency requirements for vehicles and appliances, and to the growth of local and organic agriculture movements, though much work remains to be done to create a sustainable world.

Although the rate of human population growth has slowed, the addition of another 2 billion people is projected by mid-century. It is vitally important that we prepare now for this future by doing all we can to promote resiliency in our societies and their adjacent ecosystems. It is with this preparation in mind that we have written the second edition, as who is better to carry out our unfinished work than our students? We hope that they take away from this text an enhanced understanding of how our planet operates, of how to protect themselves and their societies from natural disasters, and of how to minimize their impacts on Earth systems through wise use of resources. We also hope they take away a sense of awe and reverence for the amazing world we live in. As the first generation of people born during the Internet and social networking era, they have the potential to enact sweeping global changes, and we call upon their boundless energy to carry out a positive transformation in how we live on Earth!

CONTENT AND ORGANIZATION

The courses we teach at Franklin and Marshall College and Vassar College have evolved over the years to meet what we see as critical needs in introductory geoscience education. Given that most of our students will not take another Earth science course, we seek to impress on them the importance of understanding how Earth works, how humans affect the environment, and the characteristics of different natural hazards. Having just a little bit of knowledge on these topics can help students make informed decisions as adults about where to avoid buying a home, whether to support nuclear power or non-nuclear alternative energy resources, and how to respond as voting citizens to the threats of global climate change. In developing our course, we found no textbook that consistently weaves together basic concepts of Earth system science and environmental geoscience, though, as we noted above, such a weaving together is essential to address many of the environmental problems we face today. To compensate, we created, tested in class, and refined a set of course materials so comprehensive that they evolved, with the urging and advice of many colleagues, into the first edition of this book.

We have learned that our colleagues in the environmental geosciences want in a textbook many of the same things we wanted: information on scientific and systems thinking; information on individual systems such as soils, surface water, groundwater, atmosphere, oceans, and energy; and information on interpreting and predicting environmental change. These topics are treated thoroughly in this textbook. At the same time, we are careful to select and cover the fundamentals of physical geology that are essential to understanding the Earth system and contemporary environmental and resource issues. The second edition remains true to our original goals but has been thoroughly updated to reflect advances in knowledge made in the past 16 years.

- **Part I** presents the concept of systems in general and the various Earth systems in particular, with emphasis on reservoirs and flows, system behavior, feedbacks, and cycling of matter and energy. We also introduce the topic of human population growth, renewable versus nonrenewable resources, and the different factors that contribute to human environmental impacts. The chapter concludes with a discussion of planetary boundaries advocated by sustainability scientists, a recent idea that frames much of our writing about resources throughout the text. We come back to these boundaries throughout the book, pointing out those situations in which humans appear to have driven environmental systems toward or perhaps even beyond tipping points that are likely to lead to quite different planetary conditions in the future.

In our own teaching, we find that focusing on the evolution, processes, resources, hazards, waste problems, and policy issues for one environmental system at a time is preferable to saving all discussion of environmental topics for the end of the course. Therefore, Parts II, III, and IV concentrate on individual systems and their relationships to one another and incorporate content on hazards, resources, and relevant policy issues.

- **Part II**, Solid Earth Systems and Geologic Time, consists of five chapters covering plate tectonics, mineral and rock-forming processes, the resources and hazards of the lithosphere, and how we use the rock record to make inferences about Earth's history. We also have situated a discussion of the scientific method within the context of the development of plate tectonic theory, using the history of geological thought to bring alive the process of scientific discovery.

- **Part III**, Earth's Surface Systems, contains two chapters on the biosphere and pedosphere. While discussion of the biosphere was woven throughout the first edition, the explosion of research in biogeochemical cycles, geobiology, and geomicrobiology that has occurred since that book was released merited an expanded treatment in the current edition, and we have therefore given the biosphere a stand-alone chapter. Here we address the nitrogen, phosphorus, and carbon cycles, trophic levels and food chains, populations, and ecosystem services relevant to such topics as climate stabilization and water purification. This treatment of the solid earth and biosphere leads naturally to a chapter on the pedosphere in which we discuss weathering, soil-forming processes, soil erosion and its prevention, and the hazard of mass wasting.

- In **Part IV**, we address Earth's fluid systems, the hydrosphere and atmosphere. We begin with chapters on surface water and groundwater, first discussing the processes that distribute water between reservoirs and govern its movement. Next comes discussion of hazards such as flooding, channel instability, and drought, as well as water pollution and environmental laws and regulations designed to improve water quality. We then move to the atmosphere and the oceans, which provide the foundations for understanding the climate system and human impacts on climate in the final section of the book. In addition to addressing basic composition and circulation of these systems, we discuss hazards such as severe weather, tsunamis, and coastal erosion, as well as pollution problems such as acid rain, ozone depletion, and ocean acidification. The Montreal protocol receives attention as an example of successful international cooperation in dealing with an environmental crisis.

- **Part V,** Energy, Changing Earth, and Human-Earth Interactions, considers the driving mechanisms of all processes and change on Earth, with three chapters: energy as a system; causes and geological evidence of past environmental change; and what it means to be living in what many scientists now refer to as the Anthropocene. In the last chapter we return to the idea of planetary thresholds for stability, giving particular attention to the issue of global warming.

PEDAGOGICAL APPROACH

We are keenly aware of the need to present scientific information in a way that captures the interest of students who are taking environmental geology as their first and perhaps only college science course. Throughout all chapters, we emphasize issues relevant to everyday experience and use frequent examples and case studies. Explanations are written clearly and with a minimum of technical jargon; they are enhanced by vividly rendered and carefully labeled diagrams of structures and processes, selected tables and graphs, maps, and impressive photographs.

NEW TO THIS EDITION

We have added two categories of boxes designed to demonstrate the connections between content in chapters and current issues in the news, and to highlight state-of-the-art research in Earth sciences.

Earth Science and Public Policy Boxes

We present case studies in which research on Earth system processes has been instrumental in formulating public policy or law to deal with a variety of environmental problems. Examples include the adoption by the California legislature of the Alquist-Priolo Earthquake Fault Zoning Act, which aims to minimize earthquake damage through stringent development requirements, the gradual banning by most state legislatures of detergent phosphates to protect aquatic ecosystems, and ordinances enacted by cities in the American Southwest to conserve dwindling water supplies, among others. We also address the role of special interests and the news media in disseminating misinformation about global climate change.

BOX 3–4 EARTH SCIENCE AND PUBLIC POLICY

Earthquakes and Legislation

At 6 A.M. on February 9, 1971, a magnitude 6.6 earthquake rocked the San Fernando Valley of Southern California when a slip of about 1 m occurred on a thrust fault lying beneath the valley.[62] The event caused 65 deaths, more than 2000 injuries, and $505 million in property damage. Many of the deaths occurred when a Veterans Administration Hospital, several freeway overpasses, and part of the Olive View Hospital collapsed. Landslides triggered by the earthquake caused additional damage to buildings, railways, pipelines, and dams, and shaking was felt as far away as western Arizona.

In response to the devastation, the California legislature

In 1975, California took another step toward earthquake preparedness by creating a seismic safety commission.[64] Made up of geologists, structural engineers, seismologists, emergency management professionals, firefighters, members of social services agencies and other relevant disciplines, the commission conducts research and makes recommendations to help the state legislature and the governor in crafting laws to reduce earthquake impacts, enhance public preparedness, and monitor seismic events. Many of the building code requirements discussed in this chapter are a result of the commission's work.

Emerging Research Boxes

We profile recent and on-going research projects designed to address fundamental questions for which we presently have few or no answers. The purpose of these boxes is to make it clear to students that Earth science is a young discipline in which discoveries are made on a daily basis and to which they can contribute should they decide to continue in the discipline. Examples of projects discussed in these boxes include the San Andreas Fault Observatory at Depth, an instrumented borehole drilled into the Parkfield section of the fault to study the conditions required for and the characteristics of slip, studies of the amount of methane being released from melting permafrost as a result of global warming of the Arctic, and our evolving understanding of the Hawaiian hotspot, which has apparently migrated over time, rather than, as was assumed in the first edition of this book, being fixed in position. Awareness of this last point is requiring reinterpretation of previously agreed-on plate tectonic movements, an example of the scientific method in action!

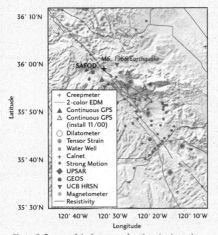

BOX 3–3 EMERGING RESEARCH

The Parkfield Earthquake and the Difficulties of Prediction

The town of Parkfield, California, lies along the San Andreas fault, midway between Los Angeles and San Francisco. Earthquakes with a magnitude of about 6 shook the area in 1857, 1881, 1901, 1922, 1934, and 1966, or on average, every 22 years (Figure 1). Mapping of offset features indicated that roughly the same area along the fault slipped in each event. These facts led the U.S. Geological Survey (USGS) to plan an experiment to determine whether earthquakes might in fact be predictable.

Beginning in 1985, USGS scientists installed equipment to measure every conceivable precursor of an earthquake and also deployed a network of seismic stations to record the next event (Figure 2).[46] On the premise that faults might slip due to changes in fluid pressure that could lubricate the fault zone, geologists installed groundwater wells that they could use to monitor water levels. Tilt meters and creep meters were installed to keep track of tiny vertical and horizontal motions that might foreshadow an impending rupture. Likewise, surveying benchmarks were installed across the fault to allow for repeated measurements between points, in an attempt to determine whether the ground moved before an earthquake. Chinese scientists had found that earthquakes are sometimes preceded by changes in electrical conductivity of the ground

Figure 2 Because of the frequency of earthquakes here, the Parkfield area has a large number of instruments to monitor ground motion, groundwater level, and other parameters.

Citations

Recognizing the importance of citing our sources in teaching science students, we now include comprehensive citations to allow students to explore further topics of interest to them.

BUILDING ON THE FIRST EDITION

This edition of *Environmental Geology* retains several features used successfully in the first edition.

Chapter Openers

Each chapter begins with a case study relating the subject of the chapter to Earth systems and environmental issues. For example, Chapter 1, "Dynamic Earth Systems," begins with an opening story about the shrinking glaciers of Mount Kilimanjaro and the worldwide causes and effects of this phenomenon on Earth's interconnected systems. A chapter outline gives the student both a preview of the content and a review device, and a list of goals establishes a context for the chapter.

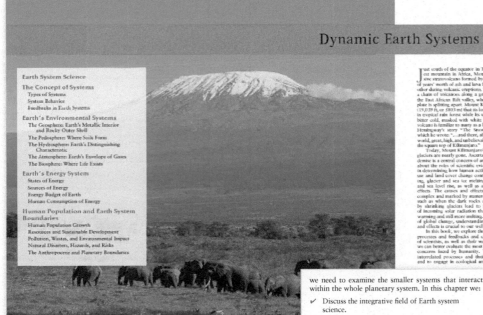

Art Program

Geologist Emily Cooper assisted in the development of the new art program. Diagrams are vividly rendered and thoroughly labeled, making geologic structures and processes clear and easily remembered.

Flow charts are constructed for clarity and visual interest, and they make concepts such as budgets, reservoirs, and fluxes easy to understand.

Color photographs and digital images provide striking illustrations, from the miniscule to the majestic, of the material presented in the text discussion.

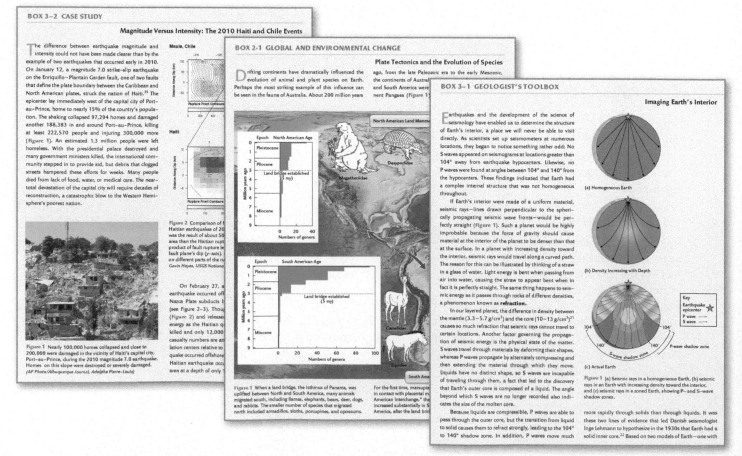

Figure 1–14 Earth's hydrologic cycle consists of the processes (and fluxes) of precipitation, evaporation, transpiration, infiltration, and runoff from one reservoir to another. Primary reservoirs are the oceans, continental water (lakes, streams, wetlands, and groundwater), and the atmosphere.

Real–World Examples

Chapters include frequent examples of how Earth scientists are solving environmental problems. For example, scientists have used the geologic record to determine the characteristics of natural streams and wetlands prior to centuries of disturbance through logging, mining, agriculture, and damming. These reconstructions are useful in restoration projects designed to enhance ecosystem services and value.

Boxes Provide Deeper Discussions

A program of boxed features enhances the text with interesting Case Studies of actual events, Geologist's Toolboxes that describe technical skills, and thought-provoking essays on Global and Environmental Change, Earth Science and Public Policy, and current-day profiles of Emerging Research.

■ New lithosphere forms along rifts at mid-ocean spreading centers on divergent boundaries, and old lithosphere is destroyed at subduction zones along convergent boundaries. The amounts of lithosphere created and destroyed are equal, keeping the lithosphere system in a steady state and ensuring that Earth's size remains the same over time.

■ Plate boundaries are zones of active rock deformation, faulting, and folding. Faulting occurs at shallow depths (generally less than 10 to 30 km), where rocks are cool and brittle, whereas folding generally occurs at greater depths, where rocks are warmer and more ductile.

■ The three basic types of faults are normal, reverse, and strike-slip. Although each may occur at all types of plate boundaries, normal faults are dominant at divergent boundaries, where crust is stretched and pulled apart; reverse faults are dominant at convergent boundaries, where crust is compressed and [...] faults are dominant at transform[...] crustal blocks slip horizontally[...]

Section Summaries

A list of the important ideas is provided at the end of each major section in a chapter to encourage students to pause, think about what they have read, and ensure that they understand the material before going on to the next section.

CLOSING THOUGHTS

We began by discussing the possible causes of rapidly shrinking glaciers atop the highest mountain in Africa, Mount Kilimanjaro. This example shows how Earth scientists investigate interconnected Earth systems. Greenhouse gases emitted from activities that include burning fossil fuels for energy lead to global warming that, in turn, can cause glacial ice masses to shrink and change the fluxes and stocks of water in Earth's hydrosphere. We also examined the ways in which scientists investigate Earth processes, using a multitude of tools and an overarching view of Earth as a dynamic system. By systematically examining each of the parts and processes within our planet's various subsystems, Earth scientists are able to determine cause and effect and to predict future outcomes.

The scale of human impact on Earth is relatively small in comparison with that of geologic processes such as volcanism, continental drift, waxing and waning ice ages, or mountain building, yet humans are by no means an insignificant force. The signatures of human activities can be detected worldwide, from extinct species to wholesale changes in land cover or the composition of our atmosphere. Whether or not the ongoing anthropogenic changes in the cycling of matter and energy on Earth result in an environment in which we can live sustainably and within planetary boundary thresholds remains to be seen.

Closing Thoughts

This unique feature is a short essay in which the authors reflect on the chapter and identify significant links to human affairs. Closing Thoughts gives the authors the opportunity to share with students their thoughts about the wide-ranging implications of an environmental problem, from the philosophic to the economic, and to contrast those implications with the scientific aspects.

Chapter Summary, Key Terms, Questions, Exercises, and Readings

Chapters end with a traditional Summary that complements the earlier end-of-section summaries; a list (with page numbers) of the Key Terms highlighted in bold-face type within the chapter; a set of Review Questions that helps students understand and remember important ideas; a set of Thought Questions that encourages students to apply what they have learned to questions not specifically addressed in the chapter; a set of Exercises that encourages students to solve quantitative problems and explore their own environment; and a list of Suggested Readings, accessible to the interested nonscientist, that enhances the major topics of the chapter.

Appendices and Glossary

At the end of the book are four helpful appendices and a glossary. Appendices include the Linnaean classification system, a periodic table of the elements, a conversion table of metric units and their English equivalents, and a list of the properties of common minerals. The comprehensive glossary includes definitions of environmental geosciences terms that may be unfamiliar to students.

ALTERNATE VERSIONS

W. H. Freeman partners with CourseSmart to provide a low-cost subscription to the text in e-Book format. CourseSmart e-Books can be read online or offline, as well as via free apps for an iPad, iPhone, iPod Touch, Android device, or Kindle Fire. Students are able to take notes, highlight important text, and search for key words. For more information, or to purchase access to the e-Book, visit www.coursesmart.com.

SUPPLEMENTS

Instructors can access resources for *Environmental Geology* at www.whfreeman.com/environmentalgeology.

- **Test Bank Questions** Fifty multiple-choice questions for each chapter are offered in the Test Bank, available for download at the *Environmental Geology* site.

- **Images** All images from the book are available for instructors to download.

- **Answers to End–of–Chapter Questions**

ACKNOWLEDGMENTS

We would like to acknowledge the contributions of the following people in helping us track down information, assisting in understanding technical details, and reading drafts: Joseph Allen, Don Anderson, Roger Anderson, Katey Walter Anthony, Wally Broecker, Joel Cohen, Mary Ann Cunningham, Mary De Jong, Kevin Dennehy, Eric Durell, Curtis Ebbesmeyer, Erle Ellis, Yehouda Enzel, David Gillikin, Flora Grabowska, David Groenfeldt, Alistair Hall, Douglas Karson, James Kasting, Lora Koenig, Joseph Krisanda, Brian McAdoo, Joe Nevins, Jim O'Connor, Roger Pielke, Jr., Bill Schlesinger, Jodi Schwarz, Jeremy Shakun, Neil Sturchio, Bill Sunda, John Taber, Jeff Walker, and Mike Wallace.

Kirsten Menking would also like to thank Nadine Reitman and Cara Hunt for their assistance in doing research.

Dorothy Merritts would like to thank the following people for their help across many areas of research: Robert Walter (F&M College), Johannes Lehmann, Alan Bussaca, Jon Major (USGS), John Ewert (Cascades Volcano Observatory), Christopher L. Sabine (Pacific Marine Environmental Laboratory), Maria Hower, Kayla Schulte, Jane Woodward (Stanford University), Karl Knapp (Stanford University), Justin Ries (Northeastern University), and Stephen Martel (University of Hawaii).

Andrew de Wet would like to thank Carol de Wet, Stan Mertzman, Chris Williams, and Rob Sternberg for answering questions related to figures, and Cameron de Wet for his editorial assistance.

The authors also wish to thank the following people who reviewed the manuscript of the second edition. Their contributions to this book have been invaluable.

Joseph L. Allen, *Concord University*
Zsuzsanna Balogh-Brunstad, *Hartwick College*
Callan Bentley, *Northern Virginia Community College*
Scott Brame, *Clemson University*
Nathalie N. Brandes, *Lonestar College*
Charles Brown, *George Washington University*

George Buchanan, *Montgomery County Community College*
W. B. Clapham, Jr., *Cleveland State University*
Jennifer Rivers Cole, *Northeastern University*
Jim Constantopoulos, *Eastern New Mexico University*
Anna Cruse, *Oklahoma State University*
Dennis C. DeMets, *University of Wisconsin–Madison*
Joseph E. Doyle, *Bridgewater State College*
Robert R. Elkin, *Miami University Hamilton*
Dennis Fries, *Somerset Community College*
Joshua C. Galster, *Montclair State University*
Tom Gardner, *Trinity University*
John B. Gates, *University of Nebraska–Lincoln*
Julian W. Green, *University of South Carolina Upstate*
Anne Larson Hall, *Emory University*
Bryce Hoppie, *Minnesota State University*
Paul F. Hudak, *University of North Texas*
Andrew C. Kurtz, *Boston University*
Julie Maxson, *Metropolitan State University*
Erich Osterberg, *Dartmouth College*
Alberto Patiño Douce, *University of Georgia*
Carol A. Prombo, *Washington University–St. Louis*
Kent Ratajeski, *University of Kentucky*
Karen L. Savage, *California State University Northridge*
Conrad Shiba, *Centre College*
Leigh Stearns, *University of Kansas*
Christine Stidham, *Stony Brook University (SUNY)*
Kevin Svitana, *Otterbein University*
Susan Swanson, *Beloit College*

Finally, the authors wish to thank the staff of W. H. Freeman: Kate Parker, publisher, Bill Minick, acquisitions editor, Brittany Murphy, development editor, Amy Thorne, media and supplements editor, Kerry O'Shaughnessy, project editor, Julia DeRosa, production manager, and Courtney Lyons, assistant editor.

ABOUT THE AUTHORS

DOROTHY MERRITTS is a geologist with expertise on streams, rivers, and the impact of humans and geologic hazards on landscape evolution. In the western United States, she conducted research on the San Andreas Fault of coastal California, and her international work focuses on fault movements in South Korea, Indonesia, Australia, and Costa Rica. Her primary research in the eastern United States is in the Appalachian Mountains and Piedmont, particularly in the mid-Atlantic region, where she is investigating the role of climate change and human activities in transforming the valley bottom landscapes and waterways of Eastern North America. Recently she partnered with other scientists and policy makers from multiple state and national government agencies to develop and test a new approach to stream and wetland restoration. She is a professor in the Department of Earth and Environment at Franklin and Marshall College in Lancaster, Pennsylvania. She is an author or co-author of more than 70 scientific articles, and the editor and contributing writer for numerous scientific books and field guides.

KIRSTEN MENKING is an environmental Earth scientist in the Department of Earth Science and Geography at Vassar College. Her research interests include using lake sediments to unravel Earth's history of climatic change, linking this history to atmospheric and hydrologic processes through a combination of numerical modeling experiments and collection of weather and stream discharge data, analyzing the evolution of landforms in response to climatic and tectonic processes, and studying the impacts of urbanization on streams. She has published journal articles documenting glacial–interglacial cycles in the Sierra Nevada mountains and adjacent Owens Valley of California, determined the climatic conditions necessary to produce a Pleistocene lake in the now-dry Estancia Basin of New Mexico, and un-covered a centuries-long mid-Holocene drought in New York's Hudson River valley. Her current research involves quantifying the amount of road salt entering the groundwater system, a topic of concern both for people dependent on well water and for aquatic ecosystems.

ANDREW DE WET is a classically trained geologist specializing in Geographic Information Systems (GIS) and remote sensing and their application to environmental problems on Earth and geological processes on Mars. He holds an honors degree in geology from the University of Natal (now the University of Kwazulu-Natal), South Africa, and a doctorate from Cambridge University, England. He has done field work in South Africa, Greece, the United Kingdom, Mongolia, Chile, Antarctica, and the United States. He teaches environmental geology, GIS and Natural Resources, and an interdisciplinary course on comparative planetology with a focus on Mars. He served as director of the Keck Geology Consortium for three years and has led Keck Geology research projects numerous times. Professor de Wet's capacity for visualizing complex systems has clarified concepts and inspired students to better understand the interconnectedness of natural systems. Through his travels across seven continents he has acquired a deep knowledge of geological and environmental conditions, which he transcribes into dynamic graphics portraying natural and anthropomorphic processes. He has published articles on geological pedagogy in the *Journal of Geological Education* and on shared faculty positions in the *Journal of Women and Minorities in Science and Engineering* and *Geotimes*. He is a member of the Geological Society of America and the American Geophysical Union. He is involved in a long-standing collaboration with researchers at NASA's Goddard Space Flight Center and publishes on environmental issues and planetary geology.

Introduction

Dynamic Earth Systems

Just south of the equator in Tanzania looms the tallest mountain in Africa, Mount Kilimanjaro, a massive stratovolcano formed by hundreds of thousands of years' worth of ash and lava flows piled one atop the other during volcanic eruptions. The peak stands among a chain of volcanoes along a great rift in Earth's crust, the East African Rift valley, where the African tectonic plate is splitting apart. Mount Kilimanjaro rises so high (19,039 ft, or 5803 m) that its lower slopes are shrouded in tropical rain forest while its upper peaks are dry and bitter cold, masked with white glaciers and snow. The volcano is familiar to many as a literary icon from Ernest Hemingway's story "The Snows of Kilimanjaro," in which he wrote: "…and there, ahead … as wide as all the world, great, high, and unbelievably white in the sun, was the square top of Kilimanjaro."

Today, Mount Kilimanjaro's "unbelievably white" glaciers are nearly gone. Ascertaining the cause of their demise is a central concern of an international dialogue about the roles of scientific evidence and investigation in determining how human activities such as fossil fuel use and land cover change contribute to global warming, glacier and sea ice melting, permafrost thawing, and sea level rise, as well as a host of other, related effects. The causes and effects of global change are complex and marked by numerous feedback processes, such as when the dark rocks and sediments exposed by shrinking glaciers lead to the greater absorption of incoming solar radiation that triggers atmospheric warming and still more melting. Despite the complexity of global change, understanding its processes, causes, and effects is crucial to our well-being as a species.

In this book, we explore the Earth system's myriad processes and feedbacks and consider the discoveries of scientists, as well as their ways of thinking, so that we can better evaluate the most pressing environmental concerns faced by humanity. To understand Earth's interrelated processes and their impacts on humans, and to engage in ecological and physical restoration,

Mount Kilimanjaro's glaciers can be seen from afar due to the stratovolcano's great height within the East African Rift valley. *(Fuse/Thinkstock)*

we need to examine the smaller systems that interact within the whole planetary system. In this chapter we:

- ✔ Discuss the integrative field of Earth system science.

- ✔ Examine the concept of systems and their components.

- ✔ Identify the forces that drive Earth's processes and examine how feedback mechanisms either amplify or regulate them.

- ✔ Consider the field of environmental geology and the scope of this book.

EARTH SYSTEM SCIENCE

Large ice masses at Mount Kilimanjaro and elsewhere existed during the last full glacial advance on Earth, which spanned the period from about 50,000 to 16,000 years ago. Modern glaciers worldwide are relics from that time, shrunken in size in response to the relatively stable warm climatic conditions of the past approximately 11,000 years. Since the early 20th century, however, Mount Kilimanjaro's glaciers have withered rapidly, shrinking laterally and thinning (Figure 1-1). Reconstructions from historic aerial photos and satellite images reveal that 85 percent of the ice cover that existed in 1912 was gone by 2011.[1] Most of the remaining glacial ice forms stagnant, isolated remnants.

In 2000, scientists drilled six ice cores in these remnants, and from the cores they were able to document nearly 12,000 years of glacial history at the summit.[2] They discovered that glacial ice had persisted in the equatorial sun without periods of prolonged melting even through centuries-long droughts. The scientists noted that the cause of recent glacial shrinking was therefore likely to be global warming due to human activity. They added that mid- to low-latitude glaciers at high altitudes are shrinking rapidly worldwide, from the Andes and Cascades in the western hemisphere to the Swiss Alps in Europe and the Himalayas in the eastern hemisphere, so the cause of their recent demise is likely to be global rather than local or regional. The scientists also suggested that the "snows of Kilimanjaro" might be gone as early as 2015.

Garnering international attention, this tentative prediction was cited in 2006 in a well-known documentary, *An Inconvenient Truth*, which called for international action to stave off global warming. By 2010, however, it looked as if the glaciers on Mount Kilimanjaro—though further shrunken—might last a few more decades. Furthermore, another group of scientists reported that deforestation and forest fires at lower elevations in the region might have affected regional weather patterns and diminished moisture at

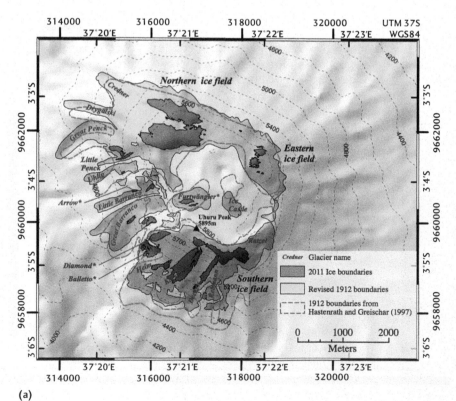

(b)

Figure 1–1 (a) Glaciers at the summit and side slopes of Mount Kilimanjaro are retreating rapidly, with about 85% of ice volume lost between 1912 and 2012. (b) Scientific investigations use data from meteorological stations and snowpits, as well as other types of evidence, in order to reconstruct the history of glaciation and determine the causes of current glacial shrinkage. *(a: Cullen, N. J., Sirguey, P., Mölg, T., Kaser, G., Winkler, M., and Fitzsimons, S. J. 2013. "A Century of Ice Retreat on Kilimanjaro: The Mapping Reloaded." The Cryosphere 7: 419–431, © 2013 the authors; b: courtesy of Douglas Hardy, University of Massachusetts)*

(a)

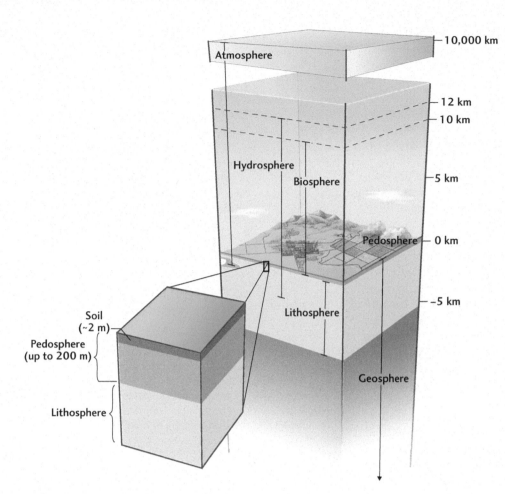

Figure 1–2 Earth's environmental systems include the geosphere, pedosphere, hydrosphere, biosphere, and atmosphere. On continents, the interface of all five of these systems is the pedosphere. The top of the pedosphere, where all Earth's systems interact, is a relatively organic rich layer of weathered rock called soil. Below the soil, weathered rock grades downward to the unweathered rock of the solid geosphere, forming an uneven layer up to 200 m thick.

the volcano's summit.[3] Based on these findings, skeptics of global warming were quick to claim that the first set of scientific studies could not verify that global warming was the primary cause of glacial shrinking. If global warming were not the cause of shrinking glaciers, the skeptics argued, it might be imprudent to make potentially costly changes in long-established habits, such as the consumption of immense amounts of fossil fuels.

As discussed in more detail in Chapter 2, science is a process, a "work in progress," in which scientists publish the results of their investigations based on the tools available and their best understanding of the data at the time. As of 2012, for example, scientists working at Mount Kilimanjaro were able to show, through a combination of climate modeling and analysis of satellite data, that regional deforestation and drying had an insignificant effect on the demise of the snows of Kilimanjaro in comparison to the effects of human-caused global warming.[4]

Why might melting glaciers matter? For Tanzania, the loss of famed glaciers might not affect tourism, as hikers will still want to reach the highest peak in Africa.[5] Nevertheless, a cherished icon will be missed. For other locales where glaciers are shrinking, as in Pakistan and India, summer meltwater from glaciers is a critical resource for farming and domestic use. The reservoirs of water stored and released by glaciers serve much the same function as a dam that holds water in a valley, and if glaciers are gone that resource will have vanished, threatening the livelihoods of billions of people. Short of a major change in climate toward cooler conditions, that loss of water resources will be permanent with respect to the timescale of human interest.

The complexity of Earth system processes such as the shrinking of glaciers at Mount Kilimanjaro is typical of the environmental problems we face today and motivates our use of a holistic approach known as **Earth system science.** In this approach, multiple fields are integrated in the study of Earth as a system. Within the whole Earth system are many smaller, interacting components that function as systems themselves but together are dynamic parts that shape the greater entity.

Earth can be viewed as consisting of five major systems that continuously interact with one another and with solar energy, gravitational energy, and the internal heat energy of the planet to produce the climates and environments we experience at Earth's surface (Figure 1-2). From the bottom up, these are the **geosphere** (Earth's rocky crust and mantle and its interior metallic core), the **pedosphere** (weathered, broken particles of rock capped with soil), the **hydrosphere** (water vapor, streams, lakes, groundwater, and ice), the **biosphere** (living organisms), and the **atmosphere** (air). These systems tend to permeate each other. Water, for example, resides not only in large pools such as oceans and rivers but also in the soil, in the air, and in living things. Similarly, living things reside on and in rocks, soils, bodies of water, and the atmosphere. Despite these overlaps, each "sphere" is an identifiable reservoir that can be viewed as an open, dynamic system into and out of which energy and matter flow. What we mean by an open, dynamic system is discussed further below.

An Earth system science approach provides not only a physical basis for understanding our world, but also the knowledge needed to attain **sustainability,** a long-term condition in which development meets the needs of the present without compromising the environment or the ability of future generations to meet their needs.[6]

Because environmental changes and geologic hazards result from the interactions of many Earth systems with one another, they need to be addressed by broadly trained Earth scientists with the ability to cross traditional disciplinary boundaries. Relevant disciplines include geology, ecology, chemistry, hydrology, soil science, atmospheric science, and climatology. As scientists learn more about the interaction of Earth systems, they can make increasingly effective recommendations for conserving resources, minimizing environmental degradation, and reducing the potential for fatalities and destruction associated with geologic hazards. They also can make better predictions, as with the timing of the demise of glaciers atop mountains in response to climate change driven by human action.

■ Glaciers on Mount Kilimanjaro, the tallest mountain in Africa, have been shrinking rapidly since the early 20th century and are nearly gone; scientists estimate they will disappear within decades as a result of human-produced global warming.

■ Earth system science integrates multiple fields in the study of Earth as a system within which there are smaller, interacting components that function as systems themselves.

■ Primary Earth systems include the geosphere, pedosphere, biosphere, hydrosphere, and atmosphere.

THE CONCEPT OF SYSTEMS

A *system* is a group of interrelated and interacting objects and phenomena consisting of reservoirs and fluxes. A **reservoir** is a container that holds an amount of a particular material. A kitchen sink, for example, holds a given volume of water that is maintained by pouring into the sink as much water as drains out (Figure 1-3). The content of a reservoir, whether matter or energy, is known as its **stock. Fluxes** are the movements of material and energy from one reservoir to another. Also called *flow rates,* fluxes are measured in amounts per unit of

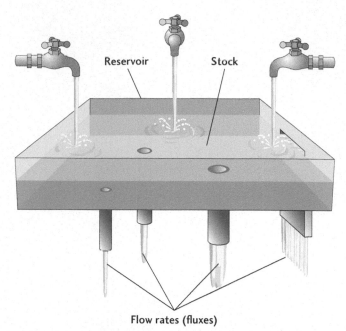

Figure 1–3 A system consists of reservoirs, which contain a particular material, such as the water in this example of a sink. The content of the reservoir is the reservoir's *stock,* and the flow of that stock from one reservoir to another is called the flow rate, or *flux.* The stock is maintained in a steady state if the inflows and the outflows are in balance.

time. The kitchen sink is a reservoir, the amount of water that it holds at any given time is its stock, and the rates at which water flows into and out of it are fluxes.

In nature, a lake operates in a manner similar to the sink. The quantity, or stock, of water held by the lake depends on the flux of water into it from streams, groundwater, and rainfall and the flux of water out of it via streams, groundwater, and evaporation. One focus of the systems approach is to learn how systems stay in balance or change over time. We do this by measuring the fluxes of energy and matter into and out of a system and from one reservoir to another.

Types of Systems

A system can be described as open, closed, or isolated, depending on how freely matter and energy can flow across its boundaries (Figure 1-4). An **open system** allows both matter and energy to flow in and out and is the most common type in nature. A lake is an open system because both matter (water molecules, sediment, organic materials, gases) and radiant energy from the Sun can enter and leave it. **Closed systems** are characterized by the ability to exchange energy but not matter across their boundaries, whereas **isolated systems** have no interactions with their surroundings and allow neither energy nor matter to cross. Although closed systems rarely occur naturally on Earth, for all practical

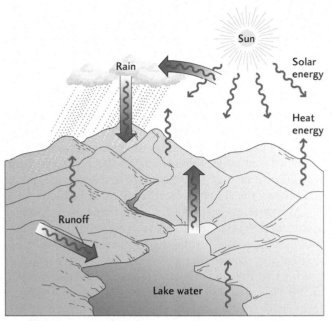

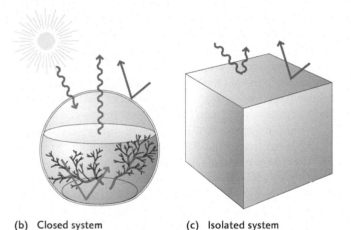

(a) Open system

(b) Closed system   (c) Isolated system

Figure 1–4 Systems are classified as open, closed, or isolated depending on the nature of their boundaries and interactions with their surroundings. (a) An open system allows matter and energy in and out. (b) A closed system, rare in nature, allows only energy in and out. (c) An isolated system blocks matter and energy from entering or leaving.

is the ability to do work, which is defined as a change brought about when a force is applied. Energy is so important to the operation of environmental systems that it is sometimes treated as a system in itself. The manner in which changes to a system occur is known as a *process;* common processes that enact change in environmental systems are the formation and subduction of tectonic plates, volcanism, mountain building, erosion, and flooding. We will define and explore these processes in later chapters.

Most Earth systems are dynamic. Life on Earth, for example, is a dynamic system in which processes such as growth and reproduction are powered primarily by energy from the Sun. The state of this system has changed with time in that the numbers and types of species on Earth have increased for several billion years and occasionally have plummeted during periods of major extinctions. In contrast, the Moon is a relatively static system in which little change occurs with time, and its appearance does not alter significantly with time.

System Behavior

In many dynamic systems, the flow of matter or energy or both into a reservoir is equal to the flow out. As a result, the stock of the material or energy remains constant with time even as the precise atoms making up that stock have changed, a condition known as **steady state.** The ocean, for example, is considered to be at approximate steady state because the total amount of water and dissolved substances it contains has changed relatively little over the past few thousand years.

If we look further in the past, however, we find that the dynamic ocean system was not always at approximate steady state. About 20,000 years ago, sea level was nearly 120 meters (m) lower than it is today because much of the world's water was frozen into extensive ice sheets during the full-glacial conditions prevalent at that time. Between 16,000 and about 6000 years ago, Earth warmed, melting all but the Greenland and Antarctic ice sheets and small remnants of once larger mountain glaciers. The ocean reservoir could not maintain its former steady state because its fluxes were no longer in balance; the inflow of water from melting ice was greater than the outflow of water through evaporation. As a result, the stock of ocean water increased with time, resulting in a rise in sea level that slowed pace about 6000 years ago. It turns out that cyclical changes in Earth's orbit around the Sun have repeatedly led to oscillations of melting and growth of ice sheets, and as a consequence, oscillatory rising and falling of sea level over tens to hundreds of thousands of years.

Sometimes the flow of material into a reservoir exceeds the flow out for an indefinite period of time. Such behavior can lead to different rates of growth, two common types of which are **linear** and **exponential growth**

purposes we can consider the planet as a whole to be closed. Aside from the occasional meteorite and small amount of cosmic dust that reach Earth's surface and the minor amounts of light elements that escape from the outer atmosphere, Earth's mass has been relatively constant for more than 4 billion years. While our planet contains numerous examples of open and closed systems, no system on Earth is truly isolated.

Systems can be further described as dynamic or static. A **dynamic system** is one in which energy inputs cause the system to change over time, whereas a **static system** is one in which no change occurs. **Energy**

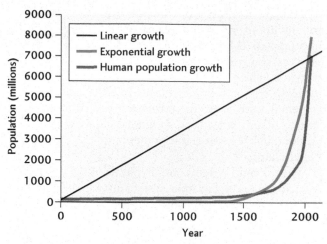

Figure 1–5 Using population as an example, if growth were linear, the number of people would increase steadily and in proportion to the amount of time elapsed. In exponential growth, on the other hand, the increase with time depends on the size of the population, such that as the stock (i.e., number of people) grows, the number of people that is added also grows over time. The actual rate of human population growth over the past 2000 years, however, has been less gradual than the rate that would have resulted if human population growth had been truly exponential since 1 CE.

(Figure 1-5). In linear growth, inflows and outflows have fixed values and the entity growing increases in proportion to time elapsed, resulting in a linear increase in the stock of the reservoir over time. In exponential growth, on the other hand, the inflow depends on the amount of material already in the reservoir, so that as the stock grows, the inflow grows with it.

The world's human population provides an example of nonlinear growth with time. The number of people alive could be viewed as a stock, with births and deaths as the fluxes into and out of this reservoir of humankind. Some aspects of human population growth are similar to exponential growth, in that the rate of growth depends on the number of people alive to reproduce. As can be seen in Figure 1-5, however, the actual rate of increase in world population differs somewhat from that of an exponential growth curve. Human population growth is discussed in more detail in Chapter 7.

Feedbacks in Earth Systems

As systems evolve over time, processes may arise that promote further change in the direction the system is moving, a phenomenon known as a **reinforcing feedback**. One example of such a process is the ice-albedo (*albedo* refers to how much incoming solar energy a surface reflects) feedback, which occurs as a result of the interaction between Earth's ice sheets and the atmo-

sphere, as shown in Figure 1-6a. A small cooling of the atmosphere can lead to the growth of ice. Unlike dark surfaces, which absorb solar energy and convert it into thermal energy, the bright white surface of clean glacial ice reflects 40 to 90 percent of the sunlight that strikes it. As a result, the atmosphere above the ice cools further, which promotes the formation of more ice, which reflects more light and leads to yet more cooling. The process of glacial ice formation thus promotes still more glacial ice formation. Reinforcing feedback processes are destabilizing in that they promote a cascade of events that propels the system toward accelerating change.

A **balancing feedback** is a process by which a change in one direction leads to events that reverse the direction of change. Balancing feedbacks are stabilizing because they counteract the effect of the initial event and help to regulate the system so that it maintains a steady state. For example, carbon dioxide is a greenhouse gas, meaning that it absorbs energy given off by Earth's surface and traps that energy in the atmosphere, which warms our planet. Increased emissions of carbon dioxide from burning coal in power plants and gasoline in automobiles would logically lead to global warming. The picture is much more complex, however, because a warmer atmosphere holds more water vapor, which can result in more cloud cover. Clouds can reflect solar energy back into space before it reaches Earth's surface and is transformed into heat. The result is cooling, as shown in Figure 1-6b. Because an initial event that causes warming results in a process that leads to cooling, this chain of events is a balancing feedback process.

Clouds also can act as reinforcing feedbacks because they can absorb the energy given off by Earth's surface. Whether the feedback effect on climate is net reinforcing or balancing depends upon the clouds' attributes (e.g., droplet size and composition, whether ice or liquid) and their elevation above Earth's surface. At present, global climate models do a relatively poor job of simulating cloud formation, so we cannot be certain that global warming will be offset by increased cloudiness.

■ A system is a group of interconnected and interacting objects and phenomena that can be separated into reservoirs of matter and energy, connected by fluxes of matter and energy from one reservoir to another.

■ In an open system, both matter and energy can flow back and forth across the system's boundaries. In a closed system, only energy can flow across the boundaries. Isolated systems exchange neither matter nor energy with their surroundings.

■ Dynamic systems are systems in motion. They display a variety of behaviors depending on the relative sizes of their inflows and outflows, whether those flows

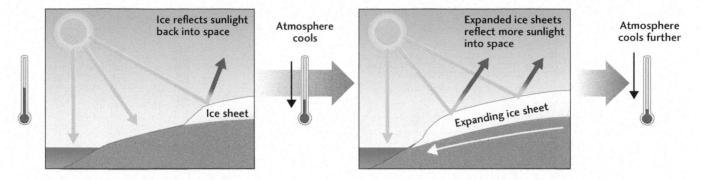

(a) REINFORCING FEEDBACK

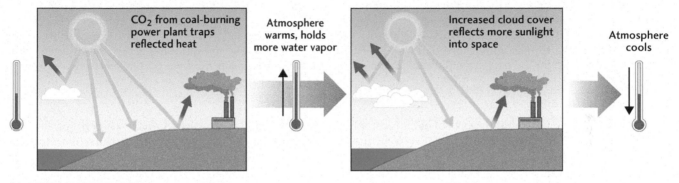

(b) BALANCING FEEDBACK

Figure 1–6 (a) An expanding ice sheet shows how reinforcing feedback is a destabilizing process. (b) A simplified model of the greenhouse effect shows how balancing feedback is a self-regulating, stabilizing process.

are constant or changing, and whether they are fixed or dependent on the reservoirs to which they are attached.

■ Most Earth systems are both open and dynamic.

■ Reinforcing and balancing feedbacks either amplify or resist changes within and among Earth systems.

EARTH'S ENVIRONMENTAL SYSTEMS

Earth's five major systems will be treated in greater detail in later chapters of this book; here, let's consider some basic information to begin exploring Earth's processes and the roles they play in environmental geology. Each of the major systems differs from the others in composition and physical properties (Figure 1-7). The outermost part of the geosphere, called the lithosphere (discussed below), is composed mostly of **silicate minerals**—solid compounds made primarily of oxygen and silicon with a few other elements including aluminum and sodium—as well as iron and nickel. The pedosphere contains elements derived from the breakdown of rock and carbon-rich organic matter and consists of solid, liquid, and gaseous compounds. The hydrosphere is mostly hydrogen and oxygen combined to form water molecules in the gaseous, liquid, and solid states. The biosphere is composed mostly of the elements hydrogen, carbon, and oxygen in the solid, liquid, and gaseous states, and the atmosphere consists chiefly of free nitrogen and oxygen in their gaseous states.

The Geosphere: Earth's Metallic Interior and Rocky Outer Shell

Earth's interior reflects the origins of the solar system 4.6 billion years ago. Based on observations of numerous galaxies, astronomers hypothesize that our Sun and its orbiting planets formed from the collapse of a massive rotating cloud of gas and dust called a nebula. Gravitational attraction drew the matter into a flattened whirling disk, with the bulk of the material at the center, where it developed into a star. Around this central star, the remaining materials in the nebula gravitated into larger bits of rock, metal, liquid, and ice that coalesced into planets.

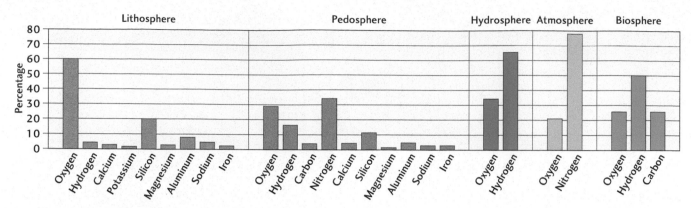

Figure 1–7 Composition of each Earth sphere in numbers of atoms relative to a total of 100. (Because some atoms, such as hydrogen, are very small and light, a comparison based on weight of atoms rather than number of atoms would yield different results.) Both the atmosphere and the pedosphere contain substantial amounts of nitrogen—vital to plant and animal life. In the hydrosphere, oxygen and hydrogen atoms combine to form water molecules, while the biosphere contains relatively large quantities of hydrogen, oxygen, and carbon, the elements from which living organisms are formed.

During the first half-billion years of Earth history, so much debris was present in the solar system that Earth was bombarded constantly by meteorites—bits of rocky and metallic matter left over from the period of planet formation. As the Sun became luminous, its radiation created solar winds, which forced much of the remaining debris out of the solar system. Since then, Earth's mass has remained essentially static, with only minor additions from cosmic dust and occasional meteorite impacts. Earth's chemical and physical characteristics had just begun to take shape, however.

Soon after it formed from accreting debris, our planet began to heat up. Heating resulted from collision of debris, sinking of dense metallic elements into Earth's interior, and radioactive decay of chemical elements. The temperature of the interior of Earth rose to about 2000°C, high enough to melt iron and nickel. Between 4.5 billion and 4.4 billion years ago, the intense heat of the primordial Earth melted much, or perhaps all, of its mass, initiating a process that formed concentric layers with different chemical compositions and physical characteristics.

Chemical Differentiation Iron and nickel, the two most abundant heavy elements, sank toward Earth's center to form the dense **core** of the hot, liquid Earth, located below about 2890 kilometers (km) (Figure 1-8). Lighter elements, particularly oxygen, silicon, aluminum, magnesium, calcium, potassium, and sodium, floated upward. When Earth's surface began to cool and solidify, these elements formed a rigid **crust,** which is Earth's thinnest and least dense rock layer, extending from the surface to a maximum depth of 80 km. Between the core and the crust lies the **mantle,** a vast layer of rock rich in oxygen, silicon, iron, and magnesium and therefore of a composition and density intermediate between the crust

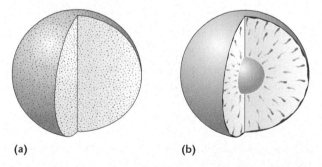

(a) (b)

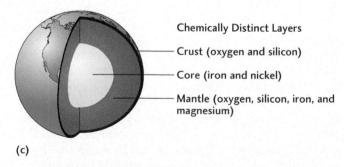

(c)

Chemically Distinct Layers

— Crust (oxygen and silicon)

— Core (iron and nickel)

— Mantle (oxygen, silicon, iron, and magnesium)

Figure 1–8 (a) Early Earth was a homogenous, rocky mixture of mostly iron, oxygen, silicon, and magnesium; there were no continents, oceans, or atmosphere. (b) During the process of differentiation that resulted from catastrophic melting of most or all of Earth's mass, dense elements sank to the center and light elements floated upward. (c) As a result, Earth is compositionally layered in concentric spheres from a dense core rich in iron and nickel, to a thick mantle of mostly oxygen, silicon, iron, and magnesium, and a light crust of mostly oxygen and silicon.

and the core. Most of the rocky matter in Earth's crust and mantle consists of silicate minerals, which are solid compounds formed largely from silicon and oxygen.

Physical Differentiation In addition to the chemical distinctions between Earth's layers, there are also physical distinctions that are the result of differences in the structure of minerals and in the way they behave under different conditions of temperature and pressure. Heating a solid makes it less dense and more malleable. Compressing a solid makes it denser and more rigid. From Earth's surface to its inner core, both pressure and temperature increase (Figure 1-9). Within Earth's crust, the geothermal temperature gradient is about 25°C per km, although the rate differs at various locations. In other words, if you descend vertically 1000 m into a mineshaft, you will feel a temperature increase of about 25°C (77°F).

Close to Earth's surface, the rocks of the crust and upper mantle are still fairly cool and behave in a brittle fashion, meaning that they can break when subjected to stresses. Below some depth, however, the temperature is high enough relative to the pressure of the overlying rock to cause the solid mantle to become somewhat ductile, or plastic, meaning that the rocks are capable of slow flowing movements. The depth at which this transition occurs varies around Earth and lies between 10 and 200 km below the surface. The overlying rigid layer of crust and upper mantle is called the **lithosphere**, and the underlying ductile layer is known as the **asthenosphere**. Since the lithosphere/asthenosphere transition occurs within the upper mantle, it is not associated with a compositional change such as that which characterizes the transition from the crust to the mantle.

The location of the base of the asthenosphere is not very well defined, but many geoscientists consider it to extend beneath the lithosphere to a depth of about 670 km. Below this level, increasing pressure results in a layer of dense, more rigid rock known as the **lower mantle** or **mantle mesosphere** (from the Greek word *mesos* for "in the middle"), which extends to about 2890 km. Between 2890 and 5150 km, the residual temperature from when Earth melted is so high that its **outer core** is still molten, despite the great pressure from the overlying rock. From a depth of 5150 km to Earth's center at 6371 km, however, the pressure is so immense that the **inner core** is solid, even though it is compositionally similar to the outer core and at least as hot.

Movements within the Geosphere The brittle lithosphere is broken into more than a dozen separate plates that

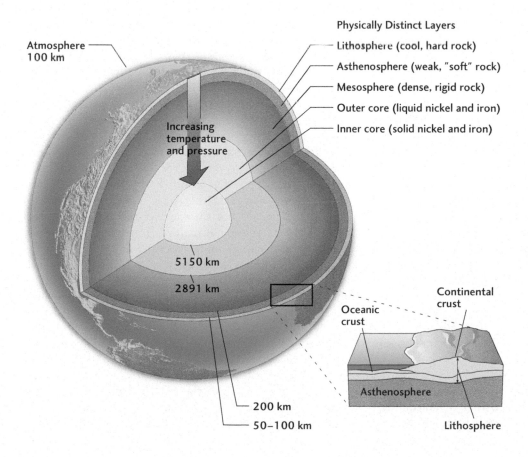

Physically Distinct Layers
- Lithosphere (cool, hard rock)
- Asthenosphere (weak, "soft" rock)
- Mesosphere (dense, rigid rock)
- Outer core (liquid nickel and iron)
- Inner core (solid nickel and iron)

Atmosphere 100 km

Increasing temperature and pressure

5150 km
2891 km
200 km
50–100 km

Oceanic crust
Continental crust
Asthenosphere
Lithosphere

Figure 1–9 Earth is zoned physically as well as chemically. The outermost 50 to 100 km, including the crust and top of the mantle, is very hard and rigid, so it can break; this is the lithosphere. Below it lies the asthenosphere, a weak zone with the consistency of stiff taffy. Under the asthenosphere lies the mesosphere, which is solid and hard, and below the mesosphere lies the liquid outer core. The inner core, although chemically similar to the outer core, is solid and rigid because of the immense pressure in the center of Earth.

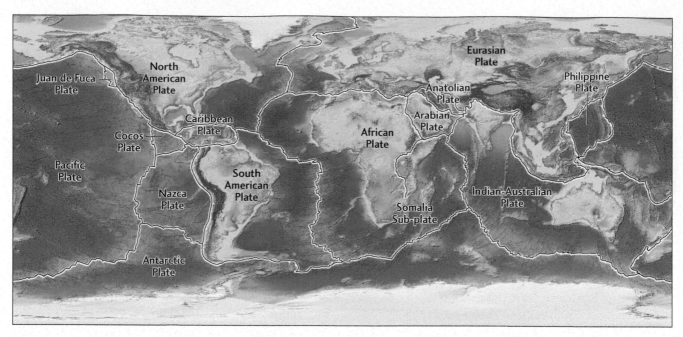

Figure 1–10 The theory of plate tectonics holds that Earth's outermost layer is broken into large, moving plates. Their motion contributes to volcanism, earthquakes, and climate change, as well as many other Earth processes.

resemble a giant spherical jigsaw puzzle (Figure 1-10). These plates slowly drift about on Earth's surface because currents of thermal energy rise from the hot interior of Earth, causing the weak, plastic asthenosphere underlying the plates to flow. The creation, jostling, and destruction of these lithospheric plates generate earthquakes and volcanoes, as we will see in Chapters 2 through 5.

Dynamic processes related to the motion, creation, and destruction of lithospheric plates are referred to as tectonism, from the Greek word *tekton,* which means "builder." Tectonic processes have been continuously building, demolishing, and rebuilding Earth's lithospheric plates and surface features for at least the past billion years, and possibly ever since Earth's crust solidified from cooling magma nearly 4 billion years ago. Since then, the plates have been moving across Earth's surface, breaking apart, colliding with one another, and continuously changing the locations of continents, ocean basins, mountain belts, and island chains. The collision of lithospheric plates typically causes one to dive beneath another into the mantle, a process called *subduction.* Catastrophic events such as earthquakes and volcanic eruptions occur in seconds or minutes, but the tectonic processes that give rise to these events in the lithosphere operate on time scales of hundreds of thousands to millions of years.

The Rock Cycle In terms of a human life span, rock seems eternal. For example, the owner of land underlain by many connected caves recently built a house and gift shop directly above one of the largest caverns, now a popular tourist site in the Appalachian Mountains. To the owner, the rock beneath his home seems permanent and unchanging, despite nearby sinkholes and the cave's evidence that over millennia rock disintegrates and changes form. Likewise, mountains seem fixed and eternal to many people, yet to the geologist they are evidence that young rocks have been pushed upward by plate tectonic processes.

The continual creation, destruction, and recycling of rock into different forms is known as the **rock cycle,** and it is driven by plate tectonics, solar energy, and gravity (Figure 1-11). The rock cycle begins when tectonic forces drive molten rock from Earth's mantle toward the surface. The hot melt is less dense than the surrounding, cooler mantle rock and rises buoyantly through it, just as hot smoke rises through the cooler air around it. As the hot melt rises, it cools and freezes into **igneous rock,** so called from the Latin *ignis,* "fire." Water, wind, biological activity, and other environmental stresses may weather the rock, dissolving and breaking it up into particles that eventually move downward and accumulate in layers of sediment (from the Latin *sedere,* "to sink down"). Under sufficient pressure, sediments harden into **sedimentary rock.** Environmental forces may again disintegrate the rock. Alternatively, tectonic forces may either drive sedimentary and other types of rock back into the mantle and remelt it into magma, or subject the rock to enough heat and pressure to transform it, without melting, into **metamorphic rock**

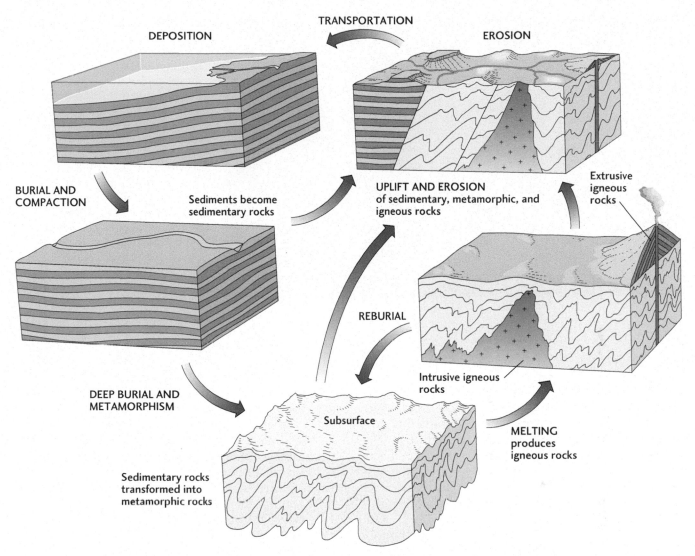

DEPOSITION

TRANSPORTATION

EROSION

BURIAL AND
COMPACTION

Sediments become
sedimentary rocks

UPLIFT AND EROSION
of sedimentary, metamorphic, and
igneous rocks

Extrusive
igneous
rocks

REBURIAL

DEEP BURIAL AND
METAMORPHISM

Subsurface

Intrusive igneous
rocks

MELTING
produces
igneous rocks

Sedimentary rocks
transformed into
metamorphic rocks

Figure 1–11 Cyclical processes link different rock types at Earth's surface. Minerals and rocks are altered by plate tectonic processes that bury rocks deep in Earth and then raise them again during mountain building, as well as by surface processes driven by solar energy, such as weathering and erosion.

(from a Greek word meaning "to change form"). All rocks on Earth can be classified as igneous, sedimentary, or metamorphic.

The Pedosphere: Where Soils Form

Atop the lithosphere lies the *pedosphere,* a layer of disaggregated and decomposed (weathered) rock debris at the surface of exposed landmasses (Figure 1-12). The prefix *ped,* derived from the Latin word for "foot," is used to refer to this layer because it lies underfoot. The composition of solid mineral matter throughout the pedosphere is similar to that of Earth's crust, consisting almost entirely of eight elements: oxygen, silicon, aluminum, iron, magnesium, calcium, sodium, and potassium. Due to its unconsolidated nature, the pedosphere is an open system into and out of which move liquids, solid particles, and gases associated with the hydrosphere, biosphere, and atmosphere. The oxygen, water, and acids abundant at Earth's surface transform rock into pedospheric materials. Minerals weathered from rocks and sediments are mixed with organic matter produced by plants and animals to produce **soil,** the topmost part of the pedosphere. However, with depth and distance from the surface, the pedosphere gradually becomes indistinguishable from the lithosphere.

Some scientists have defined a part of Earth's outermost crust as the **critical zone**. This near surface zone is the locus of complex interactions among rocks, water, the atmosphere, and living matter. The name "critical zone" derives from the fact that this zone regulates natural habitat and determines the availability of resources

Figure 1–12 Earth's terrestrial surface is subject to weathering, a suite of disintegration and decomposition processes, because the geosphere is exposed to the atmosphere, hydrosphere, and biosphere. The result is the pedosphere, of which soil forms the topmost meter or so in the zone of most intense biologic activity. This view of a road cut in southern Washington shows many buried soils, or *paleosols* (light and dark subhorizontal bands), that record about a million years of pedosphere processes during changing climatic conditions. *(Jim Richardson/Getty Images)*

needed to sustain life.[7] In a general sense, the terms *critical zone* and *pedosphere* refer to the same thing.

The total thickness of the pedosphere might be as much as 100 to 200 m in humid, tropical areas where rainfall and temperature are high and weathering processes extend deeply into the crust, or as little as 0 m in a cold, dry desert. The soil part of the pedosphere typically measures only 1 to 2 m thick, yet all terrestrial agricultural activities and food production depend on this layer for nutrients and support. Like the geosphere, soil is arranged in layers or "horizons," with the horizon poorest in organic matter at the bottom and the horizon richest in organic matter at the top. The organic matter in soil contains essential elements that are vital to plant life, such as hydrogen, carbon, nitrogen, and phosphorus.

Rates of soil-forming processes are affected by climate, rock type, organic matter, topography, and time. Most soil-forming processes operate on time scales of thousands to millions of years, and natural processes of soil erosion generally match rates of soil formation. Unfortunately, more rapid processes of soil destruction, such as erosion due to deforestation or poor farming practices, operate on much shorter time scales—years to hundreds of years. If the rate of destruction of soil exceeds that of formation, the soil stock, or reservoir, will become depleted with time, leading to lower soil fertility.

The Hydrosphere: Earth's Distinguishing Characteristic

Rain, river flow, and waterfalls are phenomena that we largely take for granted, but they do not occur on any other planet in our solar system. Earth is the only planet orbiting the Sun with abundant water and the unique combination of temperature and pressure at its surface that allows water to exist in all three states of matter: solid, liquid, and gas. Earth's abundance of water gives the planet a shimmering, blue and white marbled appearance when viewed from space (Figure 1-13).

Water permeates all of Earth's surface systems, yet its movement defines the *hydrosphere,* a zone about 10 to 20 km thick, extending from a depth of several kilometers in Earth's crust to an upper limit of about 12 km in the atmosphere. If all water on Earth (1.4 billion km³) were evenly distributed, it would form a layer nearly 3 km thick. Most of Earth's water—97 percent—occurs

Figure 1–13 Earth's abundance of water, in all three states of matter, gives the planet's surface a marbled blue and white appearance when viewed from space. This view, a photo taken from the Moon by astronaut William Anders in 1968, gave rise to the phrase "blue marble." *(NASA)*

in the oceans and is salty because runoff from the continents brings with it dissolved rock matter in the form of ions (atoms or molecules with net electric charge). Of the 3 percent of water not in oceans, 2.7 percent occurs in solid form as ice. The remaining 0.3 percent occurs as freshwater on continents (rivers, lakes, and groundwater) and as vapor in the atmosphere.

The hydrosphere started to form about 4.4 billion years ago during the period of global melting, when some oxygen and other light elements, such as hydrogen and nitrogen, floated up from the molten Earth as gases. Because of the gravitational attraction of Earth's mass, some of these gaseous elements—particularly the heavier ones—were unable to escape to space. Rather, they remained, some of them combining with one another, to form an envelope of gaseous matter around Earth. Although hydrogen is the lightest element that exists, it so readily combines with oxygen to form water (H_2O) that much of it remained in this early atmosphere. As Earth continued to cool, the water vapor in the atmosphere condensed and rained down, filling the ocean basins. Little water has been added to or lost from the surface of Earth since the first rains that fell from the newly formed atmosphere. Although volcanic eruptions still contribute some water vapor and other substances to the atmosphere and hydrosphere today, outgassing is much less pronounced than before much of Earth cooled to a solid state.

Water in its three states migrates through a vast global **hydrologic cycle** at different rates and over a range of time scales in Earth's near surface environments (Figure 1-14). As vapor in the atmosphere, water moves rapidly about the globe, on time scales of hours to days and months. The movement of water in its liquid state in streams and oceans occurs more slowly, over periods of days to thousands of years. In its frozen form, water in ice caps, ice sheets, and glaciers migrates over periods of thousands to tens of thousands of years.

Relative to other liquids, water can store vast amounts of heat energy. In areas that receive large amounts of solar radiation, water stores solar energy as heat and becomes warmer, whereas in areas that receive less solar radiation, water releases its stored heat energy and becomes cooler. As a result, the movement of water from one reservoir to another and the circulation of water within individual reservoirs such as the ocean affects local to global climatic conditions and helps to regulate Earth's surface temperature.

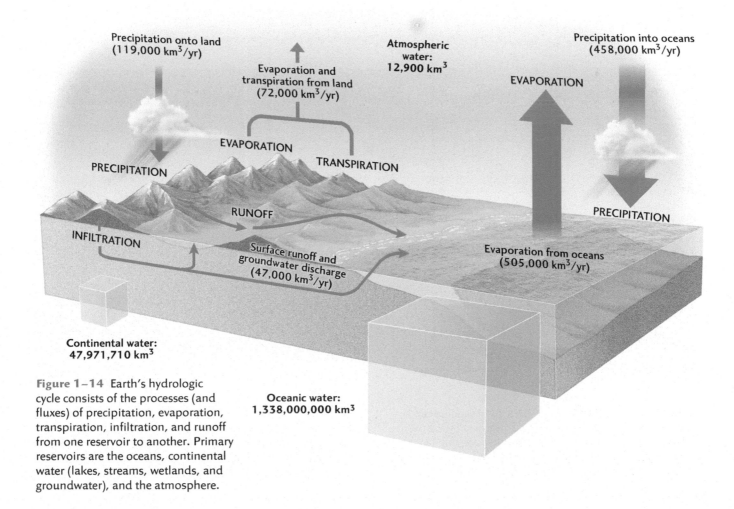

Figure 1–14 Earth's hydrologic cycle consists of the processes (and fluxes) of precipitation, evaporation, transpiration, infiltration, and runoff from one reservoir to another. Primary reservoirs are the oceans, continental water (lakes, streams, wetlands, and groundwater), and the atmosphere.

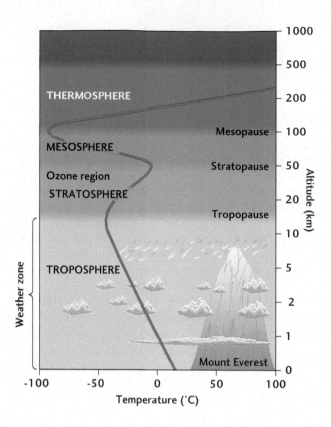

Figure 1–15 Temperature in the troposphere declines with altitude (which here is measured logarithmically) because the troposphere is heated from below by Earth's surface. Concentration of ozone (O_3) in the stratosphere allows temperatures to rise, but in the mesosphere they fall again. Molecular oxygen (O_2) in the thermosphere absorbs heat energy, so temperature rises with altitude. Clouds depict the location of water in the atmosphere.

The Atmosphere: Earth's Envelope of Gases

The Greek word *atmos* means "vapor," and provides the name for the atmosphere, an envelope of gases that surrounds Earth's surface. Part of this envelope can be seen from space as swirling clouds, large masses of air that contain tiny droplets of water and ice crystals (see Figure 1-13). The *atmosphere* actually extends from just below the planet's surface, where gases penetrate openings such as caves in the lithosphere and animal burrows in the pedosphere, to more than 10,000 km beyond Earth's surface, where gases gradually thin and become indistinguishable from the solar atmosphere (Figure 1-15).

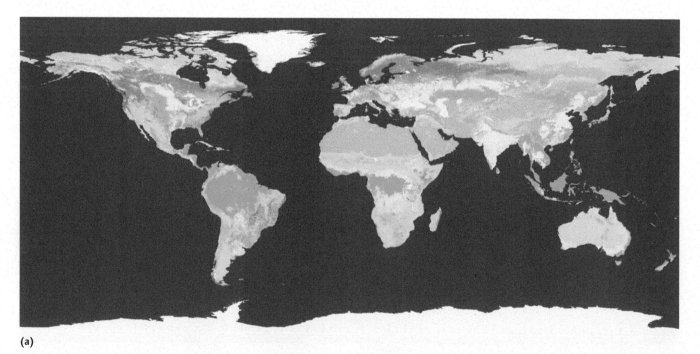

(a)

Figure 1–16 (a) Data from NASA's *Terra* satellite can be used to map land cover on Earth's surface, showing the distribution of ecosystems and land use patterns. These maps aid scientists and policy makers involved in natural resource management. In general, green colors are forests, yellows to oranges and browns are cropland, grasslands, and shrublands, white is snow and ice, and gray is barren or sparsely vegetated land. (b) The vertical extent of life on Earth is referred to as the biosphere. The area below sea level is not drawn to scale. *(a: Courtesy of Boston University and NASA GSFC)*

The clouds of moisture so obvious in Figure 1-13 actually represent only a small portion of the gases in the atmosphere as a whole (and are an example of the overlap between the hydrosphere and atmosphere). Almost all the volume of gas in the atmosphere consists of nitrogen (78 percent), oxygen (21 percent), argon (0.9 percent), and carbon dioxide (0.03 percent). Other gases, including neon, helium, nitrous oxide, methane, and ozone, occur in trace amounts. The percentage of water vapor at the base of the atmosphere varies considerably, from 0.3 percent on a cold, dry day to as much as 4 percent on a hot, wet day, but the water content decreases rapidly to nearly zero just a few kilometers above Earth's surface.

Venus and Mars are the most Earthlike planets in the solar system, yet they have atmospheres that consist almost entirely of carbon dioxide. The difference between their atmospheres and Earth's seems to be related to the existence of life, because the biological activity of plants produces oxygen. Some of Earth's oxygen combines high up in the atmosphere to form a protective layer of ozone. Other gases in the atmosphere store and transfer thermal (heat) energy at Earth's surface, modulating the planet's temperature and producing the weather systems that move water evaporated from the ocean basins over land where it can fall as rain or snow. If Earth had no atmosphere, its surface temperature would be at least 33°C cooler on average, or a chilly −18°C!

The Biosphere: Where Life Exists

The *biosphere* is a thin layer of life-forms that live in, on, and above Earth's other environmental systems (Figure 1-16). An astonishing number of different species of organisms are thought to exist on Earth—some 8 to 9 million.[8] The majority depend on *photosynthesis*, the process by which green plants convert solar energy, carbon dioxide, and water into food energy. These organisms thus are confined to a zone in which they can receive solar radiation. This zone overlaps part of the atmosphere, the land surface (the top of the lithosphere and uppermost part of the pedosphere), and the illuminated (euphotic) zone of water bodies (see Figure 1-16).

The composition of the biosphere is similar to that of Earth's other systems, but much closer to that of the hydrosphere than the lithosphere, as living cells are generally 60 to 90 percent water (see Figure 1-7). Like the atmosphere and hydrosphere, the biosphere is dominated by lighter elements. In fact, no elements with atomic numbers higher than 53 (iodine) are found in living cells except in trace amounts. By number of atoms, the biosphere consists of 99 percent hydrogen, oxygen, carbon, and nitrogen, and its high percentage of carbon distinguishes it from all the other major Earth systems.

Carbon atoms form strong bonds with one another, enabling them to develop elaborate chains of unlimited length, as well as rings and branching structures that can attach to other atoms and molecules. These complex organic (i.e., carbon-containing) macromolecules sometimes consist of thousands of atoms and form more compounds than any of the other 102 elements on Earth, which is one reason most chemistry students consider organic chemistry to be among their hardest courses! The fundamental organic molecules are carbohydrates, fats, proteins, and nucleic acids. All fossil fuels (oil, coal, and natural gas) were formed from organic molecules that originated in the biosphere but were buried by

(b)

sediments to become part of the lithosphere after the death of the organisms that produced them.

In the early history of the planet, oxygen in the atmosphere was insufficient for animal life to exist. Instead, Earth's primitive atmosphere contained large amounts of methane (CH_4), but these molecules were split apart by solar radiation, and the carbon combined with the little oxygen that existed to form carbon dioxide (CO_2). Early life-forms, including bacteria, blue-green algae, and phytoplankton, used the CO_2 for metabolism, producing oxygen as a by-product. Fossil evidence indicates that these early photosynthesizers existed by at least 3.5 billion years ago and perhaps earlier. With time, the amount of CH_4 in the atmosphere decreased, while the amounts of CO_2 and free oxygen increased. The continued evolution and spread of plants—which increased amounts of oxygen in the atmosphere—enabled oxygen-breathing animals to evolve. Some of these animals (herbivores) relied on consuming plant matter for energy, while others (carnivores) preyed on other animals to fuel their own metabolism. Increased levels of oxygen in Earth's atmosphere also allowed for the formation of ozone in the upper atmosphere, which shields organisms from ultraviolet (UV) radiation given off by the sun. UV radiation can lead to sunburn, genetic defects, and cancer, so only when the atmosphere contained enough ozone could animals evolve on land.

Earth is the only planet in the solar system known to have life. The role of water in the evolution of life is essential, in part because of the many special properties of the water molecule, such as its high heat capacity. Life began in the oceans, and those life-forms that moved onto the continents carried their ocean environments with them, in their cells. It should come as no surprise, then, that the human body is nearly two-thirds water, and the water flowing through the bloodstream and bathing the cells carries on the task of the ocean by supplying the body with nutrients and removing waste products. To stay alive, terrestrial creatures require a supply of water and, consequently, are greatly dependent on the processes that involve water on land. In this sense, the biosphere and hydrosphere are intimately connected.

■ The geosphere formed by accretion of rock and metal debris associated with the nebula that became our solar system. Subsequent melting of this material led to its chemical differentiation into the layers that became the inner and outer core, mantle, and crust.

■ The lithosphere is a rigid, rocky layer of crust and upper mantle 10 to 200 km thick that overlies a more ductile layer, the asthenosphere, within the mantle. The lithosphere is broken into plates that float above the asthenosphere and move when heat from Earth's interior induces flow in the asthenosphere. The outermost part of the lithosphere is a chemically distinct layer called the crust.

■ The pedosphere, generally less than 100 to 200 m thick, is the entire layer of weathered rock debris and organic matter at the surface of exposed landmasses. The relatively organic-rich upper meter or so of the pedosphere, called soil, provides the nutrients essential to agricultural production on land.

■ Earth's gaseous atmosphere and largely liquid hydrosphere formed by outgassing of volatile elements from the early molten Earth as it differentiated into physically and chemically distinct layers. Both atmosphere and hydrosphere serve to modulate Earth's surface temperatures, keeping them within a fairly limited range, by transporting heat absorbed from the Sun from one location to another.

■ Because of biological activity, Earth's atmosphere is rich in oxygen compared with the atmospheres of nearby planets. In turn, Earth's atmosphere shields the biosphere from harmful UV rays from the Sun by efficiently absorbing them.

■ The biosphere of Earth is unique in our solar system and formed soon after the early atmosphere and hydrosphere developed. It is an exceptionally thin layer, but it contains millions of different species that have evolved since the earliest, simple forms of life.

EARTH'S ENERGY SYSTEM

Energy can be considered a sixth environmental system. Earth generates energy in its own interior and receives energy from the Sun. Energy from these two sources, plus small amounts generated from the gravitational attraction between the Moon, the Sun, and Earth, flows through all the other Earth systems (Figure 1-17). Since energy is the ability to do work, all processes and changes would cease if Earth received no energy. Earth's access to energy and ability to use it make the planet a dynamic system.

States of Energy

Energy exists in several states, including kinetic, potential, and thermal. **Kinetic energy (KE)** is the energy of a body in motion and is defined as $KE = 0.5\, mv^2$, where m is the mass of the body and v is its velocity. **Potential energy (PE)**, in contrast, is the energy of a body that results from its position within a system. Potential energy is related to the gravitational attraction between bodies and is expressed as $PE = mgh$, where m is the mass of the body, g is acceleration of the body due to the gravitational force, and h is the relative height of the object.

Boulders perched on a hillside contain potential energy as long as they are at rest. If a process—perhaps an earthquake—causes the ground to give way, the boulders will roll downhill under the influence of gravity. Before the rockfall begins, the rocks contain potential energy proportional to their distance from Earth's center. As they roll and bounce down the slope, their potential energy is converted into kinetic energy and **thermal** (heat-related) **energy** (due to the motion of atoms). By the time the boulders settle at the bottom of the hill, all their potential energy has been converted into thermal energy as a result of frictional resistance. This example demonstrates a fundamental principle: Energy is completely conserved when work is done. While the energy has undergone a change in state (from potential to kinetic to thermal), the amount of potential energy before the work was done equals the amount of kinetic and thermal energy expended to do the work.

Energy can be transferred between bodies in the universe, as in the transfer of thermal energy. All bodies contain internal thermal energy, which arises from the random motions of their atoms. The term *heat* describes the transfer of thermal energy from one body to another. In a body receiving thermal energy, the added energy causes the atoms in the body to speed up. The temperature of a body is a measure of the average speed at which its atoms move. When heat is transferred to a body, its temperature rises because its atoms move faster.

Energy is measured in many different units, including foot-pounds, British thermal units (Btu), and electron volts, but the most common convention among Earth

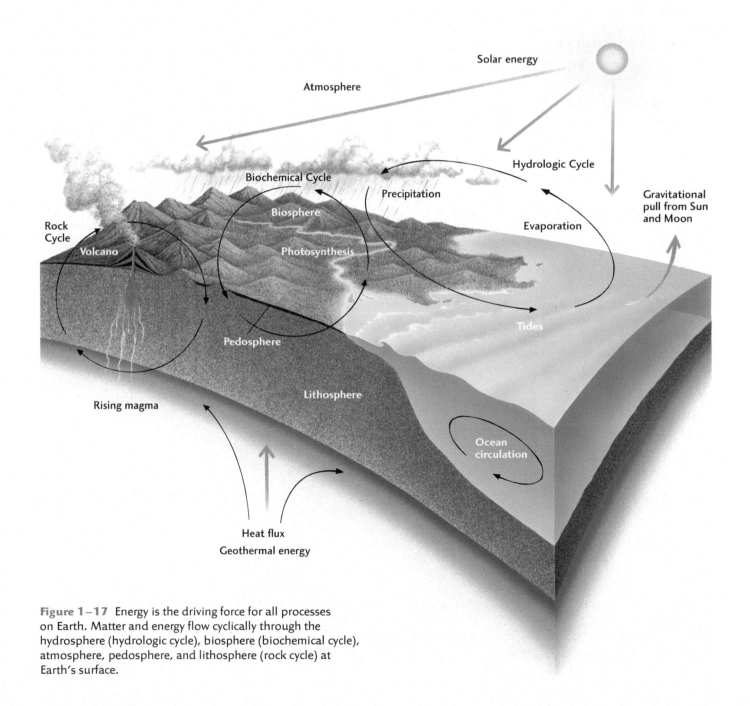

Figure 1–17 Energy is the driving force for all processes on Earth. Matter and energy flow cyclically through the hydrosphere (hydrologic cycle), biosphere (biochemical cycle), atmosphere, pedosphere, and lithosphere (rock cycle) at Earth's surface.

scientists is to measure energy in either calories or joules. One *calorie* (cal) is defined as the amount of energy needed to raise the temperature of 1 gram of liquid water by 1°C and is equivalent to 4.184 *joules* (J). When contemplating how many calories our food contains, we refer to Calories (upper case C), each of which is equivalent to 1000 cal, or 4184 J. A candy bar, for example, contains about 400 Calories, or 400,000 calories. The human body (typically 50 to 80 kg) needs to maintain a temperature of 37°C (98.6°F). Because about 70 percent of the human body is water, much of its caloric energy is used to maintain the temperature of water at 37°C. The lower the temperature of ambient air, the greater the amount of calories needed.

Sources of Energy

Three sources of energy, two internal and one external, power all the processes on Earth. Our planet receives 1.73×10^{17} watts (W; 1 W = 1 J/sec) of power (energy per unit of time) from the Sun, which amounts to 99.98 percent of its total energy budget. Smaller amounts of energy are created within Earth through spontaneous radioactive decay of some elements in rocks, and some residual heat from Earth's formation still exists. Together these amount to 32×10^{12} W of power. An even smaller amount of energy (3×10^{12} W) is supplied by the gravitational attraction within the Earth–Sun–Moon system and drives the movement of ocean water as tides.

Solar Energy In the Sun, hydrogen atoms at extremely high temperatures and pressures combine to form helium atoms in a process called fusion. This reaction releases radiant energy that is transmitted through space in all directions, and Earth intercepts some of this energy, warming the atmosphere, oceans, and land surface (Figure 1-18). The amount of sunlight that reaches Earth's surface is called **insolation,** and the amount received varies around the world as well as throughout each day and season (Figure 1-19). It is large during midsummer at midlatitudes, for example, and relatively small at the same locales in winter. During photosynthesis, plants combine solar energy with water and carbon dioxide to produce the carbohydrates on which the whole biosphere depends for food and fuel. When we burn wood in a campfire, we release this stored chemical energy, which we then can use to warm ourselves or to cook food.

Under the proper geologic conditions, some biomass may be converted to oil, coal, or natural gas. These substances are called fossil fuels because they come from plants and animals that lived many tens to hundreds of millions of years ago. When we drive cars, we are using solar energy that is many tens to hundreds of millions of years old! The amount of energy stored within fossil fuels is extremely small compared with the amount of solar energy received by Earth, because less than 1 percent of incoming solar energy is converted to chemical potential energy by plants. In less than a single month, Earth receives as much energy from the Sun as is stored in known fossil fuel reserves. Because fossil fuels that were easiest to reach and extract have diminished, solar energy is an increasingly important energy source.

Earth's Internal Energy Earth makes its own supply of internal energy through the spontaneous decay of certain elements in minerals. In this process, known as *radioactive decay,* parent atoms convert to other atoms by losing or gaining subatomic particles in their nuclei (discussed more fully in Chapter 6). This process releases relatively small amounts of thermal energy that are transferred through the surrounding rock.

Although much smaller than the influx of solar energy, it is Earth's internal heat energy that drives

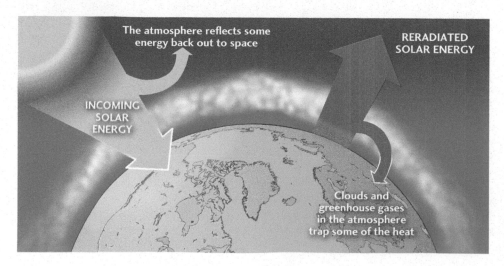

Figure 1–18 Some solar energy is reflected, and some is absorbed and reradiated to outer space. Inflow of energy is equal to outflow, keeping Earth's energy budget balanced.

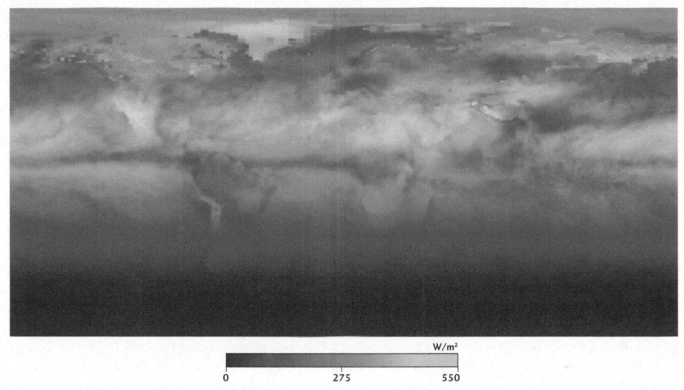

W/m²

0   275   550

Figure 1–19 Insolation is the amount of sunlight that reaches Earth's surface during a given time period, as shown here for July 1, 2012 through July 31, 2012. It commonly is measured in units of watts per square meter of land surface (W/m²). Note how little sunlight is received by the south polar region during July, which is winter for the southern hemisphere. *(NASA)*

lithospheric plate motions, which in turn result in earthquakes and volcanism. As plates move, energy can be stored in rocks, much as energy is stored in an elastic band by stretching it. When such rocks finally snap, as would a band that had been stretched too far, the amount of energy released can be quite large. The 2004 Indonesian earthquake that led to the Indian Ocean tsunami, in which more than 275,000 people perished, released about 1.1×10^{17} J of energy, equivalent to more than 1500 nuclear bombs the size of those dropped on Hiroshima, Japan, during World War II.

Gravitational Attraction Earth derives energy from gravitational attractions between itself and the Moon, the Sun, and the other bodies in the solar system. Although minor compared even with Earth's internal energy, the energy derived from gravitation causes both Earth's crust and ocean water to change shape over time in a regular fashion, resulting in rock tides as well as ocean tides.

Energy Budget of Earth

The energy system of Earth is in a steady-state condition; the amount of energy received by the whole Earth system is approximately equal to the amount of energy flowing out of the system. Because the Sun supplies all but 0.02 percent of Earth's energy, the inflow and outflow of solar energy is a close approximation of Earth's energy budget. An energy budget, like a monetary budget, is an accounting of inputs and outputs, and a steady-state condition is like a balanced financial budget.

As shown in Figure 1-18, more than half of all incoming solar radiation is returned to space before doing any work at Earth's surface. About 25 percent is reflected to space by particles or clouds in the atmosphere and another 5 percent is reflected by Earth's surface, while 25 percent is absorbed by clouds and the atmosphere. This leaves about 45 percent of incoming solar radiation to be absorbed by Earth's surface. About two-thirds of this absorbed energy (29 percent of the original incoming radiation) is cycled through Earth's environmental systems and reradiated, while one-third (16 percent of the original incoming radiation) is converted to heat (thermal energy). The wavelengths of reradiated energy are longer than those of the original (or reflected) solar radiation. Because clouds and greenhouse gases such as carbon dioxide trap energy at longer wavelengths relatively easily, reradiated energy has a strong influence on

Earth's climate. In Figure 1-18, you can see that inflow of solar energy is equal to outflow, with 30 percent of the inflow leaving Earth as reflected energy and 70 percent leaving as reradiated energy. Earth's energy budget is balanced.

Human Consumption of Energy

Humans have devised many methods to harness energy. For millennia, people have used wood for heat and wind and moving water to grind grains and process other commodities in mills; more recently, a variety of fossil fuels and other energy sources generate electricity (Figure 1-20). In Iceland, a country located on an active plate boundary with volcanism and high heat flow at the surface, the dominant energy source is geothermal. Worldwide, however, fossil fuels (largely coal, crude oil, and natural gas) have become the most important sources of energy during the past century.

The total amount of energy used has increased with time even though the dominant energy sources have varied. In 2011, humans used a staggering 487 exajoules (an exajoule equals 10^{18} joules) of energy worldwide, much of it to generate electricity. As a consequence, our planet—particularly due to its more populated areas—resembles a well-lit black marble at night (Figure 1-21).

The amount of incoming solar energy is roughly 12,000 times that of the world's current energy consumption, but relative to the amount available, little solar energy has been collected and used thus far. Most industrialized nations rely heavily on fossil fuels—oil, coal, and natural gas—to power cars, heat homes and water, and generate the electrical energy needed to run various appliances and utilities. However, Germany, Spain, and the United States currently are making significant advances in technologies in order to use more solar energy for heating and electricity.

Of all the different sources of energy available to the world, fossil fuels satisfy about 81 percent of total energy demand. However, the type of fuel used and the per capita use of energy vary markedly from country to country, as well as from region to region within countries. As of 2010, the United States used more total energy than any other country in the world except China. With only about 5 percent of the world's population, the United States consumes about 19 percent of the world's annual primary energy demand (98 quadrillion Btu in 2010). In 2010, China surpassed the United States to become the world's leading consumer of energy, using 20.3 percent of the total. Still, its far greater population means that China's per capita energy consumption remains lower than that of the United States.

■ Three sources of energy are responsible for all processes on Earth: solar energy, internal thermal energy, and gravitational energy.

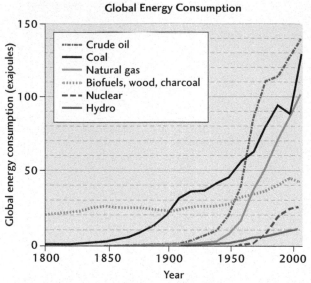

Figure 1–20 Global energy consumption has increased rapidly during the past two centuries, fueling the Industrial Revolution that continues to spread to countries worldwide. First coal and then two other fossil fuels, crude oil and natural gas, rose in prominence since the late 19th century.

■ Solar energy contributes 99.98 percent of the energy that drives processes on Earth, but Earth's internal thermal energy drives plate tectonics, while the gravitational attraction among Earth, the Moon, the Sun, and other bodies in the solar system drives the tides.

■ Earth's energy system is in a steady state; the amount of energy received by the whole Earth system is approximately equal to the amount of energy flowing out of the system.

■ Since the early 20th century, fossil fuels, largely coal, crude oil, and natural gas, have been the primary sources of energy consumed by humans.

■ World energy use is increasing over time, and was 487 exajoules in 2011.

HUMAN POPULATION AND EARTH SYSTEM BOUNDARIES

Several related issues have contributed to a growing environmental awareness since about the mid-20th century, foremost among them population growth and spreading industrialization. With population and industrial growth come ever-increasing demands for such resources as fuel, land, and clean freshwater. Environmental degradation is the nearly inevitable result

Figure 1-21 A growing population worldwide with access to electricity and high population densities in cities gives Earth a nightly glow that resembles a well-lit black marble.

(left: IM_Photo/Shutterstock; right: NASA Earth Observatory Image by Robert Simmon, using Suomi NPP VIIRS data provided courtesy of Chris Elvidge [NOAA National Geophysical Data Center])

of these pressures and demands. In addition, although people always have lived in areas prone to earthquakes, floods, and other geologic hazards, the increasing numbers of people living in these areas heightens the potential for human disaster.

Human Population Growth

As the world's population passed 7 billion sometime in late 2011 to early 2012, concern continued to mount over how many more people would be added in the future. Human population growth has been difficult to predict accurately because it depends on a large number of variables—from advances in agriculture, sanitation, and medicine to the influences of culture, religion, and medical practices. The rate of human population growth has varied throughout human history, but its most striking feature is marked growth into the billions in the last few hundred years (see Figure 1-5).

The human population has grown in a manner somewhat similar to exponential growth. It took 2 million years of human history to add the first billion people, 130 years to add the second billion (around 1927), 30 years to add the third (1960), 15 years to add the fourth (1975), and 12 years to add the fifth (1987). World population reached 6 billion in 1999 and 7 billion in 2011–2012, with each extra billion added after 12 years. Since the 1960s, annual growth rates for world population have declined from a high of roughly

2.2 percent to about 1.1 percent in 2012. As fertility and mortality rates change, population growth rates vary. The current slowdown in population growth rate is a result of falling levels of fertility.

Despite a decreasing rate since the 1960s, population growth still has considerable momentum because of the large number of people on Earth. Even at the relatively low growth rate of 1.1 percent, another 75 million people are added to Earth's population each year. Most population growth is occurring in developing rather than already industrialized countries. Various predictions of the world's future population range from 6 to 16 billion people during the next century.

What are the consequences of more than 7 billion people on Earth and a growing population with a desire for higher standards of living? To date, consequences have included increasing use of resources (e.g., minerals, fertilizers, and water), consumption of energy, and production of wastes. Unfortunately, they also have included a growing number of extinctions of plant and animal species.

More people are living in cities; in fact, more than 50 percent of people live in urban areas, and cities are increasingly larger. In the early 20th century, only 20 percent of the population lived in urban areas. As a result, population density, not just population, is increasing (Figure 1-22).

The world's first megacities, those with more than 10 million people, developed in the mid-20th century

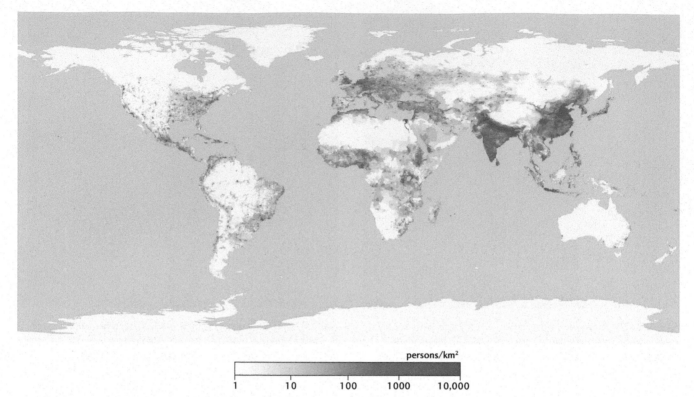

persons/km²

1   10   100   1000   10,000

Figure 1–22 World population density on average is about 50 people per km² of land area, excluding Antarctica, but can be as high as 40,000 people per km² in megacities. Note that the four most densely populated places are Singapore, Hong Kong, Bahrain, and Bangladesh. The four largest megacities are all located in Asia: Tokyo, Japan; Jakarta, Indonesia; Seoul, South Korea; and Shanghai, China. *(Image by Robert Simmon, NASA's Earth Observatory, based on data provided by the Socioeconomic Data and Applications Center [SEDAC], Columbia University)*

(New York City was the first). Today there are 28 megacities. Three are located in North America: Mexico City, New York City, and Los Angeles. In some of the world's most crowded cities, population densities are as high as 40,000 people per km² (about 100,000 per mi²). In contrast, for Earth's land area excluding Antarctica, the average population density is about 50 people per km² (130 per mi²).

Resources and Sustainable Development

Does Earth have some finite "carrying capacity," some threshold number of people beyond which it cannot sustain the human population with clean air, clean water, and adequate nourishment? If so, then at or below this population, humans could lead a sustainable existence, one in which development meets the needs of the present without compromising the environment or the ability of future generations to meet their needs as well.[6] Although perhaps impossible to determine a threshold population size with certainty, we realize that both the number of people and the resources they consume are central to evaluating the possibility of long-term sustainability.

A **resource** is anything we get from our environment that meets our needs and wants. Some essential resources, including air, water, and edible biomass (plant and animal matter), are available directly from the environment. Other resources are available largely because we have developed technologies for exploiting them. These include oil, iron, and groundwater. In general, people in affluent and highly industrialized countries use far more resources than are required for basic survival. The United States, for example, has only 4.8 percent of the world's population, yet it consumes about 33 percent of the world's processed nonrenewable energy and mineral resources.

Resources are classified into three major types according to their degree of renewability: potentially renewable, nonrenewable, and perpetual (Figure 1-23). **Potentially renewable resources** can be depleted in the short term by rapid consumption and pollution, but in the long term they usually can be replaced by natural processes. The highest rate at which a potentially renewable resource can be used, without decreasing its potential for renewal, is its *sustainable yield*. If the sustainable yield is exceeded, the base supply of the resource can shrink so much that the resource can become exhausted—used up. Soil formation, for example, occurs at rates of about 2 to 3 centimeters (cm) per thousand years, making it a potentially renewable resource. Unwise farming practices, however, can cause soil loss of 6 to 8 cm per decade. Soil cannot be renewed at this rate because its formation is dependent on plants and soil moisture, both of which are gone once the soil itself is lost. In some parts of the world, all soil has been removed and so is essentially nonrenewable.

Nonrenewable resources, such as fossil fuels and metals, are finite and exhaustible. Because they are produced only after millions of years under specific geologic conditions, they cannot be replenished on the scale of a human lifetime. The formation of oil, for example, requires that buried plant and animal matter be subjected to tens of millions of years of squeezing and heating by geologic processes, a rate of formation so slow that the resource cannot be considered renewable.

Some nonrenewable resources, such as copper and iron, can be recycled or reused to conserve the supplies. *Recycling* involves collecting, melting, and otherwise reprocessing manufactured goods. Resources commonly recycled include aluminum, paper, and iron. *Reuse* involves repeated use of manufactured goods in the same form. Glass bottles commonly are reused. Other resources, such as fossil fuels, cannot be recycled or reused because their combustion during energy production converts them to ash that falls to Earth, exhaust fumes that go into the atmosphere, and heat that ultimately leaves Earth as low-temperature radiation. Nonrenewable resources become economically depleted when so much—typically about 80 percent of the resource—is exploited that the remainder is too expensive to find, extract, and process. Oil buried in rocks deep beneath the Antarctic ice sheet, for example, would be far more expensive to find, extract, and refine than oil trapped beneath the shallow salt beds of Texas.

Perpetual resources are those that are inexhaustible on a human time scale of decades to centuries. Examples include solar energy—which has fueled numerous reactions and processes on Earth for its entire 4.6-billion-year history, heat energy from the interior of the Earth, and energy generated by Earth's surface phenomena such as wind and water flowing downhill.

One of the most significant changes in human history is the increased demand for, and use of, energy. Before the Agricultural Revolution started about 10,000 years ago, people's energy demands were primarily for food. On average, a person uses about 10 megajoules (MJ) of energy (2000 to 3000 food calories) per day, and for early humans this was the bulk of their energy needs. By 2011, however, daily per capita energy consumption in the United States was more than 900 MJ, spent mostly in industry, transportation, and temperature control. Much of this energy consumption goes toward the production and transport of food.

(a)

(b)

(c)

Figure 1–23 (a) Soil, a potentially renewable resource, is put at risk by the demands for food associated with worldwide population growth. (b) New technologies and ever–larger machines are used to mine coal, a nonrenewable resource, at increasing rates. (c) Solar energy is a perpetually available resource. Reflective panels can focus this energy on a "power tower" where it is converted into electricity. *(a: Hopsalka/Shutterstock; b: J. van der Wolf/Alamy; c: Felipe Rodriguez/Getty Images)*

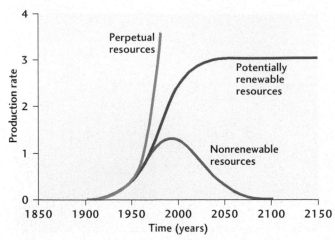

Figure 1–24 As demand for a resource increases, more of it is produced. For nonrenewable resources, production cannot increase forever, because ultimately the resource will be exhausted. Renewable resources can provide lasting supplies if the rate of production is similar to the rate of renewal. Only for perpetual resources is unlimited increase in production possible.

Production and consumption of renewable, non-renewable, and perpetual resources follow different growth curves (Figure 1-24). For potentially renewable resources, such as wood, the curve is similar to that of the growth of a biological population: Production rises exponentially for some time, and then levels off at a state of no growth that is equal to the rate of replenishment. For nonrenewable resources, such as fossil fuels and minerals, production rates rise exponentially, reach a maximum value, and then decline exponentially as supplies approach zero. For perpetual resources, such as solar energy, production can rise exponentially for essentially unlimited time periods. It is clear that humans can rely on nonrenewable energy resources only for a limited time, before turning to a perpetual source of energy.

Pollution, Wastes, and Environmental Impact

Many of the world's environmental problems are related to the wastes produced by human activities. To a greater or lesser extent, all humans produce wastes that pollute and degrade air, water, and land. **Pollution** is the contamination of a substance with another, unde-sirable material. Environmental pollution results when a pollutant degrades the quality of an environment.

Common pollutants in urban areas are sulfur diox-ide (SO_2) and oxides of nitrogen (NO_x, molecules with one or two atoms of oxygen and one atom of nitrogen), all of which are created by combustion of fossil fuels

in automobile engines and industrial power plants and released into the atmosphere. These molecules react with gases and water vapor in the atmosphere to produce smog and acidic rainwater, both of which make the air and water unhealthful, causing damage to plants, fish, land animals (including people), and even stone.

Wastes are unwanted by-products and residues left from the use or production of a resource. Early human habitation sites are often identifiable by the piles of waste left behind, including shells, bones, char-coal from fires, and broken pottery. Sometimes, early humans reused wastes in innovative ways, building huts from animal bones, for example. People tended to move on when their waste piles became substantial, although substantial amounts of human detritus can be found beneath cities that date to ancient times, such as Rome and Paris. Most waste piles made by small bands of people were biodegradable, so they did little long-term environmental damage. In contrast, modern humans in industrialized societies have much higher population densities and technologies that enable them to concentrate large amounts of waste, some of which is toxic and infectious, in small areas.

A Model of Environmental Impact Scientists have devel-oped a simple model of environmental degradation and pollution to assess the environmental impact of human populations. In this model, the extent of the environmen-tal impact depends on three variables: (1) population, (2) per capita consumption of resources, and (3) the amount of environmental degradation and pollution per unit of resource used:

Number × per capita × degradation
of people   consumption   and pollution
             of resources   per unit of
                            resource used

= environmental impact

Environmental degradation due to **overpopulation** occurs when a large group of people has insufficient resources. Although the group's per capita consump-tion of resources may be low, the amount of environ-mental degradation can be high because the group must scavenge all available resources in order to survive. Environmental protection means little in the face of starvation. In parts of northern and western Africa, for example, large areas of land have been denuded of most vegetation, and famine has occurred repeatedly in the 20th century.

Degradation due to **overconsumption** occurs when a small number of people use resources at such a high rate per person that they cause very high levels of resource depletion and pollution. Ranking first among modern examples is the United States, which has less than 5 percent of the world's population but consumes more than 33 percent of nonrenewable energy and mineral resources and produces more than a third of the world's pollution. As a result, despite its relatively small population, it has a tremendous environmental impact. In contrast, China has about 19 percent of the world's people, about four times as many as the United States, but causes less than one-tenth (or 10 percent) the environmental damage of the United States. For this reason, it sometimes is noted that the environmental impact of one person in the United States is equivalent to that of many more persons in developing nations, such as India, where resource use and pollution per person are much smaller.

As nations develop, the extent of their environmental impact due to consumption changes rapidly. In the 19th century, for example, during a rush to settle all parts of the United States, forests were destroyed to clear fields for farming, and many other sites were severely degraded by destructive mining practices. Since then, increasing environmental awareness in the United States has resulted in significant efforts to prevent such degradation in future development. Likewise, as China enters the global market of free trade and as other nations become more developed, it is likely that the nature of their environmental impacts will change.

The third component of environmental degradation is **pollution per unit of resource used**. In some cases, the amount of pollution produced per unit of resource used is so high that extreme environmental degradation occurs. For example, the increased use of coal for energy in China at a time of rapid industrial growth is resulting in substantial air pollution problems (Figure 1-25). Industrialized countries in Eastern Europe and the former Soviet Union have amassed a legacy of environmental pollution that will affect their citizens for decades to come.

Minimizing Environmental Impacts Solving the problem of environmental degradation caused by rapid population growth is essential, but the solution is not simple. Increasing access to medicine and sanitary supplies of water worldwide has reduced death rates and thus contributed to the rapid population growth of the past few centuries. As a result, reducing population will require a change in the birth rate. Solutions for reducing birth rates are related to education and standards of living, among other things. In most countries of the world, the higher the level of education and standard of living of women, the more likely women are to use birth control methods and have fewer children.

Reduced birth rates in most developed nations, as well as in China and India, caused worldwide population growth rates to drop from 2.06 percent in the late

Figure 1-25 In China, environmental pollution and degradation are extreme as a result of a large population, rapid industrialization, and heavy reliance on burning coal for power. *(Bernard Wis/Paris Match/Getty Images)*

1960s to 1.73 percent in the late 1970s, to 1.6 percent in the 1990s, and to 1.1 percent as of 2012.[9] This consistent drop in population growth rates is encouraging, but rates must drop in many more nations in order to level off global population growth.

Minimizing environmental degradation caused by consumption can be achieved through reduced resource consumption per person and coordinated efforts to recycle or reuse products. Minimizing environmental degradation caused by pollution can be achieved through environmental regulation and the use of more efficient technologies to minimize the amount of waste (much of which produces pollution) generated during industrial and resource extraction processes. Efforts to develop waste minimization procedures for different industries are now the focus of research and testing at many universities and industrial labs. Some new factories in Denmark, for example, produce nearly zero waste, designing their activities so that waste products from one stage are used as input to other stages of the manufacturing process.

As environmental regulations become more stringent, companies are developing new technologies to extract and process resources that produce fewer wastes than ever before. New methods to mine copper that require no excavation have been developed in Arizona and now are in use elsewhere (see Chapter 4). The copper is removed from sediments and rocks by solution techniques, so that heating and melting of rocks to separate precious metals are not required. The result is a substantial (or nearly complete) reduction in the production of fumes, noxious gases, and ash that once was typical of the exhaust material belched from smokestacks at mine sites.

Natural Disasters, Hazards, and Risks

In contrast to environmental issues associated with population and pollution are geologic hazards that exist regardless of human activity but can put people in harm's way as a result of human activity. For example, people who live near a volcano put their lives and property at risk in the event of a volcanic eruption. Although a volcanic eruption could occur regardless of whether or not people live nearby, the fact that they do live in the vicinity leads to a risk and potential for disaster.

A **geologic hazard** is a natural phenomenon, process, or event with the potential for negative effects on human life, property, or the environment. Examples include floods, tornados, hurricanes, volcanic eruptions, earthquakes, and landslides. Poorly consolidated sediment on a steep slope, for example, could pose a hazard to those living in a development on that slope. If slope failure, known as a landslide, occurred during a heavy rainfall event, destroying homes and killing residents, it would be a disaster.

Natural disasters typically are sudden environmental changes that happen as a result of longer-term geologic processes and have significant impacts on human life and property. The geologic hazard of earthquakes, for example, is the result of hundreds or thousands of years of accumulation of geologic deformation, but the breaking point that causes an earthquake occurs relatively suddenly.

Natural disasters can pollute the environment just as human-made disasters, such as oil spills and air pollution from factory smoke, do. Carbon dioxide gas seeping from a volcano in Cameroon, Africa, for example, accumulated at the bottom of a lake in a crater atop the volcano and suddenly belched from the lake in 1981. The geologic event suffocated and killed thousands of sleeping villagers within minutes.

Volcanic eruptions and earthquakes, geologic hazards caused by the interior processes that move Earth's tectonic plates, occur with little impact from human activities; an exception is wells that inject wastewater underground, which have been documented to cause relatively small earthquakes. Floods and landslides, in contrast, are geologic hazards that can be worsened markedly by human activities. Removing natural vegetation from hillslopes, for example, can increase the runoff of rainwater and thereby increase flooding in streams. Removing vegetation also destroys roots, which anchor the soil on the slopes. Landslides occur when loosened soil and other debris slips suddenly downhill.

Risk refers to the magnitude of potential death, injury, or loss of property due to a particular hazard. The risk of death in the United States each year by an earthquake or volcano (less than 0.1 death per million people) is far less than the risk of death by automobile accidents (120 deaths per million people) or fires (11 deaths per million people). Different regions have different risks; in California, the risk of death by earthquakes and volcanoes is greater than it is in Ohio or Kansas. In fact, much of the United States west of the Rocky Mountains is considered to be a high-risk area for earthquakes, and all active volcanoes are in western states. Nevertheless, some parts of the central and eastern United States have been devastated by large earthquakes in the past, and in some ways are even more vulnerable to damage than western areas (see Chapter 3).

The Anthropocene and Planetary Boundaries

As the human population and individual standards of living have risen, our species has become the dominant influence on many Earth systems. From increases in greenhouse gases that warm the atmosphere to changes in biogeochemical cycles or wholesale reorganizations of landscapes and water supplies, our species is leaving its mark on the surface of planet Earth. Some Earth scientists, in fact, have suggested that changes from human activities are so substantial and widespread that a new geologic era could be added to the geologic time scale, one called the Anthropocene.[10]

Decisions made now will determine whether future generations inherit a degraded planet quite different from the one we live on presently, or a planet capable of providing for human and ecosystem needs over the long term. In 2009, sustainability scientist Johan Rockström and his colleagues published a paper entitled "Planetary Boundaries: Exploring the Safe Operating Space for Humanity."[11] In it, they identified Earth system processes critical to human well-being and assessed how those processes have been altered by human hands. They noted that the past 11,000 years, a warm period called the Holocene epoch that postdates a major glacial advance and cold period, has been a time of relative stability. During that time people developed agriculture and irrigation, built permanent settlements that became cities, and made phenomenal technological advances that recently included exploring outer space and sending a remotely controlled rover to Mars.

The scientists cautioned, however, that this long-term stability might be transitory and that human activity might be capable of tipping Earth into entirely new environmental states that could be disruptive to society. They sought to identify thresholds for critical environmental processes that might lead to such *tipping points*. Operating within a safe space for humanity, within the limits of what the scientists called **planetary boundaries,** requires staying within those thresholds and not reaching those tipping points.

Rockström and his colleagues developed a list of nine factors for which exponential growth of the human population and its associated consumption of resources and generation of wastes could exceed these limits and threaten planetary stability (Figure 1-26). These nine boundaries range from local to global in their scale of process, and the threshold behavior of each ranges from slow, with no evidence of planetary scale behavior, to sharp and global in scale. The boundaries include the global cycling of major nutrients critical to life, climate, and other Earth systems (i.e., the biogeochemical cycles of nitrogen, phosphorus, and carbon); the global cycling of water; changes in land use; the loss of biodiversity that is critical to self-regulating the biosphere's resilience; aerosol loading to the atmosphere; and chemical pollution.

While the absolute values of planetary thresholds are uncertain, the idea that Earth systems are interconnected and that humans are having a marked impact on these systems is clear. Along with other topics, we address the planetary boundaries identified by Rockström and his colleagues in the chapters of this book. In carrying out our analysis we make use of the deep time perspective of Earth science, which provides us with myriad examples of former planetary states quite different from the present. In the last chapter of this book, we revisit the ideas of planetary boundaries and humanity's safe operating space.

■ As Earth entered a period of relative global warmth about 10,000 years ago, people began to develop agriculture (the Agricultural Revolution) and, since then, they have altered dramatically the composition of ecosystems on Earth. The environmental ramifications of agriculture have been substantial.

■ Since the Agricultural Revolution, the rate of human population growth has been rapid, accelerating in the years since the Industrial Revolution. Most of the recent growth is attributed to a decrease in death rates because of better sanitation and medical care.

■ With industrialization has come a marked increase in consumption of energy, in particular the use of nonrenewable sources of energy (fossil fuels).

■ Decreasing environmental degradation would require that population growth stabilize; that we

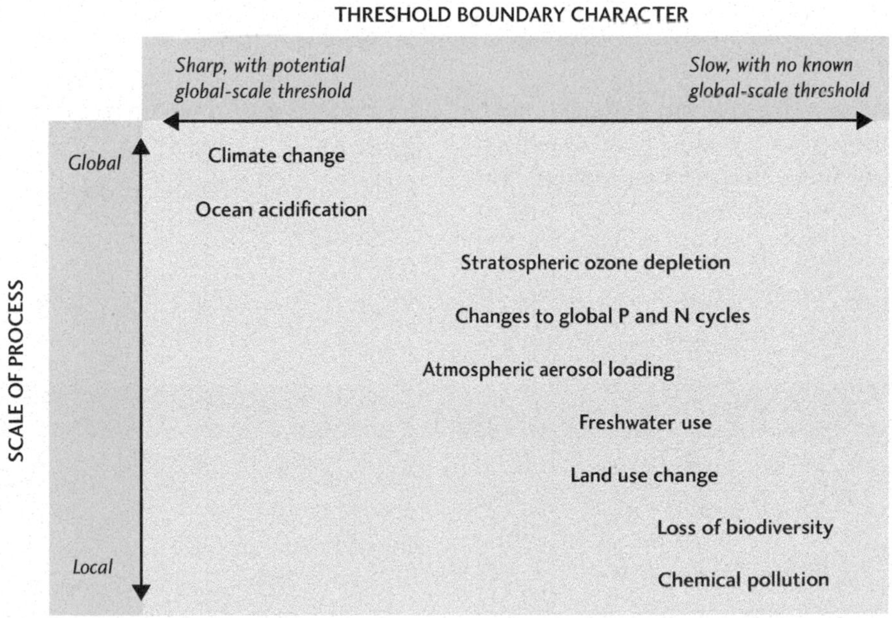

THRESHOLD BOUNDARY CHARACTER

Sharp, with potential global-scale threshold

Slow, with no known global-scale threshold

SCALE OF PROCESS

Global

Climate change

Ocean acidification

Stratospheric ozone depletion

Changes to global P and N cycles

Atmospheric aerosol loading

Freshwater use

Land use change

Loss of biodiversity

Chemical pollution

Local

Figure 1–26 Scientists have identified nine planetary boundaries that range in scale of process from local to global (bottom to top), and in scale of threshold for potential change from slow with no known global–scale threshold (right) to possibly sharp with a global–scale threshold (left).

develop perpetual sources of energy, such as solar energy; and that we decrease the use of nonrenewable resources that require extensive mining and that produce pollutants.

■ As human population increases, so does the risk associated with natural geologic hazards such as flooding, landslides, earthquakes, and volcanism. Mitigating natural disasters requires increased effort from geoscientists—to understand natural phenomena and their causes, and from all of us—to plan wisely when selecting places to live and work.

CLOSING THOUGHTS

We began by discussing the possible causes of rapidly shrinking glaciers atop the highest mountain in Africa, Mount Kilimanjaro. This example shows how Earth scientists investigate interconnected Earth systems. Greenhouse gases emitted from activities that include burning fossil fuels for energy lead to global warming that, in turn, can cause glacial ice masses to shrink and change the fluxes and stocks of water in Earth's hydrosphere. We also examined the ways in which scientists investigate Earth processes, using a multitude of tools and an overarching view of Earth as a dynamic system. By systematically examining each of the parts and processes within our planet's various subsystems, Earth scientists are able to determine cause and effect and to predict future outcomes.

The scale of human impact on Earth is relatively small in comparison with that of geologic processes such as volcanism, continental drift, waxing and waning ice ages, or mountain building, yet humans are by no means an insignificant force. The signatures of human activities can be detected worldwide, from extinct species to wholesale changes in land cover or the composition of our atmosphere. Whether or not the ongoing anthropogenic changes in the cycling of matter and energy on Earth result in an environment in which we can live sustainably and within planetary boundary thresholds remains to be seen.

SUMMARY

■ Earth is best understood from an Earth system perspective because the activities of our planet are interrelated by the cycling of matter and energy through Earth's various systems.

■ The environment can be divided into six major systems: the geosphere (rock), pedosphere (weathered rock and soil), hydrosphere (water), atmosphere (air), biosphere (life), and energy system.

■ Cycling of matter and changes in the Earth system are the result of processes that are driven by energy—solar energy, the energy from radioactive decay of minerals in Earth, and gravitational energy.

■ Geologic processes, such as plate tectonic movements or volcanism, can affect multiple Earth systems over a period of time, each change resulting in another.

■ Earth systems are characterized by both reinforcing and balancing feedback processes when they respond to changes in conditions.

KEY TERMS

Earth system science (p. 5)
geosphere (p. 5)
pedosphere (p. 5)
hydrosphere (p. 5)

biosphere (p. 5)
atmosphere (p. 5)
sustainability (p. 6)
reservoir (p. 6)
stock (p. 6)

fluxes (p. 6)
open system (p. 6)
closed systems (p. 6)
isolated systems (p. 6)

dynamic system (p. 7)
static system (p. 7)
energy (p. 7)
steady state (p. 7)
linear growth (p. 7)

exponential growth
   (p. 7)
reinforcing feedback
   (p. 8)
balancing feedback
   (p. 8)
silicate minerals
   (p. 9)
core (p. 10)
crust (p. 10)
mantle (p. 10)
lithosphere (p. 11)
asthenosphere (p. 11)

lower mantle (p. 11)
mantle mesosphere
   (p. 11)
outer core (p. 11)
inner core (p. 11)
rock cycle (p. 12)
igneous rock (p. 12)
sedimentary rock (p. 12)
metamorphic rock
   (p. 12)
soil (p. 13)
critical zone (p. 13)
hydrologic cycle (p. 15)

kinetic energy (KE)
   (p. 18)
potential energy (PE)
   (p. 18)
thermal energy (p. 19)
insolation (p. 20)
resource (p. 24)
potentially renewable
   resources (p. 24)
nonrenewable resources
   (p. 25)
perpetual resources
   (p. 25)

pollution (p. 26)
overpopulation
   (p. 26)
overconsumption
   (p. 26)
pollution per unit of
   resource used (p. 27)
geologic hazard (p. 28)
natural disasters (p. 28)
risk (p. 28)
planetary boundaries
   (p. 29)

REVIEW QUESTIONS

1. Explain how scientists do research aimed at determining the cause(s) of shrinking glaciers on mountains such as Mount Kilimanjaro.

2. Why are Earth's systems considered to be open? Why is the whole Earth system considered to be closed?

3. Is the amount of annual solar radiation (energy) reaching Earth's surface an example of a flux or a stock?

4. How and when did the geosphere, hydrosphere, pedosphere, atmosphere, and biosphere form on Earth?

5. How are the lithosphere and crust different from each other?

6. What are the major processes that operate in the lithosphere and rock cycle?

7. What are the major processes that operate in the hydrosphere?

8. What are the two most abundant elements in Earth's modern atmosphere?

9. What are the three main sources of energy on Earth?

THOUGHT QUESTIONS

1. What do you predict would happen if the flux of rock from Earth's mantle were to increase? For example, there is some evidence that periods of intensified volcanic activity have occurred during Earth's history. How would this affect the rock cycle?

2. Development of an area that was tropical forest results in removal of all vegetation and exposure of bare soil. Rainfall runs off the devegetated surface and increases the flow of water in streams. Increased runoff also has greater power to remove soil, thus causing erosion. As the amount of soil decreases, the amount of infiltration of rainfall into the soil is lessened, resulting in even more runoff, and consequently even more soil erosion. Is the preceding an example of a reinforcing or a balancing feedback? Explain your answer.

3. Why would scientists try to model Earth systems? What would be the benefits and limitations?

4. Why can't plants (and hence food or trees) be grown easily in the pedosphere if the upper meter or so of soil is eroded away?

5. Can the oceans ever contain more water than they do at present, and hence result in a rise in sea level? Can they ever contain less water than at present, and hence result in a drop in sea level? Has either of these types of change occurred in Earth's history?

SUGGESTED READINGS

Allaby, M. *The Encyclopedia of Earth: A Complete Visual Guide.* Berkeley: University of California Press, 2008.

Berner, E. K., and R. A. Berner. *Global Environment: Water, Air, and Geochemical Cycles,* 2nd ed. Princeton: Princeton University Press, 2012.

Langmuir, C. H., and W. S. Broecker. *How to Build a Habitable Planet: The Story of Earth from the Big Bang to Humankind.* Princeton: Princeton University Press, 2012.

Hazen, R. M. *The Story of Earth: The First 4.5 Billion Years, from Stardust to Living Planet.* New York: Viking, 2009.

Houtman, A., S. Karr, and J. Interlandl. *Scientific American Environmental Science for a Changing World.* New York: W. H. Freeman, 2013.

Luhr, J. F., ed. *Earth: The Definitive Visual Guide.* New York: Dorling Kindersley, 2007.

McPhee, J. *The Control of Nature.* New York: Farrar, Straus and Giroux, 1989.

McPhee, J. *Annals of the Former World.* New York: Farrar, Straus and Giroux, 2000.

Smil, V. *Cycles of Life: Civilization and the Biosphere.* New York: W. H. Freeman, 1997.

Smil, V. *The Earth's Biosphere: Evolution, Dynamics, and Change.* Cambridge, MA: The MIT Press, 2003.

Zalasiewicz, J. *The Planet in a Pebble: A Journey into Earth's Deep History.* New York: Oxford University Press, 2012.

Solid Earth Systems and Geologic Time

PART II

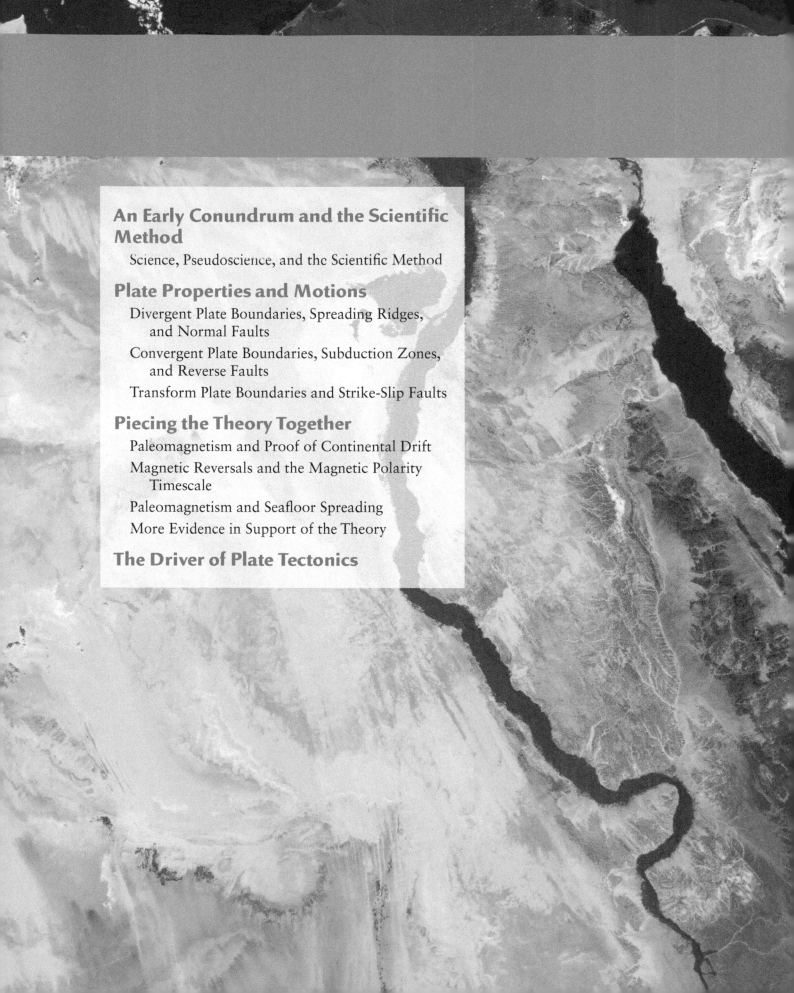

Plate Tectonics: Solid Earth in Motion

Earth is a dynamic system in which matter and energy are transferred from one reservoir to another along cyclic paths. The satellite image to the left, showing the Sinai Peninsula breaking away from northern Africa, reveals part of the plate tectonic cycle. Continental crust is splitting along a rift that deepens and becomes flooded with seawater, forming the Red Sea and its two northern fingers, the Gulf of Suez, extending northwest, and the Gulf of Aqaba, extending northeast. Magma rising from the mantle seeps upward along the rift zone and cools to form new oceanic crust. If rifting continues, an ocean basin will grow between the African and Arabian landmasses.

Earth is not expanding, so an equivalent amount of crust to that formed in the Red Sea must be destroyed somewhere else. At subduction zones, one lithospheric plate dives beneath another, incorporating crustal rocks into the asthenosphere and maintaining a constant-size Earth. Convection of heat in the mantle is one process thought to drive these plate motions, but the details of how this happens are still under investigation by scientists. The theory of plate tectonics, the overarching scientific paradigm governing our understanding of how the solid Earth works, dates only to the 1960s, and much remains to be learned.

In this chapter, we:

✔ Focus on the role of plate motions in altering Earth's surface with time.

✔ Explain how plate motions lead to geologic hazards such as earthquakes and volcanic eruptions.

✔ Explore the technological developments and cognitive leaps that promoted development of the theory of plate tectonics, and use this example to explore the practice of science more broadly.

The Sinai Peninsula is surrounded by the Mediterranean Sea to the north, the Red Sea to the south, Egypt to the west, and Saudi Arabia to the east. The Red Sea formed as the Arabian Plate split off from the African Plate and the Indian Ocean flowed in to fill the growing rift basin. The peninsula was created as rifting opened the Gulf of Suez northwest from the Red Sea, and the Gulf of Aqaba northeast. The Nile River flows north into a densely vegetated delta, emptying into the Mediterranean Sea. *(Jacques Descloitres, MODIS Land Science Team/NASA)*

INTRODUCTION

Well before satellite images such as the one that opens this chapter were possible, map views inspired speculation about what the shapes of landmasses might mean. Discovery and exploration of the Americas in the 16th and 17th centuries required new maps, and Renaissance mapmakers who were drawing in the eastern margin of the New World noticed its uncanny similarity to the western edge of the Old World, as though the two once had been joined together. Conventional wisdom, however, held that the continents were permanently fixed in place. In 1912, Austrian meteorologist Alfred Wegener (1880–1930) dared to challenge convention by proposing that the continents had been joined in the past into a supercontinent he called Pangaea, had since separated, and continued to drift apart (**Figure 2-1**).[1]

Geologists as recently as the 1960s scoffed at the idea of **continental drift**,[2] but a few were persuaded by the lines of evidence Wegener had amassed, particularly the near-perfect match of rocks and fossils of similar ages and types on opposite sides of the Atlantic Ocean. No one, however, could explain how thick, rocky continents might plow through solid ocean basins to drift about the world, and consequently Wegener's hypothesis had few supporters. He perished in 1930 on an expedition to the Greenland ice sheet, his ideas on continental drift regarded as trifling by much of the geological community.

Decades went by before Wegener was recognized as a genius ahead of his time. In 1925, the German submarine *Meteor* came upon a submerged north–south trending mountain range in the middle of the Atlantic Ocean[3] and in 1947, an American scientist, Maurice Ewing, discovered that the center of this range contains a deep, narrow rift from which lava emerges.[4] Harry Hess, another American scientist, suggested that the lava flowing out at such **mid-ocean ridges** creates new crust, which displaces the existing ocean crust on either side in a process that came to be called **seafloor spreading**.[5] As the opposing blocks of oceanic crust move apart, the continents embedded in them also move. Subsequently, long, deep trenches were discovered elsewhere in the oceans, most of them in the Pacific, and identified as places where opposing blocks of oceanic and continental crust collide and one dives beneath the other. This process of **subduction** carries crust back down into the mantle. By the end of the 1960s, most geoscientists accepted that Wegener had been right. Continents do indeed "drift": they are the above-water surfaces of lithospheric plates that move away from one another at spreading centers like the Mid-Atlantic Ridge and collide with one another at subduction zones.

This new view of the way Earth works is the theory of **plate tectonics**, which holds that the lithosphere is

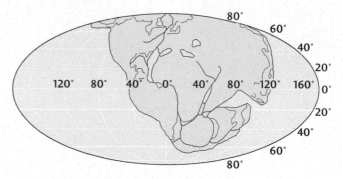

(a) Wegener's reconstruction of Pangaea

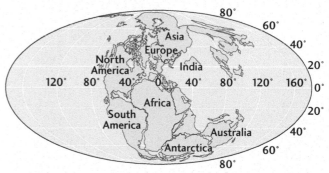

(b) Modern reconstruction of Pangaea

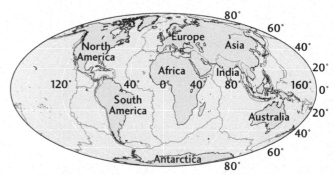

(c) Earth today

Figure 2-1 (a) In 1912, Alfred Wegener proposed that Earth's continents were once joined in a single supercontinent that he called Pangaea. As evidence, he cited matching coastlines and the existence of similar ancient rock assemblages and fossil types on modern continents on opposite sides of the Atlantic. (b) Advances in radiometric dating of rocks and evidence of plate tectonics have enabled scientists to determine that Wegener was correct. (c) About 200 million years ago, Pangaea split apart along the Mid–Atlantic Ridge, which extends the length of the Atlantic Ocean.

broken into plates that move horizontally above the ductile asthenosphere. Plate tectonics provides a consistent way to explain rock types and their origins, the making of mountains, changes in global climate, and the distribution of organisms (**Box 2-1**). Nevertheless, it was not until 1979 that the actual motion of the plates began to be measured directly, providing a test of one of the fundamental predictions of the theory.[6]

BOX 2-1 GLOBAL AND ENVIRONMENTAL CHANGE

Plate Tectonics and the Evolution of Species

Drifting continents have dramatically influenced the evolution of animal and plant species on Earth. Perhaps the most striking example of this influence can be seen in the fauna of Australia. About 200 million years ago, from the late Paleozoic era to the early Mesozoic, the continents of Australia, Antarctica, Africa, and North and South America were assembled into the supercontinent Pangaea (**Figure 1**). About 100 million years ago,

(continued)

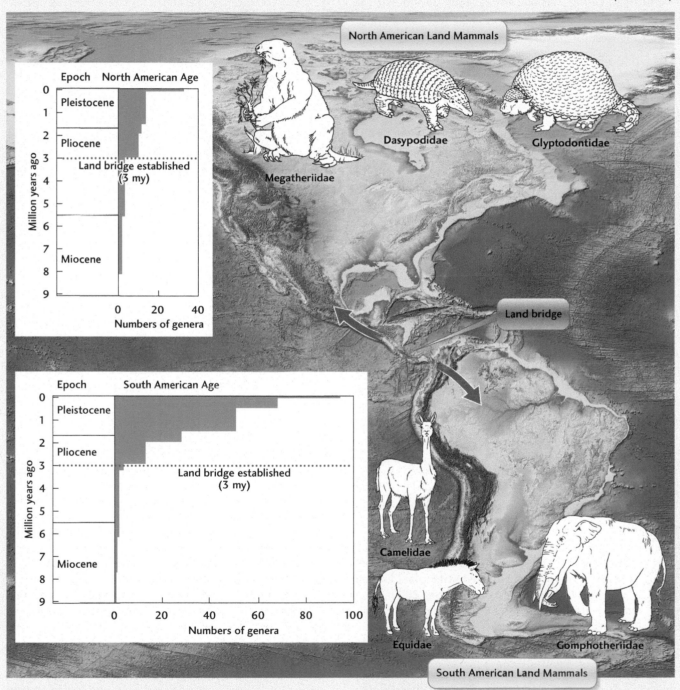

Figure 1 When a land bridge, the Isthmus of Panama, was uplifted between North and South America, many animals migrated south, including llamas, elephants, bears, deer, dogs, and rabbits. The smaller number of species that migrated north included armadillos, sloths, porcupines, and opossums. For the first time, marsupial mammals in South America came in contact with placental mammals. As a result of this "Great American Interchange," the number of genera (plural of genus) increased substantially in South America, and less so in North America, after the land bridge was established.

(continued)

marsupials evolved in what is now North America or possibly Asia, then diversified as they spread throughout the ancient continent now referred to as Pangaea. Marsupials are pouched mammals, such as kangaroos, that give birth to young that are not yet fully developed. The infants crawl up the mother's abdomen and into a pouch in which they nurse and grow. In contrast, placental mammals lack a pouch and give birth to more fully developed young.

About 70 million years ago, after Pangaea had become fragmented, placental mammals evolved on the remains of the supercontinent in the northern hemisphere. Placentals competed with marsupials for food and rapidly drove the marsupials to extinction in what became the continents of North America, Europe, and Asia. Meanwhile, marsupials thrived in South America and Australia, where there were no placentals. When South and North America became joined again approximately 3 million to 4 million years ago, placental mammals migrated south and drove to extinction all the marsupials except opossums (see Figure 1).[7] Australia, in contrast, remained isolated from the other continents and their placental mammals throughout the Cenozoic era, enabling more than 180 different marsupial species to evolve.

In recent history, humans introduced placental mammals to Australia, with adverse effects on native wildlife. From 12 breeding pairs of rabbits introduced in the mid-1800s as a game animal, the rabbit population skyrocketed to more than 20 million individuals in only 6 years. The rabbits lack any natural predators in Australia and are defoliating vast areas of grazing land vital to kangaroos and wallabies.

Beginning that year, scientists at the National Aeronautics and Space Administration (NASA) used signals sent from satellites to monitor the positions of different parts of Earth's surface. Just as Wegener had argued 75 years earlier, Earth's surface is in motion, with plates moving away from, toward, and alongside one another at rates of several centimeters per year.

AN EARLY CONUNDRUM AND THE SCIENTIFIC METHOD

Throughout the 19th century, geologists discovered that fossils of various terrestrial plants and animals could be found on continents separated by thousands of kilometers of ocean. For example, fossil leaves of a plant known as *Glossopteris,* which became extinct about 200 million years ago, are found today in South America, Africa, India, Antarctica, and Australia (Figure 2-2).[8] Likewise, fossils of the Triassic period (248 to 206 million years ago) therapsid *Cynognathus* exist in South America and Africa, and the Triassic therapsid *Lystrosaurus* is found in Africa, India, and Antarctica. Because these plants and animals lived on land, the distribution of their fossils posed a problem for early geologists, who considered it highly unlikely that terrestrial animals could have swum across thousands of kilometers of ocean to reach a new home or that plant seeds could have been dispersed such a great distance.

Prior to Alfred Wegener's ideas on continental drift, Austrian geologist Eduard Suess developed a hypothesis that explained the perplexing fossil distribution. Suess envisioned that a single crust originally covered the entire surface of Earth. He also proposed that Earth's interior shrank over time due to planetary cooling and that the crust wrinkled in response, like the skin of a piece of fruit left to dry on a kitchen counter.[9] In this way, Suess was able to explain the distribution of the fossils as well as the creation of ocean basins: the animals and plants dispersed themselves across contiguous areas of the crust and only became separated as parts subsided to form the oceans. While Suess's idea may strike us as rather quaint, in the absence of other evidence, it might seem quite reasonable. In either case, it presented a testable hypothesis, which is that the crust that underlies the ocean basins is identical to that found on the continents.

Scientists soon poked holes in Suess's "contraction theory," finding that in fact the oceans could not have formed from the same material as the continents. Surveys of Earth's gravitational field conducted in the Himalayas and the Americas showed that continental crust floated on top of a denser layer beneath.[10] If continental crust in fact floated, there was no way it could have subsided to form the oceans. Later, geologists collected the first samples of oceanic crust, thereby confirming what the gravity measurements had indicated, that oceanic crust is quite different from continental crust, but Suess's hypothesis had already been rejected as new ideas came to the fore to explain observations of natural phenomena. Steady application of the scientific method by several generations of scientists led to our present understanding of Earth's architecture and internal processes and also points toward how we will address remaining mysteries.

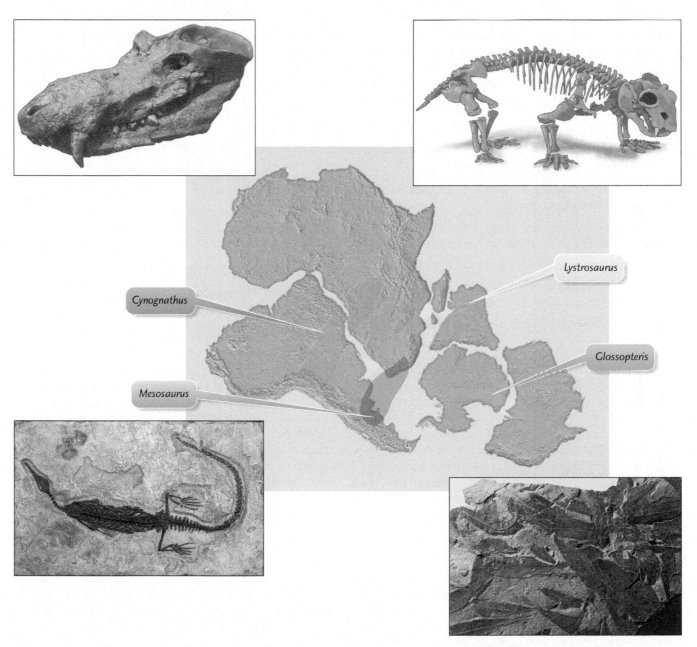

Figure 2-2 The distribution of 250- to 300-million-year-old fossils of animals from the order *Therapsida,* early reptile-like ancestors of mammals, in parts of South America, Africa, India, and Antarctica provide evidence that these continents once were connected. Such creatures include members of the genera *Lystrosaurus* ("shovel lizard") and *Cynognathus* ("dog jaw"). Other fossils with distributions indicative of a once-joined Pangaea landmass include the freshwater reptile *Mesosaurus* that is found only in South America and Africa, but could not have crossed the ocean now between the continents. Many fossils of plants, including the extinct seed fern *Glossopteris,* supplement the fossil evidence that the southern continents once were joined. *(bottom left: Courtesy of PrehistoricPlanet.com, E. Ray Garton, Curator; bottom right: ©Lloyd Homer, GNS Science)*

Science, Pseudoscience, and the Scientific Method

Science is both a body of knowledge and a logical means of understanding the universe. The word is derived from the Latin *scientia,* which means knowledge, but science is much more than a fixed body of facts and information. It is also a dynamic process of intellectual activity driven by human curiosity and historical events. The scientific approach to understanding our environment is based upon the assumption that natural phenomena have physical, and therefore ultimately knowable, causes. Scientific knowledge is acquired through the systematic and disciplined testing of ideas by careful experimental design, methodical data collection, and logical analysis, processes that together constitute the **scientific method.** The particular procedure of investigation may vary, but scientists always:

■ Observe phenomena (and review relevant scientific work on similar phenomena).

■ Develop a **hypothesis,** an explanation of the observations that is based on physical principles.

■ Predict the likely outcomes of the hypothesis.

■ Design an experiment or field investigation to test predictions based on the hypothesis.

■ Compare the real and predicted outcomes of the tests.

■ Accept, reject, or modify the hypothesis.

If testing results in the acceptance of a hypothesis, scientists will continue to test the idea under different conditions. A hypothesis that has survived repeated testing and is able to accurately explain a wide variety of phenomena is viewed as a **theory.** If such a theory explains many types of phenomena and has survived all challenges, it is considered a **universal law** or a **unifying theory.** The universal law of gravitation helps to explain why rocks fall and planets orbit their suns. The unifying theory of plate tectonics explains why volcanoes erupt and ocean basins are formed and destroyed.

Science has been described as "incomparably the most successful" way of learning about the universe, because its means of reasoning are testable, and it has been able to predict many events. The same cannot be said for nonscientific methods. An astrologer can try to predict your future based on the pattern of stars and planets in the sky at the time of your birth, but there is no observable connection between the stars and a person's life, and the predictions rarely come true. Throughout the world, dowsers, also called "water witches," are hired to help locate underground water supplies with magical sticks or rods that supposedly twitch because of supernatural or otherwise unknown forces when the dowser stands above buried water. Again, there is no observable connection between the underground water and the twitching of the sticks, and studies have shown that dowsing is no more successful at locating water than random chance.

Astrology, dowsing, and similar ways of interpreting the universe are called *pseudoscience* (false science) because they are often presented as science when in fact they are not. The way to identify pseudosciences is to recognize the ways in which they contradict scientific thinking:

■ Their fundamental assumption is that natural events can be controlled by supernatural forces.

■ Their findings are biased rather than objective; they are accepted when they support a particular idea but ignored or rejected when they do not.

■ Their ideas cannot be tested.

■ Their results cannot always be duplicated.

■ Their predictions are correct no more often than one would expect from random chance.

PLATE PROPERTIES AND MOTIONS

Over the last half-century, careful application of the scientific method has enabled us to learn that the geosphere consists of a solid inner core surrounded by a liquid outer core, which in turn is overlain by a solid rocky mantle and crust made primarily of silicate minerals. The lithosphere consists of the crust plus the uppermost part of the mantle and is broken into more than a dozen large plates that move above the asthenosphere, a part of the mantle capable of slow flow. While the chemical composition and mineral content of continental crust had been studied by geologists for more than 100 years, it was not until the 1950s that samples could be repeatedly collected from the seafloor. These samples revealed that continental and oceanic crust are quite distinct chemically. While both forms of crust are made of silicate minerals, oceanic crust is rich in iron and magnesium, whereas continental crust contains abundant potassium, calcium, sodium, and aluminum. These chemical differences cause density variations, with continental rocks averaging about 2.7 g/cm^3, while oceanic rocks average about 2.9 g/cm^3. Differences in thickness also characterize the two kinds of crust; continental crust measures 30–80 km thick, while oceanic crust is only 5–15 km thick.[11] Individual lithospheric plates may contain a mixture of oceanic and continental crust or contain only oceanic material. The composition of plates plays a role in their behavior, affecting their motions across Earth's surface.

Divergent Plate Boundaries, Spreading Ridges, and Normal Faults

Divergent boundaries are locations where two plates move away from one another. They were first recognized on the seafloor as long chains of volcanic mountains encircling Earth in an arc pattern, similar to the seams on baseballs (Figure 2-3). Because of their locations and topography, these undersea mountain chains are called *mid-ocean ridges*, and were first mapped by geophysicists Marie Tharp and Bruce Heezen. The discovery of high mountains on what was thought to be a flat ocean floor was made just after World War I, when submarines such as the German *Meteor* began roaming the world's ocean basins. Since then, numerous studies—including those that use submersible vehicles to visit the ridges first-hand—have observed long linear depressions, or *rifts*, along the centers of mid-ocean ridges, and have determined that lava erupts from them. These rifts mark the growth of new lithosphere, where new material is added as plates spread away from one another.

At mid-ocean ridges, where hot, ductile rocks from the mantle move slowly upward, the boundary between the asthenosphere and the lithosphere rises toward the surface (Figure 2-4). As warm rocks move upward, the reduced pressure of overlying rocks, combined with seawater seeping down into the rocks through cracks, lowers the temperature at which the material can melt and produce magma (molten rock). As magma seeps out, it encounters cold ocean water and solidifies, creating new oceanic crust on either side of the rift. New crust and lithosphere are created at rates of a few millimeters to as much as 15 cm per year, depending on location.[12]

While most divergent boundaries are hidden beneath the oceans, a few appear above sea level. The nation of Iceland (Figure 2-5a), straddles the Mid-Atlantic Ridge, a divergent boundary that runs north to south along the length of the Atlantic Ocean and that separates the North and South American plates to the west from the Eurasian and African plates to the east (see Figure 2-3). Volcanic eruptions occasionally wreak havoc on Iceland's population (consider, for example, the aviation chaos caused by the eruption of the Eyjafjallajökull volcano in 2010), but the people also benefit from free geothermal energy and a fascinating geyser-filled landscape that entices tourists.

Another young divergent boundary is developing in eastern Africa, rifting the continent apart to form a new ocean basin. This new plate boundary is exposed on land,

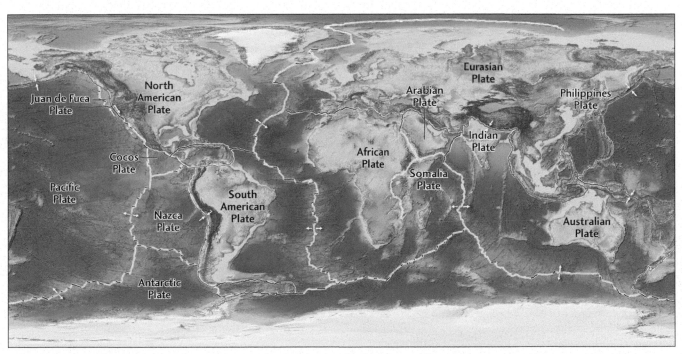

— Oceanic ridge or divergent plate boundary   ⤙ Subduction zone or convergent plate boundary   — Transform fault

Figure 2-3 Map of the mid–ocean ridge system and plates. A mosaic of about a dozen major and some smaller tectonic plates move on Earth's surface at rates up to several centimeters or more per year. The arrows illustrate the relative directions of movement between plates at a given location. Facing arrows indicate places where plates are converging. Diverging arrows indicate places where plates are separating. The Nazca Plate is converging with the South American Plate at about 10 cm per year, while the Indian–Australian and African plates are diverging at about 3 cm per year. Arrows parallel to a plate boundary, as along the San Andreas fault in western North America, indicate plates sliding past each other.

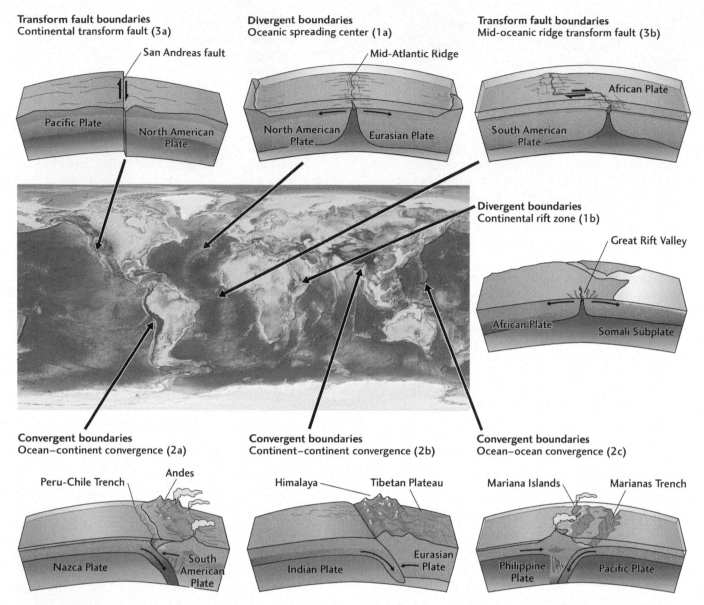

Figure 2-4 Three types of plate boundaries. There are three main types of plate boundaries. 1) **Divergent boundaries** are where plates are moving apart. They include **oceanic spreading centers** (1a), where rifting and spreading along a mid–ocean ridge create new oceanic lithosphere, and **continental rift zones** (1b), where rifting and spreading on continents are characterized by parallel rift valleys, volcanism, and earthquakes. 2) **Convergent boundaries** are where plates are moving toward each other. They include **ocean–continent convergence** (2a), where oceanic lithosphere meets continental lithosphere, the oceanic lithosphere is subducted, and a volcanic mountain belt is formed at the continental margin; **continent–continent convergence** (2b), where two continents converge and the crust crumples and thickens creating high mountains and a wide plateau; and **ocean–ocean convergence** (2c), where oceanic lithosphere meets oceanic lithosphere, one plate is subducted under the other, and a deep sea trench and a volcanic island arc are formed. 3) **Transform fault boundaries** are where plates slip past each other. They include **continental transform faults** (3a), where continental plates slip horizontally past each other, and **mid–oceanic ridge transform faults** (3b), where mid–ocean ridges are offset by transform faults.

forming the East African Rift and well-known volcanoes such as Kilimanjaro (Figure 2-6). The northern extension of this rift continues beneath the Red Sea, as shown in the satellite view at the beginning of this chapter.

Along all active spreading ridges, the lithosphere stretches and thins to the point of breaking, causing

some lithospheric blocks to slide downward relative to others along fractures known as **normal faults** (Figure 2-5b). The term normal fault comes from mining terminology. Hot fluids, rich in elements such as gold and silver, often move upward along fault planes, leaving mineral deposits in the rock cracks. Miners digging

(a)

Figure 2-5 (a) Iceland, on the Mid–Atlantic Ridge, is one of the few places on Earth where a divergent boundary can be seen above sea level. (b) Divergent plate boundaries are dominated by normal faults. (c) Rift valleys, such as the one exposed in Iceland, are created when a block of lithosphere drops down between two normal faults. *(a: John Warburton–Lee/Getty Images; c: Andrew De Wet)*

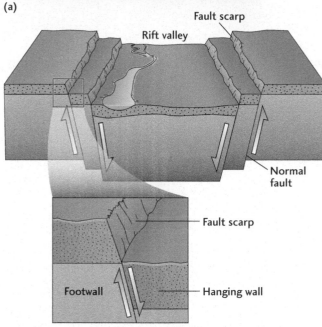

(b)

(c)

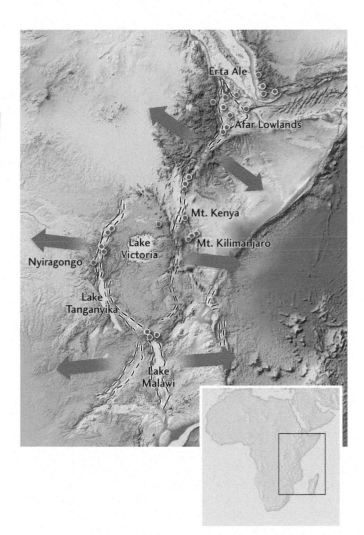

Figure 2-6 Lakes and volcanic mountains in the East African Rift, home of early hominids. A series of lakes, such as the famous Tanganyika and Malawi, lie within the sagging crust of the East African Rift basin. Ashes from the volcanoes periodically blanket the landscape. The basin is composed of a thick stack of interbedded volcanic and lacustrine (lake) sediments that have been invaluable in helping to reconstruct early hominid history.

along these planes called the block of rock above their heads the *hanging wall*, since they were able to hang lanterns from it. The block of rock beneath their feet came to be called the *foot wall*. In normal faults, the hanging wall moves downward relative to the foot wall, a motion that allows the crust to be extended.

Convergent Plate Boundaries, Subduction Zones, and Reverse Faults

New crustal rocks are forming all the time, spreading out from divergent plate boundaries at rates of millimeters to several centimeters per year, yet Earth's size remains the same. This can only mean that for Earth as a whole, **convergent boundaries,** where plates collide and lithosphere returns to the lower mantle, must be about as common as divergent ones.

Two types of interaction can result when plates collide at convergent boundaries. Subduction occurs when a denser plate collides with and sinks beneath a lower-density plate. When oceanic and continental lithospheres collide, the denser oceanic plate tends to be subducted under the continental plate. Density variations are due largely to chemical composition, as mentioned earlier, but temperature is also important. Lithosphere formed by recent volcanism generally is warmer and less dense than older rock. Therefore, when blocks of oceanic lithosphere collide, the plate bearing older, denser material will be subducted under the plate bearing newer, less dense material. In contrast, little or no subduction occurs when two bodies of continental lithosphere collide. Neither plate sinks because their densities are about equal. Instead, the edges of the two plates crumple and are pushed upward to form high mountains in a **continent–continent collision** (Figure 2-7). These behaviors explain why some continental crust dates back almost to the formation of the planet, while the oldest rocks on present-day ocean floors date back only 180 to 200 million years:[11] oceanic lithosphere is continuously recycled into the mantle.

Ongoing convergence of the Nazca and South American plates is an example of oceanic lithosphere subducting under continental lithosphere (see Figure 2-3). A long depression, or **ocean trench,** exists along the boundary between the two plates and has filled with sediment over time (see Figure 2-4). During subduction, oceanic lithosphere, overlying sediments, and seawater are dragged downward into the mantle and heated, eventually to the point of melting. Magma then rises, forming a chain of volcanoes, called a **volcanic arc,** which parallels the convergent boundary. Because the subducting plate is dipping as it sinks into the asthenosphere, the place where magma rises is inland of the subduction zone trench (see Figure 2-4). The Andes Mountains of South America and the Cascade Range of northwestern North America typify this type of convergent boundary. In

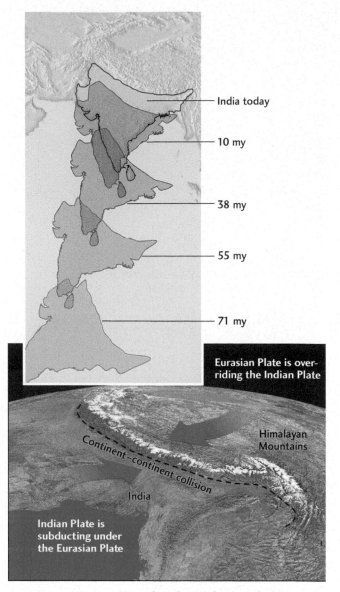

Figure 2-7 **Convergent plate boundary and mountain building.** The world's highest mountains are along the convergent margin of two continental plates, where the crust in each plate is too light to subduct. The Himalayan Mountains, along the borders of India, Pakistan, and China, are the roof of an 80–km–thick welt of folded and faulted rocks where the Indian and Eurasian plates continue to collide with each other. *(WorldSat International/Science Source)*

the Cascades, the Mount St. Helens volcano erupted catastrophically in 1980, and then was quiet before beginning another, much smaller, eruptive phase that lasted from 2004 to 2008. Things are presently quiet at the volcano, though by the time you read these words that may have changed. Plate tectonic processes are very dynamic!

The collision of the Pacific and Philippine plates is an example of older, denser oceanic lithosphere (the Pacific Plate) being subducted under a younger oceanic plate (the Philippine Plate) of lesser density (see Figure 2-3). The sediments and water dragged down with the subducting lithosphere lower its melting point, forming plumes of magma that erupt into a chain of volcanic islands, the Philippines, inland of the subduction zone. Mt. Pinatubo on the island of Luzon erupted in 1991, spewing ash and liquid aerosols into the atmosphere that blocked incoming sunlight, thereby dropping Earth's temperature by about 0.5°C over the next year.[13]

Like the Philippines, Indonesia and Japan are examples of volcanic **island arcs.** Indonesia's Tambora volcano generated the largest eruption on record in 1815, killing nearly 100,000 people and wiping the volcanic island off of the map as an estimated 30 to 33 cubic km of material was ejected from the volcano in a mere 24 hours.[14]

Continent–continent collisions are generally preceded by subduction, but when the subducting oceanic lithosphere runs out, volcanism ends, and a more gradual violence unfolds. Since both plates consist of continental rocks that are relatively light, neither is subducted. Instead, the plates remain on the surface and press against each other. This compressive force deforms large masses of rock and sediment along the edges of the plates and pushes them upward. The continental crust of the Indian and Eurasian plates has been colliding over the past 40 million years, forming the Himalayan mountain belt and raising the Tibetan Plateau to great heights (see Figures 2-3 and 2-7). Parts of the Himalayas are rising as much as a centimeter per year.[15]

Similar thick welts of deformed rock have produced great mountain belts at collision zones throughout geologic history. Between 300 and 400 million years ago, the collision of the North American, South American, Eurasian, and African plates to create Wegener's supercontinent Pangaea closed a precursor to the Atlantic Ocean, called the Iapetus Ocean, and formed the Appalachian Mountains of North America, the Scandinavian mountains of Norway and Sweden, and the highlands of Scotland. After several hundred million years of weathering and erosion, these ancient mountains still have as much as 1300 meters of relief (the difference in height between the highest and lowest points in a landscape). When Pangaea broke apart along a rift about 200 million years ago, the Atlantic Ocean was born, separating the Appalachians from their European and African counterparts (Figure 2-8).

Because the rocks tens of kilometers deep in collision zones are under conditions of high temperature and pressure, they often become intensely folded (Figure 2-9). At such high pressure and temperature, rock can behave in a ductile manner, able to flow like a liquid but remaining in the solid state. Oil and natural gas also form under conditions of high pressure and temperature. These light fluids tend to rise and accumulate under layers of impermeable rocks in the crests of folds, so pockets of these hydrocarbons are very common in folded rocks

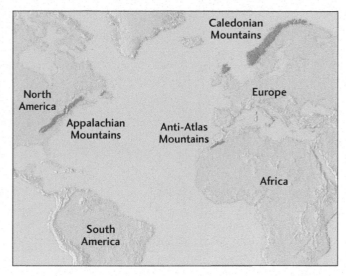

(a)

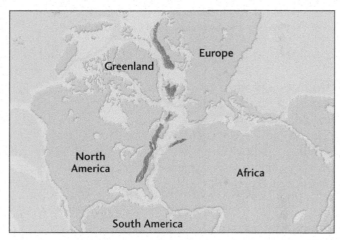

(b)

Figure 2-8 Mountain belts separated by seafloor spreading. Mountainous features equivalent to the Appalachians of North America are found in Morocco's Anti–Atlas Range in northwestern Africa. If the North American and African plates are realigned to their pre–split position, these mountains form a single chain.

in ancient convergent boundaries between continental plates. We discuss this topic in more detail in Chapter 13.

In addition to ductile deformation, brittle deformation (faulting) occurs at convergent boundaries, especially at shallow depths (less than 10 to 20 km) where rock is cool. The type of faulting generally associated with convergent plate margins is **reverse faulting.** As in normal faulting, an inclined plane of failure separates rocks on either side of the fault, but in reverse faulting, rocks in the hanging wall move upward relative to those in the foot wall (Figure 2-10). This allows rocks to stack upward upon themselves, thereby accommodating the compressional forces directed at them from the side. The angle of the failure plane can range

(b)

(a)

Figure 2-9 Rocks commonly exhibit evidence of ductile deformation along convergent plate boundaries, as in these folded sedimentary rocks in Alaska. Belts of folded rocks typically mark the locations of ancient convergent plate boundaries. They also commonly are associated with petroleum (crude oil and natural gas), which can be trapped beneath arched layers that are impermeable to the upward flow of petroleum through the rocks. *(TAO Images Limited/Alamy)*

from very shallow to very steep. In the case of angles less than about 30° from horizontal, reverse faults are called *thrust faults*.

Along a convergent margin, thrust and reverse faults at shallow levels in the continental crust typically merge at depth with a master thrust, or megathrust, that forms the actual plate boundary between the overriding and subducting plates. Some of the largest earthquakes known on Earth have occurred along great megathrusts at subduction zones. The most deadly of these in modern times occurred on December 26, 2004, when the Indian Plate slipped beneath the Sundaland Plate, generating

(a)

(b)

Figure 2-10 (a) In this view of a hillside in Northern California, the sharp break crossing the hillside is the trace of a reverse fault. The underlying crust on the upslope side of the fault has moved up relative to the crust on the downslope side. This fault has slipped several times in the past few thousand years. A nuclear reactor built over it was closed after the fault was discovered. (b) Convergent plate boundaries are characterized by reverse faults, along which rocks on the hanging wall (on the right) move upward relative to those on the footwall (on the left). *(a: Courtesy of Dorothy Merritts; b: Charles Jones)*

Subsidiary fault

Active fault

San Andreas fault

Linear valley   Offset drainage

Figure 2-11 The strike–slip fault in this view (looking east) is the San Andreas fault, a right–lateral fault that extends much of the length of California. The North American Plate is moving southward relative to the Pacific Plate. Note how streams have been offset laterally, with those on the North American Plate (background) moving to the right relative to their counterparts on the Pacific Plate (foreground). *(Sandra Schultz Burford/USGS)*

a tsunami that killed more than 225,000 people living along the margins of the Indian Ocean (we discuss tsunamis in more detail in Chapter 12). The violence of ocean–ocean subduction was also manifestly on display in March of 2011 when a massive earthquake struck Japan. In this event, the Pacific Plate slipped suddenly westward beneath the North American Plate, bringing Japan as much as 4 m (13 ft) closer to the United States and generating a tsunami that caused further destruction.[16] In all, more than 15,000 people were killed.[17]

Transform Plate Boundaries and Strike–Slip Faults

Along **transform plate boundaries,** plates are neither colliding nor moving away from one another. Instead, they slide past one another laterally. Since neither compression nor extension occurs, reverse and normal faults are typically absent. Instead, the type of faulting associated with a transform boundary is called **strike-slip faulting,** and it happens when rocks slip horizontally along a vertical or near-vertical fault plane (Figure 2-11). The orientation of the fault plane with respect to north is called its strike, and the slip of the blocks of rock on each side is parallel to the strike, hence the name strike-slip.

This type of faulting has interesting consequences. First, because the fault plane is vertical, it appears as a straight line where it intersects the ground surface, and it crosses hills and valleys with no change in direction. Second, where features such as streams, fences, and roads cross a strike-slip fault (see Figure 2-11), they are offset in a lateral direction; that is, they are displaced on either side of the fault. We assess the direction of offset

by looking across the fault to see where a disrupted feature has moved (Figure 2-12). If the feature has moved to the left relative to the position of the same feature on

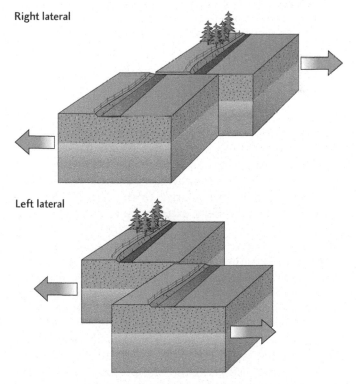

Right lateral

Left lateral

Figure 2-12 Right– and left–lateral faults. Strike–slip faults are referred to as right– or left–lateral in reference to the relative sense of motion of the crust on each side of the fault. If the crust on the block opposite that of the observer has moved to the right in a relative sense, the fault is right–lateral, and if to the left, then it is left–lateral.

Option 1: Ridge has been offset by a right lateral strike-slip fault

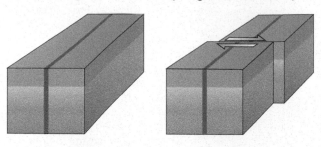

Option 2: "Transform fault" is part of a plate boundary, linking two actively spreading ridge segments

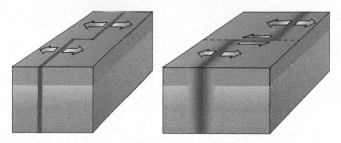

Figure 2-13 J. Tuzo Wilson's insights into the nature of oceanic transform faults. As oceanic crust moves away from spreading ridge segments, it also moves along transform faults that are roughly perpendicular to the ridges. The relative sense of offset of the ridges is opposite the actual sense of motion along the transform faults. Wilson is credited with recognizing that this apparent paradox is evidence of seafloor spreading.

the plate on which we stand, the fault is said to display *left-lateral* motion. A *right-lateral* fault displays the opposite sense of motion.

A well-known example of a transform boundary is the San Andreas fault system, where the Pacific Plate is moving northward relative to the North American Plate. The San Andreas extends more than 1500 km, from the Gulf of California in Mexico to Northern California (see Figure 2-3), and gives the region its famed earthquakes. Other transform faults on land include the North Anatolian fault in Turkey, the Alpine fault in New Zealand, and the Dead Sea transform that lies beneath the Middle East's Jordan River valley.

While a few transform plate boundaries occur on land, the majority lie beneath the waves, where they connect offset segments of the mid-ocean ridges. Geologist J. Tuzo Wilson was the first person to realize that these faults confirmed that ridges were sites of seafloor spreading.[10] Because oceanic crust moves away from ridge segments, the sense of motion (right-lateral versus left-lateral) along oceanic transform faults is opposite what it would appear to be if these underwater

mountains were simply static features offset by faults (Figure 2-13).

■ New lithosphere forms along rifts at mid-ocean spreading centers on divergent boundaries, and old lithosphere is destroyed at subduction zones along convergent boundaries. The amounts of lithosphere created and destroyed are equal, keeping the lithosphere system in a steady state and ensuring that Earth's size remains the same over time.

■ Plate boundaries are zones of active rock deformation, faulting, and folding. Faulting occurs at shallow depths (generally less than 10 to 30 km), where rocks are cool and brittle, whereas folding generally occurs at greater depths, where rocks are warmer and more ductile.

■ The three basic types of faults are normal, reverse, and strike-slip. Although each may occur at all types of plate boundaries, normal faults are dominant at divergent boundaries, where crust is stretched and pulled apart; reverse faults are dominant at convergent boundaries, where crust is compressed and shortened; and strike-slip faults are dominant at transform plate boundaries, where crustal blocks slip horizontally past each other.

PIECING THE THEORY TOGETHER

Our present understanding of how Earth works is the result of many years of careful scientific observation and analysis. In her edited book, *Plate Tectonics: An Insider's History of the Modern Theory of the Earth,* geologist and historian of science Naomi Oreskes reveals that progress also relied on seemingly unrelated historical events, on a willingness of scientists to take risks, and on the existence of an intellectual community in which scientists could try out radical new ideas.

The need to defeat Adolf Hitler's Germany during World War II led to a massive campaign to map the ocean floor and to study Earth's magnetic field. German U-boats (submarines) exacted heavy casualties on Allied forces, and the U.S. Navy realized it needed to know everything it could about its watery battlefield. The Navy also hoped to detect enemy forces by the magnetic fields associated with their ships and to hide its own vessels. This necessitated the development of the magnetometer, an instrument capable of measuring the strength of magnetic fields. Ships towing echo-sounding equipment and magnetometers crisscrossed the oceans for several years, making recordings of water depth and seafloor magnetism. These campaigns revealed that mid-ocean ridges ran through every ocean basin and that the strength of the magnetic field varied substantially with location.

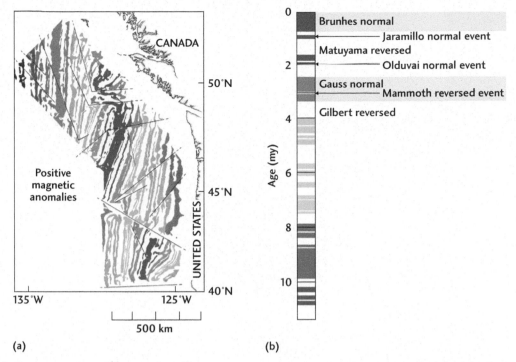

(a)

(b)

Figure 2-14 Paleomagnetic evidence of seafloor spreading. (a) As scientists gathered data from seafloor explorations, they constructed a magnetic polarity timescale that reveals a pattern similar to that of zebra stripes. These patterns were used as evidence of seafloor spreading. The youngest rocks are found at spreading ridges, and progressively older rocks are found on both sides of the ridges. (b) The magnetic polarity timescale is a record of Earth's magnetic field orientation over time. *(a: Vine, F. J., and Matthews, D. H. 1963. "Magnetic Anomalies over Oceanic Ridges." Nature 199: 947–949. © Nature Publishing Group)*

Geophysicists Arthur Raff and Ronald Mason published a map of a magnetometer survey of the Pacific Ocean just offshore of Washington and Oregon in 1961.[18] It showed an odd pattern of high and low magnetic intensity zones that looked like zebra stripes (Figure 2-14). Scientists who saw these patterns had no idea what they were looking at. Geophysicists working on land would soon provide an explanation.

Paleomagnetism and Proof of Continental Drift

Earth's magnetic field has long been a subject of interest, dating back centuries to the Chinese invention of the compass for navigation. The field is believed to result from fluid flow within Earth's molten outer core, which probably occurs because of heat-driven currents and Earth's daily rotation. Indeed, it is in part because Earth has a magnetic field that we believe the core is primarily made of iron. Like a bar magnet, Earth's field has a positive and a negative end, or a north and a south pole (Figure 2-15). Magnetic field lines emanate from the south pole, wrap around the planet, and reenter it at the north pole. Consequently, the magnetic field points straight up at the south pole, straight down at the north pole, and lies parallel to Earth's surface at the equator.

Many rocks contain magnetic minerals such as magnetite (Fe_3O_4) that become aligned parallel to Earth's magnetic field and point toward the north magnetic pole. For instance, magnetic minerals in sediments align themselves with Earth's magnetic field before the sediments are lithified (turned into rock), locking in the direction of Earth's magnetic field at the time and place of sediment deposition. Similarly, as a lava flow cools below a temperature known as the **Curie point** (570°C for magnetite), the magnetic crystals it contains lock in the orientation of the field. Lavas solidify into rock at temperatures above 900°C, so the magnetic crystals do not themselves rotate to align with Earth's field as they do in sedimentary rocks. Still, the cooling of the solid rock below the Curie point locks in the field direction. If a lava flow were to be produced at Earth's north magnetic pole, the magnetization direction of the magnetite crystals it contained would point straight down. If, on the other hand, the lava flow were to cool at the equator, the magnetization direction would be oriented parallel to Earth's surface.

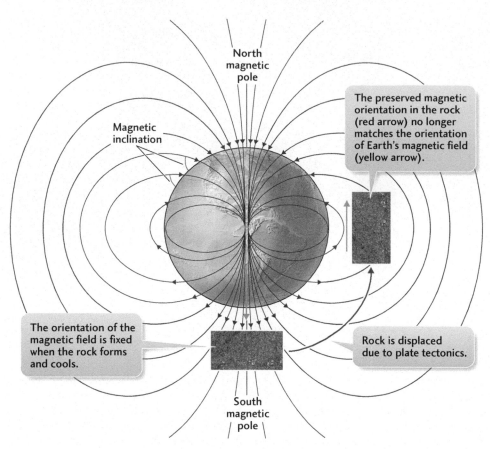

North
magnetic
pole

Magnetic
inclination

The preserved magnetic
orientation in the rock
(red arrow) no longer
matches the orientation
of Earth's magnetic field
(yellow arrow).

The orientation of the
magnetic field is fixed
when the rock forms
and cools.

Rock is displaced
due to plate tectonics.

South
magnetic
pole

Figure 2-15 Moving plates and Earth's magnetic field. In this example, volcanic rock formed near Earth's south magnetic pole travels toward the equator with plate motion. In its new location, the direction of the rock's magnetic field, fixed when the rock cooled to a solid, no longer matches that of Earth.

In the mid-1900s, geophysicists studied the magnetism of rocks of many different ages and from many different locations. They found that very young lava flows showed magnetization directions identical to Earth's magnetic field, but that older rocks displayed very different directions (Figure 2-16). There could be only three possible explanations: Either (1) rocks spontaneously remagnetized in a different direction over time due to factors such as exposure to the elements; or (2) Earth's magnetic field had changed over time, with the magnetic poles wandering across the planet's surface (Figure 2-17a); or (3) the older rocks had formed at different locations and had subsequently moved to their present places while the poles remained fixed in place. The first explanation was quickly ruled out by examining rocks of the same age in many different climatic environments on a single continent. All pointed toward the same magnetic north pole (Figure 2-17b).

British geophysicists Keith Runcorn and P.M.S. Blackett soon devised an experiment to test which of the remaining two hypotheses was correct.[10] They would examine rocks of different ages in Europe, Asia, and North America, and reconstruct where the magnetic pole had been at the time of each rock's formation. If the pole had wandered but the continents had remained fixed in position, then rocks of the same age but on different continents should point toward the same magnetic north pole (see Figure 2-17). If, on the other hand, the continents had drifted but the pole had remained fixed, rocks of the same age but on different continents should point toward different magnetic north poles. Runcorn and Blackett constructed **apparent polar wander** paths for Eurasia and North America and found that there were two different but very similar arc-shaped paths from 400 to 160 million years ago. When North America and Eurasia were brought together, the two paths merged into one; thus, the experiment's results confirmed continental drift (hypothesis 3)!

Since Runcorn and Blackett's discovery, *paleomagnetists*—scientists who study Earth's magnetic field and its history—have used the magnetic directions recorded in rocks to reconstruct the latitudes at which rocks formed for every continent on Earth. These studies have led to the conclusion that many rocks have moved very

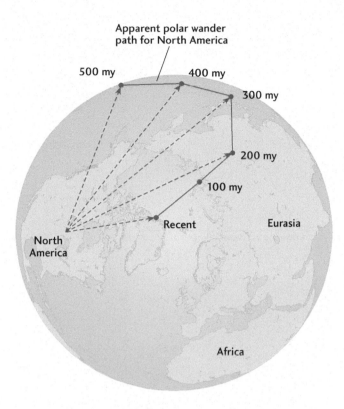

Figure 2-16 Apparent polar wander paths. Rocks of different ages, shown here for 500 to 0 million years in age, yield different apparent magnetic north poles. The paths outlined by the apparent magnetic poles for rocks of different ages are known as apparent polar wander paths.

long distances, in some cases thousands of kilometers, from the sites of their formation.

Magnetic Reversals and the Magnetic Polarity Timescale

As paleomagnetists fanned out across the globe to chart the course of the continents, they made another remarkable discovery: Earth's magnetic field reverses its polarity over geologic time. Japanese geophysicist Motonori Matuyama first encountered reversely magnetized rocks in the 1920s, a discovery that would be replicated by Jan Hospers in Iceland and Allan Cox in the United States in the 1950s.[10] At first, the reversed orientations were attributed to spontaneous remagnetization, but as evidence began to pour in from all over the world, this hypothesis was rejected.

In the late 1950s, scientists developed the potassium–argon (K–Ar) radiometric dating technique (explained in more detail in Chapter 6), which allowed them to determine the precise time when lava flows erupted and cooled on Earth's surface.[10] Cox, along with colleagues in the United States and Australia, soon found that flows of the same age showed the same magnetic polarity, a fact that would be extremely improbable if magnetic minerals were capable of spontaneous reversal.

Using the K–Ar dating method, scientists meticulously dated lava flow after lava flow, simultaneously taking samples to determine whether Earth's field had

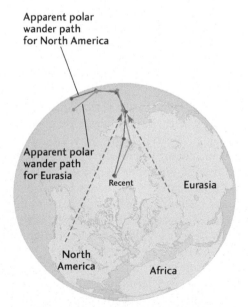

(a) Present plate configuration

(b) Reconstructed plate configuration
200 million years ago

Figure 2-17 Two possible polar wander scenarios. (a) In this scenario, the magnetic poles are assumed to have wandered while continents remained fixed in position.

(b) Here, the poles are assumed to have remained fixed while continents wandered (continental drift). The second case is consistent with data from actual continental rocks.

been in the normal or reversed orientation when the flows cooled. Gradually they built up a timeline of Earth's magnetic field orientation, which has come to be known as the **magnetic polarity timescale** (see Figure 2-14). This record now extends back over 160 million years. Major reversals in Earth's field have been named after physicists and geophysicists who have made significant contributions to the study of magnetism and paleomagnetism, with the most recent reversed interval named in honor of Matuyama.

Paleomagnetism and Seafloor Spreading

In the early 1960s, as they learned of magnetic field reversals, a pair of British scientists working together, Frederick Vine and Drummond Matthews, and the Canadian geophysicist Lawrence Morley independently realized that Raff and Mason's (see Figure 2-14) zebra-stripe seafloor magnetism map provided a test of Harry Hess's seafloor-spreading hypothesis.[19] If Hess were correct that the mid-ocean ridges were sites where new oceanic crust formed from molten magma emanating from Earth's interior, then that crust should record reversals in Earth's magnetic field over time in a pattern of stripes of lower and higher magnetic intensity. Higher-intensity stripes would reflect rocks magnetized in the same direction as Earth's present field because the two fields would be additive; lower-intensity stripes would reflect rocks magnetized during reversals, which would somewhat cancel out Earth's present magnetic field (Figure 2-18). Furthermore, if Hess's idea were correct, the pattern of seafloor stripes should be symmetrical around the mid-ocean ridges. Though Raff and Mason's map included a mid-ocean ridge, the ridge had not yet been recognized, so it was not possible to confirm Hess's idea.

Both the Vine and Matthews team and Morley quickly wrote scientific papers describing their epiphanies, and though Morley's paper was submitted for publication first, his paper was rejected by reviewers as being too speculative—a bit of bad luck for a brilliant scientist.[20] It would take nearly two decades for his discovery to be recognized, leading to what had come to be known, after successful publication of their paper, as the Vine–Matthews hypothesis, now often being referred to as the Vine–Matthews–Morley hypothesis.

Not long after Vine and Matthews published their paper, scientists at the Lamont-Doherty Earth Observatory of Columbia University confirmed their test of the seafloor-spreading hypothesis. Director of the observatory Maurice Ewing had towed a magnetometer behind the ship on every oceanographic cruise his institution had conducted, amassing huge amounts of data. Processing and examination of the

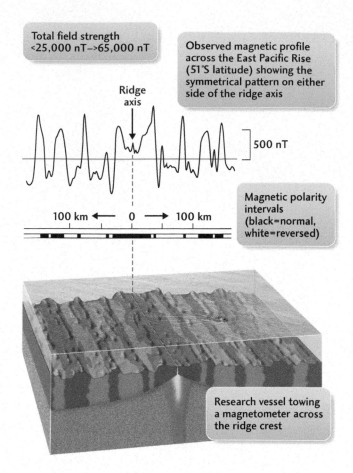

Figure 2-18 Variations in the intensity of Earth's magnetic field. Earth's magnetic field intensity is measured in units called nanoteslas (nT). Earth's present magnetic field intensity ranges from less than 25,000 nT to over 65,000 nT. The magnetic reversals represent variation of a few hundred nT superimposed on this large background field strength.

data revealed the symmetrical pattern of stripes around mid-ocean ridges.[10] The seafloor-spreading hypothesis was confirmed!

More Evidence in Support of the Theory

Further supporting evidence of seafloor spreading and the developing plate tectonic theory came with the late 1960s launch of the *Glomar Challenger,* an oceanographic research vessel designed to drill into the seafloor. Fossils in the marine sediment retrieved by the ship revealed that the age of the crust does indeed increase with distance from the mid-ocean ridges, as plate tectonic theory predicts.[21] Research cruises also found that the thickness of oceanic sediments on top of the crust increases with distance from ridges, as would be expected if the ridges were the sites of crustal formation.

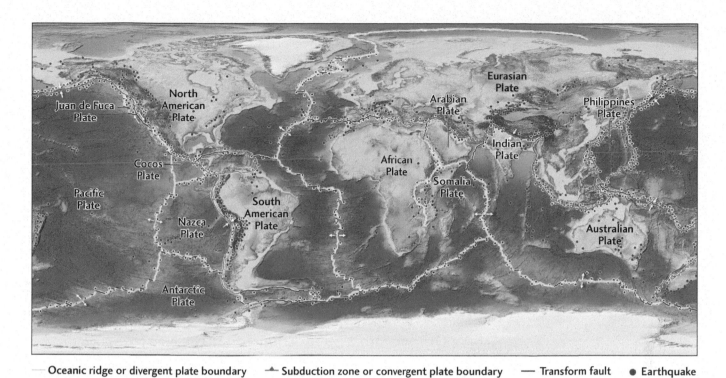

—— Oceanic ridge or divergent plate boundary   ▲— Subduction zone or convergent plate boundary   —— Transform fault   ● Earthquake
                                                                                                                                  ● Volcano

Figure 2-19 A global view of Earth and its plate boundaries illustrates the link between plate tectonics and earthquake activity. Nearly all earthquakes (purple) occur along plate margins.

Advancements in the science of seismology also contributed to the development of plate tectonic theory. Seismologists discovered how to accurately pinpoint the locations of earthquakes (described in Chapter 3) and created maps of global earthquake distributions (Figure 2-19). They found that earthquakes occur in bands that overlap the mid-ocean ridges. Earthquakes also were found along transform faults and the previously discovered deep ocean trenches, the sites where plates now are known to subduct into the asthenosphere (Box 2-2).

Seismologists Hugo Benioff and Kiyoo Wadati discovered an interesting pattern of earthquakes at deep ocean trenches.[22] The quakes lay along a dipping plane extending deep into Earth, now known as a Wadati–Benioff zone (Figure 2-20). This pattern is produced as stresses cause down-going slabs to slip in subduction

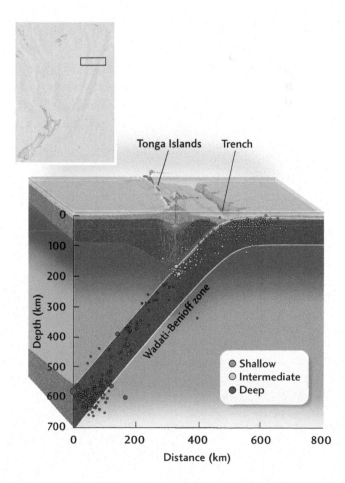

Figure 2-20 Zone of earthquakes along a subducting tectonic plate. Named for the two scientists who discovered them, the Wadati–Benioff seismic zones are zones of earthquakes along a subducting plate. Earthquakes occur from near the surface to as deep as 700 km, providing a three-dimensional portrayal of the subducting slab.

zones. Subducting slabs have been traced to depths as great as 700 km by this method.

■ The theory of plate tectonics integrates information derived from paleomagnetism, seismology, and geochronology (the study of the age of rocks) developed over several decades of scientific research.

■ Earth's magnetic field is similar to that of a large bar magnet. Magnetic field lines can be envisioned as coming out of the south magnetic pole, wrapping around Earth, and reentering it at the north magnetic pole.

■ Magnetic minerals in rocks become aligned in the direction of Earth's magnetic field at the location where they were formed.

■ Lack of congruence between the magnetic field recorded in rocks and Earth's field at any particular location indicates that the rocks have moved from their formation location to their present position.

■ The polarity of Earth's magnetic field changes over time, with the north magnetic pole becoming the south magnetic pole and vice versa.

BOX 2-2 EMERGING RESEARCH

Drilling into a Subduction Zone: The NanTroSEIZE Project

The largest earthquakes in the world are known to occur in subduction zones, but we still have only a rudimentary understanding of how they are generated. The aim of the NanTroSEIZE project is to remedy our lack of knowledge by drilling directly into a subduction zone, the Nankai Trough off the coast of Japan. Here the Philippine Plate dives beneath the Eurasian Plate at a rate of 4 cm per year, generating the volcanoes of the Southwest Japan Arc along with catastrophic earthquakes and tsunamis similar to those that struck northern Japan in March of 2011.[23] In a multi-year project that began in 2007, the Japanese research vessel *Chikyu* has bored numerous holes into the trough, bringing up sediment and rock samples that a team of scientists are studying to assess the strength and frictional properties of the subduction zone (**Figure 1**). Using technology typically employed for oil exploration, the *Chikyu* drilled more than 7000 m below the seafloor in the summer of 2010, passing all the way through the plate boundary. The hole it drilled will be fitted with instruments to record the characteristics of future earthquakes directly at their source. More than 100 scientists have been involved in the project thusfar.

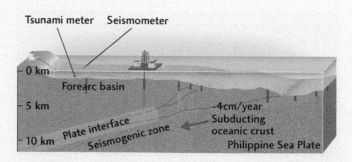

Figure 1 Drilling into a subduction zone. The Japanese scientific research vessel *Chikyu* is used by scientists to drill directly in the Nankai Trough, a subduction zone off the coast of Japan that forms the convergent boundary between the Philippine Sea (subducting) and Eurasian (overriding) Plates. Bore holes are sited across the subduction zone in order to collect samples of sediment and rock that will be investigated to determine their material properties.

■ Symmetrical patterns of normally and reversely magnetized crust surround mid-ocean ridges, confirming the seafloor-spreading hypothesis.

■ Earthquakes occur along plate boundaries. The deepest earthquakes lie along subduction zones and result from stresses in the down-going lithospheric slab.

THE DRIVER OF PLATE TECTONICS

A number of mechanisms have been proposed to explain the force that drives Earth's lithospheric plate motions. As with any dynamic process, plate tectonics depends on a source of energy, of which there are two possible candidates—thermal energy and gravitational energy.[24] Recall from Chapter 1 that at a depth of 10–200 km, a boundary separates the stiff lithosphere from the underlying plastic asthenosphere that can move slowly over long periods of time. The motion of the asthenosphere has led some scientists to propose that heat from Earth's interior causes convection in the mantle. **Convection** is the mechanism of heat transfer in which hot material rises because of its lesser density, while cool material sinks because of its greater density (Figure 2-21a). Common in liquids and gases, this is the same process that causes the boiling motion in a pot of soup simmering on a stovetop and that drives atmospheric circulation. By this hypothesis, the convective motion of the asthenosphere sets the lithospheric plates in motion and accounts for continental drift, because as

plates move, they carry along the continents embedded in them (Figure 2-21b).

The source of heat for convection may in part arise from the formation of Earth, as discussed in Chapter 1, but is also likely related to the radioactive decay of elements within the mantle and the gradual solidification of the liquid outer core. Evidence in favor of mantle convection includes the fact that Earth's interior is hot. Magma rises to the surface at mid-ocean ridges and inland of subduction zones. It also wells up in places we would not expect it to, in the interior of plates. The Hawaiian Islands (Box 2-3) are an example of such a "**hot spot.**" The generally accepted model for these features has been that they originate as plumes of molten material rising upward through the mantle from great depths (sometimes as deep as the core–mantle boundary) and melt their way through the overlying lithospheric plate. More than 100 volcanic features have been identified as hot spots in the last several decades. Recent evidence has called the mantle plume origin of hot spots into question, however, and debate has ensued over whether convection involves the entire thickness of the mantle, is restricted to shallower depths, or even whether it occurs at all.[25]

Another potential driver of plate motions is the force of gravity. Magma rising at mid-ocean ridges buoys up the seafloor, creating submerged mountain ranges at high elevations compared to the oceanic trenches where plates subduct. The potential energy associated with this elevation difference is converted to the kinetic energy of plate motion. In this model, the motion of the lithospheric plates induces flow within the underlying asthenosphere rather than the other way around. Furthermore,

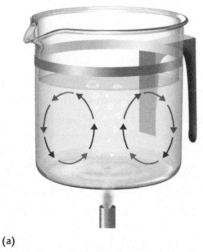

(a)

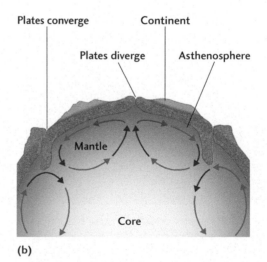

(b)

Figure 2-21 (a) As water molecules in the coffee pot absorb energy from the heat source, they expand, become less dense, and then rise. The cooler, denser molecules at the surface sink. (b) Convection currents deep in Earth's interior are thought to be the driving force behind plate movement.

As the warm asthenosphere rises under plate boundaries, it flows in opposite directions, dragging the plates along and forcing them to separate. Where the current pushes one plate into another, the cooler and denser rock sinks into the asthenosphere.

differences in subduction and crustal formation rates create stresses that can lead plates to crack, providing conduits for small pockets of magma in the asthenosphere to rise to the surface. The mantle is very close to its melting point, so a small reduction in pressure allows solid rocks to become molten.[25] Melting and magma generation are also driven by the recycling of oceanic lithosphere into the mantle, which drags with it water and sediment that can lower rock melting points.

Support for the gravitational plate tectonic driver theory comes from the fact that many of the volcanic features that have been identified as hot spots occur either along mid-ocean ridges or the fracture zones that offset them, places where pressures are reduced by stretching of the crust. In addition, for many years, imaging of Earth's interior failed to demonstrate that hot spots reside above large reservoirs of molten material, as predicted by the mantle plume hypoth-

BOX 2-3 EMERGING RESEARCH

Our Evolving Understanding of the Hawaiian Hot Spot

In addition to uncovering the nature of oceanic transform faults, geologist J. Tuzo Wilson proposed that the Hawaiian Islands were produced by upwelling of magma from a hot spot in Earth's mantle.[27] The islands extend 600 km in the central Pacific, lining up with a chain of submerged islands, called **seamounts,** that stretch all the way to the Aleutian Islands of Alaska, a total of 6000 km. A pronounced kink in the chain separates the Hawaiian Islands and seamounts from the Emperor chain (**Figure 1**). Wilson hypothesized that both the seamounts and the islands had originally lain atop the Hawaiian hot spot, which he viewed as a stationary source of magma. As the Pacific

Plate moved over the hot spot, he reasoned, magma melted its way through the lithosphere, eventually spilling out on the ocean floor. Submarine volcanic eruptions built a volcano that gradually climbed above the waves to form an island. Continued motion of the Pacific Plate eventually moved the island off of the hot spot, clearing the way for the creation of a new island. In this model, seamounts are islands that have migrated away from the hot spot and have slipped beneath the ocean surface due to a combination of wave erosion and loss of thermal buoyancy associated with the hot spot.

Wilson hypothesized that the Hawaiian Islands and their seamount extensions would increase in age away from the island of Hawaii, which is home to the chain's presently active volcanoes Mauna Loa and K'lauea. Radiometric dating of the islands confirmed that they do indeed increase in age to the northwest (**Figure 2**).[28] Each volcano stayed active for about 2 million years and was extinguished when the plate carried it past the source of magma. The youngest volcano is Lō'ihi, southeast of Hawaii. Nearly 5 km high, the seamount is within a kilometer of the sea surface. Lō'ihi will continue to grow upward and is estimated to emerge above sea level in a few hundred thousand years.

If Wilson's assumption that hot spots remain fixed in position were in fact correct, then the position of Hawaiian volcanoes and their ages could be used to determine the Pacific Plate's long-term speed and direction, an analysis that could be conducted on any plate containing a hot spot-generated volcanic island chain. Just this sort of analysis was used to determine that the plate moves at a rate of 8 cm per year in a northwesterly direction. The major

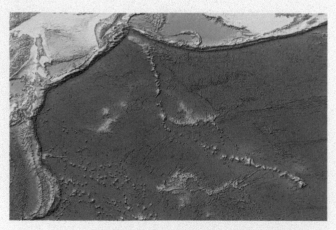

Figure 1 Chain of volcanoes created at the Hawaiian hot spot. At least 129 volcanoes form the 5800-km-long Hawaiian–Emperor chain. Only 4 are active, and some 125 or so are extinct volcanoes, dormant volcanoes, seamounts, and atolls. The bend in the chain is now thought to have resulted from movement of the hot spot location.

esis. Only in 2009 did geophysicists finally collect the seismic data necessary to establish that Hawaii is underlain by a zone of warm, plastic rocks that extends as deep as 1500 km.[26] Similar evidence has not yet been found for other hot spots. Our understanding of the mechanisms governing plate motion is continually evolving, and you may be the next young scientist to collect data to determine what really causes Earth's plates to migrate!

■ Plate tectonic motions are driven by two potential energy sources—thermal energy and gravitational energy.

■ Movement of heat within Earth's mantle may cause convection, though the details of how this happens are uncertain.

■ Differences in elevation between mid-ocean ridges and subduction zones may generate potential energy that is converted to the kinetic energy of plate motion.

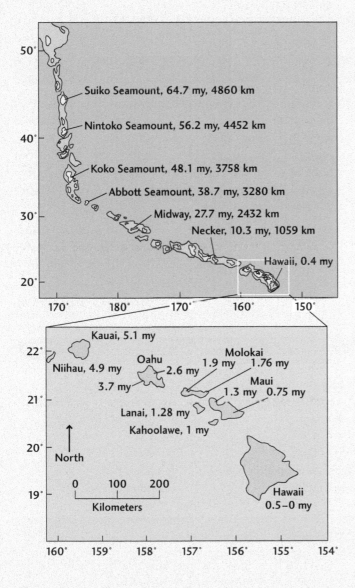

Figure 2 From southeast to northwest, the volcanoes progress in age from youngest to oldest, with the oldest—now in the northwest Pacific Ocean—about 86 million years old. The youngest volcano is Lō'ihi Seamount, located about 34 km southeast of the island of Hawaii. Still growing upward, Lō'ihi is 1000 m below the level of the ocean surface. The extinct volcanoes at the bend located approximately in the middle of the chain are 41 to 43 million years old.

bend in the chain about midway along its length was attributed to a change in plate direction from north to northwest that occurred about 45 million years ago (see Figure 1). One of the big mysteries of plate motion has been why the plate should have changed its direction so radically at that time.

Work published in 2003 has shown that the simple model of a fixed hot spot shown in every introductory geology textbook over the last few decades (including the first edition of this text) is wrong. Rather than staying in a stationary position, the Hawaiian hot spot migrated southward between 81 and 47 million years ago at speeds in excess of 4 cm per year. Evidence for this migration came from paleomagnetic measurements on rocks in the Emperor chain that proved that the seamounts had not formed at the present latitude of the Hawaiian hot spot.[29] The mysterious kink in the Hawaiian–Emperor chain was produced not by a change in plate motion, but by a change in the location of the hot spot.

Our evolving understanding of the Hawaiian hot spot perfectly illustrates the objective nature of scientific inquiry. In the light of new evidence, the hypothesis that hot spots are fixed in position has been rejected. The realization that hot spots can migrate will spawn many new investigations.

CLOSING THOUGHTS

Through careful application of the scientific method over the last century, we have learned much about how the solid Earth works. From Eduard Suess's contraction theory to the development of the plate tectonic paradigm, generations of Earth scientists have amassed evidence of seemingly fantastic processes and events in Earth history. We now know, for example, that rocks from the South Pole were once adjacent to those presently located in the northern hemisphere and that continents have repeatedly coalesced and split apart, migrating in response to plate motions. These motions have constructed mountain ranges and ocean basins, governed the distribution of plant and animal species, and caused volcanic eruptions and earthquakes.

Yet with all that we have learned, mysteries remain. The cause of lithospheric plate motion is still not fully understood. In addition, earlier assumptions about the fixity of hot spots have been overturned as new evidence has revealed that the Hawaiian hot spot migrated over time. Earth science, like the planet itself, is dynamic, and each discovery we make generates additional questions about how the planet functions. Still, the knowledge we have constructed to date has proven vital to human affairs. Our understanding of plate tectonic processes has governed the development of fossil fuel and mineral resources that contribute to our daily lives and has given us a framework for understanding geologic hazards such as earthquakes and volcanic eruptions. We explore these topics further in Chapters 3, 5, and 13.

SUMMARY

■ Plate tectonics describes the motion of Earth's lithospheric plates on top of the underlying asthenosphere, as well as the interaction of plates along their boundaries (for example, subduction, continent–continent collision, transform motion, and divergence).

■ Alfred Wegener used fossil and rock evidence to propose that all the continents on Earth had once been joined into a supercontinent that he named Pangaea, and had subsequently moved apart in a process he termed continental drift.

■ Plate tectonic theory expanded on Wegener's ideas and arose from measurements of the magnetic fields

recorded in rocks, which demonstrate that continents have migrated over time and that the seafloor is spreading apart along mid-ocean ridges. In addition, seismological information revealed the location of plate boundaries, and drilled samples of oceanic crust showed that oceanic rocks increase in age with distance from mid-ocean ridges.

■ In the cool, upper parts of the lithosphere, brittle deformation leads to faults, while in the warm, lower parts, plastic deformation leads to folding.

KEY TERMS

continental drift, p. 36
mid-ocean ridge, p. 36
seafloor spreading, p. 36
subduction, p. 36
plate tectonics, p. 36
scientific method, p. 40
hypothesis, p. 40

theory, p. 40
universal law, p. 40
unifying theory, p. 40
divergent boundary, p. 41
normal faults, p. 42
convergent boundary,
  p. 44

continent–continent collision, p. 44
ocean trench, p. 44
volcanic arc, p. 44
island arc, p. 45
reverse faulting, p. 45
transform plate boundary, p. 47
strike-slip faulting, p. 47

Curie point, p. 49
apparent polar wander, p. 50
magnetic polarity timescale,
  p. 52
convection, p. 55
hot spot, p. 55
seamount, p. 56

REVIEW QUESTIONS

1. How did Alfred Wegener determine that all continents once were connected in the single landmass that he called Pangaea?

2. What roles did World War II and new technology play in the development of the plate tectonic theory?

3. How have plate motions affected the distribution of plant and animal species on Earth?

4. What was contraction theory, and what was it used to explain?

5. What are the components of the scientific method? How does science differ from pseudoscience?

6. How do the different densities of oceanic and continental crust affect their behavior in convergent plate boundary settings?

7. What are mid-ocean ridges?

8. Which types of faults are associated with divergent boundaries? Convergent boundaries? Transform boundaries?

9. How are continental and island arcs formed?

10. What is the Curie point?

11. How do the magnetic fields contained within rocks indicate that plates have moved over time?

12. What is the magnetic polarity timescale?

13. When geophysicists discovered that rocks of different ages had different magnetic inclinations, what were the various hypotheses advanced to explain this phenomenon? What experiment was devised to test these hypotheses?

14. What evidence did oceanographic drilling provide in support of the plate tectonic theory?

15. How does the pattern of earthquakes support the plate tectonic theory?

16. What are the potential driving forces behind plate tectonics?

17. How has our understanding of the Hawaiian hot spot changed over time?

18. Along what type of plate boundary is oil most likely to be found in sedimentary rocks?

19. What is a megathrust? Where on Earth might one exist?

20. How fast do Earth's tectonic plates move? Given these speeds, how long would it take for plates to move 100 km relative to one another?

THOUGHT QUESTIONS

1. Suppose that Earth's magnetic field never reversed polarity. How might our understanding of plate tectonics differ?

2. Given that the Atlantic Ocean contains no subduction zones, what must be happening to the size of the Pacific Ocean over time?

3. Based on your answer to question 2, what will eventually happen in the Pacific Basin (provided no other changes in plate motion occur), and how soon will it happen?

4. Why do most earthquakes occur at relatively shallow depths (generally less than 20 to 30 km beneath the surface)?

EXERCISES

1. If scientists determined the ages of two extinct volcanoes along a chain in an ocean basin to be 25 and 50 million years old, respectively, and the distance between the volcanoes is 1000 km, what is the rate of motion of the oceanic crust over the hot spot that produced the volcanoes?

SUGGESTED READINGS

Anderson, D. L. *New Theory of the Earth*. New York: Cambridge University Press, 2007.

Condie, K. C. *Plate Tectonics and Crustal Evolution*, 4th ed. Oxford: Butterworth-Heinemann, 1997.

Glatzmeier, G. A., and P. Olson. "Probing the Geodynamo." *Scientific American* 292, no. 4 (2005): 50–57.

Gurnis, M. "Sculpting the Earth from Inside Out." *Scientific American* 284, no. 3 (2001): 40–47.

Haddok, E. "Birth of an Ocean." *Scientific American* 299, no. 4 (2008): 60–67.

Hsu, K. J. *Challenger at Sea: A Ship That Revolutionized Earth Science*. Princeton, NJ: Princeton University Press, 1992.

Kearey, P., K. A. Klepeis, and F. J. Vine. *Global Tectonics*, 3rd ed. West Sussex, UK: Wiley-Blackwell, 2009.

McPhee, J. *Annals of the Former World*. New York: Farrar, Straus, and Giroux, 1998.

Medawar, P. *The Limits of Science*. New York: Oxford University Press, 1988.

Oreskes, N. *The Rejection of Continental Drift: Theory and Method in American Earth Science*. New York: Oxford University Press, 1999.

Oreskes, N., ed. *Plate Tectonics: An Insider's History of the Modern Theory of the Earth*. Boulder, CO: Westview Press, 2003.

Tarduno, J. A. "Hotspots Unplugged." *Scientific American* 298, no. 1 (2008): 88–93.

Earthquakes: Their Causes, Hazards, and Risks

The ancient city of Scythopolis, now called Beit She'An, in Israel was a thriving commercial and administrative hub of the Roman and Byzantine empires.[1] A grand boulevard lined with shops, a 5000-seat theater, a hippodrome for horse racing, and a Roman bath adorned with exquisite mosaic floors all attest to the affluence of the community. Located about 30 km south of the Sea of Galilee, the city sat along an important trading route connecting the Mediterranean coast to cities east of the Jordan River, and taxes on passing caravans supported a lavish lifestyle.

In 749, CE, an earthquake destroyed most of Scythopolis and ruptured the aqueduct that supplied the city with drinking water. As archeologists excavated the site in the 1960s, they uncovered evidence of the earthquake, finding that most of the columns that once supported the roofs of buildings had toppled in the same direction. Such evidence forms the basis for the field of *archaeoseismology*, which uses archeological evidence to uncover fault activity. Based on the direction that the columns had toppled, the likely source of the earthquake that destroyed Scythopolis is the Dead Sea transform fault, which separates the African Plate to the west from the Arabian Plate to the east.

In this chapter we:

✔ Discuss how earthquakes are generated by slip along locked fault planes.

✔ Explore how seismographic stations record earthquake waves and what the pattern of waves can tell us about where earthquakes originated and how Earth's interior is structured.

✔ Learn what factors determine earthquake magnitude and intensity.

✔ Discuss the prospects for predicting earthquakes and the steps we can take to minimize damage and loss of life.

Most of the ancient Roman and Byzantine city of Scythopolis, now Beit She'An in Israel, was destroyed during an earthquake in 749 CE that also ruptured the aqueduct that supplied the city with its drinking water. *(Kirsten Menking)*

INTRODUCTION

Humans long have wondered about the origin of sudden and immense shakings of Earth that belie the phrases "solid earth" and "terra firma." The native Yurok people of the Pacific Northwest have myths that tell of a god named Yewol (in English, "Earthquake") who:

"... sank the ground.... Every little while there would be an earthquake, then another earthquake, and another earthquake....And then the water would fill those places...." (as recorded

and translated by A. L. Kroeber, Yurok Myths *[Berkeley: University of California Press, 1978]).*

The Yurok live near the Cascadia subduction zone between the Juan de Fuca and North American tectonic plates, just offshore of the western coast of North America that extends from British Columbia to Northern California (Figure 3-1). Their myth about Yewol describes a common phenomenon observed along oceanic–continental convergent boundaries (see Chapter 2): Friction along the trench prevents the oceanic plate from subducting, causing the overriding continental plate

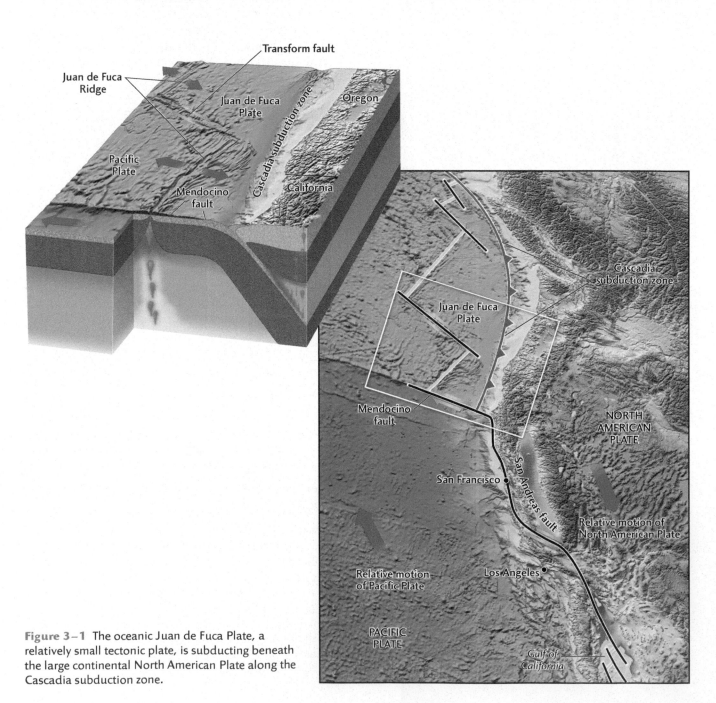

Figure 3–1 The oceanic Juan de Fuca Plate, a relatively small tectonic plate, is subducting beneath the large continental North American Plate along the Cascadia subduction zone.

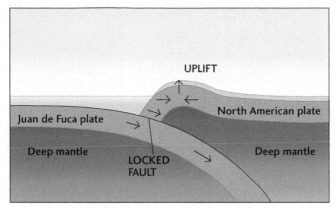

(a) Before earthquake

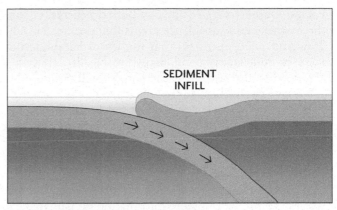

(b) After earthquake

(c)

(d)

Figure 3–2 (a) Elastic strain energy accumulates in the overriding North American Plate as it flexes upward before an earthquake. (b) After an earthquake, the outermost lip of the upper plate remains emergent above sea level, while inland, the coast subsides to form a trough that fills with water and sediment. (c) A ghostly forest of dead red cedar trees in a tidal marsh indicates that the coast of Washington subsided suddenly around the year 1700. Since the forest was drowned, a meter of mud has buried the bases of the trees. (d) Restoration Point, near Seattle, is a beach raised out of the water by a major earthquake that occurred sometime between 800 and 900 CE. *(c: Atwater, B. F., Satoko, M.–R., Kenji, S., Yoshinobu, T., Kazue, U., and Yamaguchi, D. K. 2005. The Orphan Tsunami of 1700: Japanese Clues to a Parent Earthquake in North America. Reston, VA: USGS; Seattle: University of Washington Press; d: Courtesy Washington State Department of Ecology)*

to flex upward, or bulge, as the oceanic plate pushes against it (Figure 3-2). Eventually, the bending energy stored in the continental plate exceeds the frictional resistance along the fault plane and the two plates slip past one another, producing an earthquake. Slip on the plate boundary allows the bulge in the continental plate to straighten out. The outermost lip of the plate is pushed upward in the process, resulting in coastal uplift that raises tide pools and beaches out of the ocean. Inland of the uplifted area, the bulging crust collapses, causing the land to sink, and drowning the coastline. It is the latter phenomenon that the Yurok observed, and evidence of such submergence exists along convergent plate boundaries throughout the world in the form of buried peat and dead trees killed by the exposure of their roots to saltwater or by burial in sand and mud.

Although no major earthquake has occurred along the Juan de Fuca subduction zone in recent years, geologists know from the evidence of dead trees that an event like that described by the Yurok occurred around 1700 CE, before Europeans settled in the area. The fact that no large earthquakes have occurred in several centuries does not mean that faults along the plate boundary are inactive and that no hazard exists. Rather, the period of quiet indicates that energy is storing up for the next event, a worrisome situation given that the Northwest coast is now home to major population centers such as Vancouver, British Columbia; Seattle, Washington; and Portland, Oregon. Indeed, the 1700 CE earthquake was likely one of the strongest events the plate tectonic system is capable of producing, measuring around 9.0 in magnitude and generating a tsunami that crossed the

Pacific Ocean to strike Japan.[2] Buried vegetation from three other rapid subsidence events that occurred in 400, 750, and 1000 CE reveal that the Cascadia subduction zone is quite active, and many people are concerned about when the next great earthquake will occur.[3]

The March 11, 2011 Tōhoku earthquake that struck northeastern Japan suggests what the Pacific Northwest may have in store when the Cascadia subduction zone next slips. The magnitude 9.0 tremor off the coast of Japan occurred when the Pacific Plate lurched westward beneath a part of the North American Plate that extends into Asia and generated massive tsunami waves three stories (10 m) high.[4] (More information on tsunamis can be found in Chapter 12.) Though Japan had constructed seawalls along the coast to protect against tsunamis and storms, the walls proved no match for the incoming waves, which readily overtopped them (Figure 3-3). Sweeping inland hundreds of meters, the incoming surge bore down on coastal communities, lifting homes off their foundations and smashing them to pieces. Whole villages were destroyed, with a confirmed death toll in excess of 15,000 and more than 8000 people still unaccounted for.[5]

In addition to the earthquake and tsunami, another catastrophe unfolded when several nuclear power plants along Japan's coast lost the power necessary to keep their reactor cores cool.[6] Partial meltdowns led to explosions that dispersed radioactivity across a wide area, and it is unclear when the area surrounding the damaged facilities will again be inhabitable. Given their danger, it is important to learn as much as we can about earthquakes and about how to mitigate their impacts.

THE BUILDUP AND RELEASE OF SEISMIC ENERGY

Some earthquakes are so small that humans cannot feel them, but the largest release 20 orders of magnitude[*] more energy, an amount equivalent to the annual energy consumption of the United States. What could be happening inside Earth to cause the release of such large amounts of energy in such short periods of time?

Until 1891, most scholars believed that shaking during earthquakes caused the ground to crack and form faults (Chapter 2 describes the different types of faults and their tectonic settings). In that year, a large earthquake in Japan formed a **scarp,** or steep slope, up to 6 m high and 70 to 80 km long, offsetting, or separating, many physical features as much as 5.5 m in both the horizontal and the vertical direction (Figure 3-4).[7] Examining the scarp, geologist Bunjiro Koto hypothesized that offset along the fault had caused the earthquake, rather than the other way around, as was the common thinking at the time.[8]

In 1906, Koto's idea gained wide acceptance when a large earthquake that destroyed San Francisco was linked to slip along a known fault—the San Andreas. Before the earthquake, geologists had noticed that the

---

[*]An order of magnitude is an increase of 10 times; two orders of magnitude is an increase of 100 times, and so on. An earthquake that releases 1000 times more energy than another, for example, would have three orders of magnitude more energy.

Figure 3–3 On March 11, 2011, the city of Miyako was heavily damaged by a 37.9-m-high tsunami caused by the Tōhoku earthquake. Hundreds of people were killed in the village, and only a few dozen of the nearly 1000 ships in the town's fishing fleet survived the tsunami. (©AFLO/Mainichi Newspaper/epa/Corbis)

Figure 3–4 A scarp in Honshu, Japan, formed in 1891 by a 5–6 m offset of Earth's crust along a fault segment 70–80 km long. Geologist Bunjiro Koto examined the scarp and made a revolutionary proposal: that movement on a fault had caused both the scarp and the earthquake, and that the earthquake was an effect rather than a cause. *(Bruce A. Bolt)*

rock types on each side of the fault did not match. After the earthquake, geologist Grove Karl Gilbert and others discovered that up to 9.7 m of new horizontal offset of rocks, fences, hillsides, and even walls of buildings had occurred along a 477-km-long section of the fault (Figure 3-5).[9] Similar observations along other known

Figure 3–5 As a result of slip along the San Andreas fault during the 1906 earthquake, a fence near Bolinas in Northern California was offset about 4 meters. *(G. K. Gilbert/USGS)*

faults were made soon thereafter, confirming Koto's hypothesis.

Fault Behavior

Interestingly, not all active faults generate earthquakes. In the Central California town of Hollister, slow, steady motion, called **creep,** on the Calaveras fault offsets sidewalks, curbs, fences, and even home foundations. Movement on the fault averages approximately 15 mm per year and causes structural damage to buildings and pipelines,[10] but the steady motion does not generate the intense shaking associated with earthquakes, so no one has ever been killed and no structures have been destroyed in an instant. Another site displaying creep activity is the University of California at Berkeley campus, where the football stadium lies directly above the Hayward fault. Movement of 5 mm per year threatens the structural integrity of the stadium, which is being renovated to allow it to better adjust to fault movements.[11]

Such creeping behavior is rare.[12] In contrast, most faults display **stick-slip** behavior, in which no movement along the fault plane for decades or centuries is followed by slip. To explain stick-slip events like the 1906 San Francisco earthquake, geologist Henry Fielding Reid developed the **elastic rebound model.**[13] According to this model, friction along the fault plane prevents rocks from slipping past one another, but the slow, steady motion of Earth's tectonic plates over the ductile asthenosphere continues, deforming the crust adjacent to the fault in an elastic manner similar to the deformation of a stretched rubber band or a squeezed rubber ball. The relationship between the build-up of stress and the amount of deformation produced in elastic solids was first described by physicist Robert Hooke, who published his *Discourse on Earthquakes* in 1668,[14] well before the advent of the theory of plate tectonics. Because rocks are elastic—meaning that they can recover their initial shape when the stresses applied to them are removed—they are distorted by the gradual buildup of stress across a fault. Markers on Earth's surface, including fences and roads, likewise can become distorted where they cross active faults. If a rubber band is stretched beyond its elastic limit, it will break and snap back, releasing stored energy. Similarly, with continued stress along a fault, rocks eventually fail in a brittle manner, slipping past one another along the fault plane and releasing stored elastic energy in the form of seismic waves (Figure 3-6). The greater the amount of strain (deformation) that accumulates in the

rocks before they fail, the greater the amount of energy released.

While we used to think that all stick-slip events unfold over seconds to minutes, recent research has shown that *slow earthquakes* also occur, though some seismologists prefer to refer to these events as *aseismic deformation* since they do not generate shaking.[15] The advent of Global Positioning System (GPS) satellites and sensitive strain-meter technology has allowed scientists to track the movement of markers embedded in Earth's surface with a very high degree of accuracy and, therefore, to measure movements along faults in real time. Using these technologies, the Nankai Trough subduction zone off the coast of Japan, the Cascadia subduction zone, and the San Andreas fault have been shown to undergo episodic movements that occur so slowly that they are not detected by human

(a) Original position

(b) Build-up of strain

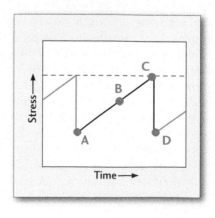

(c) Slippage

(d) Strain released

Figure 3-6 The elastic rebound model, which describes the build-up and release of stored elastic energy before and during an earthquake, explains the offset of features such as rocks, fences, and roads after an earthquake. In this example, the stream is offset in a right-lateral manner; in other words, one side of the stream moved to the right with respect to the other.

senses. Some of these earthquakes unfold over months, whereas others occur over hours or days. In some cases, the total amount of energy released is equivalent to that of a strong earthquake. In addition, slow earthquakes are occasionally followed by abrupt stick-slip events that generate significant shaking. Recent technological advances have thus made us aware that faults are capable of a spectrum of behaviors, not just rapid stick-slip events or steady creep.

What governs the kind of behavior displayed by a given fault is a topic of great interest amongst seismologists. One possibility is that high fluid pressures within a fault zone may lubricate the fault plane, reducing friction and facilitating slip by creep or slow slip. Another possibility is that lowered friction results from the pulverization of rocks along the fault zone or the growth of a slippery mineral called serpentine. Creeping or slow earthquake behavior may also result from countless microearthquakes too small to be detected. Whatever the cause of slip events, earthquakes tend to occur in the upper levels of Earth's crust, where rocks are cold and behave in a brittle fashion. At lower levels in the crust, warmer temperatures and higher pressures allow rocks to deform by flowing rather than breaking, which does not produce shaking. Only in subduction zones, where cold slabs of lithosphere descend into the asthenosphere, do earthquakes occur at depths greater than about 20 km.[16]

Seismic Sources and Waves

Several days after an earthquake struck Boston in 1755, Harvard professor John Winthrop, sitting with his feet propped on the brick hearth of his fireplace, was shaken from his repose by a small aftershock. One after the other, the bricks under his feet rose up, then dropped back down into place. Winthrop noted with surprise that it looked like "one small *wave of earth* rolling along."[17] His insight was significant: Seismic energy passing through Earth behaves similarly to sound passing though air or a ripple from a dropped stone passing through a still pond of water. In all three situations, energy is transmitted as waves, which, in the case of earthquakes, are known as **seismic waves.**

During an earthquake, failure begins at a single point on a fault plane called the **hypocenter,** or **focus;** the point directly above the hypocenter at the ground surface is called the **epicenter.** The rupture rapidly spreads out along the fault plane until most or all of the stored elastic energy is released (Figure 3-7). At the same time, seismic waves of several types propagate in all directions, each passing through Earth at a different speed. The result is that an observer at any point on Earth sees batches of different wave types arriving one after the other. Waves emanating from an earthquake hypocenter travel so fast that they sometimes reach the ground surface before the rupture itself. During a fairly large earthquake in 1983, two elk hunters driving in Idaho suddenly felt dizzy, perhaps because of the arrival of the fastest seismic waves and the rapid shaking of their vehicle.[18] Immediately afterward, they were startled to see a scarp 2 m high emerge 20 m ahead of them!

The instrument used to detect and measure seismic waves is the **seismometer.** The earliest seismometer was invented in China in the second century CE. It looked

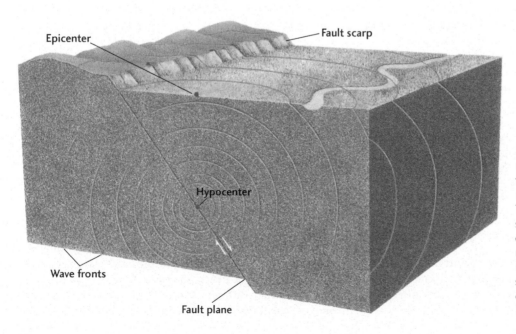

Epicenter

Fault scarp

Hypocenter

Wave fronts

Fault plane

Figure 3–7 Seismic waves radiate outward in all directions from the hypocenter, or initial rupture point, along a fault. The waves result in the ground shaking that is known as an earthquake. As the rupture moves along the fault plane, it reaches the surface and sometimes creates a steep slope, or scarp, like that shown in Figure 3–4.

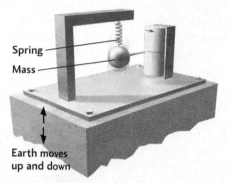

Spring
Mass

Earth moves up and down

(a) Seismograph designed to detect vertical movement

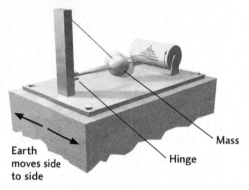

Mass

Earth moves side to side

Hinge

(b) Seismograph designed to detect horizontal movement

Figure 3–8 A seismometer measures ground motion. This scientific instrument typically consists of a dense object, or internal mass attached to a frame using a hinge, spring, or other device. The mass moves very little relative to the frame because of inertia. During an earthquake, the frame moves, and the relative motion between the frame and the internal mass is recorded. Some modern seismometers use electronic means to hold the internal mass in place.

like a vase with eight dragonheads around the outer edge, one for each cardinal direction on a compass.[19] Each head held a small ball that, when the ground shook, would drop into the mouth of one of eight toads or frogs sculpted around the base of the device, thereby allowing early Chinese seismologists to determine from which direction the earthquake waves originated. In more recent times, seismometers consisted of a pen attached to a pendulum hanging from an overlying frame (Figure 3-8). The pendulum was suspended over a rotating drum fitted with paper, and when the ground shook, the drum and paper moved with the earth while the pendulum stayed fixed in position. The pen thereby made a tracing on the paper that reflected the ground shaking.

Modern seismometers are digital and use electronic vibration detectors.[20] When the ground moves during an earthquake, seismometers monitor the direction, amplitude (height), frequency (the number of waves per

unit of time), and duration of wave motion. Scientists typically bury seismometers up to 3 m underground to minimize the noise of vibrations associated with non-seismic phenomena, such as trucks driving by or human and livestock footsteps.

Seismic waves can be plotted as a **seismogram** to show the form of the waves with time (Figure 3-9). The fastest-moving waves are called **P waves,** or **primary waves,** because they are the first to arrive at any station. Like sound waves, they compress and expand the material through which they travel, much like the shape of a Slinky toy changes if one stretches it out across a table and initiates a wave by pushing on one end. Primary waves travel at speeds ranging from about 5.5 to 14 km per second through Earth, depending on the density and rigidity of the material through which they are moving, and reach the opposite side of Earth from the earthquake source within about 35 minutes of fault rupture. **Secondary waves, or S waves,** are the next to arrive, traveling at a rate of 2.5 to 7 km per second, depending on the material they are moving through.[21] Secondary waves *shear* Earth material back and forth in a sideways motion, perpendicular to the direction in which the wave is moving, much as the shape of a rope stretched between two people changes when one of the people jerks the rope up or down. Both P and S waves are known as **body waves** because they travel through Earth's interior (Box 3-1). They were first described by French mathematician and physicist Siméon Denis Poisson in 1830.

When P and S waves reach Earth's surface, they generate lower-frequency **surface waves,** which are probably what John Winthrop observed in his brick hearth. *Rayleigh* waves move across Earth's surface in a rolling motion similar to the movement of ocean waves. *Love* waves shake the ground from side to side. Although their frequency is lower than that of body waves, the amplitude of ground motion from surface waves can be quite substantial, resulting in some of the greatest destruction during an earthquake. Objects can literally be tossed into the air in defiance of gravity as the ground rises upward, and rigid structures can be torn to pieces during the shaking. Surface waves have little impact below ground, as witnessed by miners who have survived earthquakes underground and then climbed to the surface to find widespread devastation.

Locating Earthquake Epicenters

The different velocities of P and S waves provide a convenient way to determine the locations of earthquakes. Like two runners starting a marathon, the waves arrive at the 100 m mark at nearly the same time; but as the race continues, their different speeds lead to different arrival times at the finish line, in this case the seismographic

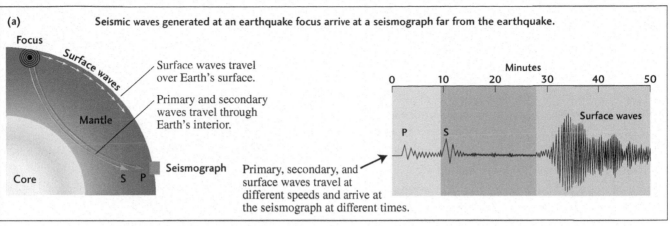

(a) Seismic waves generated at an earthquake focus arrive at a seismograph far from the earthquake.

Focus

Surface waves

Mantle

Core

Surface waves travel over Earth's surface.

Primary and secondary waves travel through Earth's interior.

Seismograph

S   P

Primary, secondary, and surface waves travel at different speeds and arrive at the seismograph at different times.

Minutes
0   10   20   30   40   50

P   S

Surface waves

(b)   Seismic waves are characterized by distinct types of ground deformation.

P-wave motion

P waves (primary waves) are compressional waves that travel quickly through rock.

Compressional-wave crest

P waves travel as a series of contractions and expansions, pushing and pulling particles in the direction of their path of travel.

The red square charts the contraction and expansion of a section of rock.

Wave direction

S-wave motion

S waves (secondary waves) travel at about half the speed of P waves.

Shear-wave crest

S waves are shear waves that push material at right angles to their path of travel.

The red square shows how a section of rock shears from a square to a parallelogram as the S wave passes.

Wave direction

Surface-wave motion

Surface waves ripple across Earth's surface, where air above the surface allows free movement. There are two types of surface waves.

In one type, the ground surface moves in a rolling, elliptical motion that decreases with depth beneath the surface (*Rayleigh* wave).

Wave direction

In the second type, the ground shakes sideways, with no vertical motion (*Love* wave).

Wave direction

Figure 3–9 Three types of seismic waves travel at different speeds and in different ways. Both primary (P waves) and secondary waves (S waves) travel from the earthquake focus through Earth's interior, whereas surface waves travel across Earth's surface. Primary waves travel about twice as fast as secondary waves. The three different wave types from the same seismic event arrive at a seismometer at different times.

BOX 3–1 GEOLOGIST'S TOOLBOX

Imaging Earth's Interior

Earthquakes and the development of the science of seismology have enabled us to determine the structure of Earth's interior, a place we will never be able to visit directly. As scientists set up seismometers at numerous locations, they began to notice something rather odd: No S waves appeared on seismograms at locations greater than 104° away from earthquake hypocenters. Likewise, no P waves were found at angles between 104° and 140° from the hypocenters. These findings indicated that Earth had a complex internal structure that was not homogeneous throughout.

If Earth's interior were made of a uniform material, seismic rays—lines drawn perpendicular to the spherically propagating seismic wave fronts—would be perfectly straight (**Figure 1**). Such a planet would be highly improbable because the force of gravity should cause material at the interior of the planet to be denser than that at the surface. In a planet with increasing density toward the interior, seismic rays would travel along a curved path. The reason for this can be illustrated by thinking of a straw in a glass of water. Light energy is bent when passing from air into water, causing the straw to appear bent when in fact it is perfectly straight. The same thing happens to seismic energy as it passes through rocks of different densities, a phenomenon known as **refraction.**

In our layered planet, the difference in density between the mantle ($3.3 - 5.7$ g/cm^3) and the core ($10 - 13$ g/cm^3)[21] causes so much refraction that seismic rays cannot travel to certain locations. Another factor governing the propagation of seismic energy is the physical state of the matter. S waves travel through materials by deforming their shapes, whereas P waves propagate by alternately compressing and then extending the material through which they move. Liquids have no distinct shape, so S waves are incapable of traveling through them, a fact that led to the discovery that Earth's outer core is composed of a liquid. The angle beyond which S waves are no longer recorded also indicates the size of the molten core.

Because liquids are compressible, P waves are able to pass through the outer core, but the transition from liquid to solid causes them to refract strongly, leading to the 104° to 140° shadow zone. In addition, P waves move much

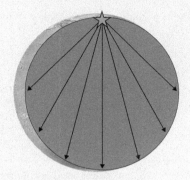

(a) Homogeneous Earth

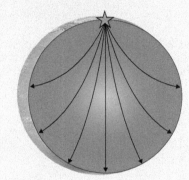

(b) Density Increasing with Depth

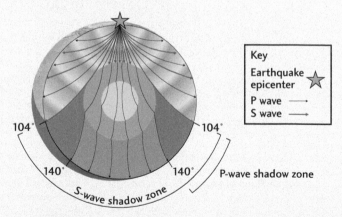

(c) Actual Earth

Key
Earthquake epicenter ☆
P wave ⟶
S wave ⟶

Figure 1 (a) Seismic rays in a homogeneous Earth, (b) seismic rays in an Earth with increasing density toward the interior, and (c) seismic rays in a zoned Earth, showing P– and S–wave shadow zones.

more rapidly through solids than through liquids. It was these two lines of evidence that led Danish seismologist Inge Lehmann to hypothesize in the 1930s that Earth had a solid inner core.[22] Based on two models of Earth—one with a mantle and a liquid core; the other with a mantle, a liquid outer core, and a solid inner core—she calculated how long it should take P waves to reach the other side of the world from their initiating earthquakes and compared her results to the actual travel times. She quickly realized that a solid inner core was necessary to match the data and calculated a radius of 1400 km for that body, a result later refined by other seismologists to the presently accepted value of approximately 1220 km.

While the gross structure of Earth's interior has been known for several decades, much remains to be learned about finer scale features. In 2007, the National Science Foundation–funded EarthScope project launched a seismographic investigation of Earth's crust underlying the United States. Called the Transportable Array, or USArray,

the experiment consists of a network of 400 portable seismometers arranged in a north–south swath extending from the U.S. border with Canada to the border with Mexico (**Figure 2**).[23] The seismometers were deployed in the western states for 2 years before being moved inland to image the next north–south swath, and all instruments in the network record incoming seismic energy from earthquakes happening across the globe. The array covers approximately a quarter of the United States at any given time and has now reached the East Coast. With a spacing of 70 km between seismometers, the array is providing detailed information about the structure and composition of Earth's crust as variations in rock type and thickness cause differences in the speeds at which seismic wave fronts propagate to reach different seismographic stations.

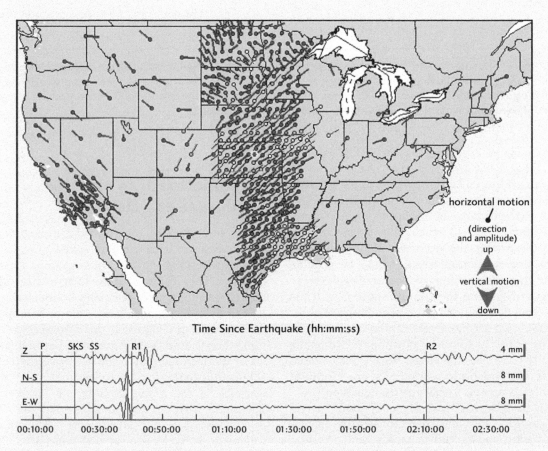

Figure 2 Seismometers for the EarthScope project's USArray recorded the incoming waves from Japan's Tōhoku earthquake on March 11, 2011. *(IRIS—Incorporated Research Institutions for Seismology. Courtesy of Manoch Bahaver, DMC)*

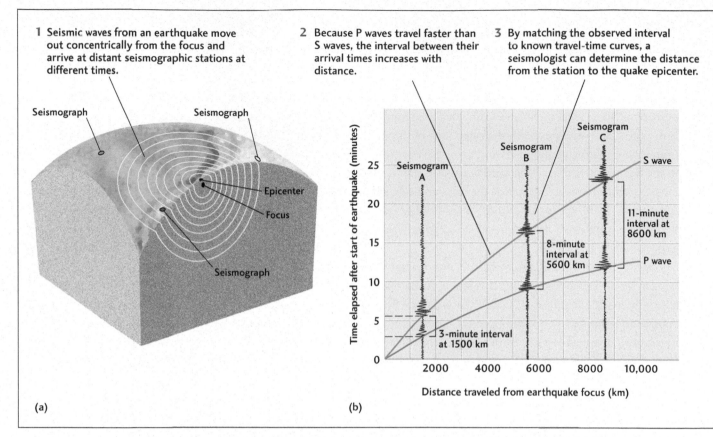

1 Seismic waves from an earthquake move out concentrically from the focus and arrive at distant seismographic stations at different times.

2 Because P waves travel faster than S waves, the interval between their arrival times increases with distance.

3 By matching the observed interval to known travel-time curves, a seismologist can determine the distance from the station to the quake epicenter.

(a)

(b)

Figure 3–10 These travel time curves, based on the speeds at which P and S waves travel, illustrate that the interval of time between the arrival of the two wave types increases with distance from the earthquake focus. This difference in arrival time can be used to calculate the radial distance from source to seismometer. If three or more such distances are calculated and plotted, the location of the source can be determined by triangulation.

station. (Figure 3-10a). The difference in time between when the two waves arrive at the seismometer is directly proportional to the distance between the seismic station and the earthquake focus, which allows us to determine where the earthquake occurred. If one only had data from a single seismometer, it would be possible to determine that the earthquake occurred somewhere along a ring surrounding the seismic station, with the radius of the ring determined by the P- and S-wave offset (Figure 3-10b). Two seismometers can be used together to narrow the potential epicenter position to two points by detecting two overlapping rings. With three or more seismometers, it is possible to pinpoint the exact location of the epicenter, a process known as **triangulation.**

Scientists have deployed networks of seismometers all over the world, allowing them to identify epicenter locations with substantial accuracy and precision within minutes after an earthquake strikes. Such rapid assessment facilitates the quick response of search-and-rescue teams and relief organizations.

Tectonic Loading and the Earthquake Cycle

Throughout the 20th century, as seismologists examined the increasingly accurate data for earthquake locations from new seismic stations, they began to see interesting patterns. They found shallow-focus earthquakes along mid-ocean ridges in all of Earth's ocean basins and shallow to deep-focus earthquakes along oceanic trenches. As discussed in Chapter 2, the developers of plate tectonic theory used these patterns to define the boundaries of the tectonic plates and to prove the existence of subduction zones. It is now clear that rather than being caused by a god, as suggested by the Yurok legend that opens this chapter, earthquakes occur as a result of lithospheric plate movements over the ductile asthenosphere. As such, we can say that faults are "tectonically loaded" by motions of the plates.

Because plates tend to move in the same direction for tens of millions to hundreds of millions of years and

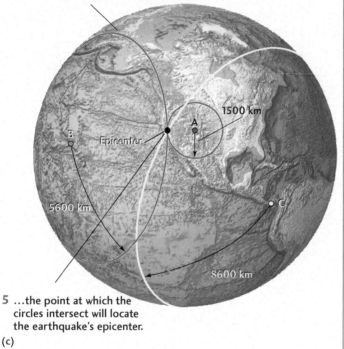

4 If the seismologist then draws a circle with a radius calculated from the travel-time curves around each seismographic station,...

5 ...the point at which the circles intersect will locate the earthquake's epicenter.

(c)

exponentially over time.[26] Due to the massive size of the Tōhoku earthquake, aftershocks may continue for years.[27] In addition to these events, aseismic deformation may take place during this *postseismic* phase as the ductile portion of the crust that was too hot to break reacts to the new state of stress by flowing. Eventually,

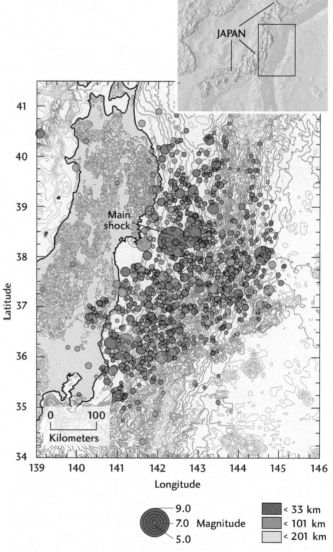

Figure 3–11 Map of the Tōhoku earthquake (magnitude 9.0) sequence in Japan, showing foreshocks, the main shock (March 11, 2011), and aftershocks. The depth of ocean water is indicated with blue contour lines; land elevation with gray contour lines; and the coastline with a thick gray line. The size of the dots indicates the relative size of each earthquake, with larger dots representing larger–magnitude events. Dot colors indicate depth for each earthquake source, from shallower (red) to deeper (progressing from orange to yellow). *(Image by C. J. Ammon, Penn State, data from the USGS)*

because friction on fault planes is finite, elastic energy builds up and is released (called *coseismic slip*) over and over again, creating what seismologists refer to as the **earthquake cycle**. Not all of the stored elastic energy is necessarily released in a single shock, however. Slip tends to be uneven, with some parts of the fault plane slipping more than others. As a result, stress may build up in new areas, and the earthquake may be followed by numerous **aftershocks** as previously stable rocks fail and release more elastic energy. Indeed, the magnitude 9.0 Tōhoku earthquake that struck the northeastern coast of Japan was preceded by four strong quakes with magnitudes between 6.0 and 7.2 that began 2 days earlier.[24] These **foreshocks** appear to have created the conditions necessary to trigger the main shock, which was then followed by hundreds of aftershocks[25] that enabled the crust to adjust to its new state of stress (Figure 3-11).

Post-earthquake adjustments can occur for months or even years after the initial rupture, but most aftershocks occur within days, their frequency falling off

the motion of the tectonic plates begins to build up strain in the rocks along the fault again (the *interseismic* phase), setting the stage for the next earthquake.

Many faults tend to fail in segments rather than along their entire length. The movement of one segment may increase the strain in adjacent segments, making them more likely to fail in the future. Turkey's North Anatolian fault, a transform boundary between the Anatolian and Eurasian plates similar to California's San Andreas fault, behaved somewhat like a zipper throughout the 20th century, rupturing from east to west in a series of events between 1939 and 1999 (Figure 3-12). Based on this progressive westward slip, scientists have expressed concern for the Turkish city of Istanbul, which lies just north of the fault and west of the last segment to rupture.[28] The 1999 Kocaeli earthquake demonstrated what could happen when the North Anatolian fault breaks near a populated area. More than 17,000 people lost their lives in the city of Izmit and its surrounding area as severe shaking collapsed numerous buildings.[29]

■ Earthquakes occur when rocks slip past one another along a fault plane, releasing elastic energy stored in the rocks. The greater the amount of rock deformed, the greater the amount of energy released and, as a consequence, the greater and longer the ground shaking during the earthquake.

■ When seismic energy is released along a fault plane, it radiates outward through surrounding rock as several types of waves, including primary, secondary, and surface waves. Each wave type travels at a different rate, with primary waves traveling most rapidly.

■ A seismometer is an instrument used to detect and record seismic waves.

■ The time offset between when P and S waves arrive at a seismometer can be used to determine where an earthquake originated.

■ The S-wave shadow zone indicates that the Earth's outer core is a liquid. The P-wave shadow zone further indicates that Earth has a solid inner core.

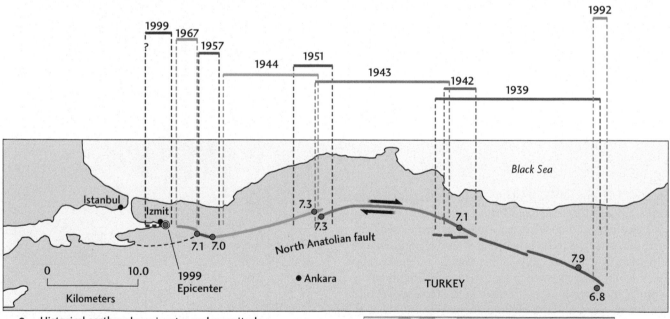

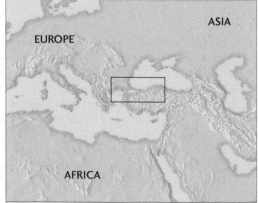

Figure 3–12 Turkey's North Anatolian fault behaved somewhat like a zipper during the past century. A series of segments that ruptured along the fault caused earthquakes between 1939 and 1999.

■ The earthquake cycle consists of the gradual buildup of deformation during an interseismic phase, release of stored elastic energy during an earthquake (coseismic phase), and postseismic adjustment of the crust to the new stress conditions.

ASSESSING EARTHQUAKE HAZARDS

The energy released during an earthquake is proportional to the size—the length and width—of the fault plane that ruptures and the amount of slip along the fault. The greater the amount of energy released during an earthquake, the greater the amplitude of seismic waves at a given distance from the earthquake source. Likewise, the greater the energy released, the worse the destruction.

Earthquake Moment Magnitude

Seeking a way to compare different earthquakes, Charles Richter developed the first magnitude scale in 1935. It was based on the amplitude of the waves recorded in seismograms, along with a correction factor that took into consideration the distance between the earthquake epicenter and the seismographic station. The correction factor was necessary because the energy of earthquake shaking dissipates with distance due to frictional losses, making the amplitude of the incoming waves decrease with distance from the epicenter. Richter's scale was in wide use until the 1970s, when seismologists developed the **moment magnitude scale** to describe the amount of energy released by earthquakes.

The *seismic moment*, M_0, is a function of the area of the fault plane that slipped (A, in square centimeters), the average amount of slip along the fault plane (u, in centimeters), and a rock property known as the shear modulus (μ, in grams per centimeter-second squared) that describes the amount of deformation a rock experiences per unit of stress applied to it; in other words, the rock's rigidity:[30]

$$M_0 = \mu A u$$

Because the range in size of earthquakes is so large, the moment magnitude scale (M_w) uses a base-10 logarithm of the seismic moment to describe each event:

$$M_w = \frac{2}{3} \log_{10}(M_0) - 10.7$$

The magnitude of an earthquake is the same no matter where it is measured, because the energy released during each earthquake is a fixed amount. The use of the base-10 logarithm means that the energy released does not increase linearly with successively higher values of magnitude. Instead, each unit on the moment magnitude scale represents a 31.6-fold increase over the preceding one. For example, the energy released by a M_w 5 event is 31.6 times more than that of a M_w 4 event, and the area of fault plane involved and the amount of slip are correspondingly larger (Figure 3-13). Events larger than magnitude 8 are considered to be "great" earthquakes. One great earthquake occurs about every year (Table 3-1).[31] More than 100 earthquakes greater than magnitude 6 or 7 occur annually. Smaller earthquakes are even more frequent, with thousands occurring each year, but those less than magnitude 3 to 4 are barely noticeable. The largest earthquake in recent history, the Indian Ocean

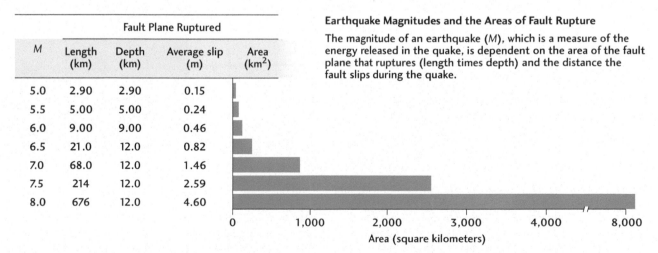

M	Fault Plane Ruptured			
	Length (km)	Depth (km)	Average slip (m)	Area (km²)
5.0	2.90	2.90	0.15	
5.5	5.00	5.00	0.24	
6.0	9.00	9.00	0.46	
6.5	21.0	12.0	0.82	
7.0	68.0	12.0	1.46	
7.5	214	12.0	2.59	
8.0	676	12.0	4.60	

Earthquake Magnitudes and the Areas of Fault Rupture

The magnitude of an earthquake (*M*), which is a measure of the energy released in the quake, is dependent on the area of the fault plane that ruptures (length times depth) and the distance the fault slips during the quake.

Figure 3–13 Earthquake magnitude is proportional to the product of the area of fault plane that slipped and the amount of slip along that plane. A magnitude 8 earthquake, for example, would be generated by a fault plane area of about 7700 km² that slipped about 5 m.

TABLE 3–1 Frequency and Occurrence of Earthquakes	
Magnitude	Average Annually
8 and higher	1
7.0–7.9	17
6.0–6.9	134
5.0–5.9	1319
4.0–4.9	13,000 (estimated)
3.0–3.9	130,000 (estimated)
2.0–2.9	1,300,000 (estimated)

From: *U.S. Geological Survey,* "Earthquake Facts and Statistics," http://earthquake.usgs.gov/earthquakes/eqarchives/year/eqstats.php.

earthquake of December 26, 2004, had a magnitude between 9.1 and 9.3, equivalent in destructive power to 23,000 Hiroshima-type atomic bombs, and spawned a tsunami that killed nearly 230,000 people[32] (we address tsunamis in more detail in Chapter 12).

Earthquake Intensity

While the moment magnitude for an individual earthquake is fixed, the amount of damage produced by the event varies considerably with location. Citizens of New York State do not feel tremors associated with earthquakes in California, for example, because seismic waves lose energy as they travel through Earth, making them noticeable only to seismometers sensitive enough to pick up tiny vibrations. In addition, earthquakes that occur in sparsely populated areas cause little damage because there are few structures that can be destroyed.

Some areas experience greater damage than others during a given earthquake because of differences in local geologic conditions. Solid rock dampens seismic waves, whereas unconsolidated sediments amplify shaking (Figure 3-14). Consider, for example, what happened in 1989 during the M_w 6.9 Loma Prieta earthquake, named for a mountain in coastal California below which the

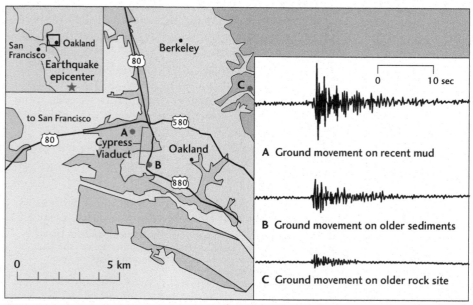

Holocene bay mud, 0–40 m thick (<10,000 years old)

Quaternary sediments, 0–50 m thick (<2 million years old)

Older rocks, mostly sandstone and shale (>60 million years old)

(a)

(b)

Figure 3–14 Impact of geologic materials on ground–shaking amplitude. About two–thirds of the deaths during the Loma Prieta earthquake in 1989 occurred on an overpass located along one stretch of Interstate 880 in Oakland, California. Much of the overpass's failure can be attributed to underlying geologic materials that did not receive necessary attention at the stage of engineering design. The overpass failed between points A and B, where it was built on young bay muds, but did not fail above older sediments and rocks. Collapse of much of the upper tier of this two–tier structure, the Cypress Street Viaduct, killed 42 people. During an aftershock, seismograms of ground motion on different geologic materials illustrated the great differences in ground shaking at these different sites. The amplitude of ground shaking was least for the older rocks and greatest for the young, unconsolidated sediments. *(H. G. Wilshire/USGS)*

(a)

(b)

(c)

Figure 3–15 Some sediments and fill can liquefy during an earthquake. (a) Liquefaction caused much damage in the historic Marina district of San Francisco during the 1989 Loma Prieta earthquake. The Marina district is underlain by man-made fill that was used to build new land in a shallow bay after the 1906 San Francisco earthquake that occurred on the San Andreas fault. (b) Liquefaction can produce "boils" when the strength of saturated sediment is reduced by a sudden change in stress, causing the soil to behave like a liquid. (c) Seismic Hazard Map showing the past and possible future occurrence of liquefaction (green) and landslides (blue) in the Marina District and surrounding areas of San Francisco. *(a: Lloyd Cluff/ CORBIS; b: Dr. Robert Kayen/USGS and UCLA; c: Courtesy California Geological Survey)*

earthquake began. Although the epicenter was nearly 75 km southwest of San Francisco, the earthquake wreaked havoc in that city and throughout the Bay Area, causing bridges, overpasses, and buildings to collapse (Figure 3-15a). In nearly all cases, these structures had been built on weak, unconsolidated, wet sediments or human-made fill that amplified ground shaking as seismic waves passed beneath them.[33] Wet sediments behave much the way Jell-O does when tapped, vibrating far more than surrounding rocks and older sediments that are more consolidated, rigid, and resistant (see Figure 3-14).

Some sediments have the potential to liquefy, or behave more like liquids than solids, during intense ground shaking. When **liquefaction** occurs, buildings and elevated freeways sink into the ground as water flows to the surface, which sometimes erupts to produce sediment-laden "volcanoes" (Figure 3-15b). In addition, if the sinking is not uniform across a structure, the structure will crack and break up in the process,

potentially leading to its collapse. During the Loma Prieta earthquake, liquefaction destroyed numerous buildings in San Francisco's historic Marina District, a neighborhood constructed on human-made fill that had been dumped into a shallow bay to create more developable land in the city after the 1906 earthquake.

Given the spatial variability of earthquake damage, an older measure of earthquake energy, known as the Mercalli intensity scale, is useful as a tool for hazard assessment. This scale, developed by Italian seismologist Giuseppe Mercalli in 1902 and since adjusted and renamed the **modified Mercalli intensity scale (MMI)**, is based on the effect of an earthquake on human structures and on how people experience the earthquake's effects (Table 3-2). Each of the 12 grades is assigned a Roman numeral, and the scale is not logarithmic. It has been carefully designed so that the increases in intensity from one step to the next are roughly equivalent. While an MMI III event feels like little more than a passing truck, an MMI VIII event can destroy masonry

TABLE 3–2 Modified Mercalli Intensity Scale

I.   Not felt except by a very few under especially favorable conditions.

II.   Felt only by a few persons at rest, especially on upper floors of buildings.

III.   Felt quite noticeably by persons indoors, especially on upper floors of buildings. Many people do not recognize it as an earthquake. Standing motorcars may rock slightly. Vibrations similar to the passing of a truck.

IV.   Felt indoors by many, outdoors by few during the day. At night, some awakened. Dishes, windows, doors disturbed; walls make cracking sound. Sensation like heavy truck striking building. Standing motorcars rocked noticeably.

V.   Felt by nearly everyone; many awakened. Some dishes, windows broken. Unstable objects overturned. Pendulum clocks may stop.

VI.   Felt by all, many frightened. Some heavy furniture moved; a few instances of fallen plaster. Damage slight.

VII.   Damage negligible in buildings of good design and construction; slight to moderate in well-built ordinary structures; considerable damage in poorly built or badly designed structures; some chimneys broken.

VIII.   Damage slight in specially designed structures; considerable damage in ordinary substantial buildings with partial collapse. Damage great in poorly built structures. Fall of chimneys, factory stacks, columns, monuments, walls. Heavy furniture overturned.

IX.   Damage considerable in specially designed structures; well-designed frame structures thrown out of plumb. Damage great in substantial buildings, with partial collapse. Buildings shifted off foundations.

X.   Some well-built wooden structures destroyed; most masonry and frame structures destroyed. Rails bent.

XI.   Few, if any (masonry) structures remain standing. Bridges destroyed. Rails bent greatly.

XII.   Damage total. Lines of sight and level are distorted. Objects thrown into the air.

U.S. Geological Survey, 2000. "The Severity of an Earthquake." U.S. Government Printing Office: 1989-288-913 http://pubs.usgs.gov/gip/earthq4/severitygip.html.

buildings, and an MMI XII event results in total destruction (Box 3-2).

One of the greatest contributions of geologists to assessing the hazards associated with earthquakes is careful mapping of faults and rock types in seismically active areas. These maps can be used to predict the intensity of earthquake shaking associated with events of various magnitudes and different geologic materials, as well as to identify locations prone to liquefaction and landsliding during shaking. Such information is valuable for planning decisions and in developing building codes to minimize loss of life and property damage. Unfortunately, while many fault planes rupture at Earth's surface, forming easily mapped scarps, others do not, meaning that their existence is unknown until an earthquake occurs.

Faults that do not rupture to Earth's surface often display reverse, or thrust, behavior (see Chapter 2), which can cause upward ground accelerations that throw objects into the air. Such was the case in the 1994 M_w 6.7 Northridge earthquake in Southern California, which caused upward ground accelerations equivalent to Earth's gravitational acceleration, killing 60 people

and injuring over 7000, as apartment buildings, parking garages, and freeway overpasses collapsed.[34]

Looking for Earthquakes in the Geologic Record

When a fault is buried beneath Earth's surface, or is visible but has not slipped within historical time, it could be difficult to determine how active it is and how much risk it poses. If we look carefully at the geologic record, however, we often find that earthquakes leave evidence of their occurrence in the landscape and geologic record. Consider, for example, the fault that destroyed Scythopolis, the ancient Roman city mentioned at the beginning of this chapter. The Dead Sea transform is a left-lateral transform fault that lies beneath the Jordan River valley, which connects the Sea of Galilee to the Dead Sea, a large inland lake between Israel and Jordan.

It turns out that lake sediments in the Dead Sea record numerous earthquakes on the transform, a fact that only became known in the last few decades as a result of changes in the lake's water balance. For the past 40 years, the level of the Dead Sea has dropped

BOX 3–2 CASE STUDY

Magnitude Versus Intensity: The 2010 Haiti and Chile Events

The difference between earthquake magnitude and intensity could not have been made clearer than by the example of two earthquakes that occurred early in 2010. On January 12, a magnitude 7.0 strike–slip earthquake on the Enriquillo–Plantain Garden fault, one of two faults that define the plate boundary between the Caribbean and North American plates, struck the nation of Haiti.[35] The epicenter lay immediately west of the capital city of Port-au-Prince, home to nearly 15% of the country's population. The shaking collapsed 97,294 homes and damaged another 188,383 in and around Port-au-Prince, killing at least 222,570 people and injuring 300,000 more (**Figure 1**). An estimated 1.3 million people were left homeless. With the presidential palace destroyed and many government ministers killed, the international community stepped in to provide aid, but debris that clogged streets hampered these efforts for weeks. Many people died from lack of food, water, or medical care. The near-total devastation of the capital city will require decades of reconstruction, a catastrophic blow to the Western Hemisphere's poorest nation.

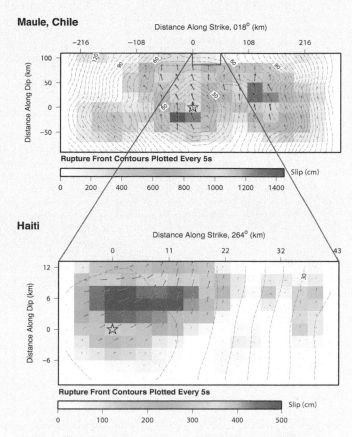

Figure 2 Comparison of fault ruptures for the Chilean and Haitian earthquakes of 2010. The Chilean rupture (M_w 8.8) was the result of about 50% more slip on a much greater fault area than the Haitian rupture (M_w 7.0). Rupture area is the product of fault rupture length (*x*–axis) and distance along the fault plane's dip (*y*–axis). Colors indicate the amount of slip on different parts of the rupture plane. *(Finite fault models by Gavin Hayes, USGS National Earthquake Information Center)*

Figure 1 Nearly 100,000 homes collapsed and close to 200,000 were damaged in the vicinity of Haiti's capital city, Port-au-Prince, during the 2010 magnitude 7.0 earthquake. Homes on this slope were destroyed or severely damaged. *(AP Photo/Albuquerque Journal, Adolphe Pierre-Louis)*

On February 27, a magnitude 8.8 subduction–zone earthquake occurred off the west coast of Chile where the Nazca Plate subducts beneath the South American Plate (see Figure 2–3). Though it ruptured a much larger area (**Figure 2**) and released more than 500 times as much energy as the Haitian quake, fewer than 600 people were killed and only 12,000 were injured.[36] In part, the lower casualty numbers are attributable to the location of population centers relative to the hypocenter. The Chilean earthquake occurred offshore and at a depth of 35 km, while the Haitian earthquake occurred directly beneath a populated area at a depth of only 13 km. The vastly different casualty

(continued)

(continued)

Estimated Population Exposed to Earthquake Shaking

Estimated population exposure (k = ×1,000)		- -*	- -*	7,090k*	6,308k	777k	749k	1,844k	710k	137k
Estimated modified Mercalli intensity		I	II–III	IV	V	VI	VII	VIII	IX	X+
Perceived shaking		Not felt	Weak	Light	Moderate	Strong	Very strong	Severe	Violent	Extreme
Potential damage	Resistant structures	None	None	None	Very light	Light	Moderate	Moderate/heavy	Heavy	Very heavy
	Vulnerable structures	None	None	None	Light	Moderate	Moderate/heavy	Heavy	Very heavy	Very heavy

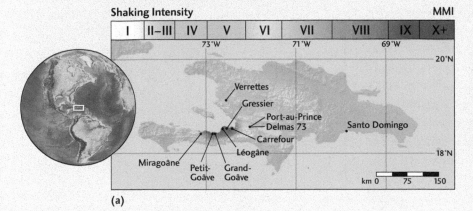

Shaking Intensity   **MMI**

| I | II–III | IV | V | VI | VII | VIII | IX | X+ |

Selected City Exposure

MMI	City	Population
X	Petit-Goâve	15k
X	Léogâne	12k
X	Grand-Goâve	5k
IX	Carrefour	442k
IX	Gressier	4k
VIII	Miragoâne	6k
VIII	Port-au-Prince	1,235k
VII	Delmas 73	383k
V	Verrettes	49k
III	Santo Domingo	2,202k

(a)

Estimated Population Exposed to Earthquake Shaking

Estimated population exposure (k = ×1,000)		- -*	- -*	454k*	1,667k*	527k*	7,578k	5,124k	0	0
Estimated modified Mercalli intensity		I	II–III	IV	V	VI	VII	VIII	IX	X+
Perceived shaking		Not felt	Weak	Light	Moderate	Strong	Very strong	Severe	Violent	Extreme
Potential damage	Resistant structures	None	None	None	Very light	Light	Moderate	Moderate/heavy	Heavy	Very heavy
	Vulnerable structures	None	None	None	Light	Moderate	Moderate/heavy	Heavy	Very heavy	Very heavy

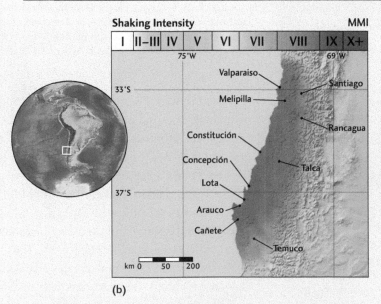

Shaking Intensity   **MMI**

| I | II–III | IV | V | VI | VII | VIII | IX | X+ |

Selected City Exposure

MMI	City	Population
VIII	Arauco	25k
VIII	Lota	50k
VIII	Constitución	38k
VIII	Concepción	215k
VIII	Cañete	20k
VII	Melipilla	63k
VII	Talca	197k
VII	Rancagua	213k
VII	Temuco	238k
VII	Santiago	4,837k
VI	Valparaiso	282k

Estimated exposure only includes population within the map area.

(b)

Figure 3 Comparison of the estimated number of people exposed to earthquake shaking in Haiti and Chile during earthquakes in 2010.

(continued)

numbers for these two events can also be understood by studying the modified Mercalli intensity values associated with each (**Figure 3**). The city of Port–au–Prince and its surroundings experienced seismic shaking in the range of severe (Mercalli VIII) to extreme (Mercalli X), and more than 1.7 million people lived in a location where the Mercalli intensity exceeded VIII. In Chile, by contrast, shaking intensity did not exceed VIII, and only about 350,000 people were exposed to this level. Thus, while the Chilean earthquake had a far larger magnitude, it had a much lower intensity. Still, an estimated 800,000 people were displaced by the damage or destruction of 370,000 homes, and the earthquake is thought to have caused total damages in excess of $30 billion. It will, therefore, take a long time for Chile to recover as well.

Besides the lower intensity of shaking than in the Haitian earthquake, the lower casualty figures in Chile reflect stringent building codes adopted by that country in the aftermath of a magnitude 9.5 earthquake that occurred in 1960, the strongest earthquake ever recorded. Port–au–Prince, on the other hand, was surrounded by sprawling shantytowns that rapidly collapsed, and lack of reinforcement in buildings in the capital city itself led many multistory structures to pancake and crush their inhabitants. We address earthquake–resistant engineering later in the chapter.

because of reduced inflow of surface water from the Jordan River, leading to the exposure of sediments that once were on the floor of the lake. Erosion by streams has carved numerous small canyons into the sediments, providing a window into the history of the lake and into past earthquake shaking (Figure 3-16a).

Thin horizontal layers of sediment are readily apparent in the canyon walls, each representing a year or perhaps a few years of deposition of sediment in the Dead Sea. Within the thinly layered horizontal sediments appear occasional thicker layers with an odd, swirling texture produced by earthquake disturbance (Figure 3-16b). These contorted sediments started out as horizontal layers located at the sediment–water interface at the bottom of the Dead Sea. There, the sediments were watery enough to slip and slide when the ground shook during an earthquake. Sediments farther below the lake floor were denser and stiffer due to compaction from the weight of the overlying sediments and so did not deform. After the earthquake disrupted the lake floor, normal sedimentation resumed, mantling the disturbed layer with thin horizontal sheets of sediment.

(a)

(b)

Figure 3–16 (a) A canyon cut into Dead Sea lake sediments (Dead Sea in background) is exposed by the erosion of streams as the level of the Dead Sea drops. (b) One of 13 seismite layers in lake sediments is evidence of ground shaking that occurred during a paleo–earthquake. Geological dating indicates that paleo–earthquakes recorded in these seismites occurred between 760 BCE and 1927 CE. *(Kirsten Menking)*

Geologists have found 13 of these disturbed *seismite* layers within the otherwise horizontally layered sediments and have used geological dating techniques (discussed in Chapter 6) to determine that their associated earthquakes occurred between 760 BCE and 1927 CE[37] Dividing the total length of time that it took to create all of the seismite layers by the total number of seismite layers yields an earthquake recurrence interval of 100 to 300 years. Based on the archeological evidence at Scythopolis and soft-sediment deformation produced by earthquakes during modern times, as well as historical records of damage at other sites, these tremors likely measured greater than 5.5 in magnitude.

Finding evidence of earthquakes in the geologic record is the subject of **paleoseismology**, a research area that has revealed important data about many faults in the United States, including, for instance, the information about the drowned coastal forests in the Pacific Northwest at the beginning of this chapter. Other locations that have received considerable attention include the San Andreas[38] and Hayward[39] faults of California and the Wasatch[40] fault of Utah. In these studies, scientists used backhoes to dig trenches across the fault lines and examined the soils exposed in the trench walls. Small offsets, or folds in soil layers, as well as fluid flow features produced by liquefaction attest to episodes of shaking. In addition, when fault motions are normal or reverse, a scarp forms, which is then subject to erosion by water and wind and the general downslope movement of mass due to gravity (see Chapter 8) (Figure 3-17). Deposition of the eroded sediment at the base of the scarp buries vegetation on the surface of the down-thrown fault block. The buried organic matter imparts a black color to the soil, making it easily distinguishable from the scarp debris. Repeated episodes of slip along the fault result in alternating sequences of buried soils and eroded scarp materials, and the age of the buried organic matter is readily determined through geological dating (see Chapter 6), making it possible to gauge when earthquakes occurred and what their likely recurrence intervals are.

The Potential for Future Earthquakes

While the geologic and archeological record provide us extremely useful information about the activity of faults, at present geoscientists know of no reliable way to use that information to *predict* a future earthquake in terms of its exact timing, location, and size (Box 3-3). *Forecasts*—which state the approximate location and probability of a future event—are possible, however, and are valuable as planning tools to help ensure that schools, hospitals, roads, dams, and other vital infrastructure components are not built on top of or near active faults.

The first step in assessing hazard potential is to examine a site in its tectonic context. It is reasonable

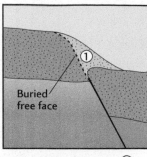
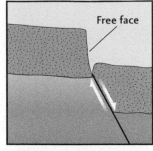
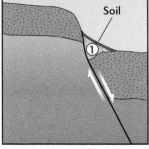
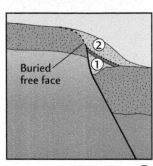

(a) First surface faulting event   (b) First colluvial wedge ①

(c) Second surface faulting event   (d) Second colluvial wedge ②

Figure 3–17 Alternating buried soils and colluvial wedges produced by normal faulting.

to forecast, for example, that future events will occur along plate margins, where brittle crust is deforming and elastic energy is accumulating along locked faults, and indeed, more than 95 percent of all earthquakes occur at the boundaries between two plates (see Figure 2-19). Most of the other 5 percent occur along ancient plate boundaries where plates have been sutured together or rifts are no longer actively spreading. These boundaries are weak spots in Earth's lithosphere. Figure 3-18 shows the relative earthquake hazard throughout the United States. The San Andreas fault in California, the Juan de Fuca subduction zone, and the Aleutian subduction zone—all plate boundaries—stand out as high hazard areas. The Great Basin, an area in eastern California, Nevada, and western Utah where Earth's crust is stretching apart along normal faults, also shows high earthquake hazard potential.

Surprisingly, the region represented by the states of Arkansas, Illinois, Missouri, Tennessee, and Kentucky shows relatively high seismic hazard, despite being in the middle of the North American tectonic plate. This area is the site of an ancient rift in Earth's crust associated with the opening of the Iapetus Ocean, a precursor to the Atlantic Ocean that formed around 542 to 488 million years ago, before the assembly of Pangaea (see Chapter 2). The rift is now buried under hundreds of meters of sediment,[41] but some faults, such as the New Madrid fault zone, occasionally slip, causing

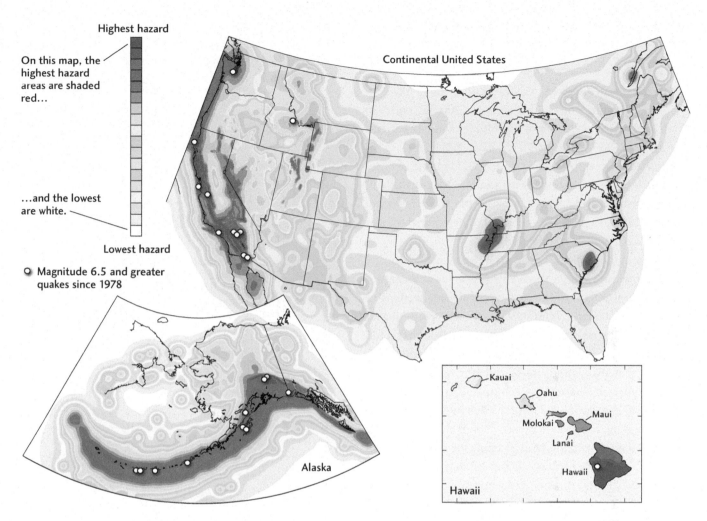

Figure 3–18 Seismic hazard potential in the United States. Regions with the highest hazard potential are located along the western coast, bounded by plate boundaries and active faults, above the hot spot located in Hawaii, and along ancient faults in the central and eastern United States. *(Adapted from U.S. Geological Survey, http://earthquake.usgs.gov/hazards/)*

earthquakes. One hypothesis is that these reactivations happen in response to the great stresses occurring at the edges of the North American Plate that are somehow transmitted into the interior of the plate.[42] Another hypothesis suggests that erosion of the landscape by the Mississippi River after Earth's last ice age reduced the amount of pressure acting downward on fault planes in the ancient rift. This downward pressure tends to keep the faults locked, whereas releasing it allows them to slip.[43] While we are not yet sure which hypothesis, if either, is correct, it is certain that the New Madrid fault zone is capable of producing strong earthquakes. In 1811 and 1812, three earthquakes with magnitudes greater than 7 struck the region.[44] Ground motions reportedly caused the Mississippi River to change its course and also generated numerous liquefaction features. The historical record of earthquakes in this location justifies its classification as an area of potential high hazard.

When attempting to determine the probability of future slip events, it is useful to know where faults are locked and where rocks are accumulating elastic energy. Faults tend to slip in segments shorter than the entire fault, and slip along one segment often releases elastic energy there but increases stresses along adjacent segments, making them more likely to fail in the future.[26] Maps of earthquake epicenters reveal those parts of faults that have slipped recently versus those on which little movement has occurred and can be used to help figure out which locations may be likely to slip in the future. GPS satellites also provide important data in that the satellites can be used to monitor small movements of Earth's crust to see how much deformation is accumulating in the rocks on either side of faults. With these pieces of information and the elastic rebound model, seismologists can estimate the probability of future earthquakes during a given time interval. The U.S. Geological Survey

BOX 3–3 EMERGING RESEARCH

The Parkfield Earthquake and the Difficulties of Prediction

The town of Parkfield, California, lies along the San Andreas fault, midway between Los Angeles and San Francisco. Earthquakes with a magnitude of about 6 shook the area in 1857, 1881, 1901, 1922, 1934, and 1966, or on average, every 22 years (**Figure 1**). Mapping of offset features indicated that roughly the same area along the fault slipped in each event. These facts led the U.S. Geological Survey (USGS) to plan an experiment to determine whether earthquakes might in fact be predictable.

Beginning in 1985, USGS scientists installed equipment to measure every conceivable precursor of an earthquake and also deployed a network of seismic stations to record the next event (**Figure 2**).[46] On the premise that faults might slip due to changes in fluid pressure that could lubricate the fault zone, geologists installed groundwater wells that they could use to monitor water levels. Tilt meters and creep meters were installed to keep track of tiny vertical and horizontal motions that might foreshadow an impending rupture. Likewise, surveying benchmarks were installed across the fault to allow for repeated measurements between points, in an attempt to determine whether the ground moved before an earthquake. Chinese scientists had found that earthquakes are sometimes preceded by changes in electrical conductivity of the ground for reasons that are not understood, so the Parkfield site was also instrumented to measure conductivity. Long tubes filled with oil and fitted with pressure sensors, called strain meters, were dropped down deep boreholes to record rock deformation. If the ground deformed, the oil would be displaced upward, causing a change in the pressure sensor reading.

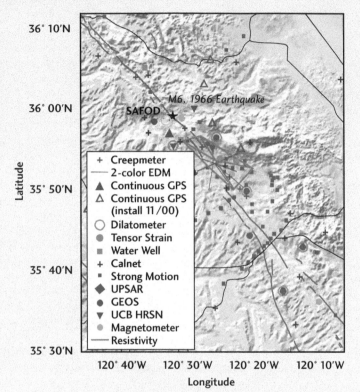

Figure 2 Because of the frequency of earthquakes here, the Parkfield area has a large number of instruments to monitor ground motion, groundwater level, and other parameters.

Given the roughly 22–year recurrence interval for the Parkfield earthquake, the USGS was hopeful that installation of the equipment would yield abundant information in the near term. However, 1988 came and went, and then 1989, and 1990. The long stretch between the previous two ruptures (1934 and 1966) of 32 years revealed that there might be a bit of a wait for the next event, but then 1998 went by, 32 years after 1966. Scientists were puzzled and began to despair that all of their efforts were for naught. Some even questioned whether the experiment should continue.

On September 28, 2004, the Parkfield segment of the San Andreas fault finally slipped,[46] generating a magnitude 6 earthquake like all of its predecessors. While the same area of the fault experienced slip as in previous earthquakes, in some ways the event was quite different. Previous earthquakes had begun in the north and ruptured southward. This event reversed direction. As far as prediction goes, analysis of the instrumental records acquired

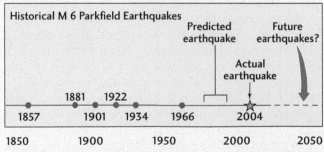

Figure 1 A plot of dates and earthquakes shows that the average recurrence interval between earthquakes in Parkfield is 22 years.

before the earthquake revealed no hint of the event to come.[47] No foreshocks preceded the main shock. The creep and strain meters showed no changes leading up to the event. No unusual electrical activity or radiowave signals were detected. Unfortunately, it appears that earthquake prediction remains an elusive goal. Still, the instrumentation at Parkfield revealed important new information about what causes faults to break in segments and about how bends in faults originate. For these reasons, the experiment was definitely worth conducting, and the site continues to be monitored in order to capture future events.

New experiments are also yielding important information about fault behavior. Between 2004 and 2007, the Parkfield experiment was supplemented by the San Andreas Fault Observatory at Depth (SAFOD).[48] Funded by the National Science Foundation's EarthScope program, scientists drilled a borehole directly into a creeping section of

the fault just north of Parkfield to study the characteristics of the fault zone itself and to install instruments to measure the stresses associated with fault movements. Examination of the rock cuttings produced during drilling revealed that the fault contains trace amounts of the minerals talc and serpentine, as well as coatings of clay minerals, well known for their slipperiness and weakness.[49] In addition, the fault zone appears to be made of fault gouge, a sediment made of pulverized rock. Pebbles within the gouge show evidence of being aligned in the direction of fault slip, and the weakness of the gouge may facilitate the steady creep of this segment. One hypothesis—that creeping behavior might be caused by high fluid pressures within the fault zone—appears to be invalid, as geologists found no evidence of high water or gas levels within the fault zone. Given these interesting findings, plans are underway for a borehole in a locked section of the fault. Stay tuned!

---

has determined, for example, that there is a greater than 99% probability of an earthquake of magnitude 6.7 occurring somewhere within the state of California in the next 30 years (Figure 3-19).[45] Higher-magnitude events have lower, but still significant, probabilities of occurring within that same time period.

While it clearly would be an advantage to have greater predictive capability, earthquake forecasts provide useful information for emergency management preparedness, development of building construction codes, and neighborhood zoning decisions.

■ The moment magnitude scale is a measure of earthquake size, and incorporates information about the area of fault plane rupture, the amount of slip during the earthquake, and the amount of deformation rocks undergo per unit of stress applied.

■ The modified Mercalli index is a measure of earthquake destructiveness and the experience of shaking. Intensity values are highest immediately adjacent to faults as well as in areas of unconsolidated sediment, where shaking is amplified.

■ Traces of earthquakes remain in the geologic record. The discipline of paleoseismology examines these clues for insights into the activity of faults.

■ Although geologists have an increasing understanding of the causes of earthquakes and where earthquakes are most likely to occur, they have not yet identified any phenomena that consistently precede an earthquake, so they are not yet able to predict the exact location and timing of earthquakes.

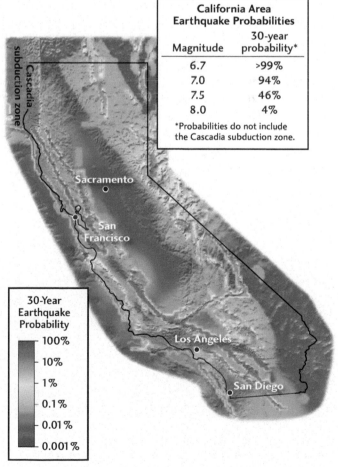

California Area Earthquake Probabilities	
Magnitude	30-year probability*
6.7	>99%
7.0	94%
7.5	46%
8.0	4%

*Probabilities do not include the Cascadia subduction zone.

30-Year Earthquake Probability

- 100%
- 10%
- 1%
- 0.1%
- 0.01%
- 0.001%

Figure 3-19 Thirty-year probabilities for different magnitude earthquakes in the state of California.

■ Though prediction is not yet possible, forecasting provides valuable information for emergency preparedness and to guide development patterns and construction codes.

MINIMIZING THE RISKS FROM EARTHQUAKES

More than 650,000 people died from earthquakes and their associated hazards in the first decade of the 21st century alone, and regrettably, many of the world's largest cities are located directly over or adjacent to plate boundaries in some of the world's poorest countries.[50] It is both possible and prudent for engineers, geoscientists, and politicians to work together to create appropriate building codes for seismically safe structures. Information needed for proper planning includes:

■ The location of possible faults, as they are the sources of earthquakes.

■ The characteristics of likely earthquakes (magnitude, duration, and type of faulting).

■ The type of geologic material between the earthquake source and structures.

Building codes in seismically active states such as California require new structures to be set back a minimum distance from active faults, and those on sediments that may liquefy must use special building techniques (Box 3-4).[51] Furthermore, California law requires that home sellers divulge to potential buyers whether their homes are in areas of seismic, landslide, or liquefaction risk, as well as to disclose any of the house's known structural weaknesses.[52] Several factors can contribute to

weakness. For example, previously, houses were simply placed on top of their foundations and frequently slipped off during earthquakes, causing structural damage or total collapse. Building codes now require structures to be bolted to their foundations to prevent this from happening, and older structures lacking these bolts can be retrofitted.[53] In earlier years, buildings on hillsides were often constructed on top of piers or pilings. When shaking occurred, these piers buckled, ripping houses apart. To solve this problem, piers can be braced, or a foundation of poured, reinforced concrete can be installed underneath the house. By simply installing cross-bracing at the corners of buildings, we can greatly improve their resistance to shaking.

One of the greatest structural problems in earthquake hazard zones is the use of masonry for construction. The mortar between bricks and concrete blocks, which consists only of mud in many developing countries, exhibits little strength when seismic waves shake the ground. As a result, brick and concrete walls are highly susceptible to collapse. In the last decade, masonry failures led to tragic consequences in many areas of the world. In 2003, for example, a magnitude 6.5 earthquake killed more than 31,000 people and demolished 85 percent of the Iranian city of Bam, previously designated a UNESCO World Heritage site for its importance as a trading center on the ancient Silk Road and for its historic buildings, including a 2000-year-old citadel (Figure 3-20).[54] Much of Bam was constructed of mud bricks, and this city of domed roofs, archways, courtyards, and narrow alleyways had an allure that drew thousands of tourists annually. Today, several years after the earthquake, recovery is ongoing as archeologists work to rebuild the citadel and new homes are constructed with steel frames that can stand up to seismic shaking.

Figure 3–20 Bam, Iran, designated a UNESCO World Heritage site for its historic buildings, before and after the M_w 6.5 earthquake that occurred in 2003. *(a: Christine Osborne Pictures/Alamy; b: Atta Kenare/AFP/Getty Images)*

BOX 3–4 EARTH SCIENCE AND PUBLIC POLICY

Earthquakes and Legislation

At 6 A.M. on February 9, 1971, a magnitude 6.6 earthquake rocked the San Fernando Valley of Southern California when a slip of about 1 m occurred on a thrust fault lying beneath the valley.[62] The event caused 65 deaths, more than 2000 injuries, and $505 million in property damage. Many of the deaths occurred when a Veterans Administration Hospital, several freeway overpasses, and part of the Olive View Hospital collapsed. Landslides triggered by the earthquake caused additional damage to buildings, railways, pipelines, and dams, and shaking was felt as far away as western Arizona.

In response to the devastation, the California legislature passed the Alquist–Priolo Earthquake Fault Zoning Act[63] in 1972 to regulate construction in seismically active areas. A companion law, the Seismic Hazards Mapping Act, was passed in 1990. These laws mandate that the California state geologist (and the staff at the California Geological Survey) create Seismic Hazard Zone maps for the entire state in order to prohibit construction on top of active faults (**Figure 1**). The maps overlay the locations of known earthquake faults and sites vulnerable to liquefaction or landsliding on top of city streets, streams, and other geographic features that are easily recognizable.

The legislation also established a permitting process for new construction. Before a structure can be built, a state-licensed geologist must inspect the site to determine whether any faults are present. If one is found, the building must be placed a minimum of 50 feet away from the surface trace of the fault to prevent the structure from being torn asunder during a slip event. Wood- or steel-frame single-family homes up to two stories in height are exempt from the state law, but local municipalities have the right to enforce stricter standards.

In 1975, California took another step toward earthquake preparedness by creating a seismic safety commission.[64] Made up of geologists, structural engineers, seismologists, emergency management professionals, firefighters, members of social services agencies and other relevant disciplines, the commission conducts research and makes recommendations to help the state legislature and the governor in crafting laws to reduce earthquake impacts, enhance public preparedness, and monitor seismic events. Many of the building code requirements discussed in this chapter are a result of the commission's work.

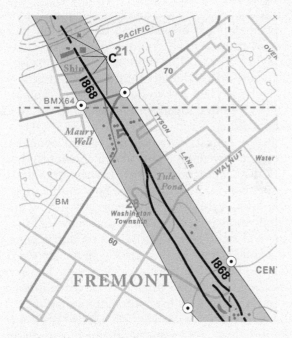

Figure 1 Seismic Hazard Zone map.

Another tragic example of masonry failure occurred when a magnitude 6.3 earthquake destroyed stone buildings in the central Italian city of L'Aquila in 2009. Ironically, while many were badly damaged, some of the city's medieval buildings fared better during the earthquake than those constructed within the last several decades. Substandard building materials and inferior construction practices were blamed for the collapse of interior walls in a hospital built as recently as the year 2000.[55] Due to the damage and to fears that the entire building might collapse, those injured in the earthquake had to be cared for in the hospital parking lot rather than in the facility itself.[56] Shoddy construction was also blamed for the collapse of schools that killed as many as 10,000 children during the 2008 M_w 7.9 earthquake in Sichuan, China. Chinese construction codes had been upgraded in the 1970s after a devastating earthquake killed nearly a quarter of a million people, so the collapse of schools led to accusations that the codes had not been enforced due to corruption.[57] Grieving parents held demonstrations for months after the event, demanding answers from local government officials.

California banned the use of unreinforced masonry for construction as early as 1933, but the state still has an estimated 30,000 such buildings, only two-thirds of which have been retrofitted to withstand seismic shaking. Still, building codes requiring the use of wood or steel framing and of reinforced concrete have greatly reduced the potential for destruction. In addition, California has instituted forward-thinking tax policies that assist people in retrofitting their homes. Realizing that retrofits increase the value of homes, the state does not increase property taxes on homeowners who take measures to enhance earthquake preparedness.[53] Additional codes require that any appliances that use natural gas, such as water heaters, stoves, and clothes dryers, be strapped to floors or walls and have flexible pipe connections to prevent the rupture of gas lines that could cause fires. The wisdom of these requirements is evident when considering the fact that the 1906 destruction of San Francisco had less to do with the earthquake than with the fires that raged across the city in the aftermath of the shaking.

Public education is another critical component of disaster preparedness in earthquake country. Televised public service announcements, newspaper articles, and websites provide information on how to shut off gas and water lines in the event of an emergency, how to create a family disaster plan, and on what supplies to put into a survival kit (Table 3-3). Education also focuses on a few simple steps that can greatly minimize the risk of injury and/or death during earthquakes. For example, dishes and canned goods can become flying missiles during earthquakes, so it is important to ensure that kitchen cabinets have secure latches. Similarly, bookcases and armoires may topple during shaking, a problem easily addressed by bolting furniture to the wall. Because glass can shatter during shaking, the Federal Emergency Management Agency recommends that beds, sofas, and other seating be kept away from windows and mirrors. Another good practice is to keep a wrench tied to any appliance that uses gas to avoid having to hunt for one after an earthquake. Gas valves can then be shut off quickly, lessening the threat of accidental fires.

Finally, earthquake drills provide emergency responders and the general public opportunities to practice what to do during an earthquake. Every year, the state of California conducts the "Great California

TABLE 3–3 Components of an Earthquake Survival Kit
• Enough nonperishable food and water for each family member for 3 days. Don't forget the pets!
• Prescription medications
• First aid supplies
• Flashlight
• Crank- or battery-operated radio and extra batteries
• Sleeping bags or blankets
• Walking shoes
• Cash, including coins
• Can opener or pocketknife with can opener attachment
• Cell phone
• Copies of important documents such as driver's licenses and homeowner's insurance policies

ShakeOut;" in 2010, more than 7.9 million people took part, about 21 percent of the state's population.[58] At 10:21 A.M. on October 21, people took cover under desks and hung on to table legs, simulating what one should do during an earthquake, and emergency responders practiced evacuation, firefighting, medical treatment, and search-and-rescue activities. Exercises such as these are essential in regions that are prone to earthquakes and may save many lives during future events.

In the not-too-distant future, Californians may also benefit from an earthquake early-warning system called the California Integrated Seismic Network ShakeAlert System.[59] Currently under development, the system will use a network of seismometers to detect incoming P waves and use their arrival times to predict when more-destructive S and surface waves will pass through an area. An automated public alert system on the Internet, television, or over phone lines may give people just enough time to dive under a table or desk, the safest places to be in the event of building collapse. The alerts could also trigger subway trains to decelerate, stop elevators and open their doors at the next floor to prevent people from becoming trapped, and issue warnings to people working with hazardous chemicals. Such a system has been in place in Japan since 2007[60] and no doubt saved many lives during the March 2011 Tōhoku earthquake. Indeed, Japan is a leader in earthquake engineering, installing steel and rubber pads under new high-rises to prevent seismic energy from traveling upward into buildings and creating a state-of-the-art shake table to study how different building designs respond to different earthquakes. The "E-defense" shake table can hold up to 1136 metric tons (2.5 million pounds) and has been used to evaluate the performance of buildings as high as seven stories.[61]

■ Masonry structures are highly vulnerable to collapse during earthquakes.

■ Building codes designed for seismically active areas can greatly diminish property damage and loss of life during earthquakes. Simple measures such as bolting homes to foundations and strapping gas appliances to walls minimize the impacts of shaking.

■ Earthquake drills allow people to practice responding to earthquakes and facilitate emergency management planning.

■ Laws in California require a minimum setback from active faults for all new development, and inspections by state-licensed geologists.

CLOSING THOUGHTS

Earth's continually moving tectonic plates present serious risks to those who live along their borders. The earthquakes they spawn are among the most destructive of natural phenomena, responsible for thousands of deaths every year. Throughout the past century, we have made incredible progress toward understanding their causes, but unfortunately, they are still not predictable. Despite the limits of our understanding, we have learned a lot about how to mitigate their impacts. Moderate-sized events in California in 1989 (Loma Prieta) and 1994 (Northridge) each caused fewer than 100 deaths because the state had developed strict regulations about where construction can be sited and what building materials can be used, and had mandated particular construction practices, such as assuring that structures be strapped to their foundations. Developers sometimes complain about these regulations because they put some lands off-limits to construction or increase costs, but such commonsense regulations are critical to reducing deaths and property damage in seismically active areas.

SUMMARY

■ Earthquakes occur if strain built up along a fault is released suddenly as seismic energy when crust on each side of the fault plane slips. The ultimate cause of the build-up of strain is motion between tectonic plates.

■ The moment magnitude scale expresses the amount of energy released during an earthquake; the modified Mercalli intensity scale reflects the amount of damage caused by the event.

■ Damage associated with an earthquake is due largely to ground shaking, which varies substantially depending on the type of geologic material through which seismic waves move. In general, loose, wet, young sediments are the most susceptible to intense ground vibration, and structures built on them are more likely to be damaged than those built on solid rock.

■ Geologists cannot yet predict the exact timing and precise location of earthquakes, but they can provide maps that locate areas where earthquakes are most likely to occur and where there are geologic materials susceptible to ground shaking and liquefaction. This information can lead to building codes that substantially reduce the risks associated with future earthquakes.

KEY TERMS

scarp, p. 64
creep p. 65
stick-slip, p. 65
elastic rebound model, p. 65
seismic wave, p. 67

hypocenter (or focus), p. 67
epicenter, p. 67
seismometer, p. 67
seismogram, p. 68
P (primary) wave, p. 68
S (secondary) wave, p. 68

body wave, p. 68
surface wave, p. 68
refraction, p. 70
triangulation, p. 72
earthquake cycle, p. 73
aftershock, p. 73

foreshock, p. 73
moment magnitude scale, p. 75
liquefaction, p. 77
modified Mercalli intensity scale (MMI), p. 77
paleoseismology, p. 82

REVIEW QUESTIONS

1. How are faults related to the cause of earthquakes?

2. What is a scarp, and how is it related to a fault plane?

3. What are the two types of movements that occur along faults? How do they differ in their destructive potential?

4. What is the purpose of a seismometer?

5. What are the different types of seismic waves, and what are their characteristics?

6. Which type of seismic wave commonly causes the greatest destruction during an earthquake?

7. What is the difference between a hypocenter and an epicenter?

8. How many seismographic stations are necessary to determine where an earthquake originated? Why?

9. What causes P- and S-wave shadow zones, and what do these zones tell us about Earth's interior?

10. How do the moment magnitude and modified Mercalli intensity scales differ from each another?

11. What kinds of evidence can geologists use to determine how active a particular fault is?

12. What is the difference between an earthquake forecast and a prediction?

13. What types of information about a given location are needed to minimize the risks associated with earthquakes?

14. What steps can be taken to minimize property damage during earthquakes?

THOUGHT QUESTIONS

1. Why do more earthquakes occur along the boundaries between tectonic plates than within plates?

2. Why do some earthquakes occur at shallow depths (several kilometers), while others occur much deeper (hundreds of kilometers) beneath Earth's surface? Would you expect earthquakes in Indonesia to be shallow or deep? Why?

3. If Earth's liquid outer core were larger than at present, what would be the impact on the P- and

S-wave shadow zones? If it were smaller than present?

4. Imagine that you are considering buying a house in Southern California. What information could you gather to assess the vulnerability of that house to damage during an earthquake?

5. Suppose that you are a member of a city council in earthquake country. What initiatives would you pursue to enhance public safety?

EXERCISES

1. If an earthquake of moment magnitude 7 occurs in San Francisco, how long will it take a P wave to reach Pittsburgh, Pennsylvania, about 4000 km away? How long will it take an S wave? Is it likely that a resident of Pittsburgh would feel this seismic wave pass through the rock underfoot? Why or why not?

2. Determine the moment magnitude of an earthquake with the following characteristics: Rupture length = 215 km, rupture depth = 12 km, total slip = 2.6 m, and shear modulus = 3×10^9 kg/ms^2. Note that you will have to conduct some unit conversions to carry out this calculation.

3. Paleoseismological analysis of the Wasatch fault in Utah has revealed numerous episodes of slip along

different segments of the fault over approximately the last 6000 years. Using the figure on the next page, determine the average recurrence interval for earthquakes along the Salt Lake City segment of the fault. What is the average recurrence interval for earthquakes along the entire length of the Wasatch fault?

4. One of the largest historic earthquakes in the world occurred along the coast of Chile in 1960. A smaller but also devastating event struck the area in 2010. Along what type of plate boundary did these events occur (refer to Figure 2-19)? Draw a vertical cross-section from west to east that shows the two plates along this boundary, and the nature of their interaction with each other.

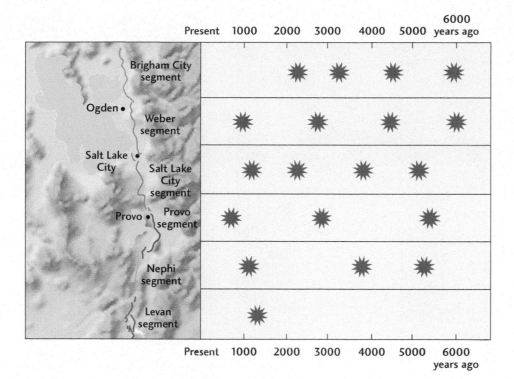

From: S. N. Eldredge and V. Clark, 1996, "The Wasatch Fault," Salt Lake City, UT: Utah Geological Survey Public Information Series, no. 40.

SUGGESTED READINGS

Bolt, B. A. *Earthquakes.* 5th ed. New York: W. H. Freeman, 2006.

California Seismic Safety Commission. "The Homeowner's Guide to Earthquake Safety." Sacramento, CA: California Seismic Safety Commission (2005). http://www.seismic.ca.gov/pub/CSSC_2005-01_HOG.pdf.

Cervelli, P. "The Threat of Silent Earthquakes." *Scientific American* 290 (March, 2004): 86–91.

da Boer, J. Z., and D. T. Sanders. *Earthquakes in Human History.* Princeton, NJ: Princeton University Press, 2005.

Fischetti, M. "Working Knowledge: Shock Absorbed." *Scientific American* 291 (October, 2004): 102–103.

Fradkin, P. L. *The Great Earthquake and Firestorms of 1906: How San Francisco Nearly Destroyed Itself.* Berkeley: University of California Press, 2005.

Geschwind, C. H. *California Earthquakes: Science, Risk and the Politics of Hazard Mitigation.* Baltimore, MD: Johns Hopkins University Press, 2001.

Hough, S. E., and R. G. Bilham. *After the Earth Quakes: Elastic Rebound on an Urban Planet.* New York: Oxford University Press, 2006.

Nur, A., and D. Burgess. *Apocalypse: Earthquakes, Archaeology, and the Wrath of God.* Princeton, NJ: Princeton University Press, 2008.

Ritchie, D., and A. E. Gates. *Encyclopedia of Earthquakes and Volcanoes,* 3rd ed. New York: Facts on File, 2006.

Schick, R. *The Little Book of Earthquakes and Volcanoes.* New York: Copernicus Books, 2002.

Shrady, N. *The Last Day: Wrath, Ruin, and Reason in the Great Lisbon Earthquake of 1755.* New York: Viking Penguin, 2008.

Stein, R. S. "Earthquake Conversations." *Scientific American* 288 (January, 2003): 72–79.

Yeats, R. S. *Living with Earthquakes in California: A Survivor's Guide.* Corvallis, OR: Oregon State University Press, 2001.

Yeats, R. S. *Living with Earthquakes in the Pacific Northwest.* Corvallis, OR: Oregon State University Press, 2004.

Yeats, R. S., K. Sieh, and C. R. Allen. *The Geology of Earthquakes.* New York: Oxford University Press, 1997.

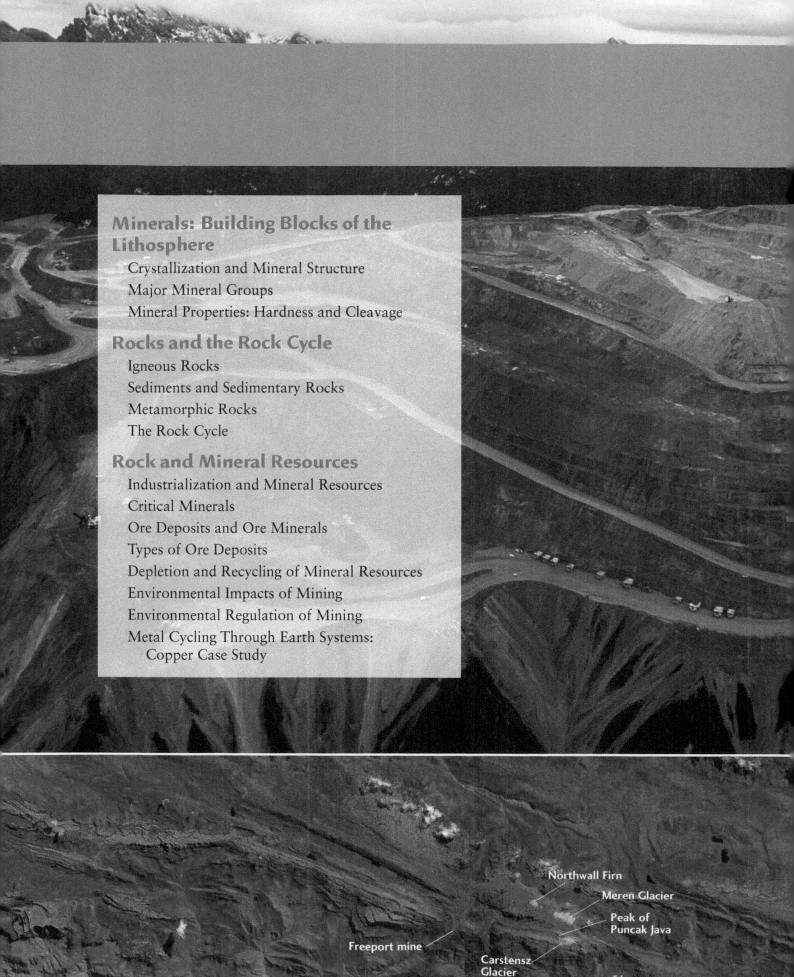

Northwall Firn

Meren Glacier

Peak of
Puncak Java

Freeport mine

Carstensz
Glacier

5 km

Earth Materials: Elements, Minerals, and Rocks

The island of New Guinea lies between the equator and Australia in the western Pacific Ocean and contains several high and actively uplifting mountain ranges. A Dutch sea captain who sailed past the island in 1623 reported glimpses of glaciers, to the scorn of his contemporaries who could not imagine finding ice in a tropical setting. Yet Captain Jan Carstensz was right. The captain had visited New Guinea during the peak of the Little Ice Age, a time when global temperatures were sufficiently low for glaciers to advance. Since then, Earth's atmosphere has warmed and many ice masses have shrunk, including rare equatorial glaciers on the island of New Guinea.

As the ice over New Guinea melted, it revealed the world's largest known deposit of gold and one of the largest of copper. A company called Freeport-McMoRan Copper & Gold uses helicopters, one of the world's longest aerial tramways, and some of the world's largest dump trucks to operate the mountaintop mine.

New Guinea's mountains were built as subduction of the Indian-Australian Plate beneath the Pacific and Philippine plates formed a volcanic arc that still shakes the island with earthquakes and volcanic eruptions. Though subduction generates geologic hazards, it also has provided the island its rich mineral deposits.

In this chapter we:

✔ Examine the origins of minerals and rocks.

✔ Discuss what ore minerals are and where they come from.

✔ Learn the uses of various Earth materials.

✔ Explore environmental issues associated with the extraction of minerals and rocks.

Freeport–McMoRan Copper & Gold, a mining company, excavates copper, gold, and silver 4000 m above sea level on the equatorial island of New Guinea. The yellowish, metal–rich rocks stand out in stark contrast to the layered gray limestone. Valley glaciers that recently covered the area have retreated higher into the mountain. (Bottom image) Small remnants of three glaciers can be seen as blue areas in this satellite image. Clouds are white; bare land, such as the rocky limestone summit ridges, is red to pink; the open pit copper mine is deep purple and gray; and the lush lowland rainforest appears green. Sediment–rich runoff into rivers appears as deep purple south of the mine. (top: George Steinmetz/Corbis; bottom: NASA Earth Observatory/Global Land Cover Facility)

INTRODUCTION

In *Naturalis historia,* the world's first encyclopedia, the Roman scholar Pliny the Elder (23–79 CE) wrote of the accidental discovery that heat could transform sand into glass. According to Pliny, a Phoenician ship with a cargo of natural soda, a rock widely used in soaps and bath salts that contains the element sodium, took harbor along the Mediterranean coast sometime in the second millennium BCE:

> *[The sailors] scattered along the shore to prepare a meal. Since, however, no stones suitable for supporting their cauldrons were forthcoming, they rested them on lumps of soda from their cargo. When these became heated and were completely mingled with the sand on the beach a strange translucent liquid flowed forth in streams; and this, it is said, was the origin of glass. (Pliny,* Natural History, *vol. X, ed. D. E. Eichholz, Loeb Classical Library 419 [Cambridge, MA: Harvard University Press, 1962], 151)*

As the Phoenicians reportedly discovered, glass can be made by melting beach sand, which typically is rich in quartz, a mineral made of the elements silicon and oxygen (Figure 4-1). During melting, some of the atomic bonds in the quartz are broken apart, enabling the material to flow as a liquid. Had the Phoenician sailors not introduced soda into the sand, they would not have produced glass. The melting temperature of quartz is about 1650°C, more than twice the temperature of a wood fire, which typically is less than 800°C. By adding sodium—in the form of "lumps of soda"—to the sand, the sailors inadvertently created a mixture with a lower melting point than quartz sand alone.

At the time that they happened upon glass, the Phoenicians were already familiar with the mysteries of melting minerals and rocks. Their cauldrons, tools, and weapons were made of bronze (an alloy of copper and tin) and iron, metals that can be extracted by smelting certain minerals at temperatures of about 900°C or more (smelting is the separation of a metal from other elements in minerals by processes that include melting). The Phoenicians were familiar, too, with the natural melting of rock, because the Mediterranean region has a long history of volcanic activity. Pliny himself died either of asphyxiation from volcanic gases or of heart failure when he ventured into the city of Pompeii to make observations during the eruption of Mount Vesuvius in the year 79 CE.

In the more than 2000 years that have elapsed since the end of the Phoenician civilization few things other than quantity have changed about our use of Earth resources. Minerals, and the rocks they are found in, touch every aspect of our lives. When we get up in the morning and flip on the coffeemaker, we use water

(a)

(b)

Figure 4–1 (a) Resistant to weathering and to the abrasion associated with transport in water, grains of quartz are common in river and beach sands. (b) Quartz sand or crushed quartz is melted in a furnace, then blown or molded—as shown here—in its molten form to produce glass articles. Glass must be worked quickly because at room temperature it freezes nearly instantaneously into a noncrystalline solid. *(a: Andrew de Wet; b: Suzanne DeChillo/The New York Times/Redux Pictures)*

transported to our taps in copper pipes derived from the mining of copper ore, such as that mined at the Freeport-McMoRan mine in Indonesia. The stainless steel faucets that dispense our water and the cast-iron pans in which we scramble eggs originated as iron ore, and the ceramic mugs from which we drink coffee started out as clay. After breakfast we brush our teeth with toothpaste made from calcium carbonate, the primary constituent of limestone. As we head out the door, we walk, ride our bicycles, or drive over asphalt and concrete made from crushed stone, and enter buildings made of stone and steel. Our cell phones and laptop computers contain rare earth elements, gold, silica, nickel, aluminum, zinc, iron, petroleum products, and many other minerals. Indeed, it is hard to imagine a world without rock and mineral resources, and they have been vital to humans since the earliest Stone Age.

According to the Minerals Information Institute, every man, woman, and child in the United States used nearly 17,000 kg (38,000 lbs, or 19 tons) of minerals and rocks in 2010 (Figure 4-2). The vast majority of this use was indirect, reflecting the labor of contractors, stonemasons, factory workers, and others who created roads, buildings, and consumer products on their behalf. When multiplied by a population just over 300 million, the per capita use rate translates into a staggering 5.3 trillion kg (11.7 trillion lbs) of minerals and rocks used by American society in 2010.

Still, the use of rocks and minerals as resources for daily living is not the only thing that makes them significant. Another important aspect of these materials is that they form the continents on which we live. The geological disciplines of *mineralogy* (the study of minerals) and *petrology* (the study of rocks) examine Earth materials with a combination of field and laboratory methods. Through these disciplines, geologists have discovered that rocks and minerals provide clues about Earth's formation, evolution, and climate over time.

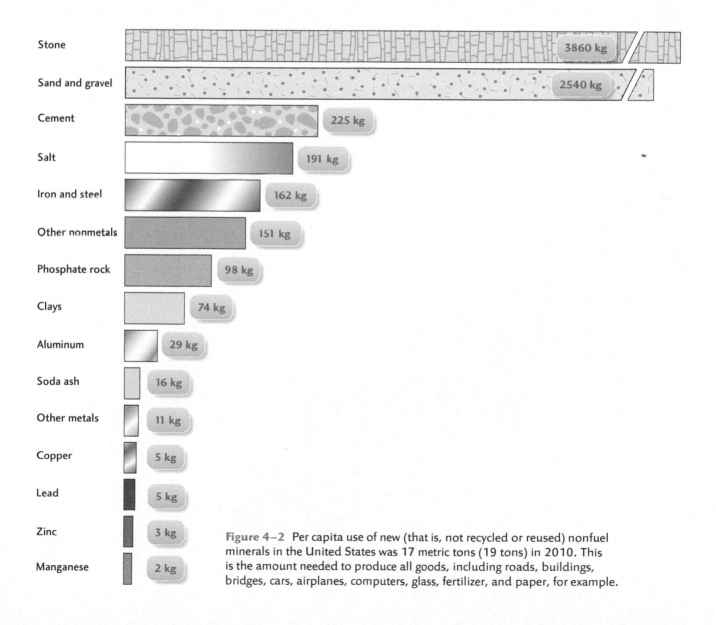

Figure 4–2 Per capita use of new (that is, not recycled or reused) nonfuel minerals in the United States was 17 metric tons (19 tons) in 2010. This is the amount needed to produce all goods, including roads, buildings, bridges, cars, airplanes, computers, glass, fertilizer, and paper, for example.

Some rocks contain information critical to geologic hazard analysis; others archive past life-forms, giving us insight into our own origins. For all of these reasons, Earth materials are worthy of study.

MINERALS: BUILDING BLOCKS OF THE LITHOSPHERE

Glowing red and gold in the Australian sunlight, majestic Uluru (also called Ayers Rock) rises nearly 350 m above a vast desert plain (Figure 4-3). It is no wonder that for thousands of years, Native Australians have used this awe-inspiring hill of bare rock as a sacred place of worship. Several different myths describe its formation, but our geological understanding is every bit as interesting. Uluru is made of sandstone, a type of rock constructed of sand grains that have been cemented together (Figure 4-4). As such, sandstone is a *sedimentary rock*, made of broken bits of other rocks. The fragments Uluru contains reveal that it originated about 540 million years ago by the erosion of nearby granitic mountains, themselves created by a continent–continent collision that produced molten material that solidified into *igneous rock*. The eroded fragments were deposited on a broad fan of sediment at the base of the mountains and eventually were cemented together

Figure 4–3 Ayers Rock in Australia is a deep–red sedimentary rock made of sand–size rock fragments. Native Australians call the rock Uluru and visit it for sacred rituals. *(John Carnemolla/Australian Picture Library/Corbis)*

to form sandstone. The relation between Uluru and its surrounding mountains illustrates part of the *rock cycle*, the continuous movement of Earth's solid materials between its surface and interior.

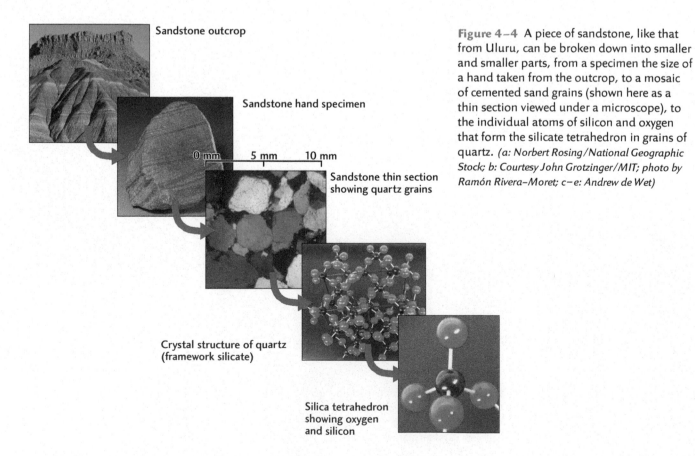

Sandstone outcrop

Sandstone hand specimen

0 mm   5 mm   10 mm

Sandstone thin section showing quartz grains

Crystal structure of quartz (framework silicate)

Silica tetrahedron showing oxygen and silicon

Figure 4–4 A piece of sandstone, like that from Uluru, can be broken down into smaller and smaller parts, from a specimen the size of a hand taken from the outcrop, to a mosaic of cemented sand grains (shown here as a thin section viewed under a microscope), to the individual atoms of silicon and oxygen that form the silicate tetrahedron in grains of quartz. *(a: Norbert Rosing/National Geographic Stock; b: Courtesy John Grotzinger/MIT; photo by Ramón Rivera–Moret; c–e: Andrew de Wet)*

If we examine rocks like those at Uluru carefully, we find that all are constructed from assemblages of **minerals**—naturally formed, inorganic, crystalline solids with specific chemical compositions or a range of compositions. Several thousand minerals have been identified on Earth. Some, called native element minerals, consist of only one element, such as gold, silver, copper, sulfur, and diamond. More commonly, however, minerals are made of two or more elements, as are quartz and feldspar, the minerals that make up most of Uluru's sand grains. These minerals comprise various mixtures of silicon, oxygen, potassium, sodium, calcium, and aluminum.

Despite the great variety of elements in the periodic table and the numerous ways in which those elements can combine to create minerals, most of Earth's crust is actually quite similar in composition. A mere 8 elements, including those from which Uluru is constructed, make up nearly 99 percent of the crust (Table 4-1), and this restricted chemical composition leads to about a dozen mineral structures that are especially common. These are the so-called *rock-forming minerals,* to which Uluru's quartz and feldspar belong.

Crystallization and Mineral Structure

Minerals form when atoms are bonded together into orderly, repeating, three-dimensional structures, a process known as **crystallization.** When crystals grow freely in open cavities, their external shapes develop unhindered and express the internal geometric order of their component atoms. These crystals have regular and planar (that is, flat) faces. Usually, however, other

TABLE 4–1 The Abundance of Elements in Earth's Crust, by Weight	
Element	Abundance (%)
Oxygen (O)	46.1
Silicon (Si)	28.2
Aluminum (Al)	8.2
Iron (Fe)	5.6
Calcium (Ca)	4.1
Sodium (Na)	2.4
Magnesium (Mg)	2.3
Potassium (K)	2.1
Other elements	~1.0

substances crystallizing around them confine growing minerals, which limits the minerals' size and external regularity (Figure 4-5).

The process of crystallization occurs in two ways—through cooling of a fluid or through precipitation of a solid from a solution. During cooling, crystallization occurs when liquids or gases lose enough thermal energy for their atoms to slow down and bond strongly to one another to form solids. Snowflakes, for example, are crystals of ice that solidify from water vapor when the air temperature falls below the freezing point of water. Similarly, minerals crystallize in magma as it cools on its way up from the hot mantle. The slower the rate of cooling, the more time the atoms have to form

(a)

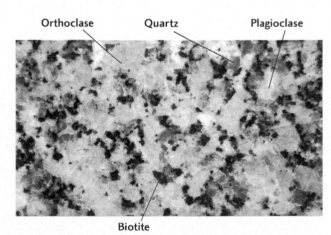

(b)

Figure 4–5 (a) During crystal growth, exquisite shapes develop if a mineral grows unhindered, as it has in these prismatic smoky quartz crystals formed in an open cavity. (b) When growth is hindered—as it was for the quartz crystals (gray) in this magnified image of a crystalline matrix of rock (granite) created from cooled magma—a solid mass of intergrown crystals forms. *(a: TH Foto-Werbung/Science Source; b: Andrew de Wet)*

Figure 4–6 Obsidian is volcanic glass, formed from magma that has cooled upon reaching Earth's surface. Because it lacks an ordered internal crystalline structure, obsidian does not break evenly along crystal faces. Rather, it breaks along curved surfaces, shaped rather like a shell. *(Gary Ombler/ Getty Images)*

extended frameworks and the larger and more perfectly formed the crystals are, whether they are snowflakes or crystals of quartz.

In contrast, instantaneous cooling freezes atoms in place before they can become arranged into orderly frameworks. The resulting solid is not considered a mineral because it is not crystalline, but instead is called a **glass,** a solid lacking a regularly repeating array of elements. Obsidian, a glass formed by rapidly cooling lava, is one such example (Figure 4-6). Obsidian can be made of silicon and oxygen, as is quartz, but it is not a mineral because its atoms have no internal symmetrical order. Lacking the planar faces of most crystalline solids, obsidian does not break into regularly shaped pieces. Instead, it breaks into razor-sharp flakes along curved, irregular surfaces. It is this characteristic that makes obsidian valuable for creating arrowheads and spear tips.

Crystallization through precipitation occurs as a liquid evaporates and atoms or ions that are left behind

bond with one another to form solids. A common example is the formation of salt crystals along the rim of a bowl of salt water. As the water evaporates, the concentration of oppositely charged ions of sodium and chloride increases and the ions are drawn closer to one another. Because ions of opposite charge attract one another, they arrange themselves so that each ion is surrounded by ions of opposite charge (Figure 4-7), forming the mineral halite (NaCl), otherwise known as table salt. In nature, halite commonly forms when sodium and chloride ions from evaporating seawater crystalize, as along the margins of a shallow bay. On beaches of hot equatorial islands, villagers often fill large clam and oyster shells with seawater so as to collect salt from the water's evaporation.

The way in which atoms are arranged into crystals is called the *structure* of a mineral. Halite, for example, has a cubic mineral structure, reflecting the 90° angles between the chemical bonds created by sodium and chlorine atoms. The arrangement of atoms, in turn, is related to the environment in which the mineral formed (Box 4-1). The pressure surrounding rocks increases with depth below Earth's surface, so atoms tend to be packed more closely together at higher pressures.

Near Earth's surface, where the pressure is low, for example, the element carbon forms the soft, low-density mineral graphite. At depths greater than 150 km, where pressure is much higher, however, carbon forms the hard, high-density mineral diamond. The cause of the observable differences between the two minerals is their crystalline structure, or the way in which their atoms are arranged.

Major Mineral Groups

Earlier we noted that the abundance of certain elements in Earth's crust (see Table 4-1) gives rise to a few common rock-forming minerals. With oxygen and silicon together making up 74 percent of the composition of the crust, it is not surprising that the most abundant minerals found there are the **silicates,** a group

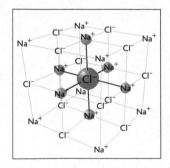
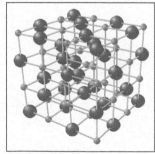

Figure 4–7 (a) and (b) Sodium and chloride ions form a cubic mineral structure in which each ion is surrounded by six ions of opposite charge. (c) The cubic shape of a halite (table salt) crystal reveals its cubic atomic structure. *(b: Roger Weller/Cochise College; c: Andrew de Wet)*

of minerals composed mostly of oxygen and silicon. In fact, at least 98 percent of Earth's crust consists of silicate minerals. The minerals that form *asbestos* are silicates, for example, and are the cause of much concern because one type of asbestos can cause cancer in humans.

The basic building block of silicates is the silica tetrahedron, a molecule constructed of four oxygen atoms arranged in a pyramidal shape around a silicon atom (Figure 4-8). Silica tetrahedra have a large negative charge (−4) which is reduced when tetrahedra link together and share oxygen atoms. For this reason, silica tetrahedra have a strong tendency to link to one another. Depending on the relative abundance of silica and oxygen, these tetrahedra can form clusters, chains, and sheets of connected molecules linked by the oxygen atoms at the corners of the tetrahedra.

Silicate minerals commonly form by crystallization of molten rock as it cools. As the temperature of the magma decreases and the silica tetrahedra move more slowly, the tetrahedra link with one another. The arrangement of linked silica tetrahedra determines which silicate structures and resultant minerals, such as olivine, pyroxene, amphibole, mica, feldspar, and quartz, are formed (see Figure 4-8).

The remaining 2 percent of Earth's crust consists of nonsilicate minerals: oxides and hydroxides, carbonates, sulfates and sulfides, halides, and native elements (Table 4-2). Each mineral group is defined by the type of anion (a negatively charged ion) in its structure. For example, oxide minerals contain oxygen anions (O^{2-}), whereas sulfides contain sulfur anions (S^{2-}). The anions combine with cations (positively charged ions) to form a variety of minerals. Iron combined with oxygen forms the oxide mineral magnetite (Fe_3O_4), known for its strong magnetic properties. If iron combines with the sulfur anion, it forms the sulfide mineral pyrite (FeS_2), also called fool's gold because its color resembles that of gold. Most metals that are mined occur in these nonsilicate groups of rock-forming minerals.

Figure 4−8 The silica tetrahedron (top) is the fundamental building block for silicate mineral structures. For example, individual tetrahedra linked by cations can form olivine. Silica tetrahedra link end to end to form single chains, as in the common mineral pyroxene, and double chains, as in amphibole. With increasing polymerization, tetrahedra can form into sheets (micas, for example), or join at all their tips to form three-dimensional frameworks, as in the minerals feldspar and quartz. *(Courtesy John Grotzinger/MIT; photos by Ramón Rivera-Moret)*

Olivine: Isolated tetrahedra

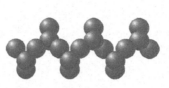

Pyroxene: Single chain of tetrahedra

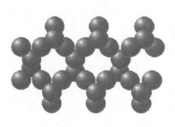

Amphibole: Double chain of tetrahedra

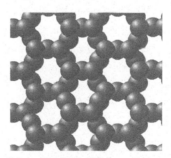

Mica: Sheet tetrahedra

Feldspar and quartz: Framework tetrahedra

BOX 4-1 GEOLOGIST'S TOOLBOX

Soft Crystals and Hard Science in the Cave of Crystals, Mexico

Selenite, a form of the soft (hardness 2) mineral gypsum with a glassy or vitreous luster, derives its centuries-old name from the Greek words for "moonstone" (selene, 'moon' + lithos, 'stone') because of its resemblance to the waxing and waning glow of moonlight, but the largest known crystals of selenite are found in dark caves deep below the groundwater table in northern Mexico's Chihuahua Desert (Figure 1). In order to extract lead, silver, and zinc ores, groundwater is pumped from tunnels and shafts below Naica Mountain. In 2000, while drilling a new tunnel about 290 m below ground, two miners discovered a cave filled with megacrystals, a veritable crystal nursery dubbed Cueva de los Cristales, or Cave of Crystals. Throughout the world, gypsum is a somewhat common mineral resource, used to make plaster, drywall, and cement, but in the Cave of Crystals it forms spectacular elongated megacrystals up to 11 m in length that are as thick around as redwood trees and weigh up to 50 metric tons (55 tons), about 5 times larger than the largest previously known crystals. The cave invokes visions of another world, an ethereal world, as noted by geologist Juan Manuel García-Ruiz, who calls it the Sistine Chapel of crystals.[1]

The discovery of the Cave of Crystals beckoned geologists from around the world. They came to determine how crystals could grow so large, when they began to form and how old they might be, and whether or not they might contain small bubbles of trapped fluid—fluid inclusions—that could provide chemical evidence of their origin. Time might be short for researchers, as the cave will flood if the mine owners shut down the pumps that drain it of groundwater. Exploring the cave is extremely difficult. The temperature is 46°C–47°C (115°F–117°F) and the humidity is 90%–100%. Before entering the cave, explorers must don special caving suits packed with ice and insulation, helmets, and masks, and are limited to only 10- to 20-minute visits at a time.

Geomicrobiologist/astrobiologist Penny Boston described her experience in the hot, humid cave as if she were trying to wade through molasses because the heat and humidity slowed her motion and ability to think clearly. In addition, the immense crystals, though soft enough to scratch with a fingernail, occur as a jumble of jagged beams jutting from the floor, walls, and ceiling of the cave, potentially as dangerous as sharp swords if an explorer falls while climbing over them.

How can such clear, flawless crystals of gypsum, a mineral that also forms in lake muds and salt flats, become so large in liquid-filled cavities hundreds of meters below a desert? Researchers found that a magma body several kilometers beneath the cave provided heat to circulate hot, metal-rich fluids through the mountain some 26 million years ago, generating hydrothermal reactions that caused sulfide minerals containing lead, zinc, and silver to precipitate along veins in the limestone bedrock above the magma. Today the Naica mine is Mexico's largest producer of lead. As the magma body cooled to a temperature below 58°C (136°F), the groundwater circulating through cavities along faults was slightly supersaturated in calcium and sulfate ions, promoting the crystallization of gypsum, a hydrated calcium sulfate mineral ($CaSO_4 \cdot 2\,H_2O$). Importantly, the groundwater stabilized and remained at this temperature for thousands to hundreds of thousands of years, enabling selenite crystals to grow to immense sizes. At temperatures above 58°C (136°F) or below 54°C (129°F), gypsum crystals cannot form.

Researchers used a sample of the crystals and mine water to reproduce conditions of crystal growth at different temperatures in their laboratory, and sophisticated microscopic techniques to detect slow growth rates. They determined that the crystals grew extremely slowly, at rates less than a thousandth of a millimeter per year.[2,3] Indeed, it took at least hundreds of thousands of years for the largest crystals to form, a time frame that is supported by radiometric dates from samples of the crystals extracted with drills.[4]

Figure 1 (Top) Enormous crystals of gypsum, a mineral so soft it can be scratched with a fingernail, jut from the floor, ceiling, and walls of Mexico's Cave of Crystals 290 m underground. The largest known crystals of gypsum in the world, they were exposed when groundwater was pumped from the cavern during lead mining operations. Note the scientists climbing on the crystals, wearing orange suits for protection from high heat and humidity in the cave. *(Carsten Peter/Speleoresearch & Films/ National Geographic/Getty Images)*

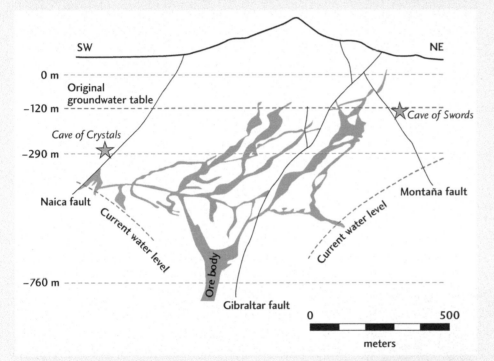

(Middle) The Cave of Crystals and a similar crystal-filled cave called the Cave of Swords are located just below the original groundwater table along faults that bound an ore body of lead- and zinc-filled fractures. Cavities and other pore spaces below the groundwater table were saturated with water, while those above it are unsaturated and contain some air. Curved dashed lines near the bottom of the illustration indicate the current shape of the groundwater table, which was drawn down by pumping to enable mining.

(Bottom) Scientists collect core samples from a gypsum crystal for uranium-thorium dating in order to determine when the mineral began to form and at what rate it grew. Crystals in this cave began growing at least hundreds of thousands of years ago, and grew extremely slowly, less than a hair's width per year. *(Carsten Peter/Speleoresearch & Films/National Geographic/Getty Images)*

TABLE 4–2 Major Categories of Rock-Forming Minerals

Class	Defining anion(s)	Examples
Silicates	Silicate ion (SiO_4^{4-})	Quartz (SiO_2), olivine (Mg_2SiO_4)
Oxides	Oxygen ion (O^{2-})	Hematite (Fe_2O_3), magnetite (Fe_3O_4), ice (H_2O), spinel ($MgAl_2O_4$)
Hydroxides	Hydroxyl ion (OH^-)	Goethite [$FeO(OH)$], gibbsite [$Al(OH)_3$]
Carbonates	Carbonate ion (CO_3^{2-})	Calcite ($CaCO_3$)
Sulfides	Sulfide ion (S^{2-})	Pyrite (FeS_2), galena (PbS)
Sulfates	Sulfate ion (SO_4^{2-})	Gypsum ($CaSO_4 \cdot 2\ H_2O$),* barite ($BaSO_4$)
Halides	Ions of chlorine (Cl^-), fluorine (F^-), bromine (Br^-), and iodine (I^-)	Halite ($NaCl$) (table salt), fluorite (CaF_2)
Native elements	None (no charged ions)	Gold (Au), copper (Cu)

*The center dot in this formula indicates that a calcium sulfate molecule is bonded with water molecules.

Mineral Properties: Hardness and Cleavage

The orientation, type, and strength of atomic bonds in crystals are important determinants of the physical properties of a mineral. In diamond, for example, the tight packing of carbon atoms leads to bonds that are equally strong in all directions; as a consequence, diamond is extremely difficult to break (Figure 4-9). In contrast, the carbon atoms in graphite are spaced farther apart and the bonds between the atoms are weaker in some directions than in others, so this mineral is soft and forms sheets that are readily pulled apart. Graphite can be broken so easily along the planes in which its carbon bonds are weakest that it leaves marks on paper. The "lead" in your pencil is actually graphite.

Hardness and the way a mineral breaks are particularly useful properties to use to identify unknown minerals. In the early 1800s, Austrian mineralogist Friedrich Mohs (1773–1839) ranked all known minerals by hardness on a scale of 1 to 10, with 10 being the hardest (Table 4-3). Later workers related the **Mohs hardness scale** to familiar materials such as steel, window glass, and fingernails. Talc, for example, has a hardness of 1 and can be scratched by harder minerals as well as by your fingernail, which has a hardness of between 2 and 3. A steel file or knife can scratch feldspar (hardness 6) but not quartz (hardness 7). Of all the minerals at Earth's surface, diamond is the hardest and can scratch all others. This is why jewelers are very careful to keep diamonds segregated from other gems.

Cleavage refers to the manner in which a mineral breaks. Some minerals, such as mica, break along parallel planes in one direction only (Figure 4-10). Others—pyroxenes, for example—break along two or more planes that are nearly perpendicular to one another. Still others, including amphiboles, break along two or more planes that are not at right angles. The planes along which a mineral breaks reflect the internal ordering of the mineral's atoms and the nature of their chemical bonds, with breakage occurring along those planes with the weakest atomic bonds. The weak ionic bonds between adjacent silicate sheets in mica break easily, so mica has perfect cleavage in one direction and poor or no cleavage in any other direction. When split thinly, mica becomes nearly transparent. For this reason, long before manufactured glass became abundant, thin sheets of mica were used to make windows for houses and lanterns.

In addition to hardness and cleavage, a number of other physical properties can be used to identify

Figure 4–9 Two different arrangements of carbon atoms result in two different minerals: graphite (bottom and left on finger), one of the softest minerals; and diamond (clear crystal on right), the hardest known mineral at Earth's surface. *(Chip Clark/Fundamental Photographs)*

Figure 4–10 Mica breaks into thin sheets along cleavage planes. The structure of mica results in pronounced cleavage along a plane between layers of silica tetrahedra. Both biotite (left) and muscovite (right), two types of mica, have the same cleavage because they are sheet silicate minerals. *(left: Andrew de Wet; right: Joel Arem/Science Source)*

minerals, such as color, density, scent (particularly useful for minerals containing sulfur), luster (whether a mineral appears glassy, metallic, waxy, etc.), and streak, the color of the mineral when powdered (Appendix 4).

■ Atoms combine in three-dimensional arrays to form minerals, which in turn combine to form rocks.

TABLE 4–3	Mohs Mineral Hardness Scale	
Hardness*	Mineral	Tests using common objects
1	Talc	Easily scratched with a fingernail
2	Gypsum	Scratched with a fingernail
3	Calcite	Scratched with a copper penny
4	Fluorite	Easily scratched with a steel knife
5	Apatite	Scratched with a steel knife or window glass
6	Orthoclase	Scratched with a steel file; (feldspar) scratches window glass
7	Quartz	Scratches a steel file and window glass
8	Topaz	Harder than most common objects
9	Corundum	Harder than most common objects
10	Diamond	Hardest mineral known (forms at depth of more than 100 km, at high pressure)

*Minerals with hardness of 8 and above are rare.

■ Although a few thousand rocks and minerals have been identified on Earth, only a few dozen of each are common, because Earth's crust is made primarily of eight elements.

■ The most common rock-forming minerals are silicates, made of silica tetrahedra that join into clusters, chains, and sheets.

■ Two key properties of minerals—their hardness and cleavage—reflect the internal ordering of their atoms and the strength of the bonds between atoms. These properties, along with other physical characteristics, make it possible to identify many minerals.

ROCKS AND THE ROCK CYCLE

Minerals combine to form rocks, and the type of rock formed depends both on the minerals and on the rock-forming processes involved. All rocks can be sorted into one of three major groups associated with their origin: igneous, sedimentary, or metamorphic. Each type can be identified from clues provided in the rock's mineralogy and texture. *Mineralogy,* which broadly means the study of minerals, refers as well to the relative proportions of each mineral in the rock. *Texture* refers to the different shapes and sizes of the minerals and the ways in which they are assembled.

Igneous Rocks

Igneous rocks (from the Latin *ignis,* "fire") form from *magma* (molten rock), and therefore consist of interlocking mosaics of crystals (see Figure 4-5b). The earliest rocks on Earth were igneous and formed 4.5 billion to 4 billion years ago as the planet differentiated from a homogeneous melt, its surface solidifying into a rigid silicon- and oxygen-rich crust. The composition of an igneous rock is a clue about where it formed, because

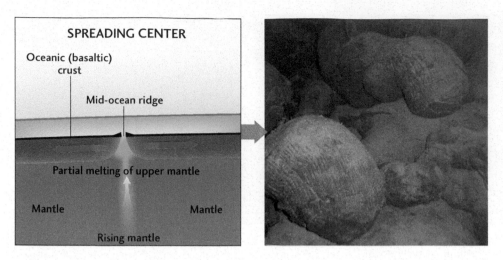

SPREADING CENTER

Oceanic (basaltic) crust

Mid-ocean ridge

Partial melting of upper mantle

Mantle   Mantle

Rising mantle

Figure 4–11 Basaltic magma forms by partial melting of rock in the asthenosphere (upper mantle), as along mid-ocean ridges at divergent plate boundaries (left) or at hot spots beneath oceanic crust, such as the island of Hawaii. Basaltic magma cools rapidly when it encounters ocean water, sometimes forming pillow shapes (right). *(OAR/National Undersea Research Program/NOAA)*

different tectonic settings tend to host different types of magmas. Likewise, the texture of the rock tells us whether it formed at or below Earth's surface. Rocks that form from *lava,* which is magma that has erupted above ground, are called **extrusive igneous rocks,** or *volcanic rocks,* and tend to have small or even microscopic crystals due to rapid cooling by the atmosphere or water bodies. Rocks that form from magma crystallizing below ground are called **intrusive igneous rocks,** or *plutonic rocks,* and have large crystals reflective of the slow cooling they experience due to being surrounded on all sides by warm, insulating rocks.

Mafic Magma at Divergent Boundaries The composition of lava emanating from mid-ocean ridges mirrors that of the igneous rock that covers nearly two-thirds of Earth's surface as oceanic crust. This lava is rich in iron and magnesium and relatively low in silicon and oxygen (about 45 to 52 percent SiO_2 by weight) and originates from the *partial melting* of Earth's upper mantle (Figure 4-11) related to the reduction of pressure and the introduction of seawater that occurs at mid-ocean ridges. The process of partial melting is similar to what happens when a chocolate chip cookie is heated in an oven: The chips begin to melt, while the rest of the cookie remains solid.

In the case of divergent boundaries, the melted materials are enriched in silicon and oxygen relative to the underlying mantle materials from which the partial melt was produced. When cooled at Earth's surface, the melt forms a dark rock called *basalt.* Basalt is known as a **mafic** rock, cooled from mafic magma and named for the chemical symbols for magnesium (Mg) and iron (Fe), important elements in this type of rock. Not all mafic magma is extruded at Earth's surface. Some intrudes into the oceanic crust and cools slowly, forming the intrusive igneous rock *gabbro.* Basalt and gabbro are chemically and mineralogically the same, but their different cooling rates lead to very different textures (Figure 4-12).

Andesitic Magma at Convergent Boundaries Igneous rocks at convergent plate boundaries are produced from magmas that are more enriched in silicon and oxygen (about 52 to 63 percent SiO_2), and correspondingly more depleted in iron and magnesium, than the mafic magmas at divergent boundaries (see Figure 4-12). The reason for the different composition probably has to do with two processes. First, magmas at convergent boundaries are produced in association with subduction of basaltic oceanic crust and marine sediments, rather than solely by partial melting of the mantle (Figure 4-13). The marine sediments contain silica-rich materials eroded from the continents and the shells of marine organisms, which provide calcium carbonate and silica to the downgoing plate. As the subducting plate moves downward into hotter parts of Earth's interior, ocean water and carbon dioxide—produced from the calcium carbonate shells—drive partial melting of the down-going plate and the overlying mantle wedge. The water and carbon dioxide act in the same manner as soda does with quartz, lowering the melting point of aluminum- and silicon-rich minerals in the overlying rocks and separating them from iron- and magnesium-rich minerals like olivine and pyroxene that have a higher melting point.

A second process involved in andesite formation may be *fractional crystallization* (Figure 4-14). As magma cools, those minerals that have the highest melting temperatures tend to form first. This partial crystallization leaves the resulting magma enriched in different elements that form a new suite of minerals as the magma continues to cool. The evolution of magma chemistry during cooling is known as **Bowen's reaction series** for the petrologist Norman Bowen, who first identified the phenomenon. Starting with a magma that has the composition of Earth's mantle, the minerals olivine, pyroxene, and calcium-rich plagioclase are the first to form from the magma as it cools. If these crystals sink

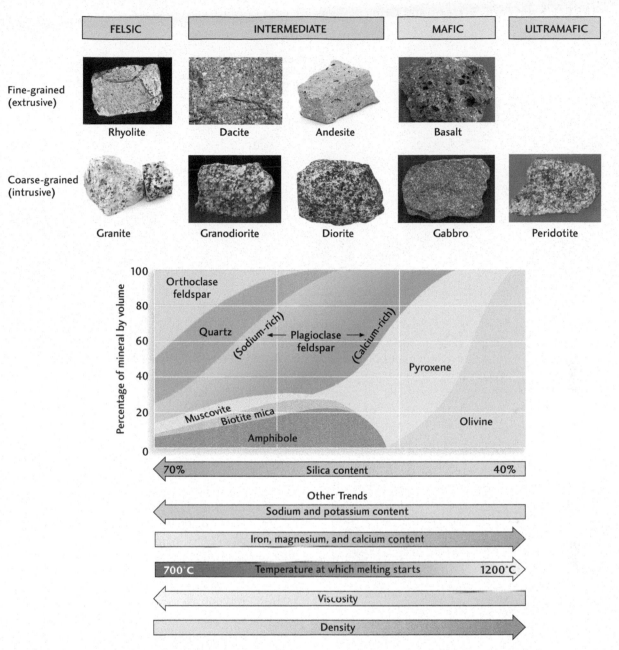

Figure 4–12 Extrusive and intrusive igneous rocks are classified according to the size of individual crystals and their mineral composition. Extrusive rocks are finer grained than intrusive rocks, for they cooled more quickly from the molten state. With increasing amounts of silica, igneous rocks contain less iron and magnesium and are lighter in color as a result. Oceanic crust consists mostly of basalt, whereas continental crust consists mostly of granite, diorite, and andesite. [(a–b: Dirk Wiersma/Science Source; c: Tyler Boyes/Shutterstock; d: Andrew de Wet; e: Tyler Boyes/Shutterstock; f: Courtesy Amir Chossrow Akhavan; g: Siim Sepp/Alamy; h: Andrew de Wet; i: Dr. Charles E. Jones, University of Pittsburgh)]

to the bottom of the magma chamber, new minerals will form as the magma continues to cool because the production of the more mafic minerals has used up much of the magnesium and iron in the process of crystallization, leaving the remaining magma more enriched in sodium, potassium, aluminum, and silicon. Together, partial melting and fractional crystallization create abundant feldspar and amphibole minerals.

The plutonic rocks at convergent plate boundaries are known as *diorite*; the associated volcanic rocks are known as *andesite*. Andesite and andesitic magmas derive their name from the Andes Mountains of South America, which are rich in this kind of igneous rock. Other subduction zones known for andesitic magma include the Cascade Range of western North America, which contains the volcanic mountains of Mount St. Helens, Mount Rainier, and Mount Hood; the volcanic mountains of Japan, including Mount Unzen and Mount Fuji; and the islands of the Philippines, where the Mount Pinatubo volcano, among others, is located. Because other magmas have even higher SiO_2 contents, andesitic magmas are said to be intermediate in composition.

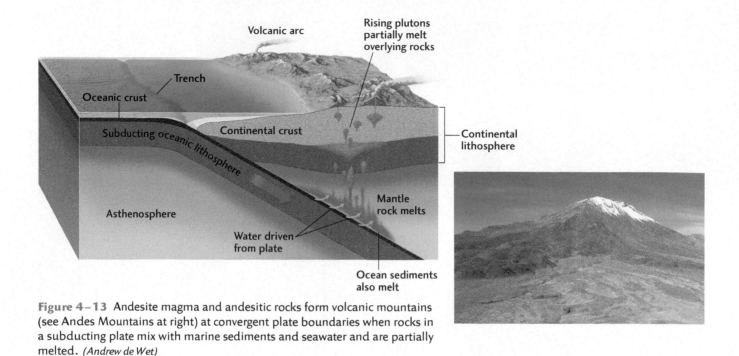

Figure 4–13 Andesite magma and andesitic rocks form volcanic mountains (see Andes Mountains at right) at convergent plate boundaries when rocks in a subducting plate mix with marine sediments and seawater and are partially melted. *(Andrew de Wet)*

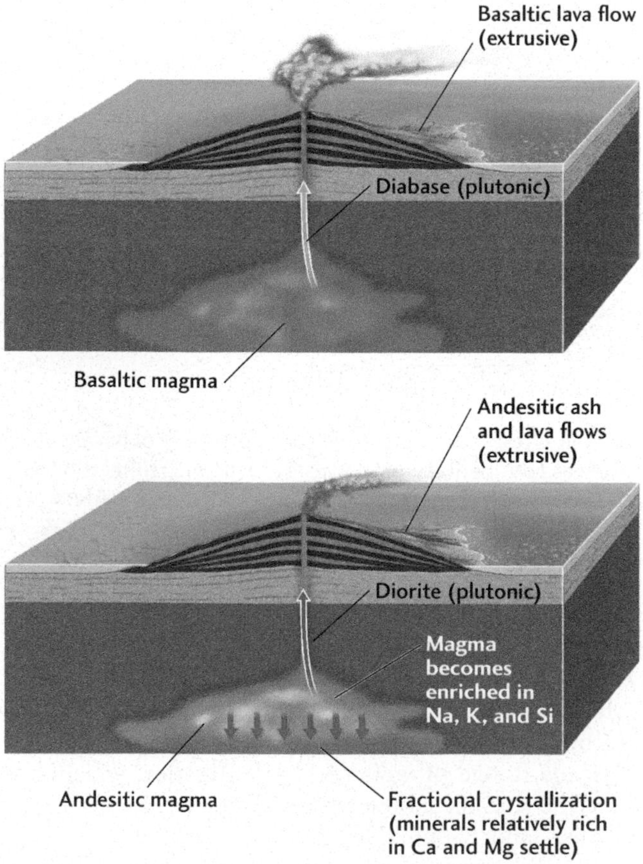

Figure 4–14 Andesite and associated igneous rocks (e.g., diorite) can form by fractional crystallization in a magma chamber (bottom). As crystals form and sink to the bottom of the chamber, the remaining magma has a different composition, in this case becoming more andesitic. In the upper diagram, by contrast, no fractional crystallization has occurred.

Felsic Magma in Continental Interiors Partial melting and fractional crystallization also are important in the formation of **felsic magmas** (*fel* for feldspar and *si* for silica) that form *granite,* an intrusive rock, and its extrusive equivalent, *rhyolite.* These rocks contain abundant quantities of quartz and mica, reflecting very high silica contents (approximately 69 percent or more). (An intermediate magma with silica content between that of andesitic and felsic magma also exists.) Felsic magmas also contain feldspars rich in sodium and potassium, elements that tend to remain in the melt as more mafic minerals crystallize out. Granites and rhyolites are common in continental interiors, such as at Yellowstone National Park in Wyoming. At this hot spot, basalt created in the upper mantle rises and heats the underbelly of the North American crust, partially melting some of it and producing magma rich in silicon and oxygen (see Chapters 2 and 5).

Another location where granite and rhyolite are found is in the cores of mountain belts that have been produced by continent–continent collision. The physical and chemical breakdown of continental rocks, or *weathering,* caused by the actions of oxygen, rain, ice, soil acids, and biota create sediments that are rich in aluminum and silicon (see Chapter 8). These sediments are transported downstream to the ocean by rivers and glaciers and deposited in marine environments. Many of these eroded sediments are trapped when continents collide and are incorporated into the growing mountain belt. Deep burial during collision may warm these materials to their melting points, and the magmas they produce are enriched in aluminum and silicon.

Sediments and Sedimentary Rocks

Sediments (from the Latin *sedere*, "to sink down") are produced by the chemical and physical breakdown of rocks into smaller particles and dissolved ions, processes collectively known as *weathering*. Once created, loose debris may be transported by *erosion*, that is, by the movement of flowing water, wind, or ice, or simply by gravity, as in a landslide. Eventually, these *clastic* sediments are laid down in **depositional environments** when transporting fluids lose energy and are unable to continue moving their sediment loads (Figure 4-15 and

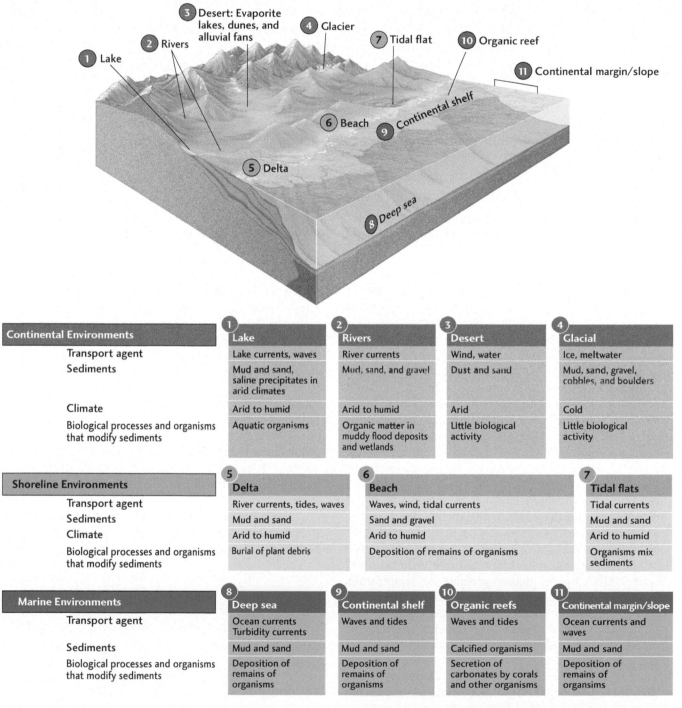

Continental Environments		1 Lake	2 Rivers	3 Desert	4 Glacial
Transport agent		Lake currents, waves	River currents	Wind, water	Ice, meltwater
Sediments		Mud and sand, saline precipitates in arid climates	Mud, sand, and gravel	Dust and sand	Mud, sand, gravel, cobbles, and boulders
Climate		Arid to humid	Arid to humid	Arid	Cold
Biological processes and organisms that modify sediments		Aquatic organisms	Organic matter in muddy flood deposits and wetlands	Little biological activity	Little biological activity

Shoreline Environments	5 Delta	6 Beach	7 Tidal flats
Transport agent	River currents, tides, waves	Waves, wind, tidal currents	Tidal currents
Sediments	Mud and sand	Sand and gravel	Mud and sand
Climate	Arid to humid	Arid to humid	Arid to humid
Biological processes and organisms that modify sediments	Burial of plant debris	Deposition of remains of organisms	Organisms mix sediments

Marine Environments	8 Deep sea	9 Continental shelf	10 Organic reefs	11 Continental margin/slope
Transport agent	Ocean currents Turbidity currents	Waves and tides	Waves and tides	Ocean currents and waves
Sediments	Mud and sand	Mud and sand	Calcified organisms	Mud and sand
Biological processes and organisms that modify sediments	Deposition of remains of organisms	Deposition of remains of organisms	Secretion of carbonates by corals and other organisms	Deposition of remains of organsims

Figure 4–15 Sedimentary rocks form in a variety of depositional environments that include continental, shoreline, and marine locations. The nature of a deposit, such as grain size and rounding, provides clues about the environment in which it was deposited. Very rounded and well sorted grains are likely to have come from sand dunes, for example, whereas much coarser and more poorly sorted ones are characteristic of glacial deposits. Because of their longer transport distances from source to sink areas (i.e., from mountains to ocean basins, respectively), clastic marine deposits tend to be finer grained (mud and sand) than many continental and shoreline clastic sedimentary deposits.

TABLE 4–4 Characteristics of Various Sedimentary Environments

Environment	Characteristics of sediments found there
Glacier	Ridges of unsorted clay- to boulder-size particles along glacial margins
Lake	Clay- and silt-size particles deposited in horizontal layers (laminations)
Desert	Well-rounded and well-sorted sands forming dunes; sand grains appear frosted from frequent impacts with one another
River	Sequences of sediments that occur in layers, from cobble to silt size; interbedded with clay and silt deposited in floodplains
Delta	Sequences of layers of well-sorted sand, silt, and clay that dip gently seaward and get finer with greater distance from the shore
Beach	Well-rounded and well-sorted sands; no frosting of grains; marine fossils (shells, corals)
Tidal flat	Fine silt and clay with mud cracks, animal burrows, and animal tracks
Continental shelf	Mixture of muds and sands from continent; shells, skeletons, and fecal pellets of marine organisms
Reef	Calcium carbonate skeletons of corals
Deep sea	Laminated fine clays and biologic oozes
Evaporite basin	Chemical sediments, deposited as salt crystals in horizontal layers in shallow water, often mixed with fine silt and clay washed from surrounding mountains

Table 4-4). Other types of sediments can be produced by chemical and biological processes. These different types of sedimentary rocks can be subdivided into three categories: clastic, chemical, and biological, as discussed below.

Deposited sediments turn into **sedimentary rocks**—a phenomenon known as **lithification**—through *compaction* and *cementation*. Compaction is a reduction in the amount of pore space between sedimentary particles that occurs as more and more sediments are deposited on top of one another and bear down on the sediments at the bottom of the pile with their great weight. Cementation occurs when groundwater percolating through the sediments precipitates dissolved solids in pore spaces. These precipitates, typically calcium carbonate, silica, or iron oxides, help to hold sedimentary particles together. Sediments deposited at Earth's surface can become lithified at depths of several kilometers and temperatures up to about 200°C, but at higher pressure or temperature, metamorphic processes begin. We will discuss this in further detail below in the section on metamorphic rocks.

Because they are made of sediments deposited at Earth's surface, sedimentary rocks contain fossils of plants and animals and thus, a record of life on Earth. Their fossil content also makes sedimentary rocks sources of carbon-rich fossil fuels (see Chapter 13). For instance, coal is produced from the burial and compaction of prehistoric swamp vegetation under layers of sediment, and oil develops from the chemical alteration (under heat and pressure) of tiny marine organisms that sink to the ocean floor upon death.

Clastic Sedimentary Rocks Clastic sediments (from the Greek word *klastos*, meaning "broken") arise largely from the action of physical weathering processes that break rocks into fragments. Ranging in size from tiny clay particles to large boulders, these fragments—or clasts—can be eroded and transported to a new location by erosional processes. If lithified, the sediments become **clastic sedimentary rocks.** Clastic sediments and sedimentary rocks are classified by the sizes of their clasts. For example, a rock formed of clay-size particles less than 4 μm (micrometers: 1 μm = 0.001mm = 0.0001 cm) in diameter is called a mudstone or a shale, while a rock formed of sand-size grains (63 μm to 2 mm) is a sandstone (Table 4-5).

The size of the grains and their shapes are clues as to the environment in which the sediment was transported and deposited. For example, the larger the grain

TABLE 4–5 Classification Scheme for Clastic Sedimentary Rocks

Particle size (diameter)	Sediment name	Rock name
<4 μm	Clay	Mudstone or shale (fine layering)
4–63 μm	Silt	Siltstone
63 μm–2 mm	Sand	Sandstone
>2 mm	Gravel	Conglomerate

size of clastic sediment, the greater the energy required to transport it. It would be impossible to move a 10-ton boulder with a trickle of water, even on a steep slope. However, a trickle may have sufficient energy to transport clay-size particles, and the boulder would provide little challenge for the moving ice in a glacier.

The size distribution of the grains, a parameter known as *sorting,* also gives important information about the environment of deposition. Beach sands, for example, typically contain a narrow range of grain sizes, reflecting the fact that waves washing sands back and forth across the beach winnow away finer silts and clays. In contrast to well-sorted beach sediments, those deposited from debris flows can include a range of grain sizes, from tiny clays to giant boulders, making them poorly sorted.

Particle shape also provides insight into the origins of clastic rocks. Some transport processes can cause individual grains to bump into one another, which knocks off sharp corners and makes the grains more rounded over time. This *rounding* is particularly pronounced in quartz sand dunes, where grains are blown back and forth by the winds. The short transport times associated with rock falls, in contrast, are insufficient for much rounding to occur, so these deposits tend to have angular clasts.

In addition to sorting and shape, additional evidence of sediment transport processes often exists in the rock record. So-called **sedimentary structures** include features such as *ripple marks* and *cross beds,* evidence of past air and water flow directions (Figure 4-16). *Sedimentologists,* the geologists who specialize in understanding sedimentary rocks, use clues provided by grain size, sorting, rounding, sedimentary structures, and the presence of marine or terrestrial fossils to figure out in which sedimentary environments different rocks were deposited. In this way, they are able to reconstruct Earth's history.

Chemical Sedimentary Rocks **Chemical sediments** originate from chemical weathering processes that partially or wholly dissolve minerals (see Chapter 8). You might have noticed that allowing tap water to dry on dishes can produce white spots. These spots are calcium and carbonate ions that had been dissolved in the water and were left behind when it evaporated. All natural water contains some dissolved ions. In freshwater, the concentration of ions is very small, in the range of just a few parts per million (ppm). In comparison, the dissolved solids content of seawater is roughly 35,000 ppm. Crystalline sedimentary deposits that form from evaporating waters are called **evaporites** and have a variety of compositions, including calcium carbonate, calcium sulfate, and sodium chloride (Figure 4-17). Most of the world's supply of table salt (sodium chloride) is mined from sediments in evaporite basins.

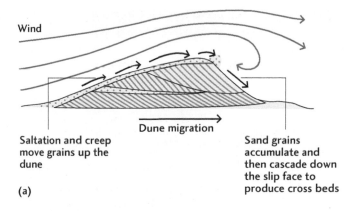

Wind

Dune migration

Saltation and creep move grains up the dune

Sand grains accumulate and then cascade down the slip face to produce cross beds

(a)

(b)

(c)

Figure 4–16 Sedimentary structures, such as cross bedding, are clues to environments of deposition. (a) Grains are blown up the stoss (upwind) side of a dune and redeposited on the lee (downwind) side of the dune, leading to distinctive layering that indicates the direction of the prevailing wind. With dune migration and shifting winds, layers with different grain orientations—known as cross beds—accumulate one atop the other. (b) The shape of these dunes reveals that the prevailing wind direction is from upper left to lower right, as this photo is oriented. The upwind side of the dune form (left) has a low slope, up which grains are blown. At the crest of the dune, the grains avalanche down the steep downwind side. (c) Sedimentary structures such as cross beds in sandstone enable scientists to better understand the environmental conditions that existed during their deposition. *(b: Stockpile/AgeFotostock; c: Marli Miller)*

Figure 4–17 Death Valley is an example of an evaporate basin, a deep basin with no streams draining out of it. Instead, streams from surrounding mountains drain water and salts into the basin, where water evaporates and salt crystallizes. Frequent wetting and drying, with the associated expansion and contraction of salt minerals, results in polygonal cracks on the desert floor. *(eye35 stock/Alamy)*

Sometimes chemical sediments crystallize even before the water in which they were dissolved has completely evaporated. Consider a lake in an enclosed basin like Death Valley. The lake continually receives stream water carrying dissolved ions. Some of the water in the lake is removed by evaporation, but the ions stay behind. If the rates of inflow of stream water and outflow by evaporation balance, the lake maintains a stable size, but the concentration of salts increases year after year. Eventually, the concentration of ions becomes so high that the water is saturated with salt and is no longer capable of holding any more in dissolved form. Minerals containing the ions will begin to form by crystallization, as they do in Death Valley, the Great Salt Lake in Utah, the Dead Sea in Israel, and elsewhere.

Biological Sedimentary Rocks   Biological sediments accumulate from the carcasses of organisms. For example, the toppling of trees and the death of small shrubs and grasses in swamps and marshes lead to the formation of peat, which may eventually turn to coal, given sufficient time and depths of burial (see Chapter 13). In addition, many organisms secrete calcium carbonate ($CaCO_3$) or silica shells to protect their soft bodies from injury and predation by other organisms. Single-celled foraminifera and coccoliths, for example, secrete millimeter-diameter shells that create thick oozes of calcium carbonate on the ocean floor when these organisms die (Figure 4-18). If the oozes are lithified, they form a rock known as a *limestone*. Limestone formed of foraminiferal oozes is also known as *chalk*, and is used to make a variety of products, including toothpaste and the chalk used to write on classroom blackboards.

(a)

(b)

Figure 4–18 (a) Layers of chalk, a type of limestone, are exposed along many kilometers of sea cliffs in Normandy, France. (b) Chalk consists of sediments rich in calcareous microorganisms called foraminifera, like the sand–size shells in this magnified image. *(a: Peresanz/Shutterstock; b: ©Alain Couette. 3339f. Thanks to Gregory Willems. http://psammophile. pagesperso-orange.fr/Page12.html)*

Another organism that secretes calcareous hard parts is coral. Coral reefs consist of large colonies of coral polyps (the fleshy animal part of the coral) and cover extensive areas along continental margins in the tropics, between latitudes of 23° north and south of the equator in today's climate. Many ancient reefs are preserved as limestones in the fossil record and, because of plate movements, can be found today at most latitudes.

Not all organisms living in the ocean secrete calcium carbonate shells. Some create skeletons or shells of silica. Single-celled microorganisms called radiolaria (animals) and diatoms (plants) create siliceous oozes when their bodies fall to the ocean floor. If lithified, these oozes turn into *diatomite* (in the case of diatoms) or *chert* (in the case of radiolaria). Sponges also secrete silica skeletons.

Many biological sediments are considered chemical in origin because the organisms that secrete carbonate and silica skeletons derive their skeletal materials from the waters in which they live. An example is the reversible chemical reaction by which mollusks, such as clams, extract bicarbonate (HCO_3^-) and calcium (Ca^{2+})

ions from water and combine the ions to form shells of calcium carbonate ($CaCO_3$):

$$2\ HCO_3^- + Ca^{2+} \leftrightarrow CaCO_3 + CO_2 + H_2O$$

For this reason, biological sedimentary rocks are sometimes called biochemical sedimentary rocks.

Metamorphic Rocks

Metamorphic rocks (from a Greek word meaning "to change form") result when either igneous, sedimentary, or other metamorphic rocks are exposed to high pressure, high temperature, or both, resulting in a change in the mineralogy of the parent material in the absence of melting. Metamorphism occurs, for example, when rocks are deeply buried, compressed, and folded during continent–continent collision (Figure 4-19). With burial, the weight of overlying rock increases, and at great depths the pressure is sufficient for minerals to react and recrystalize in a solid state. This process is aided by the fact that Earth's temperature increases about 20°C per kilometer

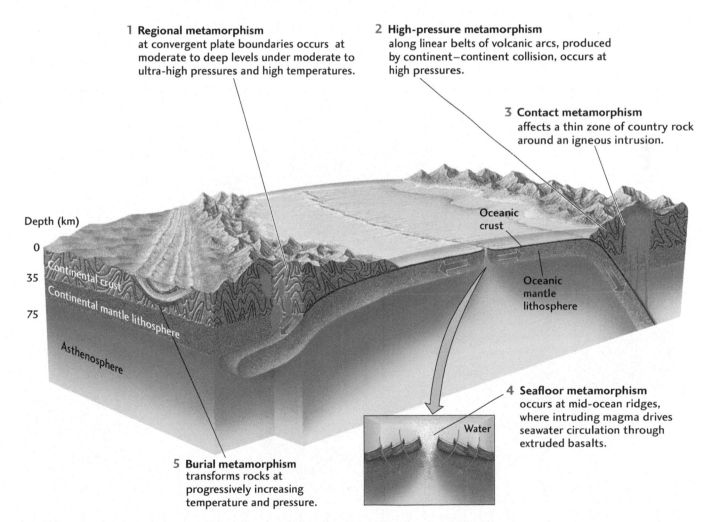

1 Regional metamorphism at convergent plate boundaries occurs at moderate to deep levels under moderate to ultra-high pressures and high temperatures.

2 High-pressure metamorphism along linear belts of volcanic arcs, produced by continent–continent collision, occurs at high pressures.

3 Contact metamorphism affects a thin zone of country rock around an igneous intrusion.

4 Seafloor metamorphism occurs at mid-ocean ridges, where intruding magma drives seawater circulation through extruded basalts.

5 Burial metamorphism transforms rocks at progressively increasing temperature and pressure.

Depth (km)
0
35
75

Continental crust
Continental mantle lithosphere
Asthenosphere
Oceanic crust
Oceanic mantle lithosphere
Water

Figure 4–19 Different types of metamorphism and metamorphic rocks occur in diverse geologic settings. [*Source: From J. Grotzinger and T. Jordan, 2010, 151, fig. 6.4*]

of depth; chemists have known for centuries that higher temperatures generally increase reaction rates.

If rocks are subducted to a depth of some 35 to 40 km, where the temperature is greater than about 650°C, igneous processes of partial melting and magma formation begin. For this reason, igneous and metamorphic rocks often are located together in mountain belts at convergent boundaries. Metamorphic rocks are also found adjacent to igneous rocks at both divergent and convergent margins because the hot magmas that produce igneous rocks expose surrounding rocks to magmatic fluids and high temperatures, which can alter the mineralogy of the rocks surrounding igneous intrusions.

Texture of Metamorphic Rocks A dominant characteristic of many metamorphic rocks is their texture, which refers to the characteristic sizes, shapes, and orientation of individual minerals in the rock. Metamorphic rocks often possess directional texture known as **foliation,** in which minerals are arranged in a parallel fashion as crystals grow under conditions of immense rock pressure (Figure 4-20). Foliation is common in metamorphic rocks that originally were fine grained and contained large amounts of clay minerals, such as the sedimentary rock known as shale. The heat and pressure of deep burial cause clays to alter to mica minerals and to grow in a parallel fashion. The new metamorphic rock, slate, is much harder than shale, and the

Figure 4-20 Metamorphic rocks often have a characteristic texture known as foliation, in which minerals are arranged in a parallel fashion. *(Andrew de Wet)*

parallel growth of flaky mica minerals makes it possible to split the rock into shingles for roofs and into large sheets for chalkboards. More intense metamorphism could produce schist, and even more intense metamorphism could produce gneiss (Figure 4-21).

In general, crystal size and degree of foliation increase with higher pressures and temperatures and continued metamorphism. With increasing degree of metamorphism, foliation is expressed in fine-grained

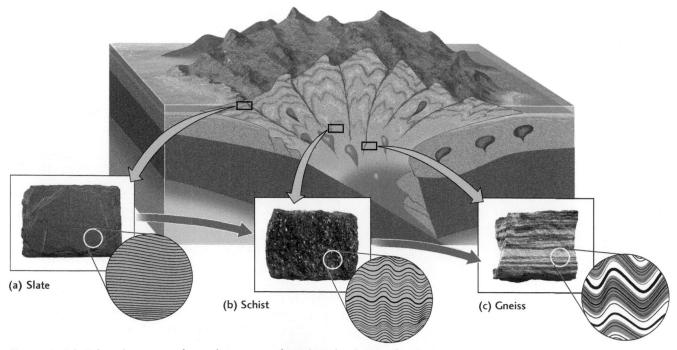

(a) Slate

(b) Schist

(c) Gneiss

Figure 4-21 Foliated metamorphic rock names are based on the degree of foliation and crystal size, which reflect the intensity of metamorphism. When shale is metamorphosed, for example, foliation occurs first as slaty cleavage in (a) slate, then as schistosity in (b) schist, and then as banding in (c) gneiss. *(Courtesy John Grotzinger/Harvard Mineralogical Museum; photo by Ramón Rivera-Moret)*

rocks first by *slaty* texture. As metamorphism continues, platy micas or needle-shaped amphiboles grow sufficiently large to be visible to the naked eye, and their parallel to sub-parallel alignment results in a texture referred to as *schistosity*. Continued metamorphism may destroy micas and produce amphibole and plagioclase. These minerals segregate into alternating bands of dark and light color, resulting in *gneissic banding*, as can be seen in Figure 4-20.

Non-foliated texture occurs in rocks that consist predominantly of one mineral—such as quartz or calcite—that does not have a tendency to form elongated or platy crystals. The most common non-foliated metamorphic rocks are *quartzite* and *marble*, which result from the metamorphism of sandstone and limestone, respectively.

Metamorphic Minerals as Geothermometers and Geobarometers The minerals that form during metamorphism reflect the conditions to which their surrounding rocks were subjected, particularly those of pressure and temperature. Recall, for example, that elemental carbon forms two different minerals—graphite and diamond—depending on the pressure the carbon is subjected to during crystallization.

In eastern North America, the continent–continent collision that formed Pangaea beginning about 500 million years ago created regionally high pressures and temperatures that metamorphosed sedimentary rocks like shales into slates, schists, and gneisses. Driving eastward from the Hudson River through New England, one moves into the heart of the collision zone and finds rocks of increasingly higher metamorphic grade, meaning rocks that experienced increasingly higher pressures and temperatures (Figure 4-22). Each change in temperature and pressure is recorded in a change in the types of minerals present in the rocks, a fact that we can use to determine what the pressure and temperature conditions were in different locations. The mineral sillimanite, for example, will not form below temperatures of about 600°C, so finding this high-grade metamorphic mineral in a rock indicates that that rock was subjected to temperatures at least this high.

The Rock Cycle

In discussing the three different rock groups, we have noted that one rock type can turn into another type through a variety of processes. As a result, the elements that make up rocks move back and forth between Earth's

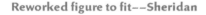

Reworked figure to fit--Sheridan

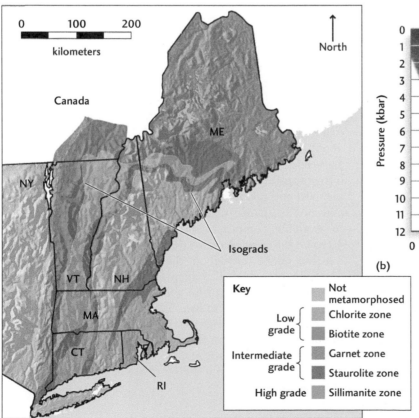

(a)

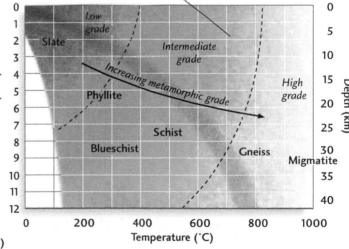

(b)

Figure 4–22 Regional metamorphic series in New England. From Maine to New Hampshire along a northeast to southwest transect, for example, the rock types change from low– to high–grade metamorphic rocks. [Source: Adapted from J. Grotzinger and T. Jordan, 2010, 157, fig. 6.9]

surface and its interior in a **rock cycle,** as discussed in Chapter 1 (see Figure 1-11). Melting of sedimentary or metamorphic rocks can produce igneous magmas. Weathering and erosion of igneous and metamorphic rocks can produce sediments. Similarly, deep burial of igneous and sedimentary rocks can produce metamorphic rocks. Plate tectonics provides a way to understand the forces that power the cycling of rock among igneous, metamorphic, and sedimentary reservoirs.

■ Magma can erupt at Earth's surface, forming volcanic (or extrusive) rocks, or can cool and solidify within the crust, forming plutonic (or intrusive) rocks.

■ The processes of partial melting and fractional crystallization generate magmas of varying silica contents.

■ Basaltic magmas are formed dominantly at divergent plate boundaries, whereas intermediate magmas are common at convergent boundaries, and felsic magmas are common in continental interiors.

■ The three types of sedimentary rocks—clastic, chemical, and biological—form in different sedimentary environments, such as floodplains and beaches (clastic), desert basins (chemical evaporites), and coral reefs (biological).

■ Sedimentary rocks are the only rocks that contain evidence of past life on Earth (as fossils) and fossil fuels such as oil, coal, and peat.

■ Metamorphic rocks form when igneous, sedimentary, or other metamorphic rocks are subjected to high pressures, temperatures, or both, which leads to the recrystallization of minerals in the rocks without melting.

■ Metamorphic rocks are either foliated or non-foliated, depending on the shape and orientation of the minerals that form during recrystallization, and the types of minerals found in the rocks provide information about the pressures and temperatures to which it was subjected.

■ The rock cycle describes the movement of Earth's solid materials among the sedimentary, igneous, and metamorphic rock reservoirs through the actions of weathering and erosion, burial and compaction, heat and pressure, and melting.

ROCK AND MINERAL RESOURCES

Long before the Phoenicians discovered how to make glass, early humans used different types of hard rock to make tools such as hand axes, and flint (fine-grained quartz) and obsidian to manufacture blades and arrowheads. The use of different Earth materials by humans through time enabled an early Danish archaeologist, Christian Jürgensen Thomsen (1788–1865), to separate millions of years of human history in Eurasia and Africa into three great technological ages: the Stone Age, which began several million years ago and ended about 6500 to 4000 years ago; the Bronze Age, which existed from about 5300 to 2600 years ago; and the Iron Age, which began about 3300 to 2600 years ago (dates vary by region).

Each age represents a period in human history during which a different Earth material was dominant in the manufacturing of various tools and weapons. Before smelting processes to separate metals from ore were developed, humans relied on hard stone for manufacturing (the Stone Age). With time and the transfer of knowledge, fostered by trade, people developed ways to mine minerals, separate elements, mix metals to make alloys, and manufacture tools from the materials they were able to work. As early metallurgists learned how to work metals, they discovered that bronze, an alloy of copper with tin, is a stronger, more durable product than copper alone; this discovery led to the Bronze Age. With time and experience, metallurgists learned how to work iron, more widely available in Earth's crust but requiring higher temperatures for smelting (the Iron Age). They also learned that an alloy of iron and small amounts of carbon produces a superior product, steel, with greater strength and rust resistance than iron alone.

Today, a wide variety of mineral resources is essential to industrial society, and the production and consumption of minerals are part of a global trade network. In 2010, the annual consumption of new (that is, not recycled or reused) nonfuel minerals and rocks used to make goods for each person in the United States was about 17,000 kg (38,000 pounds, or 19 tons).[5] Minerals are needed, among other things, to build roads and farming equipment, to manufacture trucks and trains for hauling food and goods, and to construct homes, appliances, automobiles, paper, glass, and bicycles. Some products—for example, computers and cell phones—contain some 40 to 60 different elements that are acquired from many minerals mined throughout the world.

Valuable nonfuel mineral resources are divided into two basic categories: *metals,* such as iron, copper, and aluminum; and *nonmetals,* such as sand and gravel. Metals form metallic bonds in which electrons in outermost shells can flow freely through the array of atoms. For this reason, metals are good conductors of heat and electricity, and therefore are in wide use in the modern world.

Industrialization and Mineral Resources

It is said that in the early 1800s, the French military and political leader Napoleon Bonaparte (1769–1821) pointed to China on a map and, referring to it as a sleeping giant, remarked, "When China wakes, it will shake the world."[6] Indeed, China and a number of other countries are "waking" in terms of their rapidly industrializing economies. With continued population growth and rising standards of living worldwide, the global demand for mineral resources is accelerating. Rapidly industrializing countries include China, India, Indonesia, and Brazil, and their growth rates and size are having a significant impact on the consumption and production of minerals.

The historic trends associated with industrialization include rising income, rising mineral and energy consumption, and stabilization of mineral and energy consumption at a higher level. One of the dominant characteristics of the early stages of economic development is construction of new infrastructure for housing, factories, and transportation. Such infrastructure requires large amounts of cement, steel, and glass, among other resources. The production of cement in the rapidly developing Asia–Pacific region now dominates the world cement industry (Figure 4-23).[7] Increasing consumption of cement is a key indicator of a country's economic development.[8] A main ingredient in cement is limestone, but cement can also contain gypsum, clay, and volcanic

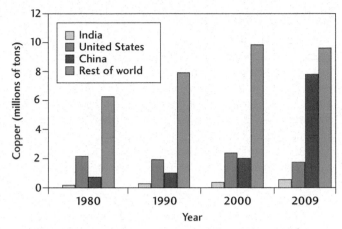

Figure 4–24 Copper consumption is a key indicator of a country's economic growth. Copper consumption in China has risen markedly since the 1990s, but decreased in the United States from 2000 to 2009. Before 2002, the United States was the world's leading copper consumer. *[Source: Data from the U.S. Geological Survey]*

ash. Cement is combined with sand and gravel or other aggregate material to make concrete. Recall from Figure 4-2 that the three nonfuel mineral resources with highest rates of per capita annual consumption in the United States are stone, sand and gravel, and cement. As countries around the world increase their standard of living and consumption, the rates at which they use these and other mineral resources are increasing as well.

With continued industrialization, subsequent stages of economic development typically include an increase in light manufacturing (for example, clothes, shoes, and furniture) and then heavy manufacturing (for example, power plants and ships). Progression through these manufacturing stages is accompanied by a rise in consumption of copper, aluminum, steel, and a variety of other industrial minerals, such as rare earth minerals. Copper is used in modern society, for example, for power generation and transmission, telecommunications systems, industrial machinery, automobiles and trucks, and wiring and plumbing for appliances and heating and cooling systems.[9] An average car might contain about 1.5 km (0.9 mi) of copper wire and 20 to 45 kg (44 to 99 lbs) of copper.[10] In recent years, China, with 1.3 billion people the largest of rapidly industrializing and growing economies, replaced the United States as the leading consumer of copper in the world (Figure 4-24).

Here we examine the geologic origin of mineral resources, the ways in which the minerals are mined, and some environmental impacts of their mining. We also examine the amounts of mineral resources thought to be available in Earth's crust and how these reserves are depleted with continued mining.

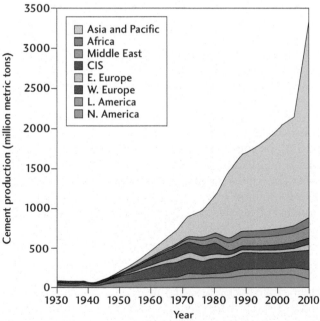

Figure 4–23 Rapidly growing Asian economies, particularly that of China, now dominate world production of cement.

Critical Minerals

Some minerals have unique properties and perform essential functions, so that few, inferior, or no satisfactory substitutes exist, and a restriction in their supply would have a substantial impact on economic activity. These are referred to as **critical minerals**.[11] Criticality encompasses the importance of a mineral's use, as well as its availability in space and time. The availability of critical minerals depends on geologic, technical, environmental, political, and economic factors, all of which can change with time. In 1973, for example, the leading producer of nonfuel minerals in the world was the United States. Today, however, the leading producer and consumer of minerals is China. From 1996 to 2009, the United States' reliance solely on imports more than doubled, from 9 to 19 nonfuel minerals.[12]

The U.S. National Research Council developed a method to characterize a mineral's criticality that is illustrated by a graph on which the impact of supply restrictions is plotted on one axis and the risk of supply disruption is plotted on the other (Figure 4-25). The degree of a mineral's criticality increases as these two measures get higher. Copper is important to modern industry, so a restriction in supply would have a fairly large impact (3 on the criticality graph), but it is widely available and the risk of disruption to its supply is low (1 on the criticality graph). On the other hand, rare earth and platinum group elements are much more critical commodities than copper because their supply risk is greater and they are essential to many manufactured products (Box 4-2).

Ore Deposits and Ore Minerals

Only three of the industrially important metals—aluminum, iron, and magnesium—occur in proportions

Figure 4–26 In order to mine economically valuable ore deposits, the overburden, or surrounding rock with no economic value, must first be removed. This waste rock, shown here in Arizona from copper mining, is referred to as "tailings." (Jim Wark/AirphotoNA)

greater than 2 percent of Earth's crust. Most others constitute less than 0.1 percent. Copper, for instance, composes only 0.006 percent, on average, of Earth's crust. Fortunately for humans, various geologic processes cause minerals to be concentrated in small areas, resulting in rich deposits from which commercially valuable minerals can be extracted profitably. These are called **ore deposits,** and the desired minerals in them are **ore minerals.** All unwanted minerals in the ore deposit are treated as waste and left at or near the mine site as spoil piles, or **tailings** (Figure 4-26).

The abundance of an element in an ore deposit compared with its abundance in the whole crust is known as the element's **economic concentration factor** (Table 4-6). Iron, for example, has a crustal abundance of about 5.6 percent, but to be economically recoverable, an iron ore must contain about 28 percent or more iron. In other words, the economic concentration factor for iron is about 5 to 7 times its crustal abundance. Copper needs to be concentrated in amounts nearly 100 times its crustal abundance in order to be economically recoverable. The global annual production of copper is more than 16 million tons, which means that 100 times that amount, or nearly 2 billion tons of rock, probably was mined in order to extract that much pure copper.

Understanding links between tectonic events and ore formation is important in mineral resource exploration and evaluation. Historical images of mining often evoke the picture of a bearded prospector and his mule working their way along a narrow trail in rugged mountains. The coincidence between mountains and ore deposits exists because most ores are associated with the high crustal temperatures found at convergent boundaries where mountain building and volcanic

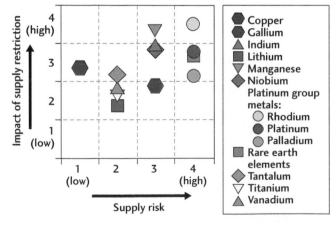

Figure 4–25 Critical minerals are those for which both the risk to supplies and the impact of supply restrictions are high. Rare earth and platinum group elements fall in this category.

TABLE 4–6 Economic Concentration Factors of Some Elements

Element	Crustal abundance (% by weight)	Concentration factor*
Aluminum (Al)	8.23	3–4
Iron (Fe)	5.63	5–7
Copper (Cu)	0.0060	100
Nickel (Ni)	0.0084	100
Zinc (Zn)	0.0070	500
Lead (Pb)	0.0014	4000
Gold (Au)	0.0000004	2000

*The concentration factor is the abundance in the deposit divided by the crustal abundance.

Sources: Data from W. M. Haynes, ed., *CRC Handbook of Chemistry and Physics*, 93rd ed. (Boca Raton, Florida: CRC Press, 2005); and W. J. Rankin, *Minerals, Metals, and Sustainability: Meeting Future Material Needs*. (Collingwood, Australia: CSIRO Publishing, 2011).

activity occur. Ore deposits along the western coasts of North and South America coincide with convergent plate boundaries and subduction zones, whereas deposits in Europe and Asia coincide with collision zones and volcanic arcs formed along the boundaries of the Indian, Australian, and Eurasian plates (Figure 4-27). Ore deposits are also being discovered along divergent boundaries, where seawater heated in cracks in basalt at mid-ocean ridge spreading centers returns to the ocean floor laden with dissolved metals.

Types of Ore Deposits

All metallic and most nonmetallic ore deposits can be classified according to one of four broad types of origin: hydrothermal, igneous, sedimentary, and weathering related. The first three are discussed here, while the fourth, which explains the origin of most aluminum ores and clay, is discussed in Chapter 8.

Hydrothermal Ore Deposits In **hydrothermal ore deposits,** mineral ores form by crystallization from extremely hot, metal-rich solutions. The term *hydrothermal* refers to the role of high-temperature seawater and groundwater in the processes of alteration and emplacement of minerals. In the 1960s, the richest underwater sulfide mineral deposits known on Earth were discovered during deep-sea drilling in the Red Sea, which marks the location of a rift between the African and Arabian plates (see Figure 2-3). Sediment cores retrieved from numerous deep basins along the rift indicate that sulfide minerals

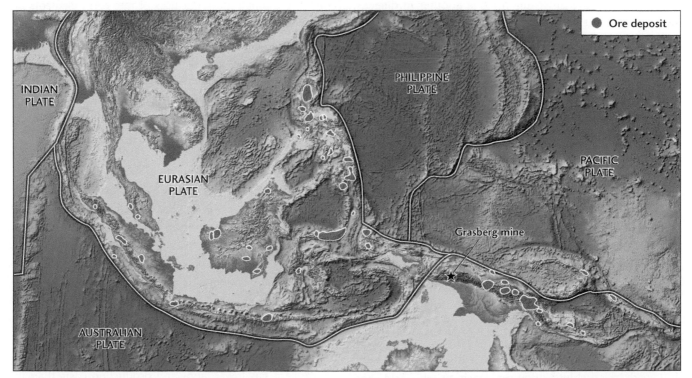

Figure 4–27 The distribution of major copper, molybdenum, and gold deposits (red dots) in the world is similar to the locations of Mesozoic- to Cenozoic-age mountain–building activities along plate boundaries, as shown here where multiple plates are converging. For example, the Grasberg mine on the island of New Guinea is the world's largest gold and third–largest copper mine.

BOX 4–2 CASE STUDY

Rare Earth Elements: Not Rare, But Hard to Get

The rare earth elements comprise 16 elements with similar properties that often are found together in geologic deposits: 15 elements in the lanthanide series from atomic numbers 57 through 71, and the element yttrium (atomic number 39) (**Figure 1**).[13] Scandium (atomic number 21) is sometimes included with the rare earth elements because it also commonly occurs with them and has similar properties. Though the minerals in which these elements—all metals—first were identified are uncommon, the rare earth elements themselves are not rare. All but one is more common than silver or mercury in Earth's crust. Unlike most industrial metals, however, the so-called rare earth elements are less likely to be *concentrated* in exploitable ore deposits. More than three-quarters of known rare earth reserves occur at a few locales in China (50%), the former Soviet republics (now some of the states in the Commonwealth of Independent States; (17%), the United States (12%), India (4%), and Australia (1%) (**Figure 2**).[14]

Rare earth elements typically occur together in geologic deposits and substitute for one another in different minerals.[15] Nearly all production of rare earth elements has come from fewer than 10 minerals, and the major economic mineral is bastnäsite, a carbonate mineral (see Figure 1). One of the main ways in which rare earth elements are thought to be concentrated in Earth's crust is by enrichment during the late stages of magma crystallization in an igneous intrusion derived from partial melting of the mantle. A second means of relative enrichment is deep weathering of minerals from igneous rocks to clays in warm and wet climatic conditions, as in the formation of laterite soils in the tropics (see Chapter 8). Clay minerals can adsorb rare earth ions into their mineral structure, creating a crust of low-grade ore forms above the unweathered parent rock. Adsorption is the adhesion, or accumulation, of a chemical substance onto the surface of solid particles.

Rare earth elements are difficult to isolate and refine to pure metal, and efficient separation processes were not developed until the 20th century. The process of extracting rare earth elements typically begins with open-pit mining followed by crushing, flotation, leaching with acid, and roasting. With typical ore grades of 5%–10%, open-pit mining of rare earth elements leaves behind large piles of tailings. In addition, ores that contain rare earth elements commonly contain radioactive substances, particularly thorium, so waste products can be radioactive.

"The Vitamins of Chemistry." Just small additions of rare earth elements to other metals can yield significant results for industrial materials and high-tech manufacturing, leading to their moniker as the vitamins of chemistry.[16] Mixing rare earth metals with iron or aluminum yields alloys that are harder and more resistant to heat and corrosion than traditional alloys. When combined with iron and boron, the rare earth element neodymium produces permanent magnets that are lighter and 10 times stronger than other magnets. Rare earth elements have enabled manufacturers to build increasingly smaller and more powerful computers and cell phones. Rare earths are also essential to the defense industry, as they are used for a range of weapons and equipment that includes night-vision goggles, precision-guided weapons, and armored vehicles.

The multiple powerful uses of rare earth elements are particularly vital to emerging "green" technologies that could help the United States wean itself from oil dependence, including compact fluorescent light bulbs, wind turbines, photovoltaic panels, and rechargeable batteries in electric cars. The hybrid Toyota Prius contains about 30 pounds of rare earth elements, and large gearless wind turbines can require as much as 2 tons.

Changing Supplies of Critical Rare Earth Minerals. The United States was largely self-sufficient in rare earth elements until the 1980s, but became nearly completely dependent upon imports from China over the past 20 years. China has established an essential monopoly on these critical mineral commodities, and whether or not it did so with geostrategic deliberation is debated by scholars. Although it hosts only about half of the world's estimated reserves of rare earth elements, China now supplies about 97% of global demand, roughly 125,000 tons a year.

China is able to produce rare earth metals at low cost because it has lower labor and production costs than the United States and, until recently, lax or nonexistent environmental regulations for rare earth mining and processing.

(continued)

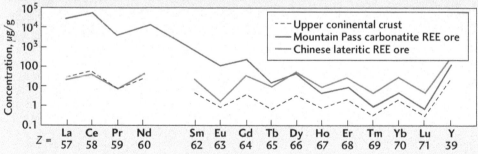

Figure 1 (Top) Bastnäsite (brown), a rare earth carbonate mineral, is one of the most important ore minerals for rare earth elements. Here it is accompanied by rhodochrosite, a manganese carbonate mineral (pink), a large crystal of feldspar (white), and a pyroxene mineral known as aegirine (dark). The sample, from a quarry in Canada, is 5.3 cm in length. *(Courtesy Tony Peterson, Geological Survey of Canada)*

(Middle) Abundances of the rare earth elements in upper continental crust (dashed line) show that the lighter rare earth elements (lower atomic number, Z), are generally more abundant than heavier rare earth elements. Rare earth ores in igneous rocks at Mountain Pass, California, are particularly enriched in lighter rare earth elements, whereas the Chinese ores in deeply weathered, clay–rich laterites are relatively enriched in heavier rare earth elements.

(Bottom) Including scandium, there are 17 rare earth elements (orange) with similar properties that often are found together in geologic deposits: 15 elements in the lanthanide series, from atomic numbers 57 through 71, and the elements yttrium (atomic number 39) and scandium (atomic number 21).

(continued)

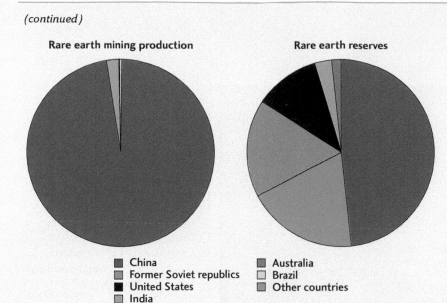

Rare earth mining production

Rare earth reserves

- ■ China
- ■ Former Soviet republics
- ■ United States
- ■ India
- ■ Australia
- □ Brazil
- ■ Other countries

Figure 2 Although Chinese reserves of rare earth elements are only about half of the world's total reserves, the country produces about 96% of the world's rare earth elements. There are other significant reserves in the Commonwealth of Independent States (some of the former Soviet republics), the United States, India, and Australia.

During the second half of the 20th century, a single open–pit mine in the United States at Mountain Pass in the eastern Mojave Desert, near the California–Nevada state border, produced large amounts of rare earth metals from rocks that contained about 9% rare earth oxides (**Figure 3**). The mine stopped operating in 2002, however, due to the

Molycorp's Mountain Pass Rare Earth Facility

Figure 3 Recently reopened, the Mountain Pass rare earth mine in southeastern California once provided the United States with much of its rare earth metals before closing in the early 21st century. *(Courtesy Molycorp, Inc.)*

lapse of a mine tailings permit, which is required for mining operations. While the permit was renewed in 2004, the then–owners of the facility (UNOCAL) elected not to re–start operations because the increased supply and low cost of rare earth elements from mines in China, in particular the Bayan Obo mining district in Inner Mongolia, northern China, made rare earth mining operations in the U.S. uneconomic. The availability of large supplies of lower–priced rare earth metals from China led to the closing of other mines worldwide as well.

The environmental and social costs of rare earth mining in China have been high, including heavily polluted air and water in the mined areas and significant health risks to workers. In recent years, the Chinese government has lowered export quotas and raised export tariffs for rare earth elements, stating that its reasons are twofold: 1) to enable increased environmental regulations at Chinese mines; and 2) to supply and boost its own domestic high–tech manufacturing. Some countries have responded with doubt, asserting instead that China is continuing to use economic tactics to maintain its control of high–tech manufacturing throughout the world.

In response to concerns about rare earth metal supplies, several federal agencies in the United States were tasked with assessing the supply of rare earth elements for the defense and energy industries.[17] Hearings on ways to reduce current dependence on Chinese rare earths were held in Congress, and lawmakers have met with scientists and mining industry representatives in the past few years. Possible solutions to the problem include exploring new sources of rare earth elements, reopening old mines, reducing dependence by finding substitutes, and recycling existing materials.

In 2011, Molycorp reopened mining operations at the Mountain Pass mine in California and began to build new processing facilities, dubbing the effort Project Phoenix. The company anticipates producing at an annual rate of 19,050 metric tons of rare earth oxides by the end of 2012. The facility's production capacity is expected to increase to up to 40,000 metric tons per year in 2013. The reopened mine operations were expanded and modernized at a cost of more than $850 million. Both energy efficiency and environmental performance were improved by new procedures that include recycling and reusing wastewater, on–site generation of high–efficiency power and steam, permanent disposal of mine tailings via advanced waste tailings technologies (which avoids the creation of a large tailings dam to store wastewater and mine tailings); and rare earth recycling capabilities.

rich in iron, zinc, and copper are crystallizing from hydrothermal brines at depths of about 1800 to 2100 m below sea level.

In 1979, geologists in a submersible vehicle exploring the East Pacific Rise spreading center had a glimpse of how such ore minerals might develop on the ocean floor. They discovered black "smokers"—tall, mineral-encrusted chimneys, up to 10 m high—venting hot metallic fluids at temperatures greater than 300°C (Figure 4-28). Based on studies of such smokers, geologists inferred that metal ores can form in the following way: Metals accumulate at rifts when seawater moves downward along fractures in the shattered rocks common at mid-ocean ridges. As it comes into contact with hot rocks or magmas, the water is heated to temperatures of several hundred degrees Celsius, making it very chemically active and able to leach metals from the surrounding rocks. The heating also causes the water to rise by convection to the seafloor, where it escapes, depositing minerals and forming chimneys as it cools.

With the discovery of active mineral deposition in the Red Sea and along the East Pacific Rise, geologists realized that some ore deposits on land must have formed this way in the past and then been pushed toward convergent boundaries as the seafloor spread. On the island of Cyprus in the Mediterranean Sea, volcanic rocks now recognized as remnants of an ancient seafloor-spreading center were thrust upward during plate collision, pushing rich deposits of copper sulfide and other ore minerals to the surface. Copper in fact takes its name from Cyprus, which holds some of the world's most ancient mines and still earns substantial revenues from its copper, chromite, and iron ores.

Hydrothermal activity also occurs at subduction zones, where metals can be remelted in the descending oceanic plate, rise as hot fluids (250°C to 500°C), and be deposited along fractures in rocks in the overlying plate. This model has been proposed to explain one of the most economically important sources of metals in the world, *disseminated porphyry metal deposits*, so called because they are widely dispersed throughout mazes of fine fractures that permeate large volumes of porphyritic igneous rocks. Porphyry is a type of igneous rock characterized by a distinctive texture of both

Figure 4–28 Black smokers and sulfide–rich mineral deposits form at seafloor vents along mid–ocean ridges. (Left) A mid–ocean hydrothermal vent is seen through the window of *Alvin,* a deep–sea submersible vehicle. The vent spews out metal sulfides that are potentially minable. (Right) Rock recovered from a hydrothermal vent chimney is rich in iron, copper, zinc, and lead. Bacteria in this environment are able to derive energy from oxidizing hydrogen sulfide rather than from sunlight. In turn, these bacteria support other bottom–dwelling organisms. *(left: Pat Hickey/WHOI; right: Tom Kleindinst/WHOI)*

large and fine crystals. Porphyritic rocks permeated with fractures containing low-grade (low concentration but minable) copper, molybdenum, silver, gold, lead, zinc, and other metals are common in the mountains of southwestern North America, western South America, the southwest Pacific, Australia, New Guinea, and Indonesia (see Figure 4-27).

An example of a tectonically controlled hydrothermal deposit in continental crust is zinc ore in the central and eastern United States. One of the explanations for the zinc's origin is that groundwater was forced to flow westward during uplift of the Appalachian Mountains to what is now the Mississippi Valley when North America collided with Europe and Africa about 300 million years ago to form Pangaea. The water was highly concentrated in zinc and lead from its interaction with igneous plutons in the mountain-building region to the east. Metals were deposited in cavities within carbonate rocks through which the groundwater flowed, leaving behind zinc and lead deposits now being mined in Missouri, Oklahoma, and Kansas.

Igneous Ore Deposits **Igneous ore deposits** are generally associated with magma chambers along the roots of island and volcanic arcs in zones of plate convergence. In magma chambers, mafic magmas can segregate into rich layers of nickel, copper, and iron sulfides, as well as chromium and platinum minerals. Segregation and layering in a magma chamber sometimes occur when dense, sulfide-rich liquids sink to the bottom before crystallization. South Africa is the source of nearly half of the world's chromium production and reserves,[18]

a strategically important metal, and much of that is mined from a layered intrusive known as the Bushveld Complex (Figure 4-29).

Nickel is found in igneous ore deposits of a type that no longer forms today. Nickel-bearing ultramafic volcanic rocks called *komatiites* formed billions of years ago when lavas from mantle magmas rich in magnesium and nickel erupted on the floors of ancient ocean basins. Komatiite is an example of a magma that no longer forms near Earth's surface because, since about 2.8 billion years ago, Earth has lost so much heat to outer space that it cannot generate a temperature

Figure 4–29 The dark layers in the layered igneous intrusive Bushveld Complex in South Africa are chrome ore. *(Spencer Titley)*

Figure 4–30 Ore minerals eroded from their source can accumulate as placer deposits at places along streams where water velocity is low. (a) One of the most common placer metals is gold, such as this nugget of quartz and gold from a stream. (b) Placer activities can be very disruptive to stream ecosystems. During many gold rushes, including those in Alaska and California in the 1800s and South America in the 1900s, miners used pans, sluice boxes, and dredging equipment to search for placer gold in streams, as this 1889 photo from the Dakota Territory shows. (c) Modern placer mining commonly uses heavy machinery, such as this mining rig, to extract placer deposits. *(a: Superstock; b: © The Protected Art Archive/Alamy; c: Michael Dunning/Getty Images)*

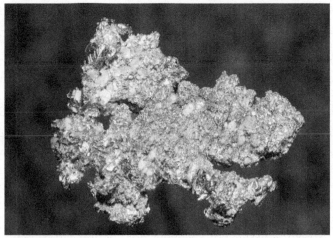

(a)

(b)

(c)

high enough to create such magmas. South Africa, Australia, and Canada are among the few countries where komatiites have been found.

The deepest magmas known to have erupted at Earth's surface are the source of diamond. Whereas most minerals crystallize from magmas at shallow depths, diamond forms at depths as great as 140 to 190 km (87 to 118 mi). By some unknown process, ultramafic lavas containing diamond crystals have been ejected to the surface at extremely high velocities, forming near-vertical pipes up to 200 m in diameter, known as *kimberlites*. Diamond-bearing kimberlite pipes exist in only a few localities. They were first found in, and named after, Kimberly, South Africa, where they continue to be mined today.

Sedimentary Ore Deposits Most sedimentary metallic ore deposits can be traced to mild hydrothermal events, usually at temperatures lower than 250°C. Examples include the heating of sedimentary rocks adjacent to igneous intrusions. These low-temperature events sometimes cause fluids rich in metals to migrate through the pores of sedimentary deposits, leaving behind cements and bands of metals that become ore deposits.

Sedimentary ore deposits not associated with thermal events include ore minerals that accumulate in streams and rivers. In India, diamonds have been found in modern stream gravels as well as in one of the rock types drained by the streams, a sedimentary conglomerate 550 million to 600 million years old. The diamonds in the conglomerate presumably were deposited after being eroded either from an even older sedimentary rock or from a kimberlite pipe. The famous blue Hope Diamond, now at the Smithsonian Natural History Museum, is thought to have been found in conglomerate in India in the 17th century. Because diamond is relatively dense (about 3.5 g/cm^3) and very resistant to abrasion (hardness of 10), it accumulates in low-velocity pools along modern streambeds. Such accumulations of mechanically segregated ore minerals are called **placers** (Figure 4-30).

A number of dense minerals, including gold, magnetite, and chromite, are also in placer deposits.

Although gold is soft, with a Mohs hardness rating of only 2.5 to 3, it is one of the densest minerals known (about 19.3 g/cm^3), and even tiny nuggets and flakes will settle from water flowing in streams. In comparison, the density of quartz, the most common mineral in stream sediments, is only 2.6 g/cm^3. Gold is also very resistant to corrosion and thus likely to be preserved in a placer deposit. The California Gold Rush of 1848 started when gold nuggets were found in placers, but miners soon traced them to their sources, hydrothermally deposited mineral veins in the Sierra Nevada that miners called the Mother Lode.

Another type of sedimentary ore deposit that is not associated with thermal events is the chemical precipitation of ore minerals in water at low temperatures (less than 80°C). For example, the floor of Death Valley, California, is covered with salt crystals precipitated from waters draining into the closed lake basin from surrounding mountains (see Figure 4-17). Some of the ions in the water combine to form the mineral borax, which contains the elements sodium and boron and is used to make cleansing products and heat-proof glass. Other commonly mined evaporite minerals are table salt (NaCl); gypsum ($CaSO_4 \cdot H_2O$), which is used for plaster and drywall; limestone ($CaCO_3$), which is used for cement and building stone; and phosphate (PO_4^{2-}), which is used for fertilizers.

Chemical sedimentary rocks known as *banded iron formations* are the most important source of iron (Figure 4-31). The deposits are so named because the iron occurs as an oxide mineral in reddish bands interbedded with layers of chert, sometimes in formations up to 1000 m thick. The deposits formed when dissolved iron (Fe^{2+}) was washed from continents to shallow ocean basins, where it oxidized to its insoluble form (Fe^{3+}) and crystallized as hematite (Fe_2O_3). Most banded iron deposits are more than 1.8 billion years old, and few equivalent deposits have formed since then. The probable reason for this is that the composition of Earth's atmosphere changed during its evolution, particularly as a result of the evolution of photosynthetic organisms,

which release oxygen. Before about 1.8 billion years ago, the atmosphere contained less oxygen than it does now, so iron that eroded from continental rocks remained in the reduced state, and thus soluble, until it reached relatively more oxygenated ocean water, where it oxidized and crystallized to form banded iron deposits. With time, iron was absorbed in the oceans and marine rocks, but as these sinks for iron were filled, excess iron became available to the atmosphere.

Depletion and Recycling of Mineral Resources

Although black smokers on the seafloor continue to produce sulfide ores, and evaporite minerals such as table salt continue to form, we have seen that some ore-forming geologic processes no longer take place because Earth systems have changed over time. Hematite no longer collects in banded iron formations because the modern atmosphere is rich in oxygen that reacts quickly with exposed iron. Nickel-bearing lavas no longer occur because Earth's interior has cooled. In general, then, mineral ores are nonrenewable resources. Even those that still can be produced by geologic processes often cannot form fast enough to match the rates at which humans extract them.

Known mineral deposits that can be mined profitably with existing technologies are called **reserves**. A **resource**, in contrast, is the entire amount of a material known to be available. The boundary between recoverable reserves and subeconomic resources—those with extraction costs greater than their market value—shifts as more efficient mining technologies are developed, existing reserves are depleted, and the price of a particular resource increases. New resources are most likely to be discovered where some mineral resources have already been found, and possible ore deposits in such locations are considered hypothetical resources.

Most ore minerals are rare, extraordinarily localized, difficult to find, and used at rapidly increasing rates worldwide. Consider, for example, the production of gold. Some 40 percent of all gold ever produced was mined from the Witwatersrand Basin of South Africa, yet the area of the basin is smaller than Los Angeles County in California. As small areas with rich concentrations of rare ores are mined, is it possible that the world will face a sudden scarcity, or will new resources be found?

Although it is not possible to predict the future with certainty, enough is known about ore minerals to indicate that a major mineral resource crisis is not likely to occur. First, Earth is made of minerals, so a source of minerals will be available as long as the planet exists. Second, the technology for locating and extracting minerals improves constantly, so that deposits once considered subeconomic now are being mined. Finally, when depletion of a mineral seems imminent, ways

Figure 4–31 Banded iron formations are sedimentary layers of light-colored siliceous chert and reddish iron oxide minerals. They are the world's largest source of minable iron ore. *(Allison Pluda/Seneca Creek Photography)*

of using substitute materials, or increasing the use of recycled materials, are developed.

Here are some examples that support these statements. Before 1988, the large gold and copper ore deposit on the island of New Guinea, shown at the beginning of this chapter, was not known to exist, and the technology needed to mine it was not developed until about the same time. Similarly, a large copper ore deposit in Chile was not discovered until 1981. At the time of these two discoveries, concern over a shortage of copper—used for telephone and electrical wiring—had already been alleviated somewhat by the development of fiber-optic cables for communication lines. Fiber-optic cables are made of quartz, and there is plenty of quartz in Earth's crust. While we are not likely to run out of minerals in spite of the escalating demand, the environmental impact of increased mining is a cause for concern.

Many mineral resources, such as copper, aluminum, and iron, can be recycled and reused as secondary materials, moving society toward more sustainable mineral production. The amount of such resources already in use can be considered a part of world reserves. With proper management, recycling and reuse can extend the use of mineral resources and minimize waste disposal. In 2008, for example, about a third of world copper consumption came from copper that was recycled.[19] In 2010, about 1.15 million metric tons of secondary lead, some 82 percent of domestic lead consumption, was recovered from previously used (postconsumer) scrap.[20] Secondary materials, primarily electric arc furnace dust, were used to recover 85,000 tons of zinc, about 41 percent of U.S. production.

Environmental Impacts of Mining

As a result of mining, elements that are found in localized concentrations in the lithosphere are *mobilized,* or released, to other Earth systems over wide areas. This is one of mining's greatest effects on the environment. Because metals have a strong tendency to form reversible bonds with many other elements (including carbon), they occur throughout Earth systems, cycling through them in a variety of forms. Elements commonly released to the hydrosphere, atmosphere, and pedosphere during mining include lead, mercury, and chromium. Lead and mercury can damage the central nervous system of humans, while chromium is a potent carcinogen. Increased concentrations of a number of elements, including lead and mercury, have been identified in sediments in lakes and bays and in layers of ice in Antarctica and Greenland. In some cases, global emissions of metals from mining and burning of fossil fuels have resulted in anthropogenic (human-induced) emission rates that are tens to hundreds of times those

of the natural background values associated with geologic processes.

In the United States, less than 1 percent of the land surface has been altered by mining minerals and coal.[21] More than that area has been affected, however, if consequences such as air pollution from smelting or water pollution from abandoned tailings and mine tunnels are included in the environmental assessment (see Box 4-2). Spoil piles and exposed ore bodies contain toxic metals and sediments, for example, that can enter streams if the piles are not carefully revegetated and stabilized.

In conventional mining, ore minerals are excavated, crushed, and then smelted with other compounds in order to produce liquid metal that can be separated from the unwanted residue. Excavation is done either underground in tunnels and shafts, or above ground in open pits. Because most metals make up less than 2 percent of the ore deposit, the amount of material excavated and left as waste is many times that of the actual metal product. The environmental impacts of conventional mining include:

- A landscape that is difficult and costly to reclaim

- Large amounts of tailings from which metal-rich, acidic waters might drain into groundwater or nearby streams

- Emission of sulfur dioxide gas into the atmosphere from smelting of sulfide ores

These impacts can be minimized, but at great cost and effort. Furthermore, unless regulations are in place (and enforced) to require such efforts, companies have little economic incentive to pay the costs for minimizing environmental degradation.

A relatively recently developed mining technique called *heap leaching* is used to extract metals from oxide and low-grade sulfide ores (Figure 4-32). In heap leaching, the ore is excavated from open-pit surface mines, crushed, spread on a pad underlain by a protective liner, and sprayed with a solvent that leaches the metal from the ore. Cyanide is the solvent used to leach gold. Sulfuric acid is used to remove copper. As sulfuric acid leaches copper from the ore, it forms a blue, copper-rich solution that is sent to a plant where the copper is extracted. This technique, like that used in conventional mining, requires that much material be excavated and left as tailings, but unlike conventional mining, heap leaching does not require heating or smelting. With care, environmental degradation off-site can be minimized by preventing leaks from heap pads, storage tanks, and pipes.

An example of environmental degradation that led to taxpayer cleanup is the Summitville mine in the Rocky Mountains of Colorado. In the 1980s, the Summitville Consolidated Mining Company excavated rock containing small amounts of gold from an open pit, crushed the rock, placed the crushed material on

(a)

(b)

Figure 4–32 During heap leaching (a), ore is excavated, crushed, spread on a protective barrier, and sprayed with solvent. This pile of crushed copper ore is being sprayed with

sulfuric acid. (b) The next step in heap leaching is to pump the dissolved copper to a plant where the solvent is removed. *(a: Gillianne Tedder/Getty Images; b: B. Brown/Shutterstock)*

leach pads, sprayed a sodium cyanide solution on the ore to leach gold from it, then collected the gold-laden solution and extracted the gold. Unfortunately, metal-laden and toxic water leaked from the system at many points, polluting streams that are used for fishing, farming, and drinking water. The chemicals include copper, cadmium, manganese, zinc, lead, nickel, aluminum, and iron. The mining company declared bankruptcy in 1992, leaving the federal government and taxpayers to foot most of the bill for cleanup.

The Summitville mine site was added to the **Superfund National Priorities List,** or **Superfund list,** the U.S. Environmental Protection Agency's (EPA) list of sites where a release or threatened release of hazardous substances may endanger public health, welfare, or the environment (Box 4-3).[22] After 1994, the EPA and the Colorado Department of Public Health and Environment engaged in a number of remedial actions at the Summitville mine site that involved removing tailings piles, filling the open pit, creating a water treatment plant to clean surface and groundwater contaminated at the site, and detoxifying and revegetating the leach pad area. Land reclamation was completed in 2002, and a new water treatment plant was completed in 2011.

A mining method that requires no excavation at all might substantially reduce the environmental impact of much mining. *In situ* (Latin for "in place") mining extracts metal from an ore deposit where it is found, with no excavation (Figure 4-33). In situ mining uses the same techniques as heap leaching, except that the solvent is injected into the ore deposit through underground wells.

In an in situ mining project in Arizona, for example, sulfuric acid injected into a deep well travels through the fractures and veins of rock containing copper oxides that occur from about 120 to 300 m below the surface.[23] The acid leaches copper from the ore-bearing rock. The

copper-bearing solution is drawn up to the surface by other wells, and the copper is extracted from the solution by chemical means in a plant. It leaves no excavation scars or tailings. The only waste is the spent solvent.

Environmental Regulation of Mining

Much mining occurs on federal land in the western United States. Since the late 19th century, laws have been enacted both to facilitate and to regulate mining on federal lands. The most important of these is the **General Mining Act of 1872.**[24] This law applies to exploring for and mining minerals on public lands, and establishes regulations for acquiring and protecting mining claims on them. About 30 percent of the land area of the United States, 2.6 million km^2 (650 million acres), is owned and managed by the federal government. In some western states, where the majority of federal lands are held and most U.S. mineral mining occurs, up to 76 percent of the total land area is public land. These lands include national parks, national forests, and national wildlife refuges. Several federal agencies are responsible for managing their natural resources. Natural resource management encompasses both the protection of resources and their use for recreation and economic growth. Agencies under the Department of the Interior that manage federal lands are the Bureau of Land Management, the National Park Service, the Bureau of Reclamation, and the Fish and Wildlife Service. The Forest Service under the U.S. Department of Agriculture also manages substantial amounts of federal land.

In 1976, the **Federal Land Policy Management Act** was implemented. It was created to revise the use of surface land at mining claims on federal land. Regulations based on this law require authorization (permits) and an

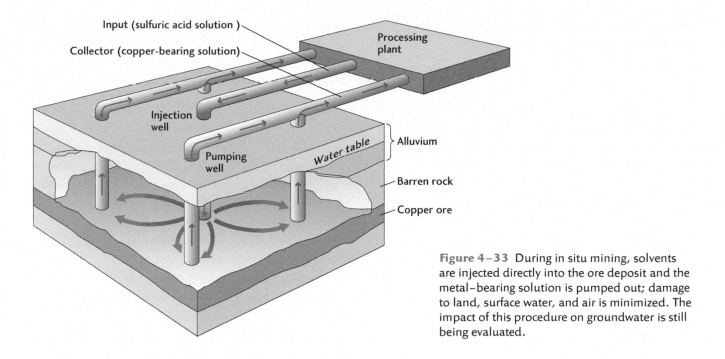

Figure 4–33 During in situ mining, solvents are injected directly into the ore deposit and the metal–bearing solution is pumped out; damage to land, surface water, and air is minimized. The impact of this procedure on groundwater is still being evaluated.

environmental assessment for activities that disturb the ground surface, and reclamation of disturbed land after mining. Reclamation includes recontouring the land, then placing topsoil on the regraded area and reseeding it with native vegetation, crops, or trees. Loading billions of tons of crushed rock back into the voids of large, open-pit mines such as the Freeport copper mine in Indonesia or the Berkeley Pit in Butte, Montana (see Box 4-3), is not likely to happen. In other cases, however, the mined area is small enough that reclamation is effective.

For the United States, the 20th century was a period of growing awareness of the need to develop new ways to mitigate, or minimize, the impact of mining, and to increase the amount of environmental legislation that applies to all industrial activities, including mining. Federal and state standards are now in place for reclaiming mined land; controlling emissions to air, soil, surface water, and groundwater; and disposing of hazardous wastes. Federal laws that serve to protect and minimize environmental degradation at mines include the following: the Clean Air Act, the Clean Water Act, the Safe Drinking Water Act, the Endangered Species Act, and the Resource Conservation and Recovery Act. On the other hand, new technologies increase humans' capacity to alter the landscape with time, as in the case of coal mining by the process of mountaintop removal (see Chapter 13).

The **Clean Water Act** of 1972 and its amendments (see Chapter 9) require that mining activities discharge close to zero pollutants into surface water. Responsible companies comply, and streams in modern mining areas generally are less impacted by modern mining practices than streams were during earlier periods of mining.

Less responsible companies, however, sometimes leave toxic sites in the hands of officials and taxpayers.

The **Clean Air Act** of 1970 and its amendments (see Chapter 15) have reduced air pollution from mining activities in the United States, but reducing emissions to zero is difficult. Many metals occur in sulfide minerals, and a by-product of separating sulfur from the metals by smelting is sulfur dioxide gas. This gas is vented through smelter stacks to the atmosphere, where it combines with water and oxygen to form sulfuric acid. The contaminated rainwater then falls as *acid rain*. To reduce atmospheric emissions, old smelters are upgraded by installing scrubbers on their stacks. These pollution-control devices contain calcium oxides and hydroxides (from limestone) that combine with and trap sulfur dioxide gas and particles in the stacks before they are released to the atmosphere.

New smelters have low emissions compared with older ones.[25] A smelter built by Kennecott Utah Copper in Salt Lake City, Utah, in 1995, for example, emits less than 2 kg of sulfur dioxide for each metric ton of copper produced, making it one of the lowest-emission smelters in the world.[26] Older smelters emitted as much as hundreds of kilograms of the gas per metric ton of copper, so the environmental regulations have had a substantial positive impact on atmospheric pollution.

Metal Cycling Through Earth Systems: Copper Case Study

Copper provides a typical example of how a metal cycles through Earth systems. Like other metals at Earth's surface, copper occurs either concentrated in ore deposits

BOX 4–3 CASE STUDY

Butte, Montana—From Boom Town to Superfund Site

Butte, Montana, is situated on what may be the richest hill on Earth, underlain as it is by sulfide ore that in places has a concentration of 50% copper. Underground veins of copper up to hundreds of meters thick, discovered in the late 1800s, were chipped away and hauled up to the surface along what eventually became almost 16,000 km of tunnels and shafts. As the world's appetite for copper to make electrical wire, pipes, motors, weaponry, and other trappings of the Industrial Age increased, the hill around Butte produced more than 9 billion kg of copper—a third of the U.S. demand and a sixth of the world's—in the first part of the 20th century.[27] The human price was high: In open tunnels where there was once ore, at least 2000 men died in accidents and another 20,000 were seriously injured or disabled. The additional number who died from mining-related lung diseases is unknown.

Anaconda Copper, one of the world's most powerful mining companies, owned the Butte mine during most of the 20th century, and as the mine's tunnels grew deeper and deeper, the company shifted to less expensive open-pit mining during the latter half of the century. As a result, the mine began to consume the city that surrounded it.

Excavation of the open–pit mine at Berkeley Pit began in 1955, and the vast hole grew to nearly 1 km wide, 1.5 km long, and 0.5 km deep, replacing one edge of the city of Butte (**Figure 1**). Pumps deep underground ran constantly to keep groundwater from filling the open pit. Surrounding the pit and city are mounds of waste rock and hills made barren by gases and particulates from the smelters.

Anaconda sold the mine and degraded landscape to an oil company, Atlantic Richfield Company (ARCO), in 1977. The mine works and pumps that kept groundwater out of the vast pit continued to operate until 1982, when world demand for copper, and hence its price, dropped. After a century of mineral extraction by human endeavor, the mine fell silent and groundwater flowing toward the low, cone-shaped depression in the water table seeped into the maze of thousands of kilometers of abandoned tunnels, flooding them and the Berkeley Pit.

With no intervention, the pit would continue filling until its water becomes level with the local groundwater surface.[28] Sometime in 2023, the water in the pit would be more than 300 m deep (**Figure 2**). It also would seep outward into the pore spaces of adjacent sediments and

(continued)

Figure 1 In 2006, an astronaut on the International Space Station photographed what was dubbed the "richest hill on Earth" for its copper, gold, and silver ores in the late 19th century, but is now composed of the deep (540 m) Berkeley Pit and several smaller pits, tailings piles, and a tailings pond. Pumps that kept groundwater out of the intricate maze of underground mine workings were turned off in 1982, and since then water from the surrounding rock basin has seeped into the pit. As of the time this photo was taken, the Berkeley Pit was about half full with water, to a depth of 275 m (900 feet). *(NASA)*

(a)

Sulfide minerals in the hills leach toxins into the water

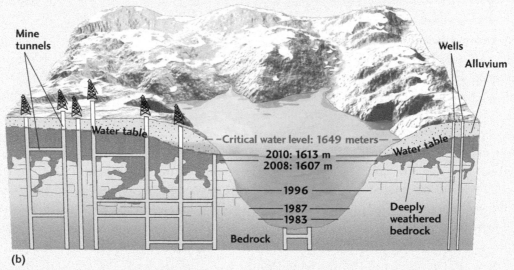

(b)

Figure 2 Mining in the Berkeley Pit ceased in 1982 and the pumps used to dewater the open pit were shut down when its bottom elevation was 1299 m (4263 ft). The pit began to fill with water from both surface runoff and groundwater. (a) The 1979 photo shows the open pit during mining, and each successive photo illustrates the rising water level, to about 1607 m (5273 ft) in 2008. (b) As of 2010, the water level was at an elevation of 1613 m (5291 ft), giving a depth of 313 m (1028 ft) of water in the pit. This is only 36 m (119 ft) below the safe, or critical, water level, which is at 1649 m (5410 ft) here. *(Montana Bureau of Mines & Geology and the Berkeley Pit Public Education Committee)*

(continued)

rocks that supply streams with some of their flow, polluting Silver Bow Creek, the headwaters of Clark Fork River. Little can be done to stop the groundwater from pouring through the old tunnels or from leaching toxins from the mineralized rock, although some water entering the pit has been diverted in an effort to decrease the rate at which the water is rising.

With the Berkeley Pit groundwater comes sulfur dioxide, cadmium, lead, and arsenic leached from sulfide minerals that permeate the hill's bedrock. After combining with oxygen from the atmosphere, the mixture becomes a toxic brew rich in sulfuric acid and poisonous metals. In the winter of 1995, the lake was so acid—with a pH of 2.3, similar to that of pure lemon juice—that it killed hundreds of Arctic snow geese on their migration to California. Autopsies revealed that tissues lining their esophaguses and digestive organs were burned and blistered.

In 1980, the U.S. government passed the Comprehensive Environmental Resource, Compensation, and Liability Act, commonly known as the Superfund law because of the large fund of money set aside to help finance the cleanup of hazardous waste sites. Under this law, the EPA has the authority to identify and clean up such sites, as well as to compel responsible parties to do the cleanup. The Silver Bow Creek/Butte area is one of the largest of more than 1297 Superfund sites identified by the EPA under this act.[29]

The EPA and the Montana Department of Environmental Quality (MDEQ) issued a record of decision (ROD) for the Butte mine flooding in 1994. This legal document describes the decisions about cleaning up the site and the scientific understanding on which they are based. One of the decisions was to pump and divert water draining into the pit from the north, in an area known as Horseshoe Bend, away from the Berkeley Pit. This diversion began in 1996 and reduced, but did not stop, the rate at which the water level in Berkeley Pit was rising. In 1998, Montana Resources, a mining company that had bought a portion of the mine works, began pumping water from Berkeley Pit in order to extract copper from it. The process involves running metal-laden water over scrap iron. Copper replaces iron, forming a recoverable copper precipitate, and iron is carried away in the water. As a result of rising electricity costs, however, Montana Resources ceased its operations in Butte in 2000.

In 2002, the EPA and MDEQ entered into a consent decree with ARCO (by then owned by British Petroleum, BP) and Montana Resources to settle for the past and future costs needed to clean up the site or to control pollution at a certain level. One aspect of this decree required BP-ARCO and Montana Resources to construct a water treatment plant at Horseshoe Bend. At a cost of $18 million to build, this plant went online in 2003, the same year that Montana Resources resumed its copper mining operations. The plant treats water from Horseshoe Bend and recycles it for use by the mine, thus slowing the rate at which water in Berkeley Pit rises. In 2004, Montana Resources also resumed its pumping of water from Berkeley Pit for copper recovery.

An important eventual function of the Horseshoe Bend water treatment plant is to pump and treat water from the Berkeley Pit in order to keep it below an elevation of 1649 m (5410 ft), the level identified as the critical level above which water from the pit could reverse flow direction and begin to flow away from the pit, into nearby groundwater and streams. Based on the long-term rates at which water has risen in the pit and information on the rates at which ground and surface water flow into it, scientists predict that the water will reach the critical level in 2023, although changes in water flow rates could affect that projection.

Water treatment at the plant involves adding lime (calcium hydroxide) to lower the acidity of the water and promote precipitation of metals. This treated water is used by the operating mine. Under the 2002 consent decree, BP-ARCO and Montana Resources will provide financial assurances for operation and maintenance costs at the Butte Superfund site (known as the Silver Bow Creek/Butte Area) in perpetuity.

(more than 100 to a few thousand parts per million) or disseminated throughout sediments and rocks in the lithosphere (Figure 4-34). From this reservoir, copper can be mobilized, largely by human activities, weathering, and volcanic activity. Most copper released to the atmosphere is the result of industrial emissions or wind erosion of weathered material; very little comes from volcanic emissions.

During weathering, oxygen in the atmosphere and hydrosphere bonds with copper to form compounds and ions that are readily transported in solution in water runoff from the landscape. Some copper is stored in carbon molecules in terrestrial systems, largely in soils and living organisms. Plants draw copper from soils to build cells, then return it to the soil when they—or the animals that eat them—die and decay.

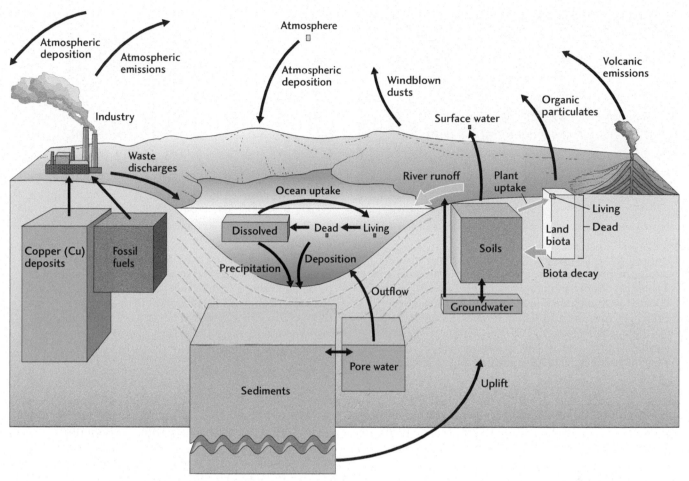

Figure 4–34 The biogeochemical cycling of a trace metal—copper—in Earth systems. Stocks (boxes) and fluxes (arrows) are drawn approximately to scale, relative to one another. The largest stock of copper is in sediment, and the next largest in copper ore deposits. Unknown fluxes are represented by question marks. Since the amounts of copper range widely, some of them cannot be represented adequately here. Small stocks of copper are represented by tiny boxes.

Once released from rock and sediment by weathering processes, most copper is transported by streams to the ocean reservoir in dissolved form or as solid particles in suspension. In the ocean, some copper remains dissolved and cycles through marine organisms, but most is buried with seafloor sediments, thus returning to the lithosphere reservoir. Eventually, because of plate tectonics activities, some of this copper returns to Earth's surface by uplift or volcanism, where it once again becomes mobilized and begins the cycle anew.

The residence time of copper in different Earth systems is important to assessing the environmental impact of mining activities. For steady-state systems, the residence time of copper in Earth system reservoirs can be calculated as:

$$\text{Residence time} = \frac{\text{Stock of reservoir}}{\text{Flow rate (flux)}}$$

The flux of the copper is the total amount flowing into or out of a reservoir, and the system is in steady state if these rates are the same. The residence time of copper in the atmosphere, for example, is small:

$$\begin{array}{c}\text{Residence time} \\ \text{of copper (Cu)} \\ \text{in atmosphere}\end{array} = \frac{26 \times 10^8 \text{ g Cu}}{709 \times 10^8 \text{ g Cu/yr}}$$

$$= 0.04 \text{ yr} \approx 13 \text{ days}$$

In contrast, copper has a much longer residence time in soils:

$$\begin{array}{c}\text{Residence time} \\ \text{of copper soils}\end{array} = \frac{6.7 \times 10^{15} \text{ g Cu}}{6.2 \times 10^{12} \text{ g Cu/yr}}$$

$$\approx 1081 \text{ years}$$

In other words, copper is cycled rapidly through the atmosphere (13 days) and slowly through soils (1081 years). An example of the environmental consequence of this difference in residence times is that even after a

(a) 1973

(b) 2012

Figure 4–35 Although the 19th century copper smelter near Copperhill, Tennessee, is gone, its environmental legacy remains. (a) Copper and other metals were emitted to the air, water, and soil from mining and smelting, leading to barren hillslopes and acid soils and water. Sulfur emissions from the smelter led to acid rain, which further damaged vegetation and degraded water quality in streams and lakes. A new smelter built in 1907 condensed and collected sulfur gases to produce sulfuric acid. Nevertheless, the legacy of deforestation, mining, and smelting hinders environmental reclamation efforts that include planting trees and grasses, as shown in these views of the area. To this day, the concentrations of copper in water and soil in the region are high. *(a: Emory Kristof/National Geographic Stock; b: Courtesy Ducktown Basin Museum)*

smelter reduces its emissions of copper to the atmosphere, it will take many years for excess copper to be removed from soils surrounding the facility (Figure 4-35).

■ The most commonly used nonfuel mineral resources are localized in concentrated ore deposits by the following geologic process: mineralization along spreading ridges, mineralization along subduction zones, density layering in magma chambers, formation of chemical sedimentary deposits, and accumulation as placers in stream channels.

■ A resource's economic concentration factor describes how much more concentrated an ore must be relative to the concentration of the resource in average crustal materials in order to be mined profitably.

■ We are unlikely to face a severe crisis in mineral availability because new ore deposits still are being discovered, new technologies diminish the need for some minerals, and new technologies increase the use of recycled material. On the other hand, many minerals are found in only a few places, so there is the potential for geopolitical instability if, for example, one country refuses to export a given mineral to other countries.

■ The total land area affected by extracting minerals from Earth's crust is fairly small, but the impact on surface water and the atmosphere has been substantial, particularly before modern environmental regulations, because mineral extraction and processing release large amounts of metals and acids into these Earth systems unless substantial efforts to remove the pollutants are made.

■ New mining methods, such as heap leaching and in situ recovery, can extract ore minerals from lower-grade ores than in the past and potentially cause less environmental degradation than past mining practices.

■ Important environmental laws related to mining and its environmental impacts include the General Mining Act of 1872, the Clean Water Act of 1972, and the Clean Air Act of 1970, and their amendments.

■ The residence time of copper in soil (1081 years) is long but short in the atmosphere (13 days); the environmental consequence is that even after reductions in atmospheric emissions, it takes many years for excess copper to be removed from soils surrounding smelters.

CLOSING THOUGHTS

Rocks contain clues to their past: the depths to which they were buried, the temperatures to which they were heated, and the places to which they have traveled. The first step in discovering these clues is identifying the rocks and the minerals within them. An experienced geoscientist can tell much from even small rock exposures. By mapping different rock types and determining their ages (see Chapter 6), geoscientists have reconstructed much of Earth's long and eventful history.

The clues in rocks and the tales they tell help us to find mineral resources, to understand earthquakes, and to predict volcanic eruptions. Rocks also contain information about past climates, atmospheres, environments, oceans, and life-forms that existed when the rocks formed. Spectacular places such as the chalk cliffs of Normandy (see Figure 4-18) and Ayers Rock in Australia (see Figure 4-3) are large rock exposures that hold a special fascination. Gold, silver, and platinum, as well as diamonds, rubies, emeralds, and other gems, have been symbols of wealth, prestige, and status for much of human history. Museums worldwide display minerals and rocks as objects of interest and beauty, and archaeologists collect stone and metal tools made by humans. It seems to be part of human nature to treasure Earth materials.

At the same time, however, humans have marred Earth's surface in order to extract the minerals they value. Landscapes scarred and contaminated by mining, such as Butte, Montana, have been described by some officials as "National Sacrifice Areas"—places sacrificed for the sake of mineral acquisition and industrialization. Mineral extraction has proceeded despite substantial risks and hazards. Beginning in the latter part of the 20th century, the environmental impacts of mining have been regulated, but then only in the most developed countries. As world population grows and developing nations become industrialized, world per capita mineral consumption rises. Mining has altered the environment for many millennia, but it remains to be seen how governments will respond to the conflict between economic interests and environmental quality in a time of rapidly growing population and industrialization.

SUMMARY

■ Minerals are naturally formed, inorganic, crystalline solids with specific chemical compositions or ranges of compositions.

■ Crystallization is the process by which atoms combine to form orderly arrays that are repeated in three dimensions.

■ The most common minerals in Earth's crust are silicates; they consist of silica tetrahedra that are either isolated or polymerized (linked in the form of clusters, chains, sheets, and three-dimensional frameworks of tetrahedra).

■ The properties of minerals reflect the internal arrangement and bonding of atoms. For example, minerals break, displaying cleavage, along planes of weak atomic bonding between layers of atoms.

■ Magma is described as mafic, intermediate, or felsic, with increasing amounts of silicon and oxygen in each form.

■ Mafic magmas result in basalt (extrusive) and gabbro (intrusive) rocks, and are most common along divergent plate boundaries. They are the dominant rock types in oceanic crust.

■ Intermediate magmas result in andesite (extrusive) and diorite (intrusive), and are most common in continental crust along subduction zones (convergent boundaries).

■ Felsic magmas result in rhyolite (extrusive) and granite (intrusive), and are the dominant rock types in continental crust.

■ Metamorphic rocks are classified as foliated or nonfoliated. Common examples of foliated metamorphic rocks are slate, schist, and gneiss; common examples of nonfoliated metamorphic rocks are quartzite and marble.

■ Sedimentary rocks are classified on the basis of their origin: clastic, chemical, or biological.

■ Sediments are deposited in different environments, which can be determined long after the environment is gone by using clues such as grain size and shape. These clues help geologists reconstruct the history of Earth's changing environments.

■ A variety of geologic processes concentrate economically valuable elements in ore minerals. The economic concentration factor describes how much greater the concentration in the ore must be relative to the average crustal concentration in order for the element to be mined profitably.

■ Critical minerals have few, inferior, or no satisfactory substitutes, and a restriction in their supply would have a substantial impact on economic activity. Rare earth elements, essential to modern technology, including new forms of green (e.g., energy-efficient) technology, currently rely on multiple rare earth elements.

■ The environmental impacts of mining include the emission of acid rain–producing gases; the creation of vast quantities of waste rock that contains toxic elements; and the introduction of toxic metals into the food chain. Strict regulations are necessary to minimize these impacts. In many newly emerging industrial economies throughout the world, environmental regulations are lax or not yet developed, yet some of these countries now produce large amounts of some global mineral resources.

KEY TERMS

minerals (p. 97)
crystallization (p. 97)
glass (p. 98)
silicates (p. 98)
Mohs hardness scale (p. 102)
cleavage (p. 102)
igneous rocks (p. 103)
extrusive igneous rocks (p. 104)
intrusive igneous rocks (p. 104)
mafic (p. 104)

Bowen's reaction series (p. 104)
felsic magmas (p. 106)
depositional environments (p. 107)
sedimentary rocks (p. 108)
lithification (p. 108)
clastic sedimentary rocks (p. 108)
sedimentary structures (p. 109)
chemical sediments (p. 109)
evaporites (p. 109)

biological sediments (p. 110)
metamorphic rocks (p. 111)
foliation (p. 112)
rock cycle (p. 114)
critical minerals (p. 116)
ore deposits (p. 116)
ore minerals (p. 116)
tailings (p. 116)
economic concentration factor (p. 116)
hydrothermal ore deposits (p. 117)

igneous ore deposits (p. 122)
placers (p. 123)
reserves (p. 124)
resource (p. 124)
Superfund National Priorities List, or Superfund list (p. 126)
General Mining Act of 1872 (p. 126)
Federal Land Policy Management Act (p. 126)
Clean Water Act (p. 127)
Clean Air Act (p. 127)

REVIEW QUESTIONS

1. Is obsidian a mineral? Why or why not?

2. What is the main mineral ingredient used to make glass? What are the two elements in this mineral?

3. What conditions favored the growth of large gypsum crystals in the Cave of Crystals in Mexico, and at what other places might such conditions have led to the slow growth of large crystals?

4. Why are silicate minerals so common in Earth's crust?

5. What are some examples of common silicate minerals formed of sheets of silica tetrahedra, and of three-dimensional frameworks of silica tetrahedra?

6. How does crystallization occur in magma? How does crystallization occur in an evaporating solution?

7. How are mineral hardness and cleavage related to the internal structural arrangement of a mineral's

atoms and the strength of the bonds between atoms?

8. What are the three major rock groups, and how are they related to one another by the rock cycle?

9. How do volcanic glasses like obsidian form? Why were they so useful for toolmaking during the Stone Age?

10. What type of sediments would you expect to find deposited along a large river? In a sand dune? In a lake?

11. How is the formation of calcareous sediments in the oceans related to the amount of carbon dioxide in the atmosphere?

12. What are the names of intrusive and extrusive igneous rocks associated with mafic, intermediate, and felsic magmas?

13. Why don't some metamorphic rocks (e.g., quartzite) display foliation in their texture?

14. When did Eurasians and Africans begin to use metals instead of stone to manufacture tools?

15. Which three nonfuel materials rank highest in terms of human consumption?

16. What are ore deposits and ore minerals, and which ones are critical minerals?

17. What are examples of geologic ore-forming processes that no longer happen on Earth?

18. Why are so many ore deposits found near present or former convergent plate boundaries?

19. What is meant by the "economic concentration factor" of an element? What is the economic concentration factor of copper?

20. How are hydrothermal ore deposits formed at spreading centers along divergent plate boundaries?

21. How did hydrothermal ore deposits of lead and zinc form along the Mississippi Valley 200 million to 300 million years ago? What did plate tectonics have to do with this process?

22. Name the four major categories of types of metallic and nonmetallic ore deposits.

23. How do diamonds form?

24. What types of minerals are likely to be found in placer deposits? Why?

25. What type of ore deposit is associated with the copper and gold mines in the mountains of the island of New Guinea? How did it form?

26. How did China acquire a near-monopoly on production of rare earth elements? In what ways might the United States become less dependent on rare earth elements from China?

27. What is the process of smelting, and what is its purpose?

28. In what ways does in situ recovery of copper reduce environmental degradation in comparison to open-pit mining and smelting?

29. What are the main environmental impacts of extracting and processing minerals?

30. What laws govern the environmental impacts of mining, and when were they implemented?

31. What is a Superfund site, and how does status as a Superfund site affect mining and cleanup activities at the Silver Bow Creek/Butte mine area in Montana?

THOUGHT QUESTIONS

1. Examine Figure 4-15 and evaluate which sedimentary environments are most likely to have well-sorted deposits. Which one(s) would have poorly sorted deposits?

2. One of the most common sedimentary rocks produced from sediments in deep ocean basins is shale. Why is that the case?

3. Start with sand deposited on a beach along a coast with an active subduction zone plate boundary, such as the coast of Chile in South America, and try to trace what might happen to that sand as it moves through the rock cycle for the next few hundred million years. For example, what rock will it become if it is incorporated with the subducting plate and moves down the subduction zone, and then what might happen to this rock?

4. List three reasons why it might be unlikely that a major mineral resource crisis will occur in the United States in the near future.

5. Examine the biogeochemical cycle for copper in Figure 4-32 and identify all the places where the copper cycle is impacted by mining, processing, and the use of metals. Can you trace the flux of copper as it moves from one reservoir to another via these human actions?

6. Urban areas contain immense amounts of used and obsolete products that can be mined for metals and stone materials. The practice of acquiring these used resources is called "urban mining." Valuable mineral resources are found in buildings that are demolished, in landfills, and in the growing numbers of used cell phones, computers, and other electronic devices. Think of the products that you've used and discarded and look around the places where you live or go to school, then try to identify the various items that might have potential for urban mining.

EXERCISES

1. Examine all the items in your kitchen, such as the refrigerator, stove, toaster, counters, etc., and make a list of the minerals used in each. For example, a refrigerator contains aluminum, copper, iron, nickel, zinc, and petroleum products. The U.S. Geological Survey maintains a website that is helpful for this exercise, at http://minerals.usgs.gov/granted.html.

2. Use a magnifying glass, hand lens, or microscope to closely examine grains of table salt (halite). These are crystals, and they are broken along cleavage planes. What type of cleavage can you see (for example, one, two, or three planes broken at right angles to one another, or at another angle)?

3. It is possible to make your own minerals by precipitating crystals from a solution. Mix a small quantity of table salt in water in a glass container. All the salt will dissolve. However, if you add more salt, the water eventually becomes saturated with sodium and chloride ions. If you try to add yet more salt, it will not dissolve but instead will accumulate as solid crystals at the bottom of the glass. If you let the solution with dissolved salt evaporate over a period of several days to weeks, the salt that is in solution will precipitate along the sides of the container. Examine this salt with a hand lens, and you will see that individual crystals have a prominent cubic shape that reflects their cubic crystal structure.

4. If a copper ore body contains 3% copper, is it economical to mine? Why or why not?

5. Examine Figure 4-25, and suggest several ways in which a critical mineral might become less critical (that is, move left and/or down on the graph) over a period of 5–10 years.

6. The newly upgraded Mountain Pass mine in California (see Box 4-2) was anticipated to produce about 19,000 metric tons of rare earth elements by 2012. What percent of the world's production of rare earth elements does this represent?

7. The U.S. Geological Survey compiles data each year on mineral commodities, including information on production, consumption, and reserve estimates. Select a metal such as copper, lead, zinc, or chromium and examine how much of the annual consumption by the United States is obtained from recycled metal. At the website for this data, http://minerals.usgs.gov/minerals, you can find statistical information by commodity, country, or state, and annual data for each commodity.

8. Use the EPA list of Superfund sites (www.epa.gov/superfund/sites/npl) to see whether any are located near your hometown or college/university. Examine the details of one of these sites to determine how hazardous waste accumulated at the site and what remedial action is to be taken or has been completed already.

SUGGESTED READINGS

Bell, F. G., and L. J. Donnelly. *Mining and Its Impact on the Environment*. New York: Taylor and Francis, 2006.

Best, M. G. *Igneous and Metamorphic Petrology*. Hoboken, NJ: Blackwell Publishing, 2003.

Boggs, S., Jr. *Principles of Sedimentology and Stratigraphy*, 5th ed. Upper Saddle River, NJ: Prentice Hall, 2011.

Boggs, S., Jr. *Petrology of Sedimentary Rocks*. Cambridge: Cambridge University Press, 2009.

Fox, T. L., F. D. Plumlee, and G. S. Hudson. *Metal Mining and the Environment*. Alexandria, VA: American Geological Institute, 1999.

Klein, C., and B. Dutrow. *Manual of Mineral Science*, 23rd ed. Hoboken, NJ: John Wiley and Sons, 2002.

Pellant, C. *Smithsonian Handbooks: Rocks and Minerals*. New York: Dorling Kindersley, 2002.

Robinson, G. W. *Minerals: An Illustrated Exploration of the Dynamic World of Minerals and Their Properties*. New York: Simon and Schuster, 1994.

Rogers, J. J. W., and P. G. Feiss. *People and the Earth: Basic Issues in the Sustainability of Resources and Environment*. Cambridge: Cambridge University Press, 1998.

Taylor, S. R., and S. M. McLennan. "The Evolution of Continental Crust." Special edition, *Scientific American* 15, no. 2 (2005): 44–49.

Wenk, H.-R., and A. Bulakh. *Minerals: Their Constitution and Origin*. Cambridge: Cambridge University Press, 2004.

Volcanoes

A volcano is an opening into Earth's hot interior, a beautiful and yet sometimes frightening glimpse through the crust to the underlying mantle. Before the development of modern volcano-monitoring techniques, hundreds of thousands of people died during eruptions of dozens of volcanoes around the world. In the immediate vicinity of eruptions, death and destruction resulted from burial under volcanic debris, high heat, or suffocating blasts of hot, dense air masses laden with ash and poisonous gases. Farther down the slopes of volcanoes, destruction often resulted from swift-flowing, ash-rich mudflows that overtook villages and their inhabitants.

In the last few decades, volcanologists have risked their lives to collect data that might help them determine the extent and timing of eruptions, and some have died in service of the public good, unfortunate casualties of the very events they hoped to predict. The dedication and sacrifice of these brave women and men have brought us greater levels of understanding and helped to prevent the deaths of many who live in the shadow of volcanoes.

In this chapter we:

✔ Learn what controls volcanic eruptions.

✔ Explore the various types of volcanoes generated by different magma compositions.

✔ Discuss some of the methods by which volcanologists monitor the activity and destructive potential of volcanoes.

✔ Consider the impact of volcanic eruptions on climate.

Scientists wearing protective masks sample the sulfurous gas seeping from Colima volcano in Mexico. When making such measurements, scientists sometimes risk their lives for the sake of understanding volcanism and predicting future eruptions. (© Roger Ressmeyer/CORBIS)

INTRODUCTION

"Vancouver, Vancouver, this is it!" shouted volcanologist David Johnston into his radio on the morning of May 18, 1980, the last words he was ever heard to say.[1] Dr. Johnston was manning an observation post on a ridge 9 km north of the summit of Mount St. Helens when the volcano produced one of the largest eruptions in the history of the United States (Figure 5-1). The volcanic mountain, located in the Cascade Range in southwestern Washington, blew apart with such force that it flattened nearly 600 km² (232 mi²) of surrounding forest, stripping the branches from towering Douglas fir trees and snapping them off at their bases (Figure 5-2).[2] Scorching clouds of volcanic ash, rock,

and gases at temperatures in excess of 300°C (572°F) moved down the north flank of the mountain at speeds as great as 1000 km (620 mi) per hour, incinerating everything in their path, including Dr. Johnston, whose body was never found. He was only 30 years old and was just beginning a career in public service.

David Johnston worked for the U.S. Geological Survey (USGS) at the time of his death. Survey scientists had descended on Mount St. Helens nearly 2 months earlier (late March) after a seismograph at the University of Washington (UW) registered unusual earthquake activity under the mountain. Within 3 days, UW and Survey geologists had installed additional seismographic stations in the vicinity of the volcano and found a worrying trend of increasing earthquake activity at

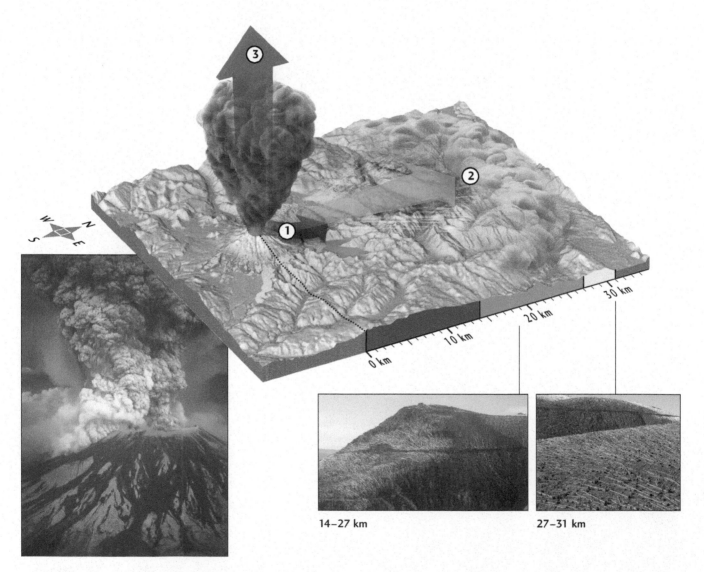

14–27 km   27–31 km

Figure 5–1 Two months after scientists detected ground shaking and began to monitor signs of volcanic activity more closely at Mount St. Helens in Washington, a massive landslide exposed the volcano's vent and the volcano erupted violently on May 18, 1980. In this diagram, 1 indicates the initial landslide, 2 indicates the resulting lateral blast, and 3 indicates the subsequent vertical eruption. *(left: Austin Post/USGS; insets: Terry Leighley, Sandia Labs/Courtesy Cascades Volcano Observatory, USGS)*

Figure 5–2 The force of the lateral blast from the eruption of Mount St. Helens in 1980 was so great that it flattened nearly 600 km² (232 mi²) of the surrounding forest of Douglas fir and other tall trees. *(Lyn Topinka/USGS)*

very shallow levels in the crust.[3] By the sixth day after the shaking began, Mount St. Helens was experiencing as many as five earthquakes greater than magnitude 4 every hour, triggering avalanches that prompted the U.S. Forest Service to close the area to the public. Only 2 days later, a small crater opened on the mountain's summit, from which eruptions of steam and ash began, precursors to the main event that was still a month and a half away. Due to the increasing threat of avalanches and mudslides produced by the melting of the volcano's ice cap, the Forest Service evacuated its personnel and their families as well as local residents and loggers— better to be safe than sorry. Eighty-three-year-old Harry Truman, owner of an inn on nearby Spirit Lake, refused to obey the evacuation order, kicking the ground and cursing the mountain every time it was rocked by an earthquake.

As news of the small eruptions spread, more scientists headed to the mountain. Little more than a week had elapsed since the UW seismographic station first revealed evidence of the volcano's renewed activity, but by this time, ash and steam were spewing from the mountain several times a day in eruptions that lasted from a few minutes to a couple of hours. Though small in comparison to the major eruption, some of these blew debris up to an altitude 3 km (10,000 ft) above the summit of Mount St. Helens.[4] Scientists flew over the volcano, collecting air samples to look for evidence of the gas *sulfur dioxide (SO₂)*, which would indicate moving magma. Others collected samples of ash to determine whether it was forming from new magma or was merely older rock blown to pieces by steam explosions; it was not yet entirely clear whether the eruptions were related to magma moving upward in the crust or

to the mountain letting off a little steam. The ash was found to be old material, but the gas measurements revealed SO_2.

The results contradictory, Survey geologists decided to install devices called *tiltmeters* to monitor the shape of the volcano. If magma were indeed moving upward, it would cause the mountain to swell, changing the angle of tilt of the surface. Over the next couple of weeks, the tiltmeters registered numerous episodes of swelling followed by collapse as small eruptions released steam and ash from the mountain. Eventually, however, the meters would record a net swelling of the volcano's northern flank. To keep track of the swelling, geologists used surveying equipment to measure the distance between specific points on the surface. Two weeks before the eruption, the swelling rate rose to 1.5 m (5 ft) per day.[4] At the same time, large cracks opened on the volcano's surface, prompting concern over catastrophic landsliding.

On the morning of the eruption, geologists Dorothy and Keith Stoffel chartered a small plane to fly around the top of the mountain and take pictures. At 8:30, they were directly above it when "the whole north side of the summit crater began to move instantaneously as one gigantic mass."[5] An immense landslide took some of the confining pressure off of superheated groundwater and magma below the surface, which exploded outward in a colossal lateral blast (see Figure 5-1). Its vent exposed, the volcano proceeded to erupt vertically, throwing ash 26 km (16 mi) into the air. The Stoffels watched in awe. In their small plane, they heard and felt nothing. Realizing the incredible danger they were in, their pilot took a steep nosedive to gain speed, and then headed south to flee the deadly ash cloud.

On the ground, David Johnston's observation post was overrun by the lateral blast within a matter of minutes. Spirit Lake and the obstinate Harry Truman disappeared, and the area where the lake existed was filled with so much debris that a new Spirit Lake formed 60 m (197 ft) above the old one. Nearly 60 people lost their lives that day, many of whom were sightseeing at the volcano in defiance of the evacuation orders. Still, the heroic work of scientists and Forest Service personnel had prevented the deaths of many more.

While both tragic and spectacular, from a geologic point of view the 1980 eruption of Mount St. Helens was just another event in the continuous evolution of a continent on our dynamic planet. Volcanism has occurred at numerous locations since a crust solidified over Earth's molten interior about 4 billion years ago, and like the other volcanoes in the Cascade Range, Mount St. Helens has erupted repeatedly over tens of thousands of years. In fact, the volcano erupted most recently from 2004 to 2008[6] and will surely do so again. Before it does, scientists at the USGS's David A. Johnston Cascades Volcano Observatory in Vancouver, Washington, will be on hand to detect the precursory evidence of activity and to issue warnings to evacuate.

VOLCANIC ERUPTIONS

More than 1300 volcanoes on Earth are active or potentially active, yet not all behave in the manner of Mount St. Helens.[7] What creates these awe-inspiring landforms, and what governs their styles of eruption? As mentioned in Chapters 2 and 4, the locations and magma compositions of most volcanoes are closely related to tectonic plate boundaries. About 80 percent of volcanoes above sea level are located along convergent margins such as the volcanic arcs that form the Ring of Fire surrounding the Pacific Ocean (Figure 5-3). There, subducting slabs of oceanic lithosphere carry rocks, marine sediments, and water into the mantle, spurring partial melting and andesitic magma formation. Most of the other 20 percent of volcanoes are located along mid-ocean spreading ridges where partial melting of the asthenosphere generates basaltic magma. A few volcanoes are found in continental interiors and have a variety of magma chemistries.

By volume, the volcanic eruptions associated with seafloor spreading ridges far surpass those of volcanic arcs and other above-sea-level eruptive vents, pouring out about 18 km^3 (4.3 mi^3) per year of new lava, compared to less than 2 km^3 (0.48 mi^3) per year that arcs release.[8] The difference, discussed below, is a result of the type of magma produced in each area.

How Volcanoes Erupt

The plumbing system for volcanoes appears to be rather straightforward: Magma from the mantle rises along narrow conduits to fill a reservoir, or **magma chamber,** in the crust (Figure 5-4). The magma rises because molten material has a lower density than the surrounding solids. As the magma chamber fills, it often swells Earth's surface, much like a balloon pumped full of water. Molten rock is squeezed into a narrow pipe above the magma chamber, called the volcano's *neck*. As the molten rock moves toward Earth's surface, the confining pressure on it decreases. As a result, **volatiles** (dissolved gases) within the magma come out of solution to form bubbles.[9] This process is identical to what happens when you open the cap on a carbonated beverage; carbon dioxide (CO_2) that had been dissolved in the liquid under pressure comes out of solution and forms the bubbles that give sodas their effervescent quality. Another source of gas bubbles in magma is partial crystallization, which can increase the concentration of volatiles in the remaining melt as the first minerals to crystallize exclude elements such as carbon, hydrogen, and sulfur.[10] Whatever the source, when the gas pressure in the bubbles exceeds the confining pressure of the molten rock, the low-density

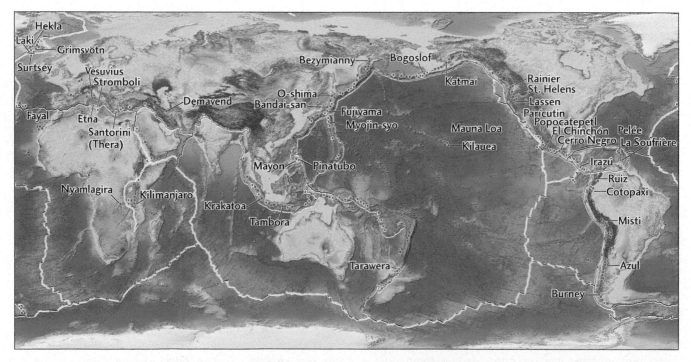

— Oceanic ridge or divergent plate boundary   ⬤ Subduction zone or convergent plate boundary   — Transform fault   ⬤ Volcano

Figure 5–3 A global view of Earth and its plate boundaries illustrates the link between plate tectonics, volcanism, and earthquake activity. Nearly all earthquakes occur along plate boundaries, and most volcanoes (red) are found along convergent and divergent plate boundaries. The few exceptions include volcanoes at hot spots, such as on the Big Island of Hawaii. The Pacific Ocean is rimmed with convergent boundaries and volcanoes.

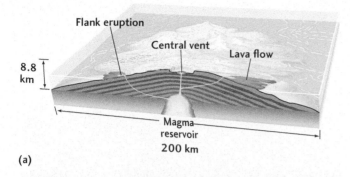

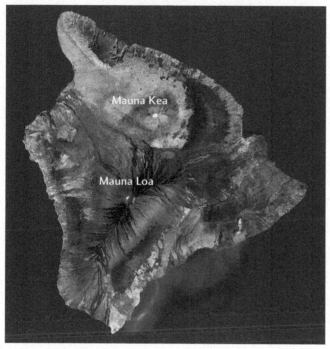

(b)

Figure 5–4 Interior architecture of a volcano. (a) A volcano forms at Earth's surface above a chamber from which magma rises along a central vent. Some magma might escape from other vents along the flanks of a volcano. The volcano in this diagram is representative of basaltic volcanoes that have constructed the island of Hawaii. (b) The broad, shield–like surface of Mauna Kea is clearly seen in this view of the island. *(b: Landsat/NASA/USGS)*

bubbles erupt, blowing some of the magma out of the volcano, just as soda may erupt out of an uncapped bottle. The loss of magma from the chamber further decreases the pressure on the remaining magma, allowing still more gas to come out of solution and driving the eruption forward.

The eruption ceases after the magma is degassed, although a small amount of magma may continue to rise to build a new *lava dome* within the eruptive crater. After the eruption, the magma chamber is partly or completely emptied and the crust above it may collapse, only to swell again as the chamber refills.

Magma Types and Eruptive Styles

The eruptive styles of volcanoes vary widely, depending on the type of magma feeding the volcano, which also affects the types of deposits that accumulate around the vent. Volcanic behavior depends largely on magma **viscosity,** a measure of its stickiness or resistance to flow. Honey, for example, has a higher viscosity than water. Differences in viscosity are related to the type and number of molecular bonds in a fluid, to its temperature, and to its dissolved gas content. Clusters and chains of linked silica tetrahedra increase the viscosity of magmas; high temperatures and dissolved gas contents decrease viscosity.

For any given magma type, the violence of a volcanic eruption typically depends on the amount of water vapor in the magma, because water vapor constitutes 60 percent or more of volcanic gases by volume.[11] Carbon dioxide, sulfur dioxide, and hydrogen sulfide account for most of the remainder. The water vapor in bubbles inside a volcano is confined at temperatures several times higher than the boiling point of water, which is 100°C (212°F). The confinement results in superheated steam that exerts the extremely high pressures that drive eruptions. Material in the 1980 Mount St. Helens eruption contained about 6 percent water by weight, making it highly explosive.

Basaltic magmas, which contain between 45 and 52 percent[12] silicon dioxide (SiO_2), have low viscosity, so they are very fluid and move upward readily along volcanic conduits. These magmas degas fairly easily and thus tend to erupt with minimal violence. Still, basaltic lavas with abundant gas may fountain tens to hundreds of meters into the sky, producing gravel- to boulder-size volcanic **bombs** that fall in a radial pattern around the vent (Figure 5-5). Many of the lava rocks (called *scoria*) used in landscaping originate in this fashion. In addition to low silica content, high temperatures contribute to the low viscosity of basaltic lavas. The temperatures of basaltic lavas are usually about 1100°C (2012°F) upon eruption and they can move as rapidly as 64 km (40 mi) per hour.[13] More typically, basaltic flows move at several kilometers per hour, making them fairly easy to outrun.

Basaltic lava flows display two primary surface textures related to the speed at which the lava moves and its viscosity. Slow-moving, lower-viscosity flows move in a ductile fashion, advancing in lobes across the surface of a volcano. These flows develop thin crusts that are deformed into a wrinkled texture by the movement of the fluid lava beneath them, a texture given the Hawaiian name **pāhoehoe** (pronounced pa-hoy-hoy), which is derived from a Hawaiian word that refers to something that is smooth and unbroken (Figure 5-6). A different texture forms when flows are forced to speed up, as happens when flows move to areas of higher slope, and/ or when flows develop higher viscosities. These factors

(a)

(b)

Figure 5–5 Volcanic fountains and bombs. Basaltic lavas have low viscosity and are very fluid, but with abundant gas can produce fountains up to hundreds of meters high and gravel– to boulder–size volcanic bombs. *(a: Westend61/ SuperStock; b: J. D. Griggs/USGS)*

cause the lava to rip or break into a series of discrete blobs, each of which quickly develops a chilled rind. These blobs tumble off the edge of the flow, forming a pile of *clinkers,* broken material, at the flow front, and over which the flow continues to advance. This lava texture is called **a'a** (pronounced ah-ah), a Hawaiian term meaning "to burn, blaze, or glow."

Felsic magmas, with as much as 70 percent SiO_2, have lower melting points than basaltic magmas and erupt at temperatures around 800°C (1472°F). Felsic magma is very sticky and flows slowly, often clogging a volcano's plumbing system as it approaches the surface and cools. Some felsic magmas cool so much while rising that they don't even reach the surface. When they do, however, their eruptions tend to be very explosive because of their higher viscosity and trapped gases. After felsic lavas erupt from a volcano, the lava shatters and cools to form debris that ranges in size from ash to building-size "bombs" (Table 5-1). Called **pyroclastic debris** (from *pyro* for "fire" and *clastos* for "particle") or **tephra** (any airborne volcanic material), some of the material ejected from these volcanoes falls back to Earth's surface near the vent and forms layers of volcanic sediments called **ashfall deposits** (Figure 5-7). The finest debris sometimes circles the globe before settling back to Earth. Ash from the Mount St. Helens eruption in

Figure 5–6 Basalt lava flow textures. Pāhoehoe (left side), is smooth, ropy, and wrinkled as a result of its low viscosity and its slow movement as it cools at the surface. The a'a flow texture (right side) is much rougher as a result of higher velocity flows (e.g., on steep slopes) and the flow front advancing over previously cooled, broken blocks of volcanic rock. *(Andrew de Wet)*

Figure 5–7 From ashes to rocks. About 300 m (1000 ft) of interbedded pyroclastic rocks and pumice from volcanic eruptions that occurred 6 to 7 million years ago has been eroded since then to create a sculpted landscape of "hoodoos," tent-shaped rocks atop which boulders protect the more readily weathered material below. This landscape is part of the Kasha–Katuwe Tent Rocks National Monument in New Mexico. Kasha–Katuwe means "white cliffs" in the traditional Pueblo Keresan language. *(Kirsten Menking)*

Washington State fell only as far as eastern Idaho, while fine ash from the much larger Mount Pinatubo eruption in the Philippines in 1991 circled Earth several times.[14]

Andesitic magmas contain 52 to 63 percent[15] SiO_2, and so are intermediate between basaltic and felsic magmas in their eruptive styles and explosiveness. Because andesitic magmas are not as viscous as felsic magmas, they reach the surface more often and sometimes flow from volcanoes as lava. However, the flows they produce move slowly, rarely traveling far from the volcano, and most of the material emitted from andesitic volcanoes consists of pyroclastic debris.

Both andesitic and felsic magmas commonly produce **pyroclastic flows,** avalanches of ash, rock fragments, and hot gases that move together like a fluid under the influence of gravity (Figure 5-8). Also called *nuée ardentes,* French for "glowing clouds," these flows can travel at speeds in excess of 300 km (186 mi) per hour, flattening everything in their path. In 1991, immense pyroclastic flows roared down the flanks of the Philippines' Mount Pinatubo during one of the largest volcanic eruptions in the past three centuries, filling deep valleys with up to 200 m (660 ft) of volcanic deposits.

Pyroclastic flows killed the victims of Mount St. Helens and have been responsible for most of the deaths attributed to volcanic eruptions throughout history. Before the advent of volcano monitoring and evacuation planning, a single pyroclastic eruption could cause thousands of deaths. Notable among these were the destruction of the Roman city of Pompeii by the volcano Vesuvius in 79 CE, which killed at least 2000 people, and of the city of St. Pierre on the Caribbean island of Martinique by the volcano Mount Pelée in 1902, which killed as many as 28,000.

Shield Volcanoes, Stratovolcanoes, and Cinder Cones

The shapes of volcanoes reflect the differences in magma types discussed earlier, as well as their gas content and explosive tendencies. Many basaltic lavas spread far and rapidly from the vent because of their low viscosity, building thin sheets of gently dipping flows (4° to 8° slope angles)[16] one atop the other. With time, a broad dome forms that resembles the convex surface of a warrior's shield. The big island of Hawaii is formed by the coalescence of several such **shield volcanoes** (see Figure 5-4), two of which, Mauna Loa and Kīlauea, are considered active. Mauna Loa last erupted in 1984, and Kīlauea has been erupting more or less continuously since 1983.[17] The source of the magma is a mantle hot spot, a site of anomalously high volcanism of controversial origin (see Chapter 2). This hot spot has poured out thousands of lava flows over the last few million

TABLE 5–1	Classification of Pyroclastic Debris by Size	
Name	Size (mm)
Bomb	> 64
Cinder	2–64
Ash	< 2

Note: All solid particles ejected from a volcano are called pyroclasts, regardless of magma composition.

Figure 5–8 Nuée ardentes (shown here from the 1991 Mount Pinatubo eruption in the Philippines) are avalanches of ash, rock fragments, and hot gases that can travel at speeds of more than 300 km (186 mi) per hour. *(Alberto Garcia/Corbis)*

years, raising the big island to an elevation nearly 10 km (6.2 mi) above the ocean floor.

In contrast to the thin, low-viscosity lava flows emitted by shield volcanoes, andesitic volcanoes emit both large amounts of pyroclastic material and thick, sticky lava flows (Figure 5-9). The high-viscosity flows are able to maintain steeper slopes (35°)[18] than are flows made of basalt, giving these volcanoes a distinctive cone shape. With time, the steep flanks of andesitic volcanoes are built up by interbedded layers of lava and pyroclasts. It is this layering that leads to the name **stratovolcano** for these features; *strata* is a geological term for layers, derived from the Latin *stratum*, meaning spread. These volcanoes are sometimes also referred to as *composite volcanoes*. Examples of stratovolcanoes include Mount St. Helens, Mount Rainier, and Mount Fuji.

Despite its low viscosity, basalt can produce steep-sided volcanic cones, and these are actually more common than either shield volcanoes or stratovolcanoes. Some basaltic magmas have such high gas content that they produce nothing but scoria pyroclasts, which rain down around the vent. These *cinders* have irregular and angular shapes that allow them to create steep slopes. If the cinders are still partially molten when deposited, they can weld together, and the strength of the welded material increases the steepness of slopes. The **cinder** or **scoria cones** these magmas create tend to be small, typically less than 300 m (984 ft) in total height, and often are

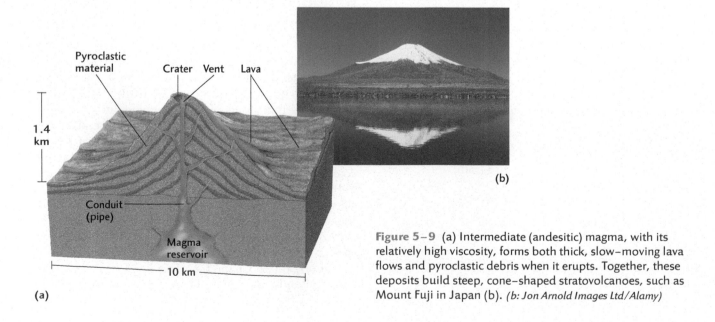

Figure 5–9 (a) Intermediate (andesitic) magma, with its relatively high viscosity, forms both thick, slow-moving lava flows and pyroclastic debris when it erupts. Together, these deposits build steep, cone-shaped stratovolcanoes, such as Mount Fuji in Japan (b). *(b: Jon Arnold Images Ltd/Alamy)*

(a)

Pyroclastic material

Crater

Central vent filled with rock fragments

(b)

Figure 5–10 Cinder cones composed of basaltic pyroclasts typically are less than 300 m (984 ft) in height and have steep slopes, like these cones in the Galápagos Islands of the Pacific Ocean. *(b: Kirsten Menking)*

found growing atop preexisting shield or stratovolcanoes (Figure 5-10). Their sides may have slopes as steep as 35°, and the cones can form extremely rapidly. Vulcan, a cinder cone that erupted in 1937 on the rim of a larger volcano in Papua New Guinea, grew to a height of 230 m (755 ft) in only 4 days![19]

Not all volcanic eruptions occur around a single circular vent. On occasion, magma comes to the surface along fissures in Earth's surface that are hundreds of meters to kilometers in length (Figure 5-11). There are numerous such fissures on the Big Island of Hawaii; they are likely produced when magma inflating the

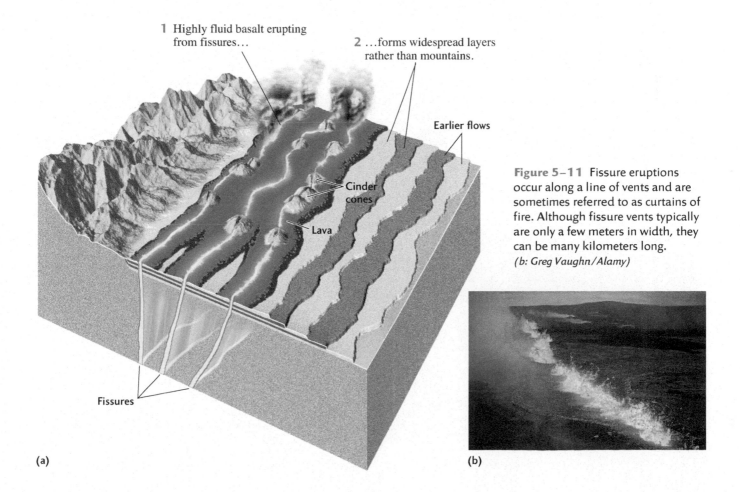

1 Highly fluid basalt erupting from fissures…

2 …forms widespread layers rather than mountains.

Earlier flows

Cinder cones

Lava

Fissures

(a)

Figure 5–11 Fissure eruptions occur along a line of vents and are sometimes referred to as curtains of fire. Although fissure vents typically are only a few meters in width, they can be many kilometers long. *(b: Greg Vaughn/Alamy)*

(b)

volcano causes the volcano's surface to stretch so much that it opens a crack into which the magma can intrude. The mid-ocean ridge system is another site where **fissure eruptions** occur. Earlier in geologic time, large-scale fissure eruptions associated with volcanic hotspots (see Chapter 2) were more common, generating voluminous deposits (up to a million cubic kilometers) called **flood basalts**.[20] The Columbia River, which forms the boundary between the states of Washington and Oregon in the United States, has cut a deep canyon through such deposits, revealing a thick sequence of stacked lava flows that erupted about 16 million years ago.

Craters, Calderas, and Climate

Craters, which are steep-sided depressions found at the summits of volcanoes, are produced by a combination of explosive removal of the mountaintop during eruptions and downward draining of magma into the emptied chamber beneath the volcano at the end of the eruption. In contrast, **calderas** form when so much molten rock has been removed from the underlying magma chamber that the chamber's roof collapses, causing the ground above it to subside (Figure 5-12). Calderas are much larger than craters, measuring from 1 to 75 km (0.62 to 47 mi) in diameter, and may form over mafic or felsic magma chambers.[21] Basaltic calderas, such as those

that exist today in Hawaii, tend to be 5 km (3.1 mi) or smaller in diameter. Larger calderas form in areas of felsic volcanism. Examples include Yellowstone in Wyoming (47 × 78 km in dimension, or 28 ×47 mi),[22] Crater Lake in Oregon (8 × 10 km in dimension, or 5 × 6 mi),[23] and Long Valley in California (17 × 32 km in dimension, or 10 × 19 mi).[24]

The most recent explosion at Yellowstone occurred 70,000 years ago, and three earlier events occurred 640,000, 1.3 million, and 2.1 million years ago.[25] Long Valley last erupted 760,000 years ago,[26] and Crater Lake erupted 7700 years ago.[27] Although these caldera-forming continental volcanoes have not erupted recently, they are not inactive. Rather, they are characterized by long periods of quiescence as the magma chamber slowly fills and reinflates. With time, the chamber might become full enough that the floor of the caldera will bulge upward sufficiently for eruption and caldera collapse to occur again. At Yellowstone, the cycle of caldera collapse and resurgence appears to be between 0.5- and 1.5-million years long, and the volume of material erupted has measured as much as 2500 km³ (600 mi³).[28] Because of its long time span of eruptive cycles and its enormous eruption volumes, Yellowstone is a "supervolcano," which is a term used to describe any volcano that generates more than 1000 km³ (240 mi³) of deposits.[29]

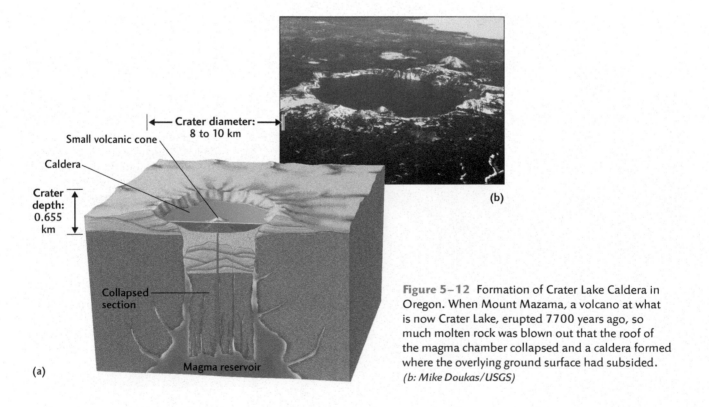

(b)

Figure 5–12 Formation of Crater Lake Caldera in Oregon. When Mount Mazama, a volcano at what is now Crater Lake, erupted 7700 years ago, so much molten rock was blown out that the roof of the magma chamber collapsed and a caldera formed where the overlying ground surface had subsided. *(b: Mike Doukas/USGS)*

A huge caldera-forming eruption in historic time was that of the Indonesian volcano Tambora in 1815. In comparison to Mount St. Helens, which erupted about 2.7 km³ (0.67 mi³) of material,[30] Tambora belched forth over 30 km³ (12 mi³) of pyroclastic debris. When it was all over, the summit of the volcano had lost a full kilometer of height and a new caldera 5 km (3.1 mi) in diameter and 1.5 km (0.93 mi) deep remained in its place.[31]

In addition to ash, Tambora erupted enormous quantities of SO_2 into the atmosphere, which combined with water vapor to form fine droplets of sulfuric acid and other sulfate aerosols. These compounds blocked so much incoming solar radiation that the planet cooled as much as 0.8°C (1.4°F) in 1816,[32] leading to what became widely known as the "year without a summer." In various locales, temperatures declined much more. In the New England region of the United States, for example, average temperatures dropped up to 2.5°C (4.5°F), and snow or frost formed even in the summer, causing crop failures that led people to migrate to warmer climates. Conditions were so dire that 8 percent of the population of Vermont moved out of the state.[33] During the same year, India and Europe were hit with famines associated with crop failures resulting from the frosty conditions.

While the situation proved dire for agriculture, ash and sulfate compounds from the Tambora eruption also generated brilliant sunsets for several years, which became a favorite subject matter for landscape painters. Indeed, artists working since the Renaissance have recorded numerous volcanic eruptions in places such as the Philippines, Indonesia, and Iceland. Though the artists might not be aware of the volcanic activity that generated the sunsets, scientists have been able to use the ratio of red hues to green hues in the paintings to estimate the *aerosol optical depth* (a measure of how much atmospheric particles and liquid droplets absorb and scatter sunlight) over time, a fascinating fusion of art history and geology![34]

The impact of volcanic eruptions on climate recorded in landscape paintings and in the strong cooling associated with the 1815 Tambora eruption and the 1991 eruption of Mt. Pinatubo has led several people to propose artificial injection of sulfate aerosols into the atmosphere as a solution to global warming (Box 5-1).

■ A majority of volcanoes are located around the Pacific Ocean rim, otherwise known as the "Ring of Fire." These volcanoes are associated with subduction zones. Others form along spreading ridges and over mantle hot spots.

■ Magma forming within Earth's mantle moves upward into the crust along narrow conduits in response to magma's lower density relative to surrounding rocks. These conduits feed magma chambers from which magmas erupt to Earth's surface.

■ Explosive eruptions occur when gases dissolved in magma come out of solution and form bubbles. Gas pressure builds in viscous magmas from which bubbles cannot readily escape. Eventually, the pressure is so great that the magma explodes.

■ Basaltic magma generally produces very fluid, fast-moving lava flows that build broad shield volcanoes. Flows typically exhibit two surface textures, pāhoehoe and a'a.

■ Intermediate (andesitic) magma generally produces explosive eruptions with slow-moving lavas and abundant pyroclastic materials that build steep-sided, cone-shaped stratovolcanoes.

■ Felsic magma generally does not reach Earth's surface, but when it does, it is associated with large calderas and very explosive eruptions of pyroclastic debris. Felsic eruptions may recur at long intervals of tens to hundreds of thousands of years.

■ Ash and sulfate aerosols injected into Earth's atmosphere reflect incoming solar radiation and lead to short-term global cooling.

VOLCANIC HAZARDS

The hazards associated with volcanoes are numerous and vary with different types of eruptions and with distance from the volcano (Figure 5-13a). For example, basaltic lava flows in Hawaii engulf buildings and cover roads, destroying vital infrastructure (Figure 5-14). Vegetation flooded by lava can produce methane gas pockets that explode violently, spraying lava bombs across the landscape and setting buildings on fire at some distance from the active flow. In areas of andesitic volcanism, ashfalls can collapse roofs of buildings, and hot or cold ash deposits can mix with rainwater to form devastating volcanic mudflows, called **lahars**. The eruption of Mount Pinatubo covered a 4000 km² (1544 mi²) area with deposits more than 5 cm (2 in.) thick, destroying many homes. The loose ash combined with rains from a coincidental typhoon (hurricane) and formed lahars that spread out as much as 40 km (25 mi) from the volcano's flanks, choking streams with debris and causing widespread flooding.

Volcanic Ash

Volcanic ash presents problems not only for people on the ground but also for those in the air. About 80,000 large aircraft a year, for example, fly over the Aleutian Islands, a chain of volcanoes along the convergent plate boundary between the North American and Pacific

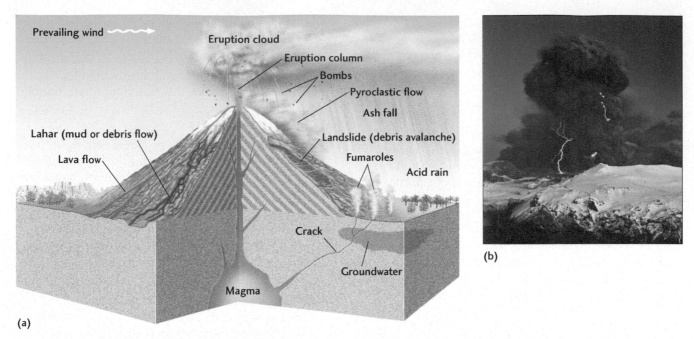

(a)

(b)

Figure 5–13 (a) The particular types of hazards a volcano poses depend upon the type of volcano and the material it erupts as well as distance from the volcano. Lava and pyroclastic flows occur near the volcanic vent, whereas ash can be carried far downwind. Lahars, which begin as melting ice and/or water mix with volcanic debris, flow downslope and along stream valleys and can happen even without an actual volcanic eruption. (b) During the April 2010 eruptions of Eyjafjallajökull volcano in Iceland, ash clouds billowed across the North Atlantic Ocean, disrupting and shutting down European air travel for weeks. Electrically charged particles in the ash caused the dramatic lightning. *(b: David Jon/NordicPhotos/Getty Images)*

plates (see Figures 2-3 and 5-3).[39] These planes carry as many as 30,000 people a day, mostly on routes between North America, Asia, and Europe, and are occasionally threatened by eruptions or by resuspension by winds of ash that has already been deposited.[40] The tiny pieces of tephra, less than 2 mm in diameter and composed mostly of fine glass shards of frozen magma, are not detected by ordinary weather radar and can damage

Figure 5–14 When Kīlauea in Hawaii erupted in May 1990, low–viscosity, basaltic magma moved rapidly down the volcano's flanks, engulfing buildings and forests. Flames along the edge of the lava tongue are burning organic matter. *(J. D. Griggs/USGS)*

BOX 5–1 EARTH SCIENCE AND PUBLIC POLICY

Creating Artificial Volcanic Eruptions to Counter Global Warming: Geoengineering or Geo–Fantasy?

Public concern over the threat of global warming has been mounting. Recent decades have brought much evidence that emissions of greenhouse gases from industrial, automotive, and other sources are trapping increasing amounts of heat in the atmosphere, warming the planet. At the same time, forests all over the world continue to be cleared for lumber, agriculture, and grazing land, destroying millions of acres of trees that could absorb CO_2, a greenhouse gas. The global temperature has generally been rising since the late 1800s and is expected to rise still higher, yet the early 1990s brought several years of global cooling.

This drop in temperature had nothing to do with human activity. It was due, instead, to the massive release of sulfate aerosols and ash from the catastrophic 1991 eruption of Mount Pinatubo in the Philippines. After more than 600 years of calm, magma collecting beneath the mountain exploded into 4 km³ (0.96 mi³) of volcanic ash and 30 billion kg (33 million tons) of SO_2 gas. The debris and resulting sulfuric acid droplets circled the globe within 22 days and persisted in the atmosphere for over 2.5 years (**Figure 1**). Nearly a year after the eruption, enough sunlight was still being blocked to cause average global temperature to decline by 0.5°C (0.9°F), as compared to a 30–year average temperature.[35] While this decline might not seem like much, consider that the difference between Earth as we know it today and the last ice age is a mere 7°C (12.6°F) drop in average temperature.

The efficiency with which volcanoes like Pinatubo and Tambora decrease global temperatures has led some scientists to propose that artificial injection of sulfate aerosols into the atmosphere could counter global warming.[36] Those who envision such a geoengineering strategy suggest that several million tons of SO_2 could be injected into the stratosphere annually through the use of balloons or missiles, or by special aircraft designed to fly at very high altitudes while burning high–sulfur fuels.

While the idea of replicating the effects of volcanic eruptions has a certain elegant simplicity, the reality is more complex. Ironically, we currently remove as much sulfur as possible from petroleum products like jet fuel

(continued)

Mount Pinatubo sulfur dioxide

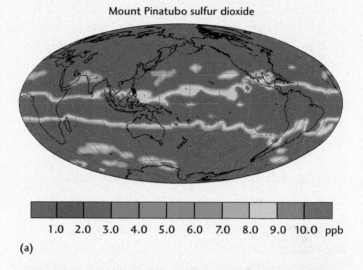

| 1.0 | 2.0 | 3.0 | 4.0 | 5.0 | 6.0 | 7.0 | 8.0 | 9.0 | 10.0 ppb |

(a)

(b)

Figure 1 After 600 years of quiet, Mount Pinatubo—a volcanic mountain in the Philippines—erupted in June of 1991. Sulfuric acid droplets circled the globe within 22 days and persisted in the atmosphere for over 2.5 years. Mount Pinatubo's plume of ash and gas affected global climate for several years. *(b: Dave Harlow/USGS)*

(continued)

because of the role of sulfur in producing acid rain and because of its damaging impact on the human respiratory system. Aside from an increased risk of acid rain, an increase in SO_2 in the stratosphere might change atmospheric circulation patterns, with consequences for rainfall. Analysis of the aftermath of the Mt. Pinatubo eruption revealed that atmospheric cooling led to a decline in rainfall in many areas of the globe and large reductions in streamflow. Geoengineering thus might cause droughts that could have a profound impact on water resources and agriculture.[37]

In addition, the chemical reactions responsible for the destruction of Earth's ozone layer, the atmospheric layer that shields the planet from the Sun's carcinogenic ultraviolet rays, are enhanced by the presence of sulfate aerosols in the stratosphere. Simulations modeled on computers of the impact of sulfate aerosol geoengineering reveal that recovery of the ozone layer over Antarctica, damaged during the 20th century by now–banned chlorofluorocarbon gases, would be delayed by 30 to 70 years.[38] In addition, a new hole may form in the ozone over the Arctic.

Clearly, more research is needed to determine the relative costs and benefits of this particular strategy. Forging ahead with the idea would merely complicate the uncontrolled experiment we're already conducting on Earth's climate system through our combustion of fossil fuels.

and decrease visibility through airplane windshields, clog engines, and even lead to engine failure.[41] In recent decades, there have been several near disasters in which planes flew unknowingly into ash clouds, lost power in all engines, dropped rapidly in altitude, and finally were able to restore engine power just before what might have been crash landings. Although there haven't been any crashes caused by aircraft encountering volcanic ash clouds, the damage to airplanes has cost hundreds of millions of dollars.

When Iceland's Eyjafjallajökull volcano (pronounced AY-uh-fyat-luh-YOE-kuutl-uh) erupted in the spring of 2010, it spewed at least 0.11 km^3 (0.024 mi^3) of ash and other volcanic tephra into the air to a height of 9 km (5.4 mi)[42] (Figure 5-13b). Prevailing winds distributed the ash cloud across European airspace. Facing prospects of catastrophic engine failures associated with aircraft engines being abraded by ash particles, European transportation ministers repeatedly grounded planes, disrupting the travel plans of millions of people and costing airlines at least $1.8 billion.[43]

The number of confirmed erupting volcanoes worldwide is about 50 to 70 each year, and scientists estimate that volcanic ash might affect global air routes for about 20 days a year.[44] More than 100 incidences of aircraft encountering airborne volcanic ash have been documented since 1973.

Volcanic Gases

Additional hazards of volcanic eruptions are associated with the poisonous gases released by volcanoes. Carbon dioxide, a common volcanic gas, is heavier than air. As a result, it tends to form layers on the ground that can be hazardous to human health. The USGS has been monitoring CO_2 releases from the Long Valley Caldera since the 1980s and has found levels high enough to cause concern.[45] Survey personnel have issued advisories warning people against camping in particular locations where high CO_2 levels have killed trees, and have also warned about the danger of gas buildup in the basements of homes.

A particularly gruesome example of the dangers of volcanic gases occurred in the African country of Cameroon in 1986. About 1700 villagers living in a valley below Lake Nyos, a volcanic crater lake, were asphyxiated in their beds when the lake suddenly belched an immense amount of CO_2 gas. Carbon dioxide given off by magma in the dormant volcano had been dissolving into the water at the bottom of the lake for probably hundreds of years. Something, perhaps a small earthquake or landslide, or perhaps simply the lake becoming saturated with CO_2, caused the water to turn over rapidly, bringing the gas to the surface where it rapidly outgassed and spilled over the crater walls into the valley below.

In addition to CO_2, fluorine in gas and ash has proven deadly. In 1783, Iceland's Lakagígar fissure erupted high amounts of the noxious gas, which is also denser than air and therefore accumulates in depressions and valleys.[46] Seventy-five percent of the island nation's livestock perished from breathing in the gas, ingesting ash-covered grasses, or from starvation when the fluorine killed the grass.[47] Similarly, hundreds of horses and cows were killed after eating grasses covered with the fluorine-rich ash erupted from Lonquimay volcano in Chile in the late 1980s.[48] Sulfur dioxide emissions have also proven problematic, combining with water vapor to make sulfuric acid that falls as acid rain. Areas downwind of Nicaragua's Masaya volcano and Costa Rica's Poás volcano have had agricultural failures and corrosion of buildings.[49]

Tsunamis from Volcanic Eruptions

Volcanic eruptions are also capable of generating devastating tsunamis. Indonesia's Krakatoa volcano (also spelled Krakatau) exploded with such force in 1883 that the eruption was heard 3500 km away in Australia. Ash clouds billowed to 80 km heights in the atmosphere as more than 21 km³ (5 mi³) of material was ejected. The resulting caldera collapse produced monstrous tsunami waves 37 m (120 ft) high that are estimated to have killed 36,000 people on the neighboring Indonesian islands of Java and Sumatra.

Active, Dormant, and Extinct Volcanoes

Most people in the United States are aware of Kīlauea and Mount St. Helens, but many are surprised to learn that the United States and its territories have 169 active volcanoes (Figure 5-15).[50] According to the USGS, an *active volcano* is one that has erupted within historic time. Of these, more than a dozen have erupted in the past few centuries.

Consider Mount Rainier, the beautiful, snow-clad volcano that rises several thousand meters above the city of Tacoma near the coast of Washington (Figure 5-16). Although it has been *dormant* (not erupting) for at least the past 100 years, it is by no means extinct. An *extinct volcano* is one that has not erupted in the geologically recent past (thousands to tens of thousands of years) and gives no signs of any future activity. Geologists mapping volcanic deposits in the Rainier area have found evidence of volcanic eruptions that occur, on average, every few hundred years. Because this is much longer than the span of human memory, most people consider it safe to live at the foot of what, to a geologist, is an active volcano that poses a clear hazard to life and property. The contrast represents the difference in human and geologic concepts of time.

The Volcanic Explosivity Index

In 1982, volcanologists Chris Newhall and Stephen Self developed an index of the explosivity of a volcanic eruption in order to provide a clear, consistent way to compare the relative magnitudes of different historical eruptions.[51] Their Volcanic Explosivity Index (VEI) integrates numerous attributes of an eruption, such as the volume of ejected material and the height to which the material is ejected, to produce a number between

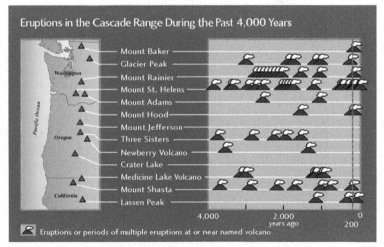

(a)

(b)

Figure 5–15 (a) Potentially hazardous volcanoes are found throughout western North America, but U.S. volcanoes have been divided into different danger classifications. Some, including Mount St. Helens, Mount Rainier, and Kīlauea, erupt every 100 to 200 years; others erupt every 1000 years or more; and some have not erupted in the past 10,000 years. The five volcano observatories in the United States are shown as red triangles. (b) Eruptions or periods of multiple eruptions at or near volcanoes in the Cascade Range of the Pacific Northwest during the past 4000 years. *(b: USGS)*

Figure 5–16 Looming behind the city of Tacoma, Washington, Mount Rainier is one of more than two dozen volcanoes in the western United States that are potentially hazardous. An active volcano located along a convergent plate margin, it erupts every few hundred years or less and ranks high in terms of cause for concern. *(Debbie Stika)*

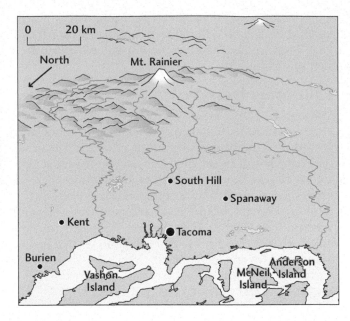

0 and 8 (Figure 5-17). Each increase of 1 in the index indicates an increase in explosivity of approximately 10 times.

A VEI of 0 represents an effusive eruption with less than 10,000 m³ of ejecta and negligible injection to the troposphere, whereas 8 represents a supercolossal eruption with greater than 1,000 km³ of ejecta and substantial injection into the troposphere and stratosphere, the lower two levels of Earth's atmosphere (see Chapter 11). From the low end to the high end of the index, eruptions are classified as gentle, explosive, severe, cataclysmic, paroxysmal, colossal, supercolossal, and mega-colossal. The 1991 Mount Pinatubo eruption had a VEI of 6, colossal, while the 1980 eruption of Mount St. Helens had a VEI of 5, paroxysmal, and the 1815 eruption of Tambora, Indonesia, had a VEI of 7, supercolossal. The frequency of eruptions is highest for those with lowest explosivity, as with the effusive eruptions at Kīlauea in Hawaii (VEI = 0). Only four eruptions with a VEI of 7 are known to have occurred in the past 10,000 years.[52]

Determining Volcanic Threat Levels

Given the variety of dangers associated with volcanoes, it is no surprise that governments around the world have invested hundreds of millions of dollars in programs to monitor volcanic activity, disaster plans, and public education campaigns designed to minimize loss of life. Geoscientists who study volcanoes provide information that helps the public understand how often eruptions might occur and how large they might be in a particular area. Scientific understanding of volcanism is acquired primarily in two ways: by monitoring active

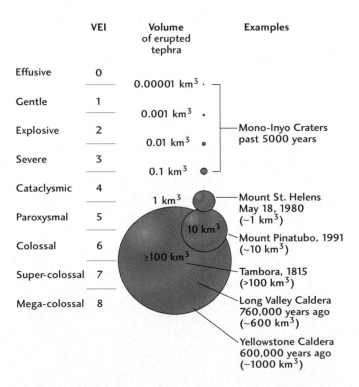

Figure 5–17 The Volcanic Explosivity Index (VEI) provides a measure of the relative magnitude of explosive volcanic eruptions. From 0 to 8, each 1–unit increase in the index represents an approximately 10–fold increase in relative size. Explosivity is estimated from a number of factors that include volume of pyroclastic material erupted, height of eruption column, and duration of eruption. Total volume of uncompacted pyroclastic material erupted is given in parentheses for selected eruptions. *(From USGS)*

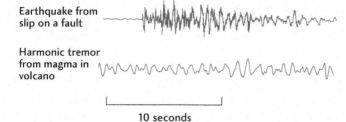

Earthquake from
slip on a fault

Harmonic tremor
from magma in
volcano

|← 10 seconds →|

Figure 5-18 Tectonic earthquakes associated with movement along faults in Earth's crust produce short–period vibrations with well–defined p-, s-, and surface wave signatures on seismograms. In contrast, magma moving through a volcano's plumbing system leads to much longer period vibrations that reflect the adjustment the overlying crust makes to the injection of magma into higher levels in the volcano and to subsequent collapse of the volcano as materials are erupted.

volcanism and by studying the remains and deposits of past volcanic eruptions.

Monitoring Volcanoes Studies of active volcanism are promoted and supported in the United States through federal funding of several volcano observatories operated by the U.S. Geological Survey (see Figure 5-15). The oldest, the Hawaiian Volcano Observatory, is located close to the eruptive center of Kīlauea on the big island of Hawaii. Other observatories have been established in Vancouver, Washington (Cascades Volcano Observatory); Anchorage, Alaska (Alaska Volcano Observatory); Menlo Park, California (California Volcano Observatory); and Yellowstone National Park (Yellowstone Volcano Observatory).

Because of studies by hundreds of scientists at these sites and elsewhere in the world, we now have a better understanding of the types of phenomena that precede eruptions. As described in the chapter opener about Mount St. Helens, for example, we now know that magma rising within Earth's crust leads both to heightened activity of a particular kind of earthquake known as a **harmonic tremor** (Figure 5-18) and to expulsions of sulfurous volcanic gases. We also have learned the benefits of using tiltmeters and surveying equipment to monitor the inflation of magma chambers.

Another recently developed technique in volcano monitoring is the use of interferometric synthetic aperture radar, InSAR for short.[53] In this technique, satellites bounce radar waves off of Earth's surface and record how much time it takes for those waves to return and be detected. The travel time is a direct function of the distance between the satellite and the surface, and makes it possible to develop detailed topographic maps with centimeter-scale resolution. Comparing maps from one time period to those at another can reveal subtle changes

in topography associated with magma movement at depth. Such work was used to document the 2004–2006 ground uplift in the caldera at Yellowstone National Park (Figure 5-19). Whether this movement will someday lead to another supervolcanic eruption cannot be determined at present, though the lack of other associated phenomena, such as earthquakes or changes in the park's geyser system, indicates no need for immediate concern.[54]

The practical applications of volcano monitoring have included a number of successful predictions of eruptions in recent years and the development of early warning systems for aircraft. Studying active volcanoes can be dangerous, however, and a number of volcanologists have been killed or injured by sudden eruptions (Figure 5-20). Those who do such work accept the risks, and the benefits of their research for humanity have been immense. Successful predictions of major eruptions at Mount St. Helens in 1980 and Mount Pinatubo in 1991 saved tens of thousands of lives.

Studying Eruptive History in Volcanic Deposits In contrast to volcano monitoring and surveillance, studying the remains of past volcanic eruptions is relatively safe. Studies of the geologic record of eruptive history also have led to a greater understanding of volcanic hazards, as the example of Mount St. Helens demonstrated.

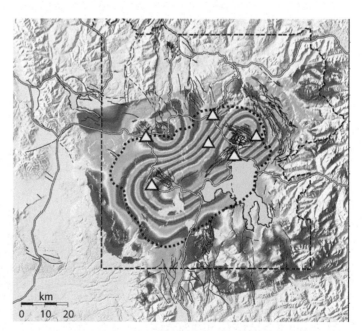

km
0   10   20

Figure 5-19 An interferogram shows ground movement in Yellowstone Caldera, as detected by an Earth–orbiting satellite. The color contour lines represent values of uplift during the interval from September 2004 to August 2006. The uplifted area is roughly elliptical and stretches from northeast to southwest across the caldera (the dashed black line). Yellow lines are roads; yellow triangles are GPS stations. Solid black lines are mapped faults. *(C. Wicks/USGS)*

Figure 5-20 Volcanologist Katia Krafft (1942-1991), photographed by her husband and colleague Maurice Krafft (1946-1991), wears heat-protective clothing while measuring properties of a lava flow. Both scientists perished when they were struck by a pyroclastic flow emanating from Japan's Mount Unzen in 1991. *(Krafft/Explorer/Science Source)*

While the exact timing was uncertain, an eruption was considered to be imminent. The public had, in fact, been alerted months earlier to the likelihood, thanks to a geologic investigation that began in the 1970s, when the USGS assigned geologists Dwight Crandell and Donal Mullineaux to evaluate the hazards of dormant but potentially active volcanoes in the Cascade Range of the Pacific Northwest.

Crandell and Mullineaux recorded evidence of past eruptions to reconstruct the history of each volcano. While hiking and driving over many thousands of square kilometers, they identified and mapped rocks and sediments stemming from lava flows, pyroclastic flows, ashfalls, and lahars. By comparing the thickness of each deposit, how much area it covered, and its position in sequence with other deposits, they were able to determine the relative size and frequency of eruptions. The oldest layers, logic dictated, lay at the bottom, and each successive upper layer was younger. Using geologic dating methods (see Chapter 6) to establish the ages of trees killed during ancient eruptions and of rocks emitted from the volcanoes, they also determined the absolute ages of some deposits. In all, they identified hundreds of deposits from numerous eruptions that had occurred in the past 5000 years.

Of all the Cascades volcanoes, Crandell and Mullineaux found Mount St. Helens to be the most active and explosive during the past 4500 years, with eruptions occurring on average every 225 years. With good reason, the name used by several Native American tribes for Mount St. Helens is *Loowit*, the Lady of Fire. Before 1980, the most recent eruptive period had been between

1831 and 1857, in the early days of the Oregon Trail, the California Gold Rush, and the arrival of the new settlers in the Pacific Northwest. In 1978, Crandell and Mullineaux warned, "In the future, Mount St. Helens probably will erupt violently and intermittently just as it has in the recent geologic past, and these future eruptions will affect human life and health, property, agriculture, and general economic welfare over a broad area.... [T]he current quiet interval will not last as long as a thousand years; instead, an eruption is more likely to occur within the next hundred years, and perhaps even before the end of this century."[55] Little did they know at the time that their warning would come true only 2 years later, when the catastrophic eruption of 1980 described at the beginning of this chapter occurred.

When Mount Pinatubo began rumbling in the spring of 1991, geologists conducted the same sort of analysis to determine the probable magnitude of an eruption. Carefully mapping volcanic deposits, they rapidly grew alarmed as they realized the enormity of earlier eruptions. Their observations, combined with instrument measurements of increasing seismicity and gas output, led scientists to issue warnings to evacuate (Box 5-2). On June 15, after more than 600 years of calm, magma reached the surface in a catastrophic eruption. Nearly 7 km^3 (1.7 mi^3) of debris were hurled 30 km (18.6 mi) into the atmosphere, resulting in the deaths of 900 people who had remained in the area. Still, the eruption destroyed the vacant homes of 200,000 others who might have been killed had warnings not been issued. Two U.S. military bases, Subic Bay Naval Base and Clark Air Base, were also damaged in the blast, leading to their eventual abandonment.

■ In addition to pyroclastic flows, lava flows, and ashfall deposits, volcanic eruptions produce lahars, poisonous gases, and fires, and can also generate tsunamis. Volcanic eruptions pose risks for aviation as aircraft engines can suck in abrasive ash particles.

■ Volcanoes that have erupted within recorded history are considered active. Those that have not erupted within recorded history are considered dormant. Extinct volcanoes have not erupted within thousands of years and show no signs of resuming activity.

■ Volcanic hazard potential can be determined by mapping and dating volcanic deposits and by actively monitoring volcanoes with seismographs, tiltmeters, interferometric synthetic aperature radar, and equipment capable of measuring gas emissions.

■ Predicting the size and timing of a volcanic eruption remains an inexact science, and volcanologists must use care to minimize unnecessary panic.

BOX 5–2 EARTH SCIENCE AND PUBLIC POLICY

The Prediction Challenge

Volcanologists must walk a tightrope between the possibility of failing to warn of a disaster and causing needless alarm. Soon after the eruption of Mount St. Helens, USGS scientists grew concerned by signs that the Long Valley Caldera in eastern California might be preparing for a new eruption. Long Valley last erupted 760,000 years ago in an event that makes the Mount St. Helens, Mt. Pinatubo, and Tambora events look puny in comparison: 600 km^3 (144 mi^3) of material exploded out of the ground,[56] creating ashfall deposits that reached all the way to eastern Kansas and Nebraska and formed a caldera 32 km (19.9 mi) long (**Figure 1**).

(continued)

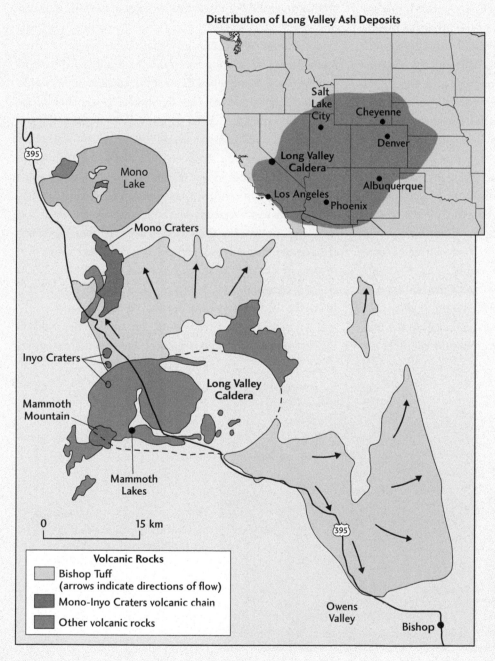

Distribution of Long Valley Ash Deposits

Volcanic Rocks
- Bishop Tuff (arrows indicate directions of flow)
- Mono-Inyo Craters volcanic chain
- Other volcanic rocks

Figure 1 The Long Valley Caldera and its resurgent dome are just east of an ancient volcanic arc that forms the rugged Sierra Nevada in California. The caldera formed after a cataclysmic eruption 760,000 years ago from which ash was carried as far east as what is now Nebraska.

(continued)

Because of a series of strong earthquakes—measuring 6 on the moment magnitude scale—in 1980 at Long Valley, scientists were concerned that the caldera might be reawakening. Over the next 2 years, there were multiple earthquake swarms, the opening of several steam vents, and surveying measurements that showed ground movements consistent with magma accumulating under the caldera.[57] Faced with this evidence in late spring of 1982, USGS scientists issued a low-level alert about the potential for future activity. A media feeding frenzy ensued, marked by misinformation and overhyping of the potential threat.

Residents of the resort town of Mammoth Lakes, which lies within the caldera, were livid because of the resulting impact on their tourism-based economy. In a short time, businesses closed and housing prices fell by about 40 percent. Several people referred pejoratively to the U.S. Geological Survey as the United States Guessing Society when Survey scientists said that they could not predict with certainty when or even if an eruption was going to occur.[57] Long Valley has in fact not yet erupted, and may not in our lifetimes, and the economy of the town eventually rebounded as people gradually calmed down. Still, the false alarm proved embarrassing to the USGS, which learned valuable lessons about how to communicate with the public.

If the Long Valley Caldera story provides an example of the perils of too much vigilance, the eruption of Nevado del Ruiz in Colombia in 1985 shows the danger of not taking hazard reports seriously enough. Nevado del Ruiz is a stratovolcano capped by glaciers. In 1984, it began showing signs of activity, including earthquakes, steam explosions, and sulfur emissions. Deep canyons eroded in its flanks revealed dozens of recent lahar deposits that led geologists to warn of the possibility of volcanic mudflows. Colombian volcanologist Marta Calvache prepared a hazard map that clearly showed the village of Armero lying in the path of a lahar if the volcano were to erupt and melt its cap of snow and ice.[58] The map was available weeks before the eruption, and a group of Italian scientists brought in to assist in threat evaluation made several simple recommendations. One was to develop an observation post that could determine when an eruption was starting and radio the information to local communities. Another was to identify local high points where people could escape from the lahars and to launch a public information campaign to let people know how to get to the high points and to foster awareness of these locations.[59] Regrettably, local officials greeted the news with skepticism, noting that the volcano was 65 km (40 mi) away and hadn't erupted in over a century. In addition, the Italian scientists were unable to say exactly when an eruption might occur and how large it might be. On the night of November 13, 1984, Nevado del Ruiz erupted, generating a lahar that swept over Armero at 30 km (18.6 mi) per hour, entombing 22,000 sleeping people in 3 m (9.8 ft) of volcanic mud (**Figure 2**).[60] It was the worst natural disaster in Colombian history.

A lesson learned from both the Long Valley caldera incident and the Nevado del Ruiz tragedy is that scientists

Figure 2 Casualty of the Nevado del Ruiz lahar. By melting the volcano's ice cap, the small volcanic eruption of Nevado del Ruiz in 1985 resulted in a flow of mud that moved rapidly down the volcano's slope to the town of Armero, Colombia, killing nearly 25,000 people. *(Jacques Langevin/ Sygma/Corbis)*

alone cannot prevent disaster. Even the best hazard maps and most careful monitoring are useless if the public is unaware of them and if officials do not use them to restrict settlement or to alert the public to imminent threats. Similarly, unless the message is strictly controlled, false alarms can lead to unnecessary panic and economic disruption. To avoid these problems, scientists must work in close coordination with social scientists, emergency management agencies, and responsible members of the media to ensure the safety of the public.

Word choice matters, particularly when human life and health are at risk. Authorities responsible for issuing warnings must take care to be as accurate and objective as possible. To that end, for hazards such as volcanic eruptions and earthquakes, scientists use different terms to describe the probability of an event. A **forecast** is an imprecise statement about the time, place, and sometimes the size of impending activity. An example of this is the forecast made by Crandell and Mullineaux that Mount St. Helens was likely to erupt in the not-too-distant future. A more precise statement that includes not only the place of an eruption but also the timing is a **prediction.** Prediction of the Mount Pinatubo event led to a large-scale evacuation that saved thousands of lives.

In 1989, the Long Valley Caldera became active once again, but this time the public was not alerted. Since the previous eruption scare, scientists had established a network of instruments in the area to monitor seismicity and ground swelling. Based on data gathered from monitoring the caldera, volcanologists concluded that an eruption is not likely to happen before the caldera rises at least another 0.5 to 1 m (1.6 to 3.3 ft). They also concluded that the volcano will probably give very strong warnings—such as large earthquakes and a substantial tilt of the ground surface—before an imminent eruption. Using this information, volcanologists will be able to issue eruption predictions in the future should the volcano awaken from its dormancy.

MINIMIZING THE RISKS FROM VOLCANIC ERUPTIONS

The energy released during major volcanic eruptions is too great to ever be easily controlled or prevented through engineering methods. However, maps can be made of volcanic deposits associated with active volcanoes such that hazardous areas are readily identifiable. In addition, because they often are preceded by precursory activity, some volcanic eruptions can be predicted to within days and hours. With appropriate monitoring, it may be possible to prevent most deaths during an eruption, although prediction does not completely alleviate destruction (Box 5-3).

In the Mount Pinatubo eruption, for example, timely prediction enabled many people to evacuate, but homes and villages were destroyed under the weight of ash, and in the end, 200,000 people were made homeless. Predictions of volcanic events also might lack important information about the direction of the eruption. In the case of Mount St. Helens, for example, while the eruption itself was forecast in advance, the lateral blast from the volcano's northern flank proved surprising to American volcanologists accustomed to vertical eruptions. In hindsight, the evacuation area could have been much larger to minimize loss of life.

The best way to minimize the risks associated with volcanoes is to prepare hazard maps for all potentially active volcanoes. Such maps have been made for about 10 percent of the world's active volcanoes. The United Nations and government agencies designated the 1990s as the International Decade for Natural Disaster Reduction, and volcanoes were one of the main concerns of their activities. Sixteen volcanoes with potentially dangerous activity and proximity to population centers were targeted for especially focused studies and monitoring, including Mount Rainier in Washington and Mount Etna on the Italian island of Sicily (Figure 5-21). Called "Decade Volcanoes," many now have volcanic hazard maps, seismic monitoring, and scientific equipment to warn of impending eruptions.

The Lahars of Mount Rainier

Crandell and Mullineaux first identified the hazards associated with Mount Rainier in the 1970s as they were assessing the threat of volcanism throughout the Cascades. Rainier is the largest of the Cascade volcanoes and looms over the cities of Seattle and Tacoma in

Figure 5–21 Sixteen volcanoes have been designated "Decade Volcanoes," those with volcanic hazards, recent geological activity, locations near populated areas, and accessibility for study. The name "Decade Volcanoes" refers to the International Decade for Natural Disaster Reduction, a United Nations–sponsored international effort to reduce loss of life, property destruction, and social and economic disruption that began in 1990. *(Source: From the Decade Volcano Project)*

Washington. Its glacier-capped summit feeds numerous rivers that flow across the Puget Lowland into Puget Sound, and careful mapping revealed that these river valleys had been repeatedly inundated with lahars during the last few thousand years (Figure 5-22). Some of these lahars were associated with volcanic eruptions,

but others were produced by landsliding in the absence of an eruption, as rocks weakened by the movement of volcanic gases and fluids crumbled and mixed with snowmelt. Mapping of the lahar deposits revealed that flows thicker than 30 m (98 ft) moved down the river valleys, probably at speeds of 70 to 80 km (43 to 50 mi) per hour. USGS scientists have determined that there is a 1 in 10 chance of a lahar traveling all the way to Puget Sound during a typical human lifespan.[62]

After a recession in the 1980s, the Puget Lowland had high growth rates related to the high-tech and aerospace industries. The spike in population was accompanied by high rates of development.[63] Between 1992 and 2000, more than 350 km² (135 mi²) of forest, grassland, and agricultural land were converted to low-density housing, shopping malls, schools, and other structures. Unfortunately, much of that development was built on top of lahar deposits, and 80,000 people now reside within the known hazard zone.[62]

The town of Orting is particularly at risk. It lies at the confluence of the Puyallup and Carbon rivers where the eastern edge of the Puget Lowland meets the foothills of the Cascades. Were a lahar to form, Orting residents would have only about 40 minutes to evacuate,[64] if they in fact knew of the danger. Given Rainier's ability to generate lahars in the absence of volcanic eruption, Survey scientists realized that catastrophe could strike without warning. For this reason, they installed

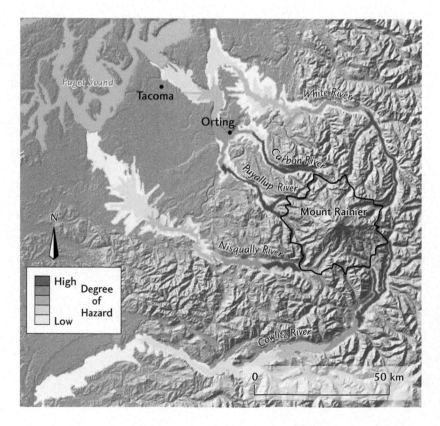

Figure 5–22 Based on geologic mapping, lahars have been identified as the greatest hazard at Mount Rainier. The largest known event, the Osceola Mudflow, occurred 5600 years ago and buried hundreds of square kilometers in mud and debris about 8 m (25 ft) deep. The town of Orting sits atop a younger lahar, the Electron Mudflow, which occurred 500 years ago. *(Source: Courtesy USGS, from Iverson et al. 1998. Geological Society of America Bulletin 110, 972–974.)*

BOX 5–3 EARTH SCIENCE AND PUBLIC POLICY

Volcano Databases and Volcanic Ash Advisory Centers

Dr. Chris Newhall, one of the scientists who served on the teams of volcanologists that forecast the deadly eruption of Mount St. Helens in 1980 and predicted the eruption of Mount Pinatubo in 1991, recently began to build a web–based system of global information for real–time monitoring of volcanoes. This global volcano database is called the World Organization of Volcano Observatories Database of Volcanic Unrest, or WOVOdat (www.wovo.org). The World Organization of Volcano Observatories (WOVO) is a sort of global volcano observatory. It coordinates communication and cooperation among nearly 80 volcano observatories and other institutions that monitor volcanoes across the globe and that warn authorities and the public when a volcano becomes restless.

In the 1990s, the United Nations' International Civil Aviation Organization (ICAO) and the World Meteorological Organization established a similar network consisting of nine regional Volcanic Ash Advisory Centers (VAAC) that coordinate and disseminate information on clouds of volcanic ash in the atmosphere.[41] Drawing primarily on information from satellites, these centers provide Volcanic Ash Advisories and use computer simulations to predict how ash clouds will move through the atmosphere. Clouds of ash are spread by wind and eventually dissipate. If ash can be identified on satellite images, however, it is assumed to pose a hazard to aircraft and should be avoided. When flying at night or in conditions of low visibility, pilots might have no indication of a volcanic ash hazard. If they are alerted in advance by reliable information from VAACs, pilots can maneuver around an ash cloud.

After the disruptions to air traffic that canceled flights and stranded tens of thousands of passengers during the 2010 eruption of the Eyjafjallajökull volcano in Iceland, the ICAO expanded the work of the VAACs to establish the International Volcanic Ash Task Force.[61] This task force is developing a roadmap that will include better means of detecting volcanic ash and of communicating with aviation controllers. A critical element of this plan is efficient communication among entities that monitor volcanoes. The WOVO developed by Dr. Newhall and his colleagues is central to this effort.

Having observed firsthand the power of scientific understanding and volcano surveillance to save lives, Dr. Newhall and other volcanologists recognized the need to develop and improve tools to forecast eruptions and for global sharing of information about volcanoes. Their work and that of other scientists is likely to have even greater consequence in the future, particularly in countries with rapidly growing populations and numerous active volcanoes, such as Indonesia.

---

five acoustic monitoring devices on the western flank of the volcano. These devices are sensitive to the kinds of ground vibrations created by huge masses of flowing debris, and transmit a warning signal to local emergency management agencies, which then can implement their disaster plans. These plans include sirens designed to alert the population to potential danger and clearly marked evacuation routes (Figure 5-23).[65]

Of course, the acoustic warning system is useful only if the local population knows what to do. In the case of Orting, emergency planners run volcano drills to help prepare the population. Schoolchildren practice loading onto buses in a calm and organized manner to be taken to higher-ground sites designated as evacuee

Figure 5–23 Volcano evacuation route near Mount Rainier, Washington. *(Zach Holmes/Alamy)*

gathering areas.[65] Unfortunately, the only major highway in and out of Orting lies directly within the path of lahars, so a speedy evacuation is critical. A faster path from the town's school complex to high ground could be afforded by trails and a pedestrian bridge over the Carbon River. The proposed "Bridge for Kids" is currently under consideration by the Washington Department of Transportation.[66]

Like that of Long Valley, Mount Rainier's hazard is real, but the timing of a specific event is uncertain. The hope is that careful mapping and zoning, extensive monitoring, and excellent communication systems will prevent the hazard from becoming a tragedy.

Stopping the Lava on Mount Etna

Twenty percent of the population of the Italian island of Sicily resides within the shadow of the stratovolcano Mount Etna, including the nearly 300,000 citizens of Catania, Sicily's second-largest city (Figure 5-24). Drawn to the volcanic soils that constitute the richest agricultural land on the island, these residents are under threat from a volcano that erupts nearly continuously.[67] Many eruptions issue from four different craters at Etna's summit, but others spew forth from fissures along its flanks. Written records of activity go all the way back to 1500 BCE, and in the last two decades, the mountain has

been quiet for a mere 4 to 5 years.[68] Given this, it should be no surprise that the Italian government and others have worked to produce volcanic hazard maps to guide development away from dangerous areas. Still, historical development patterns established before the creation of hazard maps have put people at risk.

Despite its status as a stratovolcano, Etna has erupted basaltic lava throughout the historic period of record, making its activity far less dangerous than that of Mount St. Helens or Mount Rainier, but the potential for more explosive volcanism still exists. At present, the volcanic threat is primarily from lava flows, which generally extend no further than 15 km from the volcano's summit, though a large flow in 1669 CE made it all the way to Catania, where it destroyed much of that city.[67] Beginning in December of 1991, lava nearly overran the village of Zafferana Etnea in an event that has become an interesting case study in volcanic hazard mitigation. Italian volcanologists first directed the construction of an earthen dam that contained the lava for a short time, but by April of 1992, the dam was overtopped and the lava continued its passage toward the village.[69] Three more dams were constructed farther down the mountain, but still the lava advanced.

At this point, volcanologists requested assistance from the U.S. Marine Corps. Using explosives, the volcanologists blew a hole into a **lava tube** on Etna's

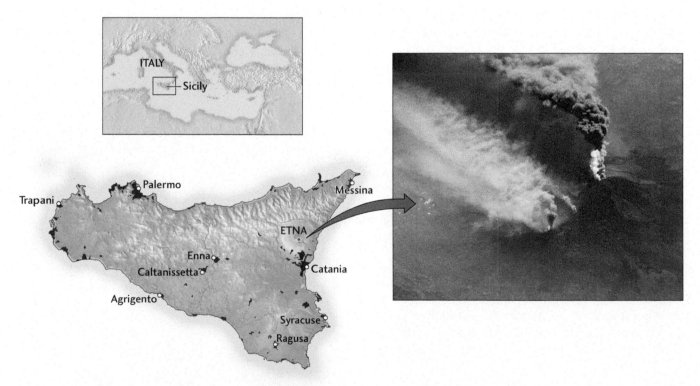

Figure 5–24 Mount Etna, on the east coast of Sicily, Italy, is the tallest (3329 m; 10,922 ft) active volcano in Europe and one of the world's most active volcanoes. *(Johnson Space Center/NASA)*

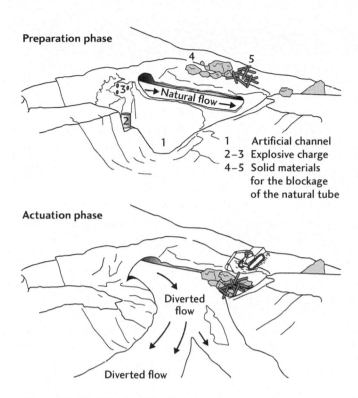

Preparation phase

Natural flow

1   Artificial channel
2–3 Explosive charge
4–5 Solid materials
    for the blockage
    of the natural tube

Actuation phase

Diverted
flow

Diverted flow

Figure 5–25 Sketch of the lava diversion carried out at Mount Etna in early 1992 to protect the village of Zafferana from being overrun with lava. *(Adapted from* Bulletin of the Global Volcanism Network, *vol. 17, no. 5)*

flanks. Such tubes form when flows develop a solid crust, which insulates the lava inside and allows it to flow farther than it might otherwise. Marine helicopters next dropped massive concrete blocks into the hole in an attempt to divert the flow.[70] While temporarily successful, the flow quickly reestablished itself. Only in late May, when the Italians dug a diversion channel and used explosives to breach the wall between that channel and the lava tube, were they finally successful in permanently diverting the flow (Figure 5-25). Lava had made it to within 850 m (0.53 mi) of Zafferana.[69] While the village avoided disaster in the early 1990s, its future is by no means assured.

■ Volcano monitoring and hazard mapping can greatly minimize loss of life from eruptions.

■ Lahars from Mount Rainier have repeatedly swept over the Puget Lowland of northwestern Washington. The USGS and local municipalities have developed hazard maps to identify likely inundation zones, have instituted evacuation drills, and have installed acoustic monitoring devices to warn of impending flows.

■ In addition to using hazard mapping, volcanic hazards can sometimes be addressed by direct mitigation. By diverting a lava flow into a constructed channel, the Sicilian town of Zafferana Etnea was saved from destruction in 1992.

CLOSING THOUGHTS

The picturesque beauty of volcanoes belies their horrific power. Over millennia, humans have been drawn to volcano's flanks to farm in the rich mineral soils produced by volcanic ash, to prospect for metal ores, and to take advantage of the streams flowing from their icy summits. In the process, we have exposed ourselves to grave danger. Throughout history, poison gases, lava flows, and pyroclastic debris have been responsible for tens of thousands of deaths as the turmoil unfolding within Earth's interior has bubbled to the surface.

Recent advances in the science of volcanology have bettered our odds of survival, as geologists have learned how to determine the likely magnitude of eruptions through careful mapping and how to monitor precursory activity such as earthquakes. Still, eruption prediction is not an exact science, and those of us who choose to live in volcano country must educate ourselves about the risks of doing so and develop contingency plans should the inevitable happen in our lifetimes.

SUMMARY

■ Earth has more than 1300 active volcanoes, most of them located along convergent (nearly 80 percent) and divergent (nearly 20 percent) plate boundaries.

■ The destructive potential of volcanoes is largely a function of magma viscosity and gas content, which together determine whether an eruption will be explosive or will produce fluid lava flows.

■ Many of the world's people live along the slopes of active volcanoes that erupt once to a few times every

few centuries, such as the residents of Tacoma, who live near the base of Mount Rainier. The recurrence time of eruptions from some volcanoes is longer than human memory, giving residents a false sense of security.

■ Geologists study past volcanic eruptions in order to understand volcanism and to prepare maps of different types of volcanic deposits. These hazard maps are used to alert officials and the public to areas of likely future volcanic activity.

KEY TERMS

magma chamber (p. 142)
volatiles (p. 142)
viscosity (p. 143)
bombs (p. 143)
pāhoehoe (p. 143)
a'a (p. 144)

pyroclastic debris (p. 144)
tephra (p. 144)
ashfall deposits (p. 144)
pyroclastic flows (p. 145)
shield volcanoes (p. 145)

stratovolcano (p. 146)
cinder or scoria cones (p. 146)
fissure eruptions (p. 148)
flood basalts (p. 148)
craters (p. 148)

calderas (p. 148)
lahars (p. 149)
harmonic tremor (p. 155)
forecast (p. 159)
prediction (p. 159)
lava tube (p. 162)

REVIEW QUESTIONS

1. What about Mount St. Helens made it clear that the volcano was reawakening after a century of slumber?

2. What techniques did geologists use to determine that the Mount St. Helens eruption was imminent?

3. How are volcanoes associated with active plate boundaries?

4. What role do volcanic gases play in eruptions?

5. Describe the interior architecture of a volcano.

6. How do silica content, temperature, and gas content affect lava viscosity?

7. Which kinds of magma are associated with the greatest loss of life, and why?

8. What types of volcanoes are formed by basaltic, intermediate (andesitic), and felsic magmas, respectively?

9. What is the difference between a crater and a caldera?

10. How can volcanic eruptions affect global climate?

11. What are the pros and cons of geoengineering projects that involve injecting sulfur dioxide into the atmosphere to combat global warming?

12. List different ways in which volcanic processes can damage property and cause death or injury.

13. What are the differences between active, dormant, and extinct volcanoes?

14. What types of phenomena commonly precede volcanic eruptions and can be used to forecast or predict an eruption?

15. What is the difference between a forecast and a prediction of a volcanic eruption?

16. Why did the USGS think that the Long Valley Caldera might be resuming activity in the early 1980s?

17. How can the risks associated with volcanic eruptions be minimized?

18. How can a map of a lahar be used to prevent loss of life?

THOUGHT QUESTIONS

1. Given the tectonic setting of the Galápagos Islands (see Figures 5-3 on page 142 and 2-3 on page 41), what style of volcanism would you expect there? What about in Alaska's Aleutian Islands?

2. Why is geological knowledge alone insufficient to deal with volcanic hazards?

3. The city of Quito, Ecuador, extends onto the flanks of the Andean stratovolcano Guagua Pichincha. What recommendations would you make to city

officials to minimize loss of life and property during future eruptions of this volcano?

4. If you were designing a volcano-alert system, what steps would you take to minimize the problem of false alarms?

5. Should volcanoes be surrounded by a zone in which development is prohibited? Why or why not?

6. Kīlauea volcano in Hawaii has such frequent activity that one always can find recent information about

recent eruptions, deformation, and other phenomena at the USGS website for the volcano: http://volcanoes.usgs.gov/hvo/activity/kilaueastatus.php.

Check this site and examine the many webcams, images, movies, maps, and updates about the volcano's activity. Write a short summary of the phenomena that have occurred during the previous month, paying particular attention to (1) the types and rates of deformation of the volcano; and (2) the types and amounts of eruptive materials.

EXERCISES

1. The rim of the crater of a volcano measures 3700 m in elevation whereas the land surrounding the volcano is at 1000 m. If the volcano has a radius of 4 km, what is the average slope angle of its sides, and given this slope angle, what was the likely silica content of the magma that erupted to form it?

2. Examine Figure 5-15b. How many times has Mount Shasta erupted in the past 4000 years? Given the value you determine, calculate the average time between eruptions. Carry out the same analysis for Newberry volcano.

SUGGESTED READINGS

Chester, D. *Volcanoes and Society.* London: Edward Arnold, 1993.

Decker, R., and B. Decker. *Volcanes,* 4th ed. New York: W. H. Freeman, 2006.

Foxworthy, B. L., and M. Hill. "Volcanic Eruptions of 1980 at Mount St. Helens: The First 100 Days." *U.S. Geological Survey Professional Paper* 1249 (1982).

Kunzig, R. "A Sunshade for Planet Earth." *Scientific American* 299 (2008): 46–55.

McPhee, J. "Cooling the Lava," in *The Control of Nature,* pp. 95–182. New York: Farrar, Straus and Giroux, 1982.

Monastersky, R. "Perils of Prediction." *Science News* 139 (1991): 376–379.

Schick, R. *The Little Book of Earthquakes and Volcanoes.* New York: Copernicus Books, 2002.

Sigurdsson, H., ed. *Encyclopedia of Volcanoes.* New York: Academic Press, 1999.

Thompson, D. *Volcano Cowboys: The Rocky Evolution of a Dangerous Science.* New York: St. Martin's Press, 2000.

Wright, T. L., and T. C. Pierson. "Living with Volcanoes: The U.S. Geological Survey's Volcano Hazards Program." *U.S. Geological Survey Circular* 1073 (1992).

Zeilinga de Boer, J., and D. T. Sanders. *Volcanoes in Human History: The Far-Reaching Effects of Major Eruptions.* Princeton: Princeton University Press, 2002.

INTRODUCTION

Modern *Homo sapiens* evolved about 100,000 to 200,000 years ago, but only the last 11,500 years, approximately, are considered the period of modern human history.[1] Called the *Holocene,* this epoch of time is a warm, interglacial stage that made the Agricultural Revolution possible. Human beings domesticated plants and animals and developed settled agricultural societies during this time. Soon after, cities were born and many of the discoveries and achievements of civilization—such as irrigation, metallurgy, and writing and record keeping—arose and gradually spread worldwide. Significant advances were made in science, technology, and medicine in recent centuries.

The past half millennium, during which a scientific revolution has occurred, is marked by rapid growth in the human population and its technological abilities. As a result, many Earth systems have been altered, some of them irreversibly. Scientific and technological advancements also have enabled us to determine that Earth experienced numerous environmental changes in the past, well before human evolution. How do we distinguish between ongoing, long-term geologic processes and those that are the result of human activity, such as current global warming? Furthermore, can natural climate changes occur abruptly, perhaps disrupting our civilization? These questions can be answered by studying Earth's history across the vast span of geologic time.

Scientists have devised ingenious ways to document climate and other changes in Earth systems over time. Detailed records of Earth's **paleoclimate**—the history of past climate changes on Earth—are gleaned, for example, from layers of ice in mountain glaciers around the world as well as from ice sheets on the continents of Antarctica and Greenland (Figure 6-1). Layers of glacial ice contain trapped bubbles of gas from ancient atmospheres, variable amounts of windblown dust that indicate wet and dry periods, fluctuations in the amounts of different atomic forms of oxygen that reveal information about past temperatures, and volcanic ash carried by wind from eruption sources.

Paleoclimate records also can be constructed from the hard parts of microscopic marine organisms buried within seafloor muds. Even tropical coral growth rings yield clues about past climates. It is the chemical composition of ice, seashells, and coral growth rings that enables scientists to extract a record of climate change from glaciers, sediment cores, and reefs (a topic explored in detail in Chapter 14), and it is the layering and datable material in these substances that enable scientists to match each record to an exact period of time. To make the most of such records, scientists scout the globe, carrying small drilling rigs up mountains

(a)

(b)

Figure 6–1 Glaciers form by the gradual burial of annual layers of snow above the approximate snow line and through the subsequent conversion of snow to ice from the weight of the overlying mass. (a) The approximate snow line is the boundary between snow at higher altitudes, where it does not melt, and the bare ground at lower elevations where it melts, such as here on the Tibetan Plateau. Over many years, snow that remains after summer melting is converted to glacial ice. (b) Each deeper layer in a glacier is older than the one above it, and the entire stack is a record of past climate. *(a: Lan Yin/ Getty Images; b: Dr. Lonnie G. Thompson)*

to glaciers in the Himalayas, drilling into the seafloor from ships, and diving into tropical coral reefs.

The records of past climates from these and other materials are provocative. They show that our planet's global climate is anything but static. Moreover, for the past few million years the most prominent ups and downs between warm and cool conditions are periodic, almost as if Earth has a climatic pacemaker. Many records of paleoclimate indicate that the past one to two centuries have been unusually warm, and the cause of this warming might be human activity. It is the task of scientists, using cores of ice and seafloor mud, growth bands of coral, layers of rock, and any other record available, to decipher the timing, causes, and effects of Earth's changing climate, both natural and anthropogenic. We are living in one part of a climate cycle, and knowing about the history and causes of such cycles might enable us to predict future climate change.

SCALES OF TIME AND EARTH SYSTEM CYCLES

Scientists describe the universe as four-dimensional, existing in space-time. Earth systems thus operate in three-dimensional space, at distances from local to global, *over time*. Time is the fourth dimension in our universe. Unlike the three spatial dimensions (length, width, and depth), the fourth dimension of time has an irreversible direction. An object in space can move in any direction from a given point, but an object existing at some point in time can only move forward. You cannot retrace your steps in history, as in a movie playing in reverse. Geoscientists refer to this unidirectional, linear nature of time as "time's arrow."[2] Layers of glacial ice, for example, are sorted by the time at which they were deposited initially as snowfall onto the ice mass. The oldest layers are at the bottom, and the youngest at the top, near the surface of the glacier. The layers of ice provide direction in time and enable researchers to reconstruct the history of glacial growth and climate change as well as variations in atmospheric composition.

Time and Earth System Cycles

While time goes steadily forward, many Earth processes are cyclical, meaning that they recur over and over again at well-defined intervals of various lengths. The daily cycle of heating and cooling on Earth, for example, results from the planet's rotation on its axis. Over the course of a day, a location on Earth's surface warms as it faces the Sun, and cools at night, when the planet has spun about and faces the opposite direction. The intensity of daily heating and cooling depends on the season. As Earth revolves around the Sun each year,

different parts of its curved surface receive more or less radiation, resulting in the seasons. In the northern hemisphere, July is often the warmest month of the year and December is among the coldest, while in the southern hemisphere, July is wintertime and summer begins in December. The daily cycles of heating and cooling are embedded in this seasonal cycling.

As a result of time's arrow, it is possible to reconstruct the history of the Earth system. It might seem as if it would be impossible to determine the history of any given part of a cycle in which matter and energy are continuously recycled. Earth system cycles occur within the overall direction of time's irreversible arrow, however, and some components of those systems retain traces of their history. The rocks that line the walls of the Grand Canyon, for example, are evidence of the cycling of rocks over more than a billion years (Box 6-1).

By any Earthly dimension of space or time, the rock cycle (see Chapters 1 and 4) that continually creates, destroys, and recycles rock into different forms is gargantuan. Driven by plate tectonics, solar energy, and gravity, it carries rocks and sediments from Earth's surface deep into its mantle, and ultimately back to the surface again as uplifted rock or as magma. From dating rocks, we know that the journey to complete the rock cycle takes, on average, some 400 million years, one of the slowest of the major Earth system cycles. Some rocks complete the cycle more rapidly, and others get caught at a standstill, such as ancient rocks in the interiors of continents that have been weathering and eroding for billions of years.

■ Earth system cycles occur within the overall direction of time's irreversible arrow, and some components of those systems retain traces of their history. Sufficient evidence of rock cycling remains to measure past time for most of Earth's history.

■ Deep incision of the Colorado River has created the Grand Canyon, which reveals rocks as old as 1.75 billion years in the bottom of the gorge and as young as 225 million years old at the rim of the canyon, representing more than a billion years of rock cycling.

MEASURING TIME: RELATIVE AND ABSOLUTE AGE DATING

Just as layers of glacial ice contain clues about the history of climate change on Earth, rocks contain evidence of tectonic cycles, earthquakes, volcanic activity, and other events throughout Earth history. Even though rock is cycled continuously through Earth systems, enough evidence remains to provide a means of measuring past time for most of Earth's history.

BOX 6–1 GLOBAL AND ENVIRONMENTAL CHANGE

One Billion Years of Rock Cycling Revealed in the Grand Canyon

Occasionally, we get deep glimpses of the earth beneath us, as in the walls of canyons and gorges. The rocks exposed in these slices contain clues to their origins and histories. One of the most fantastic views on Earth is found in the gorge carved by the Colorado River as it passes through the state of Arizona. Known as the Grand Canyon, the gorge exposes rocks to a depth of as much as 1.6 km (1 mi) over a distance of 450 km (280 mi) (**Figure 1**). From radiometric dating (discussed in this chapter), scientists have determined that the lowest rocks exposed in the bottom of the gorge are over 1.5 billion years old; at the top of the main canyon rim, the youngest rocks are 225 million years old. This deep glimpse into Earth's crust reveals more than a billion years of rock cycling.[3]

Starting at the base, with the oldest rocks, and working our way upward to the rim, we begin with a mixture of igneous and metamorphic rocks called the *Vishnu formation*. Clues in the rocks indicate that they originally formed as lava flows, volcanic ash deposits, and sediments in a sedimentary basin at Earth's surface. As a consequence of the convergence of tectonic plates, these layers were deeply buried by overlying rocks, and as they were carried to greater depths they were metamorphosed by heat and pressure, as well as intruded by magma that cooled to form igneous rocks. Still later, the entire mass was raised upward again by mountain–building processes that also deformed the rocks. Once at Earth's surface, the rocks were exposed to erosional processes that planed off a fairly smooth surface. The history of rock cycling recorded so far, just in this oldest formation exposed in the Grand Canyon, is one of sedimentation, burial, metamorphism, igneous intrusion, deformation, uplift, and erosion.

Sometime between about 1 billion and 600 million years ago, this eroded surface was submerged to become the floor of a shallow sea, and marine muds and sands were laid down upon the Vishnu formation. These sediments, too, were buried and hardened to become sedimentary rocks, but were not buried as deeply as those of the Vishnu formation had been a few hundred million years earlier. The new sedimentary rocks also were tilted and deformed, for they are no longer in their original horizontal position. Similarly to the Vishnu formation, at the top of these rocks is a prominent erosional surface that in places left small knobs of rock standing above the rest, suggesting that the rocky knobs once were islands. As a group, these layers are called the *Grand Canyon series,* and their existence adds the following events to the sequence of rock cycling: sedimentation, moderate burial, moderate deformation, uplift, and erosion.

Our examination of rocks from bottom up in the Grand Canyon brings us to more than several hundred meters of horizontal sedimentary rocks that were deposited in both marine and terrestrial environments between 600 and 225 million years ago. Yet the evidence of rock cycling is not merely one of 375 million years of continuous deposition. Instead, at least four **unconformities**—surfaces that represent discontinuities in the rock record—reveal that deposition was interrupted by prolonged periods of time in which perhaps no sediments were deposited, or erosion occurred, or some combination of the two took place. Each sequence of layered sedimentary rocks between the unconformities indicates that sediments were deposited, then buried and hardened, then uplifted again and eroded, but never was uplift associated with substantial deformation because all the rocks are still in their original horizontal positions.

Since about 225 million years ago, no sediments have been deposited above the youngest rocks exposed at the Grand Canyon's rim. The surface of erosion at the rim might become an unconformity sometime in the future, perhaps when the rocks again are carried down far enough that seas can advance across the landscape and deposit marine sediments above the youngest rocks. Were this to occur, the gorge would fill with sediment, but its existence could be inferred by a future geologist from the geometry of the sedimentary layers within it. Instead of advancing seas filling the Grand Canyon with sediment, perhaps the entire stack of rocks will be buried so deeply that it will become complexly deformed and metamorphosed. Regardless of what the future has in store, the inevitable cycling of rocks will continue, just as it always has.

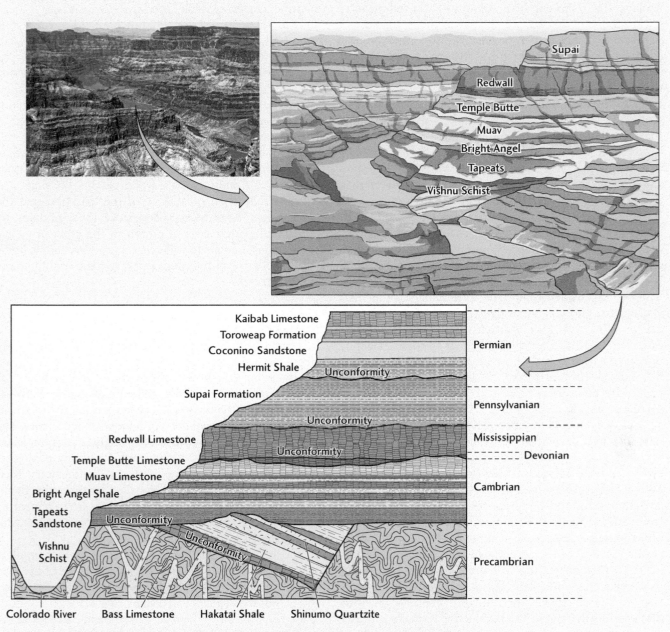

Figure 1 Rock Cycling Revealed in the Grand Canyon. The history of rock cycling is recorded by different formations exposed in the walls of the Grand Canyon. Wavy lines represent surfaces of unconformity, where a gap in the rock record exists either because no sediments were deposited, or because some were eroded away, or by a combination of the two processes. *(Inset photo: John Henshall/Alamy)*

One way that we employ the geologic record to reconstruct past events is through using fossils to learn about the evolution of life on Earth. A **fossil** is the naturally preserved remains of a plant or an animal. As layers of sediment accumulate to form sedimentary rocks, the remains of organisms or their imprints in soft sediment are sometimes incorporated and become fossils. Life forms on Earth have changed with time as a result of biological evolution, so the types of fossils in rock layers reveal the **relative ages** of the rocks. If you were to find a dinosaur leg bone jutting from a sandstone bed while hiking, for example, you would know that the rock must be quite old, because dinosaurs have been extinct for a long time.

The decay of certain unstable variants of chemical elements provides another common means of rock dating. Each **isotope** of an element has the same atomic number (that is, each isotope has the same number of protons) but a different number of neutrons. The element carbon, with an atomic number of 6, provides an example. Each carbon atom has 6 protons, but the carbon-12 isotope has 6 neutrons, whereas the carbon-14 isotope has 8 neutrons. The carbon-13 isotope has 7 neutrons. Those isotopes that spontaneously decay are called **radioactive.** Each radioactive isotope has a specific rate of decay, and the amount of that isotope remaining in a given rock reveals how long the process of decay has ensued. In contrast to yielding a relative age, this procedure provides a numerical value, or **absolute age,** for the rock.

As scientists developed new ways to date rocks over the past few centuries, their estimates of Earth's age increased by many orders of magnitude. The age estimates were refined as new scientific understanding and technologies enabled greater accuracy and precision in dating rocks and geologic events. In the 17th century, for example, Earth was thought to be about 6000 years old. Today, it is known to be about 4.54 billion years old, about 700,000 times the age estimate of the mid-1600s.

The first age-dating technique to gain widespread use relied on relative age dating of sedimentary rocks that contained fossils. In the 20th century, scientists discovered ways to use the radioactive decay of unstable isotopes in rocks, minerals, and organic matter to determine the numerical (absolute) ages of Earth's materials. Today, relative and absolute age dating usually are combined, so that different events first are placed in sequence relative to one another and then are bracketed in time by those events for which absolute dates are available.

Fossils, Evolution, and Relative Geologic Time

Among early scholars who observed Earth's rocks for clues to its age, Leonardo da Vinci (1452–1519), the Italian artist, scientist, and engineer, studied layers of sand, gravel, and fossil debris in the mountainsides of the Apennines, in central Italy.[4] Leonardo observed that the fossilized shells and other skeletal debris resembled marine creatures such as clams, crabs, and oysters, but his contemporaries did not agree. Since the time of Greek philosopher Aristotle (384–322 BCE), fossils were interpreted variously as shapes that developed in the rock itself and just happened to resemble living creatures; as discarded versions of a Creator's mistakes, which never became living creatures; and as the remains of victims of the biblical flood.

In Leonardo's time, Europeans thought that Earth's surface was stable and unchanging, having been formed during the biblical Creation and modified only once thereafter, during Noah's flood. Leonardo suggested instead that a sea once covered the land and that the land might have been raised since then, perhaps by earthquakes that were common in the area. In all his deductions, Leonardo was correct, but few people knew of his ideas until his journals were made public many years later.

Steno's Laws of Rock Layering Niels Steensen—whose name is often Latinized as Nicolaus Steno—was born almost 200 years after Leonardo and moved to Italy from his native Denmark. Independently, Steno came to the same conclusion about the origin of fossils as had Leonardo. Steno did something else as well—he recognized in the layers of rocks, or **strata,** of Tuscany a record of time, a sequence in which lies recorded a history of events. The word *strata* (singular *stratum,* Latin for "bedcovering," or "blanket") is used by geologists to describe the sheetlike nature of sedimentary rock layers. The study of strata is called **stratigraphy.** Steno inferred relative time from the spatial arrangement of rocks. This is the essence of geology, a science that did not yet exist as such during his life.

Using geometry, Steno was the first to formulate laws about the formation of rock layers. One of his laws is known as the **principle of superposition.** As explained by Steno, the sequence of events that leads to layers of rock begins with the settling of sediment particles at the bottom of a column of water. It is easy to simulate the formation of layers by mixing mud, sand, and gravel with water in a large jar, then shaking it and setting it to rest on a table. Within seconds, all the gravel and sand settle to the bottom, and over the next few hours all the clay settles as well. From coarsest to finest, the layers are ordered from bottom to top, indicating that the heavier a particle, the more rapidly it sinks.

According to Steno, subsequent deposits form one layer after another, compressing into a series of strata, as shown in Figure 6-2. The column of water might be within a river, sea, or lake. Such sediments also can be

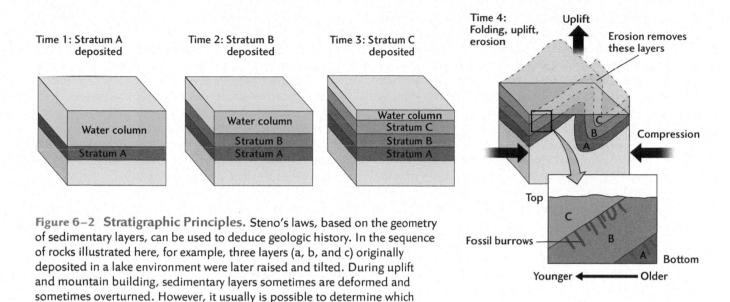

Figure 6–2 Stratigraphic Principles. Steno's laws, based on the geometry of sedimentary layers, can be used to deduce geologic history. In the sequence of rocks illustrated here, for example, three layers (a, b, and c) originally deposited in a lake environment were later raised and tilted. During uplift and mountain building, sedimentary layers sometimes are deformed and sometimes overturned. However, it usually is possible to determine which layer is the original stratigraphic top from clues such as the shapes of burrows made by worms in lake–bottom muds.

carried and deposited by wind and ice. Steno realized that in a sequence of strata, a given stratum is younger than those below it and older than those above it; hence the term "superposition."

A second law, related to the first, is the **principle of original horizontality:** Strata are deposited originally in uniform, horizontal sheets. Because plate tectonics and mountain-building processes deform rocks, strata often become tilted, as shown in Figure 6-2. Even if strata are completely overturned, the original direction of deposition often can be determined from fossil evidence and other clues. As long as the direction of deposition is known, the relative age of each layer can be determined using Steno's laws.

Fossils and the Geologic Column As powerful as Steno's principles were, they did not enable scientists, miners, and engineers to correlate rock layers from different locations. At a given site, one knows the relative age of each layer from the principle of superposition but cannot tell the relative ages of layers among different sites. For example, during the 1700s and especially after the invention of the steam engine, mining for coal and minerals increased throughout Europe, and miners began to recognize general sequences of rocks locally. In order to communicate with one another about the strata, they came up with names that referred to properties of the strata, such as their smell ("Stinking Vein") or thickness ("Ten-Inch Vein"). In coal-rich Somerset, England, coal miners knew that a Ten-Inch Vein was usually found below a "Shelly Vein." They had no way of knowing,

however, if a Stinking Vein in southeast Somerset was the same as a Stinking Vein in northwest Somerset.

The solution to the problem of correlating rocks located at different sites lay in the fossils within the rocks. An English surveyor named William Smith (1769–1839, nicknamed "Strata" Smith) studied the fossils he found while surveying canals in the area of Somerset. He realized that each layer of rock contained a particular and characteristic group of fossils.[5] When friends showed him their fossil collections, he could identify—to their amazement—exactly the stratum from which the fossils came.

Smith found that some fossils exist only in certain older strata, and never in younger ones, as if the life forms had disappeared. New forms appeared in younger strata, then they, too, disappeared from the record. Each fossil assemblage proved unique in time. Smith used this knowledge to formulate the **principle of faunal and floral succession,** which states that assemblages of plant and animal species succeed each other in time. Smith and his contemporaries in the early 1800s had no idea why such changes occurred; the theory and understanding of evolution and extinctions were not yet fully developed. Nevertheless, succession proved to be a powerful tool for correlating and mapping rock strata.

Smith and geologists after him applied the principles of superposition and faunal and floral succession to map fossil-containing rocks worldwide. For convenience in mapping, rocks were grouped into units called **formations,** which are layers that have similar properties and appearance and the same stratigraphic age. In essence, each represents a specific environment, or condition of origin. When all the known strata were combined in their relative chronological order, geologists had created a **geologic column,** a composite stratigraphic section for the world. With time, the geologic column grew longer and more complex, leading many to conclude during the 19th century that Earth must surely be more than 6000 years old.

Evolution: A Biological Timekeeper While piecing together the geologic time scale, geologists came to realize that Earth must be quite old in order for so many different life-forms to appear, flourish, and vanish. One such scientist was the British geologist Charles Lyell (1797–1875), a mentor of the famed naturalist Charles Darwin (1809–1882). When Darwin left for a 5-year round-the-world expedition on the ship HMS *Beagle* in 1831, he took along a copy of Lyell's textbook, *Principles of Geology.* During his voyage, Darwin pondered the nature of the different species he encountered and wondered if they might have changed from one form to another over time.

While visiting the Galápagos Islands off Ecuador, for example, Darwin recognized that different species of finches live on each island. After returning home and studying his notes and samples, he developed the idea that these birds had adapted themselves to their local environment by some process (or processes) that resulted in an evolutionary change from one finch species to another over time. He likened the process to that used by breeders when preferentially breeding animals with certain traits, such as cows without horns but with high milk yield. He reasoned that nature might act as a breeder by a process of **natural selection,** a process that winnows populations of those members least fit for a given environment, resulting in greater reproductive success of those best adapted for survival.

Random variation of traits within a species is quite large, Darwin realized, so he suspected that with time, natural processes of selection could result in the evolution of new species as well as the extinction of old ones. Darwin refined his hypothesis and in 1859 published his ideas in his classic book, *The Origin of Species.*[6] In this work, he recognized that both the isolation of populations and the sexual selection of individuals within those populations are key to passing on traits that allow populations to survive. From 20th-century studies of genetics, we now know that genes pass traits from one generation to the next.[7] DNA, the abbreviation for *deoxyribonucleic acid,* is a molecule that contains genetic instructions for organisms, and **genes** are segments of DNA that carry this information. Darwin's ideas preceded modern genetic discoveries, but nevertheless he realized that the processes of natural selection he envisioned would require immense spans of time to

occur. Just how immense, though, Darwin was unable to determine.

Darwin's ideas explain how Smith's principle of faunal and floral succession works. As new species evolve, they sometimes migrate from one locale to another. If something causes great environmental stress—perhaps rapid vegetation change or a catastrophic event like a meteorite impact—some species are terminated. Evolution, migration, and extinction of species are the mechanisms that lead to distinct fossil assemblages in specific strata. These distinct groupings of fossils can then be correlated from one site to another by geologists who reconstruct geologic history. Even though it was to be another half century before radioactivity was discovered and applied to age dating, Darwin's ideas about evolution made it possible to devise a relative geologic time scale from the rock record of life on Earth.

The Geologic Time Scale As the geologic column developed, geologists gave names to the specific intervals during which rocks were deposited. These relative time units were used to construct a **geologic time scale** based entirely on stratigraphy and fossils, with no notion as yet of absolute ages. Much as the year is divided into hierarchical units of months and days, minutes and seconds, geologists divided the geologic past into hierarchical units, called eons, eras, periods, and epochs (Figure 6-3).

An **eon** is the largest unit of geologic time, and only three eons are recognized. The oldest is the Archean (from the Greek for "ancient"). Sedimentary rocks of the Archean contain sparse microscopic fossils of primitive unicellular organisms. The next eon is the Proterozoic (from the Greek for "early life"). Sedimentary rocks of the Proterozoic eon also contain unicellular organisms, but these are more plentiful and complex than are those of the Archean eon. The latest eon, beginning with multicellular organisms large enough to be seen without a microscope and stretching to the present day, is the Phanerozoic (from the Greek for "visible life"). From absolute age-dating techniques developed in the last century, it is now known that each of these eons represents hundreds of millions to billions of years, as shown on the vertical scale at the left in Figure 6-3.

The Phanerozoic is the only eon divided into eras with commonly used names. **Eras** represent time spans from many tens to hundreds of millions of years. Early geologists identified eras in the Phanerozoic eon by major boundaries in the rock and fossil record, boundaries at which as many as 70 to 80 percent of all previously existing species disappeared, marking times of major **mass extinctions.** After these extinctions, many new species appear in the geologic record. Eras are divided into **periods,** which in turn are divided into smaller time units, **epochs,** which sometimes are as short as 10,000 years. As with eras, the boundaries of periods and epochs correspond with changes in marine and terrestrial fossils, but the changes are not as prominent as are those that mark the boundaries of eras.

The Phanerozoic eon is divided into three eras based on its fossil record: Paleozoic ("old life"), Mesozoic ("middle life"), and Cenozoic ("recent life"). During the Paleozoic era, fishes evolved, amphibians took to the land, and reptiles wandered early fern forests. This era ended fairly abruptly, and about 70 percent of these species disappeared, to be replaced on land by dinosaurs, conifer trees, and flowering plants during the Mesozoic era. The Mesozoic also ended fairly abruptly, when dinosaurs and many other species became extinct, making way for mammals, including humans, to become prominent in the Cenozoic era. This mass extinction most likely was caused by the impact of a large meteorite, which led to darkened skies, a toxic atmosphere, and a prolonged cold period.

The Cenozoic era is divided into two periods, the Paleogene (65.5 million to 23 million years ago) and the Neogene (23 million years ago to the present). The Neogene period is the time in which *Homo,* the genus (a group of closely related species) to which humans belong, began to spread across Africa, Asia, Australia, and Europe. It is also a time of extended, gradual global cooling that culminated, within the past few million years, in Earth's fifth major ice age. During this time, ice sheets repeatedly have advanced and retreated across North America and Eurasia, and smaller glacial masses have expanded and shrunk at high altitudes and latitudes. For these reasons, the two most recent epochs of the Neogene period, the Pleistocene and Holocene, sometimes are referred to as the "Age of Ice" and the "Age of Humans," respectively.

Radioactivity and Absolute Geologic Time

At the end of the 19th century, scientists were divided in the controversy over Earth's age. Basing their reasoning on theoretical models of the rate at which Earth might have cooled from a molten state with time, some prominent physicists and chemists argued for a young Earth of only 20 million to 100 million years. At the same time, geologists and those who supported ideas about evolution argued for an old Earth, perhaps billions of years old. What was needed to end this dispute was a means of measuring the absolute, or numerical, age of Earth.

Radioactivity: Nature's Atomic Clock With no notion about how their work would lead to a reliable method for measuring the absolute ages of Earth's strata, physicists and chemists in the late 1800s in France, Germany, and England were making numerous discoveries about a strange new energy source. In 1895, the German physicist Wilhelm Röntgen (1845–1923) called the mysterious energy "X rays."[8] French physicist Henri Becquerel (1852–1908) discovered in 1896 that such invisible rays emitted by uranium minerals left tracks on photographic plates, even if wrapped. Two scientists working together in France, Marie Skłodowska Curie (1867–1934) and her husband Pierre Curie (1859–1906), discovered that these mystery rays emanated not only from uranium but also from the elements thorium, polonium, and radium. They described the elements emitting such energy as *radioactive,* after radium.[9] Radioactivity is the propensity of certain atomic nuclei to naturally disintegrate and emit energy and/or subatomic particles. The Curies were the first to determine that it was the elements themselves that emitted the energy and particles, and for their discovery they shared the 1903 Nobel Prize for Physics with Becquerel.

Scientists busily set about identifying the properties of radioactive elements. In 1902, two physicists, Ernest Rutherford (1871–1937) and Frederick Soddy (1877–1956), isolated a radioactive gas from thorium compounds and studied changes in its activity. They called this gas *thorium X;* today, we call it *radon.* Rutherford and Soddy noted that after about 3.6 days, the activity of thorium X had decreased to half its original value, after 7.2 days to one-fourth its initial value, after 10.8 days to one-eighth its initial value, and so forth. This is because radon, or thorium X, was undergoing radioactive disintegration and diminishing over time. Rutherford and Soddy developed a hypothesis to explain radioactive decay:[10]

■ An unstable atom decays spontaneously, losing particles and energy from its nucleus, and becomes another element.

■ The number of radioactive atoms that decay in a given period of time is proportional to the number of radioactive atoms in the sample: Half will decay over a constant period of time.

This pair of statements forms the principles underlying **radiometric dating,** a quantitative method of dating based on the rate of radioactive decay of unstable atoms. Rutherford and Soddy recognized that some **parent elements,** like those in their thorium compound, decay at specific rates to produce **daughter products,** such as thorium X (radon), which in turn can decay to form other daughter products.

Radioactive Decay of Unstable Isotopes Since the work of Rutherford and Soddy, many unstable elements and chains of decay reactions have been identified and now are used as "nature's clocks" to determine the ages of

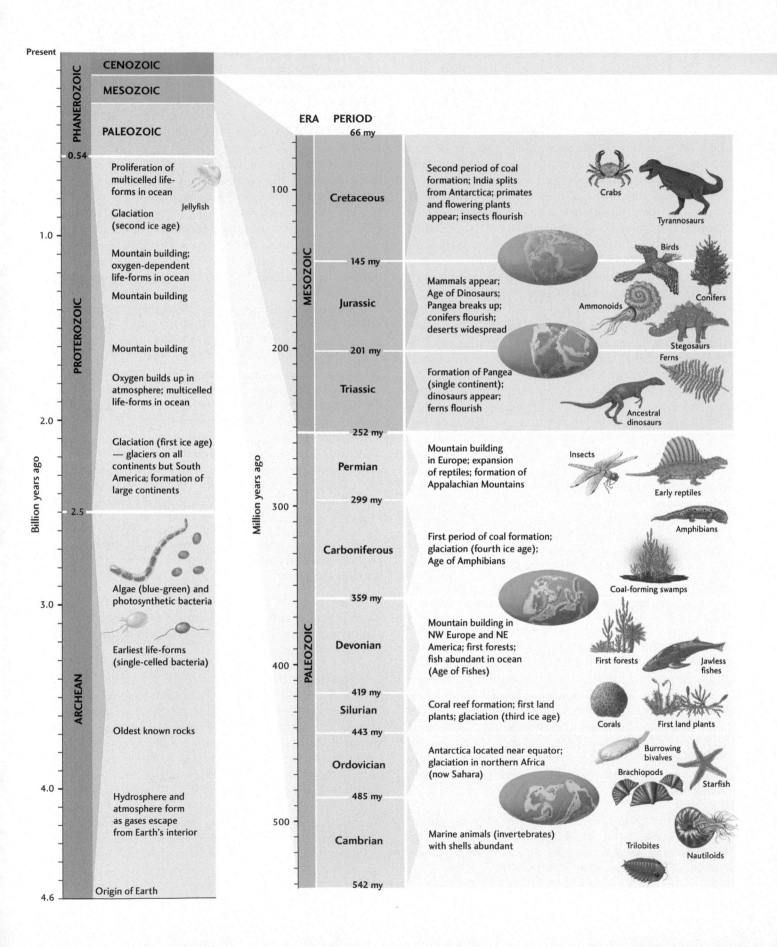

Present

PHANEROZOIC

CENOZOIC

MESOZOIC

PALEOZOIC

0.54

Proliferation of multicelled life-forms in ocean

Jellyfish

Glaciation (second ice age)

1.0

Mountain building; oxygen-dependent life-forms in ocean

Mountain building

PROTEROZOIC

Mountain building

Oxygen builds up in atmosphere; multicelled life-forms in ocean

2.0

Glaciation (first ice age) — glaciers on all continents but South America; formation of large continents

2.5

Algae (blue-green) and photosynthetic bacteria

3.0

ARCHEAN

Earliest life-forms (single-celled bacteria)

Oldest known rocks

4.0

Hydrosphere and atmosphere form as gases escape from Earth's interior

4.6

Origin of Earth

Billion years ago

ERA   PERIOD

66 my

MESOZOIC

Cretaceous

100

145 my

Jurassic

200

201 my

Triassic

252 my

Permian

299 my

Carboniferous

300

359 my

PALEOZOIC

Devonian

400

419 my

Silurian

443 my

Ordovician

485 my

500

Cambrian

542 my

Million years ago

Second period of coal formation; India splits from Antarctica; primates and flowering plants appear; insects flourish

Crabs

Tyrannosaurs

Mammals appear; Age of Dinosaurs; Pangea breaks up; conifers flourish; deserts widespread

Birds

Ammonoids

Conifers

Stegosaurs

Ferns

Formation of Pangea (single continent); dinosaurs appear; ferns flourish

Ancestral dinosaurs

Mountain building in Europe; expansion of reptiles; formation of Appalachian Mountains

Insects

Early reptiles

Amphibians

First period of coal formation; glaciation (fourth ice age); Age of Amphibians

Coal-forming swamps

Mountain building in NW Europe and NE America; first forests; fish abundant in ocean (Age of Fishes)

First forests

Jawless fishes

Coral reef formation; first land plants; glaciation (third ice age)

Corals

First land plants

Antarctica located near equator; glaciation in northern Africa (now Sahara)

Burrowing bivalves

Brachiopods

Starfish

Marine animals (invertebrates) with shells abundant

Trilobites

Nautiloids

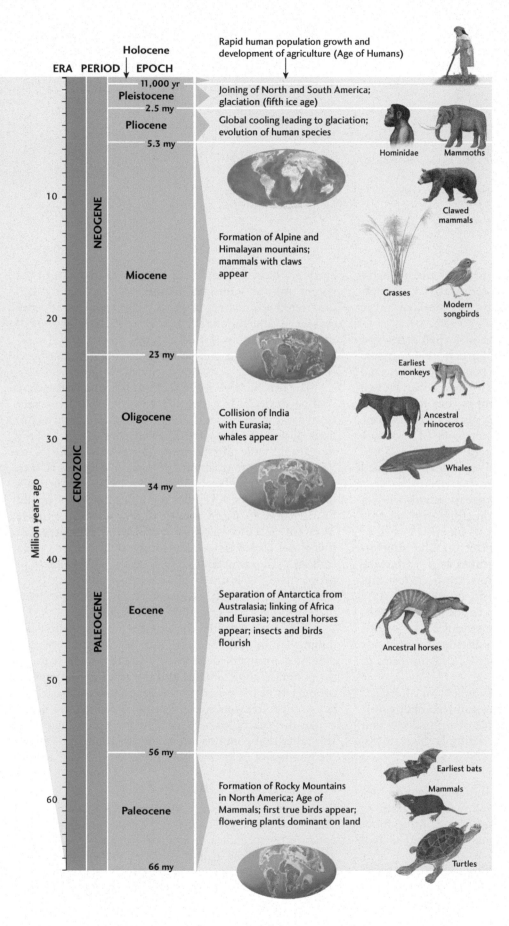

Figure 6-3 Relative Time Scale. The relative geologic time scale for Earth is constructed from stratigraphic and fossil records. Using techniques developed during the 20th century, Earth scientists have been able to use the radiometric dates of rocks to develop actual age estimates. Examples of common life forms are shown at the approximate times at which they appeared or became abundant.

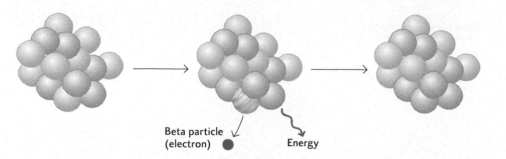

Beta particle
(electron) ●   Energy

^{14}C radioactive parent isotope
(6 protons + 8 neutrons)

Neutron becomes
a proton

^{14}N stable daughter isotope
(7 protons + 7 neutrons)

Figure 6–4 Isotope Decay. A carbon isotope, ^{14}C, decays to an isotope of nitrogen, ^{14}N, via emission of a β^- particle (electron).

fossils, minerals, and rocks. This form of chronometry is particularly valuable because radioactivity is unaffected by changes in temperature, pressure, or other environmental variables.

One form of isotope decay occurs when a neutron splits into a proton and a beta particle (an energetic electron, indicated as β^-), and the beta particle is ejected from the nucleus (Figure 6-4). This form of decay, known as beta decay, is common in isotopes that have excess neutrons, such as the ^{14}C isotope, also called *radiocarbon.* The daughter product, ^{14}N (nitrogen 14), has one less neutron than ^{14}C and one more proton. With 7 protons rather than 6, a new element—nitrogen—is formed from the ^{14}C isotope after it decays.

Another form of radioactive decay occurs when a proton transforms into a neutron in either of two ways: (1) by capturing an electron from its innermost shell of electrons, or (2) by emitting a positron (the antimatter equivalent of an electron, indicated as β^+) from the nucleus. In the nucleus, one proton has been lost and one neutron has been gained, but the atomic mass stays the same. An example of electron capture occurs when an isotope of potassium (^{40}K) transforms to an isotope of argon (^{40}Ar). An example of positron decay occurs when an unstable isotope of oxygen (^{15}O) transforms to a stable isotope of nitrogen (^{15}N).

A third type of decay occurs when an alpha (α) particle consisting of two protons and two neutrons is emitted from a nucleus. An example of alpha decay that is used commonly in radiometric dating is the transformation of ^{238}U to ^{206}Pb (lead).

Decay Rates and Half-Lives Recall that about a century ago, Rutherford and Soddy removed radon (thorium X) from thorium and observed the following: After about 3.6 days, the quantity of radon had decreased to half its original value (e.g., from 100 to 50 percent), after 7.2 days to one-fourth its initial value, after 10.8 days to one-eighth its initial value, and so forth. In other words,

every 3.6 days the isotope decayed such that half of the parent isotope vanished. Rutherford and Soddy had discovered a fundamental property of radioactive decay that is the same for all radioactive substances: *The rate of disintegration of an isotope is proportional to the number of parent atoms and occurs at a fixed rate specific to each isotope.*

The time it takes for half the atoms of a radioactive isotope to decay is called the **half-life.** Half-lives vary from as short as billionths of a second to as long as billions of years (Table 6-1). The shorter the half-life, the more rapidly the parent isotope decays. Decay of ^{14}C to ^{14}N has a half-life of about 5730 years; decay of ^{238}U to ^{206}Pb—which involves many decay steps—has a much longer half-life, of 4.47 billion years, roughly the age of Earth. Rutherford and Soddy's finding gave the half-life of radon as 3.6 days; more recent work indicates that it is closer to 3.8 days. The isotope they studied, ^{222}Ra (radon-222), is now known to pose a possible hazard to people living in houses built above specific rock types from which it seeps; exposure to its decay products increases one's risk of getting lung cancer.

With each half-life, the number of parent isotopes diminishes by half and the number of accumulating daughter products doubles. Eventually, very little of the parent isotope remains and the amount of daughter product is nearly equal to the original amount of parent isotope. In essence, the process of radioactive decay is like the ticking of a clock. The greater the age of a substance that contains radioactive isotopes, the greater the number of ticks that have occurred.

Radiometric Dating

In the general procedure used to date a sample of a mineral using a given isotope, geochronologists—scientists who specialize in geologic dating—measure the amount of a given parent isotope that remains in the mineral, and the amount of its daughter product that has accumulated in the mineral structure. Then the

TABLE 6–1 Useful Isotopes for Radiometric Dating

Radioactive parent isotope	Ultimate product of decay	Approximate half-life (years)	Useful range for dating (years)*
Rubidium-87	Strontium-87	49 billion	10 million–4.6 billion
Thorium-232	Lead-208	14.0 billion	10 million–4.6 billion
Uranium-238	Lead-206	4.5 billion	10 million–4.6 billion
Potassium-40	Argon-40	1.3 billion	2,000†–4.6 billion
Uranium-235	Lead-207	0.7 billion	10 million–4.6 billion
Carbon-14‡	Nitrogen-14	5730	~200–45,000

* In general, the effective upper limit of the dating range for a radioactive isotope is 5 to 10 times its half-life.

† Recent advances in measuring these isotopes have enabled scientists to date even younger minerals, including the ash that erupted from Mount Vesuvius and destroyed the town of Pompeii in 79 CE.

‡ This carbon isotope is useful for dating organic matter and precipitates, such as bones, wood, and shells.

duration of time that radioactive decay has been occurring is calculated from decay curves like those shown in Figure 6-5 (Box 6-2).

The usefulness of most unstable isotopes in radiometric dating is diminished after about 5 to 10 half-lives because the number of parent atoms that remain becomes too small to detect. The longer the half-life, the older the sample that can be dated (see Table 6-1). For example, with a half-life of about 5730 years, ^{14}C has a range of about 45,000 years for radiometric dating.

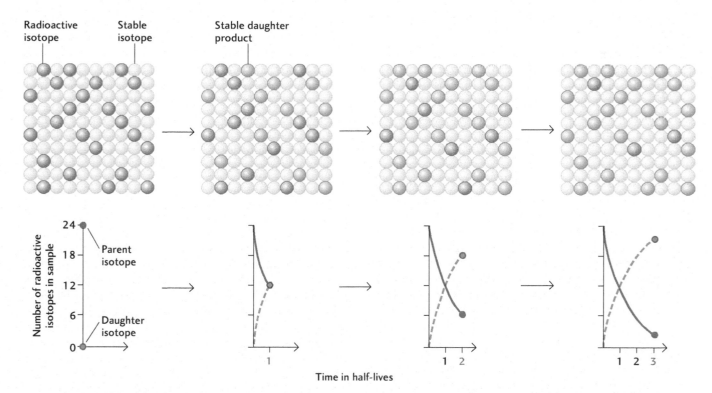

Figure 6–5 Radioactive Decay Curve. For a hypothetical group of 100 atoms that originally contained 24 radioactive atoms, after one half–life, 12 of those atoms have decayed and become stable daughter products. After two half–lives, one–half of the remaining 12, or 6, have decayed, and after three half–lives, half of the remaining 6 have decayed. The number of atoms that decays is proportional to the number of parent radioactive atoms present at any given time.

BOX 6–2 GEOLOGIST'S TOOLBOX

How Is the Rate of Radioactive Decay Used to Date Earth Materials?

Scientists use equations that represent radioactive decay in order to determine the age of something that contains a radioactive isotope. For example, if P represents the parent isotope, then P_0 is the amount of that isotope at the initial time, $t = 0$, just before decay begins, and P_t the amount at some later time, t, after decay has occurred. The half–life for an isotope is represented by $t_{1/2}$. The basic decay equation is as follows:

$$P_t = P_0 \left(\tfrac{1}{2}\right)^{t/t_{1/2}}$$

Here is an example to show how the radioactive "clock" is used. If 8000 atoms of a parent isotope exist at $t = 0$ and two half–lives have passed, then $t/t_{1/2}$ is equal to 2, and $P_t = 2000$ atoms:

$$2000 = (8000) \times \left(\tfrac{1}{2}\right)^2$$

In this example, the parent isotope has decayed from 8000 to 2000 atoms after two half–lives, and if it is a closed system (i.e., atoms cannot enter or escape), then the number of daughter isotopes would be 6000.

By measuring the number of parent and daughter isotopes in a sample, one can determine how much time has elapsed since the start of the radioactive clock, such as the time the cooling of magma takes to become crystalline rock below a certain *closure temperature.* At this temperature, the system is presumed to be closed, such that no more parent isotopes are added and no daughter isotopes can escape. The actual measurements typically are made with specialized scientific instruments, such as mass spectrometers, which can make highly precise measurements of the abundance of different isotopes.

This range is most useful for dating the evidence of events that occurred just before and during the most recent time of full glacial conditions on Earth (about 35,000 to 15,000 years ago) and the cultural and technological changes of *Homo sapiens.* In contrast, with a half-life of 4.5 billion years, the decay of uranium-238 to lead-206 is most useful for dating rocks that are tens of millions to billions of years old.

Radiometric Dating of Rocks and Sediments Radiometric dating of rocks occurred several decades before radiocarbon dating techniques were developed to date organic matter. Ernest Rutherford was the first to realize the link between his observations and the use of radioactivity to date igneous rocks. In 1905, Rutherford lectured at Yale University, where he used calculations of helium decay to show his audience that a mineral sample was 497 million years old. At the time, this value was substantially older than anyone's guess about the age of Earth.

But what does this date, and those for other igneous rocks, actually mean? What is dated is the time at which the rock cooled sufficiently both to prohibit parent isotopes from leaving the rock and to enable daughter products to be trapped in the rock's crystalline structure. As molten rock cools to form igneous rocks, isotopes become locked into the increasingly rigid crystal structures of different minerals. Once this "locking" has happened, the isotope system is set, in a sense, at the

initial time zero, or $t = 0$. From that time on, as parent isotopes decay and daughter isotopes accumulate, the radiometric clock ticks away.

Similarly, metamorphosed rocks yield ages that represent the time when the high temperatures associated with recrystallization reset the radioactive clock. Of course, one assumption in dating rocks in this manner is that neither parent isotopes nor daughter products have been able to leave the mineral structure. In other words, scientists must assume that the rock to be dated has been a closed system since the radioactive clock was set.

Even if these assumptions are valid, radiometric dating provides only part of the story of Earth's history. Processes and sequences of events of past ages are inferred from stratigraphy, not just from radiometric dating, but the two can be combined to provide a fuller picture of Earth's history. Unlike igneous rocks, sedimentary rocks do not cool from a molten state and trap radioactive decay products from possible escape, so radiometric dates cannot tell the time of origin of a sedimentary layer. If a mineral in a sedimentary rock were dated, it would yield the age of the older igneous or metamorphic rock from which it was eroded, not the age of the sedimentary layer itself. However, the ages of sedimentary rocks often can be bracketed, or put into a reasonable age range, by radiometric dating of igneous rocks and volcanic ashes that lie below and above sedimentary layers, or even

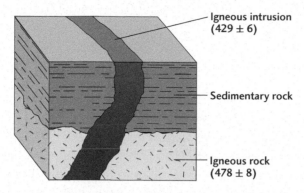

Igneous intrusion
(429 ± 6)

Sedimentary rock

Igneous rock
(478 ± 8)

Figure 6-6 Relative and Absolute Dating. This cross-section view of a wall of rock illustrates how geologists bracket the ages of sedimentary rocks. The oldest rock exposed, an igneous rock, is radiometrically dated as 478 ± 8 million years old. (The plus or minus value, with the smaller number after it, indicates the uncertainty in the age that results from possible errors in measuring the parent and daughter isotopes.) The igneous rock was eroded for some unknown time period before the deposition of sediment upon its surface. The youngest rock is an igneous intrusion that cuts across the sedimentary layer. Its radiometric age is 429 ± 6 million years old. The sedimentary layer must be older than this younger cross-cutting unit, but it also must be younger than the older, underlying igneous rock. We can bracket its age as between an oldest possible age of 478 ± 8 million years and a youngest possible age of 429 ± 6 million years.

cut across them as intrusions (Figure 6-6). In this way, stacks of interbedded volcanic ashes, river, and lake sediments—such as those in the East African Rift System—which also are rich in hominin fossils and artifacts, have been dated with potassium-argon isotopes in order to investigate the human evolution that spans the past 4 to 6 million years (Box 6-3).

Since the work of the Curies, Rutherford, and others, rocks from all over Earth, from the Moon, from Mars, and from meteorites have been dated. The oldest rocks yet found on Earth are in northern Canada and are dated at approximately 4 billion years old.[11] The oldest minerals, however, are crystals of zircon contained within sandstone, a sedimentary rock, in Australia. The crystals are about 4.35 billion years old, and were deposited in what later became sandstone about 3.06 billion years ago.[12] The oldest confirmed fossils of single-celled bacteria are found in sedimentary rocks in Australia that are about 3.4 billion years old (Figure 6-7).

Few rocks and fossils of such great ages exist because of the constant recycling of material at Earth's surface. Hundreds of rocks from the Moon have been dated, most at between 3.2 billion and 4 billion years, but some are as old as 4.5 billion years.[13] Many meteorites have been dated, and most are between 4.4 billion and 4.6 billion years old. These old rocks have

been dated with uranium-lead and rubidium-strontium isotopes (see Table 6.1). Deriving these ages has helped scientists to determine the origin and age of the solar system, which is about 4.6 billion years old.

Dating Organic Matter with Radiocarbon Methods In a standard procedure for dating a sample of organic matter, the amount of ^{14}C that remains is measured and compared to the amount thought to have existed in the atmosphere at the time of death of the organism. The age determined from radiometric dating represents the time that has elapsed since an organism died. The radioactive isotope ^{14}C accumulates naturally in living organisms because it is produced at a fairly steady rate in the atmosphere. As plants take in gases from the atmosphere and animals ingest food, the supply of ^{14}C in their tissues is constantly replenished. Once an organism dies, it stops respiring and eating, so its supply of ^{14}C is halted. The radioactive ^{14}C isotope continues to decay to ^{14}N, however, and the organic remains of the organism can be dated until about 45,000 years after its death. After that period of time, the remaining amount of ^{14}C is insufficient to use radiocarbon dating.

Radiocarbon dating differs from the dating methods described above for rocks and minerals because the daughter product of radiocarbon decay—^{14}N—is a gas that readily leaks out of the decaying material with time, so no daughter products remain to count. However, in dating both rocks and organic matter, the

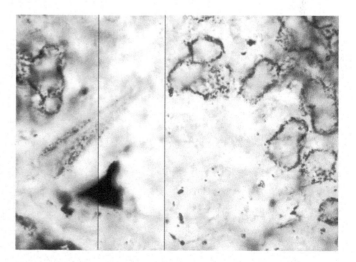

Figure 6-7 Ancient Bacteria. These fossils of bacteria, probably marine, were found in sedimentary rock in western Australia. The rocks containing these fossils have been radiometrically dated, and are 3.4 billion years old. *(Wacey, D., Kilburn, M. R., Saunders, M., Cliff, J., and Brasier, M. D. 2011. "Microfossils of Sulphur-metabolizing Cells in 3.4-billion-year-old Rocks of Western Australia." Nature Geoscience 4 (10): 698–702. © Nature Publishing Group.)*

BOX 6–3 GEOLOGIST'S TOOLBOX

Family History: Radiometric Dating and the Tempo of Human Evolution

With the advent of radiometric dating, scientists began to calibrate the evolutionary history of our own hominin ancestors, dating the times when they first walked upright, used tools, and ate meat, among other anatomical and cultural changes. Hominins include humans and early human relatives, typically associated by their bipedalism, the ability to stand and walk upright on two feet. The history of this family spans about 6 million years, from the Pliocene epoch to the present. Although fossils of human ancestors were found nearly 200 years ago, their ages were unknown before absolute dating was developed in the mid–20th century. As a consequence, it was not possible to determine the tempo of human evolution until recently.

In the 1970s, anthropologists and archaeologists discovered a remarkable trace fossil at Laetoli in Tanzania, Africa: a 27-m-long (88-ft-long) trail of about 70 footprints probably left in sediment by two or three upright-walking hominins known as *Australopithecus afarensis*. At the time they made the prints, these early hominins were walking at a gait typical of the pace we would keep if strolling through town. This fossil evidence of a distinctive feature of human behavior—upright walking on two feet—was formed 3.6 million years ago. Weathering and erosion of rocks since then exposed the hardened footprints.

How are scientists able to date such a fossil? Fortuitously, the Laetoli hominins walked barefoot through soft ash from a recent, nearby volcanic eruption; a gentle rain wet the ash, leaving raindrop impressions and making it similar to wet cement in its consistency. More ash fell, burying and preserving for millions of years the footprints of these hominins, as well as those of other animals that passed through the area. Although a print itself cannot be dated directly, minerals in volcanic ash are datable. Potassium–40 in feldspar, a common mineral in some volcanic ashes, for example, decays to argon–40 with a half-life of 1.3 billion years. The ^{40}K to ^{40}Ar decay rate is useful for dating rocks that range from as old as the solar system to as young as 2000 years, including the volcanic ashes entombing the hominin footprints at Laetoli.

East Africa has a unique combination of geologic conditions and history that makes it favorable to a relative abundance of fossils interbedded with datable volcanic ashes. This area contains the East African Rift System, a group of continental rifts that are splitting apart the African and Arabian plates (see Chapter 2). Active volcanoes line the rifts as the result of magma upwelling, and rifting itself results in deep pull-apart basins that have trapped multiple ashes and sediments in lakes, wetlands, and river floodplains for millions of years. Even as rifting pulls continents apart and forms deep basins, it tilts and flexes sedimentary layers such that modern streams cut into them, exposing what can be envisioned as a layer cake of ashes and sediment in which each deeper layer is older. This same region is the birthplace of our earliest human ancestors. Scientists scavenging the slopes of gullies and banks of streams have found tens of thousands of fossils of plants and animals, including hominins, in these layers. In addition, they have found thousands of artifacts, such as Stone Age hand axes, relatively common tools that date from about 1.8 million to 100,000 years in age.

In the same region, volcanic ashes yield ages for interbedded fossils of different hominin species, including *Australopithecus afarensis*. Several well-known examples from the Hadar Formation in Ethiopia include the skeleton of a young female nicknamed "Lucy" and 13 other hominin individuals dubbed the "First Family."[14] A nearly complete child's skeleton was found in Dikika, Ethiopia. A few dozen feldspar crystals from an ash just below the "Lucy" skeleton and above the "First Family" yielded an age of 3.18 million years, whereas the Dikika child is about 120,000 years older, at 3.3 million years.

Skeletal remains of an even older possible ancestor to humans were found more recently in Ethiopia, sandwiched between two volcanic ashes only a few meters apart that yield statistically indistinct ages of about 4.4 million years (**Figure 1**). This famous partial skeleton is of a different hominin genus, *Ardipithecus ramidus*. Nicknamed "Ardi," these fossil remains appear to push the antiquity of hominin bipedal locomotion even farther back in time.[15] Furthermore, they raise questions about what defines being human. Ardi is anatomically primitive, more closely resembling chimpanzees than humans, but has the characteristic human behavior of bipedalism. Is Ardi one of our earliest ancestors? Answering this question will require more fossil finds and dates for the critical time period of hominin evolution from about 4 to 2 million years ago.[16]

(a)

Figure 1 Thick deposits of interbedded lake, river, and wetland sediments and volcanic ashes fill deep rift basins in the East African Rift System (see inset map). These beds are exposed by tectonic uplift and stream incision, as shown here near Dikika, Ethiopia (a), where the skeleton of an *Australopithecus afarensis* child (b) was dated as 3.3 million years old. Even more common than hominin fossils are their stone tools, such as Acheulean hand axes (c) that are as old as 1.8 million years. Remarkable imprints of footsteps (d) were left as trace fossils, preserving a record of upright walking by *Australopithecus afarensis*. *(a, d: Kenneth Garrett/ National Geographic; inset: NASA/JPL/NIMA; b, c: Zeresenay Alemseged/Science Source)*

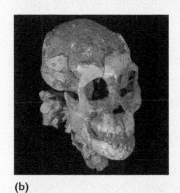

(b)

(c)

(d)

procedure relies on a decay curve like that shown in Figure 6-5.

Radiocarbon dating is particularly useful for dating human remains and associated artifacts from the past 45,000 years, an important time span that includes the spread of humans across several continents and the development of agriculture (Figure 6-8). Radiocarbon dating also has been used to date the Dead Sea Scrolls (paper and papyrus are organic matter) and to determine art forgeries (canvas is made from cotton).

■ Biological evolution, migration, and extinction of species lead to distinct fossil assemblages in specific strata that geologists correlate among sites in order to devise a relative geologic time scale from the rock record of life on Earth.

Figure 6–8 The Iceman. In 1991, hikers found the remarkably well–preserved remains of a Copper Age hunter, thought to be about 40 to 50 years old at death, just beneath the edge of a small, melting glacier in the Italian Alps, along the border between Italy and Austria. Radiocarbon dating of "The Iceman's" organic remains indicates that the man died between about 3350 and 3100 BCE, about 5050 to 5300 years ago. *(Gerhard Hinterleitner/AFP/Getty Images)*

■ The radioactive decay of unstable isotopes in minerals and organic matter is used to determine the numerical (absolute) ages of Earth's materials and can be combined with basic principles of stratigraphy (e.g., the principle of superposition) and relative age dating to reconstruct geologic history.

■ The geologic time scale is marked by major boundaries in the rock and fossil record, boundaries at which as many as 70 to 80 percent of all previously existing species disappeared during mass extinctions.

■ Radiometric dating is a quantitative method of dating based on the rate of radioactive decay of unstable isotopes.

■ The rate of disintegration of an isotope is proportional to the number of parent atoms, and it occurs at a fixed rate specific to each isotope.

■ Radiocarbon dating is particularly useful for dating human remains and associated artifacts from the past 45,000 years.

GLOBAL CHANGE OVER DIFFERENT SCALES OF TIME

Now that we have examined the ways in which scientists date different Earth materials and reconstruct geologic history, we can discuss the record of changes in the amount of carbon dioxide (CO_2) in the atmosphere and the consequent impacts on climate. This examination illustrates the different time scales of global change that are relevant to studies of the Earth system. Here we consider changes in the atmospheric concentration of CO_2 at three time scales: historic, the past several centuries, during which people have kept records of tides and temperatures; a short geologic scale spanning the past several hundred thousand years; and long geologic time periods of millions to hundreds of millions of years. Figure 6-9 shows a record of the concentration of CO_2 in the atmosphere, in parts per million (ppm) by volume, for the past 550 million years.

Global Change over Historic Time Scales

In 1958, Charles Keeling of the Scripps Institution of Oceanography in Southern California began to measure the concentration of CO_2 in the atmosphere at a station located 4300 m above the Pacific Ocean, on the volcano of Mauna Loa in Hawaii.[17] Keeling wanted to test the hypothesis that relatively recent human activity, particularly burning fossil fuels, might produce an increase in CO_2 in the atmosphere. At the time, it was unknown whether CO_2 released from fossil fuel burning—effectively excess CO_2 over a short time

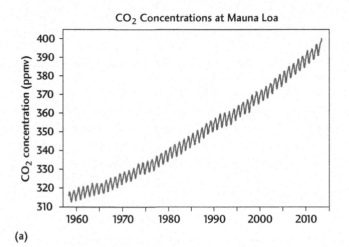

(a)

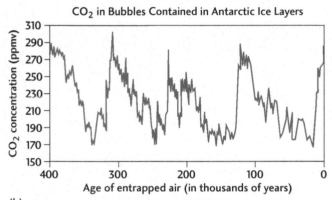

(b)

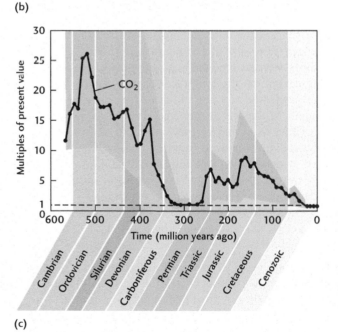

(c)

Figure 6–9 Carbon Dioxide Variations at Different Time Scales. The concentration of CO_2 in the atmosphere, in parts per million by volume (ppmv), has varied over different periods of time. The record of change illustrated in *a* is based on actual measurements at Mauna Loa, Hawaii, whereas that shown in *b* is from analyses of gas bubbles trapped in different layers of the Antarctic ice sheet. The longer–term record shown in *c* is based on a computer model, and the concentration of CO_2 is given as a multiple of the present value. Shaded area represents uncertainties in estimates. *(c: Stanley, S. M. 2008. Earth System History, 3rd ed. New York: W. H. Freeman.)*

(parts per million) in 1958 to 400 ppm in 2013 (see Figure 6-9a). Superimposed on this trend are shorter-term oscillations in the curve that reflect the annual "leafing out" of plants in the northern hemisphere, where there is more land mass and terrestrial vegetation. As leaves form and grow in the spring and summer, plants extract CO_2 from the atmosphere; after the leaves drop from plants in the fall, their decay restores CO_2 to the atmosphere.

The steady increase in CO_2 since 1850 is linked directly to industrial and agricultural activities and the use of motor vehicles; all of these activities emit CO_2 gas into the atmosphere. Furthermore, rising concentrations of atmospheric CO_2 are coincident with an increase in average temperatures at Earth's surface. Covariation between temperature and CO_2 is not surprising, because it is well established that CO_2 is a greenhouse gas. As we will discuss in Chapters 11 and 14, larger amounts of greenhouse gases in the atmosphere can lead to increased temperatures in the lower atmosphere as the gases trap outgoing heat radiated from Earth's surface.

Global Change over Short Geologic Time Scales

A record of atmospheric variation that spans several hundred thousand years was obtained from cores of ice extracted from deep in the Antarctic ice sheet, as shown in Figure 6-9b.[18] This record is similar to that revealed by ice cores from other parts of the world, including Greenland in the northern hemisphere (Box 6-4).[19] Air bubbles are trapped in snow that accumulates from year to year and eventually become permanently trapped in glaciers as the snow is compressed and converted to ice. Ice core records indicate that the rise detected by Keeling and his colleagues began at least as early as 1850, roughly coinciding with the onset of the Industrial Revolution. During the past 400,000 years, the atmospheric concentration of CO_2 has varied in periodic fashion between lower values of about 180 ppm and upper values of about 280 ppm. Low values occurred, for example, about 140,000 and 21,000 years ago, and high values occurred about 130,000 years ago. The present is also a time of high values, following a rise

scale—would be taken up by oceans and terrestrial vegetation, or whether it would accumulate in the atmosphere.

Located far from possible sources of industrial and urban pollution, the Mauna Loa site has provided a continuous, historic record of steadily increasing concentrations of CO_2 in the atmosphere, from about 315 ppm

BOX 6–4 GEOLOGIST'S TOOLBOX

How Do Scientists Date Ice Cores?

The uppermost parts of ice cores from glaciers can be dated fairly easily by counting the annual layers of accumulation of snow that gradually became ice. Sometimes each layer has slight differences in color, chemical composition, or texture that enable scientists to count back from the present, starting at the top of the glacier. Deep in a glacier, however, layers become stretched and deformed because glacial ice flows downhill under its own weight.

For some glaciers, scientists have found fossils of insects, plants, and other organic matter that were blown onto the snowpack, then buried and incorporated into the glacier as snow was converted to ice. The amount of time since these organisms died can be determined using radiocarbon dating

(discussed earlier in this chapter), but the maximum age limit for radiocarbon dating is about 45,000 years. Some thick glaciers are thought to be as much as a million years old, or even older. Other radiometric–dating techniques that are useful for exploring millions of years of Earth history can be used to date volcanic ash layers that sometimes are found within glacial ice. Volcanic ashes also can be matched to known eruptions in some cases. Finally, mathematical and theoretical models of how ice accumulates and flows down slope are used to estimate the ages of the deepest, oldest parts of thick glaciers. In general, the ages of these deeper parts are known with less certainty than upper, younger parts.

---

that began about 11,000 years ago. Note, however, that the high values of 280 to about 400 ppm of CO_2 that have existed since about 1850 CE are unprecedented in the record of the past several hundred thousand years.[20]

Independent estimates of temperature changes indicate that times of low CO_2 concentration coincide with intensely cold periods associated with advancing ice sheets, or *full glacial* conditions. Times of high CO_2 concentration coincide with relatively warm periods and shrinking ice sheets, or *interglacial* conditions. The present is a warm, interglacial period, with only two large ice masses—the Antarctic and Greenland ice sheets—remaining on Earth. During full-glacial conditions, ice sheets advanced over North America and Eurasia, as well as Antarctica and Greenland. (We'll examine the relations between periodic fluctuations in CO_2 concentrations and climatic conditions in Chapter 14).

Global Change over Long Geologic Time Scales

Finally, we examine a record of variations in atmospheric CO_2 that spans approximately the past 600 million years (see Figure 6-9c). This longer-term, geologic record is actually the result of a computer model in which the investigating scientists considered all the factors that affect concentrations of CO_2 in the atmosphere, including the types and extent of vegetation on land masses and the rates of rock weathering.

Although uncertainties are fairly large (the shaded areas around the curve in Figure 6-9c), this general

model illustrates that atmospheric CO_2 declined rapidly starting about 440 million years ago. This drop is associated with two events: first, the evolution and spread of land plants that could inhabit upland environments and accelerate the extraction of CO_2 from the atmosphere; and second, an increased rate of burial of organic carbon in extensive swamps. The second phenomenon—an increased rate of burial of organic carbon—sequestered carbon in the rock reservoir, shuffling carbon to the long-term part of the carbon cycle. As a result, the amount available for short-term replenishment to the atmospheric reservoir was depleted.

Times of global warmth and cold are referred to as **hothouse** and **icehouse climates,** respectively. Much geologic evidence indicates that global climatic conditions were warm most of the time between about 550 and 25 million years ago, with the exception of periods of glaciation that occurred about 440 million and 305 million years ago. Note that these times are marked by prominent dips in the CO_2 record (see Figure 6-9c).

■ Carbon is cycled through different Earth systems at various rates and is correlated with changing climatic conditions over geologic time.

■ The concentration of CO_2 in the atmosphere near the top of the volcano of Mauna Loa in Hawaii provides a continuous, historic record of steadily increasing concentrations of CO_2 in the atmosphere, from about 315 ppm (parts per million) in 1958 to 400 ppm in 2013.

■ Cores of ice extracted from glaciers reveal that the atmospheric concentration of CO_2 has varied in periodic fashion between lower values of about 180 ppm and upper values of about 280 ppm during the past 400,000 years. These oscillations mark changes from cold, full glacial to warm, interglacial conditions on Earth.

■ The present is a time of high values of CO_2 in the atmosphere, following a rise that began about 11,000 years ago, but the high values of 280 to about 400 ppm of CO_2 that have existed since about 1850 CE are unprecedented in the record of the past several hundred thousand years.

■ Over even longer periods of geologic time (hundreds of millions of years), atmospheric CO_2 has declined since about 440 million years ago as a result of the evolution and spread of land plants that extract CO_2 from the atmosphere and the burial of organic carbon that becomes part of the rock reservoir.

■ Geologic evidence indicates that Earth was relatively warm most of the time between about 550 and 25 million years ago, with the exception of periods of glaciation about 440 million and 305 million years ago. Both of these were times at which prominent lows are known to have occurred in the CO_2 record.

■ The past few million years were also a time of relatively cold, or icehouse, conditions marked by periodic warm interglacial intervals.

CLOSING THOUGHTS

Earth scientists study rocks, sediment, glacial ice, and fossils for clues that help to reconstruct Earth's history, to track the emergence of Earth system cycles, and to understand human origins and evolution. Inferring the sequence and rhythm of Earth's history over time spans as brief as centuries to as long as billions of years helps researchers to trace the cycling of matter and energy among Earth systems and compare the impact of human activities with those of processes that are less controllable.

For example, reconstruction of recent Earth history from ice and sediment cores reveals that the amount of carbon dioxide in the atmosphere varied—and corresponding changes in global climate occurred—for hundreds of thousands of years before the Industrial Revolution. Ever since the climate system was established more than 4 billion years ago, global climate varied between hothouse and icehouse conditions.

Nevertheless, as we will discuss in Chapter 14, these reconstructions also show that Earth's temperature is unusually high at present in comparison to geologically recent climate cycles. Furthermore, recent increases in temperature correspond to increases in greenhouse gases in the atmosphere since the start of the Industrial Revolution.

It has taken several centuries for scientists to unravel the details and timing of the history of Earth and the evolution of life on its surface. Only a little more than 100 years ago, few scientists had any notion that a method of absolute dating of rocks and fossils could actually exist. In a short time, relative to the tenure of humans on Earth, we have learned much about Earth and human history, including the facts that our home is billions of years old, that our ancestors walked upright as long as 4.4 million years ago, and that we are just one of many species that have evolved on an ever-changing Earth.

SUMMARY

■ Earth scientists use many different time-sensitive records, including cores of ice and sediment, to reconstruct Earth's climatic history.

■ The principles of superposition and original horizontality stated by Steno in the 1600s provide rational means for determining the relative ages of sequences of rocks.

■ In sedimentary rock formations, fossils and stratigraphic sequences are clues to the relative ages of the rock layers.

■ Early scientists observed that the fossils in rock records vary with time, giving rise to the principle of faunal and floral succession that Darwin later explained with his ideas of species evolution and extinction.

■ The geologic time scale is divided into hierarchical units that range from eons (hundreds of millions to billions of years) to eras, periods, and epochs (as short as 11,000 years) in duration.

■ The absolute age of an Earth material can be calculated by comparing the amount of a particular radioactive isotope in the material with the amount originally present at the time of its formation.

■ Radiometric dating of igneous rock yields an estimate of the time when the molten rock cooled to a crystalline state.

■ Because sedimentary rocks do not solidify from a molten state, their time of origin is inferred through stratigraphy and radiometric dating of igneous rock that has intruded or buried a sequence of strata.

■ A continuous historic record of steadily increasing amounts of CO_2 in the atmosphere since 1958 is available for a station atop the Mauna Loa volcano in Hawaii, and shows an increase from 315 ppm in 1958 to 400 ppm in 2013.

■ Air bubbles trapped in snow that is buried and becomes glacial ice reveal that the amount of CO_2 in the atmosphere began to increase as early as 1850, coincident with the onset of the Industrial Revolution.

■ For almost a million years, the atmospheric concentration of CO_2 has fluctuated between lower values of about 180 ppm and upper values of about 280 ppm, reflecting climatic swings between cold full-glacial and relatively warm interglacial periods, respectively.

■ Ice cores can be dated by counting annual accumulation layers for upper (younger) parts of glaciers, by radiometric dating of organic matter (for about the past 45,000 years), and by radiometric dating of volcanic ashes (for up to millions of years).

■ Computer modeling of long-term variations in atmospheric CO_2 shows a rapid decline starting about 440 million years ago as a result of two events: first, the evolution and spread of land plants that extract CO_2 from the atmosphere; and second, an increased rate of burial of organic carbon in extensive swamps.

■ Earth's climate has varied between hothouse and icehouse conditions over a period of hundreds of millions of years, with the modern Earth in a period of icehouse conditions for the past few million years.

KEY TERMS

paleoclimate (p. 168)
unconformity (p. 170)
fossil (p. 172)
relative age (of rocks, sediments, and fossils) (p. 172)
isotope (p. 172)
radioactive (isotope) (p. 172)
absolute age (p. 172)

strata (p. 172)
stratigraphy (p. 172)
principle of superposition (p. 172)
principle of original horizontality (p. 173)
principle of faunal and floral succession (p. 173)
formation (p. 173)
geologic column (p. 174)

natural selection (p. 174)
genes (p. 174)
geologic time scale (p. 174)
eon (p. 174)
era (p. 174)
mass extinction (p. 174)
period (p. 174)
epoch (p. 174)

radiometric dating (p. 175)
parent elements (p. 175)
daughter products (p. 175)
half-life (p. 178)
radiocarbon dating (p. 181)
hothouse climates (p. 186)
icehouse climates (p. 186)

REVIEW QUESTIONS

1. What are two basic types of rock dating? How do they differ from one another? In what types of situations is each useful?

2. How is the principle of faunal and floral succession related to Darwin's ideas of evolution of species?

3. How did Rutherford and Soddy discover the half-life of radon?

4. What are three common modes of radioactive decay?

5. Which isotope of carbon is useful for radiometric dating of organic materials less than about 45,000 years old?

6. Why are radioactive isotopes useful for dating over a range of only about 5 to 10 times as long as their half-lives?

7. How have concentrations of CO_2 in the atmosphere varied over historic time scales (the past 50 years), short geologic time scales (the past few hundred thousand years), and long geologic time scales (the past 550 million years)?

EXERCISES

1. Examine Table 6-1 and try to determine which isotopes would be the most useful for dating a rock that you think is about 2.4 billion years old. Explain your reasoning.

2. An isotope has a half-life of about 800,000 years. Over what time range would this isotope be useful for radiometric dating? Would it be useful to help determine the age of Earth? Why or why not?

3. Why can't Paleozoic coals be dated using radiocarbon methods?

4. Find out how old the rocks are where you live. Geologic maps have been made for almost every part of the United States. You can find such maps online or in most libraries or you can look in physical and historical geology books for less detailed maps of North America and smaller regions within the United States. If you live in a mountain belt, the rocks are likely to be older than if you live along a large river, such as the Mississippi, or in glaciated areas, such as Illinois, Minnesota, and other northern states. Think about why this is so.

5. Along the coast of Northern California, tidal platforms cut by wave action have been raised several meters out of the water during past earthquakes, and some are now covered with sand dunes. During earthquakes, all exposed, immobile marine life is killed instantly or dies soon after emergence above water. At one location, samples of fossil clamshells found in their original growth position on a platform contained only one-quarter the amount of ^{14}C in the modern atmosphere. How long ago was the earthquake that raised the tidal platform and killed the clams? A burned log left in a Native American fire pit was found in the bottom layer of dune sand covering the platform and shell. It contained half the amount of ^{14}C in the modern atmosphere. When did Native Americans camp on the dune that migrated atop the newly exposed platform? Do the radiometric ages make sense if you compare them with the sequence of events that you would have reconstructed using stratigraphic principles such as Steno's law of superposition? Why or why not?

SUGGESTED READINGS

Alley, R. B. *The Two-Mile Time Machine: Ice Cores, Abrupt Climate Change, and Our Future*. Princeton, NJ: Princeton University Press, 2002.

Coveney, P., and R. Highfield. *The Arrow of Time: A Voyage Through Science to Solve Time's Greatest Mystery*. New York: Ballantine Books, 1990.

Dalrymple, G. B. *The Age of the Earth*. Stanford, CA: Stanford University Press, 1991.

Eldredge, N. *Fossils: The Evolution and Extinction of Species*. Princeton, NJ: Princeton University Press, 1997.

Gould, S. J. *Time's Arrow, Time's Cycle*. Harmondsworth, England: Penguin, 1988.

Gradstein, F. M., J. G. Ogg, and A. G. Smith (Eds.). *A Geologic Time Scale 2004*. Cambridge: Cambridge University Press, 2005.

Hawking, S. *A Brief History of Time: A Reader's Companion*. New York: Bantam Books, 1992.

Mayr, H. *A Guide to Fossils*. Princeton, NJ: Princeton University Press, 1996.

Ogg, J. G., G. Ogg, and F. M. Gradstein. *The Concise Geologic Time Scale*. Cambridge: Cambridge University Press, 2008.

Stanley, S. M. *Exploring Earth and Life Through Time*. New York: W. H. Freeman, 1993.

Stanley, S. M. *Earth System History*, 3rd ed. New York: W. H. Freeman, 2008.

Whitrow, G. J. *Time in History: Views of Time from Prehistory to the Present Day*. New York: Oxford University Press, 1988.

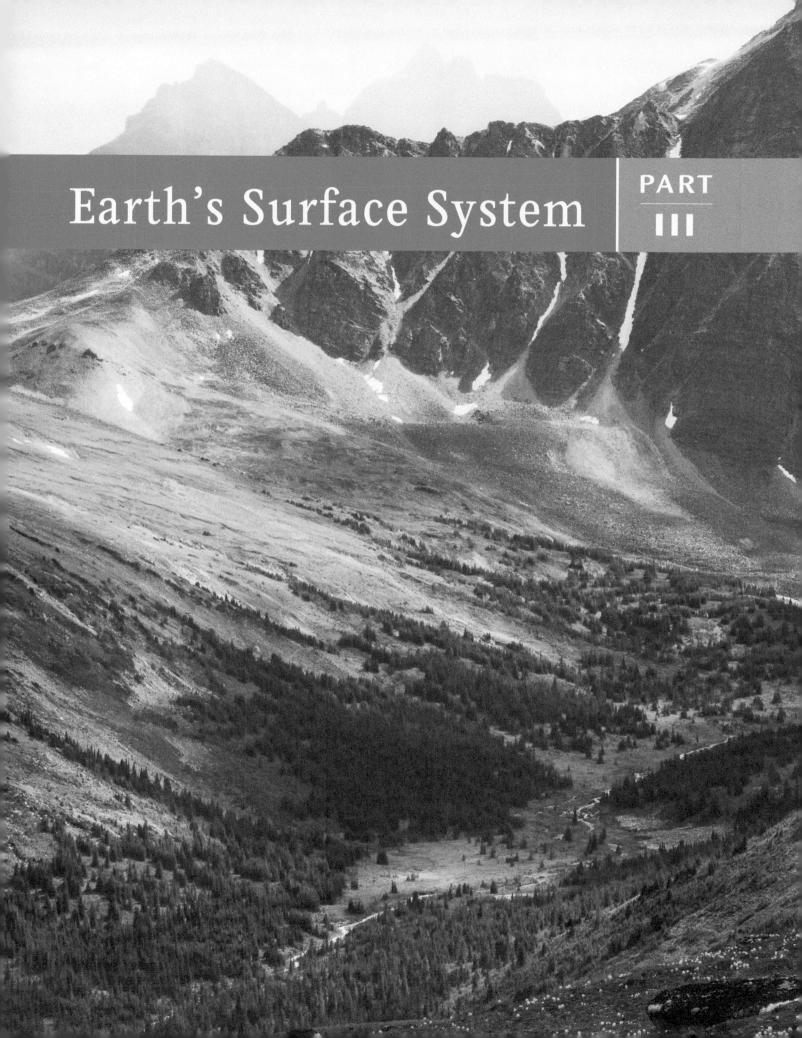

Earth's Surface System

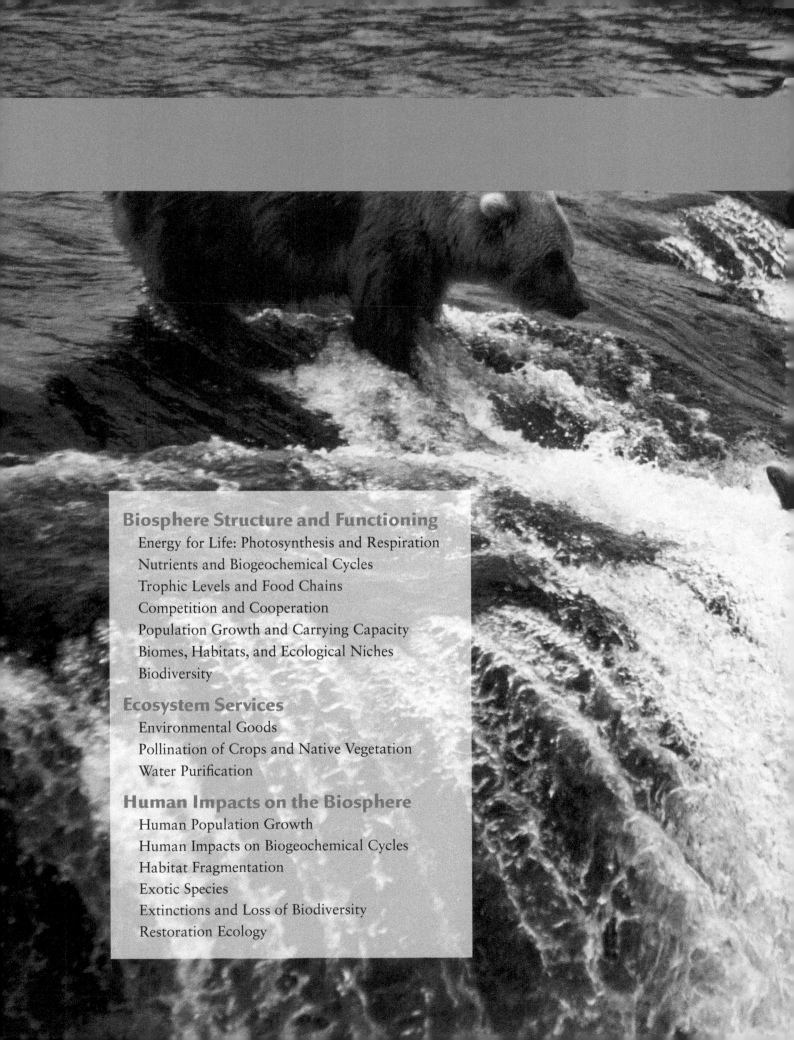

The Biosphere

Many different organisms contribute to the flow of energy and mineral nutrients within ecosystems. Upon reaching adulthood, for example, salmon swim out of the ocean and into rivers to mate, after which most die. Bears take advantage of salmon ardor, plucking them out of the water and carrying them into the forest to feed. Concerned with putting on as much weight as possible to get through winter, bears are sloppy eaters. If salmon are abundant, bears eat only those portions of the fish that are high in fat and leave the rest behind. The leftovers provide food for myriad small mammals, insects, and birds, and rotting fish carcasses release the nutrients nitrogen and phosphorus, which are critical to plants.[1] Salmon thus influence entire forest ecosystems, systems that have been disrupted by a variety of human activities, including the building of dams. While they provide hydroelectric power, dams also block fish migration and the movement of marine nutrients onto land. To deal with this problem, humans have begun to perform ecosystem services previously carried out by nature.[2] In the Pacific Northwest, the U.S. Forest Service and other agencies have distributed salmon carcasses by hand and by helicopter to areas no longer heavily populated by fish. The impacts of these experiments are not yet entirely clear, but it is hoped that this artificial fertilization will provide vital nutrients needed by forests and aquatic ecosystems.

In this chapter we:

✔ Discuss the basic structure and functioning of the biosphere.

✔ Examine how elements like carbon, phosphorus, and nitrogen are cycled between life and other components of the Earth system.

✔ Explore a variety of ecosystem services performed by nature upon which we depend.

✔ Consider the phenomenon of human population growth and its impact on the biosphere.

In Washington State, grizzly bears spread fish along streamsides, but the construction of dams blocked salmon migration, cutting off a vital source of nitrogen and phosphorus to Pacific Northwest forests. Staff from the U.S. National Forest Service now use helicopters and hand-delivery to drop dead salmon near stream banks. *(John Warden/Getty Images)*

INTRODUCTION

In September of 1991, eight people embarked on an experiment to determine whether humans might someday be able to colonize other planets, sealing themselves into a 1.28-hectare (3.15-acre) greenhouse in the desert of Arizona (Figure 7-1). Funded by a Texas billionaire and named Biosphere 2 (Planet Earth is Biosphere 1), the greenhouse was designed to be self-contained, meaning that the people living in it would have to grow all of their own food and depend on a variety of ecosystem processes to oxygenate their air and recycle their waste.[3] To ensure the greatest possibility of success, Biosphere 2's designers replicated several Earth environments in miniature, including a desert, a rainforest, a grassland, an agricultural area, a wetland, and a saltwater ocean complete with its own coral reef.

A linked energy station powered the greenhouse, with sprinkler systems used to supply rainfall and pumps used to circulate the ocean and atmosphere. Temperature, humidity, and air pressure were all carefully controlled to keep the greenhouse from overheating in the hot Arizona sun. Plants representative of Earth's different climate zones were selected from all over the world, and soils were created by mixing commercially available ingredients that had been sterilized and then inoculated with microbes representative of each environment. Chickens, pigs, goats, fish, insects, birds, and reptiles were placed into the greenhouse to provide food for the human residents, to pollinate their crops, and to create a simple food chain. In all, nearly 4000 species were put into Biosphere 2.

Shortly after the greenhouse was sealed, problems began to arise. Oxygen levels dropped while carbon dioxide (CO_2) soared to levels ten times higher than in our current atmosphere.[4] It rapidly became clear that something was awry in the photosynthesis-respiration relationship that maintains the concentrations of these gases in Earth's atmosphere. Conditions became so dire that oxygen had to be pumped into the greenhouse to ensure the survival of the crew and animals. The "biospherians" faced additional problems during their stay, including insect pests and outbreaks of plant diseases that diminished crop production; excessive levels of nutrients in their irrigation water; and interpersonal problems associated with living in cramped quarters under stressful conditions.[5] While the humans emerged hungry and irritable 2 years later, things were far worse for many of the animals in the experiment. Most of Biosphere 2's small vertebrates went extinct and every one of its insect pollinators died.[6] The news media began to report that a CO_2 scrubber system had been installed to deal with the excessively high levels of the gas, and that the crew had brought in a 3-month supply of food for themselves and a 2-year supply for their domestic animals. Cries of cheating and allegations of a lack of scientific rigor surrounded the project.

Though the original project was met with skepticism, its academic value attracted attention, and universities have operated it since the mid-1990s, using it for research and to teach about ecology and biogeochemical cycles. Perhaps the most important lesson from Biosphere 2 is a reminder that Earth is an incredibly complex system composed of numerous interacting parts through which matter and energy flow in response to physical, chemical, and biological processes. Replicating our present Earth system in miniature has proven impossible so far, but we should also remember that Earth has performed many of its own experiments in the past. Ours is a very different planet than the one that existed 4 billion, 500 million, or even 20,000 years ago, with different species and atmospheric gas contents (see Box 11-1).

While our present conditions bear little resemblance to previous eras in Earth's history, fundamental processes of evolution, photosynthesis and respiration, biogeochemical cycling, and population growth have operated throughout much of the existence of the planet and have interacted with Earth's physical, atmospheric, and hydrologic systems as well as with the energy given off by the Sun to create our modern environment. We explore these processes and how some are being altered by a particular member of the biosphere: humans.

Figure 7–1 The Biosphere 2 complex in Oracle, Arizona, replicates several Earth ecosystems in miniature, including a tropical rain forest, a desert, a savanna, a coral reef, and human agricultural lands, and was designed to be a closed system. *(ClassicStock/Alamy)*

BIOSPHERE STRUCTURE AND FUNCTIONING

As mentioned in Chapter 1, the biosphere is a thin layer of life forms living in, on, and above Earth's other environmental systems. Its chemical composition is similar to that of the other systems, particularly the hydrosphere because water makes up 60 to 90 percent of the volume of living cells. However, it also differs from the other spheres by its relative abundance of carbon. Together with elements such as nitrogen, phosphorus, iron, calcium, sulfur, and chlorine, water and carbon make up the soft tissues and skeletons of an estimated 10 million different species on Earth.[7] These organisms, whether plants, animals, fungi, or microbes, take in food, generate waste, grow, reproduce, and die in response to constantly changing environmental conditions. In the process, they also modify their environments. While the organism is the most fundamental unit of ecology, an assemblage of organisms—together with the physical and chemical environments of the assemblage—makes up an ecosystem.

Energy for Life: Photosynthesis and Respiration

All living organisms require energy to survive, and that energy is used in a variety of ways. Some goes into the basic maintenance tasks of the organism, such as replacing dying cells and ridding the organism of wastes.[8] Some is used for growth and some for reproduction. Some energy may also be used for defense. Consider, for example, the production of various chemical compounds by plants to ward off animals that might eat them, or the energy used to grow claws or to run from a predator. If an organism has abundant energy, it also might put some of that energy into storage for future use.

Most of the life forms with which we are familiar depend on photosynthesis to acquire energy. In this process, plants and a few species of bacteria use CO_2 and water in the presence of solar radiation to create organic matter and oxygen, in the process storing the Sun's energy in the form of biochemical energy in compounds such as the carbohydrate glucose. Animals and other bacteria consume the organic matter in the presence of oxygen, releasing the stored energy to carry out metabolic functions and producing CO_2 and water as by-products, a process known as respiration. Plants also respire; to maintain themselves, reproduce, and grow, they use some of the food energy they have created. They too release CO_2 and water as by-products. Simple equations demonstrate these relations:

Photosynthesis:

$6\,CO_2 + 6\,H_2O + \text{Solar energy} \rightarrow C_6H_{12}O_6 + 6\,O_2$
Carbon + Water + Solar energy → Glucose + Oxygen
dioxide

Respiration:

$C_6H_{12}O_6 + 6\,O_2 \rightarrow 6\,CO_2 + 6\,H_2O + \text{Energy}$
Glucose + Oxygen → Carbon + Water + Energy
dioxide

In the Biosphere 2 experiment, oxygen levels declined because respiration outpaced photosynthesis inside the greenhouse.[3] It turns out that the enclosure's soils contained too much organic matter, which stimulated the growth of soil microbes. As the bacteria grew, they consumed oxygen and respired CO_2. At the same time, Biosphere 2's glass roof blocked some of the radiation from the Sun, which already was reduced that year by an unusually cloudy and rainy winter. The result was that Biosphere 2's plants were unable to generate enough oxygen by photosynthesis to keep up with the increase in the amount of CO_2 produced by soil bacteria. As a result, the crew suffered from conditions similar to altitude sickness. Compounding the problem, much of the oxygen that was produced reacted with the concrete walls in the greenhouse to make calcium carbonate, further diminishing atmospheric oxygen levels.

Photosynthesis and respiration are processes that have been happening for most of Earth's history, but as in the case of Biosphere 2, they did not always produce the same composition of atmospheric gases that we live with today. We explore the impact of life on Earth's atmospheric composition throughout geologic time in Chapters 11 and 14.

Nutrients and Biogeochemical Cycles

In addition to energy, organisms require a continuous supply of essential nutrients to construct shells or skeletons and soft tissues. Growth stops if the supply of certain essential nutrients disappears. The concept of limiting nutrients is easy to understand through the analogy of baking chocolate chip cookies. Each cookie requires a number of ingredients. You can have infinite supplies of flour, sugar, salt, eggs, baking soda, vanilla extract, and butter, but if you are missing the chocolate chips, you cannot make chocolate chip cookies. Furthermore, if each cookie requires a minimum of three chocolate chips to be considered a true chocolate chip cookie, then a supply of six chips would allow a mere two cookies to be made, regardless of how large the supplies of the other ingredients.

German chemist Justus Liebig (1803–1873) is credited with originating the concept of limiting nutrients, and Liebig's *Law of the Minimum* states that growth of organisms depends on those nutrients that are in the shortest supply. The elements nitrogen and phosphorus are like the chocolate chips of the biosphere; life on Earth is impossible without them, as they are fundamental components of proteins, the genetic materials DNA and RNA (deoxyribonucleic acid and ribonucleic acid), and the compound ATP (adenosine triphosphate), which is responsible for transporting chemical energy within cells. Far more abundant in the biosphere is carbon, which rarely acts as a limiting nutrient. However, like nitrogen

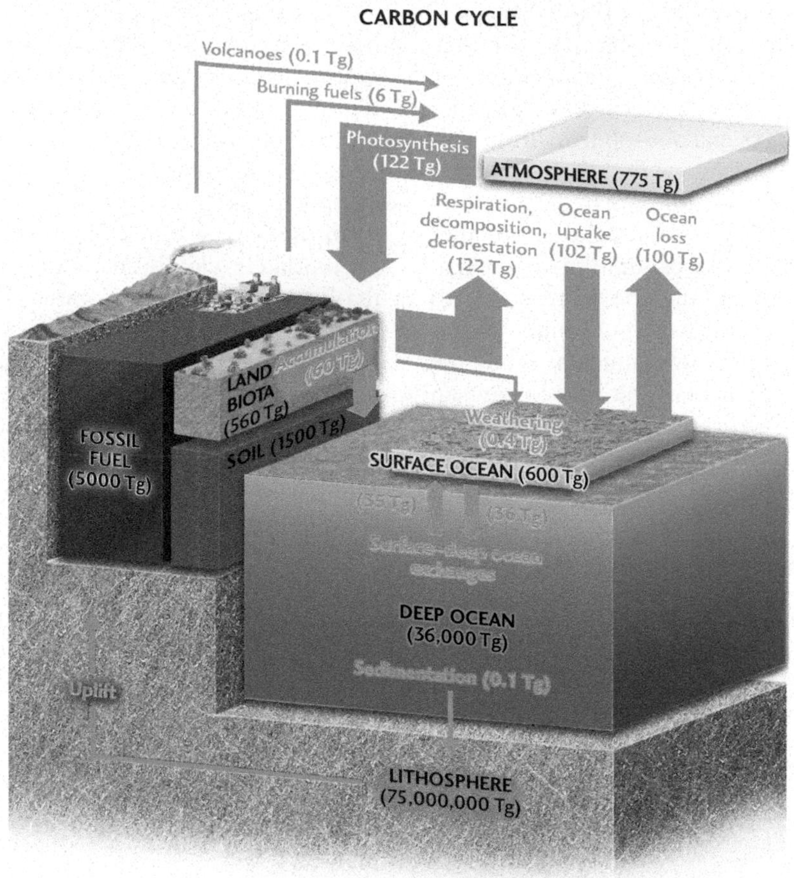

CARBON CYCLE

Volcanoes (0.1 Tg)

Burning fuels (6 Tg)

Photosynthesis (122 Tg)

ATMOSPHERE (775 Tg)

Respiration, decomposition, deforestation (122 Tg)

Ocean uptake (102 Tg)

Ocean loss (100 Tg)

LAND BIOTA (560 Tg)

Accumulation (50 Tg)

FOSSIL FUEL (5000 Tg)

SOIL (1500 Tg)

Weathering (0.4 Tg)

SURFACE OCEAN (600 Tg)

(55 Tg)   (36 Tg)

Surface-deep ocean exchanges

DEEP OCEAN (36,000 Tg)

Uplift

Sedimentation (0.1 Tg)

LITHOSPHERE (75,000,000 Tg)

Figure 7–2 The carbon cycle. Carbon cycles through all of Earth's systems, changing state along the way from a gaseous form in the atmosphere (carbon dioxide), to a dissolved form in the hydrosphere (bicarbonate and carbonate ions), to a solid in plant tissues, sediments, and rocks. Units are in Tg (teragrams, or million metric tons).

and phosphorus, carbon—a fundamental building block for life—moves through Earth's various spheres in what is known as a **biogeochemical cycle.**

The Carbon Cycle The continuous circulation of carbon among seven major reservoirs constitutes the carbon cycle (Figure 7-2). In order of decreasing size, these are the lithosphere (primarily carbonate sediments and sedimentary rocks), the deep ocean, fossil fuels, soils, the atmosphere, the surface ocean, and terrestrial organisms. Carbon moves among these reservoirs through a number of different processes, changing form along the way. In the atmosphere, it exists primarily in the form of carbon dioxide (CO_2) gas, where it plays an important role in setting Earth's temperature (discussed in more detail in Chapter 11). Carbon is removed from the atmosphere through chemical weathering of rock, diffusion into the surface ocean, and photosynthesis by plants.

Chemical weathering begins when atmospheric CO_2 combines with water vapor and falls in rain as carbonic acid (H_2CO_3). This mildly acidic rain causes some minerals to dissolve and dissociate into ions, including calcium (Ca^{2+}) and bicarbonate (HCO_3^-). These ions run off into

rivers and are transported to the oceans, where they are taken up by marine plants and animals. Additional CO_2 diffuses into surface oceans through direct contact with the atmosphere. Marine organisms process Ca^{2+} and HCO_3^- ions into hard skeletons and shells of calcium carbonate ($CaCO_3$). When these organisms die, they settle to the ocean floor as carbon-rich sediments.

Tectonic forces may cause some of these seafloor sediments to subduct into the mantle. There they melt and react with silicates to form magmas rich in dissolved CO_2. The gas returns to the atmosphere during eruptions at seafloor-spreading centers and volcanoes, completing the cycling of carbon from atmosphere to ocean to organisms and back again to the atmosphere. Carbonate sediments that are not subducted may be buried in the crust, compressed and cemented into sedimentary rocks, then uplifted and exposed during plate collision and mountain-building episodes. Chemical weathering of these rocks releases carbonate ions into solution; the ions are washed into the oceans where they are again incorporated into biological sediments.

Carbon is also removed from the atmosphere via photosynthetic production of plant biomass. In the oceans, CO_2 is used by phytoplankton (tiny floating plants), which settle to the ocean floor after they die. These phytoplankton are the dominant source of the organic substances required for the formation of petroleum (see Chapter 13), so they contribute to the store of carbon in the fossil fuel reservoir. On land, decaying vegetation and animals become incorporated into the pedosphere, where they enrich soils, or into the lithosphere, where they become fossil-rich rocks or coal. Slow weathering reintroduces the carbon they contain to the atmosphere. Natural forest fires, fossil fuel combustion, and the burning of forests and grasslands by humans also contribute CO_2 to the atmosphere. Although the annual flux of carbon through volcanic activity is small, CO_2 gas emitted from magma is the chief source of carbon in Earth's atmosphere over geologic time.

The Nitrogen Cycle Like carbon, nitrogen flows continuously through the atmosphere, biosphere, oceans, and pedosphere (Figure 7-3), but unlike carbon, it does not constitute a significant proportion of the lithosphere. Although nitrogen is the most abundant element in the atmosphere (79 percent by volume), it is an inert gas, making it largely unavailable to organisms. Only a few species of soil bacteria are able to use it directly. These organisms are valuable for their ability to "fix" nitrogen—to extract the inert gas (N_2) from the atmosphere and combine it with hydrogen into a reactive form, ammonium (NH_4^+), that plants use to construct proteins. Animals acquire their nitrogen either by consuming plants or by consuming other animals that consumed plants.

Many nitrogen-fixing bacteria live in a mutually beneficial relationship with plants, commonly in nodules attached to the roots of clovers, bean plants, and other legumes. The bacteria supply the plants with usable nitrogen and feed off the sugars and starches made by the plants. This arrangement enables the bacteria to

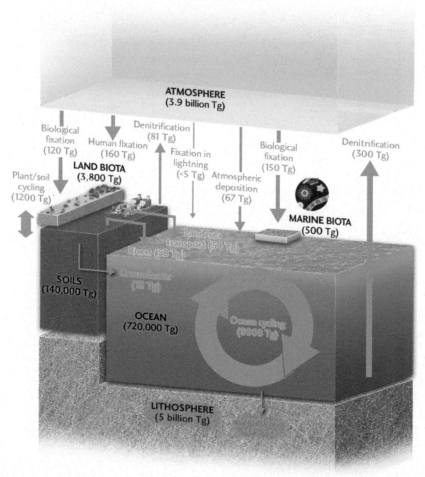

Figure 7–3 The nitrogen cycle. Nitrogen–fixing bacteria extract gaseous nitrogen from the atmosphere, turning it to ammonium, which is used by plants. Denitrification returns fixed nitrogen to the atmosphere. Units are in Tg (teragrams, or million metric tons).

survive and reproduce and, more important from the plants' point of view, to continue breaking apart N_2 molecules for the plants' use.

As clovers, legumes, and other nitrogen-fixing plants die and litter the soil, their protein molecules break down through the process of *ammonification,* releasing ammonia (NH_3), which commonly converts to ammonium. Proteins contained in other plant and animal tissues as well as in urine and feces are also converted to ammonium. Different species of bacteria oxidize part of the ammonium to nitrite (NO_2^-) and then to nitrate (NO_3^-), a process that is known as *nitrification.* Nitrate is also usable by plants, but not all nitrate cycles between the pedosphere and biosphere. Some returns to the atmosphere by *denitrification,* the natural conversion of nitrate into nitrous oxide (N_2O) and then gaseous nitrogen (N_2) through the action of another class of microbes. This entire process, from the fixing of nitrogen gas by bacteria to its release from organisms back to the atmosphere, is the nitrogen cycle (see Figure 7-3).

In Biosphere 2, systems designed to recycle sewage from the human inhabitants and their livestock into irrigation water proved inadequate, allowing inorganic nitrogen concentrations in the water to rise over time.[9] As a consequence, plants like tomatoes and sweet potatoes, which require low nitrogen concentrations in order to produce fruits and tubers, declined in productivity, reducing agricultural yields.[10] At the same time, N_2O levels in Biosphere 2's atmosphere rose significantly, showing that denitrification was not operating as efficiently

as on Earth. Fortunately, the experiment ended before those levels rose high enough to pose a serious threat to the enclosure's inhabitants, though many joked that the situation might eventually resemble a trip to the dentist— nitrous oxide is the so-called "laughing gas" used for sedation when cavities are filled or teeth are extracted![5]

The Phosphorus Cycle Like nitrogen, phosphorus (P) is a fundamental constituent of DNA, RNA, and ATP, and is therefore a limiting nutrient for life. Most phosphorus on Earth is in the +5 *valence state* (that is, it is able to accept 5 electrons), and is therefore found combined with oxygen as phosphate (PO_4^{3-}). Phosphate is found dissolved in water, as the mineral apatite in soils, rocks, teeth, and bones (nearly 95 percent of phosphorus in minerals is found in apatite), and as mineral dust in the atmosphere. Unlike nitrogen, phosphorus lacks a naturally occurring gaseous form of any significance. For this reason, the cycling of phosphorus throughout the Earth system is dependent on slow-acting geologic processes such as the weathering of rocks, the transport of sediments in streams, the decay of terrestrial plants, and the circulation of ocean water (Figure 7-4).

Plants take up phosphate ions dissolved in water or adsorbed onto soil particles, making the phosphate available to herbivores. Phosphate is released during organic matter decay and in animal urine. In bodies of water such as lakes and oceans, phosphorus cycles between the surface and depths as a result of *biological uptake* and *water column overturning.* In the summer growing

PHOSPHORUS CYCLE

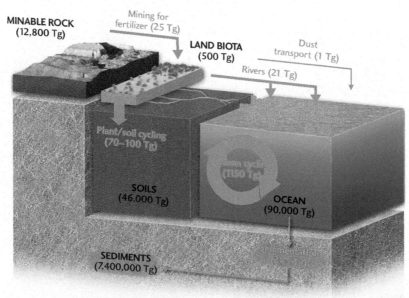

MINABLE ROCK (12,800 Tg)

Mining for fertilizer (25 Tg)

LAND BIOTA (500 Tg)

Dust transport (1 Tg)

Rivers (21 Tg)

Plant/soil cycling (70–100 Tg)

(1150 Tg)

SOILS (46,000 Tg)

OCEAN (90,000 Tg)

SEDIMENTS (7,400,000 Tg)

Figure 7-4 The phosphorus cycle. Phosphorus is found primarily in the form of phosphate ion, which exists dissolved in water and in mineral form. The lack of a significant gaseous phase means that phosphorus moves through the Earth system by slow geological processes such as subduction, mountain building, chemical weathering of rock, and stream transport. Units are in Tg (teragrams, or million metric tons).

season, phytoplankton in the shallow surface waters take up phosphate into their tissues before eventually dying and sinking to greater depths. Because surface waters are warmed by the sun, they are less dense than deeper water, and this difference inhibits vertical mixing between the surface and deeper waters. In the annual growing season, this results in the export of phosphate from the surface to the bottom waters. As the organic matter at depth decays, it gradually releases the phosphate from biological tissues and returns it to the dissolved state. In winter, the surface waters cool through contact with the chilled atmosphere, increasing in density beyond that of deep waters, and causing the surface waters to sink and the water column to overturn. Dissolved phosphate is thereby carried back to the surface waters, where it becomes available for the next growing season. Another process by which phosphate returns to surface waters is through wind-driven circulation in the open ocean and along coasts (Box 7-1).

Some small fraction of phosphate in the water column is immobilized each year by adsorption onto mineral grains or through the precipitation of minerals such as apatite. Thus trapped, the phosphate eventually subducts into Earth's interior to return to the surface in volcanic eruptions or is incorporated into rocks uplifted by continent–continent collision. In either case, chemical weathering eventually liberates it in the form of sediment particles or as ions dissolved in water, which get taken up by plants or travel in streams back to lakes and oceans.

Trophic Levels and Food Chains

Since all living organisms require energy and nutrients, the interactions among species are structured largely around food consumption. Organisms can be classified by their positions within **food chains,** which determine how matter and energy flow through biological communities (Figure 7-5). Plants and some bacteria are considered the base of the food chain because they are capable of manufacturing within their own tissues all of the food energy they need to survive. Organisms with this capability are known as **autotrophs,** or **primary producers.**

Animals, on the other hand, must eat plants or other animals to acquire the food energy necessary to grow, move, and reproduce. As such, they are known as **heterotrophs,** or **consumers.** A simple food chain might consist of a sunflower that takes in solar energy and generates seeds that are later consumed by a mouse. The mouse, in turn, is consumed by an owl. Having few predators, the owl is at the top of the food chain, and each organism— sunflower, mouse, and owl—is a member of a different *trophic level* determined by its position within the food chain.

Some of the solar energy stored in the tissues of autotrophs is lost when those organisms are consumed because some tissues are indigestible to the animal consuming them.[8] Similarly, when herbivores (animals

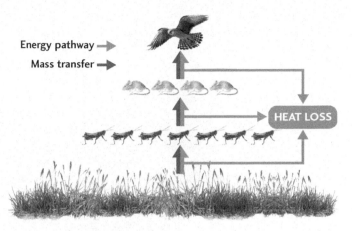

Figure 7–5 A typical food chain consisting of producers (plants) and consumers (animals). Plants fix carbon and solar energy in carbohydrates that are then consumed by herbivores, themselves consumed by omnivores or carnivores. Some 90% to 99% of the energy fixed by plants is lost in each transfer up the food chain.

that consume plants) are eaten by carnivores (animals that eat other animals), still more energy is lost since things like bones and fur are indigestible. Some portion of the energy ingested is also turned into body heat that is ultimately lost to the atmosphere through the skin. The conversion of useful energy, such as that found in food, to unusable energy, such as waste body heat, is an example of the second law of thermodynamics, which states that the universe is continually moving toward a higher degree of disorder since usable energy is necessary to keep systems ordered. As a result of the loss of energy to indigestible materials and waste heat, the number of organisms within each trophic level declines up the food chain. In our simple example, one owl would have to consume many times its weight in mice to survive for a lifetime and the mouse many times its weight in sunflower seeds. In general, 90 to 99 percent of biochemical energy is lost in the transfer from one trophic level to the next,[8] meaning that progressively fewer organisms can exist at higher trophic levels (see Figure 7-5).

While conceptually useful, the notion of the food chain is oversimplified. In reality, all life forms belong to complex *food webs* that include not only the kinds of organisms we have already described but also some that survive on the wastes produced by those organisms (Figure 7-6). **Detritivores,** for example, are animals that consume dead organic matter, such as leaves that have fallen from trees. Similarly, *scavengers* and *coprovores* feed respectively on the corpses and feces of other animals. Although not animals, fungi are also highly effective at breaking down organic tissues. Together, these *decomposer* organisms ensure that the essential nutrients for life are perpetually recycled and that wastes do not accumulate to the point of becoming toxic to the organisms that produce them.

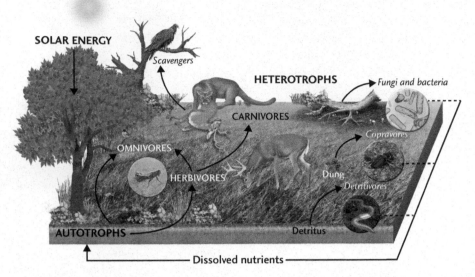

SOLAR ENERGY

Scavengers

HETEROTROPHS → *Fungi and bacteria*

CARNIVORES

Copravores

OMNIVORES

HERBIVORES

Dung
Detritivores

AUTOTROPHS

Detritus

— Dissolved nutrients —

Figure 7–6 Food chains are actually parts of more complicated food webs that contain numerous producers and consumers. Detritivores feed on the wastes and carcasses of organisms, returning their nutrients to soils and water bodies to be used again.

Competition and Cooperation

A fundamental characteristic of ecological communities is that resources are limited. As a result, organisms frequently have to compete. This competition can be *intraspecific,* that is, among individuals of the same species; or *interspecific,* among individuals of different species. After ecological disturbances, there are frequent instances of *interspecific competition* (individuals of different species competing against one another). For example, in forests, a fire, hurricane, or winter ice storm might burn or knock over a large number of trees, allowing light to penetrate to the forest floor. Sun-loving shrubs and trees quickly colonize the area, only to be shaded out later by slower-growing trees that are able to extend themselves farther upward toward sunlight. Cut off from the light, the shade-intolerant organisms eventually disappear, causing **ecological succession,** or the replacement of one community of organisms with another.

Of course, not all interactions between species are necessarily competitive. Legumes benefit from a partnership with nitrogen-fixing bacteria that supplies them with ammonium ions, while the plants provide the bacteria with sugars. Corals are part of a similar type of relationship. These colonial animals host photosynthetic algae in their tissues that supply them with sugars. At the same time, the corals provide the algae with carbon, nitrogen, and phosphorus. Species that benefit from such interactions are known as *mutualists.* If one group benefits and the other is unaffected, the interaction is described as *commensalism.* Cattle egrets, for example, feed on insects stirred up by grazing cattle and so follow cows around to get bugs. These actions benefit the egrets but do not affect the cattle. Finally, if one species is harmed by an interaction while the other benefits, the relationship is described as *amensalism.* Parasites such as the human tapeworm fall into this category.

Population Growth and Carrying Capacity

Given unlimited resources and year-round fertility, organisms reproduce in a manner that creates an exponential population growth rate over time (Figure 7-7). Such growth occurs when the population increases by a constant percentage of the individuals already in existence and is characterized by doubling during successive

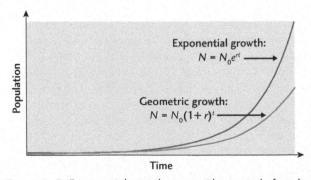

Exponential growth:
$N = N_0 e^{rt}$

Geometric growth:
$N = N_0(1 + r)^t$

Population

Time

Figure 7–7 Exponential growth starts with a period of gradual population rise followed by rapid acceleration. Such growth is characteristic of environments that have unlimited resources.

BOX 7–1 CASE STUDY

The Guano Era in Peruvian History

In the mid–19th century, the nation of Peru experienced a small economic boom. Rich deposits of guano on its Chincha Islands fueled a thriving export trade, primarily to Great Britain, which used the bird droppings for fertilizer.[11] Over a 40–year period, Peruvian miners extracted more than 20 million tons of guano, reaping more than $2 billion in profits for the economically strapped state, which nationalized the industry.[12]

Why did the nation find itself in possession of such a rich resource? Peru lies just south of the equator along the western coast of South America, an area dominated by offshore flow of winds. The so–called southeast trade winds blow from the southeast toward the northwest, pushing waters from the surface of the Pacific Ocean away from the coast and causing a flow of water from great depths toward the surface. This upwelling water is rich in the nutrients nitrogen and phosphorus needed by phytoplankton, which produce great blooms along the

coast. Zooplankton feeding on the phytoplankton support a thriving coastal fishery preyed upon by fish–eating seabirds, which nest and defecate on the Chincha Islands, transferring the marine nutrients onto land in their guano (**Figure 1**). The phytoplankton and other organisms die and sink to the seafloor, releasing the remaining nutrients into the deeper parts of the oceanic water column, from which they upwell to the surface again in a continuous cycle.

Regrettably, the supplies of guano that existed in the mid–1800s were rapidly exhausted by mining. Deposits that had taken thousands of years to accumulate were removed with pickaxes and shovels in a few decades. In the meantime, scientists in Great Britain discovered ways to manufacture fertilizers from other substances, so as Peruvian guano became scarcer and more costly, it could no longer compete in the global marketplace, causing the economic boom to go bust.

Figure 1 Seabird droppings cover Peru's Chincha Islands. The guano is rich in nitrogen and phosphorus derived from fish, which themselves are feeding on phytoplankton and zooplankton in a marine upwelling zone produced by the southeast trade winds. *(Janos Csernoch/Alamy)*

equal intervals of time. The equation that describes exponential growth is:

$$\text{(Equation 1)} \quad \frac{dN}{dt} = rN$$

where dN/dt refers to the change in the number of individuals (N) with time (t) and r is the net growth rate determined by subtracting the death rate from the birth rate. This equation shows that the change in the

number of individuals over time depends on how many individuals already exist. Solving the equation yields the exponential growth formula:

$$\text{(Equation 2)} \quad N = N_0 e^{rt}$$

where N_0 is the starting size of the population, e is a constant with the value 2.718, and the other variables are as defined above.

Exponential growth can be a challenging concept to grasp, but a simple analogy in geometric growth can help. Consider, for example, a shrewd student who negotiates the following salary arrangement for her work-study job on campus. She asks for a starting salary of 1 cent and a doubling of her wage each day. In the first week, her salary goes from 1 cent to 2 cents, 4 cents, 8 cents, 16 cents, 32 cents, and finally 64 cents, a rate of growth of 1, or 100 percent, daily. This arrangement hardly seems like a good deal, but the doubling of the salary each day yields an astounding $10.7 million daily wage by the end of the month on day 31!

The increase in the student's wage over time is an example of geometric growth. In place of the e that is used in the exponential growth equation, the geometric growth equation uses a value equivalent to 1 plus the growth rate:

$$(\text{Equation 3}) \quad N = N_0 (1 + r)^t$$

In the case of the student's salary, $r = 1$, such that her salary doubles every day. Though the geometric and exponential growth equations are slightly different, they are related and behave in essentially the same way. The geometric growth of the student's salary increase would clearly be unsustainable for her employer. Similarly, the exponential growth of any biological population is unsustainable because resources are in fact finite rather than unlimited.

Factors that limit population growth include disease and declining per capita food supplies, both of which are influenced by population density. The effect of such density dependent factors on population growth is represented by what is known as the *logistic equation*[13]:

$$(\text{Equation 4}) \quad \frac{dN}{dt} = rN \times \left(\frac{K - N}{K} \right)$$

where K is a factor known as the **carrying capacity,** the maximum population that can be sustained indefinitely without diminishing or destroying the environment's ability to produce the resources the population needs. Note that the first part of the equation looks identical to Equation 1. When the number of individuals (N) is small, the $(K - N)/K$ part of the equation has a value close to 1, such that the population grows exponentially. As the number of individuals approaches the carrying capacity, however, the $(K - N)/K$ part of the equation approaches 0 because K and N are basically the same. This causes the rate of change in the number of individuals over time (dN/dt) to also approach 0 since $dN/dt = rN \times 0$. With dN/dt equal to 0, the population stabilizes and no longer continues to increase (Figure 7-8). The result of the logistic equation is thus an S-shaped curve of population versus time, where the flat top of the S represents the carrying capacity.

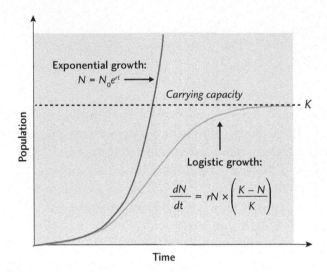

Figure 7–8 In logistic growth, populations first grow exponentially, then slow as they reach carrying capacity, the maximum population the environment can support without degradation of the resources needed to survive.

Regrettably, the progression of organisms toward carrying capacity is rarely a smooth ride. Instead, populations often grow beyond their limits, leading to environmental degradation that causes a painful population decline after the overshoot (Figure 7-9).[14] Such was the case of a reindeer herd introduced to St. Matthew Island off the coast of Alaska in the Bering Sea. Brought in by the U.S. Coast Guard to serve as an emergency food supply, the herd grew from 29 individuals in 1944 to 6000 by 1963 and rapidly consumed the island's

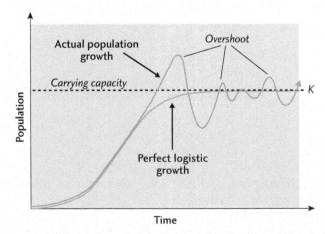

Figure 7–9 Population overshoot. Because Earth is finite, populations cannot grow exponentially forever. However, the progression to a stable population based on the carrying capacity is rarely smooth, with populations growing beyond sustainable limits and then experiencing steep declines as their resources are depleted.

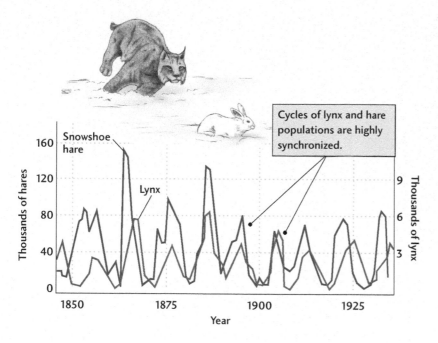

Cycles of lynx and hare populations are highly synchronized.

Figure 7–10 Lynx–hare predator–prey cycles in Canada were first observed by fur trappers in the mid– to late 1800s and early 1900s. Increasing hare populations provide more food resources for lynx, whose populations expand accordingly. As the lynx increase in number, however, they deplete the hare population and run out of food, causing their numbers to dwindle. With decreased predation, the hare population grows again to start the cycle over. *(From R. E. Ricklefs, 2008,* The Economy of Nature, *6th ed., New York: W.H. Freeman)*

lush carpet of lichen.[15] Its preferred winter food source depleted, the herd crashed to only 42 individuals during the difficult winter of 1964, only to die out completely in subsequent decades.

While environmental degradation can result in a permanent decline such as that shown by the St. Matthew Island reindeer herd, populations also oscillate in response to interactions with other species. The classic example of **predator–prey cycles** is a relationship observed in Canada between lynx and snowshoe hares.[16] When lynx populations are low, the population of snowshoe hares on which they prey increases (Figure 7-10). The larger number of hares provides more food for the lynx, whose population gradually increases. At some point, however, the higher lynx population triggers a decline in the hares, causing the lynx population to fall and the cycle to start over again. Recent biochemical research shows that physiological responses to the stress caused by the threat of predation might be as large a factor as the predation itself in causing hare populations to decline.[17] Coyotes and great horned owls appear also to play a role in the population cycles of snowshoe hares, and undergo similar cycles to those of the lynx.[18]

Factors that are not dependent on population density can also keep organisms in check. Ecological disturbances like hurricanes, ice storms, forest fires, and volcanic eruptions occur without regard to population size and kill numerous individuals. Some ecosystems are actually adapted to such disturbances and require their occurrence to generate new growth. Some jack pines (*Pinus banksiana*) in southern Canada, for example, have closed seed cones on their branches that only open

when exposed to temperatures higher than 50°C. They are able to quickly reestablish themselves after fires as their seeds drop onto the freshly exposed forest floor.[19] The ability to recover from such disturbances is known as **resilience**.[20]

Biomes, Habitats, and Ecological Niches

Organisms live in biological communities known as **biomes** that are determined largely by climate, geology, and topography but recently are also being affected by human activities (Box 7-2).[21] Seasonal variations in temperature, rainfall, and other climatic factors control the types of vegetation that can grow in a particular location, and the plants in turn determine the animal species that can live there. For example, the temperate forest biome that covers the eastern half of the United States reflects a combination of humid conditions that allow trees to grow and winter frosts that cause them to lose their leaves and go dormant. Animals living in this environment cope with the annual loss of food in different ways, including migration (birds), hibernation (bears), or starvation (deer).

In comparison, the temperate grasslands that characterize the nation's Midwest have developed in an area where evapotranspiration (a combination of evaporation from bare soils and the release of water from plant leaves) exceeds rainfall rates during the growing season; this is not a condition that supports continuous forest.[22] Occasional strong droughts lead to grass fires that also prevent trees from being established, but the deep roots

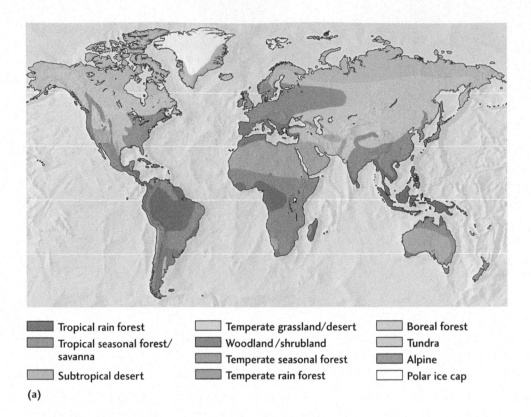

Tropical rain forest

Tropical seasonal forest/
savanna

Subtropical desert

Temperate grassland/desert

Woodland/shrubland

Temperate seasonal forest

Temperate rain forest

Boreal forest

Tundra

Alpine

Polar ice cap

(a)

Biome name		Climate zone		Vegetation
Tropical rain forest	I	Equatorial:	Always moist and lacking temperature seasonality	Evergreen tropical rain forest
Tropical seasonal forest/ savanna	II	Tropical:	Summer rainy season and "winter" dry season	Seasonal forest, scrub, or savanna
Subtropical desert	III	Subtropical (hot deserts):	Highly seasonal, arid climate	Desert vegetation with considerable exposed surface
Woodland/shrubland	IV	Mediterranean:	Winter rainy season and summer drought	Sclerophyllous (drought-adapted), frost-sensitive shrublands and woodlands
Temperate rain forest	V	Warm temperate:	Occasional frost, often with summer rainfall maximum	Temperate evergreen forest, somewhat frost-sensitive
Temperate seasonal forest	VI	Nemoral:	Moderate climate with winter freezing	Frost-resistant, deciduous, temperate forest
Temperate grassland/ desert	VII	Continental (cold deserts):	Arid, with warm or hot summers and cold winters	Grasslands and temperate deserts
Boreal forest	VIII	Boreal:	Cold temperate with cool summers and long winters	Evergreen, frost-hardy needle-leaved forest (taiga)
Tundra	IX	Polar:	Very short, cool summers and long, very cold winters	Low, evergreen vegetation, without trees, growing over permanently frozen soils

(b)

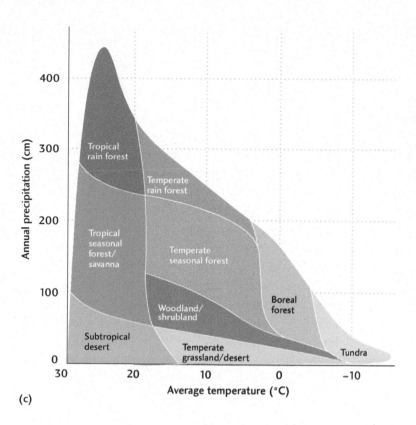

(c)

of the grasses allow rapid re-sprouting. Additional examples of biomes include temperate and tropical rain forests, tundra, savanna, and boreal forests; scientists have created various biome maps and classification schemes in the past few centuries to describe these different communities (Figure 7-11).

Within individual biomes are different **habitats,** locations where various physical characteristics determine which organisms can live there. For example, forest biomes include streams, wetlands, and lakes, each of which support different types of organisms, given variations in water flow rates, temperature, chemistry, bottom sediment, and woody debris content. The vertical structure of the forest and the presence or absence of understory shrubs and herbs makes additional habitats available. Some animals may prefer to live up in the tree canopy; others live somewhere along tree trunks; and still others prefer the forest floor. Occasional forest fires, ice storms, or other disturbances may open up meadows that host yet other organisms.

The habitat in which an organism lives is a prime determinant of its **ecological niche,** or the way in which it lives in its environment. This role is a function of the climatic conditions an organism can tolerate, its nesting behavior, the type of food it eats and its method of acquiring that food, and its interaction with other organisms. Consider, for example, the grizzly bear mentioned at the beginning of this chapter. Grizzly bears tolerate well the temperate climatic conditions of the

intermountain West and Pacific Northwest as well as the more boreal climates of western Canada and Alaska, hibernating to get through the cold winter months. During the warmer months, grizzlies feed on migrating salmon, dispersing the salmon carcasses throughout forests and bringing vital nutrients from the sea to land. When salmon are less abundant, grizzlies eat berries and other fruits of flowering plants, dispersing their seeds as they defecate in different locations. Grizzlies thus play a vitally important role for other organisms in their habitat, and their removal from that habitat could result in large changes in the environment (Box 7-3).

Biodiversity

The great diversity of climatic, topographic, and geologic conditions on Earth have led to an astonishing array of different habitats and ecological niches for organisms. Consider, for example, the diatoms—single-celled aquatic plants at the base of the food chain in many water bodies. These organisms are sensitive to water chemistry, including nutrient concentrations, pH, and salinity; to water temperature; and to the amount of light that reaches them. It sometimes seems as though there is a different species of diatom for every microhabitat within a water body. The tendency species have of evolving to fill specific and narrowly defined ecological niches contributes to biological diversity or **biodiversity,** the wealth and range of species, genes, and communities on Earth. Biodiversity

BOX 7-2 EMERGING RESEARCH

Anthropogenic Biomes: Humanity's Imprint on the Biosphere

Across-country trip today reveals that our ideas about biomes are outdated. When flying across the Midwest one doesn't see temperate grasslands. Rather, farm fields stretch as far as the eye can see. Similarly, desert and temperate forest biomes are spotted with cities and surrounding suburbs, more concrete and asphalt than vegetation. Indeed, satellite mapping shows that more than 75 percent of Earth's land surface not covered by ice has been affected by human activities that include agriculture, forestry, construction of cities and towns, and mining.[21] Impacts of these activities consist of changes in air and stream temperature associated with the heating of paved surfaces; changes in stream flow caused by the inability of rain to infiltrate into soils; increased inputs of nutrients into streams from agricultural and lawn fertilizers; contamination of soils and water with pesticides, herbicides, heavy metals, and other toxins; and changes in vegetation types from native to exotic plants.[23]

Recently, scientists have noted that we need to include *anthropogenic biomes* or *anthromes* into our thinking and models about how ecosystems may respond to future global climate change (**Figure 1**). Recognizing that nature is now contained within human systems, rather than the other way around, is critical for preserving what wild lands remain and for ensuring their viability in the future. If climate change leads to conditions no longer conducive to plants and animals trapped within a small island of wild land surrounded by human land uses, those plants and animals could go extinct. If, on the other hand, we use the concept of anthropogenic biomes to help preserve and restore degraded habitats within those biomes, species may be able to migrate to new locations as climate change occurs, ensuring their future existence.

Figure 1 Anthropogenic biomes, or anthromes, are based on the reality that biomes in many locations have been significantly transformed through human activities such as agriculture and urbanization. *(map and graph: Ellis, E. C., and Ramankutty, N. 2008. "Putting People in the Map: Anthropogenic Biomes of the World." Frontiers in Ecology and the Environment 6(8): 439–447, doi: 10.1890/070062; village and city: Thinkstock; forest: Eastcott Momatiuk/Getty Images)*

---

appears to be fostered by warm and wet environments. Some regions, like moist, tropical rainforests and coral reefs, harbor an incredibly diverse assemblage of species. Others, like some deserts and polar regions, are virtually devoid of life. Research in recent decades by ecologists and biogeographers has begun to untangle the complex questions about why these differences occur and has explored the implications of these differences, such as the possibility of predicting the response of different regions to global environmental change.

Biological diversity occurs at different levels—for example, at the genetic level and at the species level. **Genetic diversity** describes the genetic variation among

individuals within a population. Offspring receive a unique combination of genes from their parents as a result of sexual selection, and these, as well as differences in genes that arise from mutation, determine the physiology or fitness of an organism in its environment. Together, the random combination of genes from parents and the variation that is derived from mutation increase the potential for genetic variation.[28] Populations with greater genetic variation are often better able to adapt to changing environments than are those with low variation. Species that are rare have frequently lost much of their genetic variation, so they have less capacity to adapt when environmental conditions change, and

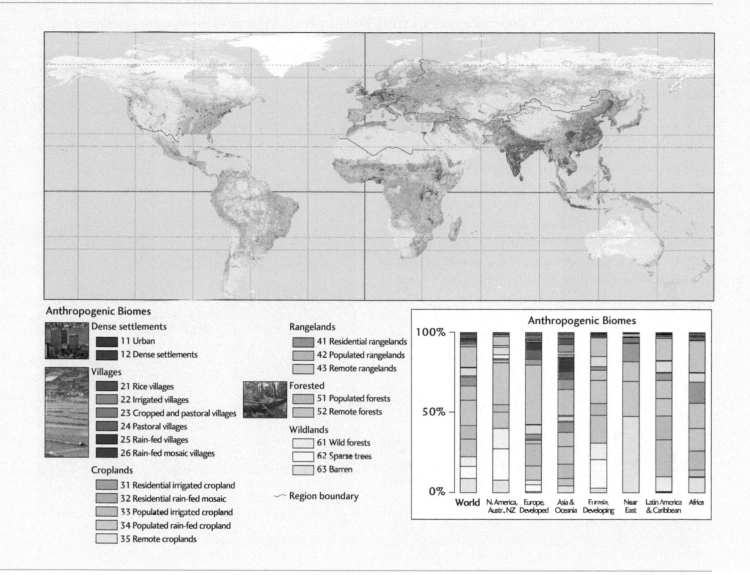

Anthropogenic Biomes

Dense settlements
- 11 Urban
- 12 Dense settlements

Villages
- 21 Rice villages
- 22 Irrigated villages
- 23 Cropped and pastoral villages
- 24 Pastoral villages
- 25 Rain-fed villages
- 26 Rain-fed mosaic villages

Croplands
- 31 Residential irrigated cropland
- 32 Residential rain-fed mosaic
- 33 Populated irrigated cropland
- 34 Populated rain-fed cropland
- 35 Remote croplands

Rangelands
- 41 Residential rangelands
- 42 Populated rangelands
- 43 Remote rangelands

Forested
- 51 Populated forests
- 52 Remote forests

Wildlands
- 61 Wild forests
- 62 Sparse trees
- 63 Barren

⌒ Region boundary

Anthropogenic Biomes

100%

50%

0%

World | N. America, Austr., NZ | Europe, Developed | Asia & Oceania | Eurasia, Developing | Near East | Latin America & Caribbean | Africa

they face increased risks of inbreeding. **Inbreeding** is the production of offspring by related individuals (e.g., brother-sister, parent-offspring, cousin matings, etc.) and reduces reproduction and survival. Studies have shown that between 80 and 95 percent of deliberately inbred populations die out after eight generations of brother-sister mating.

Species diversity describes the number and type of species on Earth. Historically, species have been described as groups of individuals that can potentially breed among themselves in the wild but do not breed with individuals of different groups. More recently, species have been defined as groups of individuals that differ morphologically (form and structure), physiologically, or biochemically from other groups. These differences can be measured by comparing DNA sequences or other molecular markers.[28]

An individual species inhabits an area either because it was dispersed to that area or because it evolved from another species. Some species are restricted, or **endemic**, to a particular location, such as a mountaintop, an island, or a continent. Regions that have been isolated from other areas for long periods of time tend to have higher numbers of endemic species. Over time, the separation and isolation of populations of a species by glaciers, drifting continents, and other barriers can result in

BOX 7–3 CASE STUDY

Consequences of Wolf Reintroduction into Yellowstone National Park

In 1995, the U.S. National Park Service decided to reintroduce gray wolves to Yellowstone National Park in Wyoming and Montana (Figure 1). While historically an important predator in the area, wolves had been exterminated from the park and surrounding areas 70 years earlier under pressure from ranchers who complained of wolf-related livestock losses.[24] In the years between, greater awareness of ecology along with the growth of the environmental movement led to calls to restore ecosystems to their former conditions. This push, combined with reductions in ranching and an increase in federal land acquisition, helped make it possible to begin the restoration efforts, which commenced with the introduction of 31 wolves brought down from Canada.

With a population now measuring in the hundreds, the consequences of wolf repatriation have been profound. After

Figure 2 Cottonwood trees in Yellowstone are regenerating as elk numbers have declined following wolf reintroduction. *(Jim Peaco/National Park Service)*

Figure 1 Wolves hunt elk in Yellowstone National Park. *(Douglas Smith/National Park Service)*

decades of decline, aspen and cottonwood trees began to show signs of recovery as the elk that had heavily browsed them began to avoid areas frequented by wolves (Figure 2). Using radio collar studies, scientists have shown that in

the evolution of new species. For example, in Australia, which began to split from Gondwanaland (one of two major continents created by the breakup of Pangaea that subsequently split into Australia, India, South America, Africa, and Antarctica) 140 million years ago and finally split from Antarctica 50 million years ago, more than 80 percent of plants and animals are found nowhere else, and the continent has more than twice the number of species as North America and Europe combined.[29] In contrast, North and South America, which have been joined for about 3 million years (see Box 2-1), have very similar types of organisms.

Species diversity can be measured in different ways. The most common measure is **species richness**, or the

areas of low wolf density, and therefore low threat of preda-tion, elk prefer to congregate in stands of aspen trees over open meadows or conifer forests, where they munch on the tender shoots of new trees. In contrast, in areas with higher wolf density, elk seek the security of conifer forests, avoid-ing the other two habitat types and allowing aspen stands to recover from their heavy grazing pressure.[25] In a similar study, scientists found that elk prefer open valley bot-toms with unobstructed views and numerous escape routes to steeper slopes containing gullies in which they might become trapped and hunted by wolves.[26] Thus the threat of wolf predation appears to be affecting elk behavior and along with it, patterns of elk foraging.

Before wolves were reintroduced to the park, the heavy grazing of elk impacted not only trees but other species of animals as well. The population of northern range beaver (*Castor canadensis*), for example, had declined steeply as young trees, the beavers' preferred food and lodge–building source, diminished in abundance. Restoring wolves to Yel-lowstone National Park has triggered a so–called *trophic cascade,* in which a *keystone predator* exerts top–down control on an ecosystem, with effects felt all the way down to the bottom of the food chain. As the trees rebound, the beaver population may eventually recover (**Figure 3**). Additional benefits of tree regrowth are expected to include stream bank stabilization, cooler stream water temperatures, and more food supplies in aquatic ecosystems related to increased leaf litter into streams.[26] All of these factors should enhance aquatic productivity, perhaps leading to a better environment for fish.

The reintroduction of wolves proved so successful that in 2008, the Bush administration relaxed rules on hunting.[27] While on the endangered species list, growing numbers of wolf packs had begun venturing out of the park and down from Canada earlier in the decade, creating problems for

Figure 3 The preferred food of beavers is young trees, so as elk numbers dwindle and aspen and cottonwood populations increase, beaver populations may eventually rebound as well, an example of a trophic cascade. *(Stephen J. Krasemann/Science Source)*

ranchers whose cattle occasionally became prey. Idaho, Wyoming, and Montana were given the authority to deal with the wolves that strayed beyond park boundaries, result-ing in 100 of the estimated 1500 wolves in the northern Rocky Mountains being killed in the first few months of the new rules. By summer of 2008, a federal judge overturned the new rules after hearing arguments from conservationists concerned about the long–term viability of the wolf popula-tion in Yellowstone. Without the ability to travel outside of the park and mix with other populations, they argued, the wolves might succumb to inbreeding and hereditary diseases that would eventually cause their demise. The court man-dated that wolves remain protected, and stated that only those animals known to have preyed upon livestock could be killed. The future of the Yellowstone wolves and the unique ecosystem they influence is secure again at the moment, but the conflict is likely to continue as wolves reestablish them-selves in their historic range.

number of species in an area, habitat, or taxonomic group.[30] The relative distribution of different species in an area can be described by the particular species present (species composition) or by indices that mea-sure *evenness*. For example, a site with 1000 species with ten individuals each would be considered more diverse than a site with 1000 species where 99 percent of the individuals belonged to the same species.[31] Spe-cies richness tends to decrease as one moves from the tropics to temperate regions in higher latitudes. Simi-larly, species richness tends to decline with increasing altitude. At geographic scales, species richness tends to be higher where energy availability is higher and where, because of varied topography and geology, environmental variation is large. This variation provides more opportunities for genetic isolation, local adaptation, and speciation to occur.[28]

On average, greater biological diversity is responsible for higher levels of productivity in plant communities, more nutrient retention in ecosystems, and better overall ecosystem stability.[32] In addition, biodiversity is important to humanity because of its continual creation of organisms of beneficial use. These include food crops, plants and animals with medicinal qualities, and a variety of other ecosystem services we explore in the next section.

■ Most organisms depend on photosynthesis to acquire the energy they need for maintenance, growth, reproduction, and defense. Photosynthetic autotrophs use carbon dioxide and water, along with incoming solar energy, to make carbohydrates.

■ In addition to energy, organisms require particular nutrients to construct shells, skeletons, and soft tissues. Nutrients such as nitrogen and phosphorus circulate throughout the whole Earth system, spending time in organisms, water bodies, the atmosphere, and rocks. This circulation is known as a biogeochemical cycle.

■ Nitrogen and phosphorus are limiting nutrients for life; that is, they are in short supply and therefore limit the growth of individuals and populations.

■ Species interactions are structured around food production and consumption. Primary producers generate their own food through photosynthesis. Heterotrophs, or consumers, eat primary producers or other heterotrophs. Between 90 and 99 percent of energy is lost from one trophic level to the next in a food chain because of the indigestibility of some materials.

■ Decomposer organisms feed on dead organic matter and feces, releasing nitrogen and phosphorus back into the environment.

■ Species both compete and cooperate with one another for resources. Competition often leads to ecological succession, or the replacement of one community of organisms with another. Succession also results from disturbances such as forest fires, hurricanes, and ice storms.

■ Given unlimited resources, populations of organisms grow exponentially over time. In reality, resources are limited, and population growth tends toward a carrying capacity that reflects the maximum number of organisms a particular environment can contain without degradation of the resources those organisms need for survival.

■ Organisms sometimes overshoot their carrying capacity, leading to population crashes as the organisms use their resources at rates that are unsustainable. Predator–prey cycles such as those exhibited by the Canadian lynx and snowshoe hare exemplify these population overshoots and crashes.

■ Organisms live in ecological communities known as biomes that are determined largely by climatic conditions. Within biomes, there are numerous habitats that host different organisms, and each organism occupies an ecological niche defined by the environmental conditions it can tolerate and its role within the food chain.

■ The tremendous variability of rock types, climatic conditions, and topography leads to a wide array of habitats for organisms and fosters biological diversity. In general, warm and wet regions on Earth host more species than do cold or dry environments.

ECOSYSTEM SERVICES

The natural functioning of the biosphere includes myriad processes upon which humans are dependent for food, water, shelter, a breathable atmosphere, and other needs. In recent years, these processes have been classified as **ecosystem services,** inasmuch as they preserve or produce environmental assets critical to human life as well as economic and spiritual well-being.[33] Consider, for example, the opening story in this chapter in which we described the role that salmon and bears play in fertilizing forests in the Pacific Northwest, an economically important region for wood products. Now that dams block salmon migration, people have had to intervene to ensure continued forest health, and federal agencies are spreading salmon carcasses by hand using helicopters. Such efforts are not inexpensive, and it's estimated that all ecosystem services on Earth are worth between $16 trillion and $54 trillion per year.[34] In the following sections we explore a sampling of these services and their importance to humanity.

Environmental Goods

While the Star Trek television and movie series allowed Vulcans, Klingons, Earthlings, and other intergalactic species to walk up to a replicator that could produce any desired object seemingly out of nothing, in reality, we are utterly dependent on the biosphere to produce the foods and fibers we use to sustain, protect, and shelter ourselves. In the United States, for example, housing is constructed primarily from timber, as are furniture, cardboard, and paper, making the country highly dependent on forests. Another globally important use of wood products is biomass energy, with wood burned for home heating and cooking as well as for industrial uses. In the first decade of the new millennium, for instance, 90 percent of wood cut in Africa was used for fuel.[35]

Like forests, grasslands are very important. Long used as grazing land for sheep, goats, and cattle, many have been converted to agricultural cropland, supplying corn, wheat, and soybeans, among other crops. Other products deriving from the biosphere include the textile fibers cotton, linen, and flax, natural rubber, herbs, spices, medicinal compounds (aspirin came from the bark of willow trees and the chemotherapy drug Taxol was extracted from the Pacific yew tree to name just two), fruits, vegetables, nuts, milk, meat, and fish. Indeed even plastic comes from the biosphere given that it is made from petroleum products, themselves originating from marine organisms.

The human value of environmental goods is illustrated nicely by the example of fish. According to the United Nations Food and Agriculture Organization, humans consumed 115 million metric tons of fish in 2008 derived both from fish farms and wild catch and valuing greater than $93.9 billion. These fish provided as much as 20 percent of the animal protein consumed by 1.5 billion people and over 15 percent of the animal protein consumed by 3.0 billion people. On top of the food provided by fisheries, these natural assets present employment opportunities. From those who catch the fish to those who process it by drying, canning, and smoking, fisheries employ an estimated 180 million people.[36]

Regrettably, like many Earth resources, wild fisheries are feeling the strain of human exploitation. Fully 53 percent of different marine food species are currently at their maximum sustainable catch levels while 32 percent are considered overexploited or totally depleted. Compounding the problems of overfishing are habitat destruction in sensitive coastal zones where many fish spawn, water pollution associated with human settlements, and dam construction that blocks fish migration. As the human population continues to grow, the viability of these and other natural resources is likely to be further threatened.

Pollination of Crops and Native Vegetation

An estimated 220,000 different species of flowering plants depend on animals for pollination, including 70 percent of human crop species.[37] Over tens to hundreds of millions of years, plants have evolved strategies to attract such species as bees, wasps, bats, and hummingbirds, typically by providing carbohydrate-packed juices. As these animals come in for a meal, they rub up against the anthers of flowers where pollen is housed, typically getting a dusting of the sperm-filled grains. Moving on to the next snack, they deposit the grains into another flower where they fertilize an egg waiting to turn into the seeds of the next generation of the plant.

The economic value of pollinators is evident in the costs required to fertilize various human food crops. In recent years, native pollinators have declined, victims of shrinking natural habitats, pesticides, and disease and parasite outbreaks. As a consequence, farmers have had to rely increasingly on industrial beekeepers to pollinate their crops. In the United States, farmers pay between $40 and $140 per hive to set up a honeybee colony near their crops, with each hive capable of pollinating about 0.4 acre.[38] Considering the acreage planted to produce all of the nation's food, the value of these pollination services reaches into the billions of dollars per year.

At Biosphere 2, the loss of every insect pollinator reveals that the experiment would likely have failed if it had continued, with the human residents no longer able to grow sufficient food for their survival and the majority of the enclosure's vegetation eventually dying out.[6] One agricultural region in China has already had to adapt to the loss of its pollinators. In southern Sichuan province, environmental degradation and overuse of pesticides killed off nearly every bee, yet the area still manages to grow a spectacular pear crop every year. How do they do it? Farmers stand on ladders or climb the trees and use feathers attached to sticks to hand pollinate every single flower on each tree in the orchard (Figure 7-12)!

Water Purification

Another important environmental service is water purification by wetlands and streamside forests. Wetlands are water-filled depressions containing vegetation that can tolerate continual submergence. These habitats are capable of trapping 80 to 90 percent of sediment particles in runoff and of taking up 70 to 90 percent of incoming nitrogen.[39] They also can filter out 20 to 60 percent of metals from water. Streamside forests are efficient water purifiers, capable of removing up to 50 percent of the phosphorus and 90 percent of the nitrogen in runoff and trapping sediment before it can reach a stream.

The importance of these services to humans is perhaps best illustrated through the example of the New York City water system. The city has the largest unfiltered water supply network in the United States, serving an estimated 9 million people. The water remains unfiltered because it meets the stringent quality criteria of the U.S. Environmental Protection Agency (EPA). Nineteen reservoirs constructed in the Hudson Highlands and Catskill Mountains between 1842 and 1964 supply more than 1 billion gallons per day through an extensive

Figure 7–12 A Chinese farmer hand pollinates peach blossoms in Zaozhuang, Shandong, China. *(TPG Top Photo Group/Newscom)*

aqueduct system that extends as far as 125 miles from the city (Figure 7-13).[40] Ninety percent of the water comes from rain falling on the Catskill/Delaware watershed, low mountains made of Devonian-age stream and delta sandstones and shales. These rocks consist primarily of quartz and feldspar and contain very low concentrations of iron, sulfur, or other impurities that can impart a bad taste or odor to drinking water. In addition, the Catskills are heavily forested with such species as paper birch, hemlock, balsam fir, spruce, red oak, and black cherry,[41] all of which serve to filter incoming rain water and prevent sediment, nutrients, or metals from reaching the reservoirs. As such, New York City is said to have the "champagne of drinking waters."[42]

In the mid-1990s, the city began to run into problems. While its drinking water was still of high quality, it became clear that development posed a threat to the water supply. At the time, the city owned a mere 3 to 4 percent of the land surrounding its Catskill/Delaware reservoirs, and as surrounding parcels were developed, they brought with them sediment running off of construction sites, nutrients from farms and sewage systems, and sewage bacteria that constituted a public health hazard. With the city's water supply threatened, the EPA stepped in, noting that there was no assurance that the city would be able to comply with the Agency's Surface Water Treatment Rule and demanding that steps be taken to either secure the watersheds or build a water treatment facility.[43] Due to an estimated cost of $5 billion to $9 billion to construct such a plant and $500 million in annual operating costs, the city hoped to avoid constructing the facility, and turned instead toward watershed protection and cleanup of existing water quality threats.

Working with the EPA, state government officials, representatives from Catskill Mountain communities, and various environmental organizations, the city reached a landmark Memorandum of Agreement in 1997 requiring it to purchase sensitive lands to protect them from development, to provide funding to upgrade the sewage treatment facilities of several Catskill communities, and to assist farmers in developing sediment and nutrient retention plans by improving streamside vegetation buffers and better managing animal manure. By 2010, the city had allocated $541 million for land purchases, nearly quadrupling its holdings to 161,000 acres or about 13 percent of the watershed, and had signed up 90 percent of farms within its watershed, paying to develop and implement agricultural best management practices. Thus far, these efforts have resulted in a 64 percent reduction in the amount of ammonia entering reservoirs along with a 36 percent reduction in particulate phosphorus and a 53 percent reduction in dissolved phosphorus.[44] With these steps, New York City has taken steps to enhance the ecosystem services that protect its drinking water and should be able to remain filtration free into the indefinite future.

■ Ecosystem services are natural processes that occur within the biosphere that create or maintain environmental assets of benefit to humans.

■ Such services include the production of environmental goods, such as lumber, food, and a breathable atmosphere, pollination of crops and native vegetation, and water purification.

■ Natural resource economists have estimated the value of all of Earth's ecosystem services at between $16 and $54 trillion annually.

■ As habitats have been degraded, humans have had to replace some natural processes. For example, some Chinese pear orchards are now pollinated by humans, and salmon carcasses are dispersed in Pacific Northwest forests from helicopters.

HUMAN IMPACTS ON THE BIOSPHERE

Unlike most animals, humans are amazingly adaptable to different climates. From the hot, humid conditions surrounding the equator to the deserts of Arabia and the frigid temperatures of the Arctic, our species has found a way to make a living in nearly every terrestrial environment on Earth. What we lack in terms of physiological fitness we more than make up for in our technological prowess, which has enabled us to pull water from deep aquifers in arid climates and to cool our bodies with air conditioning in locations where we might otherwise die

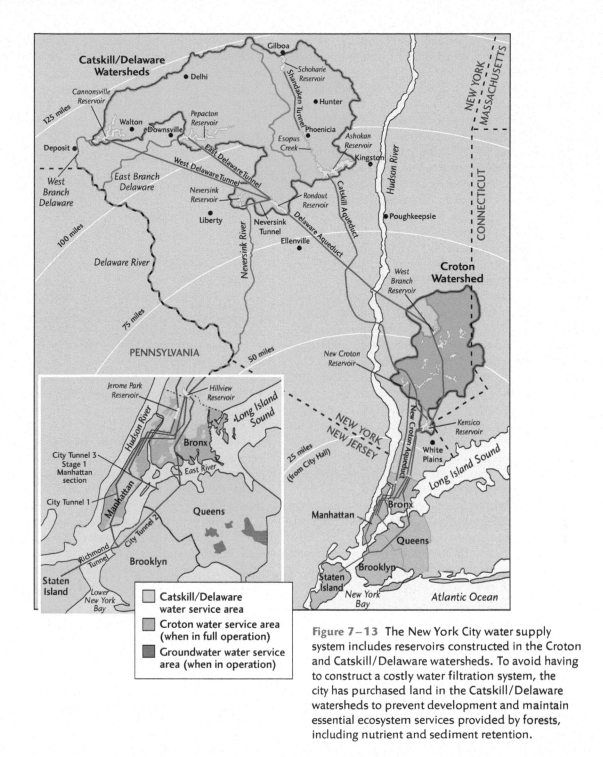

Figure 7–13 The New York City water supply system includes reservoirs constructed in the Croton and Catskill/Delaware watersheds. To avoid having to construct a costly water filtration system, the city has purchased land in the Catskill/Delaware watersheds to prevent development and maintain essential ecosystem services provided by forests, including nutrient and sediment retention.

from heat stroke. Given our ingenuity, it is perhaps no wonder that we have been highly successful in propagating our species and in transforming the planet to meet our needs. As our population has grown, however, it has become clear that what is best for us is not necessarily best for many ecosystems, nor perhaps for the planet as a whole. In this section we explore some of the impacts of humanity on the rest of the biosphere and on how the biosphere functions.

Human Population Growth

Our species, *Homo sapiens,* evolved into existence about 200,000 years ago in Africa.[45] By 10,000 BCE, our numbers had grown to between 2 and 20 million. Originally dependent on foraging and hunting for subsistence, by 8000 BCE we had developed agriculture, thereby vastly increasing our food supply and, along with it, our potential for additional population. By the year 1 CE, our population had multiplied to between 170 and 330 million, and by 1650 CE we numbered between 500 and 600 million (Figure 7-14).[46] The span of human history to about 1800 CE was needed for us to reach a population of 1 billion. Since that time, growth has been much more rapid, and our numbers surpassed 7 billion in the year 2012 (Table 7-1). Much of the recent rise can be attributed to the widespread use of fertilizer, which, along with the development of herbicides, pesticides, and mechanized agriculture, fueled the so-called "Green Revolution" that began just after World War II.

At a broad scale, human population growth appears to be occurring exponentially, the kind of growth that reflects unlimited resources. Superimposed on this larger trend are several shorter intervals of population decline, however. Genetic and archeological evidence suggests that early in our history our numbers dwindled from around 10,000 individuals to perhaps only a few hundred. Between 195,000 and 123,000 years ago, an episode of lower global temperature and glacial ice expansion led to cold, dry conditions in northern Africa that might have forced our species into limited refuges along that continent's southern coast. In these locations, small bands of people subsisted on seafood.[47]

In addition to climatic impacts on population, wars and disease outbreaks killed millions to tens of millions of people along the way toward our current numbers. Despite these events and the fact that population growth rates have fallen over the last 40 years (the r value in Equations 1 and 2 peaked at 2 percent from 1965 to

TABLE 7-1 Milestones in World Population

World population (billions)	Year when reached	Number of years it took to increase by 1 billion
1	1804	All of human history
2	1927	123
3	1960	33
4	1974	14
5	1987	13
6	1999	12
7	2012	13
8	2026	14 (projected)
9	2043	17 (projected)
9.3	2050	

Data from: United Nations Population Division, World Population Prospects: The 2000 Revision, Volume III: Analytical Report *at http://www.un.org/esa/population/publications/wpp2000/chapter5.pdf*

1970 and is now only 1.2 percent), we are on schedule to add 2 or more billion people to the world in the next 40 years (see Table 7-1).

Earlier in this chapter we noted that the interactions between organisms are structured largely around food consumption and we mentioned that the efficiency with which the solar energy fixed by plants is passed upward through the food chain is small. The result of the latter is that the number of individuals within a given species tends to decline up the food chain as energy supplies diminish. As animals who are rarely preyed upon by other organisms, humans are at the top of the food chain. Unlike other organisms, however, our omnivorous diet

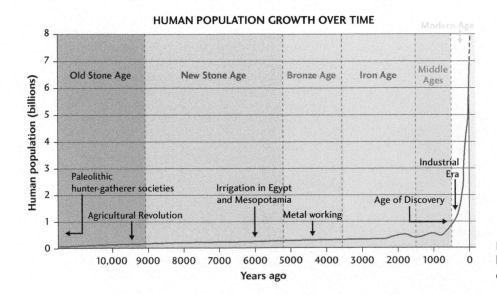

Figure 7-14 The human population has grown more or less exponentially over time.

and technological prowess have allowed us to multiply in numbers far greater than our trophic level would suggest. Studies conducted in the last few decades have estimated that our single species now consumes between 7 and 38 percent of the entire net primary production of the globe,[48] the energy and matter left over after plants have maintained themselves. Between the food crops we grow for ourselves and our domestic livestock, the fish we take from the oceans, and the trees we fell for construction, paper making, and fuel, our species has an unusually large impact on Earth. Indeed the space we use to grow crops and pasture our animals now accounts for 40 percent of the total land surface of the globe (Figure 7-15).[49] Inasmuch as each new person added represents a new competitor for resources, our impacts on the biosphere are likely to increase.

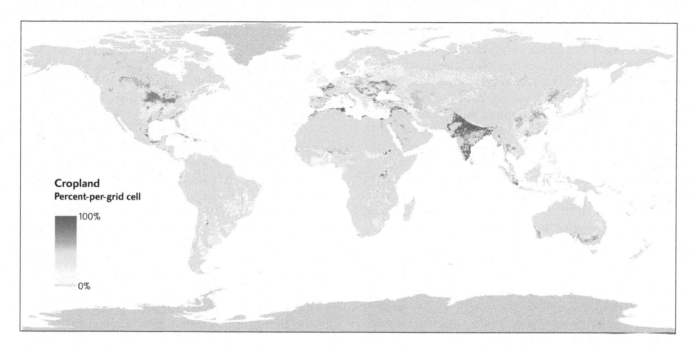

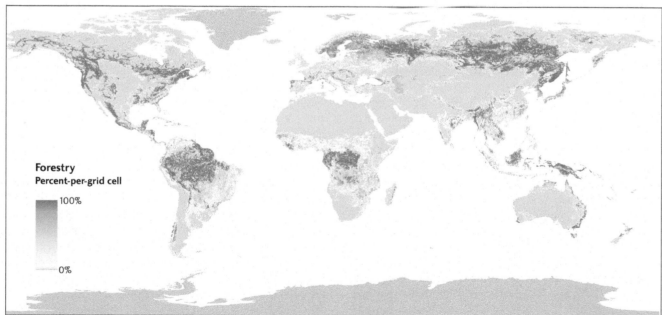

Figure 7–15 Global land–use maps with UN Food and Agriculture Organization data on cropland and forest areas at the country level. Color intensity reflects the extent of land transformation. For example, the deepest brown color on the agricultural crops map reflects a landscape in which 90% to 100% of the total area is in crop production.

Human Impacts on Biogeochemical Cycles

Since the dawn of agriculture, humans have altered the nitrogen cycle in several important ways. The first is the increasing cultivation of nitrogen-fixing legumes throughout the world. The second, known as *industrial fixation,* is the use of fossil fuels to produce synthetic fertilizer containing nitrogen. In the Haber–Bosch process, methane is reacted with steam to produce hydrogen gas and carbon monoxide. The hydrogen then combines with nitrogen at high temperature and pressure to produce ammonia, from which ammonium-based fertilizers can be made. Prior to agriculture, the background rate of nitrogen fixation likely measured between 90 and 140 Tg (teragrams, or million metric tons) per year.[50] Industrial fixation now contributes more than 80 Tg annually and that by leguminous crops an additional 40 Tg, such that nitrogen fixation now greatly exceeds denitrification. Additional alterations to Earth's nitrogen cycle have come from the combustion of fossil fuels, which results in atmospheric deposition of 20 Tg per year, and from incomplete sewage treatment. There have been similar disruptions in the phosphorus cycle (Box 7-4).

A direct result of increased nitrogen fixation has been the increased discharge of nitrogen into streams and rivers. Nitrogen is washed from farm fields and parking lots during rainstorms, running off into lakes, bays, estuaries, and the ocean. In many bodies of water, the excessive supply of nitrogen and other nutrients intensifies biological activity and produces *algal blooms,* the rapid growth of algae in response to nutrient overload. When algae and other plants die and settle to the bottoms of lakes and bays, the organisms that decompose them consume oxygen, depleting it from the water. Ultimately, nitrogen-loading results in increased oxygen depletion, which in turn destroys organisms that need oxygen, such as fish and shellfish. Because fish graze on aquatic plants, their demise worsens the problem of plant overpopulation. This overall process of nutrient-induced oxygen depletion is known as *eutrophication* and has been blamed for the deaths of many fish and shellfish in aquatic environments. We discuss this topic again in Chapter 12.

Habitat Fragmentation

Because different organisms have different physiological requirements, including dietary needs, allowable temperature ranges, and moisture tolerances, they are adapted to specific habitats around the world. Human activities like farming and urban development tend to disrupt these habitats, eliminating some and breaking others into smaller fragments. Most of the tallgrass prairie biome of North America, for instance, has been lost, replaced by agricultural crops such as wheat, corn, and soybeans. From an estimated 140 million acres before the arrival of European settlers, this unique biome has diminished to a few tens of thousands of acres, most in a national preserve in the Flint Hills of Kansas.[57] With the loss of prairie came major declines in American bison, pronghorn antelope, and prairie dogs, though many of these losses can also be attributed to the fact that these animals were hunted as part of a campaign to subdue and force American Indians onto reservations (bison) or poisoned at the behest of ranchers who claimed they competed with cattle for food resources (prairie dogs).

As habitats are transformed for human use, the species they once supported often dwindle. This is true not only of total habitat destruction, but also of habitat fragmentation, in which large habitat areas are broken into smaller parcels when roads, houses, and farm fields are constructed (Figure 7-16). These changes in the landscape reduce the size of habitat parcels, sometimes below the ability of the remaining patches to support the resident population.[58] Those creatures that are unable to migrate between patches may lose access to food or succumb to the genetic problems of inbreeding. Fragmentation also introduces edges to habitats, negatively affecting some species while favoring others. Birds, for example, commonly suffer higher levels of nest predation as the edge length of a habitat area increases in locations affected by agriculture.[59] Conversely, increased edge length favors increased species richness and abundance of those animals that prey on birds' eggs and offspring, such as raccoons and house cats.

Exotic Species

Another by-product of human population growth and technological ingenuity has been increased global trade of goods and services. While people have engaged in trading for millennia, the internal combustion engine and the discovery of fossil fuels led to a tremendous rise in trade during and after the Industrial Revolution. People can hop onto an airplane in New York City and travel to another continent in a matter of hours, bringing other species along for the ride. From cold and flu viruses to bacteria, fungal spores, and plant seeds embedded in muddy sneakers, we unintentionally transport various organisms around the globe on a daily basis. For years we have also intentionally transported species from one place to another, typically for horticultural or medicinal reasons. Many of these species have subsequently gotten out of control in their new environments, leading to serious problems. Giant hogweed (*Heracleum mantegazzianum*), for example, was imported into Great Britain and later the United States from the Caucasus region of Europe for its ornamental qualities.[60] Possessing large leaves and an impressive flower, the plant also has a

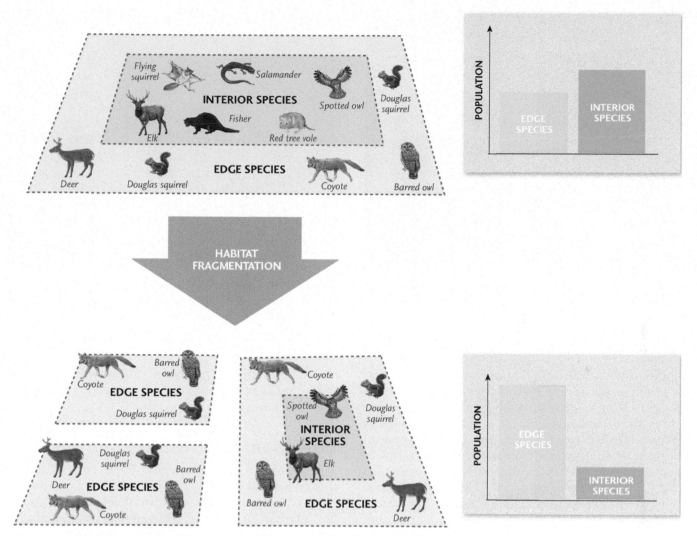

Figure 7–16 Habitat fragmentation increases the length of edges relative to interior habitats in landscapes, affecting some species adversely and others favorably. Habitat destruction is problematic for all species.

toxic sap that causes severe skin burns and can lead to blindness if introduced into the eyes (Figure 7-17). Unfortunately, it has now spread throughout New York, Pennsylvania, Washington, and Oregon and has crowded out native species as its giant leaves shade them out.[61]

Zebra mussels (*Dreissena polymorpha*) are perhaps the most well-known animal invader. Imported accidentally to the Great Lakes in the ballast water of cargo ships from Europe, these tiny mussels have no local predators.[62] The first mussels were found in the United States in 1988, and within only a couple of decades, they had spread to 25 states and two Canadian provinces. While their larvae at first swim, they eventually settle onto a hard substrate to begin the stationary, filter-feeding part of their lives. Because zebra mussels are attracted to environments with currents, they have proven particularly

problematic for water supply infrastructure since they get into pipes and form thick encrustations that block water flow (Figure 7-18). Municipal water treatment facilities, power plants, and shipping canals have all been affected by these aquatic pests, which have cost the United States an estimated $1 to $1.5 billion in damages. Not only do the mussels cause problems for human infrastructure, they also impact their local environment. As voracious filter feeders, they limit the food supplies for other organisms in Lake Erie and have also increased water clarity, which is causing more algae growth on the lake floor.

Not every exotic species introduced to a new location becomes as invasive as the giant hogweed or the zebra mussel. Indeed, most die out quickly when they reach a new environment, victims of climatic or soil

Figure 7-17 The ornamental giant hogweed is an exotic invasive species in North America that shades out native herbaceous vegetation. *(WILDLIFE GmbH/Alamy)*

Figure 7-18 Zebra mussels encrusting a pipe. Such blockages have cost the United States over a billion dollars in infrastructure repairs. *(Peter Yates/Science Source)*

conditions to which they are not adapted or subject to predation or competition for resources with other species.[63] What makes some species more likely than others to become invasive is a matter of ongoing research. Perhaps these species fill an empty ecological niche or simply proliferate as a result of a lack of predators. They may also move into a location that has recently been disturbed by a natural event or by human activities such as construction. Whatever the reason, once established they are capable of causing massive environmental changes. One of the most surprising examples may be earthworms, which most people consider benign inhabitants of soils throughout North America. In reality, earthworms are not native to the northern states or Canadian provinces, which 20,000 years ago were covered in the massive Laurentide ice sheet. Brought into these areas with transplanted soils associated with potted plants, sod, and other landscape construction materials, they have since proliferated, fundamentally changing soil structure and the cycling of nutrients in ways that may affect long-term forest health.[64] Studies are ongoing, but researchers have already found that earthworm feeding in forests significantly reduces the thickness of the leaf litter layer on the forest floor, which may make the underlying soil more prone to erosion and affect plants' roots.[65] On the other hand, earthworm feeding and burrowing also moves important nutrients downward in soils, making elements like nitrogen more readily available for root uptake. Still other studies suggest that the nitrogen moved by earthworms is readily leached out of soils when rainwater percolates through them, resulting in a

net loss of nutrient availability.[66] More work remains to be done to determine whether earthworms represent a net threat or net benefit to soils.

Extinctions and Loss of Biodiversity

In 1999, after 6 years of intensive forest surveys, scientists affiliated with the Wildlife Conservation Society in New York declared that the Miss Waldron's red colobus, a type of large West African monkey, was extinct.[67] Since that time, a photograph and a skin of a killed Miss Waldron's have been found, meaning that the species may still be clinging to life in the southeastern corner of Côte d'Ivoire in numbers so small that it is certainly on the verge of extinction, but no living specimens have been observed.[68] Five other varieties of red colobus, which are also endangered, are popular game for hunters because they are heavy and live in large, noisy social groups that are easy to track. In addition to these animals, nearly 50 percent of the world's 634 primate species are at risk of extinction within the next few decades.[69]

A species is **extinct** when there is no reasonable doubt that any individuals remain alive on earth. A species is locally extinct, or **extirpated,** when it is no longer alive in one area but can still be found elsewhere. We know from the fossil record that the number and diversity of species has increased, sometimes steadily and sometimes in sudden bursts, since life first appeared on Earth. Nevertheless, while the world presently has more species than during any other geological period, the rate of species extinction for birds, mammals, amphibians, and reptiles at the beginning of this new millennium is as

BOX 7–4 EARTH SCIENCE AND PUBLIC POLICY

Detergent Phosphates: From Good Housekeeping Necessity to Environmental Nuisance

With the chemical engineering revolution that swept the globe after World War II, manufacturers of household cleaning agents discovered that phosphate significantly increased the effectiveness of laundry and dishwashing detergents.[51] In "hard" waters, those with significant amounts of dissolved calcium and magnesium ions, phosphate ions combined with these metals, preventing them from forming greasy precipitates with the oils contained in soap. Unfortunately, the wonderful vanquisher of soap scum itself precipitated a huge environmental problem as phosphate detergents entered septic systems and municipal sewage treatment plants unequipped to deal with them. While most treatment facilities break down the organic carbon in sewage that can deplete dissolved oxygen levels and kill fish, only a small percentage of treatment plants are designed to remove other nutrient ions.[52] Since sewage treatment plants discharge all of their treated effluents to bodies of water, the resulting flow of nutrients into streams, lakes, and coastal zones led to widespread eutrophication problems. By the 1980s, millions of tons of phosphorus were entering American waterbodies each year, resulting in algal blooms in thousands of lakes.[53] In Lake Erie, for example, the floor of the coastal zone soon became carpeted in a thick green slime, and fish such as walleye and pike began to decline.

With the public clamoring for action, detergent manufacturers voluntarily cut the amount of phosphate in their products but fought against regulations. Their reductions proved insufficient for several states, including Indiana, Michigan, Minnesota, New York, and Vermont, which enacted outright bans by the mid–1980s. The bans led the industry to completely phase out phosphates in laundry detergents by 1994, given the difficulty of trying to provide different products to different states.[54] Still, phosphates continued to appear in household dishwashing detergents, but that too has recently changed. Sixteen states enacted bans in 2010, requiring that the phosphorus content be no more than 0.5 percent in these soaps.[55] Dishwashing detergent manufacturers have now reformulated their products to comply. Some people have complained that the new soaps leave spots and film on their dishes, but there is little doubt that they are better for the environment. Combined with controls on fertilizer runoff, phosphorus levels in the Great Lakes have declined significantly since manufacturers reformulated laundry detergent, reducing algal blooms in Lake Erie and allowing a gradual recovery from eutrophication.[56]

---

great as or greater than at any previous time in Earth's history.[28] The extinction crisis has been described by Edward O. Wilson, one of the most influential scientists addressing this issue, as one of our most important environmental problems.

Paleontologists generally agree that five episodes of mass extinction, defined as a loss of 75 percent or more of existing species within a time interval of 2 million years or less, have happened since the Ordovician period 500 million years ago. The largest episode took place at the end of the Permian, 250 million years ago, when an estimated 77 to 96 percent of all marine animal species went extinct, possibly within as short a time as 160,000 years. The most recent and probably best-known mass extinction occurred 65 million years ago during the Cretaceous period, when the dinosaurs disappeared and mammals emerged as the dominant species on land. The major difference between these previous extinction episodes and the current crisis, dubbed by some the "sixth extinction episode," is that the former mass extinctions were caused by natural forces such as asteroid impacts, excessive volcanism that changed atmospheric and oceanic chemistry, and tectonic activity that brought about climatic change, while the current episode is caused by human activities.[70]

Species extinction has occurred repeatedly throughout geologic history. So normal is this phenomenon that scientists have been able to determine a routine, or "background," rate of extinction for different groups of organisms. Mammals, for example, have a background rate of approximately 1.8 extinctions per million species-years, meaning nearly two mammal species go extinct every million years. During routine extinctions, the loss of species is balanced or surpassed by the appearance of new species through evolution. The evolution of new species through natural selection (see

Chapter 6) is a process that takes many generations at a scale of hundreds of thousands of years or more. Each new species typically lasts 1 to 10 million years before itself going extinct. Modern human-induced extinction rates may exceed "background" extinction rates by 100 to 1000 times, making it impossible for the evolution of new species to keep up. In this sense, human extinction rates are comparable to those in the mass extinction episodes of the geologic past.

In every part of Earth that has been colonized by humans, hunting, burning and clearing land for cultivation and other activities has resulted in large-scale changes of the landscape that have contributed to high rates of species extinctions. Large mammal populations were likely the first to feel the effects of growing human populations. For example, human colonization of Australia and North and South America is thought to have been responsible for the extinction of more than three-quarters of those mammals weighing more than 44 kg. Still, it is extremely difficult to measure extinction rates because we know so little about how many species inhabit Earth in the first place. By one estimate, 85 species of mammals (2.1 percent of known mammal species) and 113 species of birds (1.3 percent of known bird species) have gone extinct since 1600 CE. Most of these extinctions occurred in the last 150 years.[28] We know so little about insects and other life forms that extinction rates for these species are even harder to estimate.

Many species at risk of going extinct are considered endangered or critically endangered. The International Union for the Conservation of Nature defines a species as **endangered** if the extent of its occurrence is less than 5000 km², its population is less than 2500 individuals, and/or if the probability of going extinct in the wild is at least 20 percent within 20 years or five generations. A species is considered **critically endangered** if the extent of its occurrence is less than 100 km², if its population is less than 250 individuals, and/or if the probability of extinction in the wild is at least 50 percent within 10 years or three generations.

While a few individuals of a species may persist for some time, ultimately they are destined to go extinct if their populations are too small. Scientists use quantitative methods known as **population viability analysis** to study the likely future status of a population or group of populations of conservation concern.[71] Such computer-based analyses provide a quantitative estimate to determine extinction probabilities. These analyses help scientists measure the relative contributions of demographic uncertainty, environmental uncertainty, natural catastrophes, and genetic uncertainty.[72] **Demographic uncertainty** results from unexpected random effects of reproduction that can result in skewed sex ratios. For example, in 1980, only six dusky seaside sparrows remained.[73] Since all happened to be male, the species was doomed

to extinction. Populations of remnant individuals like this have been described as the "living dead."[28]

Environmental uncertainty refers to unpredictable weather events, changes in food supply, new competitors, and other unpredictable events. **Natural catastrophes** refer to extreme cases of environmental uncertainty, such as floods, fires, and storms, that are often rare and of short duration, but nonetheless can have a dramatic impact on populations. Finally, **genetic uncertainty** is particularly important for small populations. In populations with fewer than 50 individuals, genetic variability is small and the populations are prone to problems associated with inbreeding.[73]

While extinction can happen anywhere, and with virtually any species, extinction rates tend to be highest on islands. Because islands are remote, they harbor a large number of **endemic species**, those that are found in one location and nowhere else. Island species often coevolve with a relatively small number of other species, so they are highly adapted to a limited number of predators, competitors, or diseases. As a result, when people introduce new species to these island ecosystems, the consequences can be rapid and devastating. For example, all of New Zealand's land mammals, amphibians, and reptiles are endemic. When the island was first colonized by humans 700 years ago, more than 50 bird species, like the giant (3.5 m tall!) flightless moa or the 10 kg adzebill, the largest eagle in the world, went extinct.

Similarly, when the first Polynesians and their pigs and rats came to the Hawaiian islands, an estimated one-third of the islands' species, including at least 62 species of birds, went extinct. The later arrival of Europeans brought diseases and more alien species to the islands, including sheep, horses, and goats, causing the decline of another 2000 indigenous species. Today, the Hawaiian Islands have more naturalized alien plant species than native plant species. Another dramatic example of extinction on islands is the flightless Stephens Island wren. Found only on a small island near New Zealand, the entire population of wrens was killed off by one animal—the lighthouse keeper's cat.[74]

Many studies have suggested that smaller islands, both real oceanic islands and terrestrial "habitat islands," such as national parks and nature reserves that have become isolated patches of forest or grassland in a sea of the human-dominated landscape, will lose species faster and thus have fewer species than big islands.[75] Darlington suggested as a rule of thumb that a tenfold decrease in habitat area reduces the number of species by half.[76] As human-induced habitat destruction and fragmentation continues, then, we can expect increasing numbers of extinctions. A 2011 analysis of the rate and magnitude of extinctions published in the flagship scientific journal *Nature* suggests that we have not yet achieved a mass extinction on par with mass die-offs of the geologic past.[70] However, it warns that if present extinction rates continue for as little as the next 300 years, the human-caused extinction will equal or perhaps surpass those episodes.

Restoration Ecology

A new scientific discipline has emerged during the past quarter century because of the growing need to understand and respond to ecosystem degradation and the threat of species extinctions. In **restoration ecology**, an ecologist who understands ecosystem structure and function provides guidance on landscape restoration (Figure 7-19). Restoration is common in fields other than science, as in the restoration of a piece of artwork (e.g., a painting or sculpture) or an old house or car. In these cases, the goal is to restore an object to its original form, or as close to that condition as possible. Scientists identify appropriate restoration targets through their research of *paleoecology,* or ecological history. By examining soils, sediment cores, pollen, and fossil seeds, among other evidence, paleoecologists can determine which species existed at a given time and what ecosystem structure and function existed prior to environmental degradation.

One of the challenges in restoration ecology is the difficulty in determining what is "natural." Even before industrialization and European colonialism, humans modified ecosystems. Agricultural activities date to as early as the 11th century BCE in the Middle East, and Romans built bridges, aqueducts, and dams throughout the Roman Empire (from about 27 BCE to 476 CE). Furthermore, ecological change has occurred throughout Earth's history, largely in response to changing climate, landforms, and soil types. Few places on Earth have had the same terrestrial ecosystem for more than about 12,000 years, just after the end of the last full glacial episode.[77] Indeed, some late glacial (about 14,000 to 18,000 years ago) species assemblages have no modern analog.

For any given region, there is no inherent "natural" ecosystem. Species evolve and become extinct, so that different assemblages of species exist in the same place over time. Restoration ecologists understand that ecosystems are in a state of continual flux over long periods of time, and are able to provide guidance on how to foster

Figure 7–19 Before (top) and after (bottom) photographs of ecological restoration at a mine site in New Jersey. *(top: Courtesy Princeton Hydro; bottom: Patrick Verheijen/Princeton Hydro)*

biodiversity and enhance ecosystem functions in ways that maximize ecosystem function, goods, and services for a given locale. With the onslaught of anthropogenic activities that have altered and degraded many ecosystems, the work of restoration scientists is crucial. As they integrate their understanding of existing and historical ecosystems with targeted experiments, they determine which future engineered ecosystems might be stable. The viability of different levels of intervention can be determined by identifying historical states and their range of variability, as well as by determining the drift of modern systems from their historic state.

■ Human population growth has followed an exponential trend over time. We currently number around 7 billion and are expected to exceed 9 billion by 2050.

■ As the human population has grown, we have altered biogeochemical cycles through our agricultural practices, combustion of fossil fuels, and sewage.

Disruptions to the nitrogen and phosphorus cycles have led to eutrophication of water bodies.

■ Clearing land for agriculture and settlement construction has both destroyed and fragmented habitat.

■ Global trade has moved species around the globe. Some of these exotic species have become invasive, crowding out native vegetation and changing ecosystem processes.

■ Loss of habitat is leading to rates of species extinction unprecedented in geologic history.

■ The new science of restoration ecology investigates degraded or destroyed ecosystems and habitats and their responses to activities that are designed to restore or improve ecosystem structure and function.

■ The goals of ecological restoration include optimizing and maintaining ecosystem services and goods.

CLOSING THOUGHTS

Evidence from sedimentary rocks indicates that Earth's biosphere has existed for at least 3.8 billion years. During that time, evolution has produced a diverse array of organisms that have exploited every possible ecological niche, from deep-sea hydrothermal vent communities to the myriad creatures who inhabit the skies. These organisms participate in the grand cycling of matter and energy throughout the Earth system, and in the process are influenced by and affect their environments. The fossil record indicates that most life forms exist for 1 million to 10 million years before going extinct, typically casualties of climate change or some catastrophic geologic event such as an asteroid impact. Given this reality, we might ask ourselves how long humanity will last.

English economist Thomas Malthus (1766–1834) warned of the dangers of exponential growth of the

human population on a planet with finite resources and predicted that we would eventually outstrip our food supply. While this has not yet happened, we would do well to consider whether we are on a trajectory toward catastrophic overshoot or toward a sustainable carrying capacity. The question of how many people Earth can support is not easy to answer. Resource pessimists believe that finite supplies of environmental goods and services will eventually limit human population growth, whereas resource optimists think that human technology will allow us to overcome future challenges of environmental degradation. Whichever school of thought happens to be correct, it is clear that we have embarked on a global experiment whose future outcomes are quite uncertain. We would be wise to heed the cautionary tale of Biosphere 2.

SUMMARY

■ Nutrients and energy cycle through the biosphere via complicated food webs involving autotrophs, heterotrophs, and detritivores. Most energy driving the biosphere comes from the Sun and is used in photosynthesis to fix carbon. The limiting nutrients critical for life come primarily from nitrogen-fixing bacteria and from

phosphorus release by the slow geological weathering of rocks.

■ Population growth is determined by the availability of resources, which may decline or become degraded if the population grows too large. The carrying capacity of an organism is the maximum number of individuals that can exist sustainably, meaning that the environment is able to provide the population with the resources it needs and to assimilate its wastes. Populations may exhibit exponential growth followed by decline if they outstrip their resources.

■ Evolution generates biological diversity, with many species occupying narrow ecological niches determined by climatic conditions, diet, and nesting locations. Habitat destruction threatens these niches and can lead to loss of biodiversity.

■ Ecosystem services include producing environmental goods, purifying water, creating biodiversity, and pollinating crops, to name just a few, and are valued in the tens of trillions of dollars annually. Human population growth and land transformation for agriculture and settlement disrupts these services.

■ Human impacts on the biosphere include changes to biogeochemical cycles, the introduction of invasive species, habitat destruction and fragmentation, and the extinction of organisms, leading to a loss of biodiversity.

KEY TERMS

ecosystem (p. 195)

photosynthesis (p. 195)

respiration (p. 195)

limiting nutrients (p. 195)

biogeochemical cycle (p. 196)

food chains (p. 199)

autotrophs, or primary producers (p. 199)

heterotrophs, or consumers (p. 199)

detritivores (p. 199)

ecological succession (p. 200)

carrying capacity (p. 202)

predator–prey cycles (p. 203)

resilience (p. 203)

biomes (p. 203)

habitats (p. 205)

ecological niche (p. 205)

biodiversity (p. 205)

genetic diversity (p. 206)

inbreeding (p. 207)

species diversity (p. 207)

endemic (p. 209)

species richness (p. 208)

ecosystem services (p. 210)

extinct (p. 218)

extirpated (p. 218)

endangered (p. 220)

critically endangered (p. 220)

population viability analysis (p. 220)

demographic uncertainty (p. 220)

environmental uncertainty (p. 220)

natural catastrophes (p. 220)

genetic uncertainty (p. 220)

endemic species (p. 220)

restoration ecology (p. 221)

REVIEW QUESTIONS

1. What was the purpose of the Biosphere 2 project? What problems did it encounter, and what caused these problems?

2. For what purposes do organisms use energy?

3. Describe a typical food chain or trophic pyramid. Where does the energy at the base come from and how much is transmitted from one level to the next?

4. What role do detritivores play in food webs?

5. Explain Liebig's Law of the Minimum.

6. Which nutrients are most limiting for life and why?

7. Describe the similarities and differences between the carbon, nitrogen, and phosphorus cycles. Consider the forms (gaseous, liquid, solid) these elements exist in and the different reservoirs in which they are found.

8. How does carbon move from the ocean floor into the atmosphere?

9. Describe the processes that move phosphorus between the lithosphere and the biosphere.

10. What processes lead to ecological succession?

11. Explain the causes and consequences of exponential growth.

12. Can exponential growth continue forever? Why or why not?

13. How does carrying capacity influence population growth patterns?

14. What factors determine Earth's major biomes and how have humans modified these biosphere features?

15. The reintroduction of the wolf to Yellowstone National Park has triggered a "trophic cascade." Explain what this means and what the consequences have been for different species, both animal and plant.

16. What factors contribute to biodiversity? How is biodiversity measured?

17. What is an ecosystem service?

18. What factors have fostered and limited human population growth?

19. How have humans altered the global nitrogen and phosphorus cycles and what have been the impacts of these alterations on aquatic systems?

20. Are all species negatively impacted by habitat fragmentation? Why or why not?

21. Why do not all exotic species become invasive?

22. What are some mechanisms by which species go extinct?

THOUGHT QUESTIONS

1. What would happen to the carbon cycle if plate tectonic processes were to cease? How might this affect life on Earth?

2. If the winds offshore Peru transported ocean water toward that nation rather than away from it, what would be the impact on guano production on the Chincha Islands?

3. Can you think of ecosystem services provided by nature other than the ones described in this chapter?

4. What biome surrounds your hometown? What are the climatic conditions that give rise to this biome, and how might animals and plants be adapted to them? Using the map in Box 7-2, Figure 1, determine which anthropogenic biome surrounds your hometown. What are the similarities and differences between the biome and anthropogenic biome representations of the area?

5. Why do you suppose it is difficult to determine how many people Earth can support? What factors need to be considered and what uncertainties do they involve?

EXERCISES

1. Investigate the biological history of your hometown or the city where you are going to college. Which species that historically lived in this area have been extirpated? What consequences have these extirpations had on local biosphere functioning and ecosystem services?

2. Investigate the invasive species in your hometown or the city where you are going to college. Where did they come from? What conditions appear to foster their dispersal? How are the invasive species affecting native organisms? Have any control measures been instituted to control their spread?

3. In 2012, the human population exceeded 7 billion. If the population growth rate remains constant at 1.1 percent, what will the world's population be in 2050? At the end of the century?

4. In 1859, Thomas Austin released 24 European rabbits into Australia. By 1926 their population had grown to 10 billion. Rearrange Equation 2 to solve for the net growth rate required to achieve this staggering increase in the rabbit population.

SUGGESTED READINGS

Cohen, J. E. *How Many People Can the Earth Support?* New York: W.W. Norton and Company, 1995.

Cohen, J. E. "Human Population Grows Up." *Scientific American* 293 (2005): 48–55.

Daily, G. C., S. Alexander, P. R. Ehrlich, L. Goulder, J. Lubchenco, P. A. Matson, H. A. Mooney, et al. "Ecosystem Services: Benefits Supplied to Human Societies by Natural Ecosystems." *Issues in Ecology* 2 (1997).

Gende, S. M., and T. P. Quinn. "The Fish and the Forest." *Scientific American* 295 (2006): 84–89.

Harris, J. A., R. J. Hobbs, E. Higgs, and J. Aronson. "Ecological Restoration and Global Climate Change." *Restoration Ecology* 14, no. 2 (2006): 170–176.

Howarth, R., D. Anderson, J. Cloern, C. Elfring, C. Hopkinson, B. Lapointe, T. Malone, et al. "Nutrient Pollution of Coastal Rivers, Bays, and Seas." *Issues in Ecology* 7 (2000).

Jackson, S. T., and R. J. Hobbs. "Ecological Restoration in the Light of Ecological History." *Science* 31 (2009): 567–569.

Mack, R. N., D. Simberloff, W. M. Lonsdale, H. Evans, M. Clout, and F. Bazzaz. "Biotic Invasions: Causes, Epidemiology, Global Consequences and Control." *Issues in Ecology* 5 (2000).

Naeem, S., F. S. Chapin III, R. Costanza, P. R. Ehrlich, F. B. Golley, D. U. Hooper, H. J. Lawton, et al. "Biodiversity and Ecosystem Functioning: Maintaining Natural Life Support Processes." *Issues in Ecology* 4 (1999).

Ricklefs, R. E. *The Economy of Nature,* 6th ed. New York: W. H. Freeman, 2008.

Robbins, J. "Lessons from the Wolf." *Scientific American* 29, no. 6 (2004): 74–91.

van Andel, J., and A. P. Grootjans. "Restoration Ecology: The New Frontier," in J. van Andel and J. Aronson (Eds.), *Restoration Ecology*, pp. 16–28. Malden, MA: Blackwell, 2006.

Vitousek, P. M., J. Aber, R. W. Howarth, G. E. Likens, P. A. Matson, D. W. Schindler, W. H. Schlesinger, et al. "Human Alteration of the Global Nitrogen Cycle: Causes and Consequences." *Issues in Ecology* 1 (1997).

Soil and Weathering Systems

Scientists did not expect to find deep (up to 2 m) fertile soils that, for thousands of years, could be sustainably farmed in the Amazon Basin. Heavy rainfall in the tropics leaches soil, and high temperatures promote intense chemical weathering. These processes generate soils, orange to red in color, that are rich in iron and clay but are depleted in nutrients essential for crops.[1] When cleared of trees today, the thin cover of organic litter and soil beneath tropical forests quickly decomposes and the nutrient-poor subsoils bake hard in the sunlight. And yet archaeologists and soil scientists noted patches of "Terra Preta do Indio," Portuguese for "Indian Black Earth," in the Amazon as early as the late 1800s. Since then, scientists have explored fertile brown (terra mulata) to black (terra preta) soils, discovering that they cover perhaps as much as 10 percent of the Amazon and once supported sophisticated agricultural societies. In fact, these fertile soils are found only where agricultural societies once existed, supporting early Spanish explorers' accounts of an abundance of towns and people atop bluffs along Amazon rivers.[2]

Covering most of Earth's terrestrial surface, soils and weathered materials are both the products and the archival records of biogeochemical interactions among the lithosphere, atmosphere, hydrosphere, and biosphere. In the case of soils like the Amazonian dark earths, they also are a record of human interaction with the land.

In this chapter, we will:

✔ Examine the processes of weathering and soil formation that create the pedosphere and the resources in it that we use.

Scientists peer into an excavation in the Amazon basin to examine "terra preta," a carbon–rich, fertile soil both formed and farmed by humans. Studded with broken pre–Columbian pottery in this example, it was replenished with charcoal and organic waste for thousands of years, until Spanish explorers arrived and the indigenous populations were decimated by disease in the 1500s.
(Jim Richardson/National Geographic/Getty Images)

✔ Explore the causes of three hazards—soil erosion, desertification, and mass movement—and consider how they can be minimized.

✔ Investigate examples of changes in the pedosphere over human and geologic time periods.

INTRODUCTION

Soil formation generally is viewed as a slow process, one that requires thousands of years of weathering and mixing of decomposed organic and mineral matter in a relatively thin zone at Earth's terrestrial surface. Terra preta, a dark-colored anthropogenic soil rich in organic matter, and terra mulata, a lighter variety of soil that has a lower concentration of organic matter, seem to be exceptions (Figure 8-1). These soils collectively are referred to as Amazonian dark earth, and are among the few traces of pre-Columbian agricultural societies that remain readily visible in the Amazon region. Whether by chance or intentionally is unknown, but from about 7000 to 500 years ago native populations in the central Amazon region added charred residues, bones, and other organic waste to soil in and near the areas they inhabited. The exact procedure they used has not yet been determined, but the outcome was fertile patches of sustainable soil in the vicinity of early Amazonian villages that are surrounded today by infertile soils more typical of the tropics (Figure 8-2).

The high fertility of terra preta, twice that of nearby soil, has significant implications for **sustainable agriculture**.[3] Agriculture that is sustainable meets current needs without compromising those of future generations and promotes environmental health, economic profit, and social and economic equity.[4] Achieving the goal of sustainability requires that both natural and human resources be maintained, not depleted or degraded, over

the long term. The word sustain is from the Latin *sustinere—sus* (from below) and *tenere* (to hold)—meaning to maintain or support in order to keep in existence. Farming practices that maintain agricultural productivity while conserving resources and remaining commercially competitive and environmentally sound are sustainable.

The possibility of humans creating fertile soil and sustainable agriculture in the tropics portends a possible "black revolution" for soil in the developing world, in contrast to the 20th century "green revolution" for plants that relied upon hybrid seeds, pesticides, and synthetic fertilizers.[5] Terra preta has a large surface area of fine grains that absorb and retain moisture and nutrients. This engineered *super dirt* stores at least five to ten times as much carbon[6] and three times as much phosphorus and nitrogen as typical leached tropical soils.[7] It also is rich in microorganisms that break down minerals and organic matter to make more surface area and nutrients available, which explains its self-sustaining properties and resilience even in a highly leaching wet environment.

Terra preta could provide a relatively long-term sink for **carbon sequestration**, a geoengineering strategy in which carbon is deposited in a reservoir in order to reduce the amount of carbon dioxide (CO_2), a greenhouse gas, in the atmosphere (see Chapters 11 and 15). In this case, soil would be used as the reservoir to store carbon extracted from biomass by pyrolysis.[8] **Biomass** comprises any plant or animal matter, such as trees, grasses, crop residues, wood chips, or manure. **Pyrolysis** is the process of heating biomass at a low temperature (400°C–500°C) in the absence of oxygen. The product of pyrolyzed biomass, called **biochar**, contains two times more carbon and is much more durable than ordinary biomass. Whereas the carbon in plant matter decomposes readily in oxygenated soil environments, biochar carbon can persist in the soil for hundreds to even thousands of years, as in the Amazonian dark earth. Different

Figure 8–1 **Terra preta soil on the left (1 meter deep) and nearby oxisol on the right in the Amazon Basin.** The anthropogenically modified terra preta can contain 3 times as much nitrogen and phosphorus and 10 to 20 times as much carbon as surrounding soils. *(Bruno Glaser)*

Figure 8–2 Fertile soils of pre–Columbian anthropogenic origin are common in the Amazon and are sometimes associated with raised mounds characterized by high biodiversity. This "forest island" surrounded by scrub vegetation is an example of the rich vegetation that comes from these soils. *(Clark Erickson, University of Pennsylvania)*

forms of organic matter in soil, including biochar and decomposed plants, have been dated by carbon-14 radiometric dating (see Chapter 6), and show that biochar is much older than other forms of soil carbon.[9]

Scientists estimate that perhaps half of the carbon cycled annually through plants by photosynthesis and decomposition could be sequestered as biochar in soil instead (Figure 8-3). With its much slower rate of cycling and longer residence time, soil provides a better carbon sink, for example, than afforestation, the planting of trees in an unforested area. Conservative estimates for biochar in soil as a carbon sink are 1 to 2 billion metric tons of carbon each year.[8]

The implications of the sustainability and resilience of terra preta are significant for modern deforestation, leading some to call for **slash-and-char** rather than slash-and-burn approaches to logging in tropical rainforests. The complete combustion associated with slash-and-burn converts organic matter to ash and releases large amounts of carbon dioxide gas to the atmosphere. In

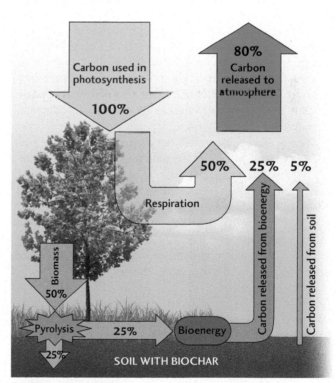

Without biochar

Net carbon withdrawal from atmosphere: **0%**

Carbon sequestration by photosynthesis:
Carbon neutral

With biochar

Net carbon withdrawal from atmosphere: **20%**

Biochar sequestration:
Carbon negative

Figure 8–3 Carbon is cycled from the atmosphere to plants by photosynthesis, and returned again to the atmosphere by plant respiration. Although about half this amount is sequestered as carbon in soil, the same amount is released to the atmosphere by decomposition and other organic processes in soil. The net result

is that no carbon is withdrawn from the atmosphere, a carbon–neutral condition. If, on the other hand, biochar is stored in soil and the energy from pyrolysis to produce biochar (bioenergy) is used to offset fossil fuel use, a net carbon withdrawal of 20% from the atmosphere is possible.

contrast, partial combustion by slash-and-char stores about 20 percent of the carbon that might be released to the atmosphere in soil.[8]

THE PEDOSPHERE: A GEOMEMBRANE TO OTHER EARTH SYSTEMS

The **pedosphere** is the entire layer of disaggregated and decomposed rock debris and organic matter at the surface of the continental Earth. Mass and energy from the lithosphere, hydrosphere, biosphere, and atmosphere are transported across the pedosphere. Because it functions much the same as a biomembrane that surrounds a living cell, the pedosphere can be thought of as the solid Earth's *geomembrane* (Figure 8-4). Dissolved ions, solid particles, and gases move back and forth across this geomembrane, altering the nature of the underlying lithosphere and creating a nutrient-rich environment for plants and animals.

The alteration of rocks and minerals at this geomembrane by physical, chemical, and biological processes is called *weathering*. Rocks in Earth's crust are strong and generally have low permeability, meaning that fluids move slowly through them. In an intricate series of feedback processes that operate at time scales of less than years to more than tens of thousands of years, however, weathered rocks are broken down and converted to particles that can form soils, hold and transmit water, and cover hillslopes. Such particles also can be eroded as sediment that eventually makes its way to floodplains, deltas, and beaches.

Weathered rocks and minerals provide essential elements to plants and animals, which in turn act to transform weathered minerals, rock fragments, and organic matter into soil, thus releasing the elements needed for life. In fact, this essential quality of soil to terrestrial life led scientists to define Earth's near-surface as the **critical zone**, in which complex interactions among "rock, soil, water, air, and living organisms regulate the natural habitat and determine the availability of life-sustaining resources."[10]

Crust—Earth's weathered face—became susceptible to weathering as soon as it formed at the surface during Earth's early evolution. Rock materials exposed to air, water, ice, wind, acidic solutions, and biological phenomena at and near Earth's surface *disintegrate* physically and *decompose* chemically. These are different weathering processes, but they generally occur in association with each other. Disintegration, or **physical weathering**, refers to the mechanical fragmentation of rocks and minerals. Decomposition, or **chemical weathering**, refers to the chemical alteration of rocks and minerals. Both physical

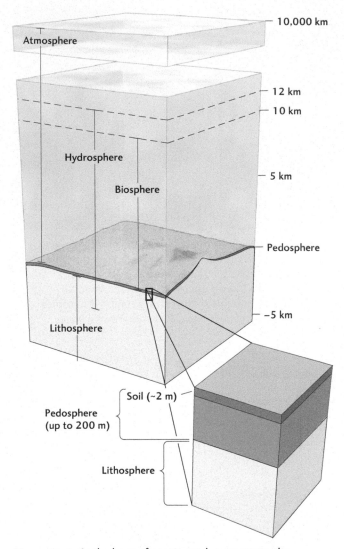

Figure 8-4 At the base of continental ecosystems, the pedosphere forms an interface between the continental ecosystem and the other earth systems. Soil, the 1- to 2-meter-thick portion that includes organic matter, forms at the top of the pedosphere, where all of Earth's systems interact. Below the soil, weathered rock grades downward to unweathered rock, forming an uneven layer up to 200 m thick.

and chemical weathering are enhanced by biological processes such as root growth along fractured rocks, or churning of soil by worms and other organisms.

Chemical reactions occur on the surfaces of rocks and minerals, so physical weathering facilitates chemical weathering by breaking apart fresh rock to expose more surface areas (Figure 8-5). The total volume of the rock remains about the same and in place, but the amount of surface area exposed to further weathering increases with the increasing number of particles.

Weathering occurs in one place; the weathered rock and mineral fragments remain at their original site. Once

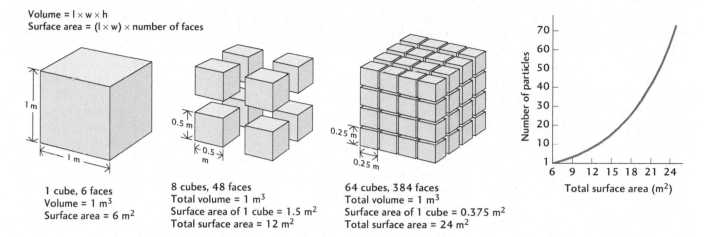

Volume = l × w × h
Surface area = (l × w) × number of faces

1 m
1 m

0.5 m
0.5 m

0.25 m
0.25 m

1 cube, 6 faces
Volume = 1 m³
Surface area = 6 m²

8 cubes, 48 faces
Total volume = 1 m³
Surface area of 1 cube = 1.5 m²
Total surface area = 12 m²

64 cubes, 384 faces
Total volume = 1 m³
Surface area of 1 cube = 0.375 m²
Total surface area = 24 m²

Number of particles
70
60
50
40
30
20
10
1

6 9 12 15 18 21 24
Total surface area (m²)

Figure 8–5 Physical weathering disaggregates rock into many small particles, increasing the amount of rock surface area exposed to chemical weathering. The total volume of the rock remains the same.

rock has been weathered, it is more easily carried away by a moving medium such as wind, water, or ice. The removal and transport of material by these processes is called **erosion.**

Energy is the capacity for doing work, which is the change effected by the application of a force. Weathering, like all processes, requires energy to do work (here, to change rock to weathered debris), and some force must act over a given distance to do that work. The force exerted on an area, in this case an area of rock, is known as **stress.** Just as rocks are deformed, or subject to *strain,* by plate tectonics processes, the actions of weathering deform Earth materials. (Stress is a measure of an applied force; strain is a measure of the resultant deformation.)

In physical weathering, stresses are created by physical forces such as the thermal expansion of minerals when heated by grassland fires. In chemical weathering, stresses are created by chemical forces such as the incorporation of water molecules into an expandable clay mineral. The greater the stress exerted on rocks and minerals, the greater the work that can be done to disintegrate and decompose them. If the stresses exceed the strengths of the chemical bonds holding the minerals together, then the minerals begin to disintegrate and decompose.

Physical Weathering

Of the many types of physical weathering processes, several are common worldwide and are responsible for most disintegration of rocks and minerals. These processes are exfoliation jointing, thermal expansion, disintegration associated with plant growth and other biologic activity, and frost weathering. Chemical weath-

ering that causes minerals to change in volume (such as through expansion) also contributes to physical weathering by increasing the amount of the surface area of the mineral that is exposed to weathering.

Exfoliation Jointing In some places, prominent fractures parallel to Earth's surface give the surface the appearance of being split into sheetlike slabs of rock. The extensive fractures are called **exfoliation joints,** or sheeting joints, as they divide the rock into multiple thin sheets (Figure 8-6).

Figure 8–6 Exfoliation weathering in Yosemite National Park, California, produces sheetlike masses of rock with large surface areas that are exposed to other weathering processes. *(© Peter L. Kresan)*

These joints are of primary importance to other weathering processes because they weaken rock and expose many fresh rock surfaces to chemical weathering by water and atmospheric gases. They also contribute substantially to groundwater flow in many places.

Scientists thought for many decades that exfoliation jointing might be caused by the gradual removal of overlying rock by erosion. A scientist working in the Sierra Nevada recently discovered, however, that it is the curvature and compressive stress parallel to Earth's surface that cause these remarkable joints.[11] Using fundamental equations based on a force balance, this research shows that when rock is under high compression parallel to Earth's surface, tension can arise perpendicular to the surface at shallow depths beneath convex topography (see Figure 8-6). Rock exposures with curvature, as along the walls of canyons, are conducive to tension in the rocks and joint formation. To test this hypothesis, the scientist used high-resolution laser mapping of Earth's surface to determine Earth's curvature at various locations. He also measured rock stress to predict where exfoliation joints would or would not occur. The results matched predictions based on the equations, strongly supporting the scientist's hypothesis. Erosion typically generates curved surfaces, so erosion plays a role, but not in the way thought by earlier scientists.

Thermal Expansion Minerals in rock expand when they are heated, and this might also contribute to disaggregation of some rock types. In the 1800s, early scientific explorers unfamiliar with hot deserts thought that much of the rubble they saw strewn about the landscapes of Africa, Australia, and North America was caused by intense heating during the day and significant cooling at night. They postulated that extreme daily fluctuations in temperature could cause repeated expansion and contraction of minerals and thereby the disintegration of rock in a process called *thermal expansion*.

One indication that thermal expansion could cause disintegration of rock is the traditional use of wood fires in quarries in India to heat rock and loosen slabs up to 15 cm thick from the quarry floor. Some scientists have proposed that natural fires are likely to reach temperatures high enough to shatter rock as well. Rock is a poor conductor of heat, so the interior of a rock mass receives much less heat than does its exposed exterior surface. As a consequence, outer parts of rock expand more than do inner parts. This variability in expansion generates substantial differences in stress throughout the mass and causes portions of the rock to break off. Recently, scientists have documented thin flakes (less than 5 cm) of rock that have spalled, or peeled, off exposed rock surfaces during range fires (Figure 8-7). Fires can occur naturally, as from lightning, and over many thousands of years the process of spalling can be quite effective in causing rock disintegration.

Daily temperature changes in some natural environments with extreme conditions might be large enough to provide the tensile stresses necessary for granular disintegration. The bonds within individual minerals, referred to as *intracrystalline* bonds, are greater than those between each crystal, referred to as *intercrystalline* or intergranular bonds. As a result, individual crystals can be released—that is, broken apart—if the tensile stresses on the minerals exceed the strength of the intercrystalline bonds. This process of grain-scale weathering is referred to as *granular disintegration*. Some scientists have proposed that the temperature must change at least 2°C per minute in order to generate the stresses needed to cause granular disintegration.[12] The technical challenge

(a)

(b)

Figure 8–7 (a) Rock surfaces can spall as a result of exposure to intense heat, as from (b) a range fire in the Sierra Nevada in California. *(a: Dorothy Merritts; b: Paul Bierman)*

Figure 8–8 Frost weathering shaped much of the angular, fragmented rock that has accumulated at the base of this slope on Belstone Tor in Devon, England. Actively forming rubble slopes are common at high latitudes and in mountainous regions at high altitudes, where diurnal and seasonal freezing conditions are common. Relict, or fossil, slopes of rock produced by frost weathering found at mid–latitudes date to the time of the last full glacial conditions on Earth, about 15,000 to 25,000 years ago. *(Miles Wolstenholme)*

in determining the necessary amount and pace of change is in measuring such small changes in temperature over short time periods (seconds to minutes) at the scale of a few grains in depth within a rock.

Some scientists recently have managed to make such measurements in laboratory and field settings, with the latter in the cold and hot dry deserts of Antarctica and Atacama, Chile, respectively.[13] The scientists measured thermal variations at the spatial scale of millimeters to centimeters, and at temporal scales of seconds. They found that the intergranular stress fields caused by fluctuations in temperature are complex and sufficient, at times, to cause granular disintegration. The absolute temperature, whether very cold or hot, is not critical; rather, it is the rate of change of temperature that is important. The amount that mineral grains heat varies, and for some minerals—such as quartz, the most common mineral—the thermal conductivity and coefficient of thermal expansion are not the same in all directions.

Frost Weathering In its frozen form, water can split rock, a notion widely considered evident by the accumulation of angular, gravel- to boulder-size debris along slope bases in high alpine and arctic areas (Figure 8-8).

The conventional model of how water can split rock during freezing was related to one of water's important properties: It expands when frozen. Recent investigations reveal that this conventional view of volumetric expansion of water leading to high stresses within rock is limited to unusual situations in which water is confined within sealed pore spaces and unable to migrate during freezing. New research proposes that water migrates toward freezing centers at temperatures well below

$0°C$ (generally $-3°C$ to $-10°C$) and causes cracks in the rock to grow.[14] As long as low temperatures are sustained and the supply of water is maintained, these cracks can grow substantially. A key difference between this and the early conventional model is that the earlier model only explained the splits that involved repeated freeze-thaw cycling; the new model accounts for splits regardless of cycling. This new model explains a broader set of environmental conditions and is now accepted as the more general explanation for the mechanism of frost weathering.

Chemical Weathering

During chemical weathering, water, ions, and oxygen react with exposed mineral surfaces. The general chemical weathering process is the reaction of unweathered minerals in crustal (mostly igneous) rocks, known as *primary minerals*, with water, acidic solutions, and oxygen from the hydrosphere, biosphere, and atmosphere. The products of these reactions are fragments of primary minerals, newly formed *secondary minerals*, and dissolved ions (Figure 8-9). Together, the residual fragments and secondary minerals released by weathering constitute *sediments*. Mechanical and chemical weathering continues to produce sediments and dissolved ions that eventually make their way to the oceans and are deposited as clastic, chemical, and biological sediments. Most chemical weathering results from exposure to acid, oxygen, or both, as expressed in the following equation:

Primary minerals + acids + oxygen →

Sediments (rock fragments + secondary minerals) + dissolved ions

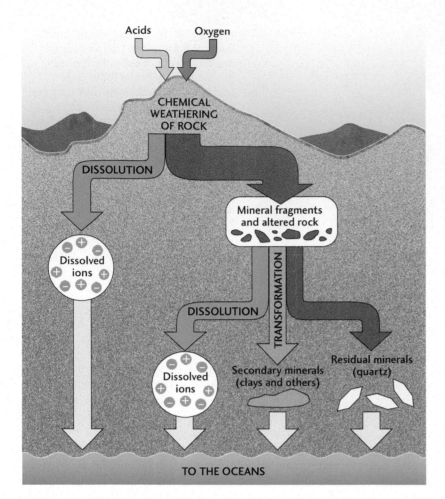

Figure 8–9 Chemical weathering decomposes primary minerals in rocks and makes the rock mass more vulnerable to fragmentation. Most chemical weathering is caused by the reaction of minerals with acidic solutions and oxygen from the atmosphere, hydrosphere, and biosphere. Products of chemical weathering include fragments of primary minerals, altered (secondary) minerals, and dissolved ions. With continued weathering, residual minerals, clays, and other secondary minerals accumulate in the pedosphere or are washed into the oceans along with dissolved ions.

Acid Formation in Soil The effect of acids is a particularly important factor in the decomposition of the upper lithosphere. Acids can partly or even wholly decompose solid mineral structures. The small, mobile hydrogen ions in acids replace cations of other elements on mineral surfaces, thus breaking apart the original mineral structure.

The more hydrogen ions in water, the greater the solution's acidity. The acidity of a solution is expressed as its pH or "potential for hydrogen." The pH scale runs from 0 to 14 and is logarithmic, so each unit of pH represents a tenfold change in the concentration of H^+ ions. A pH of 7 indicates a neutral solution; pure water has a pH of 7. A pH value below 7 indicates acidity, and a pH above 7 indicates alkalinity.

The lower the pH value, the more acidic is the solution (Figure 8-10). The pH scale is in negative powers of 10, so a pH of 1 indicates that there are 10^{-1} (or 0.1) gram of H^+ per liter of solution, a pH of 2 indicates only 10^{-2} (or 0.01) gram of H^+ per liter, and so on. Consequently, a solution with a pH of 4 is 100 times more acidic than a solution with a pH of 6 because it has that many more free hydrogen ions. While natural rainwater is slightly acidic, with an average pH of about 5 to 6, the pH of rainfall measured in areas downwind of some industrial centers has been as low as 3, an acidity comparable to that of vinegar.

One of the most common acids in the soil environment is carbonic acid, which forms when rainwater combines with CO_2 gas in the atmosphere (Table 8-1, Equation 1). Because the formation of carbonic acid removes CO_2 from the atmosphere, it is a key component of models of global climate change (Box 8-1). The carbonic acid in unpolluted rainwater gives it a pH between 5 and 6, so it is slightly more acidic than a neutral solution. In the soil zone, however, plants and decaying organic matter release additional CO_2 gas that can react with infiltrated rainwater to produce more carbonic acid. As a result, concentrations of carbonic acid are 10 to 200 times higher in soil moisture than in the atmosphere, and the pH of soils is sometimes as low as 4.

Carbonic acid can dissolve calcite at and near the Earth's surface in a chemical weathering process known as **carbonation** (see Table 8-1, Equation 2).

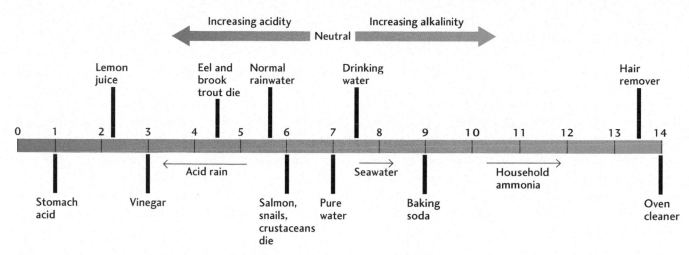

Figure 8–10 Common solutions and their acidity or alkalinity on the pH scale and the acidities at which some freshwater life-forms die.

TABLE 8–1 Common Chemical Weathering Reactions at Earth's Surface

Equation 1. Formation of carbonic acid from water and carbon dioxide:

$$H_2O + CO_2 \leftrightarrow H^+ + HCO_3^- \leftrightarrow H_2CO_3$$

Water — Carbon dioxide gas — Hydrogen ion — Bicarbonate ion — Carbonic acid

Equation 2. Carbonation of calcite in carbonic acid:

$$CaCO_3 + H_2CO_3 \leftrightarrow Ca^{+2} + 2\,HCO_3^-$$

Calcite — Carbonic acid — Calcium ion — Bicarbonate ion

Equation 3. Hydrolysis of a silicate mineral (in this example, sodium feldspar) in carbonic acid and formation of a clay mineral (kaolinite):

$$2\,NaAlSi_3O_8 + H_2CO_3 + 9\,H_2O \rightarrow 2\,Na^+ + 2\,HCO_3^- + 4\,H_4SiO_4 + Al_2Si_2O_5(OH)_4$$

Sodium feldspar — Carbonic acid — Water — Sodium ion — Bicarbonate ion — Silicic acid — Kaolinite

Equation 4. Oxidation of a mineral (in this case, pyrite) and formation of iron hydroxide and acid (common cause of acid mine drainage) occurs in a series of reactions:

$$2\,FeS_2 + 7\,O_2 + 2\,H_2O \rightarrow 2\,Fe^{2+} + 4\,SO_4^{2-} + 4\,H^+$$

Pyrite — Oxygen gas — Water — Iron ion — Sulfate ion — Hydrogen ion

The iron ion can be oxidized further:

$$4\,Fe^{2+} + O_2 + 4\,H^+ \rightarrow 4\,Fe^{3+} + 2\,H_2O$$

Iron ion — Oxygen gas — Hydrogen ion — Iron ion — Water

The iron can be precipitated as an iron hydroxide mineral:

$$Fe^{3+} + 3\,H_2O \rightarrow Fe(OH)_3 + 3\,H^+$$

Iron ion — Water — Iron hydroxide — Hydrogen ion

BOX 8–1 GLOBAL AND ENVIRONMENTAL CHANGE

Tectonic Uplift, Rock Weathering, and Long–Term Climate Change

For several decades, scientists have investigated the links among tectonic uplift and mountain building, rock weathering, and climate, leading to a plethora of studies of rocks, fossils, deep sea sediments, and isotopes.[15] A possible scenario begins with one of the most profound geologic phenomena of the Cenozoic Era: rapid uplift of the Tibetan Plateau and Himalayan Mountains of Asia to high altitudes (**Figure 1**). Tectonic uplift of this broad region during the past few tens of millions of years raised the landmass so high that the flow of air masses containing moisture that has evaporated from the Indian Ocean has been altered. During the past few million years, a period of global cooling, glaciers have formed at high altitudes in this region (and elsewhere on Earth; see Chapter 14). Moist air masses can produce intense monsoons that contribute to river flow and snow that becomes glacial ice and meltwater. Glaciers entrain sediment and cut into bedrock as they flow down valley. All of these processes promote increased rates of landscape erosion, which in turn exposes more fresh rock to weathering and erosion (**Figure 2**).

The amount of fresh rock exposed to weathering at Earth's surface affects the chemistry of the atmosphere (**Figure 3**). Water and CO_2 in the atmosphere combine to produce carbonic acid, which dissolves silicate and carbonate minerals and releases the bicarbonate ion (HCO_3^-) in solution. Through this linkage, increased exposure of fresh rock to chemical weathering can increase the amount of bicarbonate in solution. Ultimately, bicarbonate ions make their way downstream through groundwater and runoff to the ocean, where they recombine with calcium ions in the calcareous shells of marine organisms, eventually to be buried with organic matter and sediment or to be converted into limestone.

Regardless of the specific outcome, the uplift of such a great landmass affects the cycling of carbon from Earth's interior to its atmosphere and oceans, and this can affect climate. Carbon in the atmosphere originates from the outgassing that results when subsurface rocks in volcanoes melt. Carbon has a short residence time—on the order of days—in the atmosphere. As carbon from the atmosphere is used

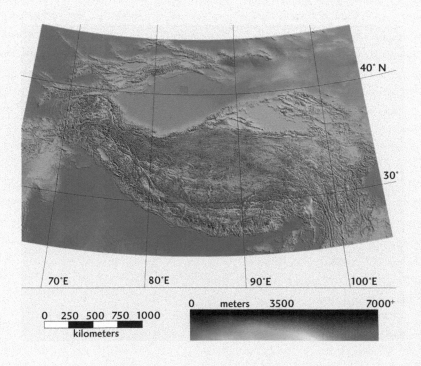

Figure 1 This topographical map of the Tibetan Plateau, bordered by the Himalayan Mountains to the south, shows the extensive area of land that has been raised many kilometers above sea level (shades of green, yellow, and red) by recent tectonic uplift.

Figure 2 The rugged Himalayan range extends thousands of kilometers across the Eurasian continent. Rivers and glacial ice carve valleys into this rapidly uplifting landmass, generating some of the greatest relief and exposures of fresh rock on Earth. This exposed rock is susceptible to physical, chemical, and biological weathering processes. *(Earth Sciences and Image Analysis Laboratory/Johnson Space Center)*

during chemical weathering processes, it can be transferred to the oceans, where its residence time is millions of years rather than days. In turn, the amount of heat-absorbing CO_2 in the atmosphere would be reduced and global climate might respond with a cooling trend.

This linkage can be envisioned as follows: Tectonic uplift leads to increased erosion, which leads to greater exposure of fresh rock to weathering, which leads to drawdown of CO_2 from Earth's atmosphere and the burial of carbon in the oceans, which leads to global cooling. Another scenario is possible though, presenting a scientific conundrum similar to that of the classic parable of "which came first, the chicken or the egg?"[16] What if global cooling leads to the formation of glaciers and higher rater of erosion, which in turn cause accelerated weathering and drawdown of CO_2 from the atmosphere, ultimately causing yet more cooling? Perhaps the removal of massive amounts of rock by weathering and erosion then causes an isostatic response of Earth's crust such that uplift occurs. In this second case, uplift is an out-

come of cooling, whereas in the previous case, tectonic uplift is the initial driver rather than the outcome, and cooling is the outcome. You can see why scientists refer to this problem as similar to that of the chicken and the egg. Scientists continue to collect samples and data and to model the possible feedbacks among uplift, weathering, and climate in order to resolve which came first and how each responds to changes in the other.

Considering all the attention being paid to global warming, you might think that this scenario of geologically recent global cooling does not make sense. The answer to this seeming contradiction is that climate change occurs over different scales of time (see Chapters 14 and 15). In fact, the global climate has been cooling for the past few tens of millions of years and, in particular, for the past few million years. Superimposed on this long-term cooling during the past approximately 2 million years are alternating fluctuations between relatively cooler (glacial) and warmer (interglacial) conditions. During the cool, full-glacial episodes,

(continued)

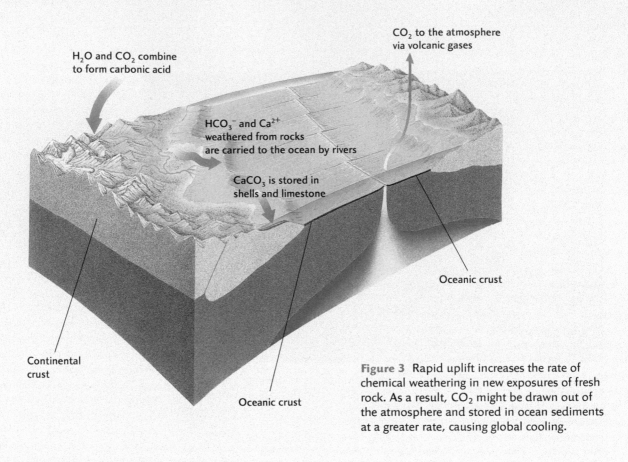

H$_2$O and CO$_2$ combine to form carbonic acid

CO$_2$ to the atmosphere via volcanic gases

HCO$_3^-$ and Ca^{2+} weathered from rocks are carried to the ocean by rivers

CaCO$_3$ is stored in shells and limestone

Continental crust

Oceanic crust

Oceanic crust

Figure 3 Rapid uplift increases the rate of chemical weathering in new exposures of fresh rock. As a result, CO$_2$ might be drawn out of the atmosphere and stored in ocean sediments at a greater rate, causing global cooling.

(continued)

massive ice sheets have waxed on continents while smaller glaciers have advanced down valleys at high latitudes and altitudes. Over the past 11,000 years, Earth has entered one of the warmer interglacial intervals. The modern warm period is cooler than the global climates of about 30 million years ago, but warmer than the cyclical full–glacial episodes that have marked the past 2 to 3 million years.

Over an even shorter time scale, that of the past century, scientists estimate that average global temperature has increased about 0.5° to 1°C, perhaps in response to the accumulation of greenhouse gases produced since the

Industrial Revolution (see Chapter 15). Some scientists point out that although the amount of warming might appear to be minor, the actual rate of warming over the past century is relatively rapid in comparison to changes in temperature over the past few million years. It is this relatively rapid, centennial–scale warming to which scientists refer when describing global warming due to human activities. This example of cooling possibly caused by the interplay between uplift and weathering illustrates the importance of specifying the timescale when referring to climate change.

Calcite (CaCO$_3$) is the mineral that forms the sedimentary rock limestone and its metamorphic equivalent, marble. Carbonation that occurs over long (geologic) spans of time results in dramatically etched landscapes and features such as sinkholes, which form when the rock structure and overlying soils collapse into caverns and solution fractures along which rock has been removed by carbonation (Figure 8-11).

Carbonic acid is not the only source of acid in soil. Organisms produce a number of organic acids. For example, plant roots release citric acid, the acid in lemons and other citrus fruits. The more organic matter in the soil, the more acidic the soil is. Acids less common in most soils are nitric and sulfuric acids, which are formed from the combination of water with nitrogen dioxide and sulfur dioxide, respectively. Soils developed on rock

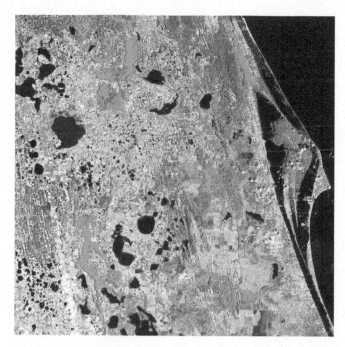

Figure 8–11 A satellite view of eastern Florida, where much of the subsurface is cavernous limestone and the surface is marked with sinkholes where cavern roofs have collapsed. Caverns and sinkholes are common in carbonate rocks, such as limestone, which are dissolved easily by carbonic acid. The point of land at right is Cape Canaveral. *(USGS/NMD EROS Data Center)*

types that contain sulfide minerals such as pyrite (an iron sulfide; see Chapter 4) sometimes have high concentrations of sulfuric acid, especially where mining activities have crushed rock and exposed large areas of sulfide mineral surfaces to oxygen (Figure 8-12).

In a reaction called **hydrolysis** because it involves a chemical reaction between a compound and water, carbonic and other acids can degrade silicate minerals (see Table 8-1, Equation 3). Most rocks in the lithosphere consist of silicate minerals, so hydrolysis is an important and common form of chemical weathering. Hydrogen ions from acidic soil solutions replace cations in silicate mineral structures, and the silicate cations in turn are released into solution. Some of the remaining elements and altered minerals form secondary **clay minerals** (Box 8-2). Clay minerals are crystalline sheet-structure silicates characterized by small particle size.

Oxidation The second basic process of chemical weathering common at Earth's surface results from the abundance of oxygen in the atmosphere. During **oxidation,** an element combines with oxygen to form oxide or hydroxide minerals (see Table 8-1, Equation 4). One of the most commonly oxidized elements is iron, which is abundant in the silicate minerals hornblende, biotite, olivine, and pyroxene. The oxidation of pyrite (FeS_2), the most abundant of sulfide minerals, alters it to an iron oxide called hematite (Fe_2O_3), and the released sulfur dissolves in water to form sulfuric acid. This corrosive acid is common in the water runoff, or drainage, from mining operations. *Acid drainage* from mines into bodies of water at the surface has caused substantial environmental degradation in some areas (see Figure 8-12).

Solubility Most elements and minerals are to some extent soluble in soil water, where typical pH values range from 4 to 9. However, solubility varies enough that some elements are more likely to be removed from the soil and carried to streams by percolating waters, while others accumulate in the soil over time. Calcium, magnesium, sodium, and potassium are highly soluble and are the most common cations found in streams,

Figure 8–12 Acid mine drainage, containing sulfuric and other corrosive acids, is created by the oxidation of sulfide minerals exposed during mining operations. At this copper mine in Michigan, with tailings piles in the background, acidic waters (red) are treated before being released into nearby streams. *(Lowell Georgia/Science Source)*

BOX 8-2 GEOLOGIST'S TOOLBOX

The Unusual Properties of Clay Minerals

Knowing the amount of clay in a soil is crucial to determining its fertility and engineering qualities. Soil consists of broken bits of primary minerals as small as sand (0.06 to 2 mm in diameter) and silt (0.004 to 0.06 mm in diameter), but its smallest mineral component—clay—makes it quite different from ground-up parent rock. Clay minerals, which are less than 0.004 mm in diameter, are secondary minerals created by the transformation of primary minerals during chemical weathering. In contrast to the larger grains of sand and silt, tiny clay minerals are highly reactive. Clays absorb and expel water, shrink and swell, attract cations, and retain *nutrients*—compounds that contain phosphorus and nitrogen that can be used directly by plants. Were it not for the capacity of clay minerals to hold nutrients in soils, land plants might not have been able to colonize the continents during the early Paleozoic era.

More than 20 different types of clay minerals are formed by the chemical weathering of mica, feldspar, olivine, and amphibole silicate minerals. Scanning electron microscopes (SEMs), with their powerful magnification, give Earth scientists a close look at the structure of clay minerals. Clay minerals are constructed from flat sheets mostly of silicon, oxygen, and aluminum that are bonded to one another by ion-sharing with cations. When arranged in stacks, the sheets produce layered clay mineral structures with extensive surface areas, much like the pages in a book (**Figure 1**). One gram of clay (about the volume of a pencil eraser) has as much as 800 square meters (about 9000 square feet) of surface area, nearly the area of an Olympic-size swimming pool![17]

The attraction and retention of cations, some of which are plant nutrients, by clay minerals (and organic matter) keeps them in the root zone; otherwise they would be washed away with water moving through the soil. This property of soils is known as *cation-exchange capacity*, the maximum quantity of cations a soil is capable of holding and that are available to be exchanged with the soil solution. The surface area of clay minerals affects their interactions with ions, other particles, and water. Negative ions—anions—can be found along the edges of clay sheets.

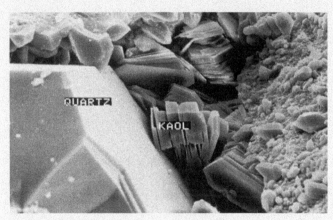

Figure 1 Kaolinite, a clay mineral formed during weathering, has grown in open pore spaces adjacent to the quartz grains seen in this scanning electron micrograph. The book-like structure of the clay mineral, a sheet silicate, is evident. *(F. J. Longstaffe, The University of Western Ontario)*

while aluminum, iron, and silica are commonly found in soils. The residual elements left in a weathered soil typically form aluminum and iron oxide minerals, clay minerals, and quartz.

Biological Processes and Their Role in Weathering

The growth and expansion of foreign substances along open spaces in rocks also can promote physical weathering. A common biological example in the upper few meters of the pedosphere is caused by the growth of tree roots (Figure 8-13). Trees wedge their roots into weak zones in rocks and split them apart as they grow, causing the ever-present bulges and cracks in sidewalks near mature trees. Plant growth in soil and weathered rocks causes **bioturbation,** the disintegration and mixing of organic and mineral matter and aeration of soil. Although roots can provide strength to soil over the short term, as in trees holding soil on a steep slope, over the longer term, roots cause disintegration and soil mixing that intensify physical weathering.

Termites also bioturbate soil by swallowing and churning through massive amounts of organic and mineral debris at Earth's surface. Partnered with microbes in their guts, they rework as many as tens of thousands

Oppositely charged particles attract each other, so the positive cations in soil solution are attracted to the clay particles. From the clay mineral surfaces, cations can move freely to the roots of plants, where they are drawn upward and distributed to stems and leaves. One teaspoonful of soil with abundant clays and organic matter might contain as many as 1.2 quintillion (1.2×10^{21}) sites where nutrients in the form of cations can be held and exchanged with plants. In general, soils with a higher cation–exchange capacity are more fertile.

Clay minerals also contribute to the properties of soil *expansion* and *plasticity,* both of which increase the soil's susceptibility to irreversible deformation. These properties play an important role in hazards, such as landslides, and are important to consider when engineering structures on different soil types. Soil expansion occurs when water enters clay minerals, causing the minerals to increase in volume. Soil plasticity refers to a soil's ability to deform in a pliable, rather than an elastic or brittle, manner. Both these properties are related to water content in the soil. Water molecules are asymmetrical in charge, with positive and negative poles, so their positive poles collect around the negatively charged surfaces of clay minerals in an oriented fashion, much the way iron filings are attracted to a magnet. Clays with weakly bonded sheets can incorporate a lot of water between the sheets, allowing the mineral structure to expand substantially during rainy seasons. Certain *expandable clays* can increase in volume by as much as 2000 percent when wet. Such soils can become

Figure 2 Soils with certain types of clay minerals are very sticky and plastic when wet and can expand, or swell, by absorbing water into their mineral structures. During dry seasons, expandable clay soils shrink and harden, which causes deep cracks. *(Frank Fennema/Shutterstock)*

too wet and sticky to till, and clay–rich dirt roads can become impassable.

Furthermore, the attraction between the individual clay particles is reduced as the mineral is wetted and swells; thus the particles become more mobile and can slip past one another more easily, resulting in plasticity (**Figure 2**). Plastic soils lose their rigidity and weaken when wet, causing roads and building foundations constructed above them to fail. During dry seasons, expandable clays expel water and shrink substantially. Although soils with expandable clays become more rigid when dry, they also can crack and become too hard to till.

of kilograms of soil in a single hectare each year. Their mounds dot the landscape in subtropical grasslands in various parts of the world.

Similarly, animals such as worms disaggregate mineral particles, bioturbating and aerating the soil, while their feces form nutrient-enriched pellets that promote soil fertility. The English naturalist Charles Darwin studied worms and wrote of their role in soil-forming processes in the late 1800s:

Worms have played a more important part in the history of the world than most persons would at first suppose. In almost all humid countries they are extraordinarily numerous, and for their size possess great muscular power...the whole of the superficial mould [topsoil]...has passed, and will again pass, every few years through the bodies of worms. The plough is one of the most ancient and most valuable of man's inventions; but long before he existed the land was in fact regularly ploughed, and still continues to be thus ploughed by earthworms.... When we behold a wide, turf-covered expanse, we should remember that its smoothness, on which so much of its beauty depends, is mainly due to all the inequalities having been slowly levelled by worms.[18]

Figure 8–13 The roots of a strangler fig tree extend throughout and assist in the crumbling of a Buddhist temple built in the late 12th to early 13th centuries in Angkor, Cambodia. *(Thinkstock)*

Following Darwin's astute observations and insights, scientists have investigated many instances of physical weathering processes by organisms, including the mound- and tunnel-building of termites, gophers, and prairie dogs.

A complex but important form of biological disintegration of minerals is caused by organisms that are too small to be seen with the unaided eye: microbes. These microscopic organisms include bacteria, archaea, fungi, and protists. In association with vascular plants, microbial communities promote physical and chemical weathering of minerals and increase the availability of nutrients derived from minerals in soils.[19]

In forest ecosystems in temperate and boreal regions, for example, the roots of trees are connected closely to the soil through an extensive network of fungal *hyphae* (the long, branching filaments of fungus) that fosters a symbiotic relation between the two types of organisms. In fact, this symbiosis produces a special environment that is called the *rhizosphere*, the soil zone that both surrounds and is influenced by the roots of plants. A thin film of fungi strands can coat the roots of trees and help the plants acquire nutrients from minerals in the soil. Experiments have shown that the fungi can apply high mechanical stresses to mineral grains, causing them to bend and even break.[20] After mechanically deforming a mineral's crystal structure, fungi contribute to dissolution and oxidation reactions that further decompose minerals. In return, the fungi receive carbohydrates from the plant they colonize. Healthy forests depend upon the ecological balance fostered by fungi and other microbes in the rhizosphere.

Some researchers have found that microorganisms are concentrated in a type of *biofilm* on the surfaces of roots, minerals, and hyphae (the long, branching filaments of fungus). This concentrate of microorganisms and the interactions between them and the roots of plants localizes and enhances weathering, contributing to the transfer of weathering-derived nutrients to water in soil and to plants. The specific processes, though complex and not yet well understood, depend upon the types of microorganisms, climate, and many other factors.

Soil Profiles and Soil–Forming Factors

Earth scientists describe **soil** as an internally organized, natural body of weathered mineral and organic constituents arranged in **soil horizons,** zones roughly parallel to Earth's land surface (Figure 8-14). The vertical arrangement of soil horizons forms a **soil profile,** which often appears as multicolored layers that are visible in road cuts and housing foundations. The total thickness of the pedosphere varies from 1 m to 200 m, with greater depths in areas where rainfall and temperature are high and weathering processes extend deeply into the crust. Horizonated soil, however, which is intimately linked to the plants and animals at the surface, is rarely more than a few meters thick.

Soil scientists label soil horizons with letters, starting at the surface. Above all is the *O horizon,* a litter of dead plants and animals over the surface. Below the O horizon, all other horizons are placed in three separate groups known as the *A, B,* and *C horizons.* In moist, temperate climates, the *A horizons* form the dark, organic-rich part of soil, also known as *topsoil.* Below are the *B horizons,* mineral-rich layers that vary in color from browns, reds, and yellows to grays and blues. Clays and other minerals, weathered from upper horizons and transported downward, accumulate in the B horizons. At even greater depths, soils grade into slightly weathered parent material with less distinctive *C horizons,* then to

Figure 8–14 The roots of these sunflowers in Kansas extend about a meter below ground, into fertile soil that is characterized by a dark, organic–rich A horizon near the surface, and browner, more mineral–rich B and C horizons at greater depth. *(Jim Richardson/National Geographic/Getty Images)*

weathered parent material with no horizons, and finally into unweathered parent material. Sometimes an E horizon, one in which *eluviation* (leaching) is significant, occurs between the A and B horizons.

Studies of soil variations throughout the world point to five major *soil-forming factors:* climate; relief of the ground surface; nature and composition of parent material; amount of time during which the soil has formed; and amount and types of living organisms. Climate may be the most significant factor in the fertility of a soil, as well as in the nature of its profile. The climate of an area is characterized by its precipitation and temperature. Precipitation affects the amount of water available for weathering processes, for transfer of material throughout a soil profile, and for removal of material from a soil. Temperature also affects chemical weathering reactions, which happen more readily at higher temperatures.

Soil profiles in Nebraska or Ethiopia are very different from those in Hawaii, Brazil, or northern Canada, because soil-forming factors vary from one location to another (Figure 8-15). In warm, wet climates, the extent of leaching and removal of iron, aluminum, and organic matter from lower A horizons sometimes is so great that a nearly white, quartz-rich *E horizon* forms in the upper part of the profile between the A and B horizons. Typically, iron and aluminum oxides accumulate lower in the soil profile, in the B horizon. The bright red soils common in Hawaii, the southeastern United States, Central and South America, and Southeast Asia owe their strong red colors to iron oxides produced by extensive weathering and oxidation (see the chapter-opening photograph). In contrast, calcium carbonate and some other salts can precipitate as crystals in dry-region soils when soil waters carrying dissolved ions are depleted as plants take up the water, as well as by evaporation. In much of the American Southwest, northern

Africa, and the Middle East, desert soils often contain nodules and even thick layers of hard calcite, known as *caliche,* that have precipitated over thousands of years.

The Interaction of Earth Systems to Form Soil

The pedosphere is an open system with fluxes of matter and energy from the lithosphere, biosphere, hydrosphere, and atmosphere. The interaction of these Earth systems contributes to weathering and the formation of soils through a variety of processes that fall into four categories: *additions, chemical transformations, transfers,* and *removals.*

The lithosphere contributes to soil formation by adding mineral matter from rocks and sediments. Soils begin as fresh rock materials, such as layers of volcanic ash and lava, or glacially deposited sands and gravels, or crystalline igneous rocks that were raised to Earth's surface by mountain-building processes. Stored in minerals are elements that are essential to organisms for growth and sustenance. These elements include phosphorus, calcium, sulfur, magnesium, potassium, and iron. Weathering processes transform primary minerals to secondary minerals, such as clays and oxides, and release ions from minerals to soil water, making the ions accessible as nutrients to plants and animals.

The biosphere adds organic matter, which enters the soil as litter from dying plants and animals. Decay transforms organic matter and releases CO_2 gas (most of which recirculates to the atmosphere) as well as nutrients that are used by soil organisms and plants. Some of the nutrients are found in large, complex, organic molecules with a high molecular weight known as **humus**. Containing carbon, nitrogen, and phosphorus, humic substances are essential to soil-forming processes: They help

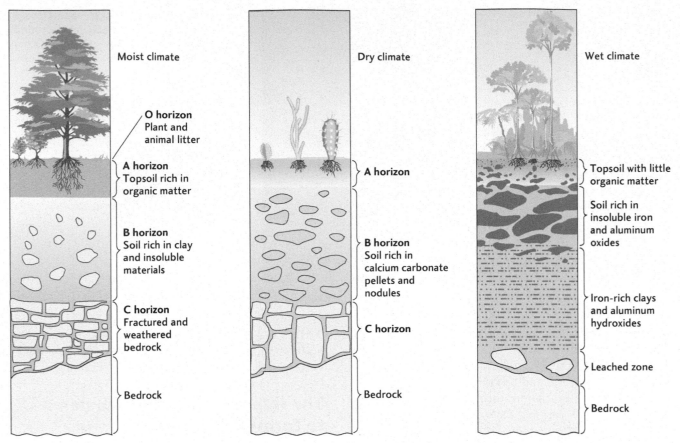

Figure 8–15 Soil profiles in moist, dry, and wet climates differ primarily in the amounts of organic matter and the thickness of their A horizons, and in the minerals that accumulate in the B horizons. In very wet climates, thin topsoil (A horizon) sometimes overlies subsoil that is so extensively leached that only layers of iron and aluminum oxides and clays remain. Technically, these are all B horizons.

to bind soil materials together, to transfer ions among soil horizons, and to retain ions in the soil. In addition, humic molecules act as soil sanitizers because their large molecular size enables them to trap pesticides, herbicides, and toxic metals such as lead.[21] Once trapped, pesticides and herbicides are broken down as the soil decomposes.

The atmosphere and hydrosphere interact at the soil surface to add many elements to soil: Sodium, potassium, magnesium, calcium, chloride, and other ions enter the soil in precipitation, particularly near coasts where sea spray is abundant. Pollutants, such as the CO_2 and sulfur dioxide emissions from power plants and automobiles, also enter the soil in precipitation. Solid particles, such as wind-blown dust, are added to the soil. One of the most important atmospheric additions is nitrogen, which enters the soil through a series of biological transformations when nitrogen cycles from the atmosphere to the soil and back.

The hydrosphere is the Earth system most responsible for the transfer of solid and dissolved substances among soil horizons and the removal of ions from soil profiles. Ions in solution can be removed if soil water continues downward to join with groundwater flow, which transports the ions to streams that drain watersheds. Some soluble substances removed from the upper part of the soil profile are precipitated deep in the soil because of the changing chemical nature of the soil horizons. Insoluble substances (clays and other minerals) added to the soil or produced during transformation can be transferred downward, primarily in water, but the depth of movement generally is limited to several meters by the decreasing size of pore spaces. For these reasons, the lower parts of most soil profiles contain layers of accumulation of clay and oxide minerals (that is, B horizons). Clay minerals contribute to the stickiness and plasticity of B horizons, and iron oxides give B horizons their typically reddish and yellowish-brown colors (see Box 8-2).

The mixture of rock and mineral fragments in soil help create the texture of the soil. Sand gives soil a gritty feel; silt has a velvety consistency: and clay is sticky and plastic. Clay minerals are an extremely important component of a soil, and together with organic matter they contribute to properties such as a soil's cohesion, structure, and ability to hold water and nutrients (see Box 8-2). *Soil cohesion* refers to the strength that is due to particles binding together. *Soil structure* refers to the arrangement of soil aggregates and pore spaces. Soil structures include *granular, blocky, prismatic,* and *platy* types. Clay minerals are crystalline sheet-structure silicates characterized by small particle size.

Cycling of Nitrogen Among the Atmosphere, Biosphere, and Pedosphere

The **nitrogen cycle**, the continuous flow of nitrogen through the atmosphere, biosphere, and pedosphere, is as essential to life as are the water and carbon cycles (see Chapter 7, Figure 7-2). Because nitrogen is a key ingredient in proteins and the nucleic acids DNA and RNA, it is a *limiting nutrient* for life—that is, life is limited by the amount of available nitrogen. Although nitrogen is the most abundant element in the atmosphere (79 percent by volume of gases in the atmosphere), it is not very reactive. Only a few organisms are able to use it directly. These organisms are valuable for their ability to "fix" nitrogen—that is, to extract the gas from the atmosphere and combine it with oxygen or hydrogen into a reactive form that is usable by plants and animals.

A few types of soil bacteria are able to fix atmospheric nitrogen. Many of these bacteria live in a mutually beneficial relationship, or symbiotically, with certain higher plants, commonly in masses of nodules attached to the roots of clovers, bean plants, and other legumes. The bacteria supply the plants with usable nitrogen and feed off the sugars and starches made by the plants. This arrangement enables the bacteria to survive and reproduce and to continue breaking apart nitrogen (N_2) molecules for the plants' use.

As organisms die and litter the soil, their protein molecules and the nitrogen they contain become available to other soil microbes. These microbes convert part of the nitrogen to ammonium (NH_4^+), a usable plant nutrient, and oxidize some of the ammonia to nitrite (NO_2^-) and nitrate (NO_3^-), both of which are also usable plant nutrients. Not all the nitrogen is recycled, however; some returns to the atmosphere by *denitrification,* the natural conversion of organic nitrates into gaseous nitrogen. This entire process, from the fixing of nitrogen gas by bacteria to its release from organisms back to the atmosphere, is the nitrogen cycle.

✔ The alteration of rock and mineral matter to broken fragments, clays, and ions in the pedosphere is called weathering.

✔ Physical weathering is the mechanical disintegration of rocks and minerals. Common processes of physical weathering are exfoliation jointing, thermal expansion, biological disintegration, and frost weathering.

✔ Chemical weathering is the chemical decomposition of rocks and minerals. Carbonation, hydrolysis, and oxidation are dominant processes of chemical weathering.

✔ Biological processes that include root growth, mound building by insects and animals, and microbial activity can intensify both physical and chemical weathering.

✔ Acids in soils contribute to the chemical weathering of minerals into secondary minerals such as clay, a major component of soil that has the ability to hold water and nutrients.

✔ Soil profiles are vertical arrangements of layered organic and mineral matter in the uppermost few meters of the pedosphere.

✔ Soil profiles vary throughout the world because of local differences in five major soil-forming factors: climate, topography (relief), rock type, types of plant and animal life, and time.

✔ The nitrogen cycle, essential to life on Earth, is the continuous flow of nitrogen through the atmosphere, biosphere, and pedosphere, from the fixing of nitrogen gas by bacteria to its release from organisms back to the atmosphere.

PEDOSPHERE RESOURCES: SOILS, CLAYS, AND MINERAL ORES

The most valuable resource of the pedosphere is fertile topsoil, the soil layer in which our food is produced. Soils are vital not only for food security; they can also help us adapt to and mitigate climate change because of their role in the carbon and nitrogen cycles. The primary terrestrial reservoir of organic carbon, nitrogen, and phosphorus is soil.[22] The weathered materials of the pedosphere also are valued as building and industrial materials. Some of these ores—such as clay and aluminum—form only as a result of weathering processes.

The Fertile Soil

Soil supports life, and the life and death of organisms are vital to soil formation. A single handful of soil contains billions of microbes that decompose plant, animal, and mineral matter and generate nutrients. Fertile soil is one of humanity's most precious resources, a medium for growing food, fiber, and wood. The more that human populations grow and standards of living are raised, the greater the demands we place on our soils. One of the great and paradoxical concerns of modern society is how to conserve soil even as we use it.

Although most undisturbed soils have **topsoil**, the relatively organic-rich upper layer of a soil that is produced by weathering processes and the accumulation of plant and animal matter (essentially the O and A horizons), not all topsoil is especially *fertile*—that is, not all of it is suitable for growing plants. The fertility of a soil depends on the amount and availability of essential elements such as nitrogen, carbon, potassium, phosphorus, and calcium. Nutrients are cycled from the roots of plants to the stems and leaves, and then returned to the soil as litter, maintaining the soil's richness. In fertile topsoil, the surface soil is dark, contains much humus, and forms stable crumbs that keep the soil structure loose and facilitate aeration. It also is porous and permeable, qualities that make for good drainage and water-storage capacities. In dry periods between rainfalls, the topsoil slowly releases water to plants.

Soils are classified according to diagnostic characteristics, many of which make them useful for particular purposes. The soil classification scheme developed by the U.S. Department of Agriculture (USDA) groups all soils in the United States into one of 12 orders (Table 8-2) and subdivides them further into suborders, great groups, subgroups, families, series, and types. Each soil order ends with the suffix "sol," from solum, the Latin word for "soil." The soil orders closely correspond to variations in precipitation and temperature, which are important soil-forming factors (Figure 8-16). The division of soil orders into increasingly finer subdivisions is based on variations in other soil-forming factors, such as topography, which affect soil properties such as horizon thickness and clay content.

Many of the soil types in the United States have been mapped on a mosaic of aerial photographs and published for each county. Soil surveys published by the USDA are especially useful for environmental planning. Tables and descriptions within published soil survey reports include information such as the depth to bedrock

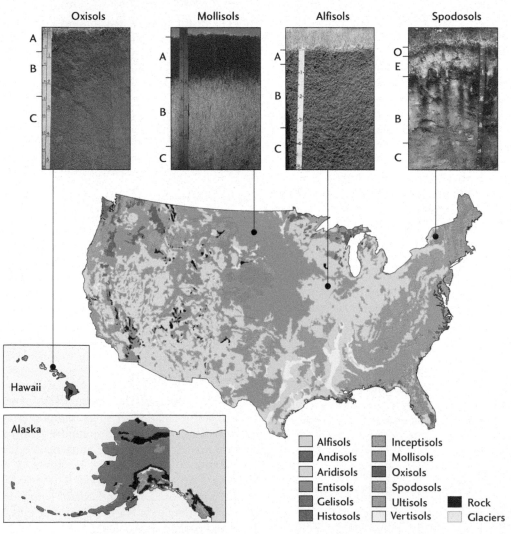

Figure 8–16 Soil orders mapped in the United States. Profiles for four soil orders, with horizon designation, are shown above the map. Note the association between geographic location, climatic conditions, and soil order. For example, aridosols are common in the dry Southwest, while mollisols are common in the wetter Midwest and ultisols are common along the humid East Coast. *(photos: Loyal A. Quandt, National Soil Survey Center, NRCS, USDA)*

TABLE 8–2 Primary Soil Orders According to USDA Classification

Soil Order (% of world total)	Source of Name	Description
Inceptisols (17)	From Latin *inceptum,* "beginning"	Relatively young soils in humid environments; horizons indicate transformation but little transfer or accumulation; usually moist, typically shallow soils
Entisols (16)	From "recent"	Mineral matter dominant; no distinct horizons; very young soils (e.g., as on floodplains)
Aridisols (12)	From Latin *aridus,* "dry"	All horizons dry for more than 6 months per year; low in organic matter; high in cations
Alfisols (10)	"Al" for aluminum; "f" (from Fe) for iron	Shallow penetration of humus; accumulation of clay in lower horizons; high in cations[†]; well-developed horizons
Gelisols (9)	From Latin *gelare,* "to freeze"	Soils on permafrost with evidence of cryoturbation (frost churning) and/or ice segregation; organic matter accumulates in dark brown to black A horizon, and no or weak B horizon present
Oxisols (8)	From French *oxyde,* "oxide"	Substantially weathered soils typically on ancient landscapes in tropical and subtropical environments; rich in kaolinite (clay), iron oxides, and sometimes humus; thick and deep B horizons
Ultisols (8)	From Latin *ultimus,* "last"	Intensely leached soils, typically in humid, warm climates; B horizons marked by substantial clay accumulation; low content of cations
Mollisols (7)	From Latin *mollis,* "soft"	Nearly black surface horizons rich in organic matter and bases; common in grasslands and forests with understory plants such as ferns
Spodosols (4)	From Greek *spodos,* "wood ash"	Soils with upper horizon that is light gray to white and extensively leached (E horizon) and overlying a reddish B horizon with substanial accumulation of iron, aluminum, and clay
Vertisols (2)	From Latin *vertere,* "to turn"	Dark, clay-rich soils that shrink and swell, forming wide cracks during dry season (also called "expansive" soils)
Andisols (~1)	From Japanese *ando,* "dark soil"	Usually in volcanic deposits; dark in color; high content of organic matter and cations (especially phosphorus)
Histosols (1)	From Greek *histos,* "tissue"	High content of organic matter (e.g., peat)

[†]Common cations in soils are calcium (Ca^{2+}), sodium (Na^+), potassium (K^+), magnesium (Mg^{2+}), aluminum (Al^{2+}), and iron (Fe^{2+} and Fe^{3+}).

Sources: Hans Jenny, *The Soil Resource* (New York: Springer-Verlag, 1980); and USDA NRCS Soil Survey Staff, *Soil Taxonomy: A Basic System of Soil Classification for Making and Interpreting Soil Surveys: Natural Resources Conservation Service* Number 436 (Washington, DC: U.S. Government Printing Office, 1999). See also USDA NRCS, "The Twelve Orders of Soil Taxonomy," at ftp://ftp-fc.sc.egov.usda.gov/NSSC/Soil_Orders/orders_hi.pdf.

BOX 8–3 EARTH SCIENCE AND PUBLIC POLICY

Population–Supporting Capacities and Soil

Vital to producing food for human consumption, soil is an essential component of any assessment of a nation's population-supporting capacity, or *carrying capacity*—the hypothetical number of people that can be supported by a given amount of land. Most (96 percent) of the world's food is derived from land resources, with the rest coming from aquatic sources such as rivers, lakes, and seas.[25]

Sometime between October 2011 and March 2012, the world's population reached 7 billion. Nearly 1 billion people were undernourished.[26] Causes of malnourishment range from political conflict to environmental degradation, raising a fundamental question with great significance to humanity: How many people can be supported by soil resources while providing sufficient food for nourishment and maintaining soil sustainability? This question has been broached before; Thomas Malthus wrote the earliest scholarly analysis in his famous 1798 essay on the feedback between population and food supply.[27]

The quantity of food that can be produced by a country depends on multiple factors, key among them the amount of arable land, the type of land (e.g., steep versus gentle slopes), climate, the intensiveness of land use, and access to technology that increases agricultural yields (e.g., fertilizer). Much of Earth's land surface is not arable because it is too cold (e.g., tundra), too dry, or too steep to farm. If a country has 100,000 hectares of arable land, for example, and this land yields sufficient food for 2 persons per hectare, then its carrying capacity is 200,000 persons (number of hectares of arable land times persons fed per hectare). The use of fertilizer, however, might increase the yield so that 4 persons per hectare are supported, increasing that nation's carrying capacity to 400,000.

In 1984, the United Nations FAO used soil maps and climatic information to estimate the carrying capacities of 117 developing nations under three different technological and economic scenarios: low optimization, which assumes traditional farming practices; medium optimization, in which some fertilizer is used; and high optimization, which includes the use of herbicides, pesticides, fertilizer, and improved crop varieties. The results for the different regions are shown in the accompanying table. For example, the value of 3.0 for Africa in 1975, given a low–optimization scenario, means that the region could support three times its population. If a country's value is less than 1, it is unable to feed its people.

The table shows that the overall average carrying capacity for different regions was cause for both optimism and concern when the FAO report was released in 1984. According to the report's projections, most regions would have sufficient resources to feed their populations in the year 2000. However, the potential population–supporting capacity diminished markedly for every region and every technological scenario between 1975 and 2000, excepting only Southeast Asia. Far more discouraging were the specific values for the individual countries within these regions. Fifty-five of the 117 countries had a value of less than 1 in 1975, indicating that those countries would have insufficient agricultural output to feed their people. By 2000, the number of countries at risk was expected to increase to 64, and much of southwestern Asia could suffer famine.

The FAO study had some shortcomings. On one hand, it did not take into account the availability of non-terrestrial food sources such as fish. On the other hand, it assumed that all the potentially cultivable land would be used to produce

---

and to the water table, soil permeability, drainage conditions, and the potential for shrinking and swelling—all factors that indicate different possible uses of land. The suitability of a soil type for cultivation, woodland, range land, wildlife, and recreation is listed, as are soil properties such as color, plasticity, and strength (the ability to support the weight of buildings, bridges, and other engineering structures).

Soil maps are valuable not only because they provide data for land and agricultural use, but also

because they indicate the number of people who can be supported by a given area of land (Box 8-3). United Nations Food and Agriculture Organization (FAO) and UNESCO maps of global soil resources became available in digital form beginning in 1991.[23] Scientists combined this digital database of global soil resources with a global climate database and a water balance model in order to compute soil moisture and temperature regimes for different parts of the world.[24] This approach provides valuable information that helps

Potential Population-Supporting Capacities Divided by 1975 and 2000 (Projected) Populations						
	1975 Ratios			2000 Ratios		
	Low	Intermediate	High	Low	Intermediate	High
Africa	3.0	11.6	33.9	1.6	5.8	16.5
Southwest Asia	0.8	1.3	2.0	0.7	0.9	1.2
South America	5.9	23.9	57.2	3.5	13.3	31.5
Central America	1.6	4.2	11.5	1.4	2.6	6.0
Southeast Asia	1.1	3.0	5.1	1.3	2.3	3.3
Average	2.0	6.9	16.6	1.6	4.2	9.3

Source: U.N. Food and Agriculture Organization (FAO). Land, Food, and People. Rome: FAO, 1984.

food rather than cash crops such as tobacco or coffee. Moreover, it did not consider the role of international trading: A nation with plentiful oil resources but little arable land might trade with another for food. In the modern global arena, a variety of resources can be exchanged for food.

Events since the 1980s have borne out some of the predictions of the FAO study, and more recent studies have produced similar results.[28] A number of nations have suffered food shortages, mass starvation, and an exodus of refugees. However, the causes have been many and not limited to the extent and fertility of available soil: Catastrophic droughts, civil strife, and political unrest that preclude the transport of food, as well as inequitable distributions of resources all have their effects. Cash crops often are grown on the best soils, while food is grown on marginal lands that are not very productive. The result of the overuse of marginal land is continued soil erosion and lowering of fertility in many parts of the world (**Figure 1**).

Figure 1 As populations increase, so does the need for food. This creates pressure to use more land for agricultural purposes. One result is that marginal land is farmed, as here in Madagascar, where farmers have cleared land to cultivate maize on a recently deforested steep slope that is subject to landslides. *(Martin Harvey/Getty Images)*

resource managers assess soil properties and population-supporting capacities.

Clay, Laterite, and Mineral Ores

Of all the mineral resources used by humans, clay ranks eighth in terms of annual per capita use in the United States (see Figure 4-2). Each person in the United States uses, on average, 74 kg (164 pounds) of clay a year. Clay is an essential ingredient in pottery, ceramics, bricks,

and tiles. When mixed with water and molded into different shapes, then baked either by the Sun or in a kiln, clay becomes a durable and waterproof material. Many early humans mixed clay with other ingredients, such as plant fibers and animal hairs, to make bricks with which to build houses. Archaeologists have found Sun-dried bricks dating to 4000 BCE in the ruins of Sumeria, and fired bricks dating to the first century CE in Roman ruins. There are also ancient brick and tile structures in Egypt, Asia, and the American Southwest.

One of the most durable brick materials comes from a type of oxisol known as a *laterite,* from *later,* the Latin word for "brick." In wet regions, some oxisols are so extensively leached that the A horizons have little organic matter and the B horizons are composed almost solely of iron and aluminum compounds. Typical minerals in the laterite subsoil are kaolinite (aluminum clay), goethite (iron hydroxide), hematite (iron oxide), gibbsite (aluminum hydroxide), and quartz. The iron oxides and hydroxides impart a striking brick-red color to the soil.

Fresh exposures of red laterite soils typically reveal a mottled, soft, earthy material that can be cut with a blade. When excavated and air dried, however, this clay-rich soil becomes extremely hard and cannot be resoftened with water. For thousands of years, people have used this type of soil to make bricks; ancient laterite temples are found in India, Thailand, and Cambodia. The very properties that make laterites desirable for building materials also make them impossible for cultivation.

The processes of chemical alteration that yield fertile soils and clays also can produce ore minerals in the pedosphere. With time and extensive leaching, weathering processes can remove all but the most insoluble elements in the pedosphere, which accumulate as *residual deposits* (see Figure 8-9). If leaching is particularly extensive, the deposits can become highly enriched in residual minerals. The least soluble minerals in the weathered zone are goethite and two aluminum hydroxide minerals, gibbsite and diaspore. The residual aluminum ores of gibbsite and diaspore are referred to as *bauxite.* Although there are iron ores in both the lithosphere and pedosphere, aluminum ore occurs only in the pedosphere. Large amounts of residual iron and aluminum ore minerals are produced in Australia and Brazil.

✔ Fertile soil, one of the world's most valuable resources, is characterized by noncompacted soil structure with good drainage, relatively large amounts of organic matter (humus), and clay that can retain water and nutrients.

✔ Weathering processes produce clay minerals that are valuable for many purposes, such as manufacturing brick and tile.

✔ Iron and aluminum are relatively insoluble and, as a result, accumulate in deeply weathered soil profiles, forming rich ore deposits in some parts of the world.

SOIL EROSION HAZARDS, LAND DEGRADATION, AND SOIL CONSERVATION

While weathering creates soils, clays, and ore minerals from primary minerals in the lithosphere, other processes in the pedosphere move the loosened debris from one place to another. *Erosion* is the general name for all processes that transport loosened Earth material downhill, down valley, or downwind; it results from the action of water, wind, glacial ice, and gravity. Erosion caused primarily by gravity is called *mass wasting.* Erosion is a natural geologic process that serves to cycle materials through Earth systems, but it also can pose serious environmental hazards to humans. Soil erosion, land degradation, desertification, and soil conservation strategies designed to minimize soil erosion are considered here; the hazard of mass wasting will be discussed in the next section.

Soil Erosion: A Quiet Crisis

When walls of dust towered above the Great Plains in the 1930s, a place dubbed the "Dust Bowl," few doubted that the accelerated erosion of the soil was due to negligent farming practices during a time of drought (Box 8-4). In 1996, a drought of equal severity destroyed much of the region's wheat crops, but it did not result in such massive soil erosion. The difference lies in conservation practices learned from experience. Elderly farmers who were children during the "dirty Thirties" have lived to see the valuable effects of planting trees as wind breaks and leaving grasses in place to provide some vegetation cover between row crops. Nevertheless, accelerated soil erosion is still a problem throughout North America as well as on every other continent on which humans practice agriculture. The economic loss from soil erosion in the United States is estimated to be about $44 billion per year.[29]

The annual global flux of sediment due to water erosion from agricultural land is estimated to be about 28 billion metric tons (28 petagrams, or Pg).[22] Plowing and wind erosion make mobile another 7 billion metric tons (7 Pg) of sediment. Not only soil, but also some of the carbon, phosphorus, and nitrogen associated with it are part of this erosional flux. Many of these sediments and nutrients are delivered to river systems, which has a substantial impact on aquatic habitat and organisms.

Humans simultaneously have increased the transport of sediment in global rivers through soil erosion (by 2.3 ± 0.6 billion metric tons per year) yet reduced the flux of sediment reaching the world's coasts (by 1.4 ± 0.3 billion metric tons per year) because of retention within reservoirs formed by dams.[30] Over 100 billion metric tons of sediment and 1 to 3 billion metric tons of carbon are sequestered in reservoirs constructed largely within the past 50 years. Even older dams, tens of thousands of which were used for water power before the advent of steam power and the use of fossil fuels, trapped sediment eroded from uplands in eastern North America for centuries after European settlement.[31]

Figure 8–17 Ruins of buildings on bare rock in Syria are a dire warning of the potential for soil to be lost unless it is managed for sustainability. This ancient city thrived until the 7th century but was one of the "Hundred Dead Cities" described by Walter Clay Lowdermilk in his report to the USDA in 1938. Lowdermilk lamented that "...the area might be repeopled again and the cities rebuilt, but now that the soils are gone, all is gone." *(Jim Richardson/National Geographic/Getty Images)*

Soil erosion is not just a modern environmental problem (Figure 8-17). One of the leaders of the Soil Conservation Service mentioned in Box 8-4, Walter Clay Lowdermilk, traveled abroad in 1938 to examine evidence of agriculture and soil erosion from ancient civilizations. A section of his subsequent report was titled "The Hundred Dead Cities," referring to places where agricultural societies once thrived but soils and people are now gone.[35] The evidence of the ancient civilizations is the ruins of buildings on bare rock, with nearly all the soil eroded away.

In Greece, geoarchaeologists have found sequences of sediments in valley bottoms that indicate periods of increased rates of upland erosion caused by changes in land use during ancient times. Layers of sediments that were deposited by streams that drained farmed hillslopes are dated to times of deforestation and agricultural activity, both of which accelerate erosion. As erosion intensified, abused sites were abandoned by prehistoric farmers and soils were able to form on the new sedimentary deposits. Over time, there were new waves of expansion and agricultural activity, repeating the sequence and resulting in additional layers of sediment and soil that buried the older deposits (Figure 8-18).

Both soil formation and soil erosion have occurred simultaneously on Earth since long before humans began using soils for agriculture. In the absence of any disturbance, natural or human-induced, the rates of soil formation and erosion generally balance each other. Typical rates of soil formation are very slow on a human time scale, ranging from only 0.0001 to 0.4 mm per year. In other words, the formation of a single meter of soil depth

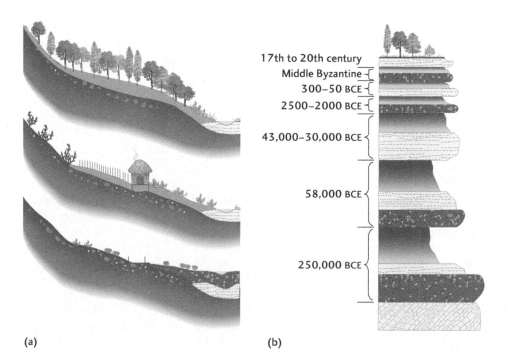

17th to 20th century
Middle Byzantine
300–50 BCE
2500–2000 BCE
43,000–30,000 BCE
58,000 BCE
250,000 BCE

(a)

(b)

Figure 8–18 (a) Hillslope topographic profiles at left show changes with time in ancient Greece as a result of historic deforestation and farming. Thick soil (top profile) is eroded from the slope (middle profile), and then the eroded sediment is deposited in the valley downslope (bottom profile). (b) Sediments and buried soils in the valley bottom record a long history of alternating landscape stability and soil erosion. Each period of soil erosion—both historically and during past climatic conditions—is represented by a downslope deposit, whereas times of landscape stability are marked by soils that form on deposits.

BOX 8-4 CASE STUDY

Droughts, Land Degradation, and the American "Dust Bowl"

On Black Sunday in 1935, a massive dust storm darkened the sky over the Great Plains and awakened the United States to the menace of soil erosion. In this semiarid, grassland region, several years of drought in the early 1930s had followed a relatively wet period during which hopeful farmers plowed up the deep-rooted sod to plant crops. The bare, pulverized soil, parched by the subsequent drought and exposed to high winds, began to blow away. On Sunday, April 14, 1935, in the worst dust storm by far, northwest winds whipped millions of tons of loose, dry topsoil into a churning wall of red-brown dust that covered the homes and fields of frightened farmers and destroyed their crops (Figure 1). Tens of thousands of impoverished families from Oklahoma, Texas, and other Great Plains states abandoned their farms and migrated west. John Steinbeck immortalized their plight in his novel *The Grapes of Wrath,* and people everywhere lamented the transformation of America's fertile "wheat heaven" into a harsh, barren "dust bowl."

The finest sediment from the Black Sunday storm was lifted high into the atmosphere and carried eastward by wind currents. In early May, the gritty haze hit the Eastern Seaboard, and even politicians in Washington, DC, found themselves sneezing, coughing, and rubbing the dust out of their eyes.[32] Congress passed the Soil Erosion Act of 1935, and President Franklin D. Roosevelt established the Soil Conservation Service (now the Natural Resources Conservation Service) and put hundreds of thousands of people to work in the Civilian Conservation Corps. The trees these workers planted and the dams they built helped to minimize future soil erosion.

Figure 1 Walls of churning dust darkened the skies over the Great Plains on Black Sunday, April 14, 1935. *(Library of Congress, Prints & Photographs Division, FSA/OWI Collection [LC-USF343- 001617-ZE])*

Two of the Soil Conservation Service's early leaders, Hugh H. Bennett and Walter Clay Lowdermilk, were instrumental in impressing on Congress the negative effects of soil erosion. Giving testimony before a congressional subcommittee, the two men silently laid a towel on the polished tabletop, poured a large glass of water onto it, and watched as the towel absorbed all the water. Then they removed the towel and poured the same amount of water onto the bare surface of the table. The water splashed, splattered, and soon ran off onto the laps of the startled committee members, demonstrating the value of healthy topsoil that absorbs moisture and protects the ground surface from erosion.

Scientists have pondered the role of human action in the Dust Bowl of the 1930s and wondered how much of

requires thousands to tens of thousands of years. In some parts of the world today, however, rates of erosion due to human activity are 10 to 100 times greater than natural rates of soil production and long-term geologic erosion.[36]

Worldwide, nations suffer billions of dollars in damage and repairs, and unestimated losses in natural resources, because of soil erosion. The threat of accelerated soil erosion associated with poor farming practices and rapid deforestation is one of the world's most

serious environmental crises. It is, however, a quiet crisis. The development and application of large amounts of synthetic fertilizers, herbicides, and pesticides, and the improved crop varieties in the 1950s—called the Green Revolution—has increased crop yields in many parts of the world, masking the problem of soil erosion and creating a false sense of security. In fact, this valuable resource is being "mined"—that is, overused—in many places. The FAO estimates, for example, that 6.3 million hectares of farmland in Africa is degraded, with

the catastrophe was the result of natural climatic factors, in this case drought, and how much a result of plowing and human-induced land degradation that altered vegetation patterns and provided a dust source. Droughts are common in the midlatitudes, modulated by sea surface temperatures (SST) in the Pacific and Atlantic oceans. Eastern tropical Pacific Ocean SSTs are lower than normal during La Niña periods, causing drier-than-normal conditions for the southwestern and southeastern United States, as was the case during the 1930s drought (see Chapter 11). Affecting nearly two-thirds of the United States as well as parts of Canada and Mexico, it was one of the worst droughts of the 20th century, but was not unusual in a geologic sense. Major droughts occur in the Great Plains about one to two times each century and even more severe droughts have lasted for decades during previous centuries (e.g., during the late 13th century and the 16th century).[33]

An unusual aspect of the 1930s drought and Dust Bowl was that the northern and central Great Plains were the areas with the highest levels of drying. Climate models (general circulation models, or GCMs, which we will discuss in Chapters 14 and 15) indicate that the drought would have been centered in southwestern North America, not the middle of the continent, if it had been a result of anomalous SSTs during that time period. Recently, researchers have found that the locus of drying shifts to the southwest if land degradation by humans is included in the climate modeling.[34] This *human forcing,* or drying as a result of degradation from land use, is associated with a reduction in vegetation cover and with crop failure, which

Figure 2 Thousands of farms like this one in Cimarron County, Oklahoma, were rendered infertile; farm families abandoned their land and sought work elsewhere, often as migrant farm workers. *(Library of Congress, Prints & Photographs Division, FSA/OWI Collection [LOC LC–USZ62–11491])*

in turn leads to a source of soil dust to the atmosphere. Feedbacks between the land surface and atmosphere shift the temperature anomaly northward and intensify drought conditions from those that would result solely from ocean–forcing mechanisms such as changes in SST. In other words, land degradation from human activities led to dust storms and amplified the drought that would have occurred anyway, producing one of the greatest environmental disasters in U.S. history (**Figure 2**).

diminished fertility and ability to hold water. In the face of a population expected to double in the next 40 years, such land degradation must be reversed in order to provide sufficient food.[37]

The environmental impact of soil erosion is felt on the slopes and fields where erosion happens, as well as downstream, where increased sediment chokes streams, rivers, and bays. On site, the loss of topsoil reduces soil productivity by removing organic matter and clay, both of which retain nutrients and moisture. Degraded

soils have lower water-holding capacities and are more susceptible to erosion, as demonstrated by Bennett and Lowdermilk to Congress in the 1930s. Downstream, sediment washed into waterways damages fish, coral reefs, reservoir-storage potential, and navigable waterways.

Soil Erosion by Water

Rainsplash and sheetwash erosion are responsible for most soil erosion by water in temperate regions. **Rainsplash**

Figure 8-19 Sheetwash erosion and rill erosion commonly occur together in places where there is a lack of vegetation cover, as on freshly harvested or plowed fields. In this example of fields in Iowa after a heavy rain, sheets of surface water accumulate more water as they flow downslope, collecting in small ephemeral channels called rills. *(Lynn Betts/USDA Natural Resources Conservation Service)*

involves raindrops hitting exposed (that is, unvegetated) soil during intense rainstorms, lifting fine particles high above the soil surface and, if there is a slope, transporting them downhill. If the soil contains much clay, the soil swells, making the surface impermeable to additional water, and continued rainfall accumulates as a nearly uniform sheet of water flowing across the ground surface. Close examination reveals that the sheet actually consists of numerous tiny rivulets that form an intricate pattern of joining and splitting "threads." Because of their rapid migration across the soil surface, over time they remove a fairly uniform layer of soil, which is why this process is called **sheetwash erosion** (Figure 8-19).

The amount of water increases toward the bottom of the slope and some tiny rivulets might increase in depth and abrasive power until small, temporary **rills** several centimeters wide and deep are gouged from the landscape. If these incised channels become so large that they cannot be removed by a plow, and become permanently entrenched features, they are called **gullies**. Gullies have dimensions of several to many meters in depth, tens to hundreds of meters in width, and lengths as great as several kilometers (Figure 8-20). Over a small, local region, rilling and gullying may be the most evident cause of soil erosion by water.

Land Degradation and Desertification

Land degradation describes a suite of environmental processes associated with human activities that result in negative change to land resources over time. It generally refers to the loss of vegetation and topsoil, a reduction in the amount of agricultural productivity that the land has, and the long-term decline of ecosystem functions.[38]

Degraded land has a diminished capacity to produce energy and food, to filter water and air, and to provide habitat. Because land degradation happens gradually over time and its early symptoms typically are subtle, by the time it is recognized it is often advanced and difficult to reverse. Common drivers of land degradation include drought, deforestation or other land clearing, sociopolitical upheaval, and poor conservation management practices.

Somalia, an arid-region African country rife with factors that have led to land degradation and tragic famines over the past few decades, provides a case study in the ways that scientists assess land degradation at a national scale.[38] A sensor on a satellite was used to collect data about the vigor and quantity of the vegetation in parts of Somalia. By gathering these data at various points in time, scientists were able to identify both the rate and the extent of land degradation. Local experts identified the types of degradation and their causes, and the results from the field sites were compared with assessments that were developed from the remote sensing data and the experts' information. This type of rapid assessment at the national scale is valuable for developing national strategies and policies to mitigate land degradation and provide relief. The assessment showed that one-third of the country of Somalia is marred by land degradation due to loss of vegetation cover and topsoil and to a decline in soil moisture. The causes of this degradation were identified as overgrazing, excessive removal of trees, and poor agricultural practices, all of which are worsened by drought, civil war, and poor governance.

One of the greatest global environmental concerns is the transformation of once-productive or marginally productive land to deteriorated land and soil that is unable to support plants and animals. Because the

Figure 8–20 Gullies range in depth from several to many meters and promote soil erosion, as in this example of land that was deforested and converted to pastures in Brazil. *(James P. Blair/National Geographic Stock)*

land becomes barren and dry, the process is described as **desertification,** although it takes place in regions other than true deserts. The United Nations Convention to Combat Desertification defines desertification as the result of "land degradation in arid, semiarid and dry subhumid areas resulting from various factors, including climatic variations and human activities."[39] Land degradation in these areas, referred to as *drylands,* results from a reduction in biological or economic productivity. With increasing population pressure, larger amounts of available land—even that which might be considered marginal for food production or habitation—are used. Much land that once was productive is over-exploited, and other land that once was considered marginally productive is converted for farming and grazing. Many marginal lands have steep slopes, are prone to high rates of erosion, and are more fragile than other lands. The result is widespread degradation and deterioration of the soil and land resource base.

Desertification occurs as a sequence of events:

■ Exposure and excessive stress dries out the soil.

■ Native plant species decline, resulting in less production of organic matter.

■ As soil fertility is reduced, the soil hardens.

■ Because less water can infiltrate the hardened soil during storms, soil drying and hardening are exacerbated in a positive feedback process.

■ During rainstorms, water runs off the landscape rather than infiltrating, scouring the soil so that rilling, gullying, and widespread erosion develop.

Depending upon how far this sequence has progressed in terms of diminished plant yield, deteriorated land is classified as slightly, moderately, severely, or very severely desertified. In cases of moderate desertification for rain-fed cropland, crop yields are reduced by 10 to 25 percent. In severely desertified land, soil is heavily eroded and crop yields can be reduced by 25 to 50 percent. In cases of very severely desertified land, all vegetation is gone, and crop yields are reduced by more than 50 percent. According to the United Nations, desertification affects some 10 to 20 percent of drylands.[40]

Assessing Soil Erosion

Since the 1940s, many researchers have gathered information on soil erosion in order to develop an equation that can be used to predict rates of erosion under different land-use or soil conservation conditions. Erosion plots are used to monitor the amount of soil removed from a given area by various factors, such as degree of slope and crop management, which are varied experimentally. From years of work on thousands of plots, scientists have discovered a relationship between erosion and four factors that control its rate: climate (rainfall intensity and duration), soil erodibility, topography (hillslope length and steepness), and land cover (erosion control and crop management). The experimentally determined relationship among these factors is called the **universal soil loss equation,** or USLE:

$$A = R \times K \times LS \times CP$$

A	R	K	LS	CP
Annual soil loss	Climatic factor	Soil erodibility	Topography	Erosion control × crop management

Using this equation, it is possible to estimate how much erosion might occur if a given crop type is planted, or if the intensity of rainstorms increases. The soil loss equation has been used in the United States to establish erosion-control and crop-management practices necessary to maintain soil erosion rates below a level of **tolerable soil loss**, which is quantified as a T value. The tolerable soil loss for a given soil is the erosion limit below which a soil can yield a high level of crop productivity indefinitely—in other words, the T value is the rate of soil loss at which the crop can be sustained. Depending on the soil type and its rate of renewal, the T value varies from about 2 to 5 tons per acre per year for different parts of the United States.[41] Erosion rates are as high as 10 tons per acre on cultivated land in some parts of the United States.[42] Small wonder, then, that the U.S. Environmental Protection Agency calls soil erosion one of the nation's five gravest environmental problems.

The USLE is an empirical model, based on the equivalent of some 10,000 years of data on runoff and soil erosion at the scale of small experimental plots.[36] Nevertheless, the actual magnitude of soil erosion at the scale of watersheds remains uncertain.[43] Comparisons of predictions from the USLE with field measurements show that the model both under- and over-predicts soil erosion in some cases, but predicts it accurately in others. Perhaps the greatest challenge is in extrapolating from the plot to the landscape scale.

In addition, the amount of soil that is eroded from its source on a hillslope is not equal to that removed from a watershed by stream flow. The path from slope to river mouth is marked by multiple possible sites for eroded sediment to be trapped and stored for different amounts of time. One of the most common places in which sediment becomes trapped is in reservoirs, in the storage areas behind dams. Even as human activities have increased the amount of soil eroded from uplands and transported as sediment by rivers (about 2.3 billion metric tons per year), the actual global flux that reaches the world's coastal areas has decreased by 1.4 billion metric tons per year because of the amount of sediment that is trapped in reservoirs behind dams.[30] Scientists estimate that more than 100 billion metric tons of sediment is stored in reservoirs that were constructed largely during the past half-century. Between 1 billion and 3 billion metric tons of carbon also is sequestered in these reservoirs.

Soil Conservation Practices

To conserve soil and water, the USDA advocates terracing, contour plowing, preserving remnant woodlands as wind barriers that create shelter belts, planting vegetation barriers along waterways, and building structures to trap water and sediment (Figure 8-21).

Terracing, still used throughout the world, is an ancient practice. There is evidence of terraces at archaeological sites thousands of years old in South America, North America, Africa, and Eurasia. Terraces conserve soil and water by reducing the length and steepness of a hillslope by dividing it into relatively short segments that trap or slow the downward movement of sediment and water. By reducing the topography component of the universal soil loss equation, the amount of soil that can be removed by erosion is reduced.

Figure 8–21 Aerial view of a watershed in Wisconsin illustrating different ways to prevent soil erosion, including contour stripping, vegetation barriers, and small reservoirs to trap water and sediment. (Kevin Horan/Getty Images)

Contour plowing is a method of cultivation in which the plows (and other farm machinery) follow contour lines, or lines of equal elevation, rather than moving up and down the farm slopes and crossing contour lines. This prevents the formation of rills and gullies along wheel ruts. Furrows with small ridges created by plowing act as mini-terraces that slow or stop the downward movement of water and soil.

Another means of minimizing soil erosion is **conservation tillage**. In conventional tillage, plows and other machines are used for cropping rows, removing weeds, harvesting crops, and other purposes. The negative effects associated with such farming practices can include compacting the soil, disrupting the soil structure and microbial communities, and drying the soil. A typical conservation tillage approach involves leaving about a third or more of the crop residue on the soil surface in order to slow the movement of water across the land surface and, as a result, to reduce soil erosion. For one type of conservation tillage system, called *no-till farming,* the soil is left undisturbed between one harvest and planting, and the planting is done along narrow seedbeds to minimize soil impact. Advantages of less frequent and more localized use of heavy machinery also include savings in labor and fuel costs, higher water infiltration rates, and storage of water and carbon in the upper parts of the soil profile. Conservation tillage was used on nearly 40 percent of U.S. cropland in 2004, a significant increase from 2.3 percent in 1965.[44]

These conservation strategies have reduced the rates of erosion on much American farmland. However, because of economic pressures and the lack of a strong national policy to protect the soil resource, sound land-use practices are not always strictly adhered to. Since the 1970s, only about 50 percent of farmers have practiced the conservation methods recommended by the USDA, and erosion rates again have been increasing in some areas. Because soil forms slowly, over thousands of years, the result of increased erosion rates will be thinner soil.

✔ Processes that transport loosened earth materials downhill, down valley, or downwind by the action of water, wind, glacial ice, and gravity are called erosion.

✔ The main processes of soil erosion by water are rainsplash, sheetwash, rilling, and gullying.

✔ In recent decades, widespread deterioration and erosion of soil associated with overgrazing and poor farming practices have resulted in desertification in nearly 40 percent of the world's agricultural lands.

✔ The tolerable soil loss for a given soil is the erosion limit below which a soil can yield a high level of crop productivity indefinitely—in other words, it is the rate of soil loss at which crop production can be sustained.

✔ Soil conservation methods that are used to reduce soil erosion, retain water, and improve soil fertility include terracing, contour plowing, wind barriers that create shelter belts, vegetation barriers along waterways, sediment-trapping structures, and conservation tillage (including no-till farming).

MASS MOVEMENT HAZARDS AND THEIR MITIGATION

The downward movement of mass on a hillslope is considered hazardous if it threatens human life, safety, property, or other features of value to humans. The hazard is substantial. In the United States, for example, mass movement causes 25 to 50 deaths and $1 billion to $2 billion in economic losses each year.[45]

Earth materials can move down a slope primarily as a result of the force of gravity. This type of erosion is called **mass wasting;** the actual process of downslope motion is referred to as **mass movement**. All loose, weathered material on hillslopes is prone to some form of gravitational movement of mass, as can be seen in the slow creep of clay-rich soil on a gentle hillside or the rapid fall of large blocks of rock from a cliff wall. Unlike other erosion processes, mass movement is viewed as a hazard more because of its catastrophic, short-term effects than because it removes soil from the landscape. Like soil erosion, mass movement is a natural process that occurs worldwide but that has been accelerated by human activities in many places (Figure 8-22). The hazard of mass movement increases as worldwide population grows and more people settle in areas where the potential for mass movement is high.

The following historical examples reveal that landslides can be triggered by both seismic and volcanic events, but as will be shown below, heavy rainfall or snowmelt also can be key factors in mass movement. In China, a large earthquake near the Tibetan border caused hillsides consisting of *loess* (windblown silt) to collapse over a wide region in 1920. Whole villages were buried, cave dwellings hollowed into the loess bluffs collapsed, and many valleys were dammed by large masses of silt and blocks of rock that cascaded down hillsides. In total, ground shaking from the earthquake, destruction associated with hundreds of landslides triggered by ground shaking, and exposure to cold after homes were destroyed caused at least 200,000 deaths, one of the greatest losses of life in recorded history from a natural disaster. The day of the event, December 16, 1920, is still referred to in China as Shan Tso-liao, or the day "when the mountains walked."[46]

In 1985, a small volcanic eruption from South America's northernmost volcano, Nevado del Ruiz in

Figure 8-22 The Brazilian state of Rio de Janeiro was subject to widespread landslides and debris flows during an unusually wet period in January 2011. Nearly 1000 people were killed, making this the second-most deadly natural disaster in Brazilian history. An equivalent of one month's rain fell during a 24-hour period from January 11-12. *(Antonio Lacerda/EPA/NEWSCOM)*

Colombia, generated a large mudflow that killed 25,000 people. The snow and ice cap over the volcano's summit melted during the eruption, providing a source of liquid to mobilize loose volcanic debris from the mountain's slopes. The torrent of mud and water emanating from the summit surged downstream at 30 km per hour, burying at least 22,000 people sleeping in the town of Armero and several thousand others nearby.

The Roles of Gravity and Water in Mass Movement

If mass movement is caused by gravity, which affects the entire Earth, why is it more likely to happen on some hillslopes than on others? And why do some mass movements occur suddenly and unexpectedly, perhaps after several days of rain, with dreadful consequences to unsuspecting residents below? The answer has to do with resistance to the forces of gravity. For a mass of earth to move, the force of gravity must be stronger than any resistance that the weathered material has. The force of gravity itself does not change substantially at Earth's surface, but the amount of resistance, or *strength*, of the material relative to the force of gravity does vary. It varies from one place to another, and it varies over time at a given location. Variation in space depends largely on the slope of the surface on which the debris rests. The steeper the slope, the more likely it is that there will be mass movement. Variation at a given location over time depends largely on the amount of moisture in the mass.

Consider a flat surface on which debris remains motionless and at rest. The force of gravity pulls the loose debris downward, holding it in place. On a sloping surface, however, the force of gravity can be separated into two components: (1) a *slope-perpendicular* component that pulls the debris in a direction perpendicular to the slope and helps to hold it in place; and (2) a *slope-parallel* component that acts to move the debris along and down the slope (Figure 8-23). The perpendicular component of gravity provides frictional resistance against mass movement, while the slope-parallel component provides a driving force that favors mass movement.

The steeper the slope, the greater the slope-parallel (shear) component of the force of gravity relative to the slope-perpendicular (normal) component. At some critical angle of slope from the horizontal, the slope becomes unstable. The critical angle above which failure occurs is called the **angle of repose,** and is about 30° to 35° in loose, dry material such as sand and gravel.

Slope steepness cannot be the only criterion for the strength of a material, because some mass remains stable on steep slopes for many thousands of years and then fails suddenly, often after prolonged, heavy rain or snowmelt. Water is a second factor in the strength of a material. If water enters the pore spaces between particles in weathered material, the material's resistance to movement changes. A certain amount of water can provide greater resistance because it holds particles together by **surface tension,** a force caused by the attraction of water molecules to one another. Along a water surface,

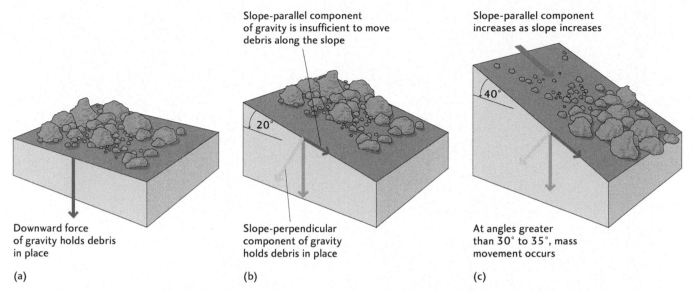

Slope-parallel component
of gravity is insufficient to move
debris along the slope

Slope-parallel component
increases as slope increases

20°

40°

Downward force
of gravity holds debris
in place

Slope-perpendicular
component of gravity
holds debris in place

At angles greater
than 30° to 35°, mass
movement occurs

(a)

(b)

(c)

Figure 8–23   The effect of increasing slope on the relative resistance, or strength, of loose debris. (a) The force of gravity is perpendicular to earth's surface and holds debris in place on a flat slope. (b) On a sloping surface, the force of gravity can be separated into slope–perpendicular (normal) and slope–parallel (shear) components. (c) On a slope steeper than about 30° to 35°, the slope–parallel force is enough to initiate mass movement for loose debris.

molecules are pulled into the water by the attraction of the molecules underneath, creating a force that can bind water and particles; this effect is seen in the cohesion of moist sand used to build a sand castle at a beach. However, if water completely fills the pore spaces surrounding the particles, then the pressure of the water in the pore spaces pushes the grains apart, causing them to move more freely. In this case, water lowers the resistance of the mass to movement, and the mass then behaves more like a liquid than a solid. In the example of a sand castle, when too much water is added, the castle collapses.

Types of Mass Movement

The phenomenon of mass movement presents a range of possibilities, from rockslides to debris flows and mudflows. With the exception of pure free fall of rock debris, the three basic mechanisms of mass movement are heave, slide, and flow (Figure 8-24). The **heave** mechanism, caused by alternating expansion and contraction of debris from freezing and thawing or wetting and drying, raises and lowers material in a direction perpendicular to the hillslope. Because of the slope and gravity, some material makes its way downhill as it is dropped back to the soil surface. **Slide** occurs when cohesive blocks of material move, or fail, along a well-defined plane, and thus the term *landslide* is seen to have a very specific meaning. **Flow** involves the debris moving like a fluid when, in contrast to sliding, there is no clear plane of failure within or below the moving mass.

All types of mass movement except rockfall can be attributed to one of these three mechanisms, but the type of debris, the amount of water, and the speed at which the debris moves can vary substantially. As a result, many different types of mass movement have been identified, including rockslides, debris slumps, debris slides, and debris flows (see Figure 8-24).

The mass movement most closely associated with heave is *soil creep,* a very slow process that involves varying amounts of water. The downward curvature of trees and gravestones and the tilt of power poles are evidence of soil creep. Because of the ubiquity of soil creep, cemeteries usually are located on flat parts of the landscape.

Rockslides, debris slumps, and debris slides can happen with little or no water, and they typically move rapidly. Intermediate between slides and a flow of stream water that carries some sediment (which is not considered mass movement) are *debris flows* and *mudflows.* Of the two, mudflows contain more water relative to sediment. Typical amounts of water in debris flows and mudflows are 10 to 30 percent of the total weight of the moving mass. Debris flows are tens to hundreds of times more viscous—resistant to flow—than honey, and can transport enormous boulders, houses, and heavy equipment. Debris flows are a common form of mass movement in the mountains that surround Los Angeles, where more than a hundred sediment-trapping basins have been constructed—at a cost of hundreds of millions of dollars—to catch debris on its way downhill, before it reaches buildings and roads.

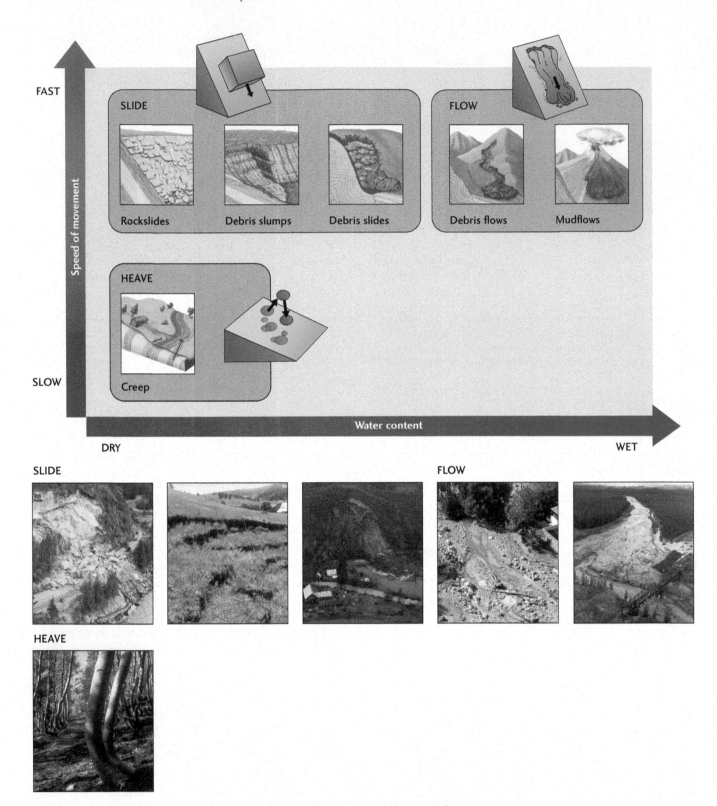

Figure 8–24 Different types of mass movement are related to the mechanism of movement (heave, slide, or flow), the nature of the mass, the amount of water, and the speed of movement. *(top, from left: AP Photo/Jonathan Hayward, The Canadian Press; NOAA/NGDC, B. Bradley, University of Colorado; Washington State Dept. of Transportation; AP Photo/Nam Y. Huh; AP Photo/NZPA, Stephen Barker; bottom: Marli Miller)*

Causes and Prevention of Mass Movement

All the causes of mass movement can be attributed to one of two factors: Either the driving forces acting on a slope are increased, or the resisting forces are decreased (that is, the strength of mass is reduced) (Table 8-3). Preventing mass movement requires that either the driving forces be reduced or the resisting forces be increased.

Road building is a common cause of both increased driving forces and decreased resisting forces along a slope (Figure 8-25). When roads are being built, hillslopes often are cut and steepened adjacent to lower-gradient benches, increasing the driving force (the slope-parallel force shown in Figure 8-23) that tends to push material down the steepened slope. In addition, excavating the material that had provided some lateral support against the weight of the material above reduces the resisting forces along a potential failure plane. Similarly, coastal erosion can steepen slopes and remove the lateral support of rock material at the base of a cliff, making mass movement common along eroding coastlines. Preventing such failures requires developers to situate roads carefully and to build structures such as benches and retaining walls along road cuts and coastal cliffs.

A common cause of reduced resistance to movement, or diminished strength, is the addition of so much water to the slope material that it begins to behave more like a liquid than a solid. For this reason, many mass movements occur immediately or soon after prolonged periods of rainfall. For example, after hurricanes in the eastern United States, there are dozens to hundreds of small slides and debris flows in hilly areas. Structures that drain water from potential slide masses can prevent or minimize failure. Sometimes the cause of increased moisture in hillslope debris is the concentration of water along newly constructed roads or the leakage of water from sewage pipes, canals, or water pipes. In all cases, mass movement can be mitigated by not permitting water to accumulate.

Because the roots of vegetation provide much strength to material on hillslopes, the removal of plants often results in mass movement. Vegetation provides a further stabilizing role in that it absorbs moisture and minimizes the accumulation of water in the soil. In logged areas, the combination of road building and the removal of vegetation leads to conditions very favorable for mass movement. Clear-cutting and associated road building were blamed for debris slides that killed five people in Oregon in late 1996. Failure is most common within a few years of vegetation removal, after the roots have fully decayed. Mitigating mass movement on devegetated slopes requires replanting as soon as possible, and taking measures, such as mulching, to help establish the plants.

Finally, the nature of the geologic material on a hillslope plays a role in mass movement. For example, if strata are inclined toward a valley bottom such that beds

TABLE 8–3 Causes of Mass Movement and Preventive Measures

Causes	Preventive measures
Factors related to increased driving forces	
Slope gradient (steeper slopes are more prone to mass movement)	Reduce slope (e.g., by constructing benches)
Factors related to decreased resisting forces (reduced strength of mass)	
Lateral support removed by erosion or construction	Reinforce base of slope with retaining walls or by grouting
Moisture content	Seal surface cracks to prevent infiltration; drain surface water from potential mass movement material with ditches; install subsurface drainage system
Vegetation	Replant slopes immediately after removing vegetation during logging or development; protect slopes with cover or mulch while seedlings become established
Nature of geologic materials (e.g., highly weathered; or rock layers inclined parallel to slope)	Construct pilings through mass; avoid building on slopes with rock layers inclined parallel to the slope (they are prone to failure)

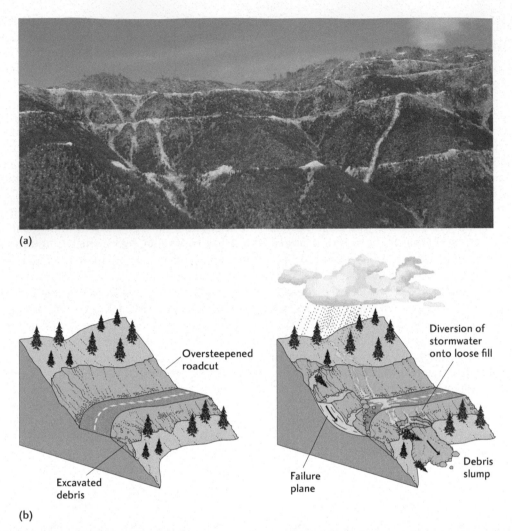

(a)

(b)

Oversteepened roadcut

Excavated debris

Diversion of stormwater onto loose fill

Failure plane

Debris slump

Figure 8–25 (a) Along roads such as these in the Cascade Mountains, which span western Canada, Washington, and Oregon, many slumps and slides (white scars) are initiated along fresh road cuts (horizontal benches), where slopes are steepened and rainwater collects and infiltrates the soil. (b) After being excavated to make the building of a road possible, the hillslope is steeper and might fail. Stormwater from rain and snow collects along roads and drains into loose, excavated debris, making that material likely to fail as well. After failure, the angle of the new slope is similar to that of the hillslope before it was excavated, close to the angle of repose. *(a: David R. Montgomery, University of Washington, Seattle)*

are parallel to the hillslope, mass movement is more likely than if beds are horizontal or dipping into the slope. The reason for this is that the contacts between each rock layer can become planes of failure themselves, and entire beds of rock can slide downhill along these contacts.

It is the job of geoscientists to map soil, sediment, rock types, deposits from previous mass movement, and areas of likely instability. Such maps, known as *slope hazard maps,* can be used as planning tools. Potential instability does not necessarily preclude development, for engineering methods to strengthen structures, stabilize slopes, and prevent their failure are known and practiced in many places. In Japan, for example, a country plagued by mass movement, a government effort to reduce landslide losses was initiated in 1958 and has markedly lowered death and property losses.

A case where geologists predicted mass movement was the 1985 mudflow at Nevado del Ruiz in Colombia. After observing a year of precursory volcanic rumblings, scientists prepared a hazard map that clearly showed the village of Armero lying in the path of a possible debris flow if the volcano erupted. The map was available weeks before the eruption in 1985; unfortunately, government officials did not urge residents to flee. The decision to ignore the scientists resulted in the deaths of nearly 25,000 people.

✔ The three mechanisms of mass movement are heave, slide, and flow, and in combination with the type of debris, amount of water, and speed of motion, these mechanisms are used to classify mass movement.

✔ Steepening hillslopes (by road building) and leaving the land bare increases the slope-parallel (shear) force on the slope and decreases the slope materials' resistance to movement, increasing the possibility of mass movement.

CLOSING THOUGHTS

Developing strategies to protect Earth's outermost, fertile layer, while the world's human population continues to grow, is one of the greatest environmental challenges of the 21st century. Throughout the world, much land is at risk of soil erosion, desertification, gullying, and catastrophic mass movement. Urbanization has been increasing since the Industrial Revolution began in about 1760. With the development of mechanized farming equipment have come much larger farms and the decline of many farming communities. Energy-intensive agricultural techniques since the Green Revolution of the 1950s have increased crop yields by nearly four times in some places, so that many more people can be supported on an increasingly precious farmland base.

Increasing crop yields, however, can lead to enough soil erosion to cause a long-term loss of the resource required to grow future crops.

In urban areas, few people have the opportunity to see much soil. Environmental concerns such as global warming and air pollution easily catch the attention of a largely urbanized society, while the quiet crises of soil erosion and the long-term degradation of soil due to diminishing fertility rarely makes front-page news. And yet, archaeological sites in the Amazon Basin reveal that farmers learned ways to make productive soil, using biochar, in a region with relatively infertile soils thousands of years ago.

SUMMARY

■ Physical weathering is the mechanical disintegration of rocks and minerals; decomposition, or chemical weathering, is the dissolution and chemical alteration of rocks and minerals.

■ The two main processes of chemical weathering are reactions of mineral surfaces with acids and with oxygen. Acids react with silicate minerals to produce clay minerals and dissolved ions, and can completely dissolve carbonate minerals. Oxygen reacts with silicates to form oxide and hydroxide minerals, and when it reacts with sulfide minerals, it also produces sulfuric acid, which leads to acid drainage from mines.

■ Biological weathering processes include not only the obvious examples of tree roots wrenching apart sidewalks, and mounds built by termites, but also the deformation and decomposition of minerals by symbiotic microbes that colonize plant root masses in the rhizosphere.

■ Clay minerals, a secondary weathering product, are a distinguishing feature of soils; they are chemically and physically different from their parent minerals, which are derived from the lithosphere.

■ All Earth systems interact at the surface of Earth to move ions, particles, and gases across the pedosphere,

resulting in a vertical arrangement of soil horizons (a soil profile) that varies in composition and appearance with depth in the pedosphere.

■ Three primary resources formed by weathering in the pedosphere are fertile soil for agriculture, clay minerals for building materials, and mineral ores (iron and aluminum oxides and hyroxides) used in industry.

■ Water can remove substantial amounts of soil if the soil surface is unvegetated or steep. Rates of soil formation are very slow in human terms, and in some parts of the world human activities have increased soil erosion to rates 10 to 100 times higher than natural rates.

■ Studies of controlled soil-erosion plots enable scientists to assess erosion and soil conservation strategies. Fairly simple, often ancient, practices that include terracing, contour plowing, and conservation tillage are effective strategies.

■ The two primary factors that determine whether or not mass movement will happen on a hillslope are hillslope angle and the amount of water in the weathered rock debris.

KEY TERMS

sustainable agriculture
   (p. 228)
carbon sequestration
   (p. 228)
biomass (p. 228)
pyrolysis (p. 228)
biochar (p. 228)
slash-and-char
   (p. 229)
pedosphere (p. 230)
critical zone (p. 230)
physical weathering
   (p. 230)

chemical weathering
   (p. 230)
erosion (p. 231)
stress (p. 231)
exfoliation joints (p. 231)
carbonation (p. 238)
hydrolysis (p. 239)
clay minerals
   (p. 239)
oxidation (p. 239)
bioturbation (p. 240)
soil (p. 242)
soil horizon (p. 242)

soil profile (p. 242)
humus (p. 243)
nitrogen cycle (p. 245)
topsoil (p. 245)
rainsplash (p. 253)
sheet wash erosion
   (p. 254)
rill (p. 254)
gully (p. 254)
desertification (p. 255)
universal soil loss
   equation (USLE)
   (p. 255)

tolerable soil loss
   (p. 256)
conservation tillage
   (p. 257)
mass wasting (p. 257)
mass movement
   (p. 257)
angle of repose
   (p. 258)
surface tension (p. 258)
heave (p. 259)
slide (p. 259)
flow (p. 259)

REVIEW QUESTIONS

1. How does the black soil called terra preta differ from surrounding soils in the Amazon basin?

2. What is sustainable agriculture in the context of soil-forming processes and soil erosion?

3. Explain physical, chemical, and biological weathering and provide an example of each.

4. How do microbes contribute to weathering?

5. How does the growth of plant roots and ice crystals cause physical weathering of rocks and minerals?

6. What types of acids are involved in chemical weathering in the pedosphere? How are they produced?

7. How has the emission of waste products into the atmosphere affected rates of chemical weathering since the Industrial Revolution?

8. How do soils in humid, tropical regions differ from soils in drier and cooler regions?

9. How can soil maps be used to (a) assess the carrying capacity of nations, and (b) make local land-use decisions?

10. What natural events and human activities caused the Dust Bowl in the Great Plains in the 1930s?

11. How have soil conservation practices changed in the United States since the Dust Bowl of the 1930s?

12. How do chemical weathering processes result in the accumulation of rich ores of iron and aluminum?

13. How can buried, ancient soils be used to reconstruct past land-use patterns?

14. What types of evidence indicate that past civilizations had problems with soil erosion?

15. Make a list of the different types of mass movement and characteristics of each.

16. Using concepts of force and resistance, explain how deforestation and road building can trigger mass movement.

THOUGHT QUESTIONS

1. In which parts of the world and types of climate might you expect physical weathering processes, such as frost weathering, to be more important than chemical weathering processes, such as oxidation?

2. Why can't plants (and thus food) be grown easily in partly weathered bedrock once the upper meter or so of soil is eroded? Could such eroded soil be amended so that it could be more productive for crop growing and, if so, how?

3. Clay minerals, smaller than silt and sand, provide stickiness and plasticity to soil. Explain what causes these properties and how they can be valuable for growing crops.

4. After heavy rains and pronounced landslide activity in tectonically active, mountainous areas with active logging, such as in the U.S. Pacific Northwest, many blame logging companies for the deaths and damage that result from the slides. Some have

called for government intervention to regulate, and even to stop, logging in such places. The logging companies and some forest officials argue that in tectonically active areas with mountainous terrain, and in particular during years of exceptionally high rainfall, landslides are likely to happen regardless of logging activities. How might scientists go about studying landslides in order to determine the role of logging in landslide occurrence?

5. Ancient soil deposits in Greece are indicative of different erosion rates over time. What types of changes are revealed in these soil sequences? Using radiometric and other dating methods, researchers studying these soils and sedimentary deposits concluded that the youngest of these deposits and soils were not the result of regional climatic changes, such as prolonged droughts, but rather of patterns of land-use change. What might have been their reasoning for this conclusion?

EXERCISES

1. Make sketches of an entisol, a soil series that might typically occur somewhere such as the Sand Hills in the Great Plains of north-central Nebraska, and a mollisol that is typical of the grassland areas outside the Sand Hills (see Figure 8-16 and Table 8-2). Use online and/or library resources to determine which soil types occur in this region and the typical thicknesses of different soil horizons (e.g., the USDA Natural Resourcse Conservation Service, at http://websoilsurvey.nrcs.usda.gov/app/HomePage.htm). Show the typical depth of A, B, and C horizons for these soil types. What might be the parent material from which each soil type formed? Think about the five different soil-forming factors and how they determine the types of soils in each area.

2. The average soil erosion rate measured on a hillslope in Kenya is about 5 mm per year, but average rates of soil formation are about 0.02 to 0.11 mm per year. Viewing soil formation as an input and soil erosion as an output, how much will the stock of soil on this hillslope reservoir decrease each year? Assuming that the soil is 2 m thick, in

how many years will the soil be completely depleted by erosion?

3. The USDA Natural Resources Conservation Service provides online access to soil maps and data for most U.S. counties and a tool called the Web Soil Survey (WSS) to search this database (http://websoilsurvey.nrcs.usda.gov/app/HomePage.htm). Use the WSS to determine what soil types are mapped for the county in which you live, and compare them to a county in another part of the United States that you think might be a nice place to live. Use the Soil Data Explorer within the WSS to determine the suitability of the soils in these places for agriculture, forestry, and other uses that you find of interest.

4. Using a pile of loose sand, make as steep a cone as possible and measure its angle of repose with a protractor. Why is the angle you measure likely to be between 30° and 35°? Then add a small amount of water to the sand, make as steep a cone as possible, and measure the new angle. Why is this angle much steeper than the previous one?

SUGGESTED READINGS

Colman, S. M., and D. P. Dethier. *Rates of Chemical Weathering of Rocks and Minerals*. New York: Academic Press, 1986.

Cornforth, D. *Landslides in Practice: Investigation, Analysis, and Remedial/Preventative Options in Soils*. Hoboken, NJ: Wiley, 2005.

Evans, A. M. *An Introduction to Ore Geology*, 2nd ed. Oxford: Blackwell Scientific Press, 1987.

Hatfield, J. L., and T. J. Sauer, eds. *Soil Management: Building a Stable Base for Agriculture*. Madison, WI: American Society of Agronomy, 2011.

Hillel, D. *Out of the Earth: Civilization and the Life of the Soil*. Berkeley: University of California Press, 1992.

Logan, W. B. *Dirt: The Ecstatic Skin of the Earth*. New York: W. W. Norton, 2007.

McLaren, R., and K. Cameron. *Soil Science: Sustainable Production and Environmental Protection*. New York: Oxford University Press, 1996.

McPhee, J. *The Control of Nature*. New York: Farrar, Straus and Giroux, 1989.

Montgomery, D. R. *Dirt: The Erosion of Civilizations*. Berkeley: University of California Press, 2008.

Morgan, R. P. C. *Soil Erosion and Conservation*, 3rd ed. Hoboken, NJ: Wiley-Blackwell, 2005.

Prothero, D. R. *Catastrophes!: Earthquakes, Tsunamis, Tornadoes, and Other Earth-Shattering Disasters*. Baltimore, MD: Johns Hopkins University Press, 2011.

Retallack, G. J. *Soils of the Past: An Introduction to Paleopedology*, 2nd ed. Oxford: Blackwell Science, 2001.

Steiner, F. R. *Soil Conservation in the United States: Policy and Planning*. Baltimore: Johns Hopkins Press, 1990.

Troeh, F., and L. M. Thompson. *Soils and Soil Fertility*, 6th ed. New York: Oxford University Press, 2005.

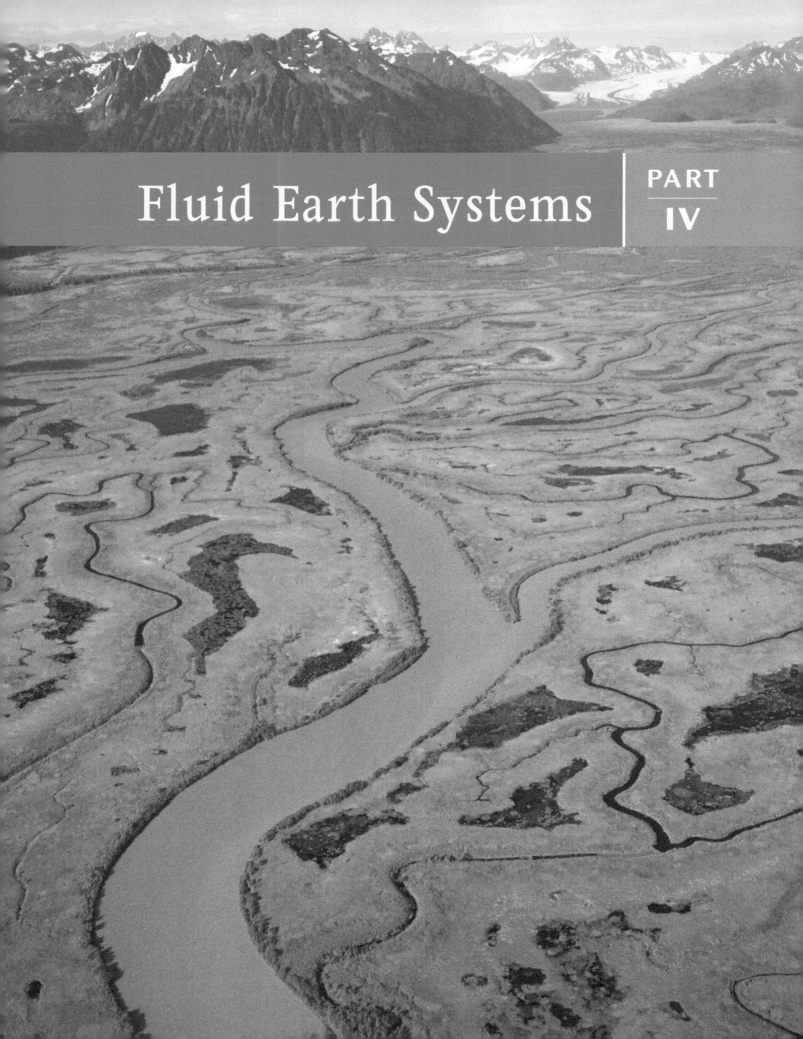

Fluid Earth Systems

✔ Examine ways in which land use change, engineering infrastructure, and climate change have altered the surface water system.

✔ Look at different approaches to ecological restoration, including dam removal, for streams and wetlands in different geographic regions.

✔ Review the ways in which surface water is used as a resource and is protected by various policies and laws.

INTRODUCTION

Floods are natural hydrologic events, part of Earth's global water cycle that is vital to the development and maintenance of floodplains and wetlands adjacent to streams and rivers. When people live and grow food along rivers, however, floods can become humanitarian disasters. In the late summer of 2010, the Secretary-General of the United Nations, Ban Ki-moon, addressed the General Assembly with grim news of a slowly unfolding catastrophe in Pakistan that he likened to a slow-motion tsunami: "Almost 20 million people need shelter, food, and emergency care. That is more than the entire population hit by the Indian Ocean tsunami, the Kashmir earthquake, Cyclone Nargis, and the earthquake in Haiti—combined." Flooding in the Indus River valley was the worst in decades, covering about a fifth of Pakistan with muddy water, ruining millions of acres of farmland, rendering tens of millions of people homeless, and killing nearly 3000 others (Figure 9-1). One in ten persons in Pakistan was affected, and the total economic impact was more than 10 billion dollars (U.S.). Recovery and reconstruction are likely to take years, if not decades, and to cost an additional 10 billion dollars. The human suffering was immense.

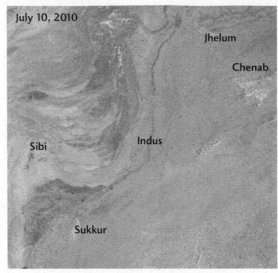

(a)

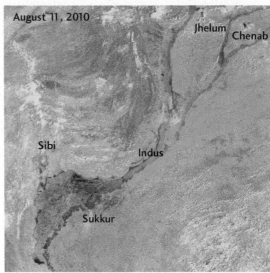

(b)

(c)

Figure 9-1 The satellite image acquired on July 10, 2010, (a) shows the Indus River region in Pakistan at the beginning of the devastating flood, which continued into August (b) and September as floodwaters moved southward from northern mountains to the river mouth in the Arabian Sea. (c) Floodwaters spread widely across lowlands along the Indus River and its tributaries, covering about 20% of Pakistan, and remaining on the landscape in some places for months. The satellite images were acquired by NASA's *Aqua* satellite with an instrument known as the Moderate Resolution Imaging Spectroradiometer (MODIS). Colors represent clouds (pale blue–green), water (bright blue to navy), bare land (pink–beige to brick red), and vegetation (bright green). The *Aqua* satellite mission collects information on all aspects of Earth's water cycle, including evaporation from oceans, clouds, precipitation, soil moisture, and snow cover. *(a, b: NASA images courtesy the MODIS Rapid Response Team at NASA GSFC; c: Paula Bronstein/Getty Images)*

Scientists and the public have wondered if extreme events such as the Pakistan flood of 2010 are occurring more often and possibly with greater intensity and duration. One line of evidence that the hydrological cycle is changing comes from analysis of changes in the annual volume of four components of the global hydrologic budget: water in the oceans (estimated from sea level data), water that falls on the oceans as precipitation, water that evaporates from the oceans, and water that runs off continents to the oceans. For the 13-year period from 1994 to 2006, satellite observations indicated that the amount of freshwater shed from continents to the world's oceans increased by about 18 percent.[1] The increase is equivalent to that of adding the outflow of one more Mississippi River each year. When scientists published the results of their analysis in 2010, they noted that 13 years of data might be too short a period of time to determine whether or not a long-term trend actually exists, but the findings thus far support the hypothesis that global warming is speeding up the world's hydrologic cycle.[2] A warming atmosphere, according to this hypothesis, leads to more evaporation from the oceans. The atmosphere can hold about 4 percent more moisture for each 1°F (about 0.5°C) increase in temperature.[3] Indeed, since 1970 the amount of water vapor measured in the atmosphere over oceans has increased about 4 percent, an increase attributed to the atmosphere's warming.[4] A moister atmosphere, in turn, leads to more clouds that can distribute rain and snow upon continents and, subsequently, greater discharge from runoff into rivers.

A second hypothesis is that anomalous atmospheric events, such as unusual patterns of a jet stream, might occur more often as a result of global warming. These events could lead to greater rainfall and flooding, as occurred with the catastrophic floods in Pakistan when the northern hemisphere jet stream remained locked in place for a week during the summer of 2010.

Yet a third hypothesis holds that the loss of natural floodplains and riparian wetlands to land use development, compounded by the building of flood protection structures such as canals and levees, yields fewer places for floodwater to spread out and be stored. As a result, river flows are greater during large hydrologic events. If this were the case, the restoration of wetlands and deliberate breaching of levees could mitigate the problem somewhat, by spreading floodwaters and reducing stream flow to downstream areas.

The cause of increased discharge of freshwater to the oceans could be a combination of all of these potential contributors. Scientists grapple with these types of questions throughout their careers. Resolving them requires methodical data collection, careful analysis, and rigorous testing of ideas with an open mind, all of which are fundamental to scientific practice (see Chapters 1 and 2).

THE HYDROLOGIC CYCLE

Water is everywhere on Earth's surface, although the total amount is small relative to the overall size of Earth (Figure 9-2). If distributed evenly as a thin film on Earth's surface, water would form a layer 3000 m deep, yet only 2 to 3 percent of this stock is freshwater; the rest is saltwater in oceans and bays (Figure 9-3). Nearly 70 percent of freshwater is locked in ice caps and glaciers, and another 30 percent is stored underground.[5] A mere 0.01 percent of all water on Earth is found in rivers and lakes, and an even smaller amount, 0.001 percent, resides in the atmosphere.

The movement of water from one reservoir to another constitutes the **hydrologic cycle**, the fundamental, unifying concept in the study of water on Earth. From dark clouds to raindrops falling on a lake, from snowflakes to ice in a glacier, or from a cold glacial meltwater stream to a chilly sea with floating icebergs, water moves easily among atmospheric, continental, and oceanic reservoirs, changing state from liquid to solid to vapor along the way.

The simplest model of the hydrologic cycle shows water evaporating from oceans to form atmospheric

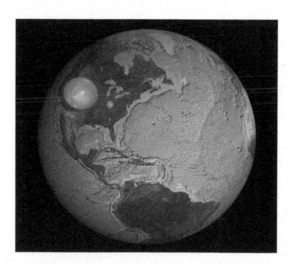

Figure 9–2 The size of each three–dimensional sphere in the image represents volume, with the largest representing Earth. Of the blue spheres, the largest represents all water on, in, and above Earth (about 332,500,000 mi³, or 1,386,000,000 km³). This includes water in the oceans, glaciers, groundwater, lakes and rivers, atmosphere, plants, and animals. The next smaller sphere, shown over the state of Kentucky, represents the liquid freshwater on Earth, which occurs in groundwater, wetlands, rivers, and lakes (2,551,000 mi³, or 10,633,450 km³). The smallest bubble, located over Georgia and barely detectable at the scale of the image, represents fresh surface water in lakes and rivers (22,339 mi³, or 93,113 km³). *(Illustration by Jack Cook © Woods Hole Oceanographic Institution)*

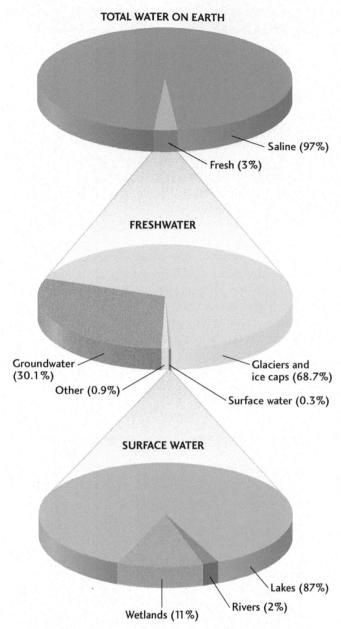

TOTAL WATER ON EARTH

Saline (97%)
Fresh (3%)

FRESHWATER

Groundwater (30.1%)
Other (0.9%)
Glaciers and ice caps (68.7%)
Surface water (0.3%)

SURFACE WATER

Lakes (87%)
Rivers (2%)
Wetlands (11%)

Figure 9–3 Pie diagrams illustrate that most of Earth's water is saline (97%), and of the approximately 3% of freshwater, the majority is frozen in ice caps and other glaciers (about 69%), or lies beneath Earth's surface as groundwater (about 30%). Of the tiny remainder (0.3%) that is freshwater at Earth's surface, the majority occurs in lakes (87%), wetlands (11%), and rivers (about 2%). All other freshwater, including water vapor in the atmosphere and moisture in plants and animals, occurs in relatively small amounts that are less than 1% of the total.

vapor that later condenses and falls as *precipitation* on continents. Once deposited, water returns to the oceans through *runoff* in streams (Figure 9-4). In reality, *evaporation* can occur anywhere on Earth, and much water returns to the atmosphere from evaporation over landmasses, lakes, and rivers. Likewise, precipitation

can occur anywhere, and much of the evaporated water ends up condensing and falling onto the ocean. Other processes involved in the movement of water are *transpiration*—the release of water vapor by plant and animal cells—and *infiltration* of water below the ground surface along openings such as pore spaces in sediments or fractures in rocks. Sometimes the processes of evaporation and transpiration are combined, in which case we use the term *evapotranspiration*. Once in the ground, water not held to particles by surface tension seeps downward until it reaches a zone in which all pore spaces are saturated. In this zone, water flows as *groundwater* toward areas of lower elevation and pressure where it discharges into streams and springs. Eventually, this water, too, returns to the atmosphere or ocean and continues to cycle.

Surface Water Distribution

The distribution of freshwater on Earth is controlled by the climate system, governed primarily by atmospheric and oceanic circulation. We will explore this topic in greater detail in Chapters 11 and 12. For now, it is sufficient to know that both precipitation and evaporation vary spatially in response to latitude, topography, and the locations of rising and descending air masses. As a result, nations and regions are endowed with differing amounts of water resources. In North America, for example, the abundance of surface water varies markedly from east to west.

By comparing precipitation rates to values of *potential evapotranspiration* (the amount of water that would evaporate and/or be transpired by vegetation each year given an unlimited water supply), it is easy to see why most water shortages and *ephemeral streams* (those that do not flow throughout the year) in the United States occur in the western states (Figure 9-5). With the exception of mountainous areas, the average annual precipitation is less than 15 to 20 in. (38 to 51 cm) nearly everywhere west of the Mississippi River, and in many areas precipitation is lower than the average annual potential evapotranspiration. In these semiarid to arid areas, the potential evapotranspiration is so great that even if more rain were to fall, it still might not be enough to create surface runoff or to recharge soil moisture and groundwater supplies. Because of these arid conditions, many who live in the western states depend heavily on surface water imported by aqueducts from locations with more abundant water resources, as well as on deep groundwater supplies.

East of the Mississippi, the average annual precipitation is greater than 20 in. (51 cm) and exceeds the average annual potential evapotranspiration. Such conditions are typical in subhumid to humid regions. Most of the water used in eastern North America comes from rivers because they are *perennial*—that is, they flow year round. Other than the melting of snow during warm

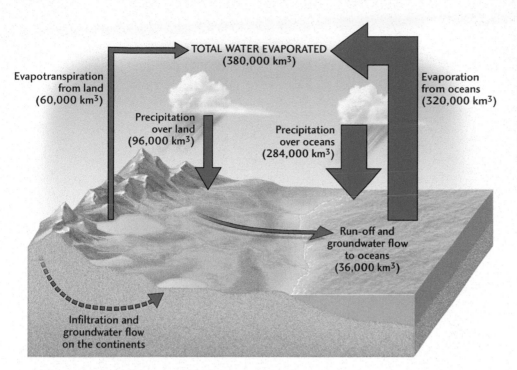

Figure 9–4 Earth's hydrologic cycle comprises the continuous movement of water through the processes of evaporation, precipitation, transpiration (from plants), infiltration, and runoff.

months, the cause of continuous flow in streams between rainfall events is the abundant storage of water in soils, rocks, and sediments. Long after rains end, this water drains into streams and lakes as base flow, the inflow of groundwater that maintains the water level in perennial streams between storms.

Hydrologic Budgets

Earth's hydrologic system is considered a *closed system*. With the exception of small amounts of water vapor that escape from occasional volcanic eruptions or geothermal vents, no new water is released at Earth's surface. Likewise, with the exception of small amounts of hydrogen that escape from the outermost atmosphere many kilometers above the surface, no water is lost from our planet. The conservation of water at any location on Earth can be expressed with a **mass balance equation:**

$$\text{Inflow} - \text{outflow} = \pm \text{ changes in stock}$$

If inflow is greater than outflow, the stock increases (positive change in stock) at a rate equal to the rate of excess, and the reservoir grows. If inflow is less than outflow, the stock decreases (negative change in stock) at a rate equal to the rate of deficit and the reservoir becomes depleted. If the system is in steady state, with inflow equal to outflow, no change in stock occurs.

Water inflows to Earth's continental surfaces are precipitation rates, and outflows are rates of evapotranspiration and surface runoff. The amount of water that infiltrates soil and becomes groundwater represents the change in stock when the inflows do not balance the outflows:

$$\text{Precipitation} - \text{evapotranspiration} - \text{surface runoff} = \pm \text{ changes in soil moisture and groundwater storage}$$

This **hydrologic budget equation** is a powerful tool for scientists, environmental planners, and engineers. It is the basis of water budget studies for lakes, watersheds, irrigation networks, reservoirs, and cities.

As an example, Figure 9-5 shows that the Las Vegas, Nevada, region receives precipitation of 4 in. (10 cm) per year while potential evapotranspiration measures 36 to 48 in. (91 to 122 cm) per year. Runoff is less than 1 in. (2.5 cm) per year. Annual change in storage of soil moisture and groundwater can be calculated with the hydrologic budget equation:

$$\begin{aligned}\text{Changes in soil moisture and groundwater storage} &= \\ \text{precipitation} - \text{evapotranspiration} - \text{runoff} &= \\ (4 \text{ in./yr}) - (36 \text{ to } 48 \text{ in./yr}) - (0 \text{ to } 1 \text{ in./yr}) &= \\ -32 \text{ to } -45 \text{ in./yr} \end{aligned}$$

As the hydrologic budget analysis for this region shows, groundwater is not being recharged significantly under current climate conditions (negative change in storage). Because evapotranspiration is so much greater than precipitation, the inhabitants of the Las Vegas region are dependent on deep groundwater and the import of Colorado River water from Lake Mead, a reservoir impounded by the Hoover Dam, for drinking water and other domestic purposes, as well as for agriculture and industry (Box 9-1).

Residents of regions that are seasonally dry or experience droughts need to maximize conservation and reduce water consumption to ensure sustainable water supplies. One strategy that has worked well in parts of the American Southwest is a financial incentive for those who convert lawns to water-efficient desert landscapes, with a rebate for each square foot of land converted. Billions of gallons of water have been conserved in the Las Vegas area, for example, through conversion of

(continued on page 277)

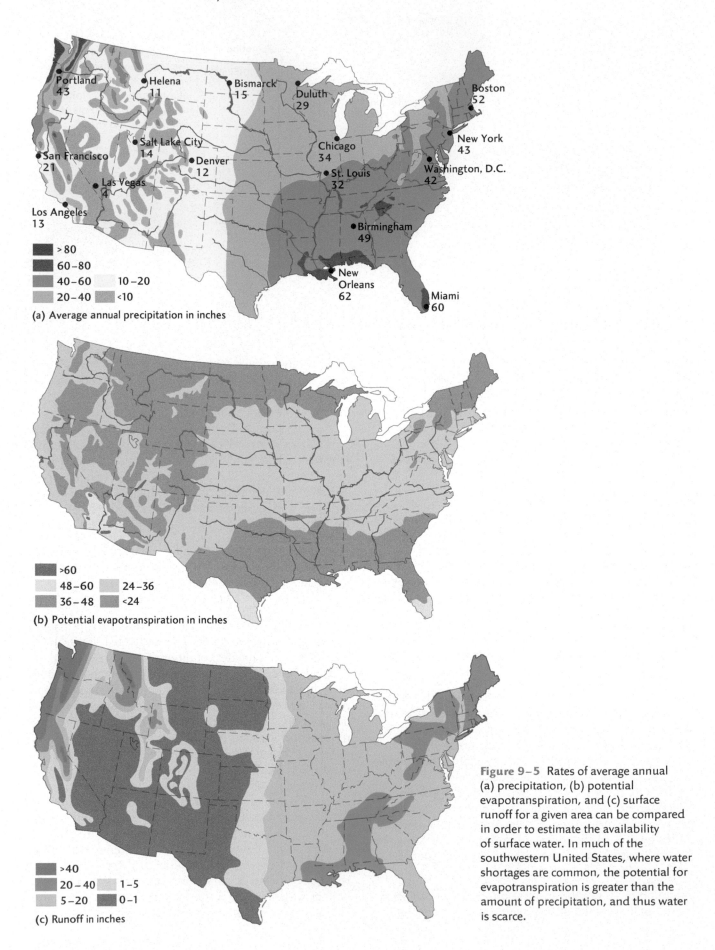

(a) Average annual precipitation in inches

Portland 43
Helena 11
Bismarck 15
Duluth 29
Boston 52
Salt Lake City 14
Chicago 34
New York 43
San Francisco 21
Denver 12
St. Louis 32
Washington, D.C. 42
Las Vegas 4
Los Angeles 13
Birmingham 49
New Orleans 62
Miami 60

>80
60–80
40–60   10–20
20–40   <10

(b) Potential evapotranspiration in inches

>60
48–60   24–36
36–48   <24

(c) Runoff in inches

>40
20–40   1–5
5–20   0–1

Figure 9–5 Rates of average annual (a) precipitation, (b) potential evapotranspiration, and (c) surface runoff for a given area can be compared in order to estimate the availability of surface water. In much of the southwestern United States, where water shortages are common, the potential for evapotranspiration is greater than the amount of precipitation, and thus water is scarce.

BOX 9–1   EARTH SCIENCE AND PUBLIC POLICY

Surface Water Use in the Colorado River Basin and Long–Term Records of Runoff from Tree Rings

Residents of the American Southwest have good reason to worry about water resource availability in the future. Analysis of tree rings indicates that the original estimate of how much water is available annually in the Colorado River was made during an unusually wet period in the early 20th century.[7] Thirty million users of runoff to the Colorado River in the semiarid to arid American Southwest, including residents of large cities such as Las Vegas, Los Angeles, and Denver, 15 Native American tribes, and farmers who produce goods from 4 million acres of irrigated land, divvy up its water according to a number of laws and treaties that date back nearly a century (**Figure 1**). The most important of these is the Colorado River Compact, negotiated in 1922 by the federal government and seven states along the Colorado River. According to this Compact, 7.5 million acre–ft (MAF) of water is allocated each year to the upper basin states (Colorado, Wyoming, Utah, and New Mexico), and another 7.5 MAF each year to lower basin states (Nevada, Arizona, and California). In 1944, a treaty signed between the United States and Mexico allocated another 1.5 MAF annually to Mexico.

At the time the Compact was negotiated, the average flow measured at one gauge station on the river (at Lees Ferry, Arizona) over a period of only 20 years was 16.4 MAF, slightly less than the amount allocated to users by the time of the 1944 treaty. Streamflow as used here is equivalent to the term "runoff." From 1906 to 2004, however, the average flow measured at the Lees Ferry gauge was only 15.1 MAF.

The Colorado River's water was over–allocated, albeit inadvertently, because its flow was apportioned based on a short record of only 20 years. Given the high degree of regulation of the river by dams and canals, and of its apportionment to multiple users, the modern Colorado River no longer flows all the way to its mouth at the Sea of Cortez; instead, it dries to a trickle upstream. Perhaps of most concern is that population growth, climate change, ecological needs, and a drought that began in 1999—the most severe in more than a century—make projections of future availability and sustainability of water supplies in the Colorado River highly uncertain.[8] In response to this situation, the U.S. Bureau of Reclamation initiated a study in 2010 to define the current and future imbalances of the

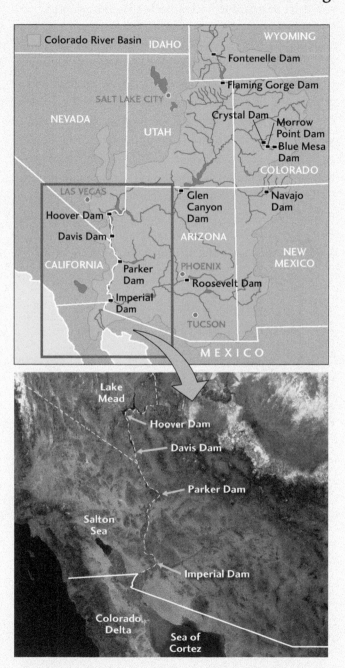

Figure 1 The Colorado River Basin drains from Wyoming to the Sea of Cortez in Mexico, providing water to 30 million residents as it flows through the American Southwest. Throughout this largely semiarid region, multiple dams, including Hoover Dam, slow and impound the Colorado's downstream flow, much of it from annual snowmelt along the western slopes of the Rocky Mountains. Reservoirs of water behind each dam can be seen in the satellite image. *(b: NASA)*

(continued)

(continued)

supply and demand of water in the Colorado River Basin for the next half century. Reports from this study include strategies to adapt to and mitigate these imbalances.[9]

Long-term information on Colorado River flow from tree-ring chronologies extends the record back many centuries, so that water managers need not rely on short records of measurement at gauge stations (**Figure 2**). Each annual growth ring of a tree reflects the environmental conditions of that year, with wide rings indicating good conditions for growth and narrow rings indicating poor ones. Because plant growth is greatly dependent on the availability of water, wide rings can indicate wet years and narrow rings, drought. Reconstructions of streamflow records for ten stream gauge stations from a regional record of tree rings in the Colorado River region indicate that wet and dry periods have been cyclic over the past twelve centuries (from 762 to 2005 CE).

The long-term average annual flow estimated from tree-ring analysis is about 14.7 MAF/year for the Lees Ferry gauge, 1.8 MAF less than the amount allocated for water use. A surprising finding from these records is evidence of a long drought that spanned many decades in the mid-1100s. All but 9 years during this 57-year drought had annual flows less than the average recorded at the gauge from 1906–2004. Two other prolonged droughts occurred in the 800s. The period from about 800 to 1300 CE was characterized by a well known climate anomaly called the Medieval Warm Period, for which other paleoclimate records indicate dry, relatively warm conditions in the American Southwest (see Chapter 14). If the Colorado River goes through a period of drought like those revealed in the tree-ring record, the tens of millions of users of its water could experience a severe water shortage for a number of years.[10]

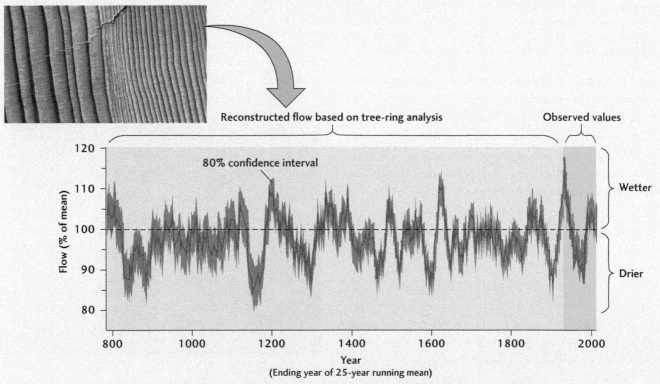

Figure 2 Streamflow reconstruction based on dated tree rings, like those shown in this graph, from several tree-ring chronology sequences for the period 762 to 2004 CE, plotted with respect to the average of values measured at the Lees Ferry, Arizona, gauge station in the Colorado River basin from 1906–2004 (18.6 billion m^3, or 15.1 MAF). Flows are shown as a 25-year running mean, so that each point actually represents a 25-year average of that year and the preceding 24 years. The shaded band about the reconstructed flows encompasses 80% of the probable estimates of flow values, and is known as a confidence interval. *(a: © Peter L. Kresan)*

155 million ft² of lawn through the Southern Nevada Water Authority's Water Smart Landscapes rebate program, and other western water authorities have implemented similarly successful programs.[6]

✔ The hydrologic cycle consists of the movement of water among the oceans, atmosphere, and continents via the processes of evapotranspiration, precipitation, infiltration, groundwater flow, and runoff.

✔ Water shortages occur in some areas because surface water is distributed unequally among and within continents.

✔ The balance among precipitation, evapotranspiration, and runoff determines how much soil and groundwater storage occurs at a given place, a relation expressed in the hydrologic budget equation.

✔ Tree-ring chronologies can extend the record of streamflow back many centuries earlier than the period of modern observation and measurement at stream gauge stations.

DRAINAGE BASINS AND STREAMS

To this point we have discussed the general operation of the hydrologic cycle and what controls water resources. We now shift our focus to examine what happens when water from the atmosphere falls onto land. Each raindrop or snowflake falls onto a catchment area, a tract of land that delivers water and sediment from hillslopes to stream channels. Earth scientists and land use planners refer to such catchments as **drainage basins** or **watersheds** (Figure 9-6). The boundaries of drainage basins are

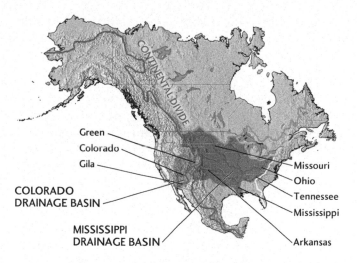

Figure 9–7 The North American continental divide in the United States is shown here in red, the divide for waters that drain to the Gulf of St. Lawrence in green and orange, and the divide for waters that drain to the eastern seaboard in yellow. Drainage networks for the Colorado and Mississippi rivers are blue and their drainage basins are orange and purple, respectively.

determined by the locations of **drainage divides**, high points in the landscape that separate waters flowing in adjacent watersheds, and each drainage basin contains many smaller subwatersheds. For example, the Ohio River watershed lies within the larger Mississippi River watershed, which means that activities within the Ohio watershed can affect the Mississippi River downstream.

Every continent contains major drainage basins separated by continental divides. In North America, a continental divide follows the crest of the Rocky Mountains of Canada and the United States and the Sierra Madre Occidental of Mexico and separates basins that drain westward, toward the Pacific Ocean, from those that drain eastward into the Gulf of Mexico (Figure 9-7). Another divide runs along the Appalachian Mountains, separating basins that drain westward into the Mississippi from those draining eastward into the Atlantic. In South America, the continental divide runs along the Andes Mountains and separates water flowing westward into the Pacific from that flowing eastward into the Atlantic.

Drainage Networks and Base–Level Controls

The entire system of stream channels in a drainage basin forms an intricately linked **drainage network**, similar in function to the human circulatory system or to veins in a leaf: This network transports fluids to or from a surface. Because many river networks are treelike in form, their patterns are called dendritic, from the Greek word *dendron*, which means tree. The major arteries of the river

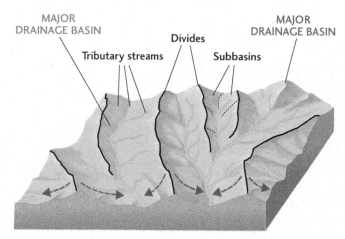

Figure 9–6 A drainage basin, or watershed, consists of drainage divides encircling tributary stream channels that drain downstream to a main stem channel. Drainage divides separate the waters between drainage basins. During a rainstorm, some waters flow on one side of a divide into a drainage basin, and some flow on the other side into another drainage basin.

channel network are called *trunk streams,* and the much finer channels that feed the trunk streams are *tributaries.*

The primary driving force for flow in a river network is gravity, and all water and sediment moves in one direction—down slope. This flowing water and entrained sediment typically is confined to a trough-like feature known as a stream channel, as it does in the Colorado River in the Grand Canyon. In other cases, however, it spreads wide across vegetated wetlands with multiple small channels, as it does in the Everglades. The lowest point toward which water and sediment can flow in a drainage basin is its **base level.** While the ocean surface is the ultimate base level for most of the world's streams, many local base levels, such as lakes or dams, temporarily slow water along stream networks. A noted exception is the Middle East's Jordan River, which flows to the Dead Sea more than 420 m below sea level, the lowest elevation of land on Earth's continents.

In a general sense, over relatively long time periods erosion and deposition gradually cause stream channels to acquire a concave-up longitudinal profile (a vertical section along the length of a stream), adjusted at the downstream end to local or global base level (Figure 9-8). If base level falls, such as during a glacial period when water is removed from the oceans and stored on land in the form of ice sheets, erosion enables the entire stream network to cut downward into the landscape to adjust to the new base level. If base level rises, such as occurs when a dam is built across a stream, sediments are deposited in the stream channel to reestablish the longitudinal profile.

What Happens When It Rains?

We have used the word runoff several times in this chapter and noted that drainage basins deliver water from hillslopes to streams, so you might have the impression that raindrops move immediately from their point of contact with the landscape into stream channels. In reality, such **overland flow** is fairly limited in natural settings. If you live in a humid environment, you will likely notice during a rainstorm that much of the rainfall is intercepted by leaves on trees and shrubs, called *canopy storage,* before eventually moving along plant stems and dropping to the ground (i.e., *throughflow* and *stemflow*) once leaf surfaces are fully wetted (Figure 9-9). Leaves, herbs, and grasses on the ground absorb additional water, reducing the amount making its way into the soil. When the rain ends, much of this stored water quickly returns to the atmosphere through evaporation; much of the rain that fell may return directly to the air. The rest infiltrates the ground, soaking into pore or open spaces in the loose material that makes up soil.

If you live in an arid environment, you may notice a different phenomenon that minimizes overland flow during rainstorms (see Figure 9-9). Arid regions lack significant vegetation cover, so canopy storage is minimal. However, the bare soils exposed between isolated shrubs and clumps of grass often show deep cracks, produced by drying of clay minerals, that shrink or swell depending on soil moisture conditions. Rain falling on these landscapes soaks into the fissures, reducing the amount of runoff. If rainfall continues, however, clay minerals may swell and close the cracks (see Chapter 8), blocking further downward movement of water into the soil. At this point, additional rain runs off as overland flow.

The key determinant of whether overland flow will be generated is the ground's **infiltration capacity**—its ability to absorb rainwater—relative to the

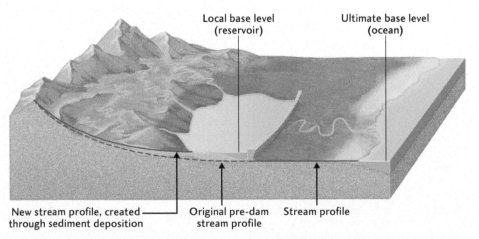

Local base level
(reservoir)

Ultimate base level
(ocean)

New stream profile, created
through sediment deposition

Original pre-dam
stream profile

Stream profile

Figure 9–8 Due to gravity, water moves downstream from higher to lower elevations, moving ultimately to the lowest elevation, or base level. Upstream of either the level of the ocean surface, the ultimate base level, or a local base level such as a dam, the flow of water in a stream cannot drop below that of the base level. Note that the original stream profile is generally concave–up, typical of many streams.

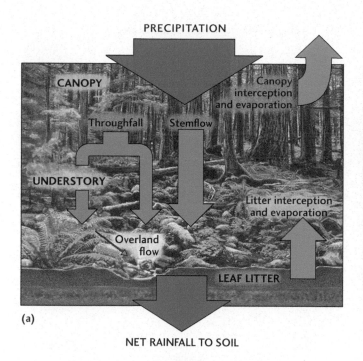

PRECIPITATION

CANOPY

Canopy interception and evaporation

Throughfall   Stemflow

UNDERSTORY

Litter interception and evaporation

Overland flow

LEAF LITTER

(a)

NET RAINFALL TO SOIL

PRECIPITATION

Evaporation

Litter interception and evaporation

Overland flow

(b)   NET RAINFALL TO SOIL

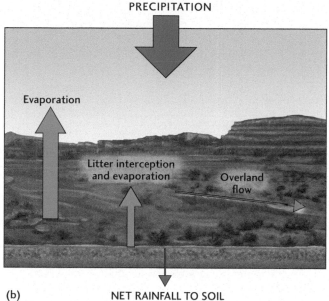

Figure 9–9 (a) Forests such as the primeval temperate rain forest of the Olympic Mountain Range in the Pacific Northwest receive up to 4 m of precipitation (snow and rain) a year. Much of it is intercepted in the lush forest canopy and evaporated or transpired from plants. (b) In arid or semiarid desert landscapes, much less rain is intercepted and, as a result, more of the precipitation becomes overland flow. *(a: William Manning/ Alamy; b: Thinkstock)*

intensity (amount of rain falling per hour) and duration of the precipitation event. Infiltration capacity depends strongly on soil characteristics. As discussed in Chapter 8, soils contain different horizons with depth. The B-horizon often contains an accumulation of clay particles that have been transported downward from the overlying A-horizon because of infiltrating rainwater. Because clay particles are small and fill the pore spaces between larger grains, the B-horizon can retard the downward movement of water, allowing the A-horizon above it to fill. Any additional rainfall runs off, because it no longer can soak into the soil.

In addition to soil texture and the presence of shrinking and swelling clay minerals, organisms influence infiltration capacity. Worms, insects, and rodents dig burrows that increase storage space and create easy pathways for water to move through soils, and the rotting roots of dead plants provide additional conduits. Another factor that determines infiltration capacity is the amount of moisture already in the ground: A soil can take in far less water if it rained yesterday than if it rained three weeks ago. Characteristics of the storm also affect the potential for runoff generation. Low intensity "misting" rains typical of the Pacific Northwest create far less overland flow than the torrential downpours common in the Southwest, because low intensity rainfall may match the rate at which water moves through slow-percolating soil horizons.

For any given storm, infiltration capacity tends to decline over time as soil storage spaces fill with water and as clay minerals expand and block pore spaces. If rainfall intensity surpasses the infiltration rate, surface water accumulates in small depressions on hillsides, and then overflows these small basins, forming rivulets that make their way toward stream channels. At the same time, water that already infiltrated the soil migrates laterally toward streams. This *subsurface stormflow* travels along the interface between soil horizons of different infiltration capacities and seeps from the ground at springs along the edges of floodplains and stream channels.

Human activities can greatly modify the potential for overland flow and subsurface stormflow generation as well as for groundwater recharge. While walking around in the rain, you are likely to notice sheets of water pouring off of roads, parking lots, and sidewalks. These *impervious surfaces* have so little infiltration capacity that even a light drizzle can produce puddles, and a prolonged or heavy rainstorm sends water rushing across them in sheets. In urban settings, curbs funnel rainwater toward storm sewers, which often consist of pipes that carry water directly to stream channels. As a result, subsurface stormflow and groundwater recharge are limited, while stormflow in stream channels increases.

Stream Stage, Discharge, and Hydrographs

A **hydrograph** is a graph of the amount of water flowing past a particular location in a drainage network; it displays either the elevation of the stream surface, called the **stage,** or the volumetric flow rate, called the **discharge,** as a function of time (Figure 9-10). Hydrographs provide valuable information to planners interested in assessing water resources and flooding potentials of different streams. Stage can be measured by observing the height of water on a staff gauge (large ruler) mounted on a culvert or bridge piling around which a stream flows. It also can be determined by measuring the height of water in a tube connected to the stream. This method can be automated by placing a pressure sensor or float in the tube that can continuously monitor changing water

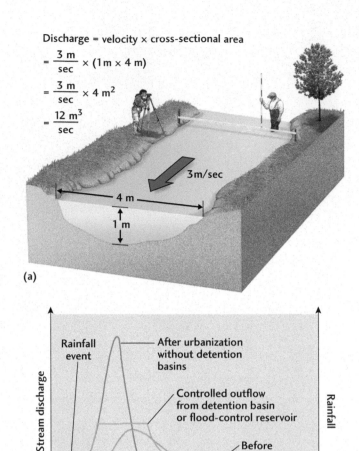

(a)

(b)

Figure 9–10 (a) Stream discharge is the velocity of the water flow multiplied by the cross–sectional area of the stream channel. Typically, scientists survey the depth and width of the stream and use flow meters submerged at different depths and points across the channel to measure velocity. Scientists measure the area of flow in the cross–sectional direction at greater resolution than shown here in approximate form. (b) The rising and falling levels of water during a rainfall or snowmelt event, converted from stage (height) to discharge (volume with respect to time), can be plotted against time to produce a hydrograph that characterizes the changing discharge. Note that the crest of the event, at peak discharge, typically occurs after the peak of the flow–producing event. Urbanization generally results in more impervious land surface and greater surface water runoff, so that peak discharge occurs sooner and with greater magnitude after a flow event. This example also illustrates the effect of detention reservoirs on minimizing the peak discharge during a flood. Without such control structures, peak discharge would be much greater.

levels. Discharge is determined by multiplying the average velocity of a stream (measured with a propeller flow meter or by timing how long it takes a floating object to travel a known distance) by the cross-sectional area of the channel. Discharge and stage can be related by means of a *rating curve* if both of these parameters are measured at varying levels of flow.

Examination of a hydrograph reveals that stage and discharge rise and fall in response to rainfall events (see Figure 9-10). During storms, only 1 to 5 percent of the water falling on a watershed lands directly on stream channels, but subsurface stormflow and overland flow move quickly toward streams, raising water levels. The *rising limb* of the hydrograph reflects this increase in discharge. As storms end, water levels decline, creating the hydrograph's *falling limb*. The rising and falling limbs typically are not symmetrical about the hydrograph peak because subsurface stormflow and groundwater leakage continue long after the period of overland flow ends. Streams thus tend to rise more quickly than they fall. Eventually, even the subsurface stormflow is depleted and only groundwater leaks out of the channel banks to provide the stream with **base flow.**

Based on your understanding of infiltration capacity gained in the last section, you may be able to hypothesize about the impact of land use development on stream hydrographs. Construction of homes, shopping malls, schools, parking lots, roads, and driveways accelerates the movement of precipitation into stream channels and lowers the amount of soil storage and groundwater recharge. As a result, for the same storm intensity, a hydrograph after land use development shows a steeper rising limb with an earlier onset than a predevelopment hydrograph (see Figure 9-10). In addition, post-development hydrographs reach a higher peak stage and discharge than predevelopment hydrographs, with important implications for flooding hazards (discussed later in this chapter). Finally, the reduced amount of subsurface stormflow and groundwater recharge decreases the base flow contribution to the stream, with consequences for stream organisms. Trout, for example, are extremely sensitive to water temperature, and reduction of cool groundwater flows can warm stream water beyond their ability to survive.

Stream Channel Patterns and Processes

Flowing water sculpts the landscape as it transports sediment eroded from continents to oceans. The higher the discharge of a stream, the larger the particles it can carry, an attribute known as *stream competence,* and the larger the volumes of sediment it can transport, an attribute known as *stream capacity.* Particles carried or deposited by flowing water in a stream or river, collectively called **alluvium,** include clay, silt, sand, and gravel. Alluvial streams have beds underlain by alluvium, whereas the channels of bedrock streams are relatively free of alluvium. The most common patterns of alluvial streams

are braided and meandering, depending on the energy of the stream system, the nature of the geologic material on which the water is flowing, and the type and volume of sediment being transported.

Braided streams have many branching channels that repeatedly divide and recombine around bars of mounded sand and gravel, giving them the appearance of a braid (Figure 9-11a). Commonly characterized by a steep stream gradient, braided streams have large widths relative to their depths—ideal characteristics for transporting material under rapidly fluctuating conditions of streamflow. During high flows, the entire width of the stream channel is submerged and water flows over the tops of the gravel bars, transporting some of the particles from these bars downstream. During lower flows, the bars are exposed and water flows among them, carrying smaller particles due to the stream's lowered competence.

Braided streams are common in locations where large volumes of coarse sediment are produced and where hydrologic conditions change rapidly, such as near glaciers and in mountainous and arid areas. They also are common in disturbed and logged landscapes, where deforestation has resulted in accelerated mass movement (and thus a greater sediment load) and increased stormwater runoff. In all of these environments, the abundance of coarse debris and frequently changing water levels inhibit the establishment of stabilizing vegetation along the channel banks, making banks easily eroded and prone to failure during floods.

Meandering streams consist of a single-thread channel that migrates back and forth across a nearly flat plain, with the channel twisting and turning across the valley bottom. The term is taken from the ancient name of the Meander River—now known as the Buyuk Menderes, in modern Turkey—which was noted for its many bends. The middle reach of the Mississippi River is a classic example of a meandering stream (Figure 9-11b). This section flows upon hundreds of feet of alluvium deposited by ancient braided streams that carried glacial meltwater during repeated full glacial cold conditions of the past few million years (see Chapters 6 and 14).[11] The modern Mississippi River channel, receiving no glacial meltwater during the present interglacial warm stage, is characterized by fine-grained sediment, a gentle gradient, and a small width relative to depth in its middle reach. Although meandering streams may carry large amounts of sediment, the sediment generally is fine grained and constant in supply. In other words, meandering streams can have a high capacity but a low competence in comparison to braided channels.

Braided streams are characterized by the bars that separate individual low-flow channels, and meandering alluvial streams are characterized by alternating deep and shallow water **pools** and **riffles,** respectively

(a)

(b)

Figure 9-11 The most common types of stream channels in terms of plan view pattern are braided and meandering. (a) The Rakaia River in New Zealand is a braided stream. During warm seasons, melting ice in its mountainous headwaters releases large amounts of water and poorly sorted sediment, whereas during cold seasons both discharge and sediment load are much lower. (b) The middle reach of the Mississippi River is meandering and highly sinuous in places, with numerous crescent–shaped oxbow lakes and ridge and swale topography formed as a result of channel migration and abandonment during the past 12,000 years. *(a: AgeFotostock/ Superstock; b: Ingram Publishing/Superstock)*

(Figure 9-12). Pools typically occur on the outer bank of a meander bend, where the energy of more swiftly flowing water is concentrated as it moves into the bend. Fast-flowing water scours this *cut bank,* sinks downward, and scours the channel floor, forming a deep depression. Continued erosion shifts the stream's position and gradually sweeps it sideways toward the outer bank and across the valley floor. At the same time, sediment is deposited opposite the pools along the inside banks of meander bends, where the water flows more slowly and has less energy. Here, silt, sand, and gravel are deposited in crescent-shaped *point bars* that migrate in the same direction as the laterally sweeping stream.

The twin processes of cut bank erosion and point bar deposition lead to the development of a nearly flat **floodplain** that exists alongside meandering streams; during high flow, floodwaters deposit sediment on this surface. In the latter process, known as **overbank deposition,** silts and clays suspended in stream water are deposited on the landscape surrounding the stream as floodwaters ebb, filling in depressions and building a level surface. Coarser sediments, such as sands and gravels, cannot be carried as far from the channel and tend to be deposited immediately adjacent to it, forming **levees**—berms that help to confine the majority of flows to the channel.

The fine-grained nature of the floodplain and the more constant flow of water in meandering streams allow grasses, trees, or shrubs, depending on local climate, to colonize the landscape. These plants can stabilize low channel banks with their roots, slowing the rate of cut bank erosion and meandering migration of the water flow. Still, erosion continues to occur and when meander bends become exceptionally curvy, streams can cut through the narrow neck of land that remains, creating a bypass and straightening the channel. Eventually, sediment deposition seals off the old meander bend and converts it to a crescent-shaped lake known as an *oxbow,* and the newly straightened channel begins to curve again as erosion and deposition continue (see Figure 9-11b).

Between meander bends, sediment scoured from the pools accumulates as riffles along straight stretches of

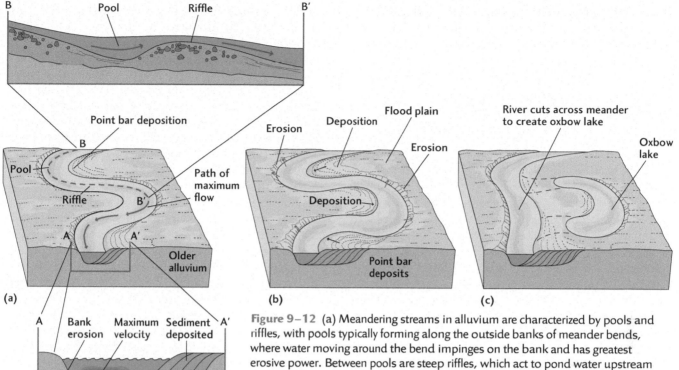

Figure 9–12 (a) Meandering streams in alluvium are characterized by pools and riffles, with pools typically forming along the outside banks of meander bends, where water moving around the bend impinges on the bank and has greatest erosive power. Between pools are steep riffles, which act to pond water upstream in the pools. On the inside banks of meander bends, where stream velocity is low, sediment is deposited as point bars. (b) Meandering streams tend to migrate across their valley floors with time, in the direction of bank erosion, so that meander bends become accentuated and horseshoe–shaped. (c) If a stream cuts across the neck of a meander bend, it abandons its former channel and creates an oxbow lake.

the stream channel. Riffles appear as humps in a longitudinal stream profile, and these humps help to hold water in pools upstream of a riffle during times of low flow (see Figure 9-12b). For most of the natural streams in the world, the distance between pools is about five to seven times the channel width. The repeating pattern of pools is critical to providing cool, deep water for many river-dwelling plants and animals, including fish such as salmon and trout, and the turbulence in riffles mixes oxygen into the water, a necessity for the health of aquatic organisms. Good anglers know that the best fishing is along streams with deep, cool pools separated by fast-flowing rapids.

With the exception of artificially channelized streams, straight single-thread channels are rare in nature. They can be created in laboratory sandboxes, but within a short period of time, the channel will alter its shape, creating pools and riffles and a meandering pattern, or developing braid bars. Rivers also change character along their lengths. For example, many streams originate in mountainous areas where they display braided behavior, but the same streams may exhibit meandering channels farther from a source of coarse debris. Thus, alluvial streams continually adjust their form in response to variations in water discharge and sediment characteristics.

✔ Surface water flows within stream networks that drain water from the land; the area of drainage—bounded from adjacent drainage areas by divides, or high spots—is called a drainage basin or watershed.

✔ Each stream network generally develops a concave-up longitudinal profile adjusted at its downstream end to base level.

✔ Canopy storage as well as plants and leaf litter on the ground can intercept much of the rainfall before it reaches the soil.

✔ The infiltration capacity of a landscape, and the intensity and duration of storms, together determine whether rain falling on a watershed will generate overland flow, subsurface stormflow, and/or groundwater recharge.

✔ The volume of water that flows in a given period of time in a stream channel is the stream's discharge, which is calculated as the area of flow in the stream channel times the water flow velocity.

✔ A hydrograph is a plot of the discharge or stage of a stream over time determined at a single point in the drainage network.

✔ Alluvial streams are described as braided, meandering, or straight, depending on the nature of their channel patterns. Braided streams typically carry large amounts of coarse sediment (i.e., high competence), whereas meandering streams carry finer silts and clays (i.e., low competence).

✔ Pools and riffles are ubiquitous features along meandering alluvial streams; they are spaced fairly uniformly along channels, with the distance between adjacent pools roughly six times the channel width.

FLOODS

The archaeological record demonstrates that people have settled along rivers and lakes for easy access to freshwater for tens of thousands of years, and for access to fertile floodplain soils and water for crops since the development of agriculture about 10,000 years ago. Unfortunately, the same process that replenishes the land for agriculture—flooding and deposition of fine sediment in floodplains—causes problems for human settlements when homes, crops, and infrastructure are damaged or destroyed during flooding. Floodwaters, erosion, and deposition all contribute to damage, loss, and fatalities, and the hazards are closely linked to land use practices throughout watersheds (Figure 9-13).

Flood Hazards

Floods cause more property damages and deaths than the sum of all damages and fatalities from tornadoes, earthquakes, and wildfires.[12] From 2002 to 2011, floods were the fourth deadliest, after heat, hurricanes, and tornadoes, of all weather-related hazards in the United States, but for the 30-year period from 1982 to 2011 floods were the most deadly (Figure 9-14).[13] Furthermore, from the 1960s to 1990s the average number of floods

increased from 394 to 2444 per year, and average damage as a result of them increased from about $46 million to $19 billion dollars per year (adjusted for inflation).[14, 15] Flood damages totaled more than $16 billion (about $28 billion in 2013 dollars) for flooding that occurred along the upper Mississippi River in 1993, the greatest flooding disaster in the history of the United States in terms of property damage, excluding Hurricane Katrina in 2005. In developing countries with high population densities, such as Bangladesh, floodplains are heavily populated and the loss of lives during flooding is far greater than in the United States.

Floodplains are a natural part of rivers and become the channels of streams during high-water stages. Floods occur when the amount of water carried by a stream exceeds the ability of the channel to convey it as a result of heavy rains, rapid snowmelt events, channel constrictions created when culverts under roads are not sized properly, or failure of natural or artificial dams that impound lakes. Considering only those events driven by weather, streams fill to the tops of their channel banks on average once every 1 to 2 years. Every 2 to 10 years or so, floodwaters spill over channel banks and spread across part of the floodplain (Figure 9-15). Every 100 to 500 years, large floods may inundate whole valley bottoms. Though such large events occur infrequently when viewed from the perspective of a human lifespan, they are in fact common when viewed from the perspective of geologic time.

The frequency, duration, and intensity of flooding may be affected by global warming. Scientists analyzing rainfall data from the period 1958 to 2007 found that the amount of rain during the heaviest events has increased in some parts of the United States.[4] In the Midwest, for example, the largest storms—those that represent 1 percent of all events—have 31 percent more rainfall. In the

(a)

(b)

Figure 9–13 Along floodplains adjacent to the Mississippi River and its tributaries in the American Midwest, repeated flooding has caused billions of dollars of property damage. Particularly devastating floods, both deemed to be 500-year events, occurred in 1993 (a, St. Louis, Missouri) and 2008 (b, Cedar Rapids, Iowa). *(a: Les Stone/Sygma/Corbis; b: David Greedy/Getty Images)*

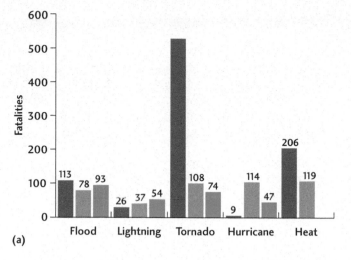

(a)

Figure 9-14 (a) In recent decades, floods were one of the deadliest of weather–related hazards. (b) The number of fatalities and injuries from flooding in the United States varies substantially each year, but in general the historical trend is of a decrease in fatalities and an increase in property damages over the past 5 decades.

■ Weather fatalities for 2011
■ 10-year average (2002–2011)
■ 30-year average (1982–2011)

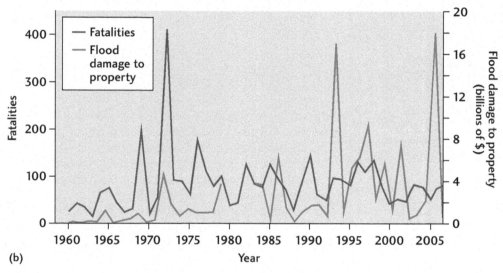

(b)

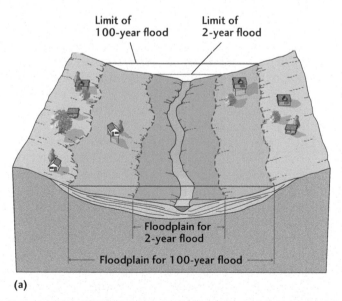

(a)

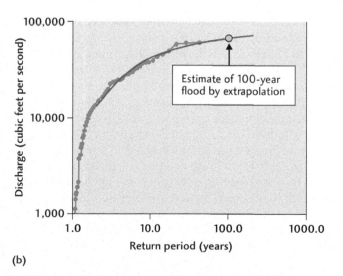

(b)

Figure 9–15 (a) The 2–year flood level coincides roughly with the active floodplain, which is built and maintained by frequent small floods. Estimated 100–year–flood levels based on flood frequency analysis can be used to identify the height and limits of flooding along the margins of a stream channel. (b) In this example of a flood frequency analysis, peak annual stream flow data (blue dots) were collected at a gauging station on the Navasota River, Texas. The red line is based on a statistical analysis of the data that can be used to estimate the discharge for floods of different recurrence intervals. The 100–year flood is estimated to have a discharge of 68,000 cfs, about five times as large as the 2–year flood (13,000 cfs) and nearly twice as large as the 10–year flood (42,000 cfs).

Southeast the increase is 20 percent. This increase can result in greater floods and might have contributed to the devastating floods on the Mississippi River in 2008.[16]

Flood Frequency Analysis

To understand how flood frequency analysis is conducted, consider first a simple analogy, the chances of winning a lottery in which six numbers between 1 and 100 will be drawn. The chance of getting one number that matches one of the six is fairly good, but the chance of getting all six numbers is small. Similarly, the chance of a small flood in a given year is large, but the chance of a large flood is small. In other words, small floods are more frequent than large floods and have a greater probability of occurring in a given year.

In **flood frequency analysis**, geologists examine streamflow records to determine the largest flow at a particular gauging station for each year of record. Once the maximum flow is determined for each year, the flows are ranked by size, with the highest flow given the rank of one and successively lower flows given increasingly higher ranks. The question addressed by statistical analysis is how often a flood of a particular rank is likely to occur or be exceeded, a value that can be calculated with the following equation:

$$\text{Probability of a flow being equaled or exceeded} = M * 100/(N + 1)$$

where M = the rank of an individual flow within a series of maximum annual flows and N = the number of years of record. The factor of 100 is used to express the probability as a percentage.

Each year of record provides the geologist with a sample, and the more sample points, the more reliable the analysis. Small floods have a high probability of occurring, perhaps a 50 percent probability of being equaled or exceeded in a given year. A 50 percent probability is a 1 in 2 chance, which is the same as the chance of heads appearing on a single toss of a coin. Large floods have a low probability of occurring, perhaps a 1 percent probability of being equaled or exceeded in a given year. A 1 percent probability is the same as a 1 in 100 chance.

The **recurrence interval** is the average period of time, in years, that elapses between events of a given size and is calculated by taking the inverse of the probability that a flood of a given size will recur or be exceeded in any given year:

$$\text{Recurrence Interval} = 100/\text{Probability} = (N+1)/M$$

The greater the probability of occurrence, the more frequent the event, and the smaller the recurrence interval. Use 50 percent probability as an example: The inverse of 50 percent is 2 (100 ÷ 50 = 2). Hence, a 2-year flood has a 50 percent probability of being equaled or exceeded

TABLE 9–1 Recurrence Intervals and Associated Probabilities of Exceedance for Floods of Different Magnitudes	
Probability that flood will be equaled or exceeded in a given year (%)	Recurrence interval of flood (years)
1	100
2	50
10	10
20	5
50	2
100	1

in any given year. In contrast, a flood with a 1 percent chance of being equaled or exceeded in any given year recurs on average every 100 years (100 ÷ 1 = 100). Note that the larger the recurrence interval, the smaller the probability of that event occurring, and the greater the discharge associated with it (Table 9-1).

The recurrence interval concept for flooding causes some confusion among the general public. Calling a particular sized flood a 100-year flood does not mean that such a flood recurs exactly every 100 years. Instead, it means that such a flood has a 1 percent chance of being equaled or exceeded in a given year. Two or more 100-year floods could occur in just a few years followed by several centuries in which no floods of this magnitude were to occur. In fact, Houston, Texas, had three such floods in 1979. Given this reality, it makes more sense to think of floods in terms of their probabilities of occurring in any given year. Still, the language of recurrence intervals continues to be widely used in the news media, and even the Federal Emergency Management Agency (FEMA) refers to the 100- and 500-year floods when determining which areas are likely to be inundated.

In a complete flood frequency analysis, either the probability that a flood will occur or be exceeded, or the recurrence interval, or both, are plotted against the corresponding discharge (see Figure 9-15). Using statistical analysis, a flood frequency curve can be developed from the data in order to estimate the discharge for different probabilities, or recurrence intervals. Such curves are essential for estimating the size of large flood events, because few gauging stations have records that span more than several decades, meaning that the discharge associated with large events must be extrapolated from the limited amount of historical data usually available. In the chart in Figure 9-15, an 85-year-long record is used to extrapolate the discharge of the 100-year (1 percent probability) flood. The discharge amount obtained for that flood then can be used to determine the stage of the water during such an event and to predict what areas will be flooded.

The U.S. National Flood Insurance Program

Flood control structures promote a false sense of security that has led to widespread development on floodplains in the United States. In the early 1960s, after decades of disastrous floods and despite billions of dollars spent on flood control, the federal government began to examine alternatives to engineering solutions. The result was the National Flood Insurance Act, under which the National Flood Insurance Program (NFIP) was created.[17] The NFIP provides insurance to any community subject to periodic flooding if the community agrees to establish and enforce standards for both land use and flood control for development in flood-prone areas. The insurance program became one of the largest of national government disaster programs and, in general, was financially sustainable until recent costly disasters that included Hurricane Katrina in 2005. As a result of the large number of claims from recent disasters, the program is now being reevaluated by FEMA and the U.S. Congress.[18]

In order to receive NFIP funds, a community first must conduct a flood frequency analysis in which it determines the odds of occurrence for floods of particular magnitudes that might damage or destroy critical infrastructure built near a stream (see Figure 9-15). Such an analysis informs zoning decisions, building construction codes, and emergency management plans in order to minimize loss of life and damage to property. The procedure relies on statistics and probability, combined with a record of past flood discharges. Historical flow data are obtained from a gauging station, and data from

prehistory sometimes can be obtained by radiocarbon dating of ancient flood deposits (Box 9-2).

Communities that apply for NFIP are sent a team of FEMA engineers and geologists who define **Special Flood Hazard Areas (SFHAs)** within a floodplain—areas that would be inundated by the 1 percent annual probability flood, for example. From this study, FEMA provides the community with a Flood Insurance Rate Map that delineates flood elevations and flood-risk zones for guidance in local zoning and building code ordinances (Figure 9-16). These maps now exist for nearly 100,000 communities, and are used by mortgage lenders to determine whether those applying for a home or business loan must purchase flood insurance.[19] Communities that fail to adopt NFIP are ineligible for federal disaster relief and low-interest loans. Incidentally, the 1 percent annual probability flood actually has a 26 percent chance of occurring within the time span of a typical 30-year mortgage, so it is clear that the 100-year flood is not as unlikely as one might think. In addition, the frequency of 100-year floods has increased in many locations throughout the time NFIP has been in existence, as urbanization has led to increased impervious surface in watersheds and higher peak flows. The current 1 percent probability flood can be much greater in discharge and stage than that of several decades ago in heavily urbanized areas. In the late 1990s, FEMA began to remap many areas, an effort that continues today.

Ideally, communities would prohibit development within SFHAs. However, this is not always possible, so FEMA requires that any new construction within these areas has its lowest floor above the elevation of the 1 percent annual probability flood. A common way of complying with this regulation is to elevate buildings on pilings, providing space for parking below and a staircase to reach the first floor. Some of the delay in rebuilding

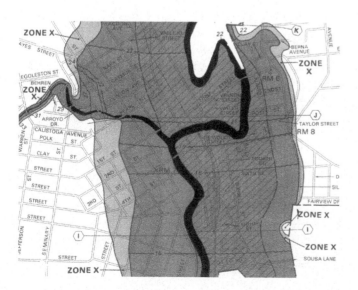

Figure 9-16 FEMA has prepared Flood Insurance Rate Maps, such as this one for Napa, California, for thousands of communities in the United States. Using statistical analysis of historical discharge data, surveyed cross-sections of stream channels and floodplains, and other information, FEMA scientists and engineers determined the area adjacent to the Napa River (heavy black line) that will be flooded by the 100-year flood (dark gray) and the 500-year flood (Zone X, light gray). Also shown as Zone X, but in white, are areas outside the 500-year floodplain. Within the 100-year flood zone, contour lines with numbers indicate the elevation to which floodwater will rise. At the intersection of Fourth and Burnell streets, for example, 100-year floodwaters will rise to about 18.5 ft. This base elevation is used as a guideline for how high to raise a house if it is built in the floodplain. The first lived-in floor of a house must be at least 1 ft above base elevation.

BOX 9–2 CASE STUDY

The Geologic Record of Flooding in South Dakota's Black Hills

During the night of June 9–10, 1972, more than 25 cm (over 1 ft) of rain fell over a 150 km² area in the Black Hills of South Dakota, triggering massive floods on several streams.[22] Flooding on Rapid Creek in Rapid City killed 238 people and injured more than 3000 as the stream rose over 4 m in just a few hours, cresting as people were asleep in their beds. More than 1300 homes and 5000 cars were destroyed, and total damages were estimated at $160 million dollars ($895 million in 2013 dollars) (Figure 1). Stream gauge records dating from the early 1900s to the present reveal that the 1972 floods were larger than any other events during the past century (the "flood of record"), making it difficult to determine their probability of recurrence. Such information is critical for making decisions about land use and to plan for future disasters, but when gauge records are too short, we can often turn to the geologic record for help.

Flooding experts at the U.S. Geological Survey have studied stream canyons throughout the Black Hills and have uncovered evidence that numerous floods like the 1972 event occurred during the late Holocene[23] (Figure 2). The Black Hills consist of a central core of metamorphic rocks, which as they weather, generate quartz–rich sands that are transported through canyons incised into the flanking carbonate rocks. Perched high above the modern streams, caves in these carbonate rocks are filled with quartz sediments deposited at times of high water levels. Further evi-

Figure 1 During flooding in South Dakota in 1972, cars swept downstream by floodwaters were strewn about and sometimes stacked one atop the other in imbricated fashion (with all but the Volkswagon beetle pointing upstream), much the same way as coarse sediment is deposited on bars during floods. *(Rapid City Journal)*

New Orleans after Hurricane Katrina arose from the fact that FEMA needed to remap the SFHA zones in that city, and residents were reluctant to rebuild their destroyed homes until they were certain that they would be in compliance with FEMA's elevation requirements and thereby able to afford flood insurance premiums.

New emphases for flood mitigation include options such as buyout or relocation for those houses caught in a cycle of "disaster, rebuild, disaster, rebuild" (see Figure 9-13). Many people have begun to question the engineering approach to flood hazards (e.g., higher levees) inasmuch as it promotes a false sense of security that has led to increased development on floodplains.[20] After the 1993 flood in the American Midwest, federal programs provided financial assistance to buy out nearly 12,000 properties in nine states and to relocate or elevate

another 500 structures. Land sold in buyouts cannot be developed, but can be used for purposes such as parks and recreation. The 2008 flood along the Mississippi and Missouri rivers led to additional voluntary buyouts, and as of 2011 FEMA had bought 36,707 properties nationwide at a cost of $2 billion.[21] On the other hand, a 2005 study documented that some 28,000 new homes were constructed on land that was inundated by the 1993 flood, counterbalancing the relocation of homes elsewhere.

✔ Stream channels are able to carry water flow most of the time, but about once every two years a stream flows over its banks and spills onto the floodplain, a nearly level alluvial surface that bounds the stream along its margins.

(b)

(a)

Figure 2 (a) Scientists visit Boxelder Creek in the Black Hills of South Dakota to look for evidence of past floods in the geologic record. Alcoves such as the small cave at lower left create conditions of low water velocity during flooding and trap fine–grained sediment (sand sized or finer), such as (b) the stratified deposits from several floods. At this site, flood slackwater deposits preserve a record of about 10 large floods. An exceptionally large flood left a single slackwater deposit in the alcove above the ladder at right. *(Courtesy James E. O'Connor, USGS)*

dence of the flooding origin for these *slackwater deposits* comes from the sedimentary structures they contain: climbing ripples form as sediment–laden waters enter these caves, decelerate, and drop their particle loads.

Organic materials such as leaves, driftwood, and charcoal are light enough to float on top of floodwaters, so they commonly are found at the top of slackwater sediments, enabling geoscientists to distinguish among individual flood deposits in the stratigraphic record. Organic matter also provides material suitable for radiocarbon dating. Through careful examination of the Black Hills deposits, the USGS scientists found evidence of flooding as far back as 3800 years ago. Based on the elevations of the caves, they also determined that several of these floods were in fact larger than the 1972 event, an ominous finding for people in the region but one of value to planners.

✔ More property damage and deaths are due to floods than from tornadoes, earthquakes, and wildfires combined.

✔ From 1958 to 2007, the amount of rain that fell during the heaviest rainfall events increased in parts of the United States; scientists attribute this increase—which causes greater flooding—to global warming.

✔ The record of streamflow events over time is used in flood frequency analysis to estimate recurrence intervals for different sized floods. Geologists use mapping, stratigraphy, and radiocarbon dating of flood deposits to extend flood records into prehistoric periods.

✔ Flood frequency analysis is useful to planning and regulatory agencies that attempt to control floods and provide flood insurance. The concept of the 100-year flood, used as a regulatory standard, is based on statistical analysis of the historical (and sometimes prehistorical) record of flooding of a stream.

✔ The 100-year flood has a 1 percent chance of being equaled or exceeded in any given year, and recurs on average about every 100 years, but changes in climate, land use, or other factors can affect the magnitude and frequency of floods over time.

✔ Geologists map Special Flood Hazard Areas (SFHAs) in order to determine which parts of floodplains will be flooded by different recurrence interval events. Such mapping is used for national flood insurance and to guide local zoning and building code ordinances.

ALTERED STREAMS

Natural streams commonly have a range of flow conditions, or *natural flow variability,* from season to season and year to year, as shown in hydrographs from gauge station data. Minimum flows are necessary for aquatic habitat, whereas less frequent high flows can remove fine sediment that accumulates in channels between large events and can replenish floodplains by overbank deposition. During their evolution, native species—such as alder or sedges along a floodplain—adapt to the natural variability and timing of natural flow conditions, coming to depend upon this range of flows during their life cycle.

A nationwide study by the U.S. Geological Survey (USGS) reported in 2010 that the flow of water in 86 percent of the streams and rivers that were assessed in the United States has been altered significantly.[24] The most common forms of alteration are the result of dams, diversion of stream water from its source, drainage of water in the shallow subsurface (e.g., tile drains), pumping of groundwater, and increased runoff from impervious surfaces such as parking lots, roads, and rooftops. Widespread throughout whole watersheds in many places, these practices contribute to aquatic ecosystem degradation as well as the loss of native species. Here we examine the impacts of three common engineering practices that alter the natural variability of streamflow: dams, channelization, and levees.

Dams

One of the most commonly employed tactics for controlling the flow of water is the construction of dams. Water can be released from the reservoir impounded upstream of a dam at a controlled rate in order to create space to capture future rain or snowmelt events as well as to prevent streams from spilling onto their floodplains downstream of the dam (Figure 9-17; see also Figure 9-8). In addition to flood control, dams are built for the purposes of storing water for irrigation, navigation, public water supply, recreation, and hydroelectric power generation.

As dam building impounds water upstream of the dam, it actually causes flooding in this area in order to diminish the risk of flooding or to provide other benefits downstream. In many cases this affects human beings, as upstream residents are displaced in order to protect downstream residents. Many thousands of people were displaced from their villages in the Catskill Mountains of New York State when dams were built in the first half of the 20th century to create the reservoirs that supply New York City with drinking water. The Three Gorges Dam on the Yangtze River of China forced the relocation of about 1.3 million people whose lands were submerged by the creation of the dam and reservoir during the late 1990s to 2006, when the main body of the dam was completed (Figure 9-18). On the other hand, Chinese

officials state that it protects from flooding the 15 million people who live downstream, and engineers claim that the dam is built to withstand floods as rare as those that occur every 10,000 years or so. Before the dam was built, 300,000 people died as a result of Yangtze River flooding and its aftermath during the 20th century.

Humans have built dams for thousands of years, but dam construction accelerated during the Industrial Revolution in response to the increasing demand for energy to run mills and forges, with tens of thousands of dams built in the United States by the Civil War.[25] Most of these were relatively small in comparison to the size of dams built during and since the 20th century, but their ubiquity had a substantial impact on valley bottom wetlands, streams, and sediment trapping. The largest dams on Earth today are on the order of hundreds of meters in height, such as the Hoover Dam (221 m) in Colorado and the Three Gorges Dam (181 m) in China (see Figures 9-17 and 9-18). The tallest dams that exist or are under construction exceed 300 m in height. The impoundments of these large dams store an amount of water equivalent to one-sixth of the total annual flow of water from rivers to the oceans.

The U.S. Army Corps of Engineers and FEMA compile information on location, size, safety evaluations, and other attributes of dams of different heights and reservoir impoundment sizes for the United States in the National Inventory of Dams.[26] As of 2009, this database included nearly 84,000 dams, about 8100 of which are considered to be "major" dams, generally those with a height of some 15 m or more. About 50,000 major dams exist worldwide, impounding and fragmenting about half of the world's large rivers. The total number of dams is much greater, thought to be as high as several million in the United States alone if small dams were included, although the actual number is unknown. Recently, scientists began collating datasets for the world's dams and reservoirs in order to produce the first Global Reservoir and Dam (GRanD) geographic database, and as of 2011 this database included 6862 dams and associated reservoirs.[27] A world map of these dams and their reservoir capacities shows vividly the great impact of dams in many countries, particularly China, the United States, India, and Japan (Figure 9-19).

Dams and their associated reservoirs have numerous environmental impacts on aquatic systems. Massive concrete walls are impassable barriers for fish to scale, blocking fish migration to their ancestral spawning grounds. Dams in the Pacific Northwest are considered largely responsible for the decline of salmon fisheries, and expensive retrofits are being made to enable fish to migrate. Retrofits include fish "ladders" that consist of zigzagging channels of concrete filled with water and are attached to the front of dams to create an upstream passage. Other dams are fitted with locks that serve as fish "elevators" to allow fish to swim into an area at the base of a dam.

Figure 9–17 Hoover Dam, the highest dam (221.4 m) in the United States, was built on the Colorado River at the Arizona–Nevada state line and began operating in 1936. Located in an arid to semiarid region that gets most of its water from snowmelt in the Rocky Mountains to the east, the dam's primary authorized purposes include regulation of river flow, improvement of navigation, and flood control. Secondary purposes include the delivery of water for irrigation and other domestic use. A hydroelectric power plant at the toe of the dam generates an average of 4 billion kWh of hydroelectric energy each year. *(Blaine Harrington III/Corbis)*

Doors close behind them and the water level is raised to the elevation behind the dam, whereupon doors open to allow the fish to continue their upstream migration. Fish are also caught, placed in tanker trucks, and rereleased upstream of a dam, a process known as "barging."

The transport of sediment is disrupted by dams that alter the flow of water in streams. Streams drop their sediment loads as they enter the still water of reservoirs, forming delta deposits as they adjust to their new local base levels. As a consequence, the storage capacity of

(a)

(b)

(c)

Figure 9–18 The main body of the Three Gorges Dam on the Yangtze River in China was constructed between 1996 and 2006. The *Landsat* satellite image from 1987 (a) predates dam building and reservoir impoundment, whereas the 2006 image (b) shows the massive reservoir created upstream of the dam. (c) Holding trillions of gallons of water, this elevated pool extends 600 km upstream, submerging 13 cities and 466 towns and forcing the displacement of more than 1.3 million people, the greatest relocation in human history. *(a, b: NASA images created by Jesse Allen using Landsat data provided by the University of Maryland's Global Land Cover Facility/NASA Landsat Program; c: Du Huaju/Xinhua Press/Corbis)*

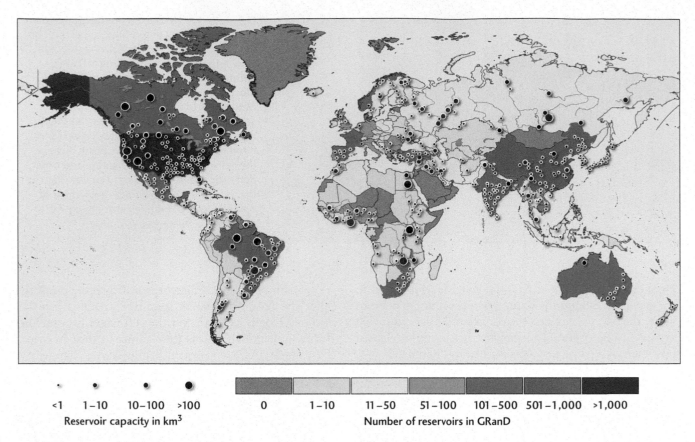

<1   1–10   10–100   >100   0   1–10   11–50   51–100   101–500   501–1,000   >1,000
Reservoir capacity in km³   **Number of reservoirs in GRanD**

Figure 9–19 Global distribution and number of reservoirs formed by dams, with associated reservoir capacity, from the Global Reservoir and Dam (GRanD) Database. Countries with the greatest number of reservoirs and amounts of reservoir capacity are China, the United States, India, Japan, Spain, and Canada.

dams tends to decline over time. The extent to which this is a problem depends on conditions within the watershed that influence the rate of sediment delivery toward streams. Deforestation and poor farming, ranching, and construction practices can expose soils to erosion, increasing the delivery of sediments into streams and filling the space behind dams.

At least 4 to 5 billion metric tons of sediment is trapped behind dams each year, about 25 percent of the total global sediment flux in rivers.[28] With sediment trapped in the reservoirs, sediment-free water spills out the downstream side of dams, promoting erosion. Such a scenario has unfolded along the Colorado River for many decades. In the past, broad sandbars lined the river within Marble Canyon and the Grand Canyon, providing streamside habitat for plants and animals and campsites for people on rafting trips. Completion of the Glen Canyon Dam in the 1960s to capture streamflow for drinking water and irrigation effectively cut off the source of sediment that supplied the sandbars, which have eroded ever since.

Beginning in the late 1990s, the USGS conducted numerous experiments to try to restore the bars by releasing controlled floods from the Glen Canyon Dam.[29] The idea behind these efforts was that sediments would be scoured from the channel bottom and redistributed to the canyon walls by the floodwaters, thereby rebuilding the bars. Thus far, these experimental floods have shown mixed results. Some sandbars grew in size as a result of the flooding but then rapidly declined again, and it appears that changes in dam releases alone will not be capable of restoring bars—more sand is needed.

Changes in flow regimes brought about by damming rivers affects geologic and biologic systems beyond sediment transport and fish habitat. Streamside vegetation is often adapted to the cycle of flooding and may even require floods to regenerate. In New Mexico, for example, dams and the introduction of nonnative plants along the Rio Grande contribute to the demise of stately cottonwoods that have lined the river for millennia. Cottonwoods drop their seeds around the time of late spring and early summer floods caused by melting snow in New Mexico's mountains, taking advantage of the floods for seed dispersal and germinating in the moistened soils. The same high flows carry away dead and dying trees and scour out underbrush, opening up patches in the tree canopy through which light can reach the flood-plain floor and allowing shade intolerant cottonwood seedlings to sprout. Damming the Rio Grande to supply water to New Mexico's cities and towns interrupted this natural cycle, and cottonwoods no longer regenerate. Nonnative invasive species such as salt cedar and Russian olive are taking over, leading to declines in the native animals dependent on the cottonwood ecosystem. Ambitious projects to restore the cottonwoods are presently underway at various points along the river. The ecosystem services previously provided by flooding now are provided by volunteers and park service staff, who plant cottonwood seedlings by hand and rip out nonnative plants. These efforts are not cheap, costing as much as $1300 per hectare of land.[30]

Dam Removal

Dam removal, particularly of small dams, has become increasingly common since the 1980s (Figure 9-20).[31] The reasons frequently cited for dam removal include aquatic and riparian habitat improvement, safety (i.e., drowning hazards), and economics. Dams fragment fluvial systems and associated aquatic and riparian ecosystems.[32] Removing dams eliminates safety hazards, restores variable hydrologic flows, and allows for unimpeded passage of fish and other aquatic organisms. For tens of thousands of obsolete low-head dams built in order to provide power for mills, forges, and other U.S. industries in the 18th through early 20th centuries, removal can be more cost-effective than continued maintenance. Despite safety, ecologic, and economic advantages, however, dam removal also can lead to channel incision, bank erosion, and increased sediment loads downstream.[33]

The U.S. Department of the Interior recently removed two early 20th century hydroelectric dams—the 33-m-high (108 ft) Elwha and the 64-m-high (210 ft) Glines Canyon dams—as part of an ecosystem restoration project on the Elwha River in the Olympic National Park, Washington, from 2011 to 2013.[34] A primary goal was to open fish passage for five species of Pacific salmon that swim upstream to spawn, because the dams were built without fish passage structures. Removal was staged over time to enable the river to flush sediment released from the reservoir and was stopped during times of fish migration. It was not only the largest dam removal project of its kind but also the largest controlled release of sediment into a river and coastal waters to date. Scientists estimated that 19 million m³ of sediment was trapped in the reservoirs behind these two dams, and that nearly half will be released as the Elwha River carves a new channel through the former reservoirs. To imagine the volume of this sediment, it would be equivalent to filling a football field to a height 11 times that of the Empire State Building.

(a)

(b)

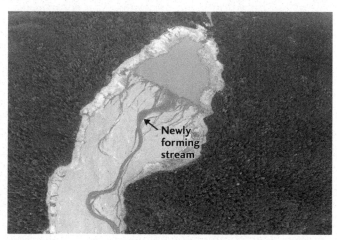
(c)

Figure 9-20 From 2011 to 2013, the U.S. Department of the Interior removed two dams on the Elwha River in the Olympic National Park, Washington. Cameras aimed at the dams document the removal process (a, Elwha Dam before removal; and b, after removal). (c) An aerial photo reveals the Elwha River cutting into reservoir sediment behind the Glines Canyon Dam as it was removed in 2012. *(a, b: National Park Service; c: Tom Roorda/Roorda Aerial)*

Artificial Levees and Floodwalls

A common flood control strategy is the construction of artificial levees along stream channels to effectively increase the height of channel banks and allow streams to convey higher flows before spilling onto their floodplains. Most artificial levees are trapezoidal in cross-section, constructed of compacted dirt, and covered in grass, rock, or concrete to prevent erosion during high flow events. In areas that flood frequently, levees may be placed at some distance from the stream channel on the floodplain to permit even higher flow volumes to be conveyed. The land enclosed by the levees is then typically left undeveloped and used as wildlife habitat or parkland. To provide additional protection where rivers pass through cities, concrete floodwalls may be set into the tops of levees. Large steel gates within the walls can be opened during low flow conditions to provide access to the river and closed during higher flows to prevent flooding.

While raising walls against floods might seem like a good idea, levees are not without problems. First, their construction can lead to a false sense of security that promotes development on floodplains.[21] This is problematic because levees and floodwalls fail from time to time. Within a decade of the devastating flooding along the Mississippi River in 1993, nearly 30,000 homes were built on land that had been underwater during the flood but was within the zone of levee protection.[35]

Another problem with levees is that they can raise the height of water in stream channels during high stage events.[36] If a levee fails, the resulting flood may be far worse than had water been able to leave the channel and spread out on a floodplain upstream. As a consequence, **levee setbacks** are under consideration in some places, and have already been implemented in others. As levees age, need repair, and sometimes fail inspections, some agencies opt to build new levees that are set back from their original location, hence the name "levee setbacks." These setbacks allow floodwaters to spread onto part of the floodplain, reducing the level of flooding downstream and hence reducing flood losses.[37]

Channelization

Channelization, the process of replacing natural stream channels with artificial culverts and ditches, is used to drain land for agriculture and development, and is another tactic commonly employed to deal with flooding. Typically, an artificial channel that is rectangular or trapezoidal in cross-section and lined wholly or in part with concrete is constructed and placed within the stream, and the former stream channel is filled in. The artificial channel is usually much straighter than the original channel, which increases the slope of the stream. Because the velocity of water is directly proportional to the steepness of the channel slope, the steeper gradient results in greater speed. Roughness of the artificial channel bed typically is lower than that of the original bed, meaning that there is less frictional resistance between the stream and its banks, another factor that causes water velocity to increase.

While speeding the passage of floodwaters reduces flooding in the artificially channelized area, its consequences are not necessarily positive for downstream communities where the stream is not channelized. Flooding can be exacerbated downstream because upstream floodwaters have not been able to disperse onto a floodplain storage area. Furthermore, the increased flow velocities hasten the arrival of peak discharge after a storm.

The first efforts at channelization took place on the Mississippi River in the 1870s, but the pace of channelization quickened after passage of the Watershed Protection and Flood Prevention Act in 1954. Subsequently, tens of thousands of kilometers of straightened channels were dredged nationwide in order to reduce flood hazards. Simultaneously, riverine wetlands adjacent to streams were drained and filled to increase the amount of land available for agriculture and development. The economic benefits to farmers and developers are undeniable, but at a cost of environmental damage to ecosystems.

✔ Aquatic organisms have adapted to a range of flow conditions (i.e., natural flow variability) in streams and rivers, with less frequent high flow events (floods) essential to removal of fine sediment from channels and replenishment of floodplains with fine sediment.

✔ Flow has been altered significantly in 86 percent of streams and rivers in the United States.

✔ Three of the most common engineering practices that alter the natural variability of streamflow are dams, channelization, and levees.

✔ Dam building that spans thousands of years of human history has impacted valley bottom streams, wetlands, and sediment storage.

✔ At least 84,000 dams exist in the United States, of which 8100 are major dams (more than 15 m high), and 50,000 major dams exist worldwide, impounding half of the world's large rivers.

✔ About one quarter of the global flux of sediment in rivers, some 4 to 5 billion metric tons, is trapped behind dams each year.

✔ Dam removal is a growing practice as aging dams become obsolete and as the ecosystem services provided by streams and aquatic habitat are better understood and valued by human society.

✔ Scientists learn from each dam removal project; these projects are, in a sense, natural experiments. In particular, scientists are studying the release of sediment from reservoirs and the ecosystem response as channels incise to their original base level.

✔ Many artificial levees have been built along streams and rivers to confine high flow events to channels rather than floodplains, effectively making that land available for use and development. This practice is problematic, however, because their presence has encouraged development in flood-prone areas and levees can fail.

✔ Channelization of streams typically leads to straighter channels with steeper slopes, less rough beds, and higher water velocity, sometimes worsening flooding downstream of channelized reaches. Various strategies have been employed to prevent flooding, including damming streams, constructing levees to contain flows, and artificial straightening of streams, known as channelization. All of these strategies have environmental consequences.

WETLANDS

Geoscientists, hydrologists, and ecologists have, in the past few decades, paid increasing attention to wetlands as their importance to hydrologic processes and ecosystems becomes apparent. To most early Americans, wetlands were foul pestholes associated with disease. In colonial Baltimore in 1766, for example, public opinion forced town officials to pass an "Act to Remove a Nuisance," giving landowners four years to wall off and fill in a large marsh along the harbor and waterfront because of its "noxious vapours and putrid effluvia."[38] By the mid-1950s more than half the area of wetlands in the contiguous United States was submerged by reservoirs behind dams, filled in, drained, farmed, paved over, or developed. Many of the nation's prominent cities, including all of Washington, D.C., and much of New York, Philadelphia, New Orleans, and San Francisco, rest above filled-in wetlands. Today, however, the word "wetlands" is more likely to invoke a peaceful image of sedges, rushes, and mosses, with aquatic birds and turtles.

With this change in attitude have come legislative attempts to protect remaining wetlands from further destruction. More than 220 million acres of wetlands are thought to have existed in the 1600s, at the time of early European settlement, in what is now the conterminous United States.[39] By 2009, that amount was estimated to be about 110 million acres. Over half of the original wetlands are gone. The rate of loss has decreased from earlier decades, yet an additional 62,300 acres of wetlands were lost in the United States between 2004 and 2009. Effective protection requires enforceable legislation, economic incentives and disincentives, and the ability to identify wetlands as well as to understand the factors necessary for their sustainability.

Characteristics and Benefits of Wetlands

What are wetlands? Scientists, environmentalists, and policy makers have developed a definition of a wetland that also is used for legislative and regulatory guidelines under the Clean Water Act (described later in this chapter).[40] This definition describes wetlands based on three factors: hydrology, vegetation, and soil type. **Wetlands** are ecosystems that are inundated or saturated by surface water or groundwater all or part of the year, and the frequency and duration of saturation support *hydrophytes,* plants that are adapted for life in such conditions. Over time, these hydrologic conditions and vegetation produce a soil typical of anaerobic, reducing conditions, known as a *hydric* soil. Wetland soil usually is dark or mottled.

Wetlands occur in a variety of places on the landscape, from drainage divides to valley bottoms (Figure 9-21). Different types of wetlands are categorized largely by their sources of water, although wetland classification among specialists is complex. In general, *bogs* are sustained by precipitation, while *fens* are sustained by the mineral-rich water from subsurface groundwater. Drainage may be impeded if the substrate—the underlying layer of soil or sediment—has a low infiltration capacity, such as the poorly sorted glacial sediments found at the border between the United States and Canada, or if a topographic feature such as a shallow depression traps water and prevents it from flowing away. One of the largest salt-marsh wetlands in the world, covering nearly 13,000 km², is the Florida Everglades, formed on a low-relief, low elevation surface consisting of limestone bedrock.

Wetlands provide a multitude of ecosystem services and goods to humans. Along with tropical rain forests, wetlands are the most productive terrestrial ecosystems on Earth. They are home to a third of threatened and endangered species in the United States. In their more commonly known forms as rice paddies, peat lands, mangrove forests, cypress swamps, and salt marshes containing rich beds of shellfish, wetlands have provided humans with food and fiber for millennia. Unless already saturated, spongy wetlands along valley bottoms act as buffers against seasonal flooding, temporarily storing floodwaters and releasing them slowly. They also act as sources, sinks, cyclers, and transformers of important chemicals and biological substances, such as carbon and nitrogen; they filter pollutants from surface water and provide a plentiful source of purified recharge to groundwater systems. Several states, including California, Maryland, and Louisiana, have learned by experience to use the nutrient filtering capacities of wetlands as alternatives to more expensive, high-technology sewage treatment plants.

A characteristic shared by all wetlands is a low concentration of oxygen in their soil, a consequence of water saturation. The result of such anaerobic conditions can

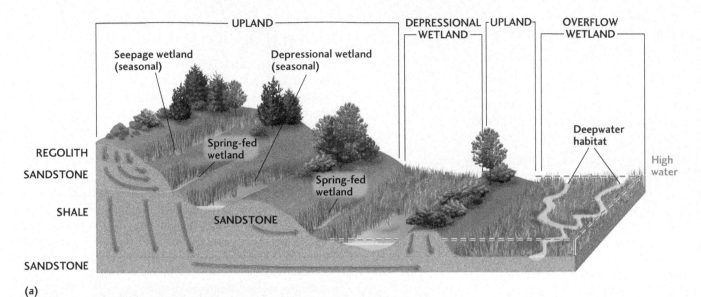

(a)

(b)

Figure 9–21 Many types of wetlands exist and are found at different locations where water is present in the landscape, from drainage divides to valley bottoms and coastlines. (a) Arrows show pathways of groundwater flow. (b) This wet meadow in Tuolumne Meadow, Yosemite National Park, is a type of marsh with a high groundwater table and soil that is saturated much of the year. Sedges and rushes are the most common types of plants in wet meadows. *(b: Russ Bishop/Alamy)*

be the accumulation of dead organic matter. As organic-rich matter accumulates it becomes thicker and the bottommost layers become compressed, forming a substrate with a low infiltration capacity.

Protecting Wetlands

Federal protection first came to wetlands in 1972 when the Clean Water Act (CWA) was passed, and later was expanded with other laws.[41] Permits are required now before solid material may be discharged into navigable waters, which are broadly defined to include wetlands. The CWA does not prohibit dumping or filling in wetlands; it merely attempts to minimize the adverse effects of filling and requires individuals to obtain a permit before placing dredged or fill material into a wetland

(or any waters of the United States). Furthermore, the CWA does not prohibit draining wetlands. In an effort to keep farmers from draining or otherwise converting wetlands to create agricultural lands, the federal government passed the Food Security Act of 1985 and 1990, also known as "Swampbuster." According to this law, farmers who destroy wetlands are ineligible for most federal farm subsidies. The law was later amended by the Food, Agriculture, Conservation, and Trade Act of 1990 to create the Wetland Reserve Program, which provides farmers with incentives to restore wetlands on their property.

Property laws give landowners the right to farm, mine, and develop their property, but the government requires developers to create new wetlands or restore former ones to compensate for those destroyed.

However, the natural systems in wetlands are complex and incompletely understood, and some wetland types are more difficult to create than others. Restoration is more likely to be successful than new wetland creation, because the conditions that can produce and sustain a wetland have existed in the past. In addition, many wetland projects call for mosquito control, typically achieved by spraying insecticides. Although many people are beginning to recognize the value of wetlands, they still prefer not to deal with the problems that made early settlers view wetlands with such disdain, problems that still accompany a wetland area.

At the international level, the Ramsar Convention on Wetlands is an intergovernmental treaty for the conservation and wise use of wetlands.[42] Adopted in 1971, the Convention now has 163 contracting parties and 2066 sites designated in its list of wetlands of international importance.

✔ Wetlands are identified by scientists on the basis of hydrology, vegetation, and soils, and are associated with hydrophytic plants adapted to wet conditions and hydric soils formed by anaerobic processes.

✔ Wetlands provide a multitude of ecosystem services and have substantial economic value to humans as a result of the food and fiber produced in them (for example, rice from paddy fields).

✔ Recent environmental understanding has altered the historical view of wetlands. Once regarded as smelly pestholes to be filled in or destroyed, they now are better protected and valued because they store floodwater, provide habitat for many organisms, and cycle carbon, nitrogen, and other elements through different Earth systems.

✔ Despite the change in attitudes toward wetlands, protecting them is difficult because of disagreements over their definition and identification and because the pressure is great to develop land for urbanization or crop production.

MANAGING AND RESTORING STREAMS AND WETLANDS

Scientific studies often provide relevant information to policy makers and engineers who strive to minimize or reverse the impacts of different land use practices on stream systems. According to studies, negative impacts can occur in upland areas as well as along waterways themselves. Upland parts of watersheds are sources of water, sediment, debris, and dissolved substances (e.g., nutrients, or pesticides) to waterways. When altered for development or agriculture, among other types of land use change, the supplies of these substances

to waterways often are increased. Deforestation, for example, commonly leads to increased rates of soil erosion and, if slopes are relatively steep, mass movement. In the valley bottoms, waterways are altered by practices such as channelization, stream channel straightening, wetland drainage, and damming. Such changes affect water flow velocity, the amount of water in streams and wetlands, the transport of sediment and nutrients, and even the recharge of surface to groundwater.

Whereas water pollution is a fairly obvious indicator of environmental degradation in a waterway, the loss of other aspects of sustainable river and wetland systems that provide a rich array of ecosystem services is not as obvious. Today, in the United States, much effort and money are spent to restore streams and wetlands, and the amount of both effort and money increases each year.

River Corridor and Wetland Restoration

The U.S. Environmental Protection Agency (EPA) as well as other government agencies and groups use **ecological restoration** as a management tool for streams in which chemical, physical, and biological habitats have been impaired. Restoration is the process of returning a degraded ecosystem to a close approximation of its condition prior to disturbance and impairment, by re-creating or repairing both the structure and ecological functions of the restored ecosystem.[43] In many cases, complete restoration is not possible, because the original condition might not be known sufficiently, some species might be extinct, or watershed conditions at present might be such that the original condition can no longer be sustained (Box 9-3).

The motives for ecological restoration are many, including environmental and economic reasons. In the Chesapeake Bay watershed, for example, states are required to reduce the loads of sediment and nutrients in their streams because the Bay is an impaired water body under the Clean Water Act. A regulatory goal, which provides a motive for the states, is reduction of sediment and nutrient loads, so *stream restoration* is a management tool for states to meet that goal. Another motive for restoration might be to re-create habitat for an endangered species, such as the bog turtle, once widespread in the mucky soils of wet meadows in the mid-Atlantic region. An economic motive might be removal of a dam in order to allow fish passage so as to improve the production of fisheries and fishing, a substantial concern in many parts of the world.

Restoration approaches and engineering practices vary geographically and depend upon the nature and history of environmental impact in each area. Restoring a stream or wetland to a condition that closely

BOX 9–3 EMERGING RESEARCH

Stream and Wetland Restoration: What Is Natural?

Efforts to restore streams prompt scientists to dig into geologic history in order to better understand how to restore them. Sometimes what they learn is surprising, in that streams were so altered during recent centuries that our concepts of what is natural are, ironically, unnatural. Restoration requires better understanding of natural conditions and geologic history as well as of causes of degradation. Skills and knowledge from a range of fields that includes geomorphology, biogeochemistry, ecology, and engineering are melded together in stream and wetland restoration projects.

In rivers of the Pacific Northwest, scientists are finding close connections between physical and biological aspects of streams and floodplains that are crucial to biodiversity and sustainability.[44] For example, fallen trees up to hundreds of feet long act as "foundation species" that affect many aspects of local habitat. These trees are part of *a floodplain large-wood cycle* in which wood jams are essential to maintaining organic-rich floodplains with multi-branching, anastomosing channel patterns (**Figure 1**). Woody-debris jams at the upstream ends of small channel segments force

Alternate stable states

Higher complexity ⟷ Lower complexity
and diversity   and diversity

1 Diversity of stable main and perennial secondary channel habitiats.

2 Abundant, high quality edge habitat.

3 Large trees entrained by river at eroding banks.

4 Stable jams at flow splits and secondary channel inlets.

5 Deep scour pools associated with stable jams.

6 Forest age and species patch diversity, including mature conifer patches on stable "hard points."

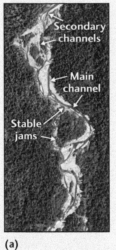

(a)   (b)

1 Braided, unstable main channel and shifting, ephemeral secondary channels.

2 Low quality of edge habitat.

3 Riparian forest recruitment limited to small wood.

4 Unstable pieces and accumulation of fluvial wood.

5 Few, shallow pools.

6 Low forest patch age and species diversity, dominantly ephemeral, young stands of pioneer tree species.

(c)

Figure 1 Streams in the state of Washington illustrate the differences between those with (a) and without (b) riparian trees and abundant, large woody debris (c) in the valley bottom. The stream on the left (Hoh River) has multiple secondary channels attached to the floodplain as a result of debris jams and high channel stability, whereas the stream on the right (Cowlitz River) is braided, with frequent shifts and high channel mobility. For both images, flow direction is from top to bottom. *(a, b: Brian D. Collins; c: Courtesy Sitka Conservation Society)*

(a)

(b)

(c)

Figure 2 (a) Geologists identified "fossil wetland landscapes" with dark hydric soils that were little changed for thousands of years, even after European settlement in the early 1700s and dredging of a straight channel in the early 20th century (far right). Radiocarbon dates and fossil seeds indicated that this sedge–dominated wet meadow in southeastern Pennsylvania existed for at least 11,000 years. (b) At many other nearby valley bottoms, geologists discovered dark soils at the level of the groundwater table, buried beneath up to several meters of historic sediment that accumulated behind early milldams. (c) State and federal agencies worked with scientists at Big Spring Run, Pennsylvania, to develop and test a new wetland–floodplain restoration approach in which historic sediment is removed down to the level of the original wet meadow and hydric soil surface. *(Photos by Cheryl Shenk)*

flow to go over banks and onto floodplains, maintaining a connection between multiple channels and the floodplain, relatively low stream velocities, and both channel and habitat stability. Much of this wood was removed during the 1800s by logging, woody–debris removal, and other practices. Removal of woody debris results in a cascade of negative impacts, including the loss of side channels and valley bottom wetlands. The ecological services provided by the natural stream channels and woody debris, such as the wetting and nutrient enrichment of soil during flooding, were unrecognized (**Figure 2**).

Recent listing of Pacific salmon under the Endangered Species Act has led to substantial interest in the scientific basis for river restoration in the Pacific Northwest.[45] Millions of dollars in state and federal funding have been allocated for habitat restoration efforts to stem the decline of salmon populations in the region. Currently at only a few percent of their estimated abundance in the early 1800s, salmon populations are thought to have declined primarily due to overfishing and habitat loss. Forested wetlands once provided diverse and vital habitat to the upstream migration of these and other *anadromous* fish that are born in freshwater

(continued)

(continued)

streams but migrate to the oceans as young adults and return to their native streams to reproduce before dying. A valuable natural as well as cultural resource, wild Pacific salmon are crucial both to Native American tribes, which have traditional fishing rights, and to the region's economy.

Policy makers and environmental agencies hope that scientific input on stream restoration might help streams to become self-sustaining and resilient and, as a result, to slow or even reverse the decline of salmon populations. Scientists recommend restoration approaches that begin with levee setbacks and riparian planting of fast- and slow-growing trees along riverbanks. As fast-growing trees mature they provide early recruitment of wood into forested wetland floodplains. Large pieces of wood might be added as engineered logjams in order to hasten forest recovery until slower growing trees also mature.

In the eastern United States, scientists have found that a different type of valley bottom wetland once connected stream channels with their floodplains, playing an equally vital role in both aquatic ecosystem diversity and stability. Geologists studying sedimentary deposits and dating fossil seeds with radiocarbon methods discovered that wet meadows, a type of wetland dominated primarily by sedges that remains wet much of the year, were widespread during the Holocene in the mid-Atlantic region.[46]

Holocene wet meadows formed over millennia at the level of the groundwater table were characterized by vegetated floodplains and anastomosing channels with low, stable banks, but unlike the forested wetlands of the Pacific Northwest, these valley bottom wetlands had few trees or woody debris jams. Trees were rare because valley bottoms consisted of wet mucky soils that could not support large trees. Instead, sedges and shrubs dominated the valley bottom (see Figure 2a). As in other places, the biodiversity and variety of ecosystem services provided by these natural stream systems were unrecognized at the time. Like the streams in the Pacific Northwest, the original wetland-floodplain landscape in the mid-Atlantic region provided crucial habitat for endangered or threatened species of plants and animals, such as the bog turtle.

When European settlers arrived, they built thousands of dams for water powered milling throughout the 18th to 20th centuries. Multiple dams that raised local base levels along valley bottoms coincided with increased upland soil erosion rates from deforestation (including for charcoaling), mining, and agricultural cultivation. Massive volumes of fine-grained upland sediment entered valleys and buried presettlement landscapes (see Figure 2b). For about 150 years, millpond reservoirs filled with fine sediment, culminating in "valley flats" that were mistaken for natural floodplains.

As water-powered milling became obsolete, dams breached (due to either dam removal and natural causes) and streams incised into the fine sediment, producing high banks that erode rapidly. Although increased stormwater runoff can lead to bank erosion, the breaching of tens of thousands of dams is the root cause of much of the stream channel incision and high banks in this region.

Scientists working with state and federal officials developed and tested a new way to restore streams and floodplains impacted by such a legacy of changes, and the Pennsylvania Department of Environmental Protection proposed this restoration approach as a new best management practice to the EPA.[47] In 2011, after years of studying a small headwater stream named Big Spring Run in southeastern Pennsylvania, scientists worked with engineers to remove most of the historic sediment that had buried the original Holocene wet meadow and its small streams, springs, and floodplain (see Figure 2c). They planted wetland species that fossil seeds revealed had existed in the buried wet meadow soil prior to European settlement. They also found remnants of unburied wet meadows at nearby locations with no milldams; these "living fossil" landscapes demonstrate biological diversity and ecosystem stability, and provide a restoration target. Continued monitoring after restoration indicates a substantial reduction in sediment and nutrient loads from the restored stream.

The difference in these two case studies of stream restoration is that one requires removal or setback of levees and tree planting, whereas the other requires removal of sediment and planting of sedges. An important similarity in both cases is the value of scientific research to guide and inform the restoration approach so as to achieve a sustainable stream and ecosystem.

approximates that prior to disturbance requires identifying the original condition and determining how each was impacted to reach its present condition. For example, if a stream was straightened, a restoration approach might involve restoring it to a closer approximation of its original channel pattern. In-stream structures using different materials such as stumps of trees with roots or clusters of boulders might be placed along a stream's bed and banks in order to provide ecological complexity and to protect the bed and banks from erosion.

If woody debris was removed from a stream corridor in order to aid navigation or reduce flooding, then trees might be planted and woody debris added to the valley bottom. If the original valley bottom was a sedge-dominated wet meadow with multiple small stream channels that later was buried with fine sediment upstream of an old low head dam, and the modern stream has incised into that sediment since dam breaching, then a different approach is needed. In this case, restoration might involve removing some of the historic sediment and planting sedges and wetland shrubs, such as alder, at the level of the exhumed water table.

Unchannelizing the Kissimmee River

An example of the problems and long-term repercussions of stream channelization is provided by the Kissimmee River of Florida. In the 1960s the U.S. government spent $30 million to straighten the stream, the main source of freshwater for the Everglades wetland system (Figure 9-22). The U.S. Army Corps of Engineers (the Corps) turned 166 km of twisting, sandy channel into a straight, concrete-lined drain 90 km in length, dubbed the C-38 Canal, in an attempt to regulate flooding and make more land available for agriculture. Soon after, the Kissimmee's aquatic ecosystem began to decline as cattle pasture replaced floodplain wetlands, and the water quality of Lake Okechobee, into which the river flows, became polluted with agricultural runoff. Fertilizers caused algal blooms that depleted dissolved oxygen when the plants died and decayed, and lack of the natural alternation between pools and riffles likewise contributed to oxygen decline. Six species of fish disappeared from the river/canal along with 92 percent of ducks, geese, and other waterfowl. Six billion shrimp died, and bald eagles and ibis that once had nested in the treetops relocated to more hospitable environments as their food supplies dwindled.

By the early 1970s, Floridians were clamoring for the project to be undone. Given the project's success at flood control, the Corps was reluctant to reverse course and argued that river restoration would provide no economic benefit. After a Florida senator crafted a bill authorizing the Corps to work on projects of environmental, not just economic or military, importance, the Corps relented. Beginning in 1999, the Corps partnered with the state to open the old stream channel and fill in the canal. Slated for completion in 2014, the results already are markedly positive. Oxygen levels in the river have increased, native fish have returned, and the

(a)

(b)

Figure 9–22 Restoration of the Kissimmee River and associated wetlands requires that the stream be reconnected with its historic floodplain. (a) From 1962 to 1970, an artificial channel known as the C–38 canal was dredged to straighten the Kissimmee River between Lake Kissimmee and Lake Okeechobee, Florida.

Two–thirds of the historic floodplain was drained, devastating the wetland ecosystem. (b) During restoration the C–38 canal was filled in and a meandering stream channel was created, with auxiliary side channels, oxbows, and wetlands. *(Courtesy South Florida Water Management District)*

numbers of wetland birds such as roseate spoonbills and great blue herons have increased substantially.[48] In contrast, the unrestored canal portion of the river continues to show signs of impairment, with much lower diversity of fish and bird species and lower oxygen levels.

The restoration project is likely to cost nearly one billion dollars by the time it is completed. The ultimate goals of restoration include greater recharge to groundwater supplies, increased nutrient storage in floodplain soils, and improved habitat for plants and animals.

✔ Ecological restoration is a management tool for streams in which chemical, physical, and biological habitats have been impaired. It is used, for example, to reduce sediment and nutrient loads in waterways, to create habitat for endangered species, and to improve fish passage.

✔ Ecological restoration is the process of returning a degraded ecosystem to a close approximation of its condition prior to disturbance, by re-creating or repairing the structure and ecological functions of the original ecosystem.

✔ Geologists use mapping, stratigraphy, radiocarbon dating, and paleoecology to determine the condition of ecosystems prior to disturbance.

✔ A common practice of stream restoration is undoing the channelization of prior decades, as in the case of the Kissimmee River, Florida. The river ecosystem is vital to the Everglades wetland system, and hundreds of millions of dollars are being spent to undo the damage of tens of millions of dollars worth of channel straightening that had been done for flood control in the 1960s.

WATER RESOURCES AND PROTECTION

Surface water provides us with most of our freshwater, a resource vital to life. The United Nations predicts that during the next few decades water will become the most strategic resource in the world, the "next oil" in terms of its relevance to global economics, power, conflict, and development. In contrast to other resources, such as various minerals or fossil fuels, water has no substitute for most of its functions and cannot be replaced by technological innovation. On a global scale, water is considered a renewable resource because it is recycled continuously through the hydrologic cycle. If water use exceeds local replenishment rates, or if water is polluted during use, freshwater can become a limiting resource, placing an ever-lowering ceiling on the number of people that can be sustained, especially in arid and highly populated regions.

Immense amounts of water are transferred from the atmosphere to Earth's surface each year, equivalent in size to nearly 10 times that of the world's third largest freshwater lake by volume, Lake Superior. The amount is easily sufficient to satisfy the needs of every person on the planet, except that it falls unevenly on Earth and nearly two thirds returns again to the atmosphere by evapotranspiration (see Figure 9-4). The remainder flows into streams, lakes, wetlands, and groundwater aquifers. Part of this amount is tapped by people for agriculture (5.1 percent), irrigation (1.4 percent), and industrial or municipal use (0.1 percent). Another 1.3 percent evaporates from open water bodies.

Freshwater Use and Virtual Water

A person needs about 5 L (1.3 gal) of water per day for fluid replenishment.[49] In the United States, an average of 1.323 trillion L (349,600 million gal) of freshwater was consumed or used in some way each day in 2005.[50] Given an estimated population of about 297 million people in 2005, that amounts to 4455 L (1177 gal) per day for every person in the country, equivalent annually to about three fifths the volume of a typical Olympic-sized swimming pool. Annual water use commonly is given in units of cubic meters, with 1 m^3 equal to about 264 gal of water, so the average annual per capita water use for a U.S. citizen is about 1628 m^3.[51] As a fully developed industrial society, the United States is one of the heaviest users of water in the world.

Water is used for much more than drinking and personal hygiene, with multiple uses that include toilet flushing, food preparation, watering lawns and gardens, agricultural irrigation, and industrial purposes (cooling power plants, processing products in factories, mining, etc.). Immense amounts of water are used to produce food. It takes 2400 L (630 gal) of water, for example, to produce a single hamburger. The total amount of freshwater a person uses, either directly or via the production of goods and services that he or she consumes, is called a **water footprint.**

In all countries, domestic needs require the smallest portion of water use; agricultural and industrial processes require far greater amounts. The use of water for agriculture is largely *consumptive,* meaning that the water is removed from further use, because much of the water used for irrigation evaporates. This is also true for some water used for industrial purposes, because it is stored in ponds or passed through towers for cooling.

Some scholars have pointed out that the global trade of water-intensive products could help to reduce water use globally. Consider water that is used to grow crops and produce food. If such commodities are exported to another country, the amount of water used is, in essence, embedded in them. This amount of water that is embedded in a product, in a virtual but not a real sense, is known as **virtual water.**[52] If a country with ample water resources produces goods that require large amounts of water, such as wheat (1000 L per kg of weight produced), and ships those goods to another country where water is less abundant, local water scarcity can be alleviated. Researchers have found that in Asia, where population and economic development are growing relatively rapidly, the import of virtual water, in soy products from Brazil, for example, has increased by more than 170 percent in recent decades.[53] This global trade is associated with an increase in water efficiency and reduction in worldwide water use. Clearly water, food, global trade, and economic development are linked closely to one another.

The United Nations and the International Water Management Institute estimate the amount of freshwater available for each country and determine whether it is experiencing water stress or scarcity with respect to its population size (Figure 9-23).[54, 55] If the annual supply of water is less than 1700 m^3 per person, then the area is experiencing *water stress.* If per capita annual supplies are less than 1000 m^3, the country faces *water scarcity.* Countries with amounts less than 500 m^3 per person are described as facing *absolute scarcity* of water.

Two types of water scarcity can be identified. One is *physical scarcity,* in which the demand for water locally exceeds its availability. Technically, the definition refers to countries that withdraw more than 75 percent of their river flow for agriculture, industry, and domestic

purposes. Examples include the American Southwest, northern Africa, northern China, and the Mideast (e.g., Iraq and Iran). One fifth of the world's population lives in areas with physical scarcity of water. The other type is *economic scarcity,* in which access to water is limited by corrupt governments, poverty, or infrastructure, even though sufficient supplies of renewable freshwater are available. One quarter of the world's population is faced with economic scarcity of water, including central Africa and parts of southeastern Asia and India. Because of water scarcity of either type, large numbers of people—the majority of them women and children—must walk farther each day to get clean and sufficient supplies of water, sometimes at great personal risk.

The United Nations estimates that nearly 2 billion people will be subject to water scarcity by the year 2025. Three factors contribute to growing scarcity: population growth, global climate change, and degradation of water quality. For the past century, the rate of water use has grown at about twice the rate of population growth. Global climate change can lead to greater aridity and flooding, both of which affect water supplies. Degraded water quality, due to waste disposal, untreated sewage, industrial pollutants, runoff from farm fields that contains fertilizers and pesticides, and the contamination of coastal aquifers with salt water all threaten water supplies worldwide.[56]

Multiple strategies are available to reduce water consumption. Conservation measures range from simple to sophisticated, from not letting faucets run while

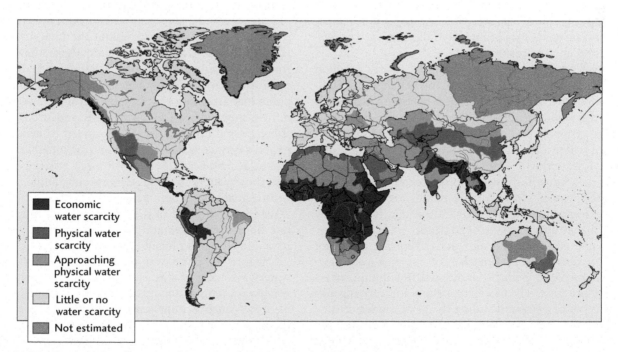

Figure 9-23 Climate, population, and economic circumstance determine freshwater availability, use, and scarcity.

washing or brushing teeth to installing water-efficient fixtures such as low-flow shower heads and replacing aging water lines that are rife with leaks. Gray water that has been used for washing can be recycled and used for purposes other than drinking. Drip irrigation systems are superior to other types, such as irrigation sprinklers, in that water seeps slowly and directly from tubing to the root zone of crops. Water can be stored underground rather than in reservoirs behind dams in order to reduce loss from evaporation in arid and semiarid regions, as is done with some "water banks" in Arizona and California, as well as in other places.

Surface Water Systems and Waste Disposal

In much of the developing world, water is unsafe to drink, because it is polluted during use and returned to its source along with sewage and other contaminants. Water pumped from a stream, for example, commonly is used by a community and then returned to the stream. If wastewater is not treated, however, disease-causing bacteria and viruses from human feces are discharged into streams and lakes. In a stream, because the same water is used again by more people as the water continues to flow downstream, the ratio of wastewater to clean water becomes larger and larger.

In most developed nations, governments require that wastewater from urban areas be treated in sewage waste treatment plants and that homes in rural areas have septic tanks or other facilities to treat waste before used water returns to groundwater or local streams. In many developing nations, however, wastewater is not treated, and periodic epidemics of intestinal diseases such as dysentery and cholera result.

The United Nations estimates that nearly 1 billion people have no access to safe drinking water, and 2.5 billion people lack access to improved sanitation facilities.[57] Sadly, the second leading cause of death for children is diarrhea, which is in large part a result of lack of access to clean water and sanitation facilities.[58] One child dies every 21 seconds, or 4100 children per day, as a result of diarrhea caused by waterborne bacteria ingested in untreated water. Most of these deaths are in Asia and Africa. Ironically, it is the dehydration that is caused by diarrhea that ultimately leads to death.

People realized long ago that rivers are good not only for transportation but also for carrying away sewage and garbage. With the Industrial Revolution, rivers became progressively more important for carrying different types of waste, including the toxic sludge created as a by-product of new manufacturing technologies. By the middle of the 20th century, many rivers in the United States and other industrialized countries transported reeking mixtures of raw sewage and industrial waste, and many were dead—devoid of aquatic life and unsafe for drinking.

As recently as 1968, St. Louis spewed raw sewage waste—as much as 300 million gal per day—directly into the Mississippi River in the hope that the river's great size would make the effects of the dumping inconsequential to those drinking from its waters downstream. In Poland, the Vistula River flows through areas of heavy industrial activity and became so polluted during the 20th century that more than half of its length was devoid of any aquatic life.[59] In both cases, however, by the late 20th century environmental regulations for disposal of sewage and industrial waste had led to measurable improvements in water quality.[60]

The Clean Water Act and Watershed Management

One of the most important U.S. laws, intended to maintain and restore the physical, biological, and chemical integrity of the nation's waters and managed by the EPA, is the **Clean Water Act (CWA)**.[61] The nation's waters include rivers, streams, lakes, and wetlands that are deemed navigable, or have a significant nexus to navigable waters, but the phrase "significant nexus" has been problematic with respect to judicial interpretation. As a result, the EPA's authority has been contested in some cases.

Point and Nonpoint Source Pollutants The CWA is intended to prevent pollution in U.S. waterways, to improve treatment of wastewater from publicly owned treatment plants, and to maintain the integrity of wetlands. **Point source pollutants** are those that can be traced to a direct source of discharge, such as a pipe from a factory or a sewage facility. **Nonpoint source pollutants** are those that emanate from broader, more diffuse sources, such as farm fields and urban streets. When passed in 1972, the CWA aimed to make all the nation's surface waters fishable and swimmable by 1983 and to achieve zero discharge of point source pollutants into those waters by 1985. Despite substantial progress, neither goal has yet been achieved. The EPA estimated in 2010 that about half of the nation's rivers and streams, one third of its lakes and ponds, and two thirds of its bays and estuaries remain impaired.[62]

Before the CWA was passed, many of the nation's waterways were in dire condition, and although some waterways remain impaired, there are notable examples of water quality improvement since then. Lake Erie, the southernmost of the Great Lakes, was described as "dying" by *Time* magazine in 1969, and the heavily polluted Cuyahoga River in Cleveland, Ohio, made

Figure 9–24 Fine sediment becomes suspended in water and is one of the leading causes of water quality impairment in streams, rivers, and lakes of the United States. *(Courtesy Scott Sanderford, City of Tuscaloosa, Alabama)*

national headlines several times when oily debris in the river caught fire. These cases were not unusual, as other American rivers in industrial areas also caught fire prior to clean-up efforts that resulted from the CWA.[63] Fish in many waters were contaminated with DDT and other pesticides, and birds that fed on these fish, such as the brown pelican, were at risk of extinction. Raw sewage was common in some coastal waters and many streams, resulting in high bacteria levels. Forty years after this landmark law was passed, the chance of a stream catching fire in an urban area due to heavy pollution or evidence of raw sewage along a coast or in a stream is unlikely.

The CWA and its amendments from 1980 brought control of point source discharges and improved wastewater treatment plants to mostly urban rivers. Under the authority of the CWA, the EPA established the National Pollutant Discharge Elimination System (NPDES) permit program to regulate point sources that discharge pollutants into U.S. waters. For example, if a factory intends to discharge a pollutant into a waterway, it must apply for a permit from the EPA or authorized state. Permits are based on the technology available to control the pollutant, as well as remaining within the limits that are needed to protect the quality of the receiving water body.

Stormwater discharge from municipal separate storm sewer systems (i.e., those stormwater systems not combined with a sewer system), construction activities, and industrial activities also are regulated by the EPA NPDES Stormwater Program. Most stormwater discharge is treated as a point source, so the entities that produce these sources apply for discharge permits.

Pollution of waterways from nonpoint source pollutants is of growing concern. Eroding stream banks and runoff from farms, animal feedlots, mines, construction sites, backyards, streets, and parking lots can pollute streams, rivers, and lakes with sediment, pesticides, herbicides, fertilizers, animal manure, oil, and household chemicals (Figure 9-24). Nonpoint sources are everywhere, and they are not as easily treated as single points of discharge.

A national water quality assessment of 28 percent of the approximately 3.5 million mi. (5.6 million km) of U.S. streams reveals that fine sediment (mostly silt and clay) is a major cause of impairment.[62] Fine sediment suspended in streams leads to high **turbidity**, a measure of the clarity of water. High turbidity (low clarity) blocks light from reaching aquatic plants, increases water temperature as particles absorb more heat, clogs fish gills, and degrades aquatic habitat.[64]

Two examples, both involving pesticides, illustrate the continuing challenge of protecting waterways from nonpoint source pollutants. In the 1960s, fish at the mouth of the Mississippi were heavily contaminated with agricultural pesticides from upstream farms. One of their predators, the brown pelican, no longer nested in Louisiana—the "pelican state"—where more than 50,000 had lived in the early 20th century. The bird had disappeared from the state completely by the late 1960s, shortly after it was officially named the state bird. Fledglings were brought from Florida to attempt repopulation. Scientists began to realize that **biological amplification** of the insecticide *endrin* was occurring. In other words, endrin had accumulated in fish to such high levels that it was lethal to animals higher on the food chain that fed on the fish (see Chapter 7). This *organochloride* (organic compound containing chlorine) has been used by countless farmers on fields far upstream from the mouth of the Mississippi River since the early 1950s. No one foresaw how quickly the toxin would make its way across farm fields with runoff during storms and then downstream along the river, to accumulate in sediment and the fatty tissue of fish and birds. With increased understanding of the environmental impacts of endrin, its use and sale in the United States were banned by the EPA in 1986. In this case, then, banning the substance contributed to reducing its presence in the environment.

Another organochloride pesticide that was found in the Missippi River, dichloro-diphenyl-trichloroethane (DDT), impaired reproduction by eggshell thinning of predatory birds, including pelicans, which feast on fish. Brittle, thin-shelled eggs are more likely to be crushed during incubation in the nest. DDT was banned for agricultural use in the United States in 1972.[65] Since then, the amount of DDT and endrin found in fish, birds, and other organisms generally has decreased in the United States. The brown pelican was listed as an endangered species by the U.S. Fish and Wildlife Service in 1970, but due in part to the banning of DDT and endrin, it had repopulated in Louisiana and elsewhere in the eastern United States by the late 1990s and was subsequently removed from the federal list of endangered species.

In the second example, a national water quality monitoring program has provided crucial information regarding the presence of the most widely and heavily used herbicide, atrazine, in U.S. streams and groundwater. The USGS began its National Water-Quality Assessment Program (NAWQA) and compilation of water quality data for 51 river basins and aquifers in 1991. Atrazine is the most commonly detected pesticide in water samples from all river basins and for all three types of land use categories: agricultural, urban, and mixed (Figure 9-25).[66] Concentrations typically are much greater for agricultural basins, especially those in the Corn Belt of the Midwest where atrazine use is heaviest. Repeat sampling over time enabled the USGS to determine that the proportion of agricultural streams in the Corn Belt with atrazine concentrations greater than the EPA benchmark for human health (0.003 mg/L, or 3 parts per billion) increased during the 1993 to 2000 measurement periods.[67] In addition, atrazine has been detected in drinking water supplies, raising further concerns about human health issues.[68]

An effective weed killer, atrazine might cause cancer in humans and is an endocrine disruptor for amphibians, fish, and mammals (including humans). **Endocrine disruptors** are synthetic chemicals that can mimic or block hormones and disrupt normal functions when absorbed into the body. Atrazine has been linked to

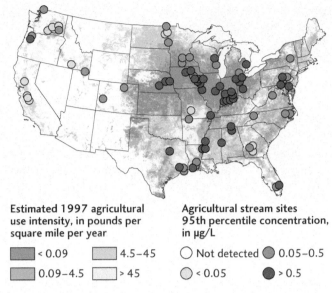

Estimated 1997 agricultural use intensity, in pounds per square mile per year

< 0.09	4.5–45
0.09–4.5	> 45

Agricultural stream sites 95th percentile concentration, in μg/L

◯ Not detected	⬤ 0.05–0.5
◯ < 0.05	⬤ > 0.5

Figure 9-25 Intensity of use of atrazine is shown in shades of green, with greatest use in agricultural regions. Colored circles show atrazine concentration in streams sampled in agricultural areas. Atrazine concentrations are given as 95th percentile values (i.e., 95% of measurements are less than this value), in micrograms per liter. *(Courtesy USGS)*

adverse reproductive effects such as male frogs with female sex characteristics. The pesticide has been banned in the European Union, and some groups in the United States urge that it be phased out or banned as well.[69] Certain farming techniques such as crop rotation and use of winter cover crops are recommended as alternative (nonchemical) means of weed control in order to reduce use of atrazine. Meanwhile, atrazine continues to accumulate in the environment, including in waterways and organisms.

Best Management Practices In 1987, in recognition that most impaired water sites are the result of runoff from uplands (that is, nonpoint source pollution) rather than direct discharge of point source toxins, the CWA was amended to incorporate nonpoint source management programs that recommend **best management practices.** These practices are designed to prevent soil erosion; reduce runoff from agricultural fields, urban areas, and construction sites; and limit the use of pesticides and fertilizers on lawns and golf courses.

Protecting water-absorbing wetlands and prohibiting development on floodplains are the first steps towards preventing degradation to stream systems. A wide variety of best management practices are used to mitigate adverse effects on water quality from land use change in the upland areas of watersheds. *Detention basins* capture stormwater runoff from paved surfaces of parking lots or whole developments, slowing its introduction into stream channels, so that the flow of water off the landscape more closely resembles predevelopment conditions (see Figure 9-10).

So-called *Better Site Design* principles also call for the narrowing of roadways and shortening of driveways to reduce the amount of pavement needed and the use of permeable paving materials wherever possible. Cluster development, in which homes are concentrated in one area and the remaining land is left as open space, also reduces the amount of pavement necessary to service a particular community, thereby decreasing the amount of stormwater runoff. Individual homes can create *rain gardens,* depressions that capture runoff from rooftops and allow it to infiltrate slowly into the ground.

To reduce sediment loading to waterways, municipalities require that contractors take steps to minimize the erosion of sediment from active construction sites. One common approach is to require the installation of *silt fences.* These textile barriers are inserted into the ground and extend up to several feet above the surface to trap eroding sediment from hill slopes. Contractors might also be required to construct *settling basins* that capture runoff from construction sites and allow sedimentary particles to settle out before the water moves onward toward the stream channel.

Measures that target nutrient runoff and channel stability include ordinances requiring that a strip of vegetated land, called a **riparian buffer,** be left along stream channels. Grasses, shrubs, and trees can absorb fertilizer and pesticide runoff from farms and lawns, trap sediments eroded from hillsides, and provide valuable wildlife habitat. They also can store rainwater and provide some root strength to channel banks.

Total Maximum Daily Loads (TMDLs) As noted earlier, despite pollution control efforts some waters still are not fishable or swimmable and do not meet water quality standards. Under the auspices of Section 303(d) of the CWA, the EPA requires states to identify and list these waters, and then to develop **Total Maximum Daily Loads (TMDLs)** in order to protect them.[70] A TMDL is an estimate of the upper limit of a pollutant that a water body can receive and yet still meet water quality standards. Once the amount of load reduction that is needed within a watershed is estimated, nonpoint source programs are established to achieve the load reductions.

New Threats to Water Quality Though many U.S. waterways have much higher water quality than prior to passage of the Clean Water Act in 1972, new risks are emerging that require innovative strategies to continue to protect water. Relatively new practices that include mountaintop removal for coal mining and "hydrofrack-

ing" for natural gas in shale pose additional threats to the nation's waterways (see Chapter 13).

Regulations to Protect Drinking Water

As scientists investigate pollutants from human activities and the effects of these substances on the environment, ecosystems, water quality, and human health, they communicate their results through various types of writings. In addition to scientific articles that are read mostly by other scientists and policy makers, scientists at times write directly to the public. One such book, *Silent Spring,* has become one of the best known books on environmental contamination and is credited with leading to greater protection of the environment through various laws and regulations.

Written in 1962 by biologist Rachel Carson of the U.S. Bureau of Fisheries (later the U.S. Fish and Wildlife Service), *Silent Spring* spurred a growing environmental movement and public clamor for environmental regulations to protect air, water, wildlife, and public health. The book's first chapter described a spring in a fictitious town where no songbirds were heard because they had vanished. Carson created the scenario to alert the public to the long-term detrimental effects of widespread poorly regulated use of pesticides, and in particular the use of DDT. Like other pesticides, DDT can be adsorbed by soil and sediment, wash off the landscape in runoff, bioaccumulate in organisms, and magnify upward through the food chain. As a result of biomagnification, DDT can have high concentrations not only in pelicans, as mentioned above, but also in large predator birds, such as eagles and condors. *Silent Spring* raised public awareness and concern regarding DDT and other pesticides in the environment, prompting U.S. President John F. Kennedy to ask his Science Advisory Committee to evaluate the book's claims. The committee released a report in 1963 that largely supported Carson's account.

In 1970, the EPA was formed in order to have one federal agency with primary responsibility for research, monitoring, standard-setting, and enforcement related to environmental protection. One of its first acts in 1972 was to ban use of DDT for agriculture, as noted earlier. A slew of other environmental legislation was enacted by Congress soon after, including two of the most important federal laws to protect U.S. waters, the Clean Water Act (1972), described above, and the **Safe Drinking Water Act** (1974).

The Safe Drinking Water Act (SDWA), amended in 1986 and again in 1996, is the primary federal law for ensuring the safety of U.S. drinking water. It provides the United States with perhaps one of the world's safest potable water supplies. Under the SDWA's mandate, the EPA sets standards for drinking water quality, oversees various suppliers of water (e.g., municipalities), supervises implementation of drinking water standards by states and localities, and requires numerous actions to protect drinking water sources. Primary drinking water standards are legally enforceable standards that limit the levels of contaminants in public drinking water systems. The standards are set below the level at which a contaminant would be expected to cause adverse human health effects after a lifetime of exposure. The EPA requires that water be tested frequently by certified laboratories and that public water systems provide customers with annual water reports that identify water sources and any evidence of contaminants.

For each of about 90 contaminants and indicators, the EPA has determined a **Maximum Contaminant Level Goal (MCLG)**, which sets the level of contaminants in drinking water at which no adverse health effects are likely to occur.[71] These goals are not, however, enforceable. An enforceable regulation was also established, called the **Maximum Contaminant Level.** This level is set as close to the MCLG as possible, but might be lower because the cost of contaminant removal to a higher goal is prohibitive, or the technology for treatment and removal to that level is not available.

Even if water is deemed safe at a water treatment plant, it might still become unsafe during transit to a domestic tap. For example, houses built before 1930 often had lead plumbing. Lead corroded from the pipes can cause severe brain damage if ingested frequently over a long period. In 1986, the use of lead was banned from new public water systems. Water suppliers now are required to test for lead and take corrective action if more than 15 parts of lead per billion parts of water are found in tap water.

Because of water regulations, tap water in the United States could be safer than some bottled water.[72] Federal standards for water quality in the United States apply only to commercially bottled water that is sold in bulk and marketed across state lines. Furthermore, the bottlers of commercial water are regulated by the Food and Drug Administration rather than the EPA and are allowed to do their own testing. Mineral waters and seltzers are considered "beverages" and do not have to meet EPA standards for drinking water. Bottled water may taste better than tap water, but more than a third of all bottled water in the United States is obtained from the same sources as tap water. At present, companies that are bottling water are not required to list the water source on their labels; in fact, they may draw it directly from a tap.

The most common reasons that bottled water tastes slightly better are the presence of dissolved minerals in bottled water obtained from springs and groundwater and the use of ozone as a disinfectant in the bottling process. Municipal water suppliers use chlorine instead of ozone, and this can leave a noticeable aftertaste. Although bottled water is not likely to be safer or different from tap water (except in taste) and although it costs about 700 to 1500 times more than tap water, the per capita consumption of bottled water is steadily increasing.

The National Wild and Scenic Rivers Act

The National Wild and Scenic Rivers (NWSR) Act was created by Congress in 1968 to preserve and protect certain rivers with "outstandingly remarkable" natural, cultural, and recreational values in a free-flowing condition, for the "benefit and enjoyment of present and future generations" (see photo of the Lewis River, Washington, at beginning of chapter; Public Law 90-542; 16 U.S.C. 1271 et seq.). As designated, Wild River Areas are either rivers or parts of rivers that are free of impoundments and generally inaccessible except by trails. Scenic River Areas are similar, except that they are accessible by roads at some places. Wild or Scenic Rivers are designated by Congress or the Secretary of the Interior.[73]

The NWSR Act protects less than a quarter of one percent of American rivers, only 12,598 miles of 203 rivers. In contrast, 17 percent of American rivers have been altered by more than 80,000 large dams that impact 600,000 miles of streams. Congress created the NWSR Act after decades of large dam construction for hydroelectricity and flood control throughout the United States, and after centuries of small dam building for water-powered mills that date to early Colonial settlement.[26] In the NWSR Act, Congress referred to the need for balance between using and protecting river resources:

The Congress declares that the established national policy of dams and other construction at appropriate sections of the rivers of the United States needs to be complemented by a policy that would preserve other selected rivers or sections thereof in their free-flowing condition to protect the water quality of such rivers and to fulfill other vital national conservation purposes.

When President Lyndon Johnson signed the NWSR Act and three other conservation acts on October 2, 1968, his remarks about their potential legacy included the following:

...today we are initiating a new national policy which will enable more Americans to get to know more rivers.... An unspoiled river is a very rare thing in this Nation today. Their flow and vitality have been harnessed by dams and too often they have been turned into open sewers by communities and by industries. It makes us all very fearful that

all rivers will go this way unless somebody acts now to try to balance our river development.[74]

✔ Freshwater is predicted to become the most strategic resource in the world, the "next oil," and has no substitute for most of its functions.

✔ Urbanization initially causes sediment fluxes toward streams to increase as soils are exposed for construction. Later, paved surfaces speed stormwater flows toward channels.

✔ Better site design principles such as reducing the width of roadways, using permeable paving materials, and directing rooftop runoff toward rain gardens can reduce the impact of urbanization on streamflows.

✔ U.S. citizens are some of the heaviest users of water in the world, using about 1600 m^3 per person annually.

✔ The primary sectors of water use are agricultural, industrial, and domestic.

✔ A water footprint is an estimate of the total amount of freshwater a person uses both directly and through the consumption of goods and services that were produced with water.

✔ Virtual water is the amount of water embedded in a product in terms of how much water was used to produce and transport the commodity, and is a useful concept because of the worldwide links among food, trade, and economic development.

✔ A country can have physical or economic water scarcity, with the former referring to cases in which local demand for water exceeds local availability, and the latter referring to cases where access to water is limited for socioeconomic reasons.

✔ The United Nations estimates that nearly 1 billion people have no access to safe drinking water at present, and nearly 2 billion people will face water scarcity by the year 2025.

✔ Laws such as the U.S. Clean Water Act (CWA) prevent pollution in waterways, improve treatment of wastewater, and maintain the integrity of wetlands.

✔ The CWA has made substantial progress controlling point source pollutants, such as discharge from a factory, but nonpoint source pollutants such as runoff from farm fields are still difficult to regulate.

✔ A major cause of stream impairment is fine sediment that can be suspended in streams, leading to high turbidity and degradation of aquatic habitat, and the sources of this sediment are largely nonpoint in origin.

✔ Long-term monitoring of water quality in the United States reveals that the pesticide atrazine, a widely used weed killer, is the most commonly detected pesticide in U.S. waterways, especially in agricultural areas.

✔ Best management practices are designed to prevent soil erosion, reduce runoff, and limit the use of pesticides and fertilizers.

✔ Means of reducing stormwater runoff include protecting wetlands, prohibiting development on floodplains, and using detention basins to slow the flow of stormwater runoff into stream channels.

✔ The most important federal law for protecting drinking water in the United States is the Safe Drinking Water Act (SDFA), which sets standards for drinking water quality and oversees water suppliers.

DROUGHTS

Types of Droughts

Climatologists measure drought in four different ways: meteorological, hydrological, agricultural, and socioeconomic (Figure 9-26). Whereas the first three of these treat drought as a phenomenon with a physical basis, the last views drought in terms of its impacts on water supply and demand.

Meteorological drought is the result of a precipitation (rain and snow) deficit that occurs in a region over an extended period of time, typically at least one season in duration. The onset of drought is determined with respect to the magnitude of departure from the long-term (e.g., 30-year) average of a climatic variable such as precipitation.[75] Comparing departures of climatic variables from a long-term average is necessary because a decrease in seasonal precipitation of just a few inches, for example, could have devastating effects in an already dry region where average annual rainfall might be only 10 inches, but might have little effect for a humid region in which annual rainfall is 40 inches or more. Although it is common to think of drought as a deficit in precipitation, other meteorological conditions such as high winds and temperature can lead to drought.

Deficits in effective precipitation (i.e., the amount not evapotranspired) can decrease surface and subsurface water supplies, including streamflow, lake levels, and groundwater, but these hydrological effects typically lag behind the actual shortfall in precipitation, that is, the meteorological drought. As the impacts develop, cascading events occur in the hydrologic cycle and a hydrological drought occurs. In addition, the deficit in precipitation leads to agricultural drought as plants receive insufficient water. Finally, socioeconomic drought occurs when the supply and demand for an economic good, such as electricity from hydropower, become imbalanced as a result of a deficit in water supply.

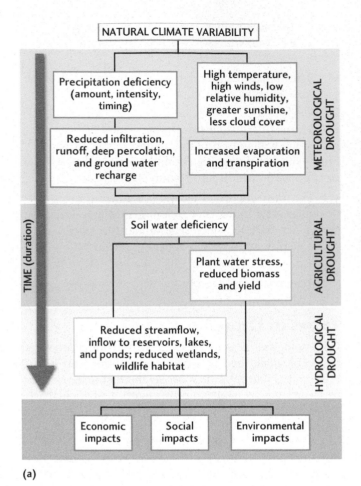

(a)

(b)

Figure 9-26 (a) Droughts begin as meteorological phenomena, but if prolonged they can become agricultural and hydrological droughts with devastating socioeconomic and environmental impacts. (b) The mega–heat wave and drought of 2010 that caused widespread wildfires and led to the deaths of 55,000 people in Russia burned this woman's house and garden, where afterward she dug for potatoes. *(b: Yuri Kochetkov/EPA/Newscom)*

If a meteorological drought is prolonged, the likelihood of agricultural, hydrological, and socioeconomic drought increases. Extended rainfall or snowfall deficits typically lead to a decrease in soil moisture, groundwater recharge, lake levels, and the flow of water from springs and in streams. Decreased soil moisture leads to stress on plants, and extreme to exceptional drought intensity can lead to widespread crop failure, as occurred during the Dust Bowl of the American Southwest in the 1930s (see Chapter 8). The increased demand for water from streams and lakes during droughts, a socioeconomic effect, can exacerbate reductions in streamflow, a hydrological effect, leading to a worsening spiral during a drought. Greater use of groundwater, particularly for agricultural needs, is another common response to drought that contributes even further to reduced flow in streams.

Impacts of Droughts on Socioeconomic and Environmental Systems

As one particularly devastating example of heat and drought, the mega-heat wave that occurred in the Russian Federation during the summer of 2010 was the most extreme heat wave recorded for that region since instrumental records began in 1880.[76] Temperatures topped 40°C (104°F), with maximum recorded temperatures as high as 44°C (111°F) breaking previous instrumental records. A large part of western Russia had temperatures at least 7°C (12.6°F) warmer than normal during July. More than 500 wildfires broke out as a result of heat and dryness, producing smoke that worsened atmospheric conditions and respiratory distress for those downwind. In Moscow alone, 11,000 deaths occurred. Crop failures in multiple areas led to a 30 percent loss in grain harvest. In response, world wheat prices soared. Cumulative impacts of the mega-heat wave and drought caused 55,000 deaths and $15 billion USD in damages in Russia.

In addition to the socioeconomic impacts, the environmental effects on aquatic life can be devastating. This has been the case for Spring Creek in southwestern Georgia (U.S.), well-known among biologists and conservationists for its rich beds of native river mussels (Figure 9-27). Biologists found as many as 14 species of mussels in one survey, three of them listed on the federal endangered species list. Once abundant, mussel populations have dwindled substantially due to land use change

Figure 9–27 (a) Scientists worked with resource managers to establish flow augmentation to Spring Creek, Georgia, from a groundwater well during drought conditions in 2006–2009. (b) This flow was vital to maintain minimum flow conditions for river mussels, including the endangered oval pigtoe mussel held in the hand of a U.S. Fish and Wildlife biologist. *(a: U.S. Fish and Wildlife Service/ Southeast; b: Tom MacKenzie/U.S. Fish and Wildlife Service)*

that includes deforestation and agriculture, as well as due to other pressures on water quality and in-stream flow. Mussels live in clean flowing water, and their filter feeding helps to maintain water quality. Water flow in Spring Creek, as well as many other streams in the southeastern United States, dropped substantially during the drought of 2006–2009, which had progressed to include hydrological, agricultural, and socioeconomic drought.[77] Native mussel populations were decimated, with layers of dead mussels found in dried mud and stagnant pools of water.

In response to the drought, a number of groups and agencies formed a partnership to fund and implement a water augmentation project at a site along Spring Creek where the oval pigtoe mussel, an endangered species, was abundant. A nearby stream gauge was established to indicate when the water drops below a threshold level, and this triggers groundwater from wells to be pumped into Spring Creek. This flow augmentation provides an additional 2 ft³ per second of water to the stream, a small amount but possibly enough to maintain mussels during times of drought.

Drought History and Global Warming

Residents of arid western states in the United States are accustomed to thinking about drought and the need for water conservation, but the same has not been true in the humid eastern states, where water supplies historically have been plentiful. Yet prehistoric records, such as those from tree-ring studies, indicate that droughts of long duration and relatively high frequency have occurred even in relatively wet areas in the past (Box 9-4). Further, global warming can lead to more severe droughts. Less water vapor is present in the atmosphere after heavy rainfall, so dry spells can become longer.[3] A warmer atmosphere in the absence of rain causes greater drying of Earth's surface. The combination of less rain and more drying contributes to more intense drought. For Russia, for example, scientists estimate from computer model experiments that the probability of a mega-heat wave such as that of 2010 occurring within the next four decades has increased due to global warming.[76]

✔ Meteorological drought, the result of a deficit in precipitation, can become a hydrological drought if soils dry, groundwater levels drop, and streamflow diminishes, and ultimately can become an agricultural and socioeconomic drought if plants become stressed, crops fail, and drinking water is in short supply.

✔ Hydrological droughts impact aquatic ecosystems, as some species cannot survive prolonged low flow conditions in streams, but geologic investigations indicate that global warming can lead to more severe droughts.

BOX 9-4 CASE STUDY

Water Battles in Wet Places: Drought and the Tri-State Water Dispute

We don't typically consider a modern city in a humid part of a developed country as a place where several million citizens might run out of water in a few months, or where states battle over the rights to water in a lake. In October of 2007, however, the governor of the state of Georgia, Sonny Perdue, declared a state of emergency as the southeastern United States sank deeper into a drought that had started the year before (**Figure 1**). Lake Lanier, a reservoir formed by damming the Chatahoochee River in 1957, is the primary drinking water supply for more than

3 million residents of metropolitan Atlanta. As 2007 ended, however, the lake's water level had fallen so low that the city would run out of water within three months if the drought continued (**Figure 2**). Restrictions on watering outdoor plants were issued, and the governor directed restaurants to serve water only upon request and urged residents to take shorter showers.

At the same time, Georgia attempted to gain an exemption from the federal Endangered Species Act, which mandates that the state provide a minimum flow of water

(continued)

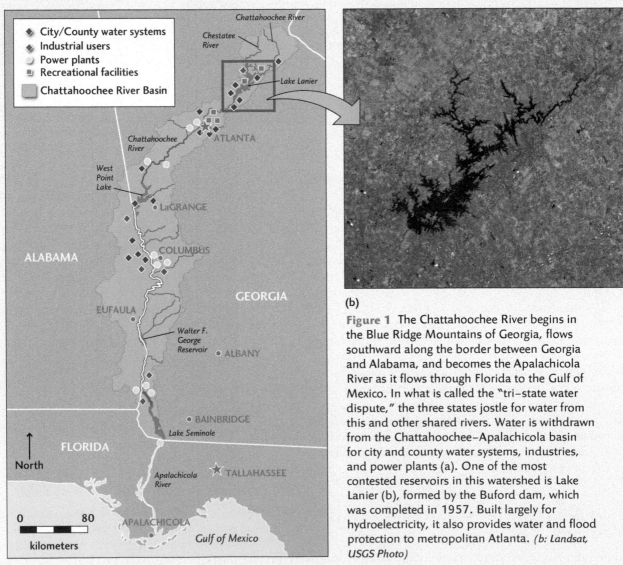

Figure 1 The Chattahoochee River begins in the Blue Ridge Mountains of Georgia, flows southward along the border between Georgia and Alabama, and becomes the Apalachicola River as it flows through Florida to the Gulf of Mexico. In what is called the "tri-state water dispute," the three states jostle for water from this and other shared rivers. Water is withdrawn from the Chattahoochee–Apalachicola basin for city and county water systems, industries, and power plants (a). One of the most contested reservoirs in this watershed is Lake Lanier (b), formed by the Buford dam, which was completed in 1957. Built largely for hydroelectricity, it also provides water and flood protection to metropolitan Atlanta. *(b: Landsat, USGS Photo)*

(a)

Figure 2 During the drought that engulfed the southeastern United States from 2006–2009, parts of the region experienced extreme to exceptional drought intensity (a). Extreme drought is associated with major crop losses and widespread water shortages or restrictions, whereas exceptional drought is marked by exceptional and widespread crop losses and shortages of water in reservoirs, streams, and wells. This map (b) from October 16, 2007, shows drought conditions as the water level in Lake Lanier dropped to its lowest level (c). *(a: Pouya Dianat/AP Photo/Atlanta Journal–Constitution)*

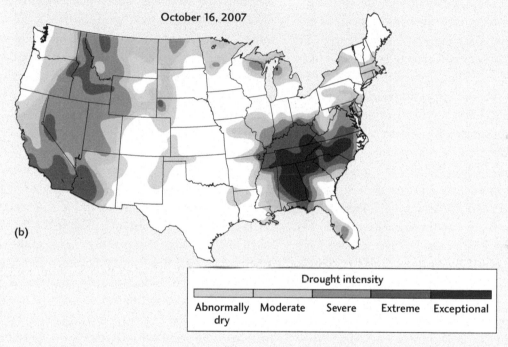

October 16, 2007

(b)

Drought intensity				
Abnormally dry	Moderate	Severe	Extreme	Exceptional

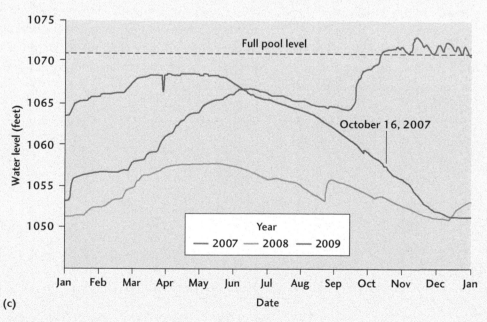

(c)

(continued)

to Alabama and Florida downstream in order to protect endangered mussel species in the Apalachicola River into which the Chatahoochee flows.[78] Residents of the two downstream states were outraged by Georgia's action. Florida governor Charlie Crist appealed to the federal government for assistance, pointing out that Florida's $134 million commercial fishing industry, supported by freshwater from the Apalachicola, would be crippled by reduced flows. Furthermore, he noted that Lake Lanier was not created for the purpose of supplying metro–Atlanta with water, but rather to control floods, generate hydropower, and aid downstream navigation. Over the years since the reservoir was constructed, the U.S. Army Corps of Engineers, which operates the reservoir, had allowed increasing amounts of water to be withdrawn for municipal purposes. In Florida's opinion, using the water this way without seeking Congressional approval constituted a violation of federal law.

In the summer of 2009, a U.S. District Court judge agreed. Ruling that use of Lake Lanier water for municipal purposes requires Congressional approval, the court issued an order to freeze water allocations at present levels for the next three years. Fortunately, the drought came to an end in 2009. The legal wrangling continues, however, and the court ruling was overturned in 2012, allowing Atlanta to withdraw water based on demand. As a result, the U.S. Army Corps of Engineers must develop a new water allocation plan.

Fortunately, many steps can be taken to reduce water consumption and thereby increase the amount of water available to streams. In response to the drought of 2006–2009, Georgia adopted a comprehensive statewide water management plan in 2008 and passed the Georgia Water Stewardship Act in 2010.[79] The goal of these actions is to create a culture of water conservation in the state, even though annual rainfall is 45–50 inches (114–127 cm) a year. The Water Stewardship Act requires local governments to adopt ordinances that restrict outdoor water use between 10 A.M. and 4 P.M. each day, when evaporation is greatest. It also requires local governments to update and enforce plumbing codes that specify high–efficiency plumbing fixtures (e.g., toilets and showers). Public water providers are required to complete annual water loss audits and reduce leakage from water systems. State agencies must enhance programs and incentives for voluntary water conservation. Already these actions have resulted in notable decreases in water usage. As water utilities in Georgia completed their audits, for example, they found that as much as 10% to 20% of the water they provide cannot be accounted for when compared to the amount billed to customers. The difference can be traced largely to leaking pipes and faulty meters, both of which can be fixed so as to reduce system–wide water losses.

Heavy rains in the Southeast restored Lake Lanier to its predrought levels as of 2010, but the long–term prognosis for Atlanta is unclear. The metro area is experiencing one of the fastest rates of population growth in the United States, and it now is obvious that growth occurred under erroneous assumptions regarding long–term water resource availability. Scientists recently studied tree–ring records that span the last 346 years for the Apalachicola, Chattahoochee, and adjacent Flint River basins. These records reveal that between 1696 and 1820, even longer droughts than that of 2006–2009 were frequent. Perhaps most ominous, these records also revealed that the time period for which water management decisions were made for the southeastern United States, the mid– to late 20th century, was one of the wettest since 1665.

CLOSING THOUGHTS

Human interaction with and use of streams, wetlands, and freshwater has grown in intensity since the time of early humans. Today, streams and rivers are enjoyed for kayaking, rafting, and swimming among other recreational activities, and their wildness and scenic value are appreciated by many.

The ecosystem services provided by streams and wetlands are valued increasingly with time, as they become better understood through scientific investigation. Wetlands are valued in particular for their biodiversity, and are widely recognized as prime habitat for bird nesting. On the other hand, streams and rivers have been dammed for irrigation and water power over the past millennia, leading to a massive reorganization of immense amounts of sediment in Earth's waterways and to barriers of passage for fish that migrate up and down streams during their life cycles. Because of base-level controls on stream gradients, the impacts of damming have rippled throughout watersheds. Meanwhile, increasing population and urbanization have led to widespread disposal of sewage in waterways, and the Industrial Revolution also brought new pollutants to streams and wetlands.

The mid-20th century saw a rise in environmental awareness and understanding, leading to new laws to protect and restore aquatic ecosystems. Ecological restoration requires, however, that we understand what existed prior to degradation in order to re-create or restore the ecosystems as well as the services they once provided. Geologists, with their focus on earth history and the use of clues in the fossil and sedimentary record, are joining with ecologists and paleoecologists to determine what once was, and are partnering with engineers and policy makers to reverse the degradation of waterways throughout the world.

SUMMARY

■ The hydrologic cycle, the fundamental, unifying concept in the study of water, is the system that controls the distribution of water resources on Earth.

■ The hydrologic budget equation shows how water resources can be unevenly distributed as the rates of precipitation, evaporation, and runoff are affected by factors that include latitude, topography, and the locations of rising and descending air masses.

■ Stream networks form as the result of overland flow, subsurface stormflow, and groundwater recharge and discharge.

■ Hydrographs display how the stage and discharge of water in a stream rises and falls in response to rainfall events, which proves helpful for land use and environmental planners.

■ Stream channel patterns and processes depend upon the geologic conditions in a watershed, the type of sediment transported within the stream, and the energy of the stream system.

■ Alluvial streams have beds and banks that are composed of sediment, and patterns that are characterized as braided or meandering.

■ To prevent the hazardous impacts of floods, planning and regulatory agencies conduct flood frequency analyses that rely upon the historical record of streamflow events to predict the recurrence intervals, magnitudes, and locations of future flooding events.

■ The frequency, duration, and intensity of flooding may be affected by global warming. Scientists analyzing rainfall data from the period of 1958 to 2007 found that the amount of rain during the heaviest events has increased in some parts of the United States.

■ Dams, channelization, and levees are widely employed to prevent flooding and increase the amount of land available for development, but each has environmental consequences that are often negative.

■ Channelization of streams sometimes leads to additional flooding downstream of channelized reaches, while damming results in some 4 to 5 billion metric tons of sediment trapped behind dams each year.

■ Artificial levees have encouraged development in flood-prone areas and can lead to disastrous flooding if they fail.

■ Wetlands are identified on the basis of hydrology, vegetation, and soil type, and are highly valued for ecosystem services that include nutrient filtering.

■ Stream restoration, the process of re-creating or repairing the structure and ecological functions of a degraded stream, is a management tool to reduce sediment and nutrient loads in waterways and to improve species habitat.

■ Although water is considered a renewable resource on a global scale, freshwater can become a limiting resource if water use exceeds local replenishment rates or if water is polluted due to improper waste disposal, untreated sewage, industrial pollutants, runoff containing fertilizers and pesticides, or the contamination of coastal aquifers with salt water.

■ The Clean Water Act (CWA) has made substantial progress controlling point source pollutants from entering waterways, improving the treatment of wastewater, and maintaining the integrity of wetlands.

■ Nonpoint source pollution continues to be problematic, with high levels of fine sediment (suspended load) and atrazine found in many streams in the United States.

■ Best management practices are intended to reduce the effects on nonpoint source water pollution through the prevention of soil erosion, reduced runoff from agricultural fields and urban areas, and limited use of pesticides and fertilizers.

■ Droughts are classified as meteorological, hydrological, agricultural, or socioeconomic, and might become more severe with global warming.

KEY TERMS

hydrologic cycle (p. 271)
mass balance equation (p. 273)
hydrologic budget equation (p. 273)
drainage basin (p. 277)
watershed (p. 277)
drainage divide (p. 277)
drainage network (p. 277)
base level (p. 278)
overland flow (p. 278)
infiltration capacity (p. 278)
hydrograph (p. 280)
stage (p. 280)
discharge (p. 280)

base flow (p. 281)
alluvium (p. 281)
braided stream (p. 281)
meandering stream (p. 281)
pool (p. 281)
riffle (p. 281)
floodplain (p. 282)
overbank deposition (p. 282)
levees (p. 282)
flood frequency analysis (p. 286)
recurrence interval (p. 286)
Special Flood Hazard Areas (p. 287)

levee setback (p. 294)
channelization (p. 294)
wetland (p. 295)
ecological restoration (p. 297)
water footprint (p. 302)
virtual water (p. 303)
Clean Water Act (CWA) (p. 304)
point source pollutant (p. 304)
nonpoint source pollutant (p. 304)
turbidity (p. 305)
biological amplification (p. 305)

endocrine disruptor (p. 306)
best management practices (p. 306)
riparian buffer (p. 307)
Total Maximum Daily Loads (TMDLs) (p. 307)
Safe Drinking Water Act (p. 307)
Maximum Contaminant Level Goal (p. 308)
Maximum Contaminant Level (p. 308)
meteorological drought (p. 309)

REVIEW QUESTIONS

1. What is the hydrologic budget equation? How can it be used to determine the availability of surface water runoff for a given area?

2. What are the main parts of a drainage basin?

3. What are the major flow paths to streams that rainwater follows after a storm?

4. How is the infiltration capacity of the ground related to the route that rainwater or snowmelt follows on its way to streams?

5. How is stream discharge determined? Why is it an important property to measure?

6. Why are braided streams so common downstream of glaciers and in logged or otherwise disturbed terrain?

7. What are pools and riffles? How do they differ from each other?

8. Compare and contrast the characteristics, benefits, and negative aspects of natural and artificial streams.

9. Why is the U.S. Army Corps of Engineers restoring the Kissimmee River in Florida to its unchannelized state?

10. How are floodplains formed by streams?

11. How are floods of different probabilities of occurrence and recurrence intervals determined from flood frequency analysis?

12. Why, in the late 1990s, did some scientists urge the U.S. government to create a flood along the Colorado River?

13. Where do wetlands occur on the landscape, and how are they related to groundwater?

14. What is a wet meadow, and what types of plants typically are found in this type of wetland?

15. Why is woody debris important in mountainous streams in the Pacific Northwest, and how has it been removed from streams?

16. What is ecological restoration, and how is it applied in cases of dam removal?

17. How much water is used annually by U.S. citizens (per capita)? How does this amount compare to water use in other countries?

18. What are the two types of water scarcity, and what countries are experiencing each of these types?

19. What is virtual water, and how is it related to a person's water footprint?

20. List the most important federal laws in the United States for the protection of water in streams and wetlands and of public supplies of drinking water.

21. What are the differences between point and nonpoint source pollutants, and what are typical examples of each?

22. What is turbidity, and how does high turbidity in a waterway contribute to the degradation of aquatic habitat?

23. What is a Maximum Contaminant Level Goal, and how does it differ from a Maximum Contaminant Level?

24. What is a Total Maximum Daily Load, and how is it related to protection of waterways and aquatic habitats?

25. What is a meteorological drought, and how is it related to hydrological, agricultural, and socioeconomic drought?

THOUGHT QUESTIONS

1. Would you predict that groundwater is or is not being recharged by infiltration of excess surface water in the area of Phoenix, Arizona? Explain your reasoning.

2. How might urbanization contribute to increased stream discharge? (Use a hydrograph in answering this question.)

3. How does the use of artificial levees and straightened stream channels result in greater flooding for areas downstream of such control structures?

4. In what ways might global climate change lead to different amounts of discharge and flooding in existing streams?

5. Has the amount of stream discharge in different areas varied in the geologic past as a result of global climate change? If so, give an example, stating when and why the change occurred.

6. Is the 100-year flood a reasonable standard for land use planners and policy makers to use for zoning and insurance practices, or would you recommend considering another recurrence interval, such as the 200-year flood? Explain your reasoning.

7. Should scientists and engineers attempt ecological restoration in a former reservoir after a dam removal, or should they just let the system adjust itself over time? Why or why not?

8. In the case of mountainous streams in the Pacific Northwest, which typically carry high sediment loads and coarse sediment (e.g., sand and gravel), restoration might involve putting woody debris in the valley bottom. In some stream systems in the eastern

United States, where relief is lower, wet meadows with multiple small streams were widespread before European settlement, but these stream wetland systems were characterized by abundant sedges and rushes with some shrubs. Woody debris was not as common in these valley bottoms. Compare and contrast these two types of stream and wetland systems, and evaluate the different ways in which they might be restored in order to return the ecosystem services to their original state.

9. Why was it easier for regulations associated with the Clean Water Act to reduce point source pollutants in waterways in contrast to nonpoint source pollutants?

10. Does the approach provided by establishing Total Maximum Daily Loads for an impaired waterway seem like an effective way to reduce the amount of nonpoint source pollutants that reach the waterway?

EXERCISES

1. If a stream has a width of 20 m, a depth of 5 m, and a velocity of 2 m/sec, what is its discharge?

2. If a natural stream has a width of 10 m, what would you predict is the average distance between each pool along its length?

3. A meandering stream flows from an elevation of 1000 m to sea level, and its channel length is 3000 m. The river is straightened for flood control, and its final length is only 1500 m. How much has the slope of the straightened channel increased from its original state?

4. What is the recurrence interval for a flood that has a 0.25 probability of being equaled or exceeded in a given year?

5. If a person uses 50 gal of water a day for domestic purposes, what is that person's annual water use in cubic meters? How does this amount compare to the average per capita annual water use of a U.S. citizen?

6. Use online resources for the National Park Service (www.nps.gov) and the United States Geological Survey (www.usgs.gov) to compare different dam removal projects. Identify differences in dam height and reservoir sediment volume, amount of sediment estimated to be released during and after dam removal, and the purposes for removing the dams. Look for evidence provided by scientists about long-term monitoring after dam removal, and evaluate the effectiveness of dam removal and restoration in achieving the original goals that were established as motives for removing the dams.

7. Use online resources from the U.S. Environmental Protection Agency (www.epa.gov) to compare Maximum Contaminant Level Goals and Maximum Contaminant Levels for the pesticides atrazine and endrin.

SUGGESTED READINGS

Barnett, T. P., and D. W. Pierce. "Sustainable Water Deliveries from the Colorado River in a Changing Climate." *Proceedings of the National Academy of Sciences* 106 (2009): 7334–7338. doi:10.1073/pnas.0812762106.

Brody, S. D., W. E. Highfield, and J. E. Kang. *Rising Waters: The Causes and Consequences of Flooding in the United States.* Cambridge, UK: Cambridge University Press, 2011.

Collier, M. P., R. H. Webb, and E. D. Andrews. "Experimental Flooding in the Grand Canyon." *Scientific American* (January 1997): 82–89.

Collier, M. P., R. H. Webb, and J. C. Schmidt. "Dams and Rivers: A Primer on the Downstream Effects of Dams." *U.S. Geological Survey Circular* 1126 (1996).

Committee on Characterization of Wetlands, National Research Council. *Wetlands: Characteristics and Boundaries.* Washington, D.C.: The National Academies Press, 1995.

Dunne, T., and L. Leopold. *Water in Environmental Planning.* New York: W. H. Freeman and Company, 1978.

Fradkin, P. L. *A River No More: The Colorado River and the West.* New York: Alfred A. Knopf, 1981.

Gleick, P. H., ed. *Water in Crisis: A Guide to the World's Fresh Water Resources.* New York: Oxford University Press, 1993.

Gleick, P. H., ed. *The World's Water, Volume 7: The Biennial Report on Freshwater Resources.* Washington, D.C.: Island Press, 2011.

Hartig, J. H. *Burning Rivers: Revival of Four Urban Industrial Rivers That Caught Fire.* Essex, UK: Multi-Science Publishing Company, 2010.

Hillel, D. *Rivers of Eden: The Struggle for Water and the Quest for Peace in the Middle East.* New York: Oxford University Press, 1994.

King, R. O. *National Flood Insurance Program: Background, Challenges, and Financial Status.* Congressional Research Service, Report for Congress, 2011.

Leopold, L. *A View of the River.* Cambridge, MA: Harvard University Press, 1994.

McCully, P. *Silenced Rivers: The Ecology and Politics of Large Dams.* London: Zed Books, International Rivers Network, and *The Ecologist,* 1996.

McPhee, J. *The Control of Nature.* New York: Farrar, Straus and Giroux, 1989.

Montgomery, D. R., ed. *Restoration of Puget Sound Rivers.* Seattle: University of Washington Press, 2003.

Mutel, C. F. *A Watershed Year: Anatomy of the Iowa Floods of 2008.* Iowa City: University of Iowa Press, 2010.

National Research Council. *Colorado River Basin Water Management: Evaluating and Adjusting to Hydroclimatic Variability.* Washington, D.C.: National Academies Press, 2007.

O'Connor, J. E. and J. E. Costa. "Large Floods in the United States: Where They Happen and Why." *U.S. Geological Survey Circular* 1245 (2003). Available online at http://pubs.usgs.gov/circ/2003/circ1245/pdf/circ1245.pdf.

Reisner, M. *Cadillac Desert: The American West and Its Disappearing Water.* Revised edition. New York: Penguin, 1993.

The Heinz Center. *Dam Removal: Science and Decision Making.* Washington, D.C.: H. John Heinz III Center for Science Economics and the Environment, 2002.

U.S. Geological Survey. *National Water Summary, 1986 to Present.* U.S.G.S. Water-Supply Paper series. Washington, D.C.: U.S. Printing Office.

Wu, M., M. Quirindongo, J. Sass, and A. Wetzler. "Still Poisoning the Well." Natural Resources Defense Council, Washington, D.C. (2010). Available online at: http://www.nrdc.org/health/atrazine/files/atrazine10.pdf.

The Groundwater System

Our lives depend on freshwater. The largest supply of liquid freshwater—32 percent of the total supply on Earth—is underground, flowing through spaces between sedimentary grains and cracks in rocks. This groundwater discharges at Earth's surface into springs, wetlands, streams, rivers, lakes, and ultimately into the world's oceans. Because the growing human population demands ever-increasing amounts of clean freshwater for agriculture, industry, and domestic use, groundwater is now more heavily used than it has been at any time in the past. Yet human stewardship of groundwater has been careless at times, leaving this essential resource exposed to pollution from sewage disposal, leaking underground chemical storage tanks, and chemical spills. As global dependence on groundwater increases, nations worldwide are making efforts to protect and manage groundwater resources and to clean up contaminated sites.

To illuminate the issues surrounding groundwater, in this chapter we:

✔ Examine how groundwater gets underground and flows from one place to another.

✔ Consider groundwater use and the environmental changes that have affected its quality and quantity in recent years.

✔ Investigate hazards associated with the extraction of groundwater, such as sinkholes, subsidence, and contamination by seawater.

✔ Examine how groundwater can be cleaned if it becomes contaminated.

A spring–fed oasis in a seemingly dry landscape reveals the presence of underlying groundwater. Rocks and sediments of the lithosphere and pedosphere naturally filter many impurities out of the groundwater and add some minerals that enhance taste, making springwater desirable as drinking water. *(Frank Krahmer/Digital Vision/Getty Images)*

INTRODUCTION

"Water below the surface can hide from the naked eye, but not from GRACE."[1] NASA scientist Matt Rodell made this statement in 2009. GRACE is the acronym for Gravity Recovery and Climate Experiment, a joint mission between NASA and the German Aerospace Center. A pair of satellites that orbit Earth from 310 miles (500 km) above its surface measure changes in Earth's gravity caused by differences in mass. These measurements provide valuable information about water and climate. Areas with large amounts of groundwater or glaciers, for example, exert a stronger gravitational pull than other areas, so relative positions of the twin satellites change as they move over places with differences in water content. Scientists using GRACE data have studied changing sea level, the accumulation and melting of snow packs and ice sheets, the severity of droughts, and depletion of groundwater resources.

An example of the last of these types of studies was done in northwestern India, a region known as a *bread basket* because of its rich soils, mild climate, and agriculture surplus. Farmers depend heavily on groundwater to irrigate many of the crops, including wheat and rice, that are grown in this region. GRACE data and hydrologic modeling revealed, however, that the level of the water table dropped nearly a foot (30.5 cm) each year from 2002 to 2008, indicating that more water is being pumped from the ground than is being replenished (Figure 10-1).[2] Water tables can vary seasonally and annually, but when groundwater is pumped faster over a prolonged period than the rate at which it is recharged, the use of groundwater becomes unsustainable (Figure 10-2).

The primary sources of groundwater recharge are rainfall and infiltration. Researchers studied annual rainfall data for northern India and found that it was close

Figure 10–2 A man irrigates his field in northern India by pumping water from the ground. *(AP Photo/Anupam Nath)*

to normal during the period from 2002 to 2008, so they concluded that the cause of groundwater depletion must be more discharge from the aquifer. The primary source of such discharge in northern India is overpumping to provide water for irrigation. The amount of groundwater depletion in northern India is substantial, equivalent to two times the capacity of the country's largest surface water reservoir, and three times that of the largest human-made reservoir in the United States.

The GRACE satellites are important for many reasons, among them the ability to monitor hydrologic data from space, especially for places where data might otherwise be sparse or difficult to obtain. Data from GRACE have been used to monitor groundwater storage in Texas, California, Africa, and elsewhere, and have revealed that depletion of aquifers as a result of human consumption of water for irrigation and other purposes is a problem not only in India. By determining

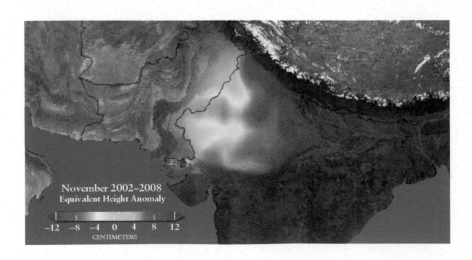

November 2002–2008
Equivalent Height Anomaly

-12 -8 -4 0 4 8 12
CENTIMETERS

Figure 10–1 An analysis of gravity data from GRACE satellites showed that there was a substantial loss of groundwater and lowering of the groundwater table in northern India from 2002 to 2008. The losses were the result of water being pumped out to irrigate agricultural crops. Deep reds indicate negative values of water table change (i.e., lowering) and blues indicate positive values relative to the mean for the period. *(NASA/Trent Schindler and Matt Rodel)*

how much water there is in the ground, how much is needed both for humans and for the environment, and how both the amount available and the amount needed change with time, scientists and resource managers will be better able to guide sustainable use of water resources in the future.[3]

WATER IN THE GROUND

Water in the ground, like water at Earth's surface and water vapor in the atmosphere, is in motion because it has energy. Raindrops and snowmelt entering the ground seep downward because of gravity and the porous nature of soil, rock, and sediment (Figure 10-3). Water continues to move underground, flowing to valley bottoms and eventually to the ocean, the ultimate base level. From the ocean and the continental surface, water molecules evaporate into the atmosphere and contribute to rainfall. Water is always entering and leaving the groundwater system, forming an essential link in the global hydrologic cycle.

The Water Table

Rainfall and snowmelt infiltrate the subsurface and drain downward through any underlying unsaturated material along interconnected pore spaces. At the point where all the voids are saturated, infiltrating water becomes part of the groundwater system (see Figure 10-3). The boundary between the saturated and unsaturated zones forms a surface called the **water table**. Technically, all water below Earth's surface is *underground water*, but only the water below the water table is considered **groundwater**. Unsaturated rock or sediment above the water table provides an avenue along which surface water can recharge groundwater through gravity drainage.

In areas where the water table rises so high that it intersects with Earth's surface, groundwater is exposed in springs. Likewise, low places in Earth's topography, such as streams, lakes, and wetlands, expose the water table, and groundwater discharges into these surface water systems (Figure 10-4). Recall from Chapter 9 that groundwater discharge provides the base flow contained in streams between episodes of rainfall or snowmelt. In addition to supplying surface water features, groundwater discharges into the oceans along coastal areas. In some coastal areas of the Mediterranean Sea and the state of Hawaii, large subsurface channels discharge springwater at such high pressures that it rises above the surface of the sea. A poem by the Roman philosopher Lucretius (first century BCE) describes such a fountain in the Mediterranean:

At Aradus there is a spring within the sea.

The water that this spring pours forth is fresh,

It parts the salty water all around.

And elsewhere too the level sea gives help to thirsty mariners,

By pouring out fresh water mid the salty waves.

(De rerum natura, *Book VI, translated by Alban Dewes Winspear, New York: S. A. Russell, Harbor Press, 1956)*

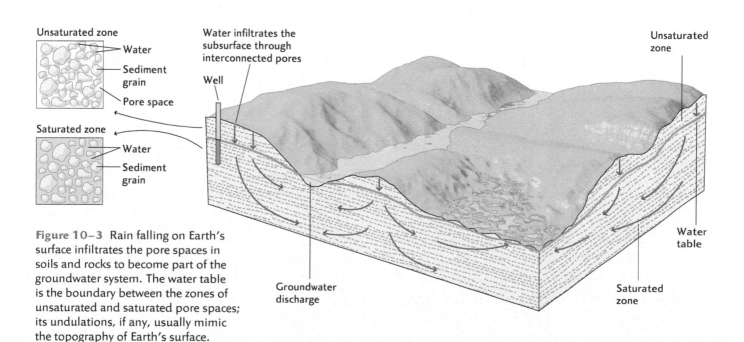

Figure 10–3 Rain falling on Earth's surface infiltrates the pore spaces in soils and rocks to become part of the groundwater system. The water table is the boundary between the zones of unsaturated and saturated pore spaces; its undulations, if any, usually mimic the topography of Earth's surface.

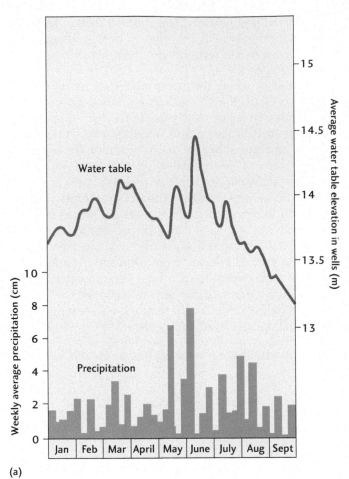

(a)

(b)

Figure 10-4 (a) Monthly trends of precipitation and water table levels in Maryland from January through September. The water table rises shortly after each rainfall. Over the longer term, it rises in the cool, wet spring and drops in the warm, dry summer. (b) The level of the water table varies in response to precipitation and evapotranspiration at Earth's surface. If the groundwater zone is thought of as a reservoir, the input is precipitation and the outputs are discharge from evaporation, springs, streams, and wells.

The level of the water table at any location changes in response to seasonal and yearly weather patterns. In a humid, temperate region in the northern hemisphere, for example, snow melts at relatively low temperatures during the spring, when the potential for evapotranspiration is low. This meltwater percolates downward into the unsaturated zone of the soil and can cause the water table to rise. Conversely, during hot summer months, high temperatures may result in significant evapotranspiration and the depletion of soil moisture, causing the water table to fall (see Figure 10-4). During prolonged droughts, water tables may fall over large areas, causing wells, springs, and streams to go dry.

Porosity and Groundwater Storage

It is surprising how much water can be stored underground in the tiny pore spaces between grains of sediment or along cracks in rocks. If all pore spaces that are filled with water in the subsurface of the United States were contiguous to one another, they would form one large water-filled cavern, at least several tens of meters high, underlying the entire country. **Porosity** is the ratio, usually stated as a percentage, of the volume of pore space between grains to the total volume of rock or sediment in a segment of Earth. The more porous the rock or sediment, the more groundwater it can store.

Some pores originate as spaces between adjacent sedimentary particles or as vesicles—small cavities—that form in igneous rocks as gases escape from cooling lava (Figure 10-5a). This type of porosity is called *primary porosity* because it develops at the same time the rock itself is formed. In sediments that are compacted and cemented into sedimentary rock, the precipitation of minerals—the cement—into the pore spaces reduces

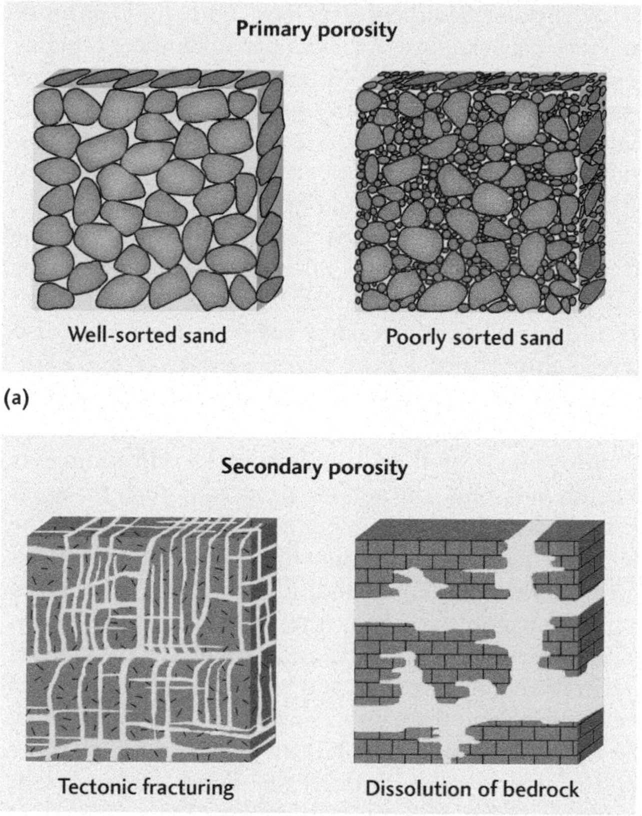

Primary porosity

Well-sorted sand   Poorly sorted sand

(a)

Secondary porosity

Tectonic fracturing   Dissolution of bedrock

(b)

Figure 10–5 (a) Primary porosity consists of void spaces between sediment grains in sedimentary rocks and vesicles in lava flows. The amount of porosity in sedimentary rocks is a function both of the original sorting of the sediment and of the amount of cement holding the grains together. Poorly sorted sediments have lower porosities than well–sorted sediments because small grains fill the spaces between larger grains. (b) Secondary porosity results from tectonic fracturing and/or dissolution of bedrock.

the original porosity of the deposit, but because the cementation process is part of rock formation, the remaining pore space continues to be considered primary porosity. *Secondary porosity* develops after a rock is formed. Fractures, caused by tectonism, uplift, and erosion, contribute to secondary porosity (see Figure 10-5b). In limestones, these fractures may be enlarged by dissolution of the rock, leading to an increase in secondary porosity over time.

In general, primary porosity is highest in sediments and sedimentary rocks and lowest in unweathered crystalline igneous and metamorphic rocks (Table 10-1). Likewise, primary porosity is higher in volcanic rocks than in plutonic rocks. Some volcanic rocks contain lava tubes, which form when molten material continues to flow beneath cooled, hard crusts during volcanic

eruptions. In addition, as extrusive igneous rocks cool, gases escape and the rocks crack, so the tops of individual flows within a stack of lava flows often are vesicular and fragmented, further increasing the primary porosity of volcanic rocks. In metamorphic rocks, most pore spaces either were squeezed shut by the immense pressures to which the rocks were subjected or filled with crystals that precipitated from hot fluids deep in the lithosphere. Because porosity generally decreases with depth, most groundwater is located within several kilometers of the surface, although water-filled fissures have been found at the bottom of the deepest drill hole on Earth, the Kola Hole in Siberia,[4] which was drilled to 12.3 km.

In clastic sedimentary rocks with grains that are silt-size or larger, porosity is unaffected by grain size but is strongly influenced by the degree of particle sorting and the style of grain packing. Well-sorted spherical grains typically have porosities between 26 and 48 percent, depending on how closely the grains are packed together. Poorly sorted sediments have lower porosities because small grains fill the pore spaces between larger

TABLE 10–1 Typical Values of Porosity and Specific Yield for Common Geologic Materials

Material	Porosity (%)	Specific yield (%)
Sedimentary		
Gravel, coarse	24–36	23
Gravel, fine	25–38	25
Sand, coarse	31–46	27
Sand, fine	26–53	23
Silt	34–61	8
Clay	34–60	3
Sedimentary rocks		
Sandstone	5–30	21–27
Siltstone	21–41	15–25
Limestone	0–50	5–45
Shale	0–10	0–3
Crystalline rocks		
Fractured schist	30–38	26
Fractured crystalline rocks	0–10	5
Dense crystalline rocks	0–5	0.1
Basalt	3–35	8
Weathered granite	34–57	40–45
Weathered gabbro	42–45	35–40

grains (see Figure 10-5); thus the sedimentary environment in which grains are deposited strongly influences their future groundwater-storage potential. Well-sorted fluvial channel sands make far better groundwater reservoirs than poorly sorted glacial till, for example. Clay-size particles exhibit a very large range of porosities (see Table 10-1). Clay minerals are sheet silicates like micas, so they tend to form small platy particles. In recently deposited sediments, these particles may form structures, similar to a house of cards, that hold large amounts of water. Shaking or compaction causes these structures to collapse, leading to the low porosities typical of shale and mudstone.

Although grain size does not affect porosity for particles that are silt-size and larger, and thus does not affect the amount of water that can fill the pore spaces, it does affect the amount of water the grains can hold onto by **surface tension**—the attractive force between the molecules at the surfaces of grains and those at the surface of the enclosed water. Consequently, grain size affects the quantity of water that can flow through pore spaces. Smaller grains and fractures have more solid surface area per unit volume of sediment and rock than do larger grains and fractures. As a result, the smaller the grains or fractures, the more water that adheres to the sediment or rock when water is drained out or is pumped from it.

The amount of groundwater that can be gravity drained from a porous rock or sediment is called the **specific yield,** expressed as the ratio of water drained to the total volume of rock or sediment. Because of surface tension, it is not possible to extract all the water from an underground reservoir, just as it is never possible to get all the oil out of the ground by pumping on wells, so specific yield is always lower than porosity (see Table 10-1). The ratio of the amount of groundwater retained by surface tension to the total volume of water is called **specific retention.** The greater the specific retention, the harder it is to remove every bit of a contaminant from groundwater by flushing and pumping the contaminated area.

Permeability and Groundwater Flow

Note that some of the geologic materials listed in Table 10-1, such as clay, have high porosity and low specific yield, meaning that little of the water in their pores can move. The ability of rock or sediment to transmit water through its pores is called its **permeability.** Materials with high permeability are those in which pore spaces are relatively large and interconnected. Highly permeable materials include well-sorted, coarse sand and gravel, as well as crystalline rocks, such as granite, with extensive fractures. Rocks and sediments with low permeability include crystalline rocks with few or no fractures and very poorly sorted sediments in which the finer particles block pore spaces.

A rock or sediment can have both high porosity and low permeability. An example is pumice, a glassy volcanic rock formed by the rapid cooling of gas-rich lava. Pumice contains many vesicles created by magmatic gases, but they are largely unconnected to one another. For this reason, pumice is not likely to contain much water in its pores, and even after a soaking rain, the water left in the pores cannot flow through the rock. Pumice has so few interconnected pore spaces that it will float when placed in water, like a natural Styrofoam. The fine-grained sediment clay can also have high porosity, but its pores are narrow and elongate, hindering the flow of water so effectively that its permeability is very low.

Movement of fluids through rocks and sediments depends not only on aquifer properties such as grain size and degree of sorting, but also on properties of the fluids themselves. To account for differences in movement between water and crude oil, for example, scientists define a factor known as the **hydraulic conductivity,** which depends on both the permeability of the porous material and the density and viscosity—stickiness—of the fluid moving through it. The more viscous the fluid, the lower the hydraulic conductivity and the slower the fluid is able to move through the material. Conversely, increased density speeds up fluid flow (Table 10-2).

Because permeability and the related property of hydraulic conductivity help determine the rate at which water—and pollutants—can flow through rocks and sediments, they are major concerns in the siting of facilities for hazardous and radioactive wastes. Perhaps the most serious danger in storing nuclear waste or any other hazardous substance underground is that the contaminant could leak into the surrounding host rock or sediment and, ultimately, into groundwater. Once a contaminant reaches groundwater, it may flow with the water toward springs, wells, wetlands, streams, and lakes. For this reason, hazardous waste-disposal facilities are sited where the water table is deep and the host rock has low hydraulic conductivity. Because igneous rocks typically have low values of porosity, permeability, and hydraulic conductivity, igneous formations often have been considered potential sites for hazardous waste disposal.

Aquifers

Rocks and sediments that are porous, permeable and contain water are called **aquifers,** from the Latin words meaning "water" and "carry." The most efficient aquifers include gravel, sand, sandstone, fractured limestone, and fractured basalt. Low-permeability rocks and sediments adjacent to aquifers are called *confining layers* or *aquicludes* because they restrict the flow of water between adjacent aquifers. The most common and effective confining layers are clay and shale, though unfractured crystalline rocks also make good aquicludes.

TABLE 10–2 Typical Values of Hydraulic Conductivity for Common Geologic Materials

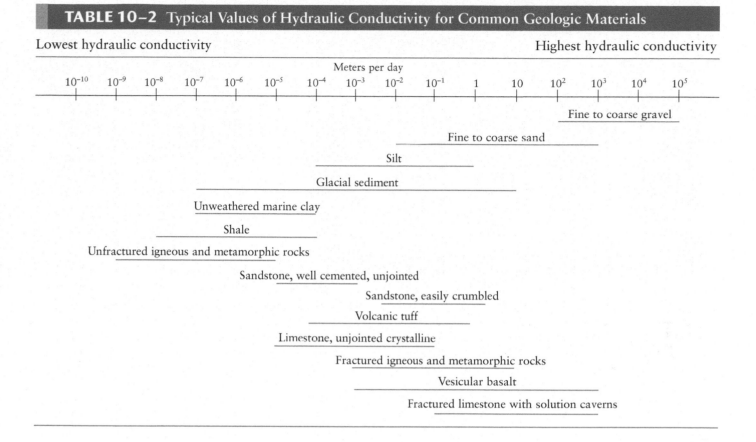

Lowest hydraulic conductivity   Highest hydraulic conductivity

Meters per day

10^{-10} 10^{-9} 10^{-8} 10^{-7} 10^{-6} 10^{-5} 10^{-4} 10^{-3} 10^{-2} 10^{-1} 1 10 10^2 10^3 10^4 10^5

Fine to coarse gravel

Fine to coarse sand

Silt

Glacial sediment

Unweathered marine clay

Shale

Unfractured igneous and metamorphic rocks

Sandstone, well cemented, unjointed

Sandstone, easily crumbled

Volcanic tuff

Limestone, unjointed crystalline

Fractured igneous and metamorphic rocks

Vesicular basalt

Fractured limestone with solution caverns

Aquifers are either confined or unconfined, depending on the subsurface geometry of sediments and rocks. **Confined aquifers** lie sandwiched between layers of relatively impermeable rock or sediment. Because of this impermeability, rainwater and snowmelt are impeded from percolating downward into the layers and thus don't reach the aquifer except in a narrow recharge zone where the aquifer intersects Earth's surface (Figure 10-6). This narrow recharge zone has the benefit of inhibiting the movement of contaminants into the aquifer. Thus, confined aquifers tend to have high water quality and are desirable water sources. In **unconfined aquifers,** porous and permeable sediments or rocks are open to the ground surface along their entire lengths, which means that these aquifers can be recharged everywhere, making them more vulnerable to contamination (see Figure 10-6).

In addition to their differing contamination potentials, confined and unconfined aquifers also differ somewhat in their flow behavior. The concept of the water table really applies only to unconfined aquifers. In these reservoirs, the shape of the water table mimics that of the overlying topography, rising underneath hills and falling toward stream valleys, and the elevation of the water table in any particular location changes over time depending on rainfall and evapotranspiration rates

(see Figures 10-3 and 10-4). The **hydraulic gradient,** or the slope of the surface along which water flows from an area of higher to lower potential energy, is determined by the differences in elevation of the water table from one place to another. It is this hydraulic gradient that drives water from higher elevation recharge areas toward lower elevation discharge areas, thereby supplying streams with base flow.

In a confined aquifer, in contrast, all pore spaces are perpetually filled with water, and one might think that the water table would mark the boundary between the top of the aquifer and the overlying confining layer. Such a designation would tell us little about flow within the aquifer, however, because even though that boundary might slope, if the aquifer is completely confined, there is no flow in response to gravity. For these aquifers, it makes more sense to speak of the **potentiometric surface,** a plane to which water would rise within wells drilled into the aquifer (see Figure 10-6). This surface always lies above the top of the confined aquifer because the weight of the overlying rocks and soils exerts downward pressure on the water it contains. If water can somehow escape from the confining layers—for instance, through a fracture—it will rise and perhaps even flow out at the ground surface as an **artesian spring.** Similarly, wells that tap water under high pressure in confined aquifers and thus need little or

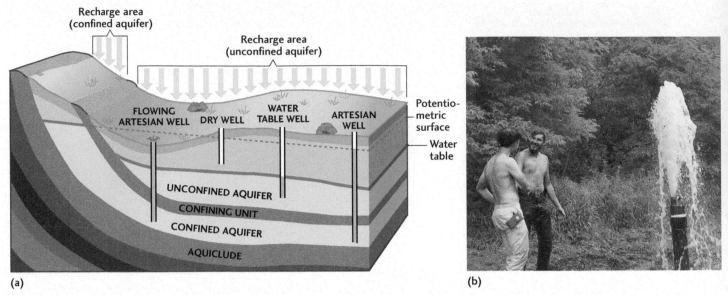

Figure 10-6 (a) Confined aquifers are capped by low permeability layers and have limited recharge areas, whereas unconfined aquifers are open to the atmosphere throughout most of their extent. Water in a confined aquifer is under so much pressure that it can rise to the *potentiometric surface*—a level that can occur well above the top of the aquifer. When the elevation of the top of a well is lower than the potentiometric surface, water flows out of the ground spontaneously, making a flowing artesian well. For wells drilled in unconfined aquifers, water rises to the elevation of the water table. (b) A geologist (left) and a driller are drenched but delighted after drilling into an artesian aquifer in glacial sands and gravels in Massachusetts. *(b: Paul Bierman, Geology Department, University of Vermont)*

no pumping are called **artesian wells** (see Figure 10-6), from the name of the Artois region in France, where many wells of this type were drilled in the 18th century.

The role of pressure in groundwater flow causes groundwater to behave quite differently from surface water—groundwater actually can flow upward, in seeming defiance of the force of gravity. This is true regardless of whether water resides in a confined or an unconfined aquifer. At the water table, the water pressure in pore spaces is equal to the overlying atmospheric pressure, just as it is at the surface of standing water in a lake or a drinking glass, and it cannot rise any higher. With increasing depth, however, the water pressure in pore spaces becomes greater than atmospheric because of the weight of the overlying water column.

In humid regions streams are recharged by groundwater. Deep under the stream channel, groundwater flows upward because water pressure at the stream is atmospheric, and therefore relatively lower than the pressure at depth below the stream (see Figure 10-3). This upward flow contributes to the base flow of streams in humid environments. Streams into which groundwater discharges are called *gaining streams*. In arid environments, the water table may be so far below the stream that the opposite happens: Surface water in streams infiltrates into the subsurface to recharge the groundwater system (Figure 10-7). We say that such a stream is a *losing stream*, because it loses water along its length.

Wells Wells are tubes sunk into aquifers for the purpose of extracting groundwater. The permeability of an aquifer and the hydraulic gradient of the groundwater flowing through it affect the flow rate of the aquifer's water and, therefore, the amount of water a well can yield. This relationship was first identified by Henry Darcy, a French engineer, who expressed it in a formula for groundwater velocity that came to be known as **Darcy's law** (Box 10-1). When a pump is inserted into a well and begins operating, its suction creates an area of low pressure in the well hole. As water is removed from the well, the elevation and pressure of the water in its vicinity are lowered in an unconfined aquifer, and the pressure and elevation of the potentiometric surface are lowered in a confined aquifer. Since water flows from areas of high to areas of low elevation and pressure, water flows from the aquifer toward the well. As more water is withdrawn from the well, replenishment comes from farther out in the aquifer. According to Darcy's law, the velocity of flow to the well is proportional to the aquifer's hydraulic conductivity and the hydraulic gradient. This means that more permeable and higher-pressure aquifers can sustain larger pumping rates and have higher-yielding wells than those with lower permeability and pressure.

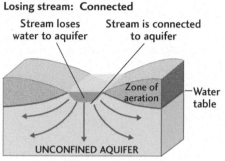

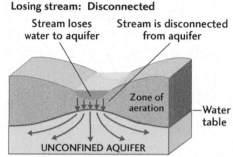

Figure 10–7 Gaining streams acquire water from an adjacent unconfined aquifer; losing streams lose water to the subsurface. The former are typical of humid environments where the water table is very near Earth's surface; the latter are common in arid environments. Depending on how dry the environment is, losing streams may either be connected or disconnected from the water table in their surrounding aquifers.

If the rate of pumped discharge is to remain constant in an aquifer, water must flow along steeper and steeper hydraulic gradients as it nears a well, because the area through which it flows becomes smaller and smaller. Eventually, a funnel-shaped area of depressed water level stabilizes around the well in an unconfined aquifer (Figure 10-8). If one could look down from above and see this depressed area, it would appear circular, with the well located in the center of the circle, as in a bull's-eye target. From the side, the depressed surface has the shape of a downward-pointing cone, with the tip of the cone at the well intake. This cone-shaped feature is known as the well's **cone of depression** (sometimes called the *cone of drawdown*), and it exists only while the water is being pumped out. After the pump is turned off, the cone refills and the original

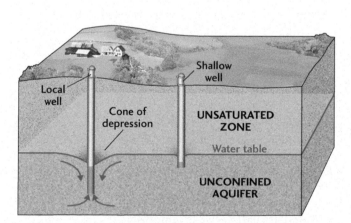

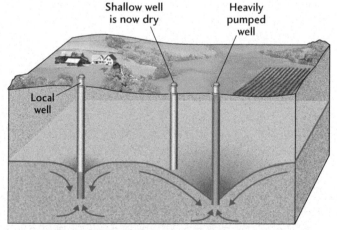

Figure 10–8 A three–dimensional view of the cone of depression of groundwater around a pumping well. Nearby shallow wells in the vicinity of a well that pumps large amounts of water can go dry if the water table drops below their intake level.

BOX 10–1 GEOLOGIST'S TOOLBOX

Darcy's Law and the Flow of Water and Contaminants in Rocks and Sediments

In the mid–1800s, engineer Henry Darcy was hired by the city of Dijon, France, to help its residents obtain a reliable and clean source of water. Using tubes of sand as water filters, Darcy tried, as he put it, "to determine the law of the flow of water through filters." Darcy's goal was related to a practical objective: to determine the most efficient way to pipe water to the city from a large spring nearly 10 km away and to have it flow by gravity to different supply points. Darcy was successful, and he is credited with giving Dijon a reliable and clean water supply for the first time in its long history.

In the course of his experiments on the flow of water between two points, Darcy discovered that the greater the change in *hydraulic head*—a quantity combining elevation and the water pressure at that elevation—the greater the volumetric flow of water per unit area from his tubes. The difference in hydraulic head (h) between two points divided by the distance (d) between them is the *hydraulic gradient* $[(h_1 - h_2)/d]$, and volumetric flow (Q) per cross-sectional area of the tube (A) is proportional to hydraulic gradient:

$$\frac{Q}{A} \propto \frac{h_1 - h_2}{d}$$

Since volumetric flow per cross–sectional area is the same as velocity (v), this proportionality may be written (**Figure 1**).

$$v \propto \frac{h_1 - h_2}{d}$$

Darcy also found that water flows more quickly through coarse sands than through fine sands, silts, or

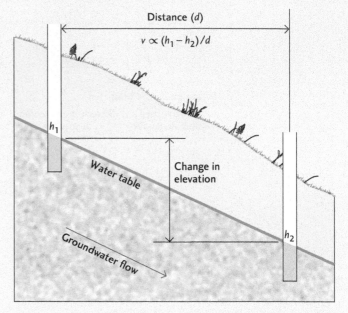

Figure 1 According to Darcy's law, the amount of groundwater flow per unit of time is proportional to the drop in elevation divided by the horizontal distance between two points in an aquifer, and proportional to the hydraulic conductivity of the material through which it is flowing. This law can be used to estimate the travel time of a contaminant from its source to points farther down gradient.

clays because the larger pore spaces make the material more permeable, enabling water to pass more readily. Water velocity is thus proportional not only to hydraulic gradient but also to hydraulic conductivity (K), a factor that depends on the permeability of sediments and rocks as well as the properties of the fluid flowing through them.

water table level gradually reestablishes. The size of the cone of depression depends both on the hydraulic conductivity of the aquifer and on the rate of pumping. The higher the pump rate, the larger the cone; the greater the hydraulic conductivity of the aquifer, the smaller the cone.

If a shallow well lies within the cone of depression of a deeper and larger well, it will go dry when the water level is drawn down below its base (see Figure 10-8). If there is more than one well, a cone of depression will develop around each and, at any one location,

the total drawdown is equal to the sum of all of the drawdowns from every pumping well in the area. The more wells in an area and the more water each well pumps, the greater the chance that wells will go dry unless they are deepened. Large cones of depression up to tens of kilometers in diameter are common in heavily irrigated areas that use groundwater.

Another result of a cone of depression is that groundwater flow directions may reverse locally while pumping is happening. This reversal could have adverse impacts on water quality, if, as shown in Figure 10-9,

$$v \propto K$$

$$\text{where} \quad K = \frac{k\rho_w g}{\mu} \quad \text{and}$$

k = permeability
ρ_w = fluid density
g = gravitational acceleration, and
μ = fluid viscosity

Putting the variables of hydraulic conductivity and hydraulic gradient together gives the relationship that is known as Darcy's law:

$$v = K \times \frac{h_1 - h_2}{d}$$

Darcy's law has become the foundation of the science of *hydrogeology*—the study of groundwater flow, contamination, and well hydraulics.

Darcy calculated water flow velocity assuming that water moved through the entire cross-sectional area of his experimental tubes, when in fact most of that cross-sectional area contained sediment. The modern version of Darcy's law goes a step further by taking into account the tortuous path that water molecules take as they move from pore to pore between grains. This *pore velocity* is determined by dividing Darcy's original equation by the porosity (n) of the rock or sediment in question:

$$v = K \times \frac{h_1 - h_2}{d \times n}$$

This equation can be used to estimate how long it would take water to travel from one site to another. If its physical characteristics are known, the travel time of a pollutant can also be estimated by Darcy's law. For example, assume that a chemical leaks into an aquifer from an underground storage tank and that the chemical moves at the same velocity as groundwater. If the leak is into a coarse sand aquifer with a hydraulic conductivity of 60 m per day, a porosity of 30 percent (0.3), and a hydraulic gradient of 1 m per 1000 m, the chemical will travel into the aquifer at the following velocity:

$$v = K \times \frac{h_1 - h_2}{d \times n}$$

$$= (60 \text{ m/day}) \times \frac{1 \text{ m}}{(1000 \text{ m} \times 0.3)}$$

$$= 0.2 \text{ m per day, or 73 m per year}$$

The direction of travel is from an area of high to one of low potential energy, or high to low hydraulic head. If the leak is into a fine-grained clay with a hydraulic conductivity of 0.0001 (or 1×10^{-4}) m per day, a porosity of 20 percent (0.2), and a hydraulic gradient of 1 m per 10 m, the chemical will travel much more slowly than it would in a coarse sand aquifer:

$$v = K \times \frac{h_1 - h_2}{d \times n}$$

$$= (10^{-4} \text{ m/day}) \times \frac{1 \text{ m}}{(10 \text{ m} \times 0.2)}$$

$$= 5 \times 10^{-5} \text{ m per day, or 0.02 m per year}$$

Contaminants can move much more rapidly in materials with high hydraulic conductivity than in those with low hydraulic conductivity; thus, high hydraulic conductivity in an aquifer can pose a considerable threat to drinking water supplies. The faster a contaminant moves through a groundwater system, the larger the amount of the contaminant that will appear in the drinking water supply per unit of time, and the more extensive the area likely to be affected.

pumping draws household sewage from a septic leach field into a well. To minimize the possibility of this happening, many developers hire geological consultants to conduct pumping tests to determine the size of the cone of depression for a typical household pumping rate; they then space wells and septic systems appropriately.

Springs and Geysers In contrast to wells, which are artificial groundwater retrieval systems, *springs* are natural mechanisms through which groundwater reaches the surface. The larger the groundwater recharge area for a spring and the greater the rate of water recharge to the aquifer, the greater the water pressure in the aquifer and the greater the discharge from the spring.

Some springs spout hot water and steam above the ground. This phenomenon is caused by superheated steam that builds up pressure in deep subsurface conduits. Such springs are known as *geysers*, from the Icelandic word *geysir*, which means to "gush" or "rage." Geysers are common in areas where the heat flow from the crust is high—for instance, above magma chambers filled with molten rock. Geysers form when groundwater

SMALL, CLEAN WELL

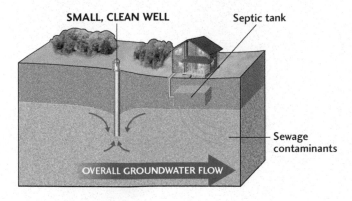

Septic tank

Sewage contaminants

OVERALL GROUNDWATER FLOW

HEAVILY PUMPED WELL

Small well is now contaminated by sewage

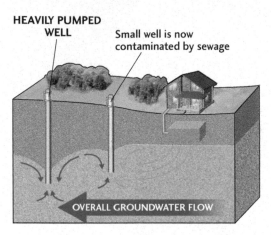

OVERALL GROUNDWATER FLOW

Figure 10–9 The cone of depression around a pumping well can reverse local groundwater flow directions. This can present a problem if the well is situated too close to a septic leach field, as sewage may be drawn into the well.

flows downward to depths of 3000 m or more, where it is heated to extremely high temperatures (Figure 10-10). Because of the great rock and fluid pressures at such depths, the water can be raised to temperatures much higher than its surface boiling point (100°C) before it boils and begins turning into steam. Once steam is formed, it expands and pushes the remaining liquid water upward, where pressures are lower. At these reduced pressures, water boils at lower temperatures, so even more steam is produced and more water is forced upward to erupt at the surface. The process continues until the pressure is dissipated and the groundwater cavity begins to refill with water and steam. Such eruptions happen on a periodic schedule at Old Faithful, a well-known geyser at Yellowstone National Park in Wyoming.

Natural Groundwater Chemistry

Rain and snow, the sources of groundwater, contain very little dissolved matter. Human activities such as the burning of fossil fuels or wood, and geological events such as the release of volcanic gases, emit compounds into the atmosphere that are later dissolved in falling

rainwater, but generally, rainwater contains less than 20 mg of dissolved substances per liter of solution.[5] By the time precipitation has percolated through soil, sediments, and rocks to replenish groundwater, it has accumulated many times those amounts of dissolved substances. Just as table salt dissolves in water, the minerals in rocks and sediments are soluble in water. Each mineral has a different solubility in pure water, and solubility varies with the acidity of the water and the amount of oxygen dissolved in it. Groundwater readily dissolves the most soluble minerals in rocks. With increasing depth, the amounts and types of materials dissolved in groundwater vary, with deeper waters

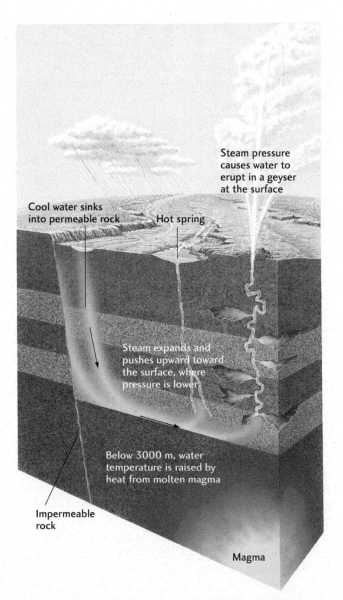

Steam pressure causes water to erupt in a geyser at the surface

Cool water sinks into permeable rock

Hot spring

Steam expands and pushes upward toward the surface, where pressure is lower

Below 3000 m, water temperature is raised by heat from molten magma

Impermeable rock

Magma

Figure 10–10 When groundwater is heated to high temperatures, steam and water erupt periodically with great force, forming geysers. Hot water may also seep from the system and collect at the surface as a hot spring.

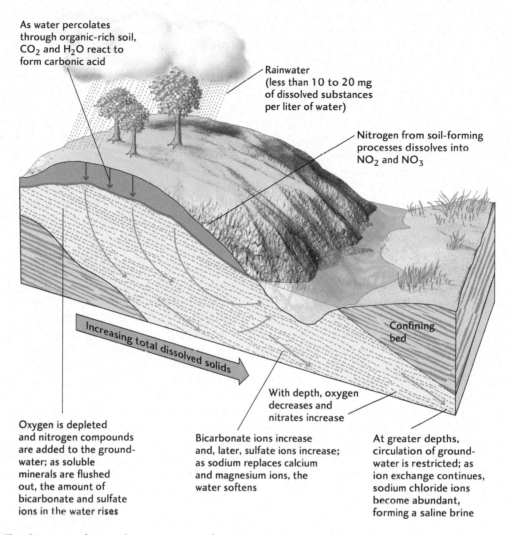

As water percolates through organic-rich soil, CO_2 and H_2O react to form carbonic acid

Rainwater (less than 10 to 20 mg of dissolved substances per liter of water)

Nitrogen from soil-forming processes dissolves into NO_2 and NO_3

Increasing total dissolved solids

Confining bed

With depth, oxygen decreases and nitrates increase

Oxygen is depleted and nitrogen compounds are added to the groundwater; as soluble minerals are flushed out, the amount of bicarbonate and sulfate ions in the water rises

Bicarbonate ions increase and, later, sulfate ions increase; as sodium replaces calcium and magnesium ions, the water softens

At greater depths, circulation of groundwater is restricted; as ion exchange continues, sodium chloride ions become abundant, forming a saline brine

Figure 10–11 The chemistry of groundwater in an aquifer and its evolution with time and depth. Groundwater acquires various ions through reactions with rock, soil, and organic matter particles.

generally having higher concentrations of salts and less oxygen (Figure 10-11).

The composition of rainwater begins to change soon after it infiltrates the soil and becomes underground water. Biological processes remove much oxygen from groundwater; for example, microbes use oxygen to decompose plant and animal matter. Soil-forming processes result in the production of the nitrogen compounds nitrite (NO_2) and nitrate (NO_3). Nitrogen enhances plant growth and is abundant in manure and artificial fertilizers, both of which farmers and gardeners add to soils in large quantities. The use of these fertilizers has caused the level of nitrates in groundwater in many farm areas to increase rapidly in recent decades. While nitrogen is an essential plant nutrient, high levels of nitrate in groundwater are toxic to infants, resulting in a serious illness commonly known as *blue baby syndrome*, so these rising levels are

a cause for concern. The U.S. Environmental Protection Agency's (EPA) standard for drinking water requires that water have no more than 10 mg of nitrate per liter.[6] Ion exchange or reverse osmosis filters are used to reduce the amount of nitrate to acceptable levels, but a better strategy is to prevent nutrient contamination in the first place by, for example, using fertilizers more wisely. Soil testing can show how much fertilizer is necessary, and loss of nutrients to groundwater and surface water can be avoided by precisely timing fertilizer applications based on plant growth schedules.

As water moves through organic-rich soil, much carbon dioxide gas (CO_2) in unsaturated pore spaces dissolves into the water. The reaction of CO_2 and H_2O forms carbonic acid (H_2CO_3), a weak acid, which in turn reacts with carbonate and silicate rocks. Ions that are common in groundwater from dissolution of minerals are calcium, magnesium, potassium, iron, and sulfur.

Together, calcium and magnesium ions contribute to the quality known as **water hardness.** Hard water contains 150 mg/L or more of dissolved calcium and magnesium ions per liter of water.[5] These ions can crystallize to form mineral deposits that leave rings in bathtubs and a scaly deposit in water pipes. The ions in hard water also change its flavor and react with soap so that suds cannot readily form. Soft water, with low amounts of these ions, has a much better taste and forms a better lather when soap is added. Many people living in areas underlain by limestone, a calcium carbonate rock, complain about their hard water and often treat it with *water softeners,* chemicals that react with calcium and magnesium to reduce the amounts of these free, dissolved cations.

Iron and sulfur are common constituents of groundwater that sometimes also cause trouble for households dependent on wells. In the low oxygen environment typical of aquifers, iron is highly soluble so groundwater sometimes has a relatively high concentration of dissolved iron. At Earth's surface, however, iron reacts with oxygen to form insoluble iron oxide and hydroxide compounds. These can produce orange-colored sediments that stain porcelain and impart an unpleasant flavor to water. Dissolution of minerals such as pyrite and gypsum introduce sulfur into groundwater. In the low oxygen environment of aquifers, reduced sulfur combines with hydrogen to make hydrogen sulfide, a gas with a distinctive rotten egg scent. Various filtration systems have been devised to deal with these issues if homeowners find them intolerable.

While many dissolved substances are mere nuisances in that they affect smell or taste of water, some, like nitrate, are dangerous in high quantities. Another example of such a substance is arsenic, which is a common, naturally occurring constituent of groundwater derived from rock weathering. Chronic exposure to arsenic is known to cause skin, bladder, lung, and kidney cancer, as well as vascular disease. Bangladesh is at particularly high risk of arsenic-related problems because of the arsenic-rich rocks in the subsurface and the increased reliance on wells to combat diarrheal diseases spread through contaminated surface water resources. The World Health Organization reports that 200,000 to 270,000 people will die there from cancers contracted from arsenic poisoning.[7] The U.S. EPA has placed a limit of 10 parts per billion (ppb) on arsenic in domestic water supplies. Dilution and/or filtration can be used to make water with higher concentrations potable. Between a half and two-thirds of Bangladesh's population of 140 million people drink water that exceeds 50 ppb arsenic,[8] so a low-cost filtration method is desperately needed. Fortunately, arsenic readily sticks to iron particles, so simple filters can be made out of buckets containing iron filings sandwiched between layers of sand. More than 90,000 of these simple filters have been distributed to villages throughout Bangladesh

and neighboring Nepal, but many more will be needed to solve Bangladesh's arsenic crisis.

The process of dissolution continues as water moves deeper into the subsurface. The total amount of dissolved solids rises, and there is more exchange of ions between water and mineral surfaces (see Figure 10-11). Clay minerals, common in many rocks and sediments, generally are associated with large amounts of sodium during deposition in offshore marine environments, so cation exchange between calcium and magnesium from the water and sodium on exposed surfaces of clay minerals results in water rich in sodium bicarbonate.

The age of groundwater also increases with depth: Radiometric dating of carbon in groundwater has shown that its age can be at least 40,000 years[9] in the deeper parts of some aquifers.[10] The general pattern of groundwater evolution with depth and age is a transformation from bicarbonate-rich water to sulfate (SO_4^{2-})-rich water, then finally to brine, a water solution rich in sodium chloride. Because circulation of water at great depths usually is much more restricted than at shallow levels, sodium chloride—which is highly soluble—is not flushed from groundwater. Although some early hydrogeologists thought that the brines pumped from deep wells were ancient marine waters trapped in sediments as they were laid down, it is now clear that groundwater can evolve to become more saline with depth and age even if it started out fresh.

About half of all groundwater is deep and saline and thus not nearly as likely to be used as the freshwater found at shallower depths. Still, in water-scarce regions such as the American Southwest, all water is valuable, even when salty. For many years, the city of Santa Fe, New Mexico, considered pumping salty groundwater from the adjacent Estancia Basin, which would have required constructing a desalination plant that would be used to reduce the salt in the water. Political issues and the cost of a pipeline and water treatment facility eventually killed this proposal, and Santa Fe residents began to implement water conservation practices that obviated, at least for now, the need for Estancia Basin water, reducing their per capita water use by an astonishing 40 percent since 1995 (Box 10-2).

■ Groundwater recharge occurs at Earth's surface from precipitation that drains downward through the unsaturated zone to the saturated zone. The water table rises as the amount of saturation increases.

■ Groundwater is discharged to Earth's surface at springs, wetlands, streams, lakes, and ultimately into the oceans.

■ Elevation and water pressure are the main determinants of the direction in which groundwater flows. In general, groundwater flows from areas of

BOX 10–2 EARTH SCIENCE AND PUBLIC POLICY

Saving Water in the Desert

In the United States, the city of Santa Fe, New Mexico, is a leader in water conservation. The average withdrawal of municipal water in the U.S. is 220 m³ per person per year, or about 600 liters per day.[11] In Santa Fe, however, average withdrawal is about 380 liters per day.[12] Located in the high desert at the base of the Sangre de Cristo Mountains, Santa Fe is situated in a region in which evaporation routinely exceeds rainfall, making the city highly vulnerable to water shortages. Santa Fe gets its water from a mixture of groundwater pumping (30 percent), surface water stored in reservoirs in the Sangre de Cristos and released to the Santa Fe River (40 percent), and diversions of water from outside the watershed (30 percent).[13]

Historically, poor management of water resources led to the drying up of the Santa Fe River within the city. Excessive groundwater pumping and river withdrawals for domestic consumption left little water behind to support stream flow. With the exception of spring snowmelt and thunderstorm periods, this part of the river no longer contains sufficient water for fish, despite historic records showing that it once was a trout stream. Measurements of depth to groundwater taken from wells show that the decline in water in the river has been matched by a decline in the water table of 100 to 200 feet over the past half century as pumping has exceeded recharge.

To ensure adequate supplies of water for the future of the city and to restore some flow to the Santa Fe River, the city of Santa Fe has implemented aggressive water conservation programs,[14] including a rebate program in which citizens can receive $100 to exchange water-guzzling clothes washers for water-efficient models. The city also gives away low-flow toilets and provides rebates to encourage its citizens to install rain barrels, which capture rainfall from rooftops and store it so that it can be used for outdoor irrigation. Santa Fe also instituted a number of ordinances designed to curtail water usage. These include prohibiting outdoor irrigation between 10 A.M. and 6 P.M., when evaporation is at its peak; restricting car washing at private residences to no more than once a month; banning the use of water to clean sidewalks and driveways; requiring that swimming pools be covered when not in use; and requiring that no more than 25 percent of lawn turf be Kentucky bluegrass, a variety that needs heavy watering. Restaurants have been directed to provide water and hotels told to change sheets and towels daily only upon request. Violations of these restrictions carry fines of $50–$200. Since these programs were implemented in 1995, per capita water usage has fallen from 635 liters per day to 380 liters per day, and though the city's population has grown 15 percent since that time, total water use has dropped by nearly 25 percent.

Despite Santa Fe's substantial progress toward curtailing its water use, future climate change and population growth mean that the city is likely to remain vulnerable to water shortages into the indefinite future. Groundwater mining continues. Furthermore, the Santa Fe River still does not have sustainable flows and is dry about 9 months of the year. As a result, the city must remain vigilant about water use. The city already has plans in the works to squeeze more value out of every drop consumed by using treated wastewater for irrigation, a strategy that has proven effective in other arid regions.

---

high elevation to those of low elevation and from areas of high pressure to those of low pressure.

- Darcy's law states that the rate of flow of groundwater is proportional to the change in elevation and pressure between two points and the hydraulic conductivity of the geologic material through which the water flows.

- Groundwater dissolves minerals during its travels through rocks and sediments. Water that contains relatively large amounts of calcium and magnesium is known as hard water.

GROUNDWATER AS A RESOURCE

Groundwater is not as accessible as surface water, so it became a widely used resource only as new technologies to exploit it were developed. Whereas earlier societies were dependent on hand-dug wells, drilling rigs with diesel engines now can bore hundreds of meters into Earth's crust to draw out groundwater. The stock of groundwater, though not very accessible, is much larger than the stock of surface water—at any given time nearly 120 times as much water is stored underground as at the surface[15] in the conterminous United States alone. Despite its seeming abundance, our use of groundwater has skyrocketed to a point where withdrawals have become unsustainable in many areas. In addition, toxic chemical spills at Earth's surface prior to the advent of environmental regulations have left a legacy of groundwater contamination. Our stewardship of water resources must improve significantly to ensure that we have adequate and safe water into the future.

Global Distribution and Use

The United Nations Educational, Scientific and Cultural Organization (UNESCO), in partnership with a number of other nonprofit international organizations, recently launched the World-wide Hydrogeological Mapping and Assessment Programme (WHYMAP). The goal of WHYMAP is to determine the extent and quality of groundwater resources across the globe in order to facilitate better planning and more coordination among nations in managing groundwater resources. As Figure 10-12 shows, groundwater basins frequently cross national boundaries (there are at least 273 transnational aquifers), requiring that governments cooperate with one another to ensure that water resources are equitably and sustainably used. These maps also show that groundwater resources are not evenly distributed: Some nations sit on top of major groundwater basins that easily can meet their water resource needs, while others are atop small aquifers of only local importance.

Like distribution, water quality varies widely: Many coastal and some inland areas have only saline groundwater supplies. Given the spatial variability in water quality and quantity, groundwater use has the potential to cause conflict or to foster cooperation among nations.

For arid countries such as Saudi Arabia, groundwater may provide nearly 100 percent of water resources for irrigation, industry, and domestic use.[11] For other, wetter locations like Brazil, groundwater may be used secondarily to surface water. Likewise, the uses of the water vary, depending on the regional climate. In Germany, for example, an estimated 82 percent of the water extracted from the ground and from surface resources is used in industrial production; in Chad, that same percentage is used in agriculture. Chad, located in the Sahara, needs more irrigation water than does humid Germany, and the variance in the two countries' water usage also reflects differences in the economic development status of each nation. Information about the climate conditions and groundwater recharge potential of different regions can be used to promote sustainable water use and to determine how to allocate water resources.

Groundwater Resource Management

In contrast to oil and gas, which are *nonrenewable* at timescales of human generations, groundwater has the potential to be a *renewable* resource, one for which the replenishment rate equals or exceeds the rate of

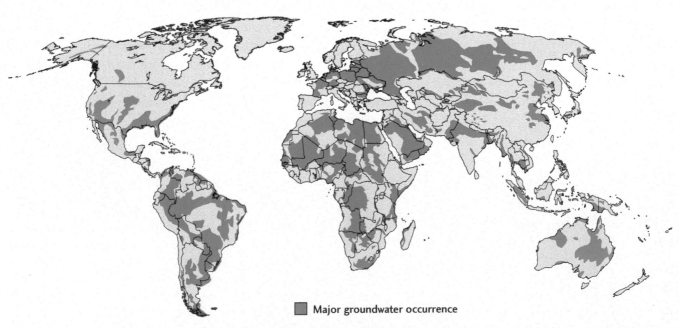

■ Major groundwater occurrence

Figure 10–12 Groundwater resources of the world are not distributed evenly and do not follow national boundaries. Managing the resources requires intergovernmental cooperation.

(b)

(a)

Figure 10–13 (a) Subsidence of the ground in this San Joaquin Valley neighborhood has exposed a groundwater well. The top of the well used to be flush with the ground surface, but groundwater withdrawals have caused the land to sink. (b) Land subsidence also can lead to fissures at the ground surface, as here in Arizona. *(a: Richard Ireland, photographer/ USGS, courtesy Devlin Galloway; b: Brian Conway/Courtesy Arizona Department of Water Resources)*

consumption. In some areas, however, groundwater is withdrawn at rates that exceed recharge, a phenomenon known as *groundwater mining*. Even if there were to be no further pumping in these locations, aquifers might not recover their previous water storage potential. The reason is that as water is withdrawn, sedimentary particles in aquifers tend to shift and compact, leading to a decline in porosity and permanent loss of aquifer storage capacity. Compaction also leads to a phenomenon known as **land subsidence** (Figure 10-13). The effect is similar to that of air loss in a bicycle tire. When air leaks out, the tire deflates, or decompresses, because the pressure of air on the inner surface of the tire is reduced. Similarly, when water is removed from an aquifer, pressure on the inner walls of the pores decreases and the porous structure deflates. Subsidence is an issue—sometimes on a very big scale—throughout

the world where groundwater is mined: One instance in California has resulted in nearly 9 m of sinking.[5]

Another example of groundwater mining is in aquifers that no longer receive any recharge, or in those that receive only very small amounts under present climatic conditions. For example, the nation of Libya discovered vast quantities of groundwater in sedimentary rocks buried beneath the sands of the Sahara while prospecting for oil in the 1950s. The Nubian Aquifer extends for 2 million km^2 beneath parts of Libya, Egypt, Chad, and Sudan, and is estimated to contain more than 375,000 km^3 of water.[16] Radiometric dating using krypton-81 in the groundwater indicates that it is up to a million years old.[17] Other isotopes in the water show that the water originated during cyclical glacial episodes when Earth's climate was cooler and North Africa received abundant rainfall derived from the Atlantic Ocean.[18]

Though the Nubian groundwater reservoir is a nonrenewable resource, Libya made the decision to tap its "fossil water," drilling more than 1000 wells and constructing 5000 km of pipelines to deliver 6.5 million m³ per day of water from the southern part of the country to the population centers along its northern coast (Figure 10-14). The project, called the Great Man-Made River, began in 1984 and is projected to cost $20 billion by the time it is completed. Already more than 100,000 hectares of new farmland have been brought into production through irrigation. If its estimated size is correct, the Nubian Aquifer should be able to supply Libya with water for hundreds of years at present rates of consumption.

Other fossil waters may be depleted far more rapidly. The High Plains Aquifer (also called the Ogallala Aquifer) of the central United States is made up of sediments that were shed off of the Rocky Mountains and extends 450,000 km² beneath the states of South Dakota, Nebraska, Kansas, Oklahoma, Texas, Wyoming, Colorado, and New Mexico (Figure 10-15).[19] Dating back from the late Pleistocene through the middle Holocene epochs, the fossil groundwater the aquifer contains is being pumped at a rate of 66,200 million liters per day.[20] Ninety-seven percent of the withdrawals are used to irrigate wheat, sorghum, corn, and other crops. In fact, the High Plains Aquifer provides 30 percent of the irrigation water of the United States. Along with these withdrawals have come steep declines of the water table, with some areas falling more than 200 feet (see Figure 10-15). The resulting compaction has led to an average decline of 7 percent in the total aquifer storage capacity, with local declines of as much as 18 percent. At current rates of pumping, parts of the

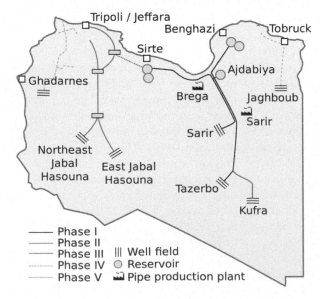

Figure 10–14 Libya's Great Man–Made River Project transports water drawn from the Nubian Aquifer beneath the Sahara to coastal communities via pipelines. *(a: Thomas Hartwell/Time & Life Pictures/Getty Images; b: Dan Michael O. Heggo)*

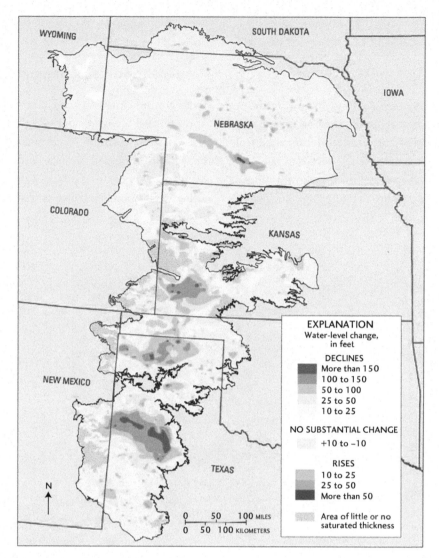

Figure 10-15 The U.S. High Plains Aquifer supplies most of the water used for irrigation in the region that has been dubbed "the bread basket of the world" for its production of wheat and other cereals. Water is being withdrawn faster than it is recharging in much of this region, leading to large declines in the elevation of the water table. *(McGuire, V. L. 2007. "Water-level changes in the High Plains aquifer, Predevelopment to 2005 and 2003 to 2005." U.S. Geological Survey Scientific Investigations Report 2006–5324, 7 p. Courtesy USGS.)*

aquifer could go dry within a few decades, leaving the breadbasket of America dried out.

Artificial Recharge If natural recharge rates are low, groundwater management may include artificially adding water to the groundwater system. Recharging an aquifer can reduce the impact of pumping. For example, in areas of natural recharge of an unconfined aquifer, some precipitation may not infiltrate into the unsaturated zone and later to the water table if the surface becomes saturated during a storm or a period of snowmelt. In such cases, the water becomes surface runoff, makes its way downslope to streams, and leaves the basin. So that it is not entirely lost, runoff and streamflow can be collected by diversion and then spread over a large land area to infiltrate in places where the water table is not close to the surface. This practice, known as **water spreading,** is common in many arid and semiarid regions where there is a substantial zone of permeable, unsaturated sediments above the water table, and it allows surface water to be captured for future use as groundwater (Box 10-3).

Other means of resupplying aquifers are through using recharge basins and injection wells. **Recharge basins** are less extensive in area than are water-spreading grounds, but can hold a greater depth of water; consequently, a large hydraulic gradient is created that, in keeping with Darcy's law, helps increase the rate of recharge.

BOX 10–3 CASE STUDY

Water Resource Management in Los Angeles

The metropolitan area of Los Angeles, in Southern California, is built on a broad, gently sloping coastal plain made of marine sediments and eroded debris shed from mountains that surround the plain. These deposits have accumulated in the Los Angeles alluvial basin for millions of years, and the upper strata contain large supplies of

fresh groundwater. As the climate of the region is semi-arid—rainfall measures only 23 cm per year while evapotranspiration is about 76 cm per year—Los Angeles relies heavily on imported surface water and on groundwater from its own basin. Surface water is imported through aqueducts (tunnels or culverts that carry water) from other

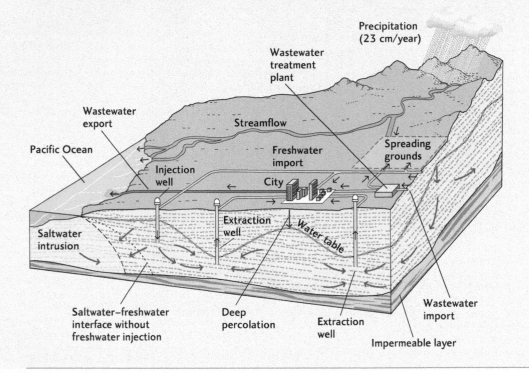

Figure 1 Spreading basins are used to spread water above ground in order to recharge aquifers under Los Angeles.

The water added to recharge basins includes stormwater runoff from paved urban areas, reclaimed wastewater, and water diverted from streamflow. **Injection wells** are wells through which water is pumped into the ground to maintain the pressure of a reservoir. They are used to recharge water in deep, confined aquifers. Because injection wells have been used for hazardous waste disposal, at least one state—Wisconsin—prohibits their use for any purpose. Hazardous waste injection wells are legal in other states and are regulated by the EPA to minimize the risk of contaminating drinking water supplies.

Siting and Spacing Wells A comprehensive groundwater management plan includes siting and spacing wells so that there is minimal lowering of the water table and

little impact on surface water resources. Wells should be located near areas of natural recharge, because well discharge and recharge are more likely to balance in these areas—even under adverse environmental conditions or intense use—than they are in locations that are far from recharge areas. If wells are sited near areas of surface water runoff, such as streams, wetlands, and springs, then surface water may be drawn back into the aquifer, a process known as *induced recharge*. This presents a problem for surface water ecosystems, which are sensitive to fluctuations in water amounts, and downstream water users may no longer get sufficient water from the surface water source.[21] Many environmental lawsuits have been brought after supplies of surface water were depleted by the excessive pumping of groundwater.

Figure 2 Aqueducts like this one, crossing the desert of Owens Valley, California, transport water to the Los Angeles metropolitan area. *(Ron Chapple Studios/Thinkstock)*

watersheds, including the Owens Valley 200 km north of Los Angeles and the Colorado River 200 km east of the city.

Groundwater supplies about half the area's water needs. Intensive pumping, however, resulted in falling water tables that permitted saltwater along the coast to flow into the aquifers under the city. In recent decades, the region has had a number of severe droughts that have alternated with extremely wet periods, which often cause floods. In response to the drought years, the LA Metropolitan Water District developed and maintains a coordinated management plan that makes the most efficient use of what little surface water is available, uses reclaimed wastewater, and minimizes negative impacts on groundwater aquifers. The elaborate network of devices to obtain, treat, and transmit water throughout the basin includes spreading basins that are used to recharge aquifers with surface water, and injection wells along the coast that maintain a wall of freshwater to prevent the intrusion of saltwater into the groundwater system (**Figures 1** and **2**).

Clearly, prudent management of groundwater and surface water resources requires knowing the maximum stable yield of water that can be obtained from a basin. The *safe yield* is the maximum possible pumping rate that will neither exceed average annual rates of recharge nor lower the water table to such an extent that pumping becomes too costly, the aquifer quality is degraded, or the ecosystem dependent on groundwater discharges is affected. While the concept of safe yield is straightforward, determining its value for an individual aquifer can be challenging and is based on detailed knowledge of groundwater recharge and discharge rates in the absence of pumping, aquifer properties such as hydraulic conductivity, potential climatic changes such as droughts, levels of human use, and ecosystem requirements.

Wellhead Protection Ensuring safe and abundant groundwater resources into the future requires maximizing groundwater recharge. Water managers define the *wellhead protection area* for each well, which is all of the watershed area that recharges groundwater to a particular well. Communities dependent on groundwater supplies benefit from mapping these areas and restricting activities within them. For example, wellhead protection areas are often off-limits to development because constructing buildings, roads, and parking lots creates impervious surfaces that block groundwater recharge. Likewise, development brings with it both point and nonpoint source pollution, jeopardizing groundwater quality, so requiring that these areas remain as undeveloped forest or grassland can protect water supplies.

- Groundwater resources are spatially variable and often cross governmental boundaries, leading to potential conflicts as well as cooperation.

- Overextracting groundwater can cause permanent loss of storage capacity as well as subsidence from compaction.

- Groundwater is considered a renewable resource if rates of recharge are the same as or greater than rates of discharge.

- In areas where rates of discharge exceed natural rates of recharge, water management techniques can be used to restore a balance. These include using water spreading grounds, recharge basins, and injection wells.

- The goal of creating wellhead protection areas is to maximize groundwater recharge and minimize groundwater pollution. In such areas, planners and developers avoid the use of impervious surfaces that would reduce recharge. Use of pesticides and disposal of pollutants are restricted.

GROUNDWATER HAZARDS

In addition to providing a valuable resource, the groundwater system can present hazards, both natural and induced. One such hazard is sinkholes; others include land subsidence and the intrusion of saltwater into aquifers because of overpumping.

Solution Caverns and Sinkholes

An aquifer undergoes physical as well as chemical changes as water migrates through it. In rocks composed of very soluble minerals, such as limestone, the carbonic acid in groundwater enlarges pore spaces through dissolution. Since much of the pore space in limestones is secondary porosity related to fractures, mazes of interconnected caverns and tunnels form as the fractures widen by dissolution of surrounding rock with time. The terrain in areas of fractured limestone is highly irregular, characterized by tall spires and pinnacles separated by deep channels. Such topography is called *karst,* after the Karst region in southwestern Slovenia and northeastern Italy, which is dominated by such features. In karst areas, water and its contaminants can move swiftly because of the high hydraulic conductivity of soluble rocks (see Table 10-2). Furthermore, the direction of groundwater flow is hard to predict in cavernous aquifers.

As dissolution continues, the roofs of caverns can become quite thin, eventually collapsing as the weight of the overlying soil becomes too much for the rock ceiling (Box 10-4). Such collapsed areas are called **sinkholes;** they often are aligned in gridlike patterns that reflect the underlying perpendicular fracture zones in the rocks. In some places, the formation of sinkholes coincides with lowering of the water table. In South Africa, groundwater was pumped from an excavation during mining and within 2 to 3 years, eight large sinkholes, more than 50 m in diameter and 30 m deep, had formed. Evidently the water pressing against the cavern ceilings had kept the cavern roofs from collapsing.

Land Subsidence from Groundwater Mining

We already mentioned that groundwater withdrawals in excess of recharge rates cause compaction of aquifer sediments and permanent loss of storage capacity, along with ground subsidence (see Figure 10-13). When subsidence happens in a farm field, it may not cause more than localized problems. Underneath cities, however, subsidence can wreak havoc.

Mexico City is all too familiar with this problem. The city is located in a topographic bowl surrounded on all sides by mountains. The bottom of the bowl contained a shallow lake at the time of the Aztec empire. When Spanish conquistadors arrived in the early 1600s, the era of European settlement began, and the settlers began trying to drain the lake and its surrounding wetlands so they could use the land for farming. Drainage was a perennial problem, and the city continued to flood frequently. For this reason, in the early 1900s, a great tunnel called the Grand Canal was built through the mountains to better drain the city (Figure 10-16).[26]

As the population grew, surface water resources from the surrounding mountains were exhausted, so Mexico City inhabitants began to dig wells to extract groundwater. They withdrew more and more groundwater and built more and more buildings. The soft lake sediments beneath the city began to collapse, causing it to subside by more than 7 m. Not all areas subsided the same amount, which caused water and sewer pipes to break and building foundations to crack. Today, the National Cathedral lists to one side because of differential compaction, and engineers have had to install scaffolding to prevent stones from falling from the ceiling. Likewise, the Palace of Fine Arts has sunk 3 m below street level (Figure 10-17). So many buildings have been damaged that the city has banned groundwater pumping from the historic core. Still, subsidence of up to an inch a year continues in the city center, and is at much higher rates on the periphery, where water continues to be withdrawn.

As the city grew and subsidence progressed, the ability of the Grand Canal to drain the city diminished (see Figure 10-16). With a slope of 19 cm/km in the early 1900s, the canal ably conveyed water out of the city, but the slope had declined so much by the 1950s that water began to be trapped in the city. In response, Mexico City built a new, deep drainage system, the

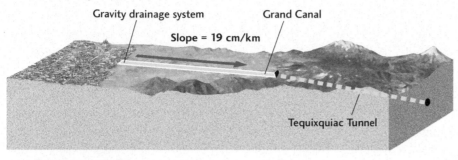

Mexico City: 1900

Gravity drainage system

Grand Canal

Slope = 19 cm/km

Tequixquiac Tunnel

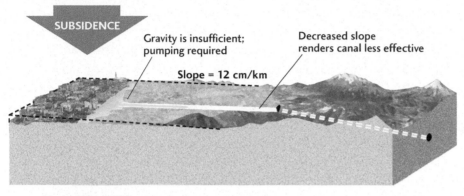

Mexico City: 1950

SUBSIDENCE

Gravity is insufficient; pumping required

Decreased slope renders canal less effective

Slope = 12 cm/km

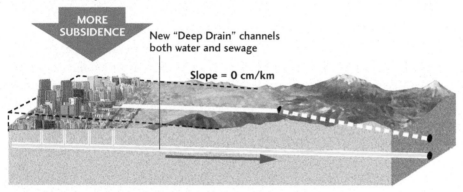

Mexico City: 1990

MORE SUBSIDENCE

New "Deep Drain" channels both water and sewage

Slope = 0 cm/km

Figure 10–16 Subsidence of Mexico City due to groundwater withdrawals rendered the city unable to rid itself of storm water and sewage. The Deep Drain that has been serving the city for the last few decades is in poor repair and is currently being replaced with a new wastewater tunnel.

Deep Drain, to export water. In time, the Deep Drain was used not only to remove excess rain and surface runoff from the city, but also to carry away sewage. Unfortunately, the drain was not constructed to handle the corrosive compounds associated with raw sewage, and the concrete began to crumble. Scientists familiar with the problem feared that the Deep Drain might someday collapse, trapping water and sewage in the city and causing widespread flooding,

Figure 10–17 Subsidence due to groundwater withdrawal is causing buildings to sink in Mexico City, as with this building that is lower than the surrounding pavement. *(Courtesy Professor Ikuo Towhata)*

BOX 10–4 CASE STUDY

Dead Sea Sinkholes

In the spring of 2009, a hiker walking along the shores of the Dead Sea in Israel was critically injured when the ground beneath him gave way and he plummeted into a newly formed sinkhole. Such occurrences unfortunately are not isolated. More than 2000 sinkholes have formed along the Dead Sea coast in the last few years, the unintended consequences of water resource use by Israel and Jordan (**Figure 1**).[22] The Dead Sea receives water from the Jordan River, which begins in Lebanon and enters the Golan Heights north of the Sea of Galilee (also called Lake Kinneret and Lake Tiberias). As the lowest place on Earth, the Dead Sea is a terminal lake from which water is removed only by evaporation; its salinity is ten times more than that of the world's oceans.

In 1964, Israel finished the construction of the National Water Carrier, a 130 km pipeline and canal system used to convey 500 million m^3 of water annually from the Sea of Galilee southward for agriculture, industry, and domestic uses.[23] The peace treaty signed by Israel and Jordan in 1994 likewise guaranteed Jordanians access to Jordan River water, and the kingdom receives 50 million m^3 annually. The reductions in flow in the Jordan River have contributed to a more than 25-m decline in the level of the Dead Sea over

the last 45 years (**Figure 2**),[24] which has fallen to a record elevation of 423 m below sea level and become ever more saline. Another factor contributing to the decline is industry, which pumps Dead Sea brine into shallow evaporation basins on both the Israeli and Jordanian sides of the lake. Carnallite, a mineral made of potassium and magnesium chloride, precipitates in the evaporation basins and is mined as a source of these elements for fertilizer manufacture. Nearly a third of the annual loss of Dead Sea water results from this process.

As lake level dropped and sinkholes began to form, Israeli geophysicists started searching for explanations. Using borehole drilling and seismic refraction, a technique in which human-made vibrations are used to image the subsurface, they soon discovered that the cavities are forming in a layer of rock salt buried at a depth of 20 to 50 m below the surface.[25] Before water diversions in the Jordan began, the Dead Sea was higher in elevation, and extremely salty water from the lake flowed into surrounding aquifers, bathing the rock salt in brine. As the sea's surface elevation dropped, so too did the water table in the surrounding hills and the elevation of the sloping interface between Dead Sea

(a)

(b)

Figure 1 (a) Sinkholes form when underground limestone or salt caverns become hollowed out by water erosion. Limestone and salt dissolve in water, and the dissolved matter is carried away to nearby bodies of water. The ceilings of caverns become thin and ultimately collapse. (b) More than 2000 sinkholes have opened along the margins of the Dead Sea in Israel, with at least a dozen in this view showing how they pockmark the landscape. *(Kirsten Menking)*

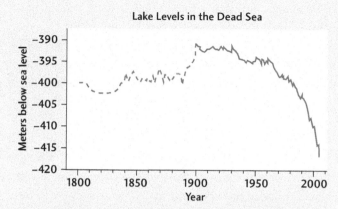

Lake Levels in the Dead Sea

Figure 2 Lake levels in the Dead Sea have declined over the last century as Jordan River inflow that previously fed the sea has been diverted for agriculture and municipal water supplies in Israel and Jordan.

brine and overlying fresh groundwater (**Figure 3**). Unlike the brine, which is supersaturated and incapable of holding any additional salt, the freshwater readily dissolves the rock salt,

creating massive cavities analogous to caves in limestone terrains

The sinkholes pose a major threat to infrastructure and human beings in the Dead Sea region. In addition to the injured hiker, sinkholes have caused roads to collapse and threaten numerous large resort hotels along the sea-shore that cater to tourists who want to bathe in the brine. Unfortunately, the problems are projected to become much worse in coming decades. Diversion of Jordan River water is expected to continue or even increase, and the sea will continue to recede, eventually reaching a steady state eleva-tion of about 550 m below sea level. The potentially mas-sive infrastructure problems have led some to propose the construction of a canal to bring water from the adjacent Red Sea to the Dead Sea to raise its elevation to pre-diversion levels, thereby raising the elevation of the freshwater–salt-water interface in the groundwater system and stopping the salt dissolution process. The so-called "Red–Dead canal"

(continued)

Dead Sea Level: 1960

Dead Sea Level: 2006

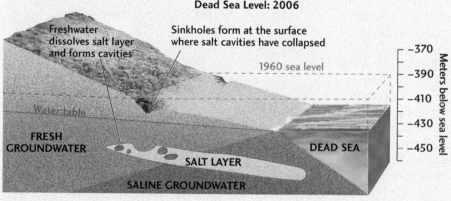

Figure 3 As the lake level declines, subsurface salts are exposed to and dissolved by fresh groundwater.

(continued)

would run along the border of Israel and Jordan and would contain several hydroelectric power plants, taking advantage of the elevation drop between the two water bodies to generate electricity. The power would be used not only for electrification projects in the two countries but also to desalinate Red Sea water to supply arid Jordan with additional freshwater resources. An alternative, the "Med–Dead canal," has also been proposed to supply the Dead Sea with water from the Mediterranean.

Before either of these projects can go forward, much research must be done to evaluate potential environmental impacts. The salinity of the Dead Sea is currently so high that only a small amount of algae are capable of living within the lake. Introducing Mediterranean or Red Sea water into

the Dead Sea could potentially shift the brine toward a new composition that would stimulate algal growth, possibly causing the entire sea to turn red or purple. Likewise, introducing Red or Mediterranean Sea water might cause precipitation of the mineral gypsum in surface waters, turning the Dead Sea white. While neither of these scenarios constitutes a dire ecological consequence since nothing but algae is capable of living in the Dead Sea, the consequences for the tourism industry could be quite profound. The Dead Sea has a gorgeous azure color that beckons to travelers from all over the world. No one is sure whether tourists would continue to flock to the area if the water were to turn purple or white, and at a projected cost of $15 billion, the economic viability of the project is in question.

---

potentially displacing millions of people. To avoid this problem, Mexico City began the construction of a new, nearly $1 billion drainage tunnel in 2008 that was scheduled to be completed in 2012.[27] This tunnel will convey stormwater and the wastes of the city's 20 million residents and will hopefully put an end to the city's drainage woes.

Intrusion of Salt Water

Lowered groundwater levels in coastal areas allow salty seawater to migrate inland into aquifers, a process known as **saline intrusion,** which makes groundwater nonpotable. In the 1930s, saline intrusion destroyed the aquifers beneath the borough of Brooklyn in New York City when the water table was lowered as much as 15 m, and forced the city to look elsewhere for water.

The causes of saline intrusion are related to the difference in density between freshwater and saltwater and to changes in the volume of freshwater when the aquifer is pumped. Seawater is denser than freshwater because it contains more dissolved ions. Freshwater therefore tends to float above seawater as the latter moves inland along coasts (Figure 10-18). At the boundary zone between the overlying freshwater and the salty seawater underneath, the two mix. Surface water drains into the sea in a volume usually large enough to push away the denser saltwater. When groundwater is pumped, however, the volume of freshwater above the boundary zone is diminished, and the boundary migrates inland and upward. Wells that once pumped groundwater begin to yield useless seawater.

Saline intrusion can be prevented or reversed by artificial recharge of the groundwater and by constructing

barriers that block the inland flow of seawater. Such efforts were made recently to prevent saline intrusion to the Biscayne Aquifer, the sole source of groundwater for

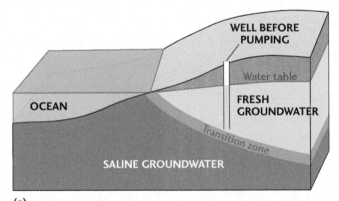

(a)

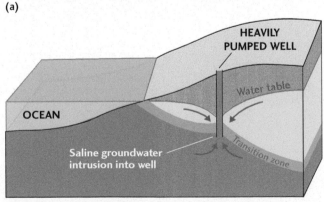

(b)

Figure 10–18 (a) The boundary between saline and freshwater zones in a coastal area before groundwater pumping. (b) After prolonged pumping, the water table is lowered and saline intrusion occurs.

Miami, Florida.[28] Today, fresh surface water from the Everglades is impounded behind levees and then carried by canals to the coast, where it is used to recharge groundwater and push seawater away.

■ Sinkholes form above soluble rocks, such as limestone and rock salt, where groundwater has dissolved large amounts of the rock ceiling and caused the roof to collapse.

■ Land subsidence happens where aquifers and confining layers settle because of withdrawal of groundwater. Subsidence under cities disrupts infrastructure such as water mains and sewers and can also damage buildings.

■ Saline intrusion into aquifers—the replacement of fresh groundwater by seawater—takes place when excessive amounts of groundwater are pumped along coastal areas.

GROUNDWATER POLLUTION AND ITS CLEANUP

Many pollutants that seep into the ground do not dissipate harmlessly. Rather, they travel with groundwater and can emerge in streams and wells far from the site of initial contamination. To understand how to protect and clean groundwater, it is important to know the types of pollutants that reach groundwater and how they interact with it.

Types and Sources of Groundwater Pollution

Groundwater can be contaminated by biological pollutants, such as bacteria, protozoans, or viruses, as well as by industrial chemicals, such as trichloroethylene (TCE) and benzene. Until the Industrial Revolution, biological contamination was the biggest threat to drinking water. Major outbreaks of bacterial diseases—among others, bubonic plague in 17th-century London and cholera in 19th-century New York—were associated with wells contaminated by human sewage. In general, however, groundwater is less susceptible to biological pollution than surface water because many pathogens are similar in size to grains of silt and sand and thus can be trapped in pore spaces in aquifers or in the unsaturated zone above the water table (Figure 10-19). For this reason, knowledgeable campers and backpackers filter any water they collect from streams through fine carbon granules before drinking it.

Since the Industrial Revolution, the number of chemicals used for manufacturing, farming, and other activities has increased rapidly. More than 80,000 chemicals are used worldwide today, many of them distributed

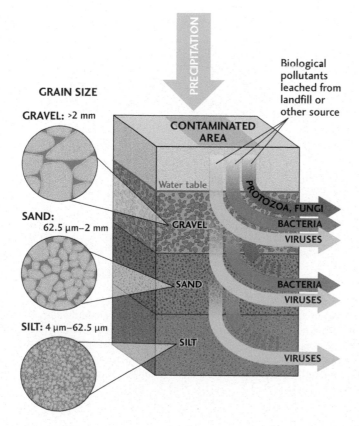

Figure 10–19 The smaller the size of the grains, the smaller the pores between the grains and the smaller the microorganisms that can be filtered from water flowing through pore spaces. At 0.02 to 0.25 μm, viruses are too small to be filtered by silt; bacteria are large enough to be filtered by silt but too small to be filtered by sand.

throughout the environment, in air, water, and soil, and the effects of many thousands of these chemicals on human and environmental health are not known. Each year another 700 or 800 new chemicals are produced. The EPA estimates that 55 million tons of hazardous chemical wastes are produced in the United States each year. Three of the most abundant compounds identified in groundwater at industrial waste disposal sites are TCE, benzene, and vinyl chloride, all of which are known human carcinogens and also can cause liver damage, brain disorders, lesions, and nervous system dysfunction. TCE may be especially prevalent. This toxic chemical is used mainly as a solvent by metal industries and dry cleaners to remove grease. It is also used to extract caffeine from coffee and is sometimes involved in the manufacture of pesticides, tars, paints, and varnishes. The recommended maximum limit of TCE in drinking water is 5 ppb (parts per billion).[29] Unfortunately, because TCE is so commonly used, drinking water wells in some states, including Pennsylvania, New Jersey, New York, and Massachusetts, have been found to contain as much as 14,000 to 27,300 ppb of TCE.[30]

Until the 1970s, no laws in the United States prevented any company or individual from disposing of hazardous wastes in underground storage tanks, injection wells, or surface impoundments (also referred to as waste lagoons or ponds) like those at the Massachusetts Military Reservation on Cape Cod (Box 10-5). In fact, after the Clean Air and Clean Water acts were passed in the 1960s and 1970s, the disposal of wastes into surface water and the atmosphere was regulated so strictly that even more waste was disposed of underground. Unfortunately, a prevailing misconception was that wastes stored underground could not travel far, if at all. Instead, because the ground is porous, buried contaminants migrate along with the water draining downward in the unsaturated zone or flow down the hydraulic gradient with groundwater in the saturated zone.

A recent assessment of the causes of groundwater contamination in all states identified leaking underground storage tanks and drainage from septic tanks as the top two sources of pollution. Other common sources include landfills, surface impoundments, infiltration of pesticides and fertilizers from agricultural fields, and accidental spills from trucks and trains (Figure 10-20).

Many chemicals, including the fuel stored at gas stations, factories, and other businesses, are stored underground before, as well as after, becoming waste products. Nearly 600,000 commercial underground storage tanks (UST) are known to exist in the United

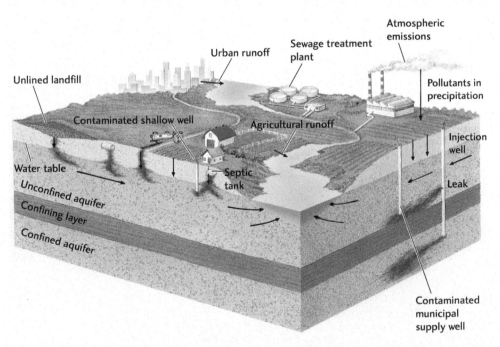

Figure 10–20 Corroded and leaking underground storage tanks are one of the main causes of groundwater pollution from industrial wastes, while leaking septic tanks are a main residential cause. Pesticides and fertilizers from farm fields seep underground and can reach the water table, as can metals and other pollutants that move through waste stored in landfills. Depending on the type of waste disposed in it, a landfill may contain metals and chemicals such as solvents and fuel oils that can enter the groundwater. Today, landfills in the United States are designed carefully to collect and treat all groundwater before it leaves the site. Although most wells pump water from underground, some are used to inject hazardous waste into the subsurface.

States.[31] With time, metal containers can corrode and supply lines can develop cracks and leaks. In 1986, Congress created the Leaking Underground Storage Tank (LUST) Trust Fund to clean up contaminated sites whose owners could not be identified and to oversee and enforce cleanup when responsible parties could be found. Every year, $70 million is allocated to dig up contaminated soils and pump and treat polluted groundwater (Figure 10-21). States and the EPA have cleaned up about 83 percent of sites, but there are still some 85,000 UST releases, and the annual number of cleanups completed nationally has declined steadily since fiscal year 2000.[32]

Precipitation or runoff infiltrating downward through landfills and surface impoundments can dissolve metals and chemicals and transport them, along with bacteria, to saturated zones. The sometimes oily, discolored product is known as **leachate**, and can be seen seeping from the ground near waste sites that are improperly lined and lack leachate collection systems (Figure 10-22).

Migration of Groundwater Pollutants

Substances carried by groundwater move at rates that vary with particle or molecular size, solubility, chemical

Figure 10–21 A leaking underground storage tank being pulled out of the ground near Detroit for disposal at a hazardous waste site. *(Courtesy Kevin Svitana)*

activity, density, and viscosity. If oil spills from a truck and seeps downward to the water table, the oil—which does not dissolve in water—floats at the water table above the zone of saturated rock or sediment. It moves along the direction of groundwater flow, giving off petroleum vapors and contaminating wells. In contrast,

(a)

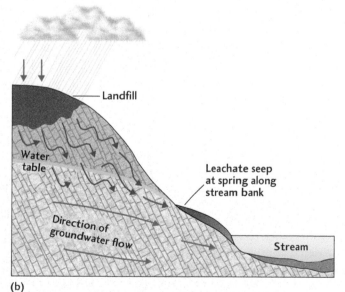

(b)

Figure 10–22 (a) Photograph and (b) diagram of leachate seeping from an abandoned landfill. Infiltrating water flows through the waste and leaches out contaminants. These contribute to the formation of leachate, a liquid that usually is relatively rich in metals and sometimes contains chemicals

such as solvents. The leachate flows downward until it reaches the water table. In this instance, the leachate emanates from the hillside at a spring and then flows downhill into a nearby stream. *(a: Courtesy Bill Wolfe, http://www.wolfenotes. com/2010/06/dep-calls-toxic-landfill-leachate-natural/)*

BOX 10-5 CASE STUDY

Contaminant Plume at Cape Cod, Massachusetts

Residents of Cape Cod, Massachusetts, rely on groundwater for their water supply. When they complained about a soapy taste in their water, scientists investigating the problem drilled a well and found the water thick with bubbly detergent lather (Figure 1). After drilling many more wells, they traced the source to a long plume of soapy water emanating from the Massachusetts Military Reservation (MMR), a combined Army National Guard, Air National Guard, and Coast Guard training facility. From 1936 to 1995, wastewater and treated sewage from base housing were disposed of in a shallow pit on the reservation grounds,[33] and until the mid-1960s the personnel at MMR, like everyone else in the United States, used detergents that do not *biodegrade,* or break down naturally in the environment by biological processes. In addition to sewage and detergent wastes, industrial solvents and jet fuel were purposefully placed on the ground and used for fire-fighting training exercises. These and numerous unintentional spills led to serious groundwater contamination problems as the wastes seeped downward and outward through sand and gravel, heading south toward the Atlantic Ocean and passing through the groundwater systems of several communities, including Falmouth, some 5 km away.

Residents of Falmouth first began to notice water quality problems in 1973, but it took 20 years for their problems to begin to be addressed. Since 1993, the EPA, the

Figure 1 Groundwater pumped from an aquifer on Cape Cod contains foaming household detergents. This well is located nearly 2 km downslope from the Massachusetts Military Reservation, which used to dispose of its sewage through a septic leach field that ended up contaminating the aquifer. *(Doris LeBlanc/USGS)*

United States Geological Survey (USGS), and the military have engaged in a massive campaign to locate and remove sources of contamination and to remediate affected groundwater supplies. These agencies identified 80 source areas of contamination and began digging up polluted soils to stop further introduction of toxic materials into the groundwater system. They also installed tens of thousands of groundwater

if salt seeps downward from roads that are salted during winter, it dissolves and flows into the saturated zone. Both oil and salt are major contaminants of groundwater. In fact, the common use of as much as 2 to 11 metric tons of salt per kilometer of single-lane road to deice roads in the course of a single winter has caused extensive environmental degradation of groundwater in northern climates. Salt enters the groundwater system when snowplows shove snow toward road margins and by being kicked up by car tires. Once it lands at the roadside, salt dissolves in snowmelt or rainwater and infiltrates the groundwater system along with the water. The stream base flow chemistry has had a steady increase in chloride concentrations over the last several decades in streams such as the Mohawk River of New

York State.[34] In addition, several wells in New York's Hudson River valley are already so contaminated with road salt that they have become undrinkable without being diluted by fresher water.

Miscible contaminants such as salt flow in groundwater as dissolved constituents, or *solutes,* mixing with the groundwater to form a single liquid. Because most groundwater flows at a rate of several meters per year or less and is subject to little turbulence, solutes migrate very slowly from the source of contamination and tend not to become diluted. In fact, recent monitoring of groundwater wells shows that some contaminants are in far higher concentrations underground than in the localities where contaminated surface water—which moves relatively rapidly—is known to be.

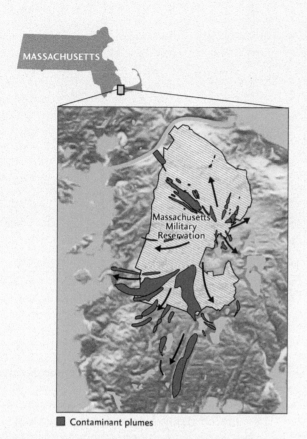

MASSACHUSETTS

Massachusetts
Military
Reservation

■ Contaminant plumes

Figure 2 Numerous contaminant plumes emanate from the Massachusetts Military Reservation. Contaminants include household detergents, jet fuel, industrial solvents, and chemicals associated with military explosives.

sampling wells in order to map the distribution of contaminants in the subsurface, eventually identifying more than 20 separate contaminant plumes (**Figure 2**). Cleaning up this pollution required installing numerous pumping wells that pull contaminated water from the ground, pass it through filtration systems, and then pump the treated water back into the ground. Thus far, more than 120 billion L of polluted water have been treated, removing over 3045 kg of toxic substances.

Affected residents of Falmouth were initially provided bottled water to drink and eventually were supplied with individual home filtration systems or with municipal water piped in from uncontaminated areas, but though their problems were solved, contamination still threatened the environment and other residents as the groundwater continued to move, necessitating cleanup. In the end, remediation is expected to cost more than $1 billion, and the work is projected to continue through the year 2055.

Thanks to environmental regulations and better awareness, intentional dumping such as that which took place at MMR is largely a thing of the past, but groundwater continues to be under threat both from earlier dumping and from present-day activities such as road salting in winter. Protecting groundwater supplies into the future requires an understanding of where groundwater comes from, how it flows beneath Earth's surface, and how contaminants get introduced into and move with groundwater.

Miscible contaminants are diluted by streams and rivers, but in groundwater they tend to spread outward in long plumes through the process of *dispersion* (Figure 10-23). In dispersion—the mechanical combining of contaminated and uncontaminated water—the contaminated water remains at close to full strength as a discrete flow, but is increasingly mixed with uncontaminated water in the downgradient direction. Much of the mixing happens when contaminants flow around individual grains in paths that cause the contaminated water to diverge from the main flow direction. Dispersion occurs in aquifers because the material through which groundwater flows is not completely homogeneous. Water flows faster through large pores than small pores because frictional resistance with rock

surfaces is lower. Also, the water flows more quickly in the centers of the pores than along the pore walls, where frictional resistance to flow is greater. The result is that although the direction of the contaminant flow is that of the regional hydraulic gradient, the shape of the contaminant flow changes. It becomes elongated, and it also may spread out perpendicular to the direction of the groundwater flow as the water follows branching pathways. Dispersion is the primary cause of the characteristic plume-like shape of contaminants in groundwater. Contaminant plumes have been identified in many aquifers for all types of miscible pollutants.

Immiscible contaminants are insoluble, and their fate in groundwater depends partly on their density. Those that are less dense than water float above the

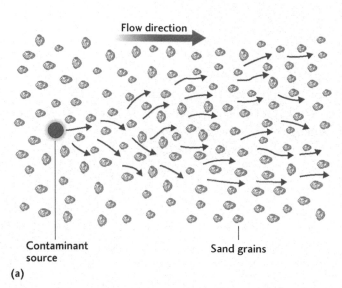

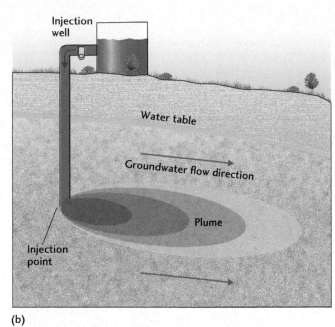

(a)

(b)

Figure 10-23 (a) A primary cause of dispersion is the separation of flow paths around grains. (b) With increasing distance from the source of contamination, such as an injection well at which wastes are injected into the ground, contaminated water is mixed with uncontaminated water. The result is that the contaminant assumes a plume–like shape along the direction of groundwater flow.

water table and move down the hydraulic gradient along the water table's surface (Figure 10-24). Gasoline is a common example of an immiscible material found floating above groundwater. Those immiscible contaminants that are denser than water, such as trichloroethylene (TCE), tend to sink to the bottoms of aquifers. Depending on the geometry of the bottom of the aquifer, the contaminant plume may or may not move in the same direction as the regional groundwater flow. In the case of light immiscible substances, some vapors may be given off that seep into the unfilled pore spaces of the unsaturated zone.

U.S. Laws Governing the Quality of Water Resources

Although groundwater is an essential national resource susceptible to contamination from many common sources, there are no federal laws in the United States solely for the protection of groundwater quality. Many aquifers have been seriously polluted, and in some cases the damage is irreparable. In the absence of a specific law, a number of federal laws designed to address other environmental problems have been invoked to protect groundwater. Some of these laws were created to make certain that public water supplies meet established health standards (see the discussion of standards for drinking water in Chapter 9) or to regulate the production, use,

and disposal of pesticides, which can affect the quality of underground water.

In 2006, the EPA published an amendment to the 1996 Safe Drinking Water Act that is known as the Ground Water Rule. Its purpose is to increase the protection of public water systems using groundwater against microbial pathogens (for example, *E. coli* and salmonella). These disease-causing microorganisms can come from sources such as fecal matter from leaking septic systems. The Ground Water Rule uses a risk-targeting strategy that involves surveying public water systems; monitoring source water when a pathogen is identified; taking corrective actions for those systems with a significant deficiency in order to reduce risk of contamination; and performing compliance monitoring to ensure effective treatment.

One of the most important laws passed for the remediation of polluted groundwater is the Comprehensive Environmental Response, Compensation, and Liability Act of 1980 (CERCLA) and its amendments, commonly referred to as **Superfund**. Many contaminated sites have been targeted for cleanup under the guidelines of this legislation. If those responsible for the pollution do not do the cleanup, it is done by the EPA and billed to the polluters if they can be determined. In 2010, the National Priorities List—those sites that have been found to warrant remediation—contained 1280 Superfund sites; 62 others were proposed for

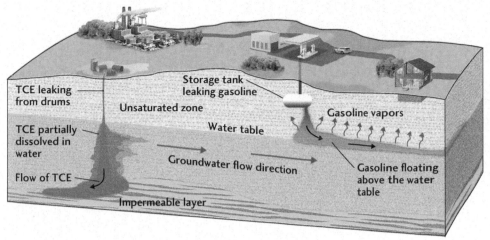

Figure 10-24 Light and heavy immiscible contaminants: Gasoline from a leaking underground storage tank floats atop the water table; trichloroethylene (TCE, at left) spilled from drums sinks to the bottom of an aquifer.

consideration. As of the same year, 347 sites already had been cleaned up and were no longer on the list.

The costs of cleanup are so high that the EPA does not have the funds necessary to remediate more than a few of the sites on its national priority list, let alone the many hundreds of other contaminated sites that dot the country. At a Superfund site in Indiana, for example, the owners of a company that used organic solvents, including TCE, went bankrupt in the 1970s, abandoning 98 tanks and 50,000 drums filled with chemicals. For years, many of these containers leaked chemicals into the ground. Cleanup under the supervision of the EPA began in 1988 and will cost between $25 million and $40 million over the next few decades.

Groundwater and Aquifer Restoration

Remediating contaminated groundwater systems is a growing business that involves hydrogeologists, engineers, lawyers, and policy makers. New technologies to treat myriad situations of pollution at hundreds of thousands of sites across the United States are continually being developed and tested. Broadly, cleanup activities can be divided into three major categories: containment, containment withdrawal, and in situ treatment of contaminants.

Containment, or isolation, is meant to prevent contaminants from leaving their source area. New solid waste landfills, for example, are underlain by impermeable liners to prevent groundwater from entering the waste. Likewise, an impermeable cap is placed on top of the waste to prevent rainwater infiltration. If water does make its way into the waste, it is captured by a drainage-collection system and the leachate treated.

Containment withdrawal is an enhanced method of containment that uses one or more wells to pump contaminated water from a site (Figure 10-25). In general, the pumped contaminated water is treated onsite or may be transported offsite. Containment withdrawal often is used at industrial sites where underground storage tanks have leaked over a period of years. First the tanks themselves are removed, then the groundwater is withdrawn for treatment, with the treated water then pumped back into the ground. Sometimes remaining soil and sediment must also be excavated and disposed of because surface tension makes it difficult to remove all contaminants from the material.

In situ treatment is used to clean contaminated water in the subsurface without removing the water. This treatment may be either chemical or microbiological in nature. One of the primary environmental problems associated with the disposal of sewage at the MMR, for example, was high levels of phosphorus from detergents (see Box 10-5). When phosphorus travels to surface water bodies, such as ponds, it can cause eutrophication by stimulating algal blooms that consume dissolved oxygen when they decay. To prevent such a scenario from destroying Ashumet Pond, in 2004 scientists made use of the fact that phosphorus binds tightly to iron; they dug a trench at the edge of the pond where the detergent plume was entering and filled it with iron-rich sediment. Thus far, the iron barrier has reduced by 95 percent the amount of phosphorus able to reach the pond.

Bioremediation of groundwater pollutants relies on the fact that microbes are often capable of breaking down toxic compounds into harmless by-products. For example, in the aftermath of oil spills, scientists have identified bacteria that consume hydrocarbons and convert them to nontoxic carbon dioxide and

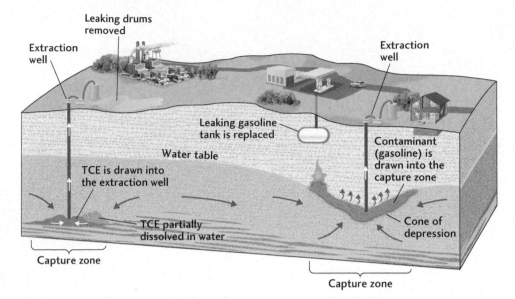

Figure 10–25 Extraction wells can be used to remove some contaminants from groundwater. In the two examples here of substances with low solubility in water, the gasoline from a leaking storage tank (at right) is less dense than water and floats on the water table, whereas the trichloroethylene (TCE, at left) is denser than water and is found lower in aquifers. Near the extraction wells, pumping causes the groundwater and contaminants to flow toward the well, from which contaminated water can be pumped to the surface and treated.

water. Bacteria that consume TCE were found in soils around dry cleaners. To speed up the rate at which naturally occurring bacteria break down pollutants, hydrogeologists can add nutrients to contaminated aquifers through injection wells. Likewise, compounds that generate oxygen to stimulate aerobic bacteria or that create reducing conditions to stimulate anaerobic bacteria may be added, depending on the contaminant that is being remediated (Figure 10-26).

■ Widening plumes of miscible contaminants form in groundwater because the contaminants are dispersed (that is, mechanically combined) with the surrounding water during transport. The amount of mixing increases with greater distance from the source.

■ Light immiscible contaminants tend to float above the water table, whereas dense immiscible contaminants sink to the bottoms of aquifers.

■ The federal Superfund program was created by Congress to clean up contaminated sites. The program is administered by the EPA and bills polluters for the cost of cleanup.

■ Groundwater contaminants may be biological or chemical in nature.

■ Those involved in cleanup use various techniques to remediate polluted groundwater, including excavating the source of contamination, pumping and treating contaminated groundwater, and bioremediation.

Figure 10–26 Just beneath the asphalt surface where a gasoline station once stood, microbes injected into the ground through tubes consume gasoline that leaked from an underground storage tank. Bioremediation can happen naturally, but does so at a very slow rate: It would take 50 to 100 years to clean up a site such as this one near Philadelphia. Intervention shortens the process to less than a year. Scientists are doing extensive research to identify microbes that break down specific contaminants, and they recently identified one type that decomposes TCE. This discovery is significant for those attempting to restore aquifers. *(Dan Oleski, Franklin and Marshall College, Lancaster, Pennsylvania)*

CLOSING THOUGHTS

Although the use and contamination of groundwater have resulted in a flurry of scientific activity to understand this important resource and to find ways to restore water quality, there is still much to learn about what happens underground. While some water flows in streams through underground caverns, the great majority of groundwater flows through small pores and thin fractures along circuitous pathways. Therefore, the speed at which groundwater flows is much less than that of surface water, except in underground caverns. How contaminants move in groundwater is not well understood, because not all flow at the same rate as their water host. In addition, there are so many different contaminants that fully understanding every one is challenging.

Preventing groundwater contamination is far more prudent than attempting to clean up after pollution has reached an aquifer. Expensive federal programs such as Superfund deal with wastes that already exist, but for the future it would be wiser to prevent contaminants from entering underground flow paths. The attention to new technologies to prevent movement of leachate from landfills to groundwater, for example, provides encouraging evidence that progress is being made in avoiding further environmental degradation of aquifers and groundwater.

SUMMARY

■ The boundary between the unsaturated and saturated zones of unconfined underground water is a surface called the water table, which mimics Earth's topography and rises and falls in response to changing amounts of recharge and discharge.

■ In uncemented, unconsolidated, well-sorted sand, porosity can be as high as 48 percent. In lithified sedimentary rocks and crystalline (metamorphic and igneous) rocks, it generally is much lower (less than 10 percent) unless the rocks are fractured.

■ Rocks and sediments that are both porous and permeable make good aquifers, while those that are neither porous nor permeable act as confining layers.

■ Groundwater contains salts because it flows slowly around minerals, dissolving them and picking up ions in the process. With depth below Earth's surface, groundwater generally becomes increasingly saline.

■ Environmental hazards associated with groundwater include sinkholes, land subsidence, groundwater mining, saline intrusion in coastal areas, and groundwater pollution.

■ Major sources of groundwater pollution are leaking underground storage tanks and septic tanks, leachate from landfills and unlined waste pits, pesticides and fertilizers from agricultural fields, and accidental spills from trucks and trains.

■ The flow of contaminants in groundwater depends on whether they are miscible (soluble) or immiscible (insoluble) in water, and if they are immiscible, on their density relative to groundwater.

■ One of the most important laws passed for remediation of polluted groundwater is the Comprehensive Environmental Response, Compensation, and Liability Act (CERCLA) and its amendments, commonly referred to as Superfund.

■ The three main ways in which contaminated groundwater is cleaned are containment, containment withdrawal, and in situ treatment of contaminants using bioremediation.

KEY TERMS

water table (p. 323)
groundwater (p. 323)
porosity (p. 324)
surface tension (p. 326)
specific yield (p. 326)
specific retention (p. 326)
permeability (p. 326)
hydraulic conductivity
 (p. 326)

aquifers (p. 326)
confined aquifers (p. 327)
unconfined aquifers
 (p. 327)
hydraulic gradient (p. 327)
potentiometric surface
 p. 327)
artesian spring (p. 327)
artesian well (p. 328)

Darcy's law (p. 328)
cone of depression
 (p. 329)
water hardness (p. 334)
land subsidence (p. 337)
water spreading
 (p. 339)
recharge basin (p. 339)
injection well (p. 340)

sinkholes (p. 342)
saline intrusion (p. 346)
leachate (p. 349)
miscible contaminant
 (p. 350)
immiscible contaminant
 (p. 351)
Superfund (p. 352)
bioremediation (p. 353)

REVIEW QUESTIONS

1. List examples of sources of recharge and discharge for groundwater.

2. What are typical values of porosity for sand, sandstone, and fractured limestone?

3. What is Darcy's law? How is it related to the flow of pollutants in groundwater?

4. Would a pollutant migrate more quickly through a well-sorted, coarse, unconsolidated sand or a poorly sorted, fine-grained unconsolidated sand? Why?

5. Why do some springs discharge so much more water than others?

6. How can nitrates from agricultural areas affect groundwater?

7. How is the groundwater that comes out of geysers at Yellowstone converted to steam below ground?

8. What does "water hardness" mean? Why is some water harder than other water?

9. List common ways that pollutants get into groundwater.

10. What is leachate?

11. In choosing underground disposal sites for nuclear waste, why is the permeability of the rock in which the waste will be stored a concern?

12. How does dispersion contribute to the formation of contaminant plumes in groundwater?

13. What happens to gasoline that is spilled on the ground or leaks from a tank into the water table?

14. What happens to salt spread on roads that seeps to the water table?

15. What are the contaminants in the Cape Cod plumes?

16. Explain how Superfund is being used to deal with the problem of groundwater pollution.

THOUGHT QUESTIONS

1. Which material is likely to have the greater permeability, unconsolidated sand or sandstone? Why?

2. Which of the following is likely to be the most efficient aquifer at yielding water to a well: unconsolidated sandstone, sandstone with cement filling all pore spaces, or crystalline granite with no fractures? Why?

3. Is groundwater in most cases a renewable or a nonrenewable resource? If there is a prolonged drought in a region, why might local groundwater change from a renewable to a nonrenewable resource?

4. What is a cone of depression? If your neighbor installs a well that is deeper than yours and pumps large volumes of water each day, might your well be affected? (Assume that both of you have wells in the same unconfined aquifer.) Why?

5. Suppose you are a hydrogeologist called in to investigate a gasoline smell in someone's house. Gas company personnel already have determined that there is no leak from their equipment to the house. What possibilities would you suspect? How would you investigate them?

EXERCISES

1. If you live in a humid region, you can learn something about the depth to the water table by examining topographic maps. Points along perennially flowing streams, lakes, and springs are outcrops of the water table surface and their elevations are equal to the hydraulic head of the water table. With the elevation of the water table at various points, it is possible to make a contour map of the water table, from which you can estimate values of hydraulic head (discussed in Box 10-1),

the difference in water table elevation from one point to another. Acquire a topographic map for your area. Once you construct a map of the shape of the water table surface, determine how deep the water table is where you live and calculate values of hydraulic head in a few locations.

2. You can find out how deep groundwater is in your area, and how much has been yielded from wells drilled there, from state officials. Well drillers are

required by law to file information with a state environmental agency about each well they drill. This information, available to the public, records the depth of each well, the depth to groundwater, the nature of the aquifer and its porosity, and the well yield. Contact your state environmental agency and ask them how to access this information. Compare and contrast different wells. Do yield values correlate to rock type, to depth, or to some other variable?

SUGGESTED READINGS

Alley, W. M., T. E. Reilly, and O. L. Franke. "Sustainability of Ground-Water Resources." *U.S. Geological Survey Circular* 1186 (1999). http://pubs.usgs.gov/circ/circ1186/pdf/circ1186.pdf.

Clark, R., and J. King. *The Water Atlas*. New York: The New Press, 2004.

DeSimone, L. A., P. A. Hamilton, and R. J. Gillion. "Quality of Water from Domestic Wells in Principal Aquifers of the United States, 1991–2004." *U.S. Geological Survey Circular* 1332 (2009). http://pubs.usgs.gov/circ/circ1332/includes/Circular1332.pdf.

Fetter, C. W. *Applied Hydrogeology,* 4th ed. New York: Prentice-Hall, 2001.

The Freshwater Society promotes conservation, protection, and restoration of freshwater resources, and has numerous publications and resources related to groundwater and surface water. Available at http://freshwater.org/about-the-society/.

Laws, E. A. *Aquatic Pollution,* 3rd ed. New York: John Wiley, 2000.

Reilly, T. E., K. F. Dennehy, W. M. Alley, and W. L. Cunningham. "Ground-Water Availability in the United States." *U.S. Geological Survey Circular* 1323 (2008). http://pubs.usgs.gov/circ/1323.

U.S. Environmental Protection Agency. "Drinking Water from Household Wells." EPA Publication EPA 816-K-02-003 (2002). http://www.epa.gov/safewater/privatewells/pdfs/household_wells.pdf.

Winter, T. C., J. W. Harvey, O. L. Franke, and W. M. Alley. "Ground Water and Surface Water: A Single Resource." *U.S. Geological Survey Circular* 1139 (1998). http://pubs.usgs.gov/circ/circ1139/pdf/circ1139.pdf.

Zekster, I. S., and L. G. Everett. *Groundwater Resources of the World and Their Use*. Paris: UN Educational, Scientific, and Cultural Organization, 2004. http://unesdoc.unesco.org/images/0013/001344/134433e.pdf

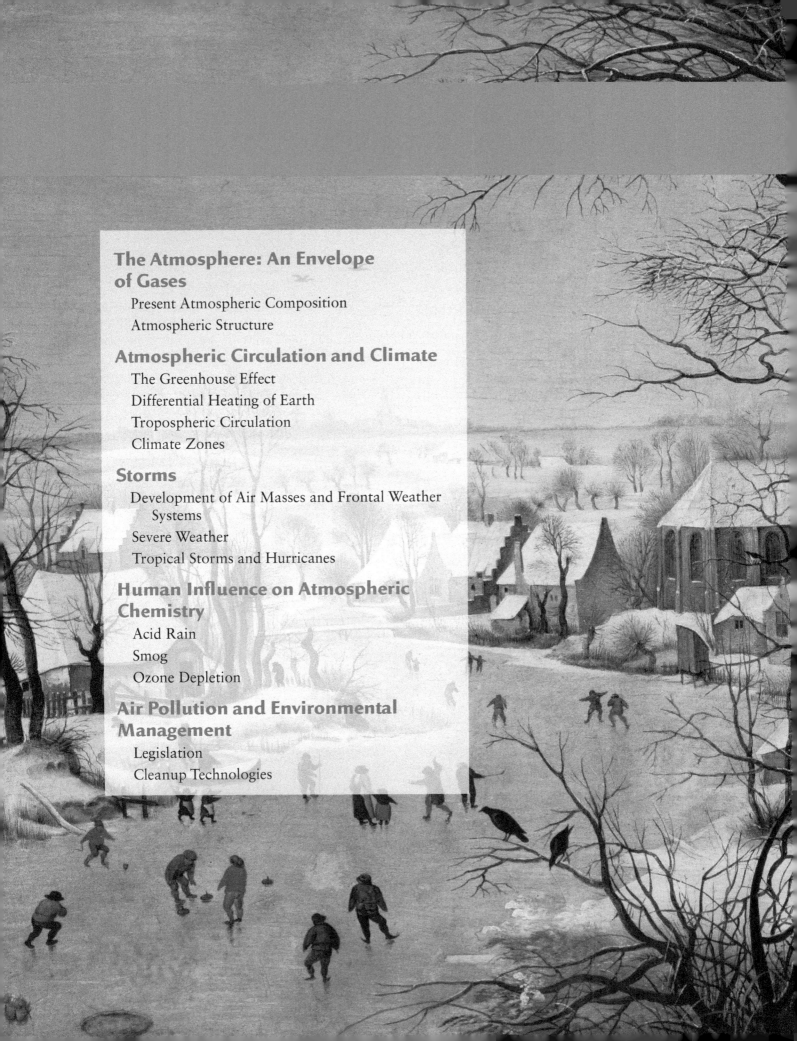

The Atmospheric System

Paintings by the 16th century Flemish artist Pieter Bruegel depict a very different Netherlands than exists today. During Bruegel's lifetime, parts of the world were experiencing a cold snap that began around 1400 and lasted until 1900, the exact timing varying with location.[1] In winter, the Dutch canals froze over so solidly that people could skate around town. Elsewhere in Europe the Baltic Sea froze, and glaciers expanded in the Alps. In North America, ice fields grew throughout the Canadian Arctic, and George Washington crossed the ice-filled Delaware River with his soldiers, mounting a surprise attack that scored a decisive victory in the American Revolutionary War. The Delaware froze over frequently in Washington's time. This interval of cooling has come to be known as the "Little Ice Age." Hypotheses advanced to explain it include the frequent eruption of volcanoes that emitted Sun-blocking sulfur dioxide and ash into the stratosphere and the persistence of a period of decreased solar radiation output related to a decline in sunspot activity. Because glacial accumulation is a destabilizing, reinforcing-feedback process, some geologists have suggested that the Little Ice Age might have been a precursor to still colder conditions had it not been interrupted by the massive outpouring of greenhouse gases into the atmosphere by the Industrial Revolution.[2]

In this chapter we:

✔ Investigate the impact of the atmosphere's composition and structure on weather and climate.

✔ Examine the links between atmospheric circulation, regional distribution of moisture, and storms.

✔ Discuss the role of human-induced atmospheric pollution in environmental change.

Winter Landscape with a Bird Trap, painted in 1565 by Pieter Bruegel the Elder, depicts a village in the Netherlands during the Little Ice Age (1450–1890). *(Pieter Bruegel the Elder,* Winter Landscape, *1601. Kunsthistorisches Museum, Vienna. Erich Lessing/Art Resource)*

INTRODUCTION

Storyteller Garrison Keillor, in his radio show *A Prairie Home Companion*, often quips that winter is nature's attempt to kill off the inhabitants of Minnesota. While citizens of the Gulf states bask in subtropical warmth, Minnesotans and their neighbors in the Great Lakes region routinely experience subfreezing temperatures, windchills well below 0°C (32°F), and snow measurable in feet. When they feel they can no longer bear it, those who can afford the time and the money head south to spend a few days sunning themselves on the beach, recharging their bodies with vitamin D and their minds with thoughts of spring. Imagine their disappointment when, in early January 2010, these escaping "snowbirds" found "old man winter" hot on their tails. Anomalously high air pressure over the Arctic pushed polar air southward, bringing temperatures reminiscent of the Little Ice Age and giving Floridians and their southern neighbors a taste of what Minnesotans and Canadians annually endure.

While the snowbirds may have felt right at home, the frigid temperatures were no laughing matter for southern states unaccustomed to such conditions. Indeed, temperatures up to 17°C (30°F) lower than normal proved downright destructive.[3] Dozens of water mains ruptured in Atlanta as the water they carried began to freeze and expand in volume.[4] As the pipes broke, water that had not yet frozen poured onto city streets and formed an icy glaze, causing traffic accidents. In Florida, citrus growers used sprinkler systems to cover their oranges and grapefruits with a protective coating of ice, hoping that by doing so, they could prevent the fruits themselves from freezing (Figure 11-1). On the stock market, orange juice futures rallied as traders anticipated losses from the state's $9.3 billion citrus industry.[5] Florida's $400 million strawberry crop was also threatened.[6]

The frigid temperatures proved dangerous not only for crops but also for people and animals. Homeless shelters throughout the South struggled to find enough beds for the thousands left out in the cold, and several people died of exposure in Nashville, Tennessee.[7] Hypothermic sea turtles, too cold to swim, washed up on the beaches of North Carolina. Florida's $33 million tropical fish industry suffered devastating losses as pond temperatures plummeted, killing 50 percent of their inhabitants, and in Riviera Beach, 400 chilled manatees holed up next to the city's power plant, warming themselves in the hot effluent of the plant's cooling water (Figure 11-2).[8] The cold weather even caused iguanas to rain from the skies! The cold-blooded animals live in trees, and as their body temperatures dropped, they were no longer able to hang onto branches.[9]

Figure 11–1 Florida citrus growers covered their fruit in a protective coating of ice to try to prevent the fruit from freezing when a cold snap hit the state in January of 2010. *(Chris O'Meara/AP Photo)*

A mysterious weather pattern called the Arctic Oscillation (AO, also called the Northern Annular Mode, NAM, and the North Atlantic Oscillation, NAO)[10] was responsible for all the misery. The AO is characterized by the difference in air pressure over the Arctic Ocean and over a latitudinal ring surrounding the Arctic at about 45°N, and it seesaws back and forth between positive and negative values for reasons that are not yet known.[11] During its positive phase, lower than normal pressure over the Arctic and higher than normal pressure over the midlatitudes tightens a strong upper atmosphere current known as the polar vortex. This current prevents polar air masses from moving southward and allows Florida and the Gulf states to bask in the subtropical warmth for which they are well known (Figure 11-3a). When the AO moves into its negative phase, however, as happened in winter 2009–2010, higher than normal pressures over the Arctic and lower than normal pressures in the midlatitudes relax the polar vortex and push polar air masses

Figure 11–2 Manatees congregate around the effluent pipes of the Riviera Beach, Florida, power plant, attempting to stay warm. *(Michael Patrick O'Neill/Alamy)*

southward, causing temperatures in normally balmy areas to plummet (Figure 11-3b).[12]

Ironically, the same factors that bring unusually cold temperatures to southern regions cause parts of the Arctic to warm, as the negative phase AO causes gyrations in an atmospheric current known as the polar jet stream that bring warmer air northward (see Figure 11-3b). Warmth associated with the negative phase AO may add to a long-term trend of Arctic warming that has unfolded over the last several decades. Indeed, December 2009 and January 2010 showed the fourth lowest wintertime Arctic sea ice extents since satellite measurements began in the late 1970s.[13] Arctic warming is understood to be largely a

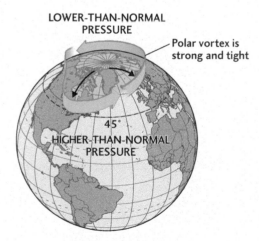

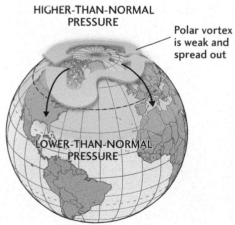

Figure 11–3 The Arctic Oscillation (AO) reflects changing pressure conditions between the Arctic Ocean and a latitudinal band at 45°N. When the AO is in its positive phase, air pressure over the Arctic is lower than normal, while air pressure over the midlatitudes is higher than normal, which prevents cold Arctic air from spilling southward. When the AO is in its negative phase, air pressure conditions are reversed, allowing cold air from the Arctic to sweep southward. Negative phase AO conditions brought frigid temperatures to the southern states in January 2010.

by-product of human activities (such as power production, the use of motor vehicles, rice paddy agriculture, and cattle ranching) that release greenhouse gases to the atmosphere. Melting of the Arctic Ocean's sea ice threatens the survival of Arctic species like polar bears and may cause major shifts in atmospheric circulation patterns that exacerbate droughts and flooding in the midlatitudes. The potential magnitude of the ecologic and economic disruption associated with long-term climate change as well as with shorter-term weather phenomena like the AO necessitates that we learn all we can about Earth's atmosphere and how human activities affect atmospheric processes.

THE ATMOSPHERE: AN ENVELOPE OF GASES

An **atmosphere** is an envelope of gases that surrounds a planet. Earth's atmosphere extends from less than a few meters below the surface, where gases penetrate pore spaces in the pedosphere and lithosphere, to more than 10,000 km (6000 mi) beyond the surface, where gases gradually thin and become indistinguishable from the composition of interstellar space (Figure 11-4).

Figure 11–4 The atmosphere seen from the space shuttle *Atlantis.* The layer at the base of the atmosphere contains ash and sulfuric acid droplets, which scatter red light. These particles, photographed in August 1992, resulted from the June 1991 eruption of Mount Pinatubo (see Chapter 5). The blue layer results from the scattering of blue light by atmospheric gases. *(Johnson Space Center/NASA)*

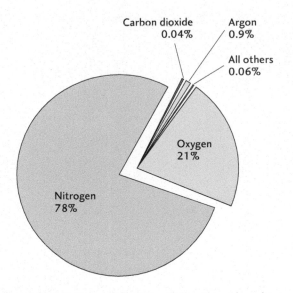

Figure 11–5 The composition of the atmosphere by volume. Water vapor is not listed because the amount in the atmosphere varies with air temperature. The small concentrations of gases such as carbon dioxide and ozone belie their importance in making Earth suitable for life.

Earth's atmosphere formed early in our planet's history. The global meltdown that occurred about 4 to 4.5 billion years ago (see Chapter 1) released light elements such as hydrogen and helium. Because of the gravitational attraction of Earth's mass, some of the gases did not escape to outer space but remained near the surface to form an atmosphere. In contrast, the Moon and Mercury, with less mass and lower gravitational attraction, were unable to hold onto much gas and so have almost no atmosphere. After Earth solidified, volcanic eruptions continued to add water vapor, nitrogen, carbon dioxide, and other gases to the atmosphere. The evolution of life also strongly influenced its composition, most notably by adding oxygen. Indeed, the atmosphere owes its present composition primarily to volcanic eruptions and to the evolution of organisms such as cyanobacteria and plants.

Present Atmospheric Composition

By volume, today's atmosphere is about 78 percent nitrogen, 21 percent oxygen, 0.9 percent argon, and 0.04 percent carbon dioxide, with neon, helium, nitrous oxide, methane, ozone, and other gases in trace amounts (Figure 11-5).[14] The percentage of water vapor varies with temperature, from about 0 percent in cold air to nearly 4 percent in hot air.

The small absolute amounts of carbon dioxide, water vapor, and methane in the atmosphere belie their importance. These greenhouse gases are critical for

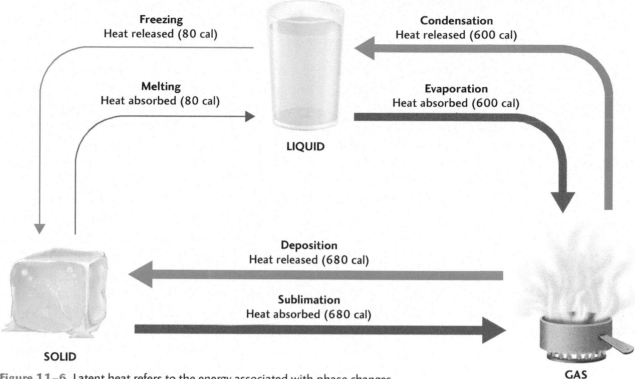

Figure 11-6 Latent heat refers to the energy associated with phase changes between solid, liquid, and gaseous water. Water evaporated from the equatorial oceans carries latent heat to higher latitudes where it may be released when the water vapor condenses back to a liquid.

regulating Earth's temperature and keeping it within a range that is able to sustain life. Without them, Earth would have an average surface temperature of −18°C (0°F), too cold for life as we know it to have originated and evolved.[15] Water vapor is particularly important because it, along with ocean currents, is responsible for transporting heat from the equator to higher latitudes. Evaporation of water from ocean basins occurs when water absorbs solar energy and changes phase from a liquid to a gas. The energy absorbed, known as *latent heat energy,* is stored in the water vapor and released when the vapor condenses back to water droplets, forming clouds (Figure 11-6). Energy that water picks up near the equator is released when the vapor reaches higher latitudes and condenses, making the higher latitudes warmer than they otherwise would be. Likewise, removal of heat from the tropics, through evaporation and other processes, keeps them from becoming unbearably warm. Evaporation and condensation also bring vital precipitation to land and to terrestrial organisms.

Like greenhouse gases, ozone also is found only in trace amounts in the atmosphere but is critical to keeping Earth habitable. Ozone at high altitudes absorbs much of the incoming ultraviolet (UV) radiation from the Sun, the same radiation responsible for sunburns, various genetic mutations, and skin cancer. Without ozone, incoming UV rays would destroy all life on land; the fossil record suggests that the terrestrial part of early Earth indeed remained lifeless until sufficient ozone accumulated in the atmosphere to block the harmful radiation (Box 11-1).

In addition to gases, the lower parts of the atmosphere contain some suspended liquid and solid particles, such as acid droplets and dust, which are referred to as **aerosols.** After volcanic eruptions, the concentration of aerosols in the atmosphere increases but then diminishes with time as the particles settle back to the surface. Aerosols play an important role in the hydrologic cycle because they provide a surface on which water vapor can condense to form water droplets to make clouds. However, some aerosols, introduced into the atmosphere both by natural processes and human activities, are also pollutants responsible for the production of acid rain. (The human impacts on atmospheric chemistry will be discussed later in this chapter.)

BOX 11–1 GLOBAL AND ENVIRONMENTAL CHANGE

The Intertwined Histories of Atmospheric Oxygen and Life

Information on the early composition and evolution of Earth's atmosphere is scant, but it is widely accepted that initially there was little free oxygen. Instead, the atmosphere probably was composed of gases such as carbon dioxide, water vapor, hydrogen, methane, ammonia, and helium. Hydrogen and helium were lost quickly due to their low masses and therefore weak gravitational attraction to the planet. Most of the other gases were stripped off the planet by solar winds, high-energy particles emanating from the Sun. When the solid Earth acquired sufficient mass, gravitational collapse of the heavier elements into its core promoted melting that released enormous volumes of carbon dioxide, water vapor, and sulfur dioxide, some of which condensed to form the oceans, and the rest of which remained to form an atmosphere. Ultraviolet rays from the Sun decomposed some of the water and carbon dioxide in the atmosphere into hydrogen (H^+), hydroxyl radical (OH^-), carbon monoxide (CO) and atomic oxygen (O):

$$H_2O + \text{solar energy} \rightarrow H^+ + OH^-$$

$$CO_2 + \text{solar energy} \rightarrow CO + O$$

The newly-liberated atomic oxygen combined with the hydroxyl radical to produce small amounts of molecular oxygen (O_2), which in turn combined with atomic oxygen to produce a very thin layer of ozone (O_3):

$$O + OH \rightarrow O_2 + H$$

$$O_2 + O \rightarrow O_3$$

Still, atmospheric concentrations of oxygen were very low, certainly less than 0.1 percent of present-day levels and possibly as little as 0.0001 percent.[16] One line of evidence for this low oxygen environment comes from the presence of ancient soils that lost their iron during weathering. As mentioned in Chapter 4, iron is extremely insoluble and, therefore, immobile when oxidized to minerals such as hematite (Fe_2O_3) and limonite ($FeO(OH) \cdot nH_2O$). Only when iron is in its reduced state can it be leached from soils and rocks, so the lack of iron in soils older than 2.5 billion years implies a low oxygen atmosphere.[17]

Early life was adapted to this world without oxygen and developed processes to acquire the energy it needed to survive without the gas. Methanogens, microorganisms belonging to the domain Archaea, reduce carbon dioxide in the presence of hydrogen to generate energy and produce methane (CH_4) as a by-product. This *anaerobic* respiration pathway thereby differs from the *aerobic* respiration described in Chapter 7 in which organisms use oxygen in combination with organic matter to generate energy. While we still don't know exactly when methanogens evolved, research published in 2006 found that microbially produced methane exists in tiny fluid pockets inside of minerals dating to more than 3.46 billion years in age, providing a minimum estimate of their antiquity. Their prodigious production of methane, a potent greenhouse gas, was partially responsible for making Earth warmer than it is today, a subject we address further in Chapter 14.

Fossil and geochemical evidence suggests that by 3.8 to 3.45 billion years ago the first photosynthesizing cells, cyanobacteria that produce microbial mats known as stromatolites,[18] had evolved. A by-product of photosynthesis is oxygen (Chapter 7), a gas toxic to methanogens. Early in Earth's history, the rate of oxygen production by cyanobacteria was far out-paced by oxygen consumption in chemical reactions involving elements such as iron, keeping atmospheric oxygen levels low and producing the banded iron formations discussed in Chapter 4. Methanogens therefore continued to be abundant.

Around 2.45 billion years ago, atmospheric oxygen levels climbed dramatically in an episode that has come to be known as the Great Oxidation Event (GOE) (Figure 1). Evidence for this event includes the cessation of banded iron deposition in the oceans and the presence, thereafter, of iron oxide minerals in soils. Based on this evidence, oxygen levels must have risen to at least 1 percent of modern levels, and possibly as high as 40 percent. What caused the GOE? Scientists are still looking for answers. It's possible that oxygen produced by cyanobacteria began to accumulate once all the iron at Earth's surface finally oxidized.

Another hypothesis calls on a catastrophic decline of methanogens brought about by changes in volcanic eruption styles.[19] Methanogens use the metallic element nickel in many of their metabolic processes. Recent evidence suggests that the amount of nickel brought to Earth's

(continued on page 366)

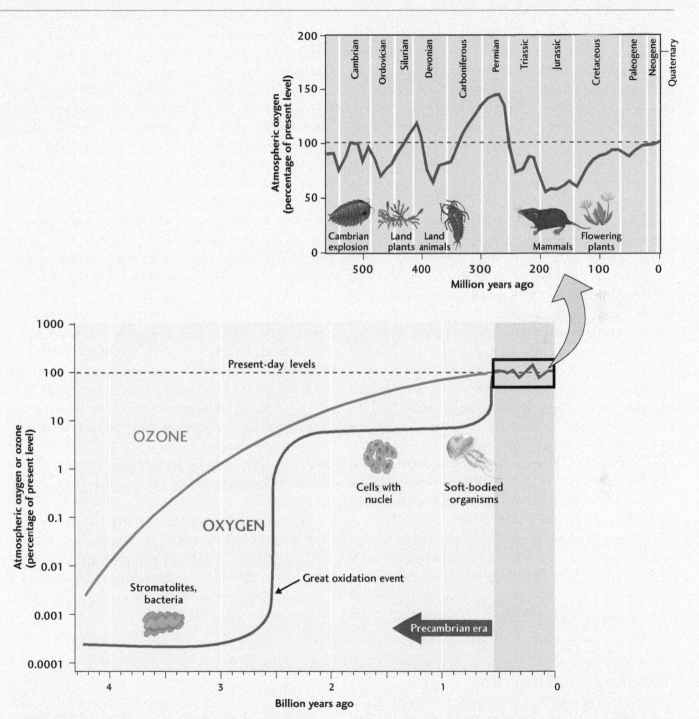

Figure 1 The evolution of atmospheric oxygen and ozone was driven by the evolution of photosynthesizing organisms. Early in Earth's history, oxygen produced by cyanobacteria oxidized iron, keeping oxygen levels in the atmosphere very low. These bacteria left a record of their existence in stromatolites, algal structures that date back to 3.5 byr in age. Eventually all the iron in contact with the atmosphere was fully oxidized, allowing oxygen to accumulate in the atmosphere and producing the so–called Great Oxidation Event. Oxygen levels climbed again around 600 mya in response to an increase in carbon burial rates. As oxygen increased, so too did atmospheric ozone, which blocked enough ultraviolet radiation so that plants could evolve on land. Thereafter, atmospheric oxygen and ozone levels rose to near their present values.

(continued)

surface via volcanic eruptions declined between 2.7 and 2.5 billion years ago as the planet cooled and magma production shifted from deeper in Earth's mantle where nickel is abundant to higher up in the crust where it is not. Declining nickel would have cut off a critical element for methanogens, decreasing their production of methane. Since methane oxidizes to carbon dioxide in the presence of oxygen, reduced levels of methane would have allowed oxygen to accumulate in Earth's atmosphere. Then again, perhaps the increasing prevalence of toxic oxygen in the atmosphere itself led to the decline of methanogens.

Whatever the cause of the GOE, it was followed by a second rise in atmospheric oxygen between 635 and 510 million years ago that paved the way for the evolution of multicellular life. A number of ideas have been advanced to explain this rise, all of which call on an increase in the rate of organic matter burial to counteract the usual respiration reactions that oxidize organic carbon to carbon dioxide (Chapter 7). For example, since zooplankton produce fecal pellets that are filled with organic matter and that sink to the bottom of the ocean, perhaps their evolution allowed for a net transport of carbon from the atmosphere into the surface ocean and then to burial in the deep ocean, promoting the atmospheric oxygen rise. While an interesting idea, this hypothesis came under fire when it was determined that oxygen levels began to rise well before the evolution of zooplankton large enough to create fecal pellets big enough to sink quickly.

An idea proposed in 2010 to explain the increased burial of carbon that drove the second oxidation event is the tectonic coalescence of the Gondwana supercontinent around 620 million years ago and the uplift of the Transgondwanan Supermountains by continent–continent collision. According to this hypothesis, the uplift of high mountains would have increased erosion rates that raised the fluxes of nutrients and sediments to the ocean basins. Increased nutrient fluxes would have stimulated biological productivity, which would have created more organic matter and oxygen. The accompanying increased sedimentation rates would have buried the organic matter, isolating it from microbial respiration that would otherwise return its carbon to the atmosphere.[20] By this mechanism, oxygen levels could have climbed to within 60–80 percent of their modern atmospheric levels.

Time will tell whether this hypothesis turns out to be correct, but whatever the cause, as oxygen levels in the atmosphere climbed, the ozone that continues to protect Earth's surface from ultraviolet radiation began to be produced in the stratosphere. Such protection permitted life to emerge near ocean surfaces and ultimately on land and enabled many invertebrate species to evolve, including soft-bodied sponges and worms and hard-shelled mollusks. About 400 million years ago, land plants evolved and diversified. As a result, atmospheric oxygen levels soared, reaching values perhaps as high as 160 percent of modern levels (or in absolute amounts, up to 35 percent of atmospheric gas volume) by 300 million years ago. This was a time marked by widespread equatorial coal swamps and very high organic carbon burial rates. Subsequent drift of continents away from the equator reduced the number of swamps and, along with them, the production of oxygen and burial of organic carbon. Nonetheless, high concentrations of charcoal in some sedimentary rocks indicate that oxygen has been at levels in excess of 60 percent of modern values, the level required to sustain fire, for at least the last several hundred million years.

Atmospheric Structure

Like Earth's interior, the atmosphere system is organized into layers with different chemical and physical properties (Figure 11-7).

The Troposphere The lowest layer of the atmosphere, the **troposphere**, extends from Earth's surface to heights between 10 and 16 km, depending on location; the elevation is highest at the equator and lowest at the poles.[21] On average, Earth's surface is warm and provides heat to the bottom of the troposphere, but the temperature drops to between –55°C and –80°C (–67°F and –112°F) at the troposphere's upper boundary, referred to as the *tropopause*. Consequently, air temperature declines with elevation at a rate of 4°C–9.8°C per km (11.5°F–28.2°F per mi) depending on location.[22] You have probably experienced this decrease in temperature with elevation, called the **environmental lapse rate,** if you have ever driven upward into a mountain range. In the summer, shorts and T-shirt weather at the base of a mountain can give way to sweater, hat, and mitten temperatures at higher elevations, and the changing temperature profoundly influences the type of vegetation that can grow with increasing altitude.

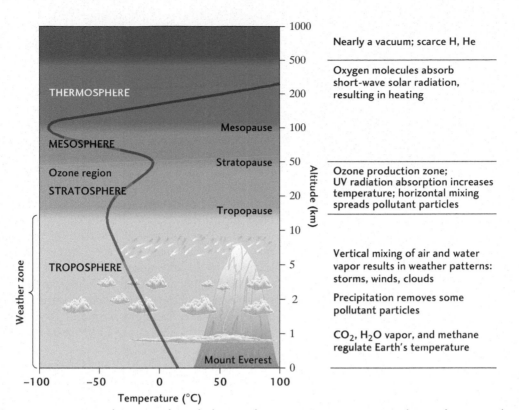

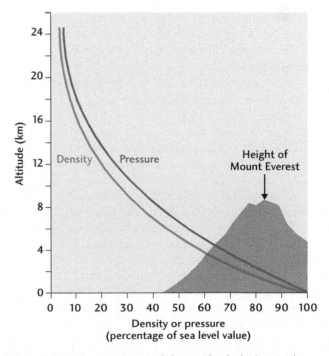

Figure 11–7 Temperature in the troposphere declines with altitude (which here is measured logarithmically) because the troposphere is heated from below by Earth's surface. Concentration of ozone (O_3) in the stratosphere allows temperatures to rise, but in the mesosphere they fall again. Molecular oxygen (O_2) in the thermosphere absorbs heat energy, so temperature rises with altitude. Clouds depict the location of water in the atmosphere.

In addition to the fact that the troposphere is heated from below by its contact with Earth's surface, much of the temperature decline with elevation results from the decline of atmospheric pressure (force per unit area) with altitude (Figure 11-8). The pressure (P) at any level within the atmosphere can be represented mathematically as:

$$P = \rho g h$$

where ρ is the density of air, g is the gravitational acceleration at Earth's surface (equivalent to 9.8 m/s^2 or 32 ft/s^2), and h is the thickness of the overlying air column. As air moves to higher elevations with less overlying atmospheric mass (i.e., lower values of h and ρ), the pressure exerted on it declines, allowing it to expand. To expand, the gas molecules must expend energy, resulting in a cooling of the air, a phenomenon known as **adiabatic cooling** (conversely, compression of air results in warming). The amount of water vapor that air can hold, called the **saturation vapor pressure**, decreases as temperature declines (Figure 11-9), and as air cools, the vapor condenses around dust particles and other aerosols. The condensation process releases latent heat energy to the surrounding air, thus the range of values for the environmental lapse rate arises from the varying moisture

Figure 11–8 Air pressure and density both decline with altitude. Because air is compressible, these parameters decrease nonlinearly.

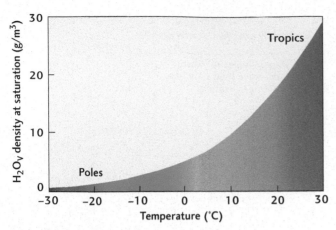

Figure 11–9 The water vapor content of air with temperature. Warm air can hold much more water in vapor form than can cold air.

contents of the atmosphere in different locations. Air parcels fully saturated with water vapor, such as might exist over an ocean or a large lake, experience a lower decline in temperature with elevation than do dry air parcels over a desert.

If condensing water droplets grow large enough that they can no longer be held aloft by turbulence in the air, they will fall as rain, thereby cleansing the air of the aerosol particles around which the droplets nucleated. This process allows many natural and human-produced compounds to be removed from the lower and middle troposphere and is one of the key processes involved in the production of acid rain. The extremely cold temperatures at the top of the troposphere cause the volume of water vapor there to approach zero and limit the amount of water that can enter higher levels in the atmosphere.

The troposphere is marked by strong mixing because of declining density (mass per unit volume) and temperature with elevation. Density at ground level, where the air is compressed by the weight of the overlying atmosphere, is much greater than density at high levels in the atmosphere, which have less overlying mass (see Figure 11-8). However, when a thin layer of air is warmed by contact with Earth's surface, it expands, and its density decreases markedly. The layer will therefore rise until it reaches a level in the atmosphere with the same low density. If a parcel of air rises in one place, *conservation of mass* says that another parcel must descend in another place, leading to vertical and horizontal mixing. The mixing of masses of air and moisture in the troposphere is manifested at Earth's surface as changing weather patterns, storms, clouds, and wind.

The Stratosphere Above the troposphere, at about 10 to 50 km (6 to 31 mi), lies the **stratosphere**, a prominent zone of ozone (O_3) production. Here UV rays from the Sun break O_2 molecules apart and recombine the O atoms with other O_2 molecules to form O_3.[23] Because ozone is a strong absorber of UV radiation and UV rays enter from the top of the atmosphere, the temperature of the stratosphere increases with altitude.[26] On its own, cooler, denser air cannot rise above warmer, thinner air, so no appreciable vertical mixing occurs in the stratosphere. The stratosphere also contains about 1000 times less water vapor than does the troposphere. This, combined with the lack of vertical mixing, allows volcanic ash and human-made pollutants to remain there for many years once introduced. Horizontal mixing still takes place, however, allowing these pollutants to be distributed over the globe. Recall from Chapter 5, for example, that the eruptions of the volcanoes Tambora and Mount Pinatubo generated significant global cooling as horizontal mixing spread Sun-blocking ash and sulfate aerosols throughout the stratosphere. Tambora is one of the eruptions that may have contributed to the Little Ice Age.

The Upper Atmospheric Layers About 99.9 percent of the total mass of the atmosphere is found in the troposphere and stratosphere,[21] so most atmospheric circulation of consequence to weather systems occurs below the *mesosphere,* a second layer of mixing that exists at altitudes of about 50 to 85 km (31 to 53 mi) above Earth. It is in the mesosphere that most incoming meteors burn up, generating the shooting stars we enjoy watching at night. Above the mesosphere, oxygen molecules (O_2) absorb short-wave solar radiation (UV light), resulting in heating and increased temperature to about 160 km (96 mi) in a zone known as the *thermosphere.* Above the thermosphere are layers primarily of helium and hydrogen, the lightest elements in the universe.

In 2007, the National Aeronautics and Space Administration (NASA) launched a satellite designed to study the mesosphere after atmospheric scientists noticed an increasing frequency of clouds of ice crystals in that layer.[25] These *noctilucent clouds,* so-called because they appear to glow at night, are thought to form from the tiny amount of water vapor that makes it into the mesosphere and nucleates on the dust of burned up meteorites, and they form primarily at high latitudes during the summer months. At present, little is known about these clouds, but it is possible that their increasing frequency is related to global climate change.[26] Many years of satellite measurement will be necessary in order to understand these interesting phenomena.

In summary, the upper atmosphere shields us from excess radiation and meteor bombardment, while the lower atmosphere determines our weather.

■ Trace gases in the atmosphere, such as carbon dioxide, ozone, and water vapor, are responsible for keeping Earth warm enough to sustain life. They also protect us from UV radiation.

■ Atmospheric oxygen levels have risen over geologic time in response to the evolution of photosynthetic organisms.

■ Vertical and horizontal mixing in the troposphere produces winds, storms, and clouds.

■ Layering of the atmosphere inhibits the vertical transport of material between the troposphere and stratosphere.

■ The concentration of water vapor in the atmosphere declines rapidly with elevation, and almost all water is confined to the troposphere.

■ Pollutants found in the troposphere are easily removed by raindrops, while stratospheric pollutants may linger for longer periods of time.

ATMOSPHERIC CIRCULATION AND CLIMATE

Circulation of the troposphere produces climate and weather. **Climate** refers to the long-term atmospheric and surface conditions that characterize a particular region. **Weather** refers to daily fluctuations in temperature, air pressure, wind speed, and precipitation. We can think of climate as the long-term average of daily weather events.

Earth's climate system determines where humans can live for long periods of time unaided by external resources. The climate of Antarctica, for example,

with its frigid temperatures and high wind speeds, is extremely inhospitable to humans, and farming on ice sheets is out of the question. Although many scientists live on Antarctica for months at a time, they are completely dependent on supplies of food and fuel flown in from lower latitudes.

Weather events may be severe, but they do not in and of themselves restrict human habitation. The winter 2009–2010 negative phase Arctic Oscillation described at the beginning of this chapter, for instance, did not force people to move away from Florida; it was a singular event from which the population readily recovered. If, however, a change in climate caused the negative phase AO to become a constantly recurring event, those who moved to Florida to escape cold winters might look elsewhere for housing.

A number of different factors are responsible for making Earth habitable for humans and other life-forms.

The Greenhouse Effect

Every year, Earth's surface receives 1.73×10^{17} watts of energy from the Sun. This is about 15,000 times as much energy as the entire human population uses annually, and it drives the photosynthetic reactions carried out by plants that we depend upon for food (see Chapter 7). Energy emitted by the Sun consists primarily of UV radiation, visible light, and infrared radiation—that is, heat (Box 11-2). About 25 percent of this incoming energy is reflected back to outer space by clouds and atmospheric gases and another 5 percent by Earth's ground surface, leaving about 70 percent to be absorbed (Figure 11-10).

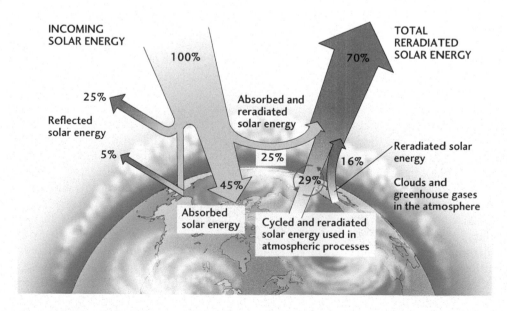

Figure 11–10 Incoming solar radiation takes several pathways once striking Earth. Some energy is reflected directly to outer space off of cloud tops and atmospheric gases, some is absorbed within the atmosphere, and some is absorbed by Earth's surface.

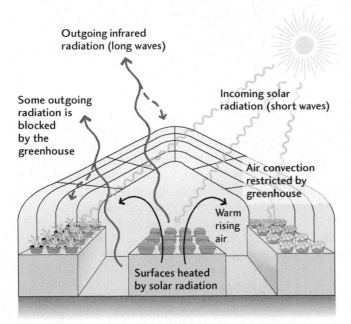

(a) A greenhouse stays warm by allowing in radiation and trapping heated air

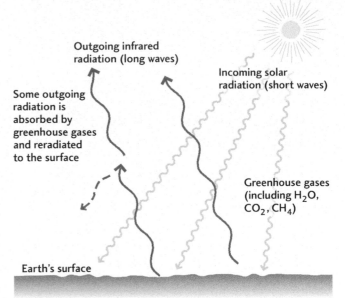

(b) Earth's atmosphere stays warm because greenhouse gases absorb outgoing infrared radiation from Earth's surface and reradiate it

Figure 11–11 (a) In a greenhouse, sunlight passes through the glass and is absorbed by plants and the ground within. The ground and plants then emit the energy as infrared radiation that heats the greenhouse air. The glass holds in all the heated air and a bit of the escaping radiation. (b) In the atmosphere, greenhouse gases absorb some of the outgoing radiation given off by Earth and emit part of the energy back toward Earth's surface, thus keeping the planet warm.

Of this remaining energy, a little more than a third is absorbed within the atmosphere (for example, by ozone, as mentioned above), leaving only 45 percent of the original solar beam to strike the ground.

As Earth's surface absorbs this energy, it heats up and then reemits the energy as infrared radiation. All of this heat would escape into space if Earth had no atmosphere. Instead, some outgoing energy is trapped by atmospheric gases. Nitrogen and oxygen, the primary constituents of the atmosphere, absorb very little infrared energy. Carbon dioxide (CO_2), methane (CH_4), water vapor, and nitrous oxide (NO_2) are present in only trace amounts but are very efficient at absorbing this energy and redirecting it, some of it to outer space and some back to Earth's surface. This process is called the **greenhouse effect** by partial analogy to the way greenhouses maintain heat (Figure 11-11a). In a greenhouse, incoming solar energy is absorbed by plants, the ground, and other objects. These objects then reemit the energy, thereby warming the air. The glass walls trap the heated air, which circulates within their confines. In the larger atmosphere, infrared radiation given off by Earth's surface is trapped by greenhouse gases, which then reradiate some of the heat energy back to Earth's surface, causing it to warm (Figure 11-11b). In the absence of the greenhouse effect, Earth's average surface temperature would be about 33°C (59°F) colder than it is today, and the entire surface of the planet would be frozen.[27]

Differential Heating of Earth

If Earth were a flat surface oriented perpendicular to incoming solar rays, its area would be heated uniformly. Because Earth is a sphere, however, the energy is not evenly distributed (Figure 11-12a). Polar locations are farther from the Sun than equatorial locations; hence polar areas receive less energy than the equator. The difference in distance to the Sun is practically negligible, however, when compared to the Earth–Sun distance itself, so the tiny difference in location between the poles and equator is of little consequence to climate.

Of far greater importance to surface temperature is the angle at which solar rays strike Earth's surface, because the angle of incidence determines the area over which the incoming radiation is spread. Due to Earth's nearly spherical shape, the incoming radiation striking near the poles is spread over a much greater area

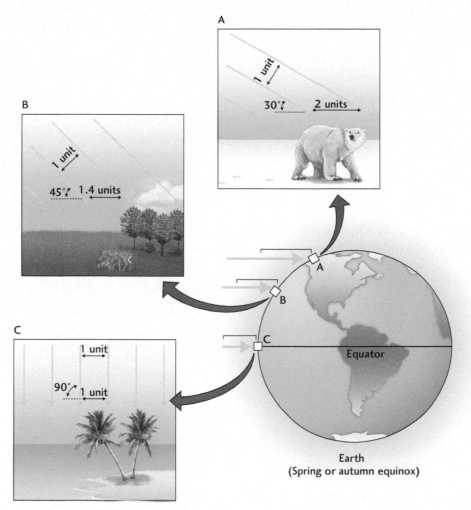

A

1 unit

30° 2 units

B

1 unit

45° 1.4 units

C

1 unit

90° 1 unit

A

B

C

Equator

Earth
(Spring or autumn equinox)

Figure 11–12 Radiation striking a flat area is distributed evenly over the surface; radiation hitting a sphere, however, is not. At high latitudes (a), the route taken through the atmosphere by the radiation is longer than at low latitudes (b), causing more energy to be reflected to outer space. In addition, solar rays hitting higher latitudes make a more oblique angle with the surface and spread out over a much greater area than those at lower latitudes. For these reasons, less energy is absorbed at high latitudes than at low latitudes.

than the same amount of radiation hitting the equator (Figure 11-12b), causing the poles to be colder. Earth's spherical shape also affects the thickness of the atmosphere through which incoming radiation travels to reach the surface. As already mentioned, atmospheric gases and clouds reflect a lot of the incoming radiation back to space. More energy is reflected by the atmosphere at high latitudes, because energy arriving at high latitudes takes a longer route through the atmosphere before reaching the ground than energy hitting the low latitudes does.

Not only does Earth's spherical shape influence how much solar energy different regions receive during the year, so too does the tilt of Earth's axis of rotation.

Summer occurs in a hemisphere when it is tilted toward the Sun, winter when it is pointed away from the Sun (Figure 11-13). This explains why one hemisphere experiences winter while the other experiences summer. The tilted axis also explains why the polar latitudes spend several months each year in total darkness. In contrast, the equatorial regions are illuminated year round, another reason why the poles are relatively colder than the equator. Still another factor influencing solar radiation receipts is the reflectivity, called the **albedo**, of Earth's surface. The polar latitudes are covered with snow and ice, which reflect 40 to 90 percent of incoming solar energy straight back to space. In contrast, the equatorial latitudes are covered in dark

BOX 11–2 GEOLOGIST'S TOOLBOX

Electromagnetic Radiation

All bodies in the universe give off *electromagnetic radiation* (Figure 1), energy transmitted in the form of waves. The wavelength (λ)—the distance from the crest of one wave to the crest of the next—of the radiation emitted is inversely related to the temperature of the body:

$$\lambda = \frac{2897}{T}$$

(T, in degrees Kelvin, K. 0 K = −273°C, and there is a one-to-one relationship between degree changes in the two temperature scales.)

Bodies at high temperature (high internal heat energy) radiate at short wavelengths (small λ), and colder objects radiate at longer wavelengths (larger λ). The range of wavelengths is called the *electromagnetic spectrum,* on which radio waves are the longest and gamma rays are the shortest. Much of the Sun's radiation, with wavelengths ranging from 0.4 to 0.7 μm, is in the visible part of the spectrum. Earth, because it is much cooler than the Sun, radiates at a longer wavelength, in the infrared part of the spectrum. We cannot see infrared radiation, but we can feel it as heat.

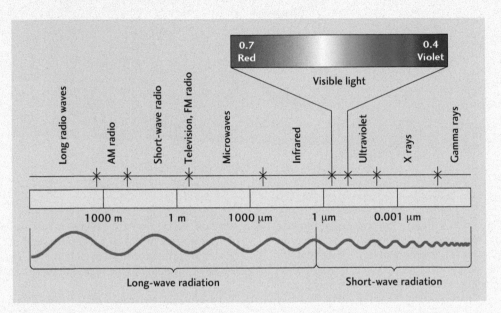

Figure 1 The electromagnetic spectrum. The Sun radiates energy primarily in the visible, infrared, and ultraviolet parts of the spectrum; Earth radiates in the infrared.

blue ocean water and dark green tropical forests, both of which reflect little incoming energy.

Averaged over Earth's entire surface, incoming solar radiation is balanced by outgoing planetary radiation—heat energy from the radioactive decay of elements in the interior as well as from solar energy absorbed by the surface and reradiated to outer space. Without this balance, our planet as a whole would not have a steady temperature but would either heat up or cool down. Despite this overall energy balance, however, individual latitudes

are not in balance. The poles lose more energy than they receive, whereas the equator receives more energy than it loses. Only at about 40° latitude north and south (the latitude of Denver, Philadelphia, Madrid, and Beijing in the northern hemisphere and Wellington, New Zealand, in the southern hemisphere) does the amount of solar energy gained equal the amount of planetary energy given off (Figure 11-14).

If the poles did not receive an inflow of energy from lower latitudes, they would become colder every

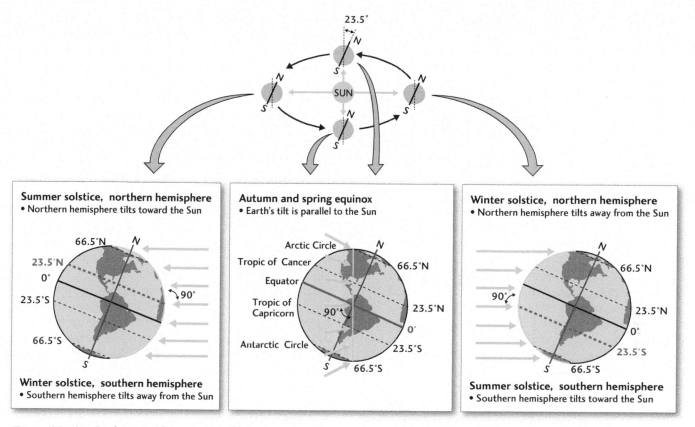

Figure 11–13 Earth's axis of rotation is tilted. The red lines in each of the globes represent the latitudes at which the Sun's rays are directly overhead. Summer occurs when a hemisphere is tilted toward the Sun. Winter occurs when a hemisphere is tilted away from the Sun. The poles are in complete darkness during winter, while the equator is illuminated year–round.

year. Likewise, because the equator receives more solar energy than it loses to space, it would continuously heat up if there was no way of removing heat energy. These effects are arrested because the atmosphere and oceans redistribute heat energy through the winds and ocean currents, carrying warm air and water from the equator to the high latitudes and colder air and water from the poles to the equator. This redistribution makes all parts of Earth more tolerable for life than they would be in the absence of circulation.

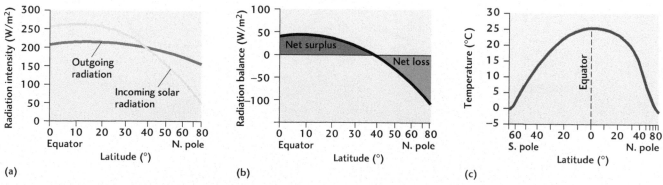

Figure 11–14 (a) Incoming and outgoing radiation varies with latitude. (b) The poles lose more energy than they receive, whereas the equatorial regions gain more energy than they lose. Energy gain and loss are equal at about 40° latitude north and south. (c) Temperature varies with latitude. The mean annual temperature of the polar regions is far lower than that of the equatorial region.

Tropospheric Circulation

The mechanism of energy transfer between the equator and the poles is *convection*, the same process you see on your stovetop when you boil a pot of water, a process that is also a possible driver of tectonic plate motions (see Chapter 2). Convection in the troposphere occurs because the equatorial regions experience a net gain of solar energy every year, heating the air so that it becomes less dense and rises. As the heated air moves poleward, it radiates heat to outer space and becomes colder and therefore denser. (It also cools off because of adiabatic expansion.) This cold, dense air sinks at the poles. Because air has mass, it is gravitationally attracted to Earth and therefore exerts pressure on the surface. Conversely, when air rises, some of the mass is pulled away from Earth's surface, leading to a lower pressure than would otherwise exist. The rising and sinking air masses lead to pressure gradients (changes in pressure over distance) in the troposphere (Figure 11-15). Air flows along these pressure gradients, producing winds. The steeper the pressure gradient—that is, the greater the pressure difference between two places—the faster the winds flow.

The Coriolis Effect and Hadley Cell Circulation If Earth did not rotate, atmospheric circulation would take the form of two huge convection cells, one for each hemisphere. Earth does rotate, however, making large convection cells unstable. Air and water currents are deflected to the right in the northern hemisphere and to the left in the southern hemisphere. This phenomenon is known as the **Coriolis effect**. To understand its cause, begin by imagining firing a rocket from the North Pole toward Mount Kilimanjaro, near the equator (Figure 11-16). The rocket is fired due south, but by the time it arrives at the equator, Mount Kilimanjaro has rotated to the east along with the rest of Earth, and the rocket misses its mark. Relative to its original trajectory from the North Pole, the rocket appears to have been deflected to the right, or west.

Earth completes a full rotation in one day, so all points on its surface come back to their starting position in 24 hours. Because of Earth's spherical shape, not all points on its surface travel at the same rotational velocity, however. Velocity varies with latitude, from zero at the poles, where Earth's circumference measured along lines of latitude is zero, to a maximum value at the equator, where the globe is widest (see Figure 11-16). Mount Kilimanjaro, near the equator, moves eastward rapidly, while the pole doesn't move at all, so that by the time the rocket reaches the equator, the mountain has moved far to the east and the rocket, if its course is uncorrected, lands west of it.

Now imagine firing a rocket from the equator toward the North Pole. The rocket has a northward velocity from the launch, but it also has an eastward

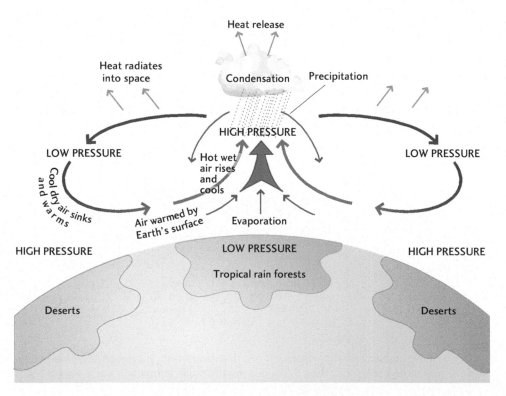

Figure 11–15 Heated air rises, producing low pressure at Earth's surface and high pressure in the atmosphere. As it rises, the air cools and water vapor condenses to water droplets. The cooled air then returns to Earth's surface, producing high pressure at the surface and low pressure at higher altitudes. Air flows along the pressure gradients from high to low pressure.

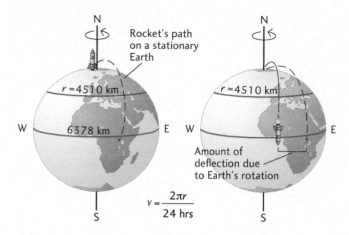

Figure 11–16 Coriolis deflection due to Earth's rotation would cause a rocket launched from the North Pole directly at Mount Kilimanjaro (marked by the red x) to land west of its target. Note that *v* is Earth's rotational velocity in kilometers per hour where *r* is the radius of Earth relative to the spin axis.

velocity because it was originally affixed to the rotating Earth. As it travels northward, the rocket passes over land moving at slower and slower eastward rotational velocities. Consequently, the rocket "overtakes" the land and again is deflected to the right (this time, east) of its intended trajectory.

Just like the rocket, air currents are influenced by the Coriolis effect. Air flowing poleward from the equator is deflected by the Coriolis effect so that by the time it reaches 30° north or south latitude it is flowing due east (Figure 11-17). Having lost much of its heat energy

by radiation to space and by expansion, the air also has cooled and become denser. The accumulation of cold, dense air at 30° is drawn downward. Some of the sinking air travels back toward the equator, creating a circulation loop known as a **Hadley cell** (Figure 11-18). The rest of the air flows poleward along the surface of Earth. At 60° latitude north or south, this poleward-flowing air converges with air traveling toward the equator from the poles. The convergence results in uplift of air at 60° and creation of another circulation cell between 60° and 90° latitude, called the **polar cell** (see Figure 11-18). The latitudinal belt between 30° and 60°, sometimes referred to as the **Ferrel cell,** is a zone of transition between the warm, tropical Hadley cell and the cold, polar cell, and has a more complicated circulation marked by a series of high and low pressure zones that change location with the seasons, particularly in the northern hemisphere (more on the influence of land and sea on tropospheric circulation below).[28]

Jet Streams The large temperature gradients associated with the boundaries of the vertical circulation cells generate large localized pressure gradients that create **jet streams,** narrow (a few hundred kilometers wide), high speed currents of air (up to 90 m/s or 200 mi/h) in the upper troposphere that flow from west to east and can encircle the entire globe. Jet streams got their names during World War II when military pilots discovered the high velocity currents of air. Today, aviation takes advantage of these winds to speed up the travel between the continents of the northern hemisphere, and if you have ever traveled between these locations, you may

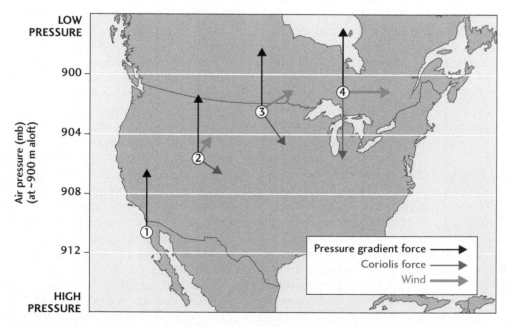

Figure 11–17 Air moving down a pressure gradient is deflected by the Coriolis effect such that the air ends up flowing parallel to that gradient.

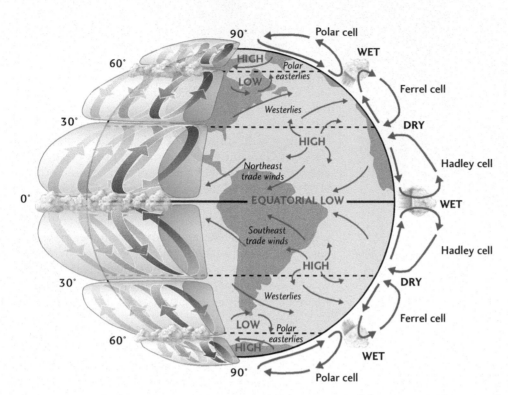

Figure 11–18 An idealization of atmospheric circulation on a rotating Earth. Circulation is broken into three large convection cells called Hadley cells. Coriolis deflection of the surface winds results in easterly winds from 0° to 30° and from 60° to 90° latitude, and westerlies from 30° to 60° latitude, north and south of the equator. Rising warm air at the equator and at 60° causes condensation, cloud formation, and precipitation at these latitudes. Sinking cold air at 30° and at the poles causes deserts to form.

have noticed that it takes about an hour longer to travel from east to west than from west to east.

Each hemisphere contains two jet streams: a subtropical jet located at roughly 30°–40° latitude and a polar front jet whose position varies markedly with season. During the summer months, the polar front jet is found at around 50°–60° latitude, but during the winter, it migrates toward the equator, developing pronounced meanders that carry parts of it toward 30° latitude. In winter the polar front jet also increases in velocity because the temperature gradient, and therefore the pressure gradient, that produces it is more extreme during the winter months than in the summer. Because the polar front jet marks the site of convergence between air moving equatorward from the pole and poleward from the equator and because converging air at the surface causes uplift, cooling, and condensation, the position of the polar front jet tends to reflect where storms are produced.

Surface Winds Like the air flowing in the upper troposphere between the equator and 30° latitude, air flowing at Earth's surface is deflected by the Coriolis force. In the warm Hadley cell, air therefore moves from the northeast toward the southwest in the northern hemisphere, and from the southeast toward the northwest in the southern hemisphere, to form belts of wind known as the **trade winds** (see Figure 11-18). Sailors traveling from Europe or Africa toward the New World quickly learned to make use of these winds to shorten their travel time. Coriolis deflection in the latitudes between

30° and 60° causes surface winds to blow from the west toward the east, and these winds are given the name **westerlies**. At the polar latitudes, surface winds flow from the pole toward the equator and are deflected to become **polar easterlies**, so called because they flow from the east toward the west.

Tricellular Circulation and Climate The tricellular model of tropospheric circulation is a simplification of the actual circulation in the atmosphere but is useful in explaining Earth's distribution of deserts and wet regions. Hot air can hold a lot of water vapor (see Figure 11-9). As that hot air rises to higher elevations and cools, the vapor condenses into clouds and then falls as rain when the droplets become sufficiently large. This process accounts for the enormous quantities of rain that fall on the equatorial regions where the trade winds meet in the **Intertropical Convergence Zone (ITCZ)** and are forced aloft. Having lost much of its water vapor, the now drier air moves poleward, sinks at 30° latitude, and warms up from adiabatic compression. This warming increases the air's saturation vapor pressure, limiting condensation and rainfall and increasing evaporation at the surface. It is no surprise then, that Earth's major deserts are found in latitude belts centered about 30° north and south (Figure 11-19).

Although they don't look like conventional deserts, the poles are also considered arid lands because they receive very little moisture. The colder air gets, the less moisture it can hold, so air masses moving toward the

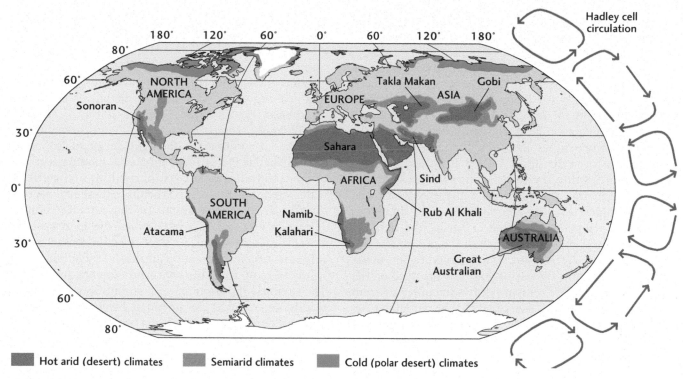

Hadley cell circulation

Hot arid (desert) climates **Semiarid climates** **Cold (polar desert) climates**

Figure 11–19 Most deserts are found in belts centered at about 30° latitude north and south. Deserts are also located at the poles and in the "rain shadows" of great mountain ranges such as the Andes, Rockies, and Himalayas.

poles are unlikely to drop much precipitation at high latitudes. Furthermore, air sinking at the poles warms as it sinks, increasing its ability to hold moisture in vapor form.

Influence of Land and Sea on Tropospheric Circulation Landmasses complicate the simple Hadley cell model for a variety of reasons and influence local climate

and weather patterns. First, air may be forced to flow around or up and over obstructions such as mountain ranges and plateaus. The air cools as it rises, causing any vapor to condense and rain out. As the now dry air flows over a mountain and down the other side, it becomes progressively warmer due to compression and can therefore hold more moisture, increasing evaporation (Figure 11-20). This is why mountain ranges

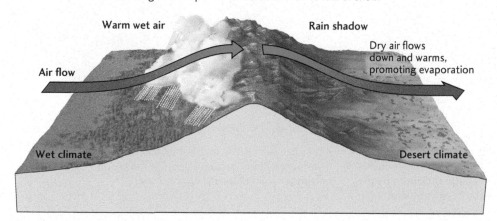

As air is forced upward over the mountains, it cools, causing water vapor to condense and fall as rain or snow.

Warm wet air

Rain shadow

Air flow

Dry air flows down and warms, promoting evaporation

Wet climate

Desert climate

Figure 11–20 Moist air on the windward side of a mountain is forced upward, where it cools and condenses. Precipitation strips the air of moisture. On the leeward side of the mountain, the air subsides and warms. The dry, warm air evaporates water from the ground surface. This process creates a "rain shadow" and a very arid climate in the lee of the mountain.

commonly have a "rainy" side and a "dry" side, and the dry side is often described as being in the "rain shadow" of the mountains.

Second, the very different thermal properties of oceans and landmasses lead to the generation of sea and land breezes. Heat from the Sun is easily carried deep (about 100 m) into ocean water because water is transparent, but on land, the Sun's heat is absorbed by only a thin surface layer. Furthermore, heating a unit volume of water by 1°C (1.8°F) takes far more energy than heating the same volume of rock or soil. As a consequence, the surface temperature of oceans fluctuates much less between day and night than does the surface temperature of land. During the day, the land heats and becomes hotter than the ocean. The hot land heats the overlying air, which becomes less dense and rises, creating a zone of low air pressure at the land surface. The low pressure draws in cooler air from over the sea, creating a "sea breeze" that moderates the land temperature (Figure 11-21). At night, the pattern of circulation is opposite: The land cools rapidly, while the ocean maintains a temperature close to that experienced during the day. The ocean is therefore warmer than the land, and air rises over it. Cooler air is then drawn from the land toward the sea in a "land breeze."

A similar though much larger-scale phenomenon produces **monsoons,** seasonally reversing winds found over large continental landmasses. India and the southwestern United States are two locations with particularly good examples of this phenomenon. In the summer, the high heating of the Tibetan and Colorado plateaus lowers the surface air pressure in each location and draws air into the continental interiors from the oceans (Figure 11-22). This air is full of moisture and is quickly forced to higher, colder altitudes by the presence of mountains—the Himalayas to the south of the Tibetan Plateau and the many mountain ranges of the Colorado Plateau. As the air rises and cools, its water vapor condenses, and therefore these regions tend to receive abundant rainfall in the summer months. The reverse is true in winter, when the land becomes extremely cold. Warm air rises over the oceans while cold air sinks over the plateaus, causing cold, dry winds to blow from the continents toward the oceans. The seasonal flip-flopping of high and low pressure so readily apparent in the northern hemisphere is much less pronounced in the southern hemisphere due to the configuration of the continents and ocean basins and the lesser amount of land there (see Figure 11-22).

Climate Zones

The general circulation of the atmosphere in conjunction with topography and seasonal variations in solar

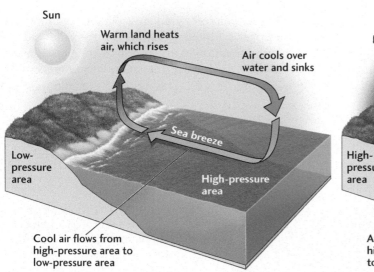

(a) Daytime sea breeze

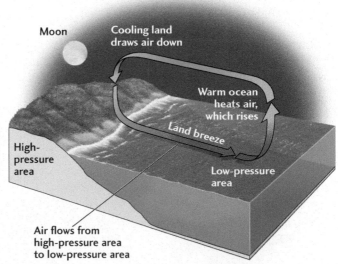

(b) Nighttime land breeze

Figure 11–21 (a) During the day, the ground heats to a warmer temperature than does the sea. Air over the ground warms, expands, and rises. The low pressure produced by the rising air pulls cooler air in from over the ocean and creates a "sea breeze." (b) At night, the breeze reverses direction, becoming a "land breeze," because the ground quickly becomes cooler than the adjacent sea.

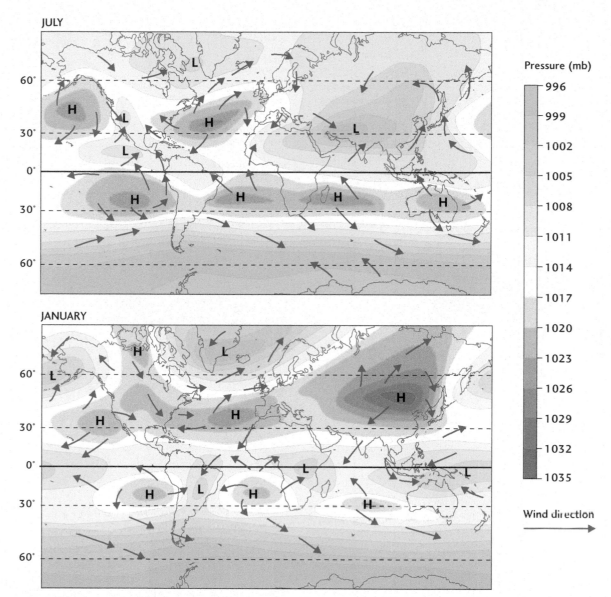

Figure 11–22 Monsoons are seasonally reversing winds driven by pressure gradients created by the thermal contrast between a landmass and its surrounding ocean. In summer, the land heats, causing the air above it to warm and rise, and lowering the air pressure at the land surface. Cooler, moisture-rich air flows into the continental interior from the coasts and may be forced upward over mountains or plateaus, where it cools and releases abundant rainfall.

In winter, the land becomes extremely cold, causing the air above it to cool and sink, and creating a high pressure zone. Cold, dry winds blow outward from the center of the continent toward the coast. The seasonal change in atmospheric pressure over the land and ocean is more pronounced in the northern hemisphere due to both the greater size of landmasses and their latitudinal distribution in that hemisphere.

radiation amounts determine the climatic conditions experienced by Earth's different regions. Starting in the late 1800s, scientists developed climate classification schemes to describe these regions. These schemes generally focused on monthly temperature and precipitation patterns, which together govern water resource potential and the types of vegetation that can grow in a particular area. One of the most popular classifications, first developed by German climatologist Wladimir Köeppen and later modified by him and others, is shown in Figure 11-23. At least 18 different zones appear on this map, and the influences of latitude and

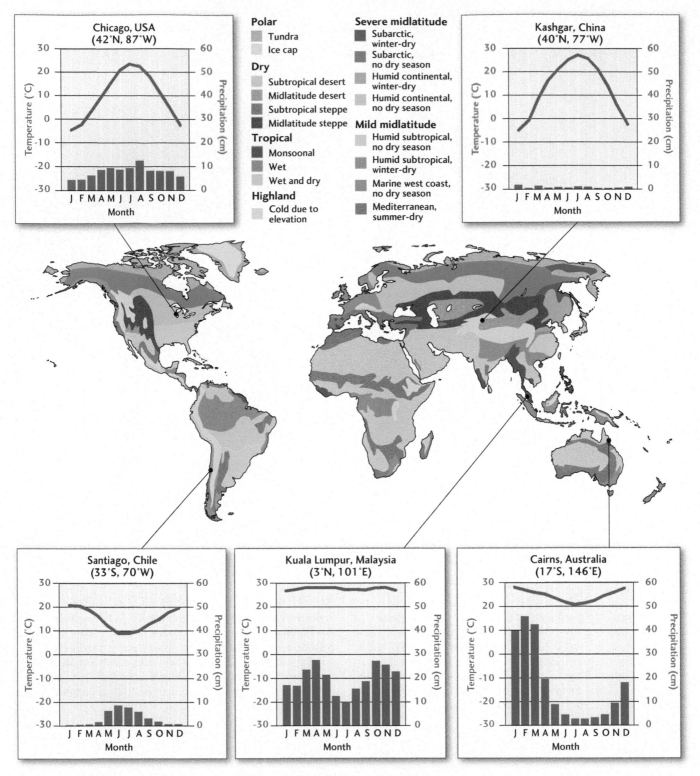

Figure 11–23 Global climate zones identified by German scientist Wladimir Köeppen and based on seasonal variations in temperature and precipitation amounts.

topography are immediately apparent in the location of deserts and wetter areas.

The long-term average distributions of temperature and precipitation are shown for several of these zones in order to illustrate the differences among climate types. For example, the city of Kashgar in northwestern China receives less than 10 cm of precipitation annually and experiences a large annual temperature range typical of a midlatitude continental interior, making its climate classification *midlatitude desert*. In contrast, Chicago, Illinois experiences nearly the same temperature cycle as northwestern China, but receives abundant rainfall distributed fairly evenly throughout the year, giving it a *humid continental* classification. Like Chicago, Kuala Lumpur, Malaysia, is watered abundantly throughout the year, but its equatorial location where the trade winds converge results in higher rainfall amounts and little range in temperature from month to month. These characteristics are typical of *tropical wet* climates.

The seasonal variations in temperature and precipitation experienced by Earth's different climate zones profoundly affect water resources, flooding potential, and the types of soils and ecosystems that develop in different locations. The summer monsoons in India, for example, bring rainfall needed for agriculture, but also cause floods that drown thousands in low-lying Bangladesh to the east. In contrast, Las Vegas, Nevada, like northwestern China, lies within the midlatitude desert climate zone and receives so little annual precipitation that it must rely entirely on groundwater pumped from below Earth's surface and surface water diverted from the Colorado River. Flooding in this climate occurs very rarely.

■ Earth's climate and weather are governed by solar heating of the atmosphere combined with Earth's rotation and gravity.

■ The poles receive less solar energy than the equator, leading to the establishment of convection in the troposphere.

■ The Coriolis effect causes deflection of air currents to the right in the northern hemisphere and to the left in the southern hemisphere. Tropospheric circulation, thus deflected, breaks into smaller convection cells: the Hadley, Ferrel, and polar cells.

■ The tricellular model of circulation explains the occurrence of deserts at 30° north and south latitudes and in the polar regions.

■ Narrow, high-speed atmospheric currents, called jet streams, form at the boundaries of the circulation cells; the polar front jets are associated with stormy weather.

■ Interaction of the atmosphere with the land and ocean influences local and regional climate by producing wet and dry mountainsides, sea and land breezes, and monsoons.

■ Earth's land surface can be divided into different climatic regions based upon the monthly pattern of surface temperature and distribution of rainfall. These regions vary markedly in their water resource and flooding potential.

STORMS

Thus far we have discussed the factors that determine Earth's overall temperature, atmospheric circulation patterns, and the different climates experienced at the surface. Shorter time- and spatial-scale phenomena like storms warrant our attention as well because they are vital in bringing precipitation to land and because they cost so many lives and so many billions of dollars in damage every year. Storms may be produced anywhere that an upward movement of warm, moist air is initiated, and they may be regional in scale—extending hundreds or even thousands of kilometers—or more local in extent.

Development of Air Masses and Frontal Weather Systems

Frequently, a large volume of air becomes a distinctive **air mass,** a parcel of air of fairly homogeneous temperature and humidity that moves as an entity. Air masses form over oceans or over land characterized by fairly uniform topography and are favored by stagnant or very slow atmospheric circulation. These conditions are met in four different latitude bands in each hemisphere: the Arctic and Antarctic, polar (55° to 65°), tropical (10° to 30°), and equatorial latitudes. Thus, air masses are classified as maritime polar, continental tropical, and so on (Figure 11-24). Air masses formed over ocean basins are typically more humid than those formed over land. Likewise, Arctic and polar air masses are much colder and therefore drier than tropical and equatorial air masses. Cold, dry air masses have higher densities than warm, humid air masses because colder temperatures cause gases to contract and because water vapor is less dense than dry air. In North America, the middle latitudes do not favor the formation of air masses because the extreme topographic variability of the western United States leads to highly variable temperature and moisture conditions in the overlying air. In addition, in the central and eastern states no topographic barrier stands in the way to block the convergence of maritime tropical air masses from the south with continental polar air masses from the north. As a result, these latitudes experience atmospheric instability that inhibits air mass formation.

Once formed, air masses eventually leave their source regions and move in response to the general circulation of the atmosphere. As a result, they come into contact with other air masses of different properties. The boundary between any two air masses is known as a **front,** and the interaction of air masses along fronts is analogous to the interaction of lithospheric plates along faults. The latter causes earthquakes; the former causes storms. If a cold air mass moves into an area occupied by a warm air

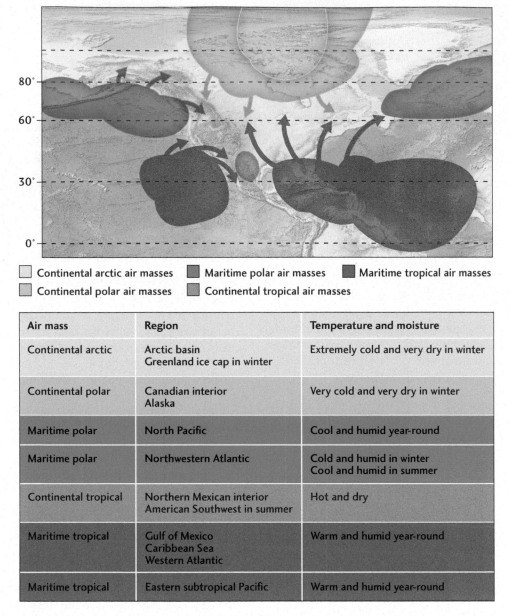

Air mass	Region	Temperature and moisture
Continental arctic	Arctic basin Greenland ice cap in winter	Extremely cold and very dry in winter
Continental polar	Canadian interior Alaska	Very cold and very dry in winter
Maritime polar	North Pacific	Cool and humid year-round
Maritime polar	Northwestern Atlantic	Cold and humid in winter Cool and humid in summer
Continental tropical	Northern Mexican interior American Southwest in summer	Hot and dry
Maritime tropical	Gulf of Mexico Caribbean Sea Western Atlantic	Warm and humid year-round
Maritime tropical	Eastern subtropical Pacific	Warm and humid year-round

Legend:
☐ Continental arctic air masses ■ Maritime polar air masses ■ Maritime tropical air masses
☐ Continental polar air masses ■ Continental tropical air masses

Figure 11–24 Source areas and trajectories of air masses that affect North American weather. Note that continental polar and continental arctic air masses arising over Canada expand their influence in winter, whereas maritime tropical air masses originating over the Gulf of Mexico and the eastern tropical Pacific are more prominent in summer.

mass, a *cold front* is said to have arrived (Figure 11-25). Conversely, a warm air mass that moves in to replace cold air is called a *warm front*. Storms occur at fronts because the collision of air masses lifts warm, low-density air over cold, high-density air. As the warm air rises, it cools, causing condensation of whatever moisture is present. Condensation releases stored latent heat energy that warms the uplifting air mass, causing its density to fall further and the *updraft* of air to continue.

Condensation also results in the formation of clouds (Box 11-3). Initially, updrafts are capable of keeping the cloud droplets suspended. However, as the droplets

grow bigger, they eventually exceed the capability of the updrafts to support them, and they begin to fall through the cloud. The falling raindrops push air down in front of them and cold air from above rushes down to replace it. The resulting *downdraft* pulls air and precipitation down with it (Figure 11-26). The cold air in the downdraft causes a drop in surrounding air temperature as it moves down toward the ground and cuts off the supply of warm air to the updraft. Eventually, the updraft loses energy and dissipates, resulting in the "death" of the storm.

The steep slopes of advancing cold fronts (see Figure 11-25) lead to rapid and large upward motions of

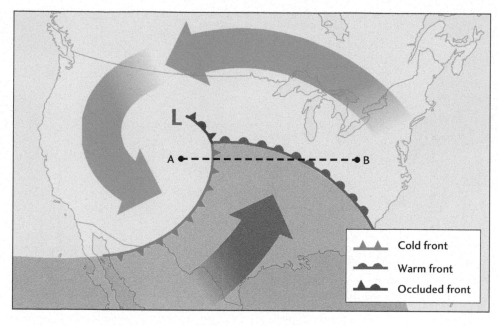

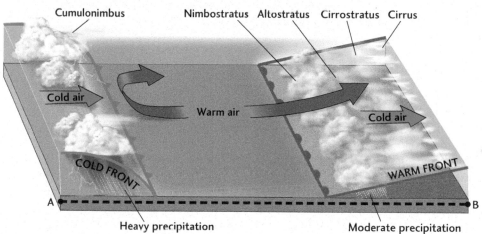

Figure 11–25 Fronts mark the location of contact between air masses of different temperature and humidity characteristics. Warm fronts, which are denoted on weather maps by red lines with red semicircles, are locations where warm maritime tropical air rides up and over cooler drier air coming from higher latitudes. Cold fronts are locations where cold dry continental polar air masses move in to displace warmer air and are denoted on weather maps with blue lines and triangles.

the warmer air mass, which tends to generate localized towering cumulonimbus clouds that produce lightning and thunder, strong winds, and heavy rains or even hail. In comparison, the gentler slope of warm fronts produces more broadly distributed nimbostratus clouds and steadier rainfall. In addition to these frontal storms with regional extents, more local storms can occur when moist air is forced over a mountain range or during the production of sea breezes.

Severe Weather

Updrafts and downdrafts can spawn atmospheric phenomena of extreme violence. Although these phenomena can be very destructive, sometimes, as in the case of lightning, they also play a beneficial role.

Lightning and Thunder While we don't yet entirely understand what produces lightning, we do know that it arises from electrical charge separation within clouds (Figure 11-27). Experiments show that as water droplets freeze, they develop an icy rind that becomes positively charged. Because water is electrically neutral, the remaining unfrozen water inside the drop acquires a negative charge. Water is unusual in the fact that its maximum density occurs in the liquid phase. Thus, as the water droplet continues to freeze from the outside inward, the interior of the drop expands, shattering the exterior icy

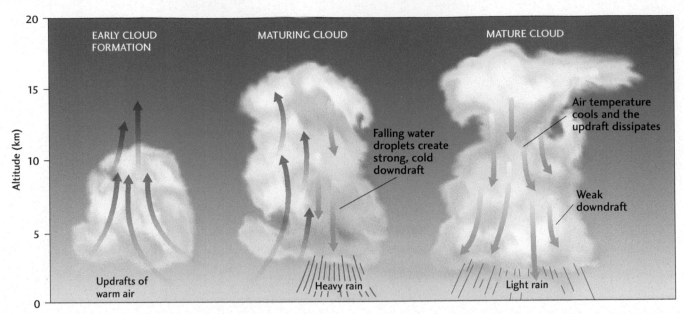

Figure 11–26 Updrafts within a storm cloud keep droplets suspended until they are too heavy to be supported. As they fall, they produce a downdraft that pulls cold, high–altitude air down to the ground surface, cutting off the supply of warm air needed to sustain the storm.

rind. In many cases, there may still be unfrozen water at the core of the drop. In clouds then, it is possible that the tiny fragments of positively charged ice are swept upward by updrafts while the negatively charged water droplet continues to fall toward the cloud base.[29]

The accumulation of negative charges at the base of a cloud tends to attract positive charges on the ground below. The charges at Earth's surface "shadow" the charges within the cloud, moving as the cloud moves. Air is a very efficient insulator, so it inhibits the flow of electrons from the cloud to the ground until the electrical potential energy is large enough to overcome the natural resistance of the air to the flow of electricity. At this point, a spark heads from the cloud to the ground, blazing a trail for the negative charges to follow. A massive amount of electrical energy, equivalent to a million times the voltage used in an average household, streaks down to Earth. As it travels through the air, the lightning instantaneously heats the surrounding air to temperatures as high as 20,000°C (36,000°F).[30] This intense heating causes rapid expansion of the air, producing a sound wave that our ears hear as thunder. Lightning discharges also occur within clouds. In fact, only about 20 percent of lightning bolts ever reach the ground.

Lightning, which can cause great damage, serves a very beneficial purpose. Lightning splits nitrogen gas (N_2), which quickly oxidizes to nitrate (NO_3^-) in the presence of atmospheric oxygen. This nitrate is then deposited on the ground with the rain and is thereby made available to plants, for which it is an essential nutrient. Every year, 3 to 5 million metric tons (3.3 to 5.5 million short tons) of nitrogen are delivered to the terrestrial biosphere in this manner.[31]

Supercell Thunderstorms The combination of an updraft and downdraft is known as a circulation *cell*, and supercell thunderstorms get their names from the enormous sizes of the cells with which they are associated. Measuring as much as 50 km across and 20 km high, these cells extend upward through the atmosphere to layers experiencing different wind velocities and directions. Closer to the ground, winds may be moving in a northeasterly direction, whereas high above Earth's surface, winds may be blowing from the west. Frictional drag with the surface causes winds nearer the ground to move more slowly than those aloft. The changing direction and velocity of the winds aloft is known as **wind shear** and causes air to rotate about an axis parallel to Earth's surface. If an updraft forms, it bends this horizontally rotating tube of air into a semi-vertical vortex known as a **mesocyclone** with dimensions of 2 to 10 km across. Updraft velocities commonly measure 240 to 280 km (150 to 175 mi) per hour, and the adiabatic cooling that occurs as the air rushes upward results in the formation of a towering cumulonimbus cloud.

Because the updraft rises along a slant, the rain, hail, and cold air in the associated downdraft falls in a different location from where the updraft originated (Figure 11-28). As a result, these storms lack the typical mechanism of dissipation that occurs when cold, subsiding air cuts off the source of an updraft. In addition, the cold air flowing out of the downdraft forms a wedge that forces warmer air adjacent to it

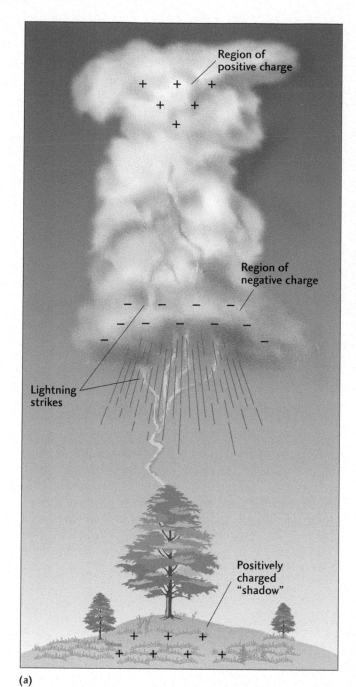

(a)

(b)

Figure 11-27 (a) In a cloud, electrical charges separate and are "shadowed" by charges at the surface of Earth. A spark to the ground leads to a massive release of electrical energy, seen as a lightning bolt (b). *(Keith Kent/Science Photo Library/Science Source)*

upward, feeding the updraft. This combination of processes allows supercell storms to last for many hours. Another factor favoring their persistence appears to be the development of a temperature inversion a few kilometers above Earth's surface. In an inversion layer, temperature increases with elevation rather than the usual decrease with elevation. As in the stratosphere, this situation leads to vertical stability and limits the altitude to which updrafts originating at Earth's surface

can rise. Since the air cannot rise beyond the inversion, it cannot mix with colder air aloft, yet heating at the surface continues. These conditions cause the air near the surface to become still hotter, and the higher the air temperature, the greater the amount of moisture it can hold. Eventually the air near the surface becomes so unstable that it forces its way upward through the inversion layer to form a massive cumulonimbus cloud. Rapid downdrafts associated with the supercell clocked at 130 km (80 mi) per hour lead to damaging winds at Earth's surface, and the high altitudes that these storms reach lead to very large raindrops and hail.

Tornadoes Some of the most destructive storms in the world occur in the Great Plains and southern states of the United States where air masses from the Arctic or from the intermountain West clash with air masses moving northward out of the Gulf of Mexico.[32] The drier, and therefore denser, air from the west or northwest flows over warm moist air traveling northward, creating a situation that is extremely unstable and producing a mesocyclone. Occasionally, cold dry air from aloft subsides along the backside (the side opposite the direction of motion of the storm) of the mesocyclone, forming a so-called *rear flank downdraft*. This downdraft pulls the mesocyclone downward, producing a slowly rotating *wall cloud* that extends below the base of the cumulonimbus cloud that hosts the mesocyclone.

While we still don't fully understand the process for tornado formation, the rear flank downdraft appears to confine the rising air in the updraft, strengthening

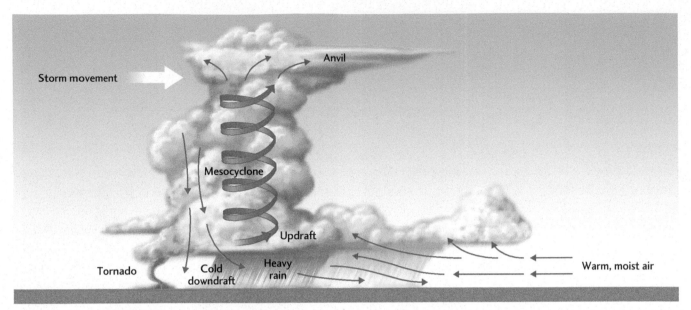

Figure 11–28 Supercell thunderstorms originate when wind shear causes air to begin rotating around a horizontal axis. When an updraft forms, it bends this rotational cell upward, creating a mesocyclone that may spawn funnel clouds and tornados.

the updraft and increasing its velocity. The intensified updraft decreases the atmospheric pressure at the ground surface, which draws the wall cloud downward. As the air subsides, it accelerates, causing the wall cloud to narrow to a funnel, which may ultimately become a **tornado** if it touches the ground. Condensation of water is initially what makes the tornado visible. Its visibility increases as the vortex picks up debris, which may include tree limbs, fence posts, cars, and sections of roof. The speed of the wind is what causes tornadoes to be so destructive (Box 11-4).

Tropical Storms and Hurricanes

Rotating over the ocean as tornadoes spin over land, tropical storms and hurricanes cause death and destruction from a combination of very high winds, torrential rains, and storm surges (Figure 11-29). **Tropical storms** are the "gentler" of the two, with somewhat slower winds; **hurricanes** (also called *tropical cyclones* and *typhoons*) have winds of at least 120 km (75 mi) per hour (Table 11-1).[43] Both of these storm systems tend to be much larger in scale (hundreds of kilometers

Figure 11–29 This satellite image of Hurricane Katrina was taken on August 29, 2005, the day the storm made landfall on the Gulf Coast as a Category 3 storm. *(GOES 12 Satellite/NASA/NOAA)*

BOX 11–3 GEOLOGIST'S TOOLBOX

Clouds: Thor's Anvil and Castles in the Air

Most of us have at some point in our lives lain on our backs on a warm summer day and stared up at the sky, imagining that we see dinosaurs, lambs, or puppies in the puffy masses of condensed water drifting high above Earth. Most of us have also welcomed the cooling effect of clouds passing in front of the Sun and have been caught without an umbrella in a downpour from an unexpected thunderstorm. What processes lead to cloud formation, and why do some portend fair weather while others spawn lightning or drop hailstones the size of baseballs?

All clouds form when water vapor condenses to water droplets as a result of atmospheric cooling. Dust particles, pollen grains, soot from smokestacks and forest fires, volcanic ash, and tiny salt crystals derived from sea spray act as surfaces on which cloud droplets nucleate. These droplets are initially so small that they are held aloft in the atmo-

sphere where they coalesce into different forms characterized by morphology and altitude of formation.

Cirrus clouds (**Figure 1**) are thin wispy clouds found at altitudes higher than 6000 m (20,000 ft) and are sometimes associated with warm fronts (see Figure 11–25). White in color, these clouds often display an arc shape (*cirrus* is the Latin word for curl) that points in the direction of air mass movement. The low water content of the atmosphere at the altitude where cirrus clouds form generally precludes their development into storm clouds that drop rain or snow. In addition, the cold temperatures associated with the high elevations at which cirrus clouds form means that they consist entirely of ice crystals.

Stratus clouds are much denser than cirrus clouds and often form in sheets that stretch across the sky from horizon to horizon. These clouds are usually made of water droplets,

(continued)

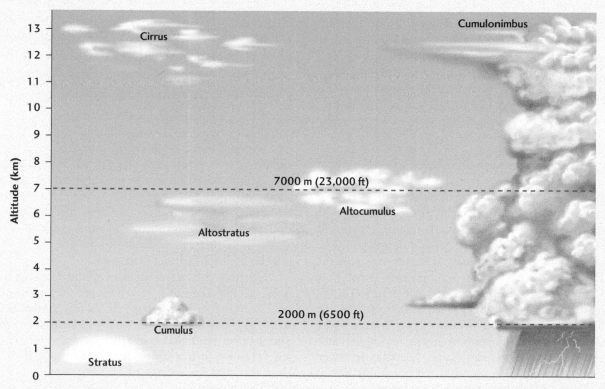

Figure 1 Clouds are classified based on their form and their altitudes of formation. Those found between 2000 and 7000 m are known as midlevel clouds and are given the Latin prefix "alto." The Latin word "nimbus" means rain cloud.

(continued)

reflecting a lower altitude of formation (below 6000 m) and giving them a grayish color. Occasional holes allow sunbeams through, but individual clouds cannot be distinguished. Stratus clouds commonly form at warm fronts (see Figure 11-25) as warmer, more humid air flows up and over cooler, drier air. The gentle slope of warm fronts gives rise to their vast lateral extent. They produce light to moderate rainfall that can linger for many hours to several days.

Cumulus clouds are the classic cotton balls commonly found in children's art. They are characterized by distinct edges, a fairly flat base, and dome or tower-shaped tops. The base of these clouds often lies at low altitudes (below 2000 m), but they can extend upward thousands of meters. As a result, they are often referred to as *vertically developed clouds*. Fair-weather cumulus clouds form from uneven heating of Earth's surface. Warm air rises in small parcels, eventually reaching elevations cool enough to cause condensation that generates the puffy shapes interspersed with blue sky. The elevation of the tops of these clouds is limited by the amount of energy released during condensation, since that energy produces the thermal updrafts that keep the cloud droplets suspended. Given their origins, fair-weather cumulus clouds never drop rain or snow.

In contrast, towering *cumulonimbus clouds* form as a result of frontal lifting, often in association with cold fronts (see Figure 11-25). Strong convective updrafts generated by energy release upon condensation allow these clouds to extend to elevations of 12 km or more above Earth's surface. As a result of their extreme vertical range, these clouds are made up of water droplets at low elevations and ice crystals at high elevations. Strong winds in the upper levels of the troposphere stretch the upper part of the cloud into a distinctive anvil shape, and electrical charges associated with the updrafts generate lightning. In ancient Norse mythology, these clouds were associated with Thor, god of thunder, whose hammer produced a lightning bolt each time it struck the anvil.

The strong updrafts associated with cumulonimbus clouds allow falling ice crystals and raindrops to repeatedly rise upward in the atmosphere, which promotes their growth in size as they collide with other particles again and again. As a result, individual raindrops are much larger when they hit the ground than those falling from other cloud types, and cumulonimbus clouds can also produce large hailstones. The large size of the precipitation emanating from cumulonimbus clouds reduces the influence of atmospheric drag, meaning that the rain and ice that fall from these clouds has a higher terminal velocity than that from other clouds. This, in combination with its large size, increases the erosive power of the precipitation when it hits the ground. The light and almost misting rain typical of stratus clouds, by comparison, hits the ground with little force.

---

across) than tornadoes, and therefore more extensive in their impact.

Tropical storms and hurricanes are born near the equator, where solar heating of the surface ocean results in heating and uplift of the overlying air, creating a low-pressure zone known as a *tropical depression*. The low-pressure zone may intensify if water vapor in the uplifting air condenses aloft and releases latent heat energy that drives more lifting. Air rushes in to fill the low-pressure zone and is very weakly deflected by the Coriolis effect. If the low-pressure zone moves away from the equator and begins to travel to higher latitudes, however, Coriolis deflection becomes stronger. The low thus becomes a spiraling vortex of air that is continually fed by new warm air from below.

The initiation of hurricanes is aided by conditions of low wind shear with elevation in the atmosphere.[44] What this means is that there is little difference in horizontal speed or direction of the winds with altitude. Such a situation allows the growing storm to build vertically as latent heat released by the uplifting air is spread over a small area (Figure 11-30). In contrast, high wind shear forces the growing thunderstorm to slant downwind, causing the latent heat release to be spread over a much larger area and lessening the updraft in any one location. This minimizes the drop in air pressure that can occur at the surface, thereby limiting the ability to form a hurricane.

Hurricanes always form over the tropical oceans because they require a steady supply of warm, moist air. Over land, temperatures rapidly fall below the 26°C to 27°C (79°F to 81°F) necessary to sustain the storm, and air over land is also usually not as humid as ocean air. Likewise, as hurricanes move poleward, they move over seas that are much colder. This, too, tends to shut off their circulation.

Average wind speeds in hurricanes exceed 100 km (62 mi) per hour, and winds in excess of 300 km (186 mi) per hour have been clocked. As with tornadoes, such winds rip roofs off of houses and turn tree limbs and other loose debris into deadly missiles. They also contribute to coastal flooding. The low-pressure zone at the

TABLE 11−1 Hurricane Classification		
Category	Wind Speed (km per hour)	Damage
Tropical depression	37–61	Minor
Tropical storm	62–118	Moderate
1	119–153	Damaging
2	154–177	Widespread
3	178–209	Extensive
4	210–249	Devastating
5	>250	Catastrophic

eye of the storm pulls the sea surface upward like water being sucked up a soda straw, and the strong winds blowing into the center of the low-pressure zone stack up water against the shore (Figure 11-31). Additional flooding occurs when the torrential rains produced by hurricanes overwhelm the capacity of the ground to soak them up. Bangladesh, with much of its land lying within a meter of sea level, has been particularly hard hit by hurricanes throughout its history. Storm surges of several meters have occurred many times in the last 40 years, resulting in flooding and the drowning of more than 1 million people.

In May 2008, the deadliest storm in recent history struck the nation of Myanmar, formerly known as Burma. Around 140,000 people were killed and 450,000 homes destroyed in the Irrawaddy River delta region when Typhoon Nargis made landfall with winds up to 240 km (149 mi) per hour and a storm surge of greater than 3 m (9.8 ft).[45] The storm also flooded rice paddies with saltwater and killed 120,000 animals used for plowing,[46] destroying the area's agricultural base. A year after the storm hit, a million people remained dependent on international food aid.[47]

In the United States, damage to life is generally far less severe because of sophisticated weather forecasting, emergency response planning, and evacuation of low-lying areas. Nevertheless, hurricanes in the United States have caused staggering property losses in the tens of billions of dollars. The deadliest hurricane to strike the United States this century was Hurricane Katrina, which made landfall as a Category 3 storm on August 28, 2005. An estimated 1570 people drowned as levees designed to protect the city of New Orleans failed catastrophically, allowing water to flood areas of the city that lie below sea level.[48] More than 275,000 homes were destroyed along with commercial properties, costing at least $81 billion in property damage.[49] While Hurricane Katrina ranks as the most expensive disaster in U.S. history, even greater casualties resulted from a 1900 hurricane that struck Galveston, Texas, well before radar equipment and satellites were used to track storms. More than 6000 people lost their lives as the storm took residents by surprise.[50]

■ Air masses interact along fronts, which are often the sites of storm activity.

■ Water vapor in warm air updrafts condenses as the air travels upward to higher, colder altitudes. Condensation releases stored heat energy that allows the updraft to continue and more condensation to

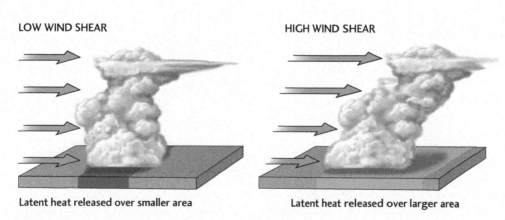

LOW WIND SHEAR

HIGH WIND SHEAR

Latent heat released over smaller area

Latent heat released over larger area

Figure 11−30 Wind shear describes the change in velocity of the winds with elevation in the atmosphere. High wind shear inhibits hurricane formation.

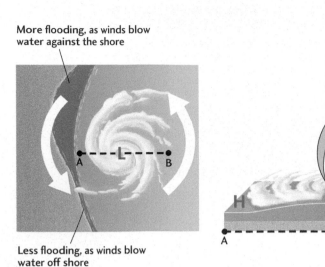

More flooding, as winds blow water against the shore

Less flooding, as winds blow water off shore

Figure 11–31 The low pressure associated with hurricanes raises the elevation of the sea surface. This, combined with strong winds, generates storm surge that can reach many meters and inundate low–lying coastal areas.

occur. The resulting water droplets fall to the ground as rain, ice, or snow.

■ Positive and negative electrical charges develop and separate in clouds when gases, ice crystals, water droplets, and dust particles rub against one another. These charges lead to lightning bolts when the resistance of air to the flow of electricity is overcome by the electrical potential energy developed in the cloud.

■ Tornadoes, rotating storms known for their high winds and destructive ability, form at the intersection of two or more air masses that have very different physical properties.

■ Hurricanes and tropical storms are produced and sustained by a supply of warm, moist air found only over low-latitude oceans.

■ Hurricane destruction results from a combination of high winds, heavy rains that produce river flooding, and coastal flooding brought about by storm surge.

HUMAN INFLUENCE ON ATMOSPHERIC CHEMISTRY

The natural processes of tropospheric circulation and the climate and weather they produce have operated for millions or even billions of years. Since the advent of the Industrial Revolution, however, human activities have begun to affect these processes. Humans introduce a wide variety of pollutants into the atmosphere through activities such as industry, agriculture, motor vehicle use, and war (Figure 11-32). Many of these pollutants are hazardous to the health of animals and plants, while others have the potential to change climate or to destroy Earth's protective layer of ozone in the stratosphere. Amazingly, these hazards are produced by

Figure 11–32 Oil wells in Kuwait were set ablaze by the retreating Iraqi army during the Persian Gulf War in 1991. The soot from the burning oil was so thick that headlights were needed even for daytime driving. Many children and older individuals required hospitalization for respiratory problems caused by the smoke and toxic gases released from the burning wells. *(Wesley Bocxe/Science Source)*

substances with concentrations lower than 0.1 percent of the total volume of the atmosphere.

The dispersal of pollutants is driven by atmospheric circulation and the reactions of pollutants with various compounds in the atmosphere. Some pollutants in the troposphere, such as sulfur dioxide and nitrogen oxides, react with water and are quickly removed through precipitation that falls to the ground. Other pollutants, such as chlorofluorocarbons (CFCs), are chemically more inert and remain in the atmosphere for long periods of time. The residence time of pollutants in the atmosphere is one of the factors that determines their environmental impact. Pollutants with short residence times tend to have a local or, at most, a regional impact, whereas those with long residence times may have a global impact. Acid rain and smog are examples of the former; ozone depletion and global warming are examples of the latter. Due to the complexity of the Earth system interactions associated with global warming, we defer the discussion of this example until Chapter 15.

Acid Rain

The acidity of a solution is a measure of its hydrogen ion concentration ($[H^+]$) and is normally expressed in terms of pH, where:

$$pH = -\log_{10} [H^+]$$

The pH scale runs from 0 to 14; a value of 7 (i.e., $[H^+]$ = 10^{-7} moles per liter) is considered neutral, lower values (higher $[H^+]$) are acidic, and higher values (lower $[H^+]$) are basic. Ordinary rainfall is slightly acidic (pH between 5 and 6) because as it falls, it reacts with carbon dioxide to make carbonic acid ($H_2O + CO_2 \rightarrow H_2CO_3$), a tiny amount of which dissociates into hydrogen and bicarbonate ions: $H_2CO_3 \rightarrow H^+ + HCO_3^-$.

As a result of new industrial processes introduced in the Industrial Revolution, the acidity of rain increased. Combustion of fossil fuels by power plants and motor vehicles, as well as industrial processes such as metal smelting, release sulfur and nitrogen oxides into the atmosphere. These gases combine with the water in raindrops to make sulfuric and nitric acids:

$$2\,SO_2 \quad + \quad O_2 \quad \rightarrow \quad 2SO_3$$
Sulfur dioxide + Oxygen → Sulfur trioxide

$$SO_3 \quad + \quad H_2O \quad \rightarrow \quad H_2SO_4$$
Sulfur trioxide + Water → Sulfuric acid

$$2NO_2 \quad + \quad H_2O \quad \rightarrow \quad HNO_2 \quad + \quad HNO_3$$
Nitrogen dioxide + Water → Nitrous acid + Nitric acid

Like carbonic acid, these acids also dissociate, but because they are "stronger" acids, they liberate more hydrogen

ion. As a result, **acid rain** often has a pH between 4 and 5, and values as low as 1.5 have been recorded.[51] Because of the short residence time of water in the atmosphere, most deposition occurs within 1000 km of the source of the acidity.

The impact of acid deposition on ecosystems varies. Acid rain falling on land underlain by carbonate rocks is readily neutralized. For instance, sulfuric acid can react with limestone to form the sulfate mineral anhydrite and carbonic acid:

$$H_2SO_4 + CaCO_3 \rightarrow CaSO_4 + H_2CO_3$$

The carbonic acid, in turn, dissociates into water and carbon dioxide gas:

$$H_2CO_3 \quad \rightarrow \quad H_2O \quad + \quad CO_2\,(g)$$

Nitric acid is similarly neutralized. Unfortunately, landscapes underlain by other rock types, such as granite or gneiss, often lack this buffering capacity. When acids react with noncarbonate minerals in rocks and soils, they may leach toxic metals or strip nutrients from these materials. Aluminum, for example, is readily released and transported by acidic waters, and it both damages the roots of plants and interferes with their calcium absorption.[52] As a result, plants growing in these soils that have been acidified become sickly or die. Telltale signs of acid-related nutrient deficiency are death of the upper levels, or *crown*, of a tree, yellow leaves or needles, and excessive dropping of needles (Figure 11-33). Acid-induced stress also causes trees to be more susceptible to pests, viruses, and fungi, causing further damage.

Toxic metals released by acid rain may also be transported by groundwater and surface water into lakes and streams. Aluminum presents a particular problem for fish, causing changes to gill tissue that

Figure 11-33 The Black Forest of Germany has been severely damaged by acid rain. Note the yellowing of the needles on these Norway spruce trees. *(Tom McHugh/Science Source)*

BOX 11–4 CASE STUDY

The Destructive Power of Tornados

In the late afternoon of May 4, 2007, the Kansas sky turned dark greenish purple. A roiling mass of clouds began to form and lightning flashed between them, signaling the onset of a severe thunderstorm. At 9:45 P.M., the clouds spawned a devastating tornado that struck the small town of Greensburg, Kansas, population 1574. With wind speeds reaching 330 km (205 mi) per hour,[33] the 2.7-km- (1.7-mi)-wide twister was among the largest the atmosphere is capable of producing, measuring an EF–5 on the enhanced Fujita scale of tornado strength. EF–5 tornadoes have wind speeds in excess of 322 km (200 mi) per hour and are exceptionally rare, constituting fewer than 0.1 percent of all tornadoes.[34] At these wind speeds, buildings are blown off their foundations, smashed to pieces, and dispersed for miles. Power lines are flattened and trees are snapped in half, all of their bark stripped away. At speeds of 330 km per hour, even blades of grass become deadly projectiles and have been observed protruding from bricks, like darts from a target, after tornadoes subside (Figure 1).

On the ground for about 30 minutes and covering 35 km (22 mi) of territory, the Greensburg tornado killed 10 people, injured dozens more, and destroyed more than 800 homes (Figure 2). All of the town's schools, the police and fire department, ten of eleven churches, and the only stoplight in all of Kiowa County lay in ruin.[35] Nearly the entire population of Greensburg was left homeless and jobless as 97 percent of buildings were totally destroyed or damaged beyond repair. That more people were not killed is testament to a population prepared for such disasters. Most homes in Kansas are built with basements that can be used as shelter during tornadoes, and community sirens warn of impending danger. Schoolchildren participate in emergency drills, and local television stations break from normal programming to continuous storm coverage until the threat has subsided.

On that fateful May night, the news station meteorologists were kept very busy issuing forecasts and warnings. The Greensburg tornado was only one of four tornadoes that struck southwestern Kansas that evening. They were spawned by a storm that formed when warm, moist air moved quickly northward out of Texas toward Oklahoma and Kansas and collided with cooler and drier air flowing from the Rocky Mountains to the west. The frequency with which such conditions occur has earned the American Midwest the nickname "Tornado Alley," and indeed the region experiences more tornadoes per square kilometer of ground than any other place on Earth. Still, tornadoes can occur in any state, and during any month of the year.

Figure 1 At speeds of 330 km per hour, commonplace objects become deadly projectiles. (Grant Hindsley/ZUMAPRESS.com/Alamy)

Figure 2 The Greensburg tornado destroyed 97% of buildings in this Kansas town. *(Greg Henshall/FEMA)*

More than 1000 of these destructive phenomena are reported annually throughout the United States, though few have the force of the Greensburg tornado. Despite the devastation of that event, the loss of life was minimal. In comparison, on so-called "Terrible Tuesday," April 3, 1974, 148 tornadoes touched down over a 12-state area extending from Alabama to Michigan, killing 309 people, injuring thousands more, and causing property damage in the hundreds of millions of dollars.[36] The month of April proved very destructive again in 2011, earning the dubious distinction of most active month for tornados on record. As many as 875 tornados occurred,[37] including more than 163 funnels spotted in only one day on April 27.[38] Citizens of Arkansas, Georgia, Virginia, Kentucky, Tennessee, and Mississippi lost their lives, though the greatest devastation was in Alabama, where a mile-wide tornado tore through the city of Tuscaloosa, killing over 200 people. Less than a month later (May 22, 2011), an EF-5 tornado struck the city of Joplin, Missouri, killing more than 150 people. In May of 2013, two massive EF-5 tornados devastated suburbs of Oklahoma City. In Moore, Oklahoma, 24 people died on May 23 as more than 1100 homes were destroyed. Just over a week later (May 31), 22 people died as the widest tornado ever recorded (4.2 km, or 2.6 mi) struck the towns of El Reno and Union City, Oklahoma. Though the destruction from these storms was enormous, the worst single event occurred on March 18, 1925, when the "tristate tornado" killed 695 people as it moved 352 km (219 mi) from Missouri to Illinois to Indiana.[39] Most of the deaths were caused by flying debris and collapsed walls and roofs from the 15,000 homes that blew to pieces.

Today, Greensburg is rebuilding. The near total devastation allowed those residents who decided to return to start over from scratch, and the town decided to rebuild in a manner worthy of its name. Town-owned buildings are being designed to qualify for LEED (Leadership in Energy and Environmental Design) Platinum certification, which will allow them to use 42 percent less energy than standard construction.[40] In addition, buildings are being constructed of recycled bricks and wood, and in 2009, the town, in partnership with two electrical power utilities, broke ground on a wind farm that will eventually supply all of its electricity.[41] Already known for being the home of the world's largest hand-dug well and the world's largest pallasite meteorite (a very rare meteorite made of silicate crystals in a matrix of iron and nickel), Greensburg will soon become a destination for those interested in learning how to create ecologically sustainable communities.[42]

TABLE 11–2 Effects of Increasing Acidity on Fish

pH	Effects on fish
9.0–6.5	harmless to most fish
6.5–6.0	significant reductions in both egg hatchings and growth in brook trout under continued exposure
6.0–5.0	rainbow trout do not occur; small populations of relatively few fish species are found; declines in salmon can be expected; if high aluminum concentrations are present, fish will die; mollusks are rare
5.0–4.5	harmful to salmon eggs and fry, and to common carp
4.5–4.0	harmful to adult salmon, goldfish, and common carp; resistance increases with age
4.0–3.5	lethal to salmon; roach, trench, perch, and pike survive
3.5–3.0	toxic to most fish; some plants and invertebrates survive

Data source: Adapted from D. M. Elsom, *Atmospheric Pollution: A Global Problem*, 2nd ed., New York: Blackwell, 1992.

promote asphyxiation[53] and interfering with the calcium uptake necessary for skeletal growth.[54] In addition, increased acidity of the water inhibits reproductive success; Table 11-2 shows the relationship between pH and toxicity to fish. Perhaps one of the largest problems of toxic metals and low pH is loss of the aquatic invertebrates upon which fish are dependent for food. This loss ripples upward through the food chain to affect the waterfowl that eat the fish.

The federal government agency Environment Canada (EC) defines a "critical load" (8–20 kg per hectare or 7–18 lb per acre) of acid deposition—that is, the maximum amount of acid a watershed can neutralize in a year without adverse impacts on its ecosystems. EC estimates that an 800,000-km^2 (308,882-mi^2) region spanning Quebec, Ontario, and the Maritime Provinces exceeds this critical load due to industrial activities and power production in the Great Lakes region, which adversely impacts nearly 95,000 lakes that lie on gneissic and granitic rocks.[55] Half of the acid deposition originates from power plants and factories in the Ohio River valley region of the United States; unfortunately, atmospheric circulation pays no attention to political boundaries. Though the United States and Canada have worked together to reduce emissions of acid rain–producing chemicals, another 75 percent decrease below 2010 emission levels will have to be achieved before all of Canada's watersheds can fall below the critical load threshold. Achieving such large reductions will be challenging, but it will also benefit parts of the northeastern United States. Studies of lakes in New York's Adirondack Mountains revealed that 23 percent of fish population losses were due to acid deposition. Losses occurred either because of metal toxicity or because of destruction of the food chain upon which fish populations were dependent.[56]

In addition to Earth's natural ecosystems, urban environments are also under attack from acid rain. In Sweden, increased acidity in well water has led to higher concentrations of the toxic metals copper, lead, and cadmium and to loss of the vital nutrients calcium, potassium, and magnesium in drinking water in some areas.[57] In some cases, the metals may be stripped from rocks or sediment particles in the groundwater system, and in others they may come from the attack of acidic water on plumbing. In addition to changing the constituents of drinking water, acid rain is slowly dissolving away the histories of many countries by attacking limestone and marble monuments, temples such as the Parthenon in Athens, and grave markers (Figure 11-34).

In an early response to local acid rain, power plants and factories built taller smokestacks. The idea was to emit noxious gases at higher levels in the atmosphere, where they would then be kept aloft by air currents and swept out of the local environment. The problem with this approach is that the acids were simply transported to new areas, where they still caused environmental damage.

Smog

Industrial activities and automobiles contribute not only to acid rain but also to the production of smog, a common sight in cities. A complex mixture of ozone, nitrogen oxides, and hydrocarbons, **smog** causes respiratory problems, reduced visibility, and damage to vegetation. Nitrous oxide gas makes a smoggy atmosphere appear brown because the molecules strongly absorb blue light and scatter yellow and red wavelengths (Figure 11-35).

Figure 11–34 Acid rain attacks historic monuments and buildings and literally dissolves them, as shown in this view of York Minster stonework. The central head used to be in the direct path of acidic rainfall, whereas the heads to either side have always been more sheltered. *(Martin Bond/Science Photo Library/Science Source)*

All the components of smog have existed in Earth's atmosphere for hundreds of millions of years. Plants produce some hydrocarbons, ozone is manufactured in the stratosphere, and lightning and forest fires generate the heat necessary to produce nitrogen oxides. Industrial activities have added significantly to the hydrocarbons and ozone in the troposphere, however,

and the intense heat of automobile cylinders produces a steady supply of nitrogen oxides. Solar energy dissociates nitrogen oxides into other nitrogen compounds and free oxygen (O). The free oxygen readily combines with molecular oxygen (O_2) to form ozone (O_3):

$$NO_2 + h\nu \rightarrow NO + O$$

Nitrogen dioxide   Solar energy   Nitrogen oxide   Free oxygen

$$O + O_2 \rightarrow O_3$$

Free oxygen   Molecular oxygen   Ozone

(Solar energy can be described as the wave frequency (ν) times h, a universal constant known as Planck's constant.)

Although in the stratosphere ozone is beneficial, blocking the transmission of harmful UV rays from the Sun, ozone in the troposphere can cause respiratory irritation and eye damage, may lead to lung cancer, and kills plant tissues. In forests around the Los Angeles basin and in Europe,[58] large stands of trees have been damaged or killed by ozone pollution in nearby cities, and the United States loses at least $1 billion to $2 billion annually in ozone-related crop damage.[59]

Smog accumulates in urban areas partly because of the large amounts of atmospheric pollutants generated by motor vehicles, power plants, and industry, and partly because atmospheric temperature can keep the pollutants from escaping. Because temperature generally decreases with altitude above Earth's surface in the troposphere, if a parcel of air at the surface becomes strongly heated during the day, its density declines and the parcel rises until it reaches a level in the atmosphere

(a)

(b)

Figure 11–35 Shanghai, China, (a) on a clear day and (b) on a smoggy day. The city's iconic skyscrapers are enshrouded by smog, a thick haze consisting of nitrogen oxides and uncombusted hydrocarbons from motor vehicles, along with ozone. *(Thinkstock)*

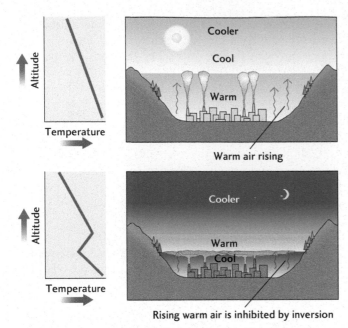

Figure 11-36 Atmospheric temperature inversions, which occur at night and in the winter, inhibit vertical mixing of the lower troposphere and lead to high pollutant concentrations.

with the same low density. The upward motion of the parcel of air allows pollutants to disperse. During the night or in winter when surface temperatures are very low, however, air at the surface may be colder and denser than the overlying air and unable to move upward. At these times of temperature **inversion**, pollutants remain concentrated at the surface instead of being carried aloft (Figure 11-36). This problem may be exacerbated by geography: If the city is in a basin surrounded by mountain ranges, the dense, cold, pollutant-rich air will not be swept away by ground-level winds. Such is the situation of Los Angeles, where the San Gabriel and San Bernardino mountains block winds and trap pollutants.

Ozone Depletion

Ironically, while some urban areas experienced increasing smog and excessive amounts of ozone near the ground, stratospheric ozone levels declined throughout the second half of the 20th century. As mentioned earlier in this chapter, ozone strongly absorbs harmful UV radiation from the Sun and limits its penetration to Earth's surface. In 1974, scientists Mario Molina and F. Sherwood Rowland warned that a class of human-made gases known as **chlorofluorocarbons (CFCs)** might destroy ozone in the atmosphere, with grave consequences for humans and other organisms.[60] Only 11 years later, satellite measurements revealed a thinning, or "hole," in the ozone layer over Antarctica (Figure 11-37).[61] In this region, ozone concentrations had dropped to as low as 5 percent of their former

levels. Extremely low levels of ozone were then discovered over other parts of the southern hemisphere, such as Australia and New Zealand, and then in the northern hemisphere, over the North Sea. Indeed, ozone levels declined globally until the late 1990s; they have subsequently stabilized,[62] though 2006 saw the largest ozone hole over Antarctica since satellite measurements began,[63] and 2011 saw the first formation of an ozone hole over the Arctic, the magnitude of which is similar to the one over Antarctica.[64]

CFCs, which were formerly widely used as refrigerants, as propellants for aerosol sprays, and in the production of Styrofoam, are broken down by sunlight to form chlorine monoxide and chlorine. The chlorine reacts with ozone to produce gaseous oxygen and chlorine monoxide, while the chlorine monoxide reacts with oxygen to form gaseous oxygen and chlorine:

$$Cl + O_3 \rightarrow ClO + O_2$$

Chlorine + Ozone → Chlorine monoxide + Molecular oxygen

$$ClO + O \rightarrow Cl + O_2$$

Chlorine monoxide + Atomic oxygen → Chlorine + Molecular oxygen

Because chlorine is recycled by these reactions, its destructive effect on stratospheric ozone is nearly continuous. The mechanism for removing chlorine from the atmosphere involves a reaction with hydrogen ions to form hydrochloric acid (HCl), which is then incorporated in rainfall. Because the concentration of water vapor in the stratosphere is extremely low, however, a single chlorine atom may reside there long enough to destroy as many as 100,000 ozone molecules.[65]

In 1989, 13 industrial nations, including the United States, agreed to stop all production of CFCs by the year 2000. The treaty they signed, called the Montreal Protocol, was subsequently ratified by 183 additional countries, such that nearly every country in the world is now involved.[66] The agreement has been successful, but it will take several decades, until the years 2060–2075,[62] before stratospheric ozone returns to its pre-1950s concentration because CFCs introduced into the stratosphere are long-lived and because additional ozone depleting chemicals are only now being phased out.[67] Hydrochlorofluorocarbons, or HCFCs, are not quite as destructive as CFCs and so have been used in the interim to replace CFCs but will themselves be eliminated by 2030 under amendments made to the original Montreal Protocol treaty.

■ Combustion of fossil fuels, metal smelting, and other industrial processes release to the atmosphere sulfur and nitrogen oxides, which combine with cloud droplets to form acid rain.

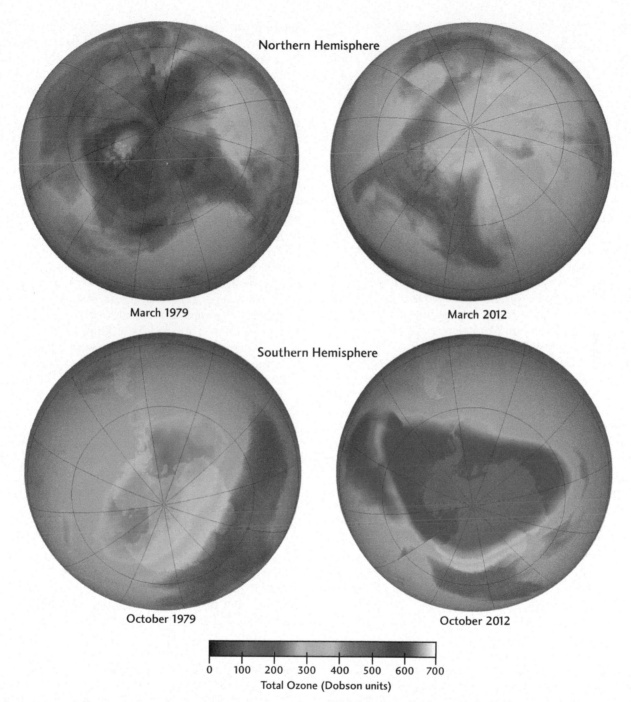

Northern Hemisphere

March 1979

March 2012

Southern Hemisphere

October 1979

October 2012

0   100   200   300   400   500   600   700
Total Ozone (Dobson units)

Figure 11–37 Ozone concentrations have been declining globally for at least the last two decades, as can be seen in these 1979 images produced by the Total Ozone Mapping Spectrometer (TOMS), and the 2012 images from the Ozone Monitoring Instrument (OMI) on board the *Aura* satellite. Declines have been most dramatic over the poles and high latitudes. Concentrations are measured in Dobson units (1 DU equals an ozone concentration of 1 part per billion), named for George Dobson, an English scientist who designed the first ozone measuring instrument. *(GSFC/NASA Ozone Hole Watch)*

■ Smog is produced by chemical reactions involving nitrogen oxides and hydrocarbons mixing with sunlight. Some of these reactions produce tropospheric ozone.

■ The effects of urban smog are exacerbated by the temperature inversion that exists above land at night and during winter.

■ Concentration of ozone in the stratosphere declined for several decades because of human-made gases known as chlorofluorocarbons (CFCs). An international agreement banning the production and use of these gases will eventually allow the ozone layer to recover.

AIR POLLUTION AND ENVIRONMENTAL MANAGEMENT

Clean air became a threatened resource, especially in urban areas, with the Industrial Revolution. For decades, smoke and chemical fumes belched from factory smokestacks, power plants, and automobile tailpipes, and combined with pollutants issuing from individual home heating systems, making life miserable for people with respiratory problems and occasionally leading to premature death.

The winter of 1952 showed just how dangerous air pollutants can be.[68] The city of London was experiencing a particularly cold December, leading to more household coal combustion than normal. On top of the usual industrial emissions, an estimated one million coal stoves used for home heating belched thousands of tons of sulfur dioxide and fine particulate matter into the sky, which became trapped near the surface by a temperature inversion in the overlying atmosphere. Combining with fog, these pollutants formed a thick chemical smog that reduced visibility in some parts of the city to as low as 10 m (33 ft) on the night of December 5. Over the next few days, an estimated 4000 people died from respiratory and cardiac failure.

Legislation

The London Smog of 1952 prompted the British government to finally enact legislation to reduce air pollution. In 1956, Great Britain passed its first Clean Air Act, which allowed municipal governments to provide funds to homeowners so that they might switch their heating systems from coal to cleaner fuels, such as electricity or natural gas. This law was followed by a second Clean Air Act that mandated the raising of smokestacks. Since wind speed increases with elevation above Earth's surface, the idea was to increase the rate at which pollutants were removed from the local environment.

Around the world, other countries also began to act. In the United States, California became the first state to enact air pollution legislation in 1947. Los Angeles had long been plagued by smog, but its source was unknown. The city government shut down a rubber-making factory, but the smog persisted. Eventually officials realized that the primary sources of Los Angeles' smog were motor vehicles, and they passed a law in 1960 mandating the installation of pollution-control devices in automobiles.[69] Today, California has the strictest automobile emissions standards in the country, related to ongoing pollution problems.

At the national level, the U.S. federal government passed the Air Pollution Control Act of 1955, which provided funds for research into the health effects of air pollution. Unfortunately the act left regulation of air quality to the states, which did little to improve it.[70] As air quality continued to worsen, Congress passed the Clean Air Act in 1963, giving authority to the Secretary of the Department of Health, Education, and Welfare to establish air quality standards and to make funds available to state and local governments to create pollution control agencies.[71] In 1970, the federal government amended the Clean Air Act, charging the newly created Environmental Protection Agency (EPA) with developing and enforcing **national ambient air quality standards (NAAQS)**. Standards were set for maximum allowable concentrations of particular pollutants (Table 11-3), as measured for a specific period of time, called the *averaging time*, at a given distance from the source.

Despite these early efforts, pollution continued to worsen in urban areas as population and, therefore, vehicle use increased. In addition, hazardous substances continued to be exported across state and national boundaries, threatening the health of humans and ecosystems. In New York and New England, for example, 20 to 50 percent of red spruce trees were killed through a combination of acid precipitation, tropospheric ozone, climatic stresses, and insect infestations.[72] In response to these continuing threats, the U.S. Congress amended the Clean Air Act again in 1990 to require that annual emissions of sulfur dioxide be reduced by 10 million tons and nitrogen oxides by 2 million tons over 1980 levels by the year 2000.[73] In addition, the amendments called for decreased production of CFCs and other gases that contribute to ozone depletion and for reductions in mercury emissions from power plants.

To ensure progress toward national air quality goals, the EPA instituted a permitting program, known as New Source Review, which requires any potential new (or existing, but undergoing renovation) source of air pollution to acquire a permit specifying the allowable size of emissions and mandating that pollution control devices be installed.[74] If the potential source lies within a region that has not yet attained the NAAQS standards, it must be constructed with the best demonstrated pollution control technology available at the time of construction to make emissions as low as possible. Furthermore, the EPA instituted a cap and trade program for sulfur dioxide. Each of the nation's coal-fired power plants was provided emissions allowances that they could sell to one another. Those plants that were able to upgrade their pollution control technology at low cost rapidly did so, selling their allowances to those that couldn't. To incentivize continuing progress at higher polluting plants, the EPA has reduced the emissions allowed over time and requires plants to pay $2000 for every ton of emissions above their allowances.

The results of the Clean Air Act and its amendments have been significant, with a 71 percent reduction in annual emissions of sulfur dioxide, a 92 percent reduction in atmospheric lead, a 46 percent reduction in nitrogen dioxide, and a 25 percent reduction in tropospheric ozone. Nitrogen oxide emissions from new cars have decreased substantially since 1967 (by 76–96 percent), but the number of people and of cars has increased, so the total declines in emissions and production of ozone have not been as large as for sulfur dioxide and lead. Nevertheless, air quality probably would have been far worse by now without the emission controls.

Another area in which little progress has been made since the 1990 Clean Air Act amendments has been in regulation of mercury. While the EPA was given regulatory authority over mercury emissions, the agency only developed a national standard for such emissions in 2011. Over the next several years, coal-burning power plants, which are responsible for 99 percent of atmospheric mercury emissions produced by the electric power generation industry, will be required to install pollution control devices. The reduction in these emissions along with those of arsenic and other toxic pollutants are projected to prevent 11,000 premature deaths, 130,000 asthma attacks, and 4700 heart attacks annually.[75]

In addition to the United States, other industrialized nations have reduced their atmospheric emissions in the past few decades. Some of the greatest successes have come in Eastern Europe, where investments in air pollution control technologies have increased since the dissolution of the Soviet Union in 1991. Emissions of SO_2 in the Czech Republic, for example, fell by 86 percent between 1990 and 2001.[76] In contrast, air pollution in China grew steadily worse throughout the end of the 20th century due to rapid industrialization and a heavy reliance on high-sulfur coal. Though some progress was made to clean up the air, for example by replacing coal stoves with natural gas, air quality remained low due to increasing automobile usage. In 1991, China's population of 1 billion people owned a mere 710,000 cars. By 2006, the number of cars had topped 20 million and by 2011 exceeded 100 million.[77]

During the 2008 summer Olympics in Beijing, the government famously placed restrictions on automobile use in an attempt to clean up the air for athletes and visiting tourists. Some modified restrictions remain in place. For example, one in five cars are banned from driving on any given day, forcing people to use subways, buses, or bicycles. In addition, the government ordered a 90 percent reduction in sulfur content of gasoline in 2008.[78] Together with an expansion of the city's subway network and conversion of public buses to cleaner-burning natural gas, these governmental actions promise to improve Beijing's air quality markedly in the future. At present, though, it does not achieve standards set by the World Health Organization.

Unfortunately, little information on air quality is available from developing countries, yet these are the very nations where rapid urban growth is occurring, air quality is severely threatened, and little progress has been

TABLE 11–3 National Ambient Air Quality Standards		
Primary standards		
Pollutant	Level	Averaging Time
Carbon Monoxide	9 ppm (10 mg/m³)	8-hour
	35 ppm (40 mg/m³)	1-hour
Lead	0.15 μg/m³	Rolling 3-Month Average
	1.5 μg/m³	Quarterly Average
Nitrogen Dioxide	0.053 ppm (100 μg/m³)	Annual (Arithmetic Mean)
	0.100 ppm	1-hour
Particulate Matter (PM₁₀)*	150 μg/m³	24-hour
Particulate Matter (PM₂.₅)	15.0 μg/m³	Annual (Arithmetic Mean)
	35 μg/m³	24-hour
Ozone	0.075 ppm	8-hour
	0.12 ppm	1-hour
Sulfur Dioxide	0.03 ppm	Annual (Arithmetic Mean)
	0.14 ppm	24-hour

*PM_{10} refers to particulates 10 micrometers or smaller, whereas $PM_{2.5}$ refers to particulates 2.5 micrometers or smaller.

Modified from National Ambient Air Quality Standards (NAAQS), Air and Radiation Criteria, http://www.epa.gov/air/criteria.html.

made to establish pollution control. Regrettably, a United Nations Environment Programme initiative to establish a Global Environmental Monitoring System (GEMS) for air quality failed, the victim of lack of participation by United Nations member states and budget cuts.[79]

Of course, passing laws is not ultimately what reduces air pollution. Cutting down on emissions of various pollutants requires technological innovations and less use of pollution-generating equipment through conservation or increased efficiency.

Cleanup Technologies

The various types and sources of air pollution require different cleanup strategies. A **catalytic converter,** part of the exhaust system of a motor vehicle, turns smog-producing waste products of gasoline combustion into relatively benign gases. The device consists of a steel cylinder filled with porous ceramic that is coated with platinum or palladium. The cylinder resides between the engine block and the exhaust pipe and receives carbon monoxide, unburned hydrocarbons, and nitrogen oxides given off by combustion. The platinum or palladium coating acts as a *catalyst* in a chemical reaction that converts carbon monoxide, unburned hydrocarbons, and oxygen into water and carbon dioxide. (A catalyst is a substance that accelerates a chemical reaction but does not otherwise change it.) The catalytic converter also turns nitrogen oxides into nitrogen gas, the most abundant gas in the atmosphere.

California was the first state to mandate the use of catalytic converters, and because the state has such a large share of the automobile market in the country, its action eventually led to all new cars in the country being equipped with this technology. The success of catalytic converters is readily apparent when examining emissions data over the last few decades. Between 1982 and 2001, the state's population grew by 40 percent, resulting in a 97 percent increase in miles traveled by car. At the same time, ground level ozone concentrations fell by 55 percent.[80]

Power plant emissions have also declined over the years. To get rid of the sulfur dioxide gas that results from fossil fuel combustion, calcium oxide or calcium carbonate particles suspended in water, known as scrubbers, are sprayed into the stream of gases coming away from the combustion chamber. The calcium oxide or carbonate combines with the SO_2 to make calcium sulfite or sulfate. Calcium sulfate, commonly known as the mineral gypsum, is used in the manufacture of cement, plaster, and wallboard. Japan, which has little natural gypsum, recycles the gypsum produced by scrubbers. Other technologies turn SO_2 gas into sulfuric acid or into elemental sulfur, both of which can be captured and used for other purposes. Elemental sulfur, for instance, is used in matches.

Another strategy to cut down on the amount of SO_2 gas produced by power plants is the use of coal and oil that are naturally low in sulfur content. This is not always practical, however, so technologies have been devised to remove sulfur from oil and coal. For coal, the technique consists of crushing followed by washing in water to remove the iron sulfide mineral pyrite. Pyrite is much denser than coal and can easily be removed by washing.

Removal of nitrogen oxides from power plant combustion is generally achieved by burning the fuel at a low temperature (700°C to 900°C; 1290°F to 1650°F) and by limiting the amount of oxygen allowed for the combustion. This keeps the organic nitrogen in the fossil fuel from oxidizing and can also inhibit the reaction of atmospheric nitrogen gas to nitrogen oxides.

■ Enactment of legislation since the late 1940s has led to significant reductions in atmospheric pollution in the developed world. Statistics for the developing world are poorly known.

■ Cleanup strategies depend on the type of pollutant produced and the source of the pollution.

■ Reduction of hazardous emissions is typically achieved through a chemical modification process such as occurs in the catalytic converters of cars or in the scrubbers of power plants.

CLOSING THOUGHTS

The lives of all of us—those who move south to Florida or Arizona to escape harsh northern winters, farmers dependent on summer rainfall to grow the crops we eat, people suffering from asthma and other respiratory ailments, and residents of "Tornado Alley" and hurricane ravaged coasts—are intimately linked to atmospheric phenomena. The atmosphere, like Earth's other systems, is in a constant state of change. It is an open system that perpetually exchanges gases and particles with the lithosphere, hydrosphere, pedosphere, and biosphere, and that transports received solar energy around the globe. In the process, it creates deserts and wet areas, air masses and fronts, severe weather and gentle rains, and it also disperses a variety of chemical compounds. Despite its significance in our lives, when skies are clear, it is easy to forget that the atmosphere exists.

For many hundreds of thousands of years, the size of the human population remained sufficiently small that our activities had little impact on the overall atmospheric composition. In the last two centuries, however, population has exploded (see Chapter 7), and this increase, combined with technological advances, has led to unprecedented changes in atmospheric chemistry. While we previously took for granted that the atmosphere would dilute and dispose of our wastes, today we are faced with the legacy of decades of environmental carelessness in acidification of forests and watersheds and lowered levels of the stratosphere's protective ozone. Fortunately, we are beginning to wake up to these problems. International agreements and actions of individual states have begun to reduce emissions of some harmful substances. Still, much work remains to be done before the chemistry of the atmosphere can approach its preindustrial composition.

SUMMARY

■ Earth's average surface temperature is maintained at a value suitable for life by the presence in the atmosphere of water vapor and trace gases such as carbon dioxide, methane, and nitrous oxide.

■ Evaporation of water in the equatorial latitudes and condensation in the higher latitudes transports heat from the equator to the poles.

■ Earth's atmosphere is organized into layers identifiable by temperature variations. Temperature reflects the distribution of gases with different capabilities to absorb solar radiation and infrared radiation from Earth's surface.

■ Air circulation, and therefore weather, occurs primarily in the troposphere.

■ Tropospheric pollutants are easily removed by reaction with water droplets. Stratospheric pollutants have longer residence times.

■ Differential heating of Earth's surface produces pressure gradients in the atmosphere along which winds flow. Those winds are deflected by the Coriolis effect, which causes air circulation to take on a tricellular form in each hemisphere and thus explains the distribution of deserts and wetter environments.

■ Air masses with homogeneous temperatures, humidities, and densities interact with one another along fronts. Frontal lifting of warm air over cold air often initiates updrafts that lead to storms.

■ Human activities have produced pollutants in the atmosphere that cause environmental change.

■ The impact of a pollutant is determined partly by its residence time in the atmosphere. Acid rain is primarily a local problem because the sulfates and nitrates that produce it react quickly with atmospheric water and fall as rain within a few thousand kilometers of the source. Stratospheric ozone depletion, on the other hand, is occurring globally because CFCs are less reactive and can therefore be mixed throughout the atmosphere before they are removed.

■ A combination of legislation and technology has led to vast reductions in pollutant emissions to the atmosphere.

KEY TERMS

atmosphere (p. 362)
aerosols (p. 363)
troposphere (p. 366)
environmental lapse rate (p. 366)
adiabatic cooling (p. 367)
saturation vapor pressure (p. 367)
stratosphere (p. 368)
climate (p. 369)

weather (p. 369)
greenhouse effect (p. 370)
albedo (p. 371)
Coriolis effect (p. 374)
Hadley cell (p. 375)
polar cell (p. 375)
Ferrel cell (p. 375)
jet streams (p. 375)
trade winds (p. 376)

westerlies (p. 376)
polar easterlies (p. 376)
Intertropical Convergence Zone (ITCZ) (p. 376)
monsoons (p. 378)
air mass (p. 381)
front (p. 381)
wind shear (p. 384)
mesocyclone (p. 384)
tornado (p. 386)
tropical storms (p. 386)

hurricanes (p. 386)
acid rain (p. 391)
smog (p. 394)
inversion (p. 396)
chlorofluorocarbons (CFCs) (p. 396)
national ambient air quality standards (NAAQS) (p. 398)
catalytic converter (p. 399)

REVIEW QUESTIONS

1. What atmospheric phenomenon plunged the southeastern United States into a deep freeze during winter 2009–2010?

2. What role is life thought to have played in the evolution of Earth's atmosphere? How did non-biological factors influence atmospheric chemistry?

3. Why are trace gases in the atmosphere so critical to the existence of life on Earth?

4. What causes adiabatic cooling?

5. What are the two reasons that temperature declines with altitude in the troposphere?

6. Explain the vertical structure of the atmosphere.

7. Why does vertical mixing occur in the troposphere? Why is it important?

8. How does temperature affect the ability of air to hold moisture?

9. What is the difference between weather and climate?

10. What happens to solar radiation as it hits Earth?

11. Explain why the Sun radiates in the infrared, visible, and ultraviolet parts of the electromagnetic spectrum while Earth radiates only in the infrared.

12. Why do the poles receive less solar energy than does the equator?

13. What keeps the poles from getting colder, and the equator from getting hotter, every year?

14. What is the Coriolis effect? What impact does it have on atmospheric circulation?

15. What causes many of Earth's deserts to lie in a latitude belt near 30°?

16. What causes a "rain shadow"?

17. Explain the origin of sea breezes, land breezes, and monsoons. In what ways are these phenomena similar?

18. What is the Intertropical Convergence Zone (ITCZ), and why is there so much rain there?

19. What impact might climate have on water resources?

20. What are air masses?

21. Why do storms occur along fronts?

22. What processes generate lightning and thunder?

23. What causes tornadoes to form?

24. What factors are necessary for hurricane formation?

25. Why do hurricanes lose their strength when they move over land or poleward in the ocean?

26. Why does acid rain have only a local or regional impact whereas production of chlorofluorocarbons (CFCs) has a global impact?

27. Ozone depletion is occurring in the stratosphere, while too much ozone is forming near Earth's surface in the troposphere. Explain why this is a problem, and discuss the causes of ozone depletion and tropospheric pollution.

28. What technologies are used to decrease the amount of acid rain produced in the atmosphere?

29. What steps have been taken to restore Earth's protective layer of ozone?

THOUGHT QUESTIONS

1. Imagine an Earth shaped like a cylinder rather than a sphere. Would the Coriolis effect operate on this planet? Why or why not? How would the atmospheric circulation differ from that which develops on a rotating sphere?

2. Draw a picture of Earth with a rocket at the south pole. If the rocket were fired toward the equator, the rocket would appear to be deflected to the west, or left, relative to an observer in the southern hemisphere; explain this effect.

3. Suppose Earth's axis of rotation were not tilted. Would we experience seasons? Why or why not? Would all parts of Earth receive the same amount of solar energy? Why or why not?

4. Would you expect hurricanes to be more or less frequent if Earth's temperature warmed because of increased concentrations of greenhouse gases in the atmosphere?

5. Suppose that the average time a molecule of SO_2 spends in the atmosphere before reacting with water

and falling to the ground as acid rain is 4 days. If SO_2 is emitted from a factory smokestack when atmospheric winds are blowing at 15 km per hour, how many miles downwind of the factory will the acid precipitation be felt?

6. Suppose that you are the minister of public health in a developing country that is experiencing rapid urbanization. What actions might you carry out to prevent air quality deterioration? Consider all sectors that contribute to air quality problems, such as transportation, housing, and industry.

EXERCISES

1. Construct a barometer to keep track of atmospheric pressure changes. Attach a piece of tubing to a water bottle. Fill the bottle halfway with water and turn it upside down while holding the tubing up so that water doesn't flow out. Water will flow partway up the tube. Tape the tubing to the water bottle and tape a small ruler next to the tubing. Record the level of the water in the tube every day. What do you observe when storms pass through your area? What causes these effects? Hint: How does the air pressure inside the barometer compare with that outside?

2. To learn more about why vertical mixing occurs in the troposphere but not the stratosphere, fill two clear glasses with lukewarm tap water. Then fill one cake pan with ice and another with hot water. Put one water-filled glass into each cake pan, making sure that the glass is resting on the bottom of the pan. Which glass represents the troposphere, and which the stratosphere? Why? Wait a few minutes for water motions to stop and then drop a little food coloring in each of the glasses. What do you observe? Hint: Think about convection.

3. Determine the pH of rainfall in different local areas by collecting rainwater in a jar or cup and testing it with a piece of pH paper (ask your college chemistry stockroom if they will supply you with a few pieces of pH paper). Try to collect rain from locations near fossil fuel–burning power plants. Does the pH at different locations vary? Do the samples taken near the power plant differ from the others? How?

4. Earth's average surface temperature is 15°C. Given this value, calculate the wavelength of electromagnetic radiation given off by our planet.

5. From the information presented in Figure 11-16, what is the rotational velocity of Earth at the equator? At 45° latitude?

SUGGESTED READINGS

Baron-Faust, R. "Under the Weather." *Scientific American* 11 (2000): 90–96.

Emmanuel, K. *Divine Wind: The History and Science of Hurricanes.* New York: Oxford University Press, 2005.

Henson, R. "Billion-Dollar Twister." *Scientific American* 11 (2000): 32–39.

Lutgens, F. K., E. J. Tarbuck, and D. Tasa. *The Atmosphere: An Introduction to Meteorology,* 11th ed. New York: Prentice Hall, 2010.

Mathis, M. *Storm Warning: The Story of a Killer Tornado.* New York: Touchstone, 2007.

Metcalfe, S., and D. Derwent. *Atmospheric Pollution and Environmental Change.* New York: Oxford University Press, 2005.

Monastersky, R. "Forecasting Is No Picnic." *Scientific American* 11 (2000): 12–19.

Rohli, R. V., and A. J. Vega. *Climatology.* Sudbury, MA: Jones and Bartlett, 2008.

U.S. Environmental Protection Agency. "Achievements in Stratospheric Ozone Protection—Progress Report." Report EPA-430-R-07-001. Washington, D.C.: U.S. Environmental Protection Agency Office of Air and Radiation, 2007.

Vallera, D. *Fundamentals of Air Pollution,* 4th ed. Burlington, MA: Academic Press, 2008.

Wise, W. *Killer Smog: The World's Worst Air Pollution Disaster.* Lincoln, NE: iUniverse.com, 2001.

The Ocean and Coastal System

The oceans form a dynamic system. Immense currents rise to the surface at some locations and sink to great depths at others, carrying water, nutrients, dissolved gases, and heat from one part of the globe to another. Where currents rich in nutrients well up along coasts, as off of Alaska and Peru, some of the world's most productive fisheries are found. Indeed, life on Earth is thought to have begun in the oceans, and the seas have nourished humans for more than a hundred millennia with abundant harvests of fish and shellfish. Unfortunately, our growing population exacts an increasing toll on ocean resources. Since the late 1800s, some marine populations have declined so much due to overfishing and pollution that they are losing their potential for renewal.[1]

While we draw much of our food from the sea, some coastal communities also use it as a cesspool for sewage and other wastes, and currents that convey water and nutrients along coastlines and out to the open ocean transport contaminants as well. Additional pollution originates from global trade when oil tankers run aground or cargo ships spill their loads during storms. An overabundance of nutrients, bacteria, viruses, pieces of plastic, and chemicals degrade marine habitats and affect species at all levels of the food chain. Moreover, as greenhouse gas levels climb, Earth's warming and increasingly acidic oceans are becoming less hospitable to many species,[2] and climate change–related sea level rise threatens the 40 percent of people who live within 60 km of coasts.[3]

Intertidal oyster reefs in temperate bays and estuaries, such as this one in the Chesapeake Bay, today comprise less than 15% of their historical abundance worldwide, representing losses far greater than those of coral reefs, mangrove wetlands, and other habitats in shallow water. With a role similar to that of coral reefs in tropical marine environments, native oyster reefs create complex, structured habitats for hundreds of species of plants and animals. The "ecosystem engineers" of bays and estuaries, these bivalve shellfish provide further benefits by filtering water. Causes of reef loss are multiple, including overharvesting, loss and degradation of habitat, disease, and the increased runoff of freshwater, nutrients, and sediment from land. Efforts to protect, conserve, and restore oyster reefs have shown some signs of success and provide models for future management strategies. *(Courtesy Virginia Institute of Marine Science, Eastern Shore Laboratory)*

To understand ocean resources, pollution, hazards, and the role of the sea in global climate and environmental change, in this chapter we:

✔ Examine the nature of the ocean basins and their shoreline boundaries.

✔ Analyze ocean circulation and its links to other Earth systems.

✔ Investigate the causes of coastal erosion, flooding, and sea level rise.

✔ Evaluate the effects of humans on the seas and efforts to conserve marine resources.

INTRODUCTION

On January 10, 1992, a cargo ship loaded with goods manufactured in Asia and bound for retail stores in the United States encountered rough winter weather in the northern Pacific.[4] As strong winds pushed against the ocean surface, 40-foot waves tossed the ship from side to side, tipping it as much as 55° from horizontal. Cargo lashed to the deck broke loose and more than 7000 cardboard boxes, each containing a set of children's plastic bathtub toys, fell into the frigid saltwater. As glue holding the boxes together dissolved, a bright yellow duck, red beaver, green frog, and blue turtle emerged from each box (Figure 12-1). Thus introduced into the ocean, the 28,800 toys began an odyssey that would float them around the globe over the next few decades.

Within months, thousands of the toys washed up on beaches surrounding Sitka, Alaska. They had hitched a ride on the northern Pacific's Subarctic Gyre,

an ocean current that circulates in a great oval between Alaska and Siberia.[5] Many more traveled even farther. By 1994, some of the toys had floated through the Bering Strait into the Arctic Ocean. Embedded in sea ice, these polar travelers eventually made their way to the North Atlantic, where they got caught up in the Labrador and Gulf Stream currents and washed up on the shores of Maine and Scotland. Other toys continue to float in the Pacific, making a trip around the Subarctic Gyre approximately every 3 years.[6]

The oceans' major currents have been recognized for as long as humans have conducted oceanic voyages, but accidental spills of plastic toys and other items, along with intentional releases of sophisticated oceanographic equipment, provide details that illuminate the role of ocean circulation in Earth's climate and in the distribution of sea life. We now know, for example, that the Gulf Stream current, which flows along the east coast of the United States before crossing the North Atlantic, brings warmth from tropical latitudes to northern Europe, greatly moderating northern Europe's winter temperatures. Similarly, we have learned that some of the most productive fisheries on Earth lie atop upwellings of nutrient-rich water from the deep sea.

One of the most ambitious projects to study ocean circulation was launched in 2000. The Argo program uses more than 3000 probes to keep track of water temperature, salinity, and current velocity in the upper 2000 m of the ocean and can detect changes in circulation associated with global climate change (Figure 12-2).[7] The probes are programmed to drift with ocean currents for 10 days at a time before ascending to the sea surface. During their climb, they take water temperature and salinity measurements that are relayed to satellites once at the surface. In conjunction with location coordinates, these measurements provide nearly real-time information about the continuously changing physical properties of the seas.

While ocean circulation can be studied with temperature, salinity, and velocity information, monitoring the health of the oceans requires additional techniques, including surveys of marine biota and measurements of ocean water chemistry. Research cruises and studies of fishing records reveal that populations of some fish have fallen as much as 90 percent over the last several decades.[1] Much of this decline is related to overfishing, but pollution also has been an important factor. Analyses of water quality reveal that many coastal waters have low oxygen levels, chemical toxins, and microbiological pathogens associated with sewage disposal by coastal cities and agricultural runoff from rural areas.

Global trade also plays a role in oceanic pollution. The 1992 tub toy spill was neither the first nor the last to gain attention. Two years earlier, a cargo ship lost 80,000 sneakers en route from South Korea to the

FIGURE 12–1 Rubber ducks (similar to this one) floating on global ocean currents continue to wash up on beaches decades after they were spilled from a cargo ship in 1992. *(Steve West/ Getty Images)*

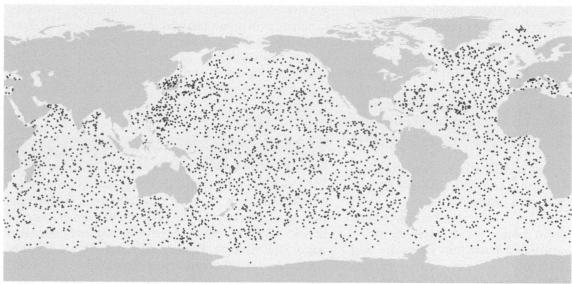

(a) 3555 Active Floats   April 2013

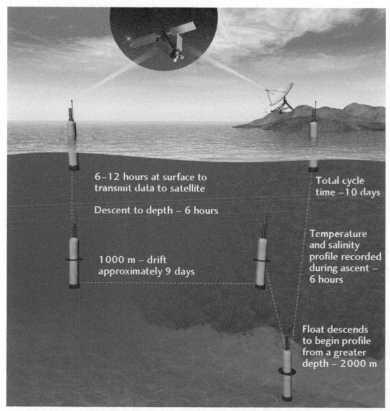

6–12 hours at surface to
transmit data to satellite

Total cycle
time –10 days

Descent to depth – 6 hours

Temperature
and salinity
profile recorded
during ascent –
6 hours

1000 m – drift
approximately 9 days

Float descends
to begin profile
from a greater
depth – 2000 m

(b)

FIGURE 12–2 (a) More than 3500 probes drifting in the oceans worldwide provide data on ocean conditions. (b) The probes descend to a depth of 1000 to 2000 m, drift for about 10 days, and then ascend, measuring temperature and salinity as they rise. Once at the surface, they communicate with satellites that record their positions and collect the data before they descend again. *(Courtesy the Argo Program, http://www.argo.net)*

United States, and in 2002, another 33,000 sneakers fell overboard.[8] In 1997, five million Legos spilled into the Atlantic just off the shore of England, and thousands of computer monitors spilled into the Pacific in 2000.[9] During the 1990s, there were as many as 10,000 cargo spills annually,[10] and at the beginning of the 21st century it is estimated that tens of millions of pieces of marine debris are produced daily,[11] resulting in the creation of great oceanic garbage patches. Trash that blows off shore or is transported to the oceans by rivers also contributes to the flotsam.

Though the image of rubber ducks floating in the seas is quaint, trash washing up on beaches is not, and oceanic spills have caused ecological damage. Marine mammals, large fish, and seabirds swallow debris, confusing it with food, and can suffer acute intestinal blockages, interference with food absorption, and/or exposure to toxic compounds.[12] As marine debris continues to circulate, it succumbs to the elements, breaking down into ever smaller pieces of plastic and heavy metals that are ingested by creatures at lower levels of the food chain. Animals also become entangled in old pieces of fishing net, ropes, and six-pack beverage rings. Recently, marine debris has been implicated in the dispersal of invasive species, which attach to it and float to new locations.[13]

While the activities of humans have significantly affected the global ocean, the ocean's impact on our lives is likewise profound. Its great mass of water absorbs the Sun's heat and moderates Earth's climate. Its energetic circulation delivers moisture to the atmosphere, resulting in rain and snow that water the land and enable us to grow crops. In a vast system of moving water, runoff from land then replenishes the oceans at the mouths of rivers and along coasts. Slow movements of the seafloor over geologic time shape the continents

we live on, and rapid movements during earthquakes generate tsunamis that have killed hundreds of thousands of people in recent years. The margins of deep ocean basins supply us with petroleum and other resources. To understand the land we live on and the processes that shape it therefore requires understanding the sea.

THE OCEAN BASINS

Viewed from outer space, Earth appears as a vast mass of water surrounding a few relatively small landmasses. Indeed, 71 percent of our planet's surface is covered in saltwater that resides in five major, interconnected basins: the Atlantic, Pacific, Indian, Arctic, and Southern (or Antarctic) **oceans**. Also connected to this world ocean, but largely enclosed by land, are smaller bodies called **seas**. Among these are the Caribbean, Red, North, Baltic, and Mediterranean seas.

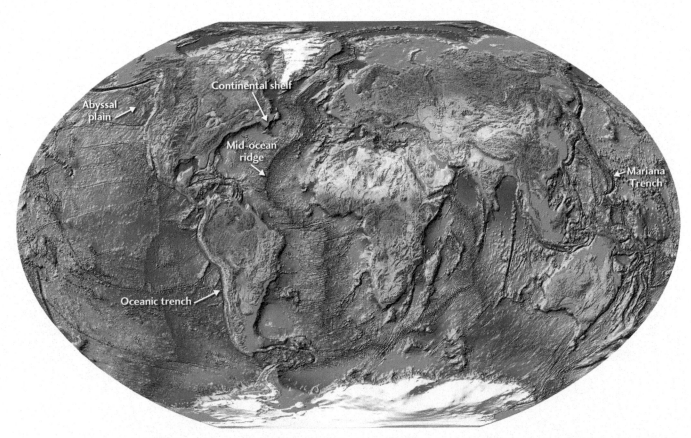

FIGURE 12–3 A digital shaded–relief map of Earth's surface. Shading is used to make apparent such features as continental shelves, abyssal plains, mid–ocean ridges, and oceanic trenches. Broad continental shelves can be seen, for example, along the east coast of North America, and deep oceanic trenches along the west coast of South America. The relatively smooth areas are deep abyssal plains, and there is a high mid–ocean ridge midway between the Americas, Europe, and Africa. Land elevations are taken from topographic maps and ocean–floor elevations from depth data acquired by the U.S. Navy and other groups. *(Michael Schmeling/Alamy)*

Features of the Seafloor

The submarine weaponry of World War II gave nations ample incentive to investigate the topography of ocean basins (see Chapter 2). **Bathymetry** (from the Greek word *bathys,* meaning "deep") is the science of inferring seafloor topography by measuring the depth of the oceans. Bathymetric studies, using techniques that range from the ancient line-and-sinker method to modern sonar depth measurement, reveal that the topography of the ocean floor varies substantially from place to place. Nevertheless, several consistent undersea landforms can be identified worldwide. These include the continental shelf, continental slope, continental rise, abyssal plain, mid-ocean ridge, and deep-sea trench (Figure 12-3).

Surrounding every continent is a **continental shelf,** a nearly flat underwater plain that slopes less than 0.5° and is covered with up to thousands of meters of layered sand and mud eroded from land and transported to the shelf by rivers and glaciers. Continental shelves provide diverse habitats for millions of marine species and vary in width from a few kilometers along the Pacific coasts of North and South America to more than 1000 km in the Arctic Ocean. At the outer edges of continental shelves, water depth is as much as 200 m. From this point, the nearly flat shelf gives way to a relatively steep drop-off called the **continental slope,** which is angled as much as 6° and reaches depths of 3000 m or more.

Like continental shelves, continental slopes are covered with thick layers of mud and sand eroded from the continent, and both features are deeply incised by *submarine canyons* that, in some places, dwarf the Grand Canyon in size (Figure 12-4). Sediments are transferred through these submarine canyons to deep-ocean basins. Sometimes the sediments are spread over a broad, gently sloping apron that forms a transition zone between the continental slope and the deep ocean. Such a transition zone, which slopes at about 1°, is called a **continental rise.**

Continental rises typically level out imperceptibly into a wide, flat **abyssal plain** that forms most of the world's ocean floor and is submerged beneath 3 to 5 km of water. The abyssal plain is mantled with hundreds of meters of sediment composed of the shells of microscopic floating organisms that sank and buried the underlying basaltic ocean crust. Although abyssal plains are the flattest features on Earth, they are not completely so. The ocean floors are also home to volcanic mid-ocean ridges, where Earth's crustal plates are splitting apart, and to deep-sea trenches at subduction zones, where plates converge (see Chapter 2). The bottom of the deepest trench on Earth, the Mariana Trench in the western Pacific Ocean, is 11 km below sea level (see Figure 12-3).[14]

The Changing Ocean Basins

Just as plate tectonics influences seafloor topography, it changes the shapes of ocean basins over time. The basin of the Atlantic Ocean has been expanding for about 200 million years. Tectonic forces tear the lithosphere apart along the Mid-Atlantic Ridge, and magma from the mantle rises to form new crust. At the same time, the Pacific Ocean has been shrinking along a vast ring of subduction zones. Although there are spreading centers in the Pacific Ocean, the Pacific Plate is being consumed considerably faster than new crust is being created, so the width of the Pacific Ocean is getting smaller at a rate of a few centimeters per year. The Mediterranean Sea also is shrinking as the African and Eurasian plates converge and the seafloor between them is subducted.

FIGURE 12–4 Multibeam sonar surveys provide detailed images of the seafloor. The continental shelf along coastal California, shown here in shades of gray to pink, is especially wide near San Francisco. The continental slope, shown in shades of red to yellow, slopes seaward, in some places more steeply than others, and grades into the continental rise, shown in shades of green to blue. The continental rise merges imperceptibly with the abyssal plain, which is nearly flat except for occasional seamounts (submerged volcanic islands) and deep canyons. Rumpled features on the abyssal plain are sediments deposited on extensive undersea fans, places where mass movement has occurred off the continental slope. *(Lincoln F. Pratson, University of Colorado, and William F. Haxby, Columbia University)*

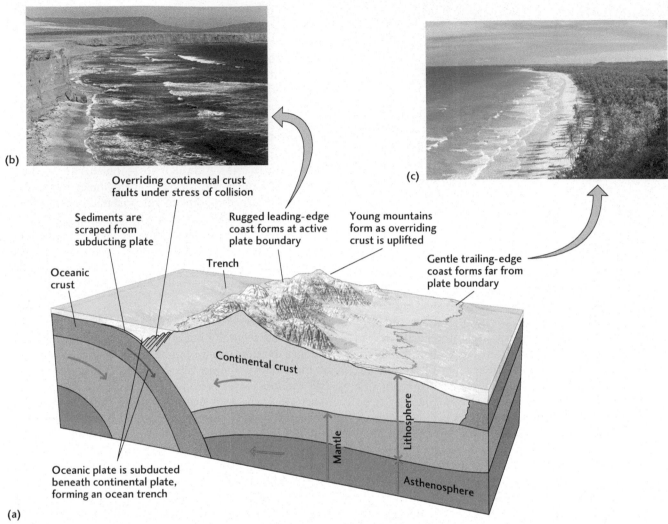

(b)

(c)

Overriding continental crust faults under stress of collision

Sediments are scraped from subducting plate

Rugged leading-edge coast forms at active plate boundary

Young mountains form as overriding crust is uplifted

Gentle trailing-edge coast forms far from plate boundary

Trench

Oceanic crust

Continental crust

Lithosphere

Mantle

Asthenosphere

Oceanic plate is subducted beneath continental plate, forming an ocean trench

(a)

FIGURE 12–5 (a) Coasts at tectonically active plate margins are sometimes referred to as leading–edge coasts because they lie at the front, or leading, edge of a moving plate where it comes in contact with other plates. Coasts located away from plate margins, embedded within and riding along with a tectonic plate, are referred to as trailing–edge coasts. (b) The coast of Peru is a leading–edge coast. Situated on the western margin of the South American Plate, where it overrides the subducting Nazca Plate, it is characterized by steep slopes and a high–energy beach. Steplike benches are remnants of past shores planed by wave action and since raised out of the water by tectonism. (c) The coast of Brazil, along the eastern margin of South America, is a trailing–edge coast, characterized by a broad coastal plain, shallow water, and gentle waves. *(b: Gerard Lacz Images/SuperStock. c: Elizabeth Whiting & Associates/Alamy)*

Juvenile ocean basins are beginning to open in the African Plate at the East African Rift, in the North American Plate at the Gulf of California where Baja California is separating from the Mexican mainland, and between the African and Arabian plates at the Red Sea (see the photograph that opens Chapter 2). The Atlantic Ocean basin probably resembled these young rifts when it began to open 200 million years ago.

Coastlines

As ocean basins change shape with time, so do their coastlines (Figure 12-5). So-called **active margin** coasts (also known as *leading-edge coasts*) form at convergent boundaries where one plate, typically continental, overrides a denser oceanic plate. Mountain building and uplift are common, raising the coastline out of the water and generating rocky cliffs (see Figure 12-5b). Uplift also causes streams to incise deep, narrow canyons. Because they typically are bounded by subduction zones and deep-sea trenches, active margins have narrow continental shelves and little coastal plain. Sediment eroded from stream canyons and cliffs is deposited in deep-sea trenches and offshore basins formed by folding and faulting. Deep water close to shore allows waves to maintain their size as they approach shore and to strike the coastline with tremendous energy and erosive power. With little shallow nearshore area available to store sediment, and with such high-energy waves to remove any sediments that do accumulate, large deltas cannot form.

Passive margin coasts (also known as *trailing-edge coasts*) are located along continental margins away from plate boundaries (see Figure 12-5c). These coasts are associated with broad, low-relief coastal plains and gently sloping, meandering streams. Sediment stored in river floodplains and beaches is reworked into deltas, sand dunes, and barrier islands. The continental shelves of passive margins typically are broad and extensive, and shallow water reduces the size of waves as they approach shore. These conditions are conducive to delta formation, so most of the world's large deltas occur along passive margins.

Both active and passive margins are found along the coasts of North America. The gently sloping east coast of the continent is a passive margin in the middle of the North American Plate. The craggy western edge borders active plate margins: the Cascadia and Aleutian subduction zones in the north and the San Andreas fault to the south (see Figure 2-19).

Sea Level Changes

Coastlines change over time in response to climatic and tectonic factors. Global cooling, such as that which happens when Earth moves into an ice age, traps water in glaciers and permafrost on land, lowering sea level and causing shorelines to advance seaward. Conversely, global warming melts ice masses and causes thermal expansion of seawater, raising sea level and causing shorelines to move inland. Tectonic uplift and subsidence of landmasses have similar effects on relative sea level, and both types of change alter the nature of sediment deposition and transport. When sea level falls, parts of continental shelves, and sometimes even continental slopes, are exposed, and rivers cut gorges across the surface on their way to the ocean basin. When sea level rises, as is currently happening, the lower reaches of rivers are drowned by ocean water, forming estuaries and fjords that are especially prominent on coasts that have little tectonism and uplift of land.

Estuaries are bodies of ocean water that have encroached inland over the mouth of a stream (Figure 12-6). The inflowing river water mixes with the salty ocean water to form *brackish* water, intermediate between fresh and salty. The salinity gradient from fully fresh to fully saline creates a great diversity of habitats for aquatic organisms, so estuaries are known for their high biological productivity. On the U.S. Atlantic coast, Chesapeake Bay and Delaware Bay are examples of estuaries once renowned for their oysters before they were subject to overfishing, habitat degradation, and water pollution.[15]

Fjords are drowned coastal valleys carved by glaciers rather than by rivers. Glacial valleys tend to be deeper and more U-shaped than those carved by rivers, causing fjords to have very steep and sometimes vertical walls. The coasts of Norway, Greenland, Alaska, British Columbia, and Chile all have fjords produced by glaciers that were more extensive during Earth's last glacial

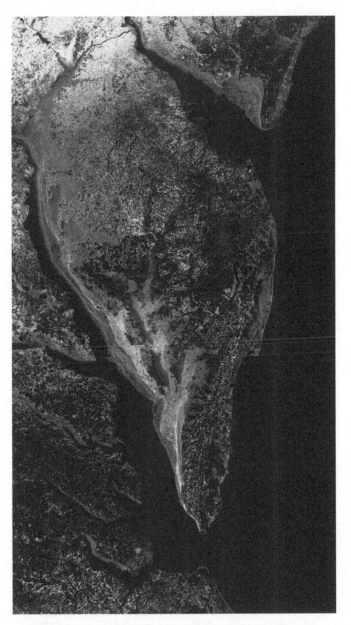

FIGURE 12–6 Chesapeake Bay (the larger body of water to the south) and Delaware Bay (to the north) are estuaries that formed about 6000 years ago when the rising sea level flooded the lower reaches of many rivers that drain into these coastal indentations. The indentation at the southern end of the Chesapeake Bay formed about 25 million years ago, when a meteorite crashed into the continental shelf. *(Landsat/NASA)*

period, which ended approximately 16,000 years ago. As the ice melted and the glaciers retreated, sea level rose, filling the valleys with ocean water.

■ From ocean bathymetry, scientists have learned that the ocean floor has substantial topographic relief and consists of six dominant landforms: continental shelves, continental slopes, continental rises, abyssal plains, mid-ocean ridges, and deep-sea trenches.

■ Plate tectonics influences the topography of the seafloor and the shape and size of ocean basins over time.

■ Coastline characteristics vary depending on tectonic setting. An active margin coast typically forms where a continental plate overrides an oceanic plate at a convergent boundary, and is characterized by steep cliffs, high wave energy, and coastal erosion. A passive margin coast forms along continental margins away from plate boundaries and is marked by broad coastal plains and sediment accumulation in deltas, beaches, and barrier islands.

■ Sea level changes caused by climatic oscillations, as well as apparent (relative) sea level changes associated with tectonic movements, cause coastlines to migrate laterally and either expose continental shelves to erosion or drown coasts and their associated river valleys with seawater.

SEAWATER CHEMISTRY

Having studied something of the ocean basins that hold the seas, we turn our attention now to the properties and circulation of the water itself. Ocean water tastes salty because it contains a large number of dissolved ions. Many of these, especially cations like calcium and sodium, originate from chemical weathering of Earth's crust. They are delivered to the seas by rivers and through the reaction of seawater with the oceanic crust. Others, particularly anions like sulfur and chlorine, are introduced when magmatic gases (sulfur dioxide, hydrogen chloride, hydrogen sulfide) are erupted along with molten rock along mid-ocean ridges.

TABLE 12–1 Major Ions in Seawater and River Water

Constituent ions	Concentration in seawater (mg/kg)*	Percentage of ions in seawater	Concentration in river water (mg/kg)*	Percentage of ions in river water	Mean residence time in ocean (million yrs)
Chloride (Cl^-)	19,350	55.05	5.75	6.44	120
Sodium (Na^+)	10,760	30.61	5.15	5.77	75
Sulfate (SO_4^{2-})	2712	7.72	8.25	9.24	12
Magnesium (Mg^{2+})	1294	3.68	3.35	3.75	14
Calcium (Ca^{2+})	412	1.17	13.4	15.02	1.1
Potassium (K^+)	399	1.14	1.3	1.46	11
Bicarbonate (HCO_3^-)	145	0.41	52	58.27	0.10
Bromide (Br^-)	67	0.19	0.02	0.02	100
Boron (B^{3+})	4.6	0.01	0.01	0.01	10.0
Strontium (Sr^{2+})	7.9	0.02	0.03	0.03	12
Fluoride (F^-)	4.6	0.01	0.10	0.11	0.5
Silicon (Si)	3.0[†]	0.01	13.1[†]	10.8[†]	0.0004[‡]
Total	35,148	100.00	89.24	100.00	

* Milligrams per kilogram is the same as parts per million by weight and can be expressed as a percentage. For example, the total of 35,148 mg of ions in 1 kg (1,000,000 mg) of seawater can be expressed as 35,148 parts per million, or about 3.5 percent.

[†] From Pinet, P. R., *Invitation to Oceanography*, 4th ed., (Boston: Jones and Bartlett, 2006), 146.

[‡] From Treguer, P., Nelson, D. M., Van Bennekom, A. J., DeMaster, D. J., Leynaert, A., and Queguiner, B., "The Silica Balance in the World Ocean: A Reestimate," *Science* 268 (1995): 375–379.

The Salinity of Seawater

On average, seawater contains about 35,000 mg of ions per kg (1,000,000 mg) of water (Table 12-1), so its salinity is roughly 3.5 percent:

Salinity = (35,000 mg/1,000,000 mg) × 100 = 3.5%

The actual value varies a bit with geographic location and ranges from less than 2.5 percent to 4.2 percent,[16] but mostly falls between 3.3 and 3.7 percent.[17] Lower values are associated with large freshwater inflows, such as at the mouth of the Amazon River; higher values are found in climatologically arid areas where abundant evaporation of water from the ocean concentrates salts, as in the Red Sea. The percentage of dissolved solids may seem small, but is enough to make seawater toxic to many living things.

Of the ions in seawater, the two most abundant are chloride and sodium, with average concentrations of about 19,000 mg per kg (1.90 percent) and 10,700 mg per kg (1.07 percent) respectively (see Table 12-1). Sodium chloride is thus the most abundant salt in the sea, accounting for most (85 percent) of the 3.5 percent of dissolved solids. In contrast to seawater, river water contains only about 0.01 percent dissolved solids, and rainwater contains virtually none. Furthermore, the ionic composition of river water is weighted heavily toward bicarbonate and calcium rather than chloride and sodium.

If the ocean is fed by rain and rivers, why is it so saline and why is its ionic composition so different from freshwater? The answer lies in the processes that introduce and remove ions to and from the ocean reservoir (Figure 12-7).

First, the evaporation of ocean water to make atmospheric water vapor leaves dissolved ions behind, raising the concentration of dissolved ions in the ocean relative to that in river water. Some of these ions are removed through sea spray blown onto land, through deposition and burial in oceanic sediments, and through reaction with the volcanic rocks of the seafloor, but the rate of removal is slow. Chloride and sodium, for example, have very low chemical and biochemical reactivities and are among the most soluble substances in seawater. Only when large evaporite basins form as a result of tectonic and climatic changes can water become sufficiently concentrated with these ions to precipitate halite (a mineral formed of sodium and chloride ions) out of solution, and that has happened relatively infrequently in geologic history.

Second, various processes remove different ions from the water column. Unlike sodium and chloride, calcium and bicarbonate (converted to carbonate) are used extensively by marine organisms to construct shells of biogenic calcium carbonate. Mollusks, corals,

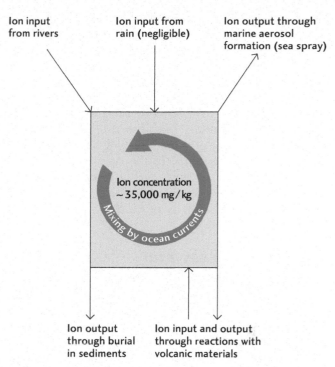

FIGURE 12-7 The ocean can be viewed as a reservoir into and out of which flow dissolved ions and sediment. This model includes only the major constituents of seawater, among them chloride and sodium ions, for which inputs and outputs are approximately balanced over thousands to millions of years. Because the world ocean has no outlet, the only ways that ions can be removed are through transfer into the atmosphere by formation of aerosols—suspensions of particles in a gas—or through transfer into the lithosphere, which takes place via burial in sediments and reaction with volcanic materials on the seafloor.

foraminifera, and other creatures rapidly deplete the levels of these ions relative to others. On a percentage basis, silica is also far less abundant in the ocean than in river water due to its uptake by diatoms, sponges, and other organisms that use it to construct their skeletons.

The makeup of ancient marine sedimentary rocks shows that seawater has had a composition similar to today's oceans for about 1.5 billion years,[18] evidence that the amount of the most abundant dissolved substances in the ocean reservoir is in a long-term steady state. The approximate residence times of the major constituents can be estimated from the amount of each element in ocean water divided by the rate at which it is added or removed. Those ions that are depleted rapidly by biological processes, such as calcium, bicarbonate, and silica, have relatively brief residence times (see Table 12-1). Those that do not participate in biological activity tend to have longer residence times. For example, the residence times of chloride and sodium are respectively 120 million and 75 million years. Because these values are much longer than the time periods over which the ions are mixed by ocean currents (1000 to 2000 years or less), their amounts are nearly the same in all parts of the ocean.

The effect of the ocean currents is similar to the homogeneity created by a beater that mixes new ingredients being added to a bowl. If the beater mixes the ingredients more rapidly than the rate at which they are added, all parts of the bowl will have a uniform distribution of ingredients. In the oceans, the circulation of water via immense currents mixes dissolved ions at a faster rate than that by which they are added to seawater, and the salinity of most seawater is about 3.5 percent.

Nutrients in Seawater

Seawater contains dissolved substances other than salts. As described in Chapter 7, nutrients are elements that organisms require in order to grow and reproduce. In the oceans, nutrients nourish life, beginning with **phytoplankton**—from the Greek words *phyto* (plant) and *plankton* (made to wander or drift)—microscopic plants that float near the surface of the water column but lie at the bottom of the oceanic food chain. Through the process of photosynthesis, phytoplankton use nutrients to produce organic compounds and in turn nourish organisms higher on the food chain that eat phytoplankton; these include *zooplankton*, microscopic animals that also float at the ocean surface (Figure 12-8). The main nutrients in seawater include nitrogen (chiefly as nitrate, NO_3^-) and phosphorus (as phosphate, PO_4^{3-}).

Unlike the salts in seawater, which tend to be mixed uniformly, nutrient levels vary markedly in different parts of the ocean. Carried in runoff from continents, nutrients are abundant along coastlines, especially near the mouths of rivers along continental shelves (Figure 12-9). Their abundance also varies markedly between the upper and lower parts of the ocean reservoir. The layer nearest the surface, called the **photic zone**, or **euphotic zone**, because it receives the most sunlight, typically is almost completely depleted of nutrients because of the high biological activity in this layer. Nutrients are partially recycled within the euphotic zone but are also steadily lost to the deep ocean as organisms die and their organic material sinks to the ocean floor. Nutrients are recycled from depth to the surface layers where currents well up at coastlines (discussed below). These areas have high rates of biological activity and are important for fisheries.

Dissolved Gases and pH in Seawater

In addition to salts and nutrients, seawater contains gases, chiefly nitrogen (N_2), oxygen (O_2), carbon dioxide (CO_2), and inert, or "noble," gases (argon, neon, and helium). (Noble gases are so called because they "stand alone" and are rarely reactive.) Each of these can be exchanged between the atmosphere and ocean across the air–sea interface. Nitrogen can be fixed (see Chapter 7) by oceanic bacteria and some species of phytoplankton, which convert it to forms that can be used by other organisms.[19] This is likely an important source of nitrate in parts of the ocean far from coastlines.

As with nitrate, concentrations of oxygen and carbon dioxide vary substantially in space and time because of

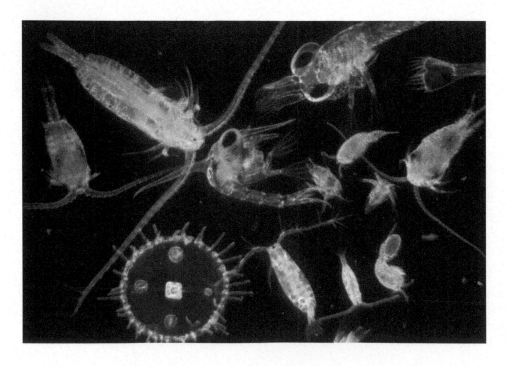

FIGURE 12–8 Passively drifting or weakly swimming microscopic organisms in the oceans include plant forms called phytoplankton and animal forms called zooplankton. *(CLPA/SuperStock)*

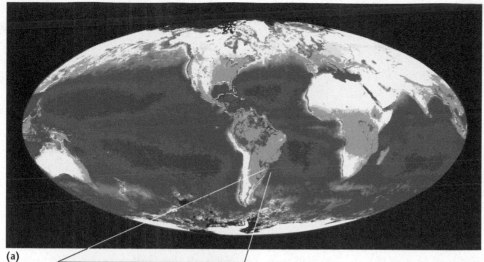

(a)

Argentina

(b)

FIGURE 12-9 (a) This false-color satellite image shows the distribution of vegetation on land and phytoplankton in the oceans. Colors represent the density of chlorophyll, from dark green (most dense) to pale yellow (least dense) on land, and from red (most dense) through yellow and blue to pink (least dense) in the oceans. Areas rich in phytoplankton are associated with broad continental shelves and upwelling currents. (b) There was a massive bloom of phytoplankton, shown here as colorful swirls in the Atlantic Ocean, on the continental shelf off the coast of Patagonia during December 2010, when cold water from the south met with warmer, relatively saltier water from the subtropics to the north and east. As the currents collided, they produced eddies of turbulent water that vertically mixed nutrients from the deep ocean with surface water. From the north, runoff of freshwater laden with nitrates and iron-rich sediment came in from the Rio de la Plata, a large river that drains much of Argentina. The result was a mass of floating nutrients and phytoplankton known as a "bloom," a rich feast for the zooplankton and other organisms higher up the food chain. *(a: Dr. Gene Feldman/NASA–GSFC/Science Photo Library/Science Source; b: NASA image created by Norman Kuring, Ocean Color Web; NASA images courtesy the MODIS Rapid Response Team at NASA GSFC)*

differing levels of biological activity. In the photic zone, phytoplankton photosynthesis draws down the concentration of carbon dioxide and raises the concentration of oxygen (Figure 12-10). Some oxygen also diffuses into ocean water from the overlying atmosphere. At depth, photosynthesis ceases where light no longer can penetrate the water, but organisms continue to respire, and dead organic matter decays as it sinks through the water column. As a result, oxygen levels generally decline and carbon dioxide levels increase below the photic zone. Depending on the rate at which vertical circulation of the oceans can deliver oxygenated water to the seafloor, some parts of the deep ocean can actually become *anoxic,* or lacking in oxygen, due to respiration and decay.

The typical pH of seawater measures between 7.5 and 8.5.[20] The concentration of carbon dioxide in ocean water strongly controls this property, as shown in the following reactions:

$$CO_2 + H_2O \leftrightarrow H_2CO_3 \leftrightarrow H^+ + HCO_3^- \leftrightarrow$$
$$2\,H^+ + CO_3^{2-} \quad \text{(Equation 1)}$$

$$H^+ + CaCO_3 \leftrightarrow Ca^{2+} + HCO_3^- \quad \text{(Equation 2)}$$

Carbon dioxide (CO_2) dissolves into ocean water (H_2O) to make carbonic acid (H_2CO_3). The carbonic acid dissociates into hydrogen (H^+), bicarbonate (HCO_3^-), and carbonate (CO_3^{2-}) ions, with the hydrogen ion concentration determining the pH. Reaction of some of the hydrogen ions with calcium carbonate ($CaCO_3$) creates dissolved calcium (Ca^{2+}) and shifts the pH of ocean water toward values more basic than that of pure water (that is, some of the hydrogen ions are consumed).

■ Seawater contains many dissolved ions, making it saline. The salts are derived from surface water runoff in rivers and outgassing of magma at mid-ocean ridges and accumulate very slowly over geologically long periods of time. Over a period of thousands to millions of years, the amounts of the major salt ions in seawater—the two most abundant of which are sodium and chloride—are roughly constant.

■ The residence times of the most abundant salt ions in seawater are on the order of millions of years, whereas the amount of time needed for ocean currents to mix seawater and its dissolved substances is on the order of thousands of years. As a consequence, the proportions of dissolved salts in seawater are similar throughout the world, on the order of 35 g of dissolved ions per kg of water (35 parts per thousand), or 3.5 percent.

■ The most important nutrients in seawater—nitrates and phosphates—are derived primarily from surface water runoff. Their concentrations vary markedly from one part of the ocean to another because they are depleted from the surface layer by biological activity, accumulate in the deep ocean by sedimentation, and are recycled to the surface by upwelling currents from the deep ocean.

■ The pH of ocean water is determined by the amount of carbon dioxide dissolved in the water as well as by the presence of calcium carbonate.

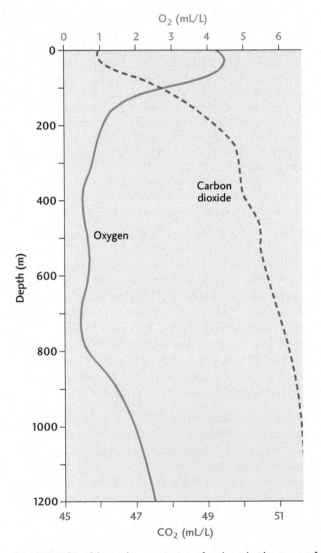

FIGURE 12–10 In the uppermost few hundred meters of ocean water, phytoplankton deplete water of carbon dioxide (CO_2) and enrich it with respect to oxygen (O_2) through the process of photosynthesis. The concentrations of O_2 and CO_2 are given in milliliters per liter, or mL/L.

OCEAN STRUCTURE AND CIRCULATION

Great currents flow through the oceans at rates of 0.1 to 2 m/sec and carry matter and energy from one part of Earth's surface to another.[21] Some are immense, relative to masses of continental water. For example, all the rivers flowing into the Atlantic Ocean carry a total discharge of about 0.6 million m³ per sec, but some of the smallest ocean currents, such as that flowing out of the Mediterranean Sea, carry about 1 to 3 million m³ per sec, about two to five times as much water.[22] Large ocean currents, such as the Gulf Stream, carry many times as much, transporting up to 150 million m³ per sec, hundreds of times the average amount of the largest river on Earth, the Amazon.[23] When ocean currents are compared in size with the largest rivers on Earth, it is easy to see why the oceans dominate the global hydrologic cycle.

Ocean currents belong to two separate but linked circulation systems, one driven by the winds and the other by variations in water density. In order to understand these systems, it is necessary to discuss the vertical structure of the oceans, which promotes their separation.

Vertical Structure of the Oceans

Mixed by winds and heated by the Sun, the **surface mixed layer** extends to about 150 m depth and is relatively vertically homogenous in temperature, salinity, and density at any given moment (Figure 12-11). Laterally, the surface mixed layer exhibits gradations in these properties. For instance, temperatures decline from the equator to the poles due to variations in solar heating (discussed in Chapter 11). Likewise, near the equator, high rates of rainfall associated with the atmosphere's Intertropical Convergence Zone reduce ocean salinity, but at about 30° north and south latitude, in the desert belt, evaporation outpaces precipitation, increasing the surface layer's salt concentration. The depth and temperature of the mixed layer also vary seasonally as solar radiation amounts change and as wind strength waxes and wanes. In winter, the mixed layer is colder and thicker because of lower solar energy inputs and stronger wind mixing. In summer, it is warmer and thinner. No matter what the location or the season, the surface mixed layer has the lowest density of all ocean water, at roughly 1.02 to 1.025 g/cm³.

Beneath the surface mixed layer, from approximately 150 to 1000 m, lies a zone where temperature, salinity, and density change markedly with depth. Called the **thermocline, halocline,** or **pycnocline,** depending on the physical or chemical property of interest, this zone prevents the surface mixed layer from interacting freely

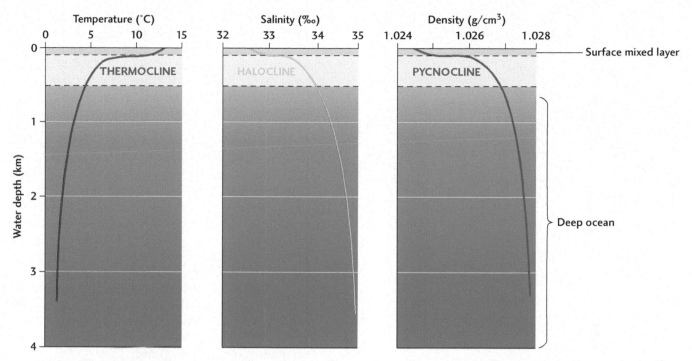

FIGURE 12-11 At mid- to low latitudes, three properties of seawater change markedly between about 150 and 1000 m water depth. The temperature of seawater decreases with depth, whereas the salinity and density increase with depth.

The zone of change is referred to as the thermocline, halocline, or pycnocline, in reference to changes in temperature, salinity, and density, respectively.

with the **deep ocean,** which lies below 1000 m. Because the deep ocean is untouched by surface phenomena (such as solar heating and winds) and cannot readily mix with the surface layer, it is, overall, colder, more saline, and denser. On average, it has a temperature of 3°C to 4°C and a salinity of 3.5 percent;[24] however, it also contains many different water masses, each with distinctive temperatures and salinities produced by different formation histories that are described below.

Circulation in the Surface Mixed Layer

Water in the surface mixed layer moves in response to the winds, which exert a drag on the top of the water column, pushing it along. The water in direct contact with the atmosphere moves at the highest speed, and forward velocity decreases with depth until eventually, no further forward motion takes place. The loss of velocity with depth is caused by the friction of water molecules moving past one another, which diminishes the amount of energy each successively deeper layer can transmit downward.

Because the surface layer of the ocean is driven forward by the winds, one might expect that ocean currents would travel in exactly the same direction

as the prevailing winds, but they do not. The great Arctic explorer and statesman Fridtjof Nansen of Norway (1861–1930) observed that icebergs floating in the North Atlantic travel at a 20° to 40° angle to the prevailing wind direction. Nansen had set out on an expedition to reach the North Pole in a specially designed ship, the *Fram*, which he allowed to become frozen into the Arctic sea ice. Every time the wind blew, both ship and ice were carried in an oblique direction. Later work by the Swedish oceanographer V. Walfrid Ekman (1874–1954) demonstrated that ocean currents are deflected by the Coriolis effect (see Chapter 11) in the same way that atmospheric currents are deflected.[25] Each successively deeper layer of water—moving more slowly than the one above it—is oriented a bit farther away from the prevailing wind direction, producing what is known as the *Ekman spiral*. Averaged over the entire column of water that responds to the winds, the net water movement direction is 90° to the prevailing winds, a phenomenon called the **Ekman transport.** In the northern hemisphere, the net motion is 90° to the right of the winds (Figure 12-12); in the southern hemisphere, net motion is 90° to the left of the winds.

The Ekman transport of water plays an important role in the distribution of nutrients and oxygen within the oceans. When winds in the northern hemisphere

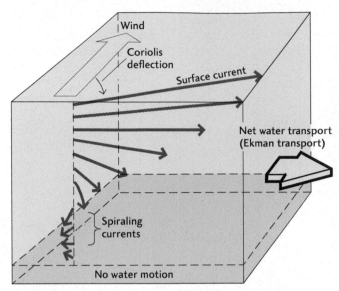

FIGURE 12–12 Ekman transport, shown for the northern hemisphere, results from the interaction of wind friction and Coriolis deflection. Here, the Coriolis deflection causes the surface current to flow to the right of the direction in which the wind is moving. Moving downward, each layer of water in the current is deflected a bit to the right of the layer above it, producing a spiraling current called the Ekman spiral. Averaged over the entire layer influenced by the winds, the water moves 90° to the right of the direction of the generating wind. In the southern hemisphere, Ekman transport is to the left of the direction of the generating wind.

the net water transport is onshore (Figure 12-13). In the southern hemisphere, the opposite is true.

The differences between these two scenarios are important to the locations of the world's fisheries and places of marine biodiversity. When net water transport is offshore, cold water from the deep ocean rushes up to the surface to replace the displaced surface water, in a process called coastal **upwelling** (see Figure 12-9b). The deep water brings with it dissolved nutrients produced by the decay of sunken organic matter and replenishes the surface mixed layer, which is typically very low in nutrients because of photosynthesis. Upwelling also occurs near Earth's equator. There the action of the northeast and southeast trade winds leads to Ekman transport away from the equator, causing upwelling from the deep ocean to replace the water that has moved away at Earth's surface (Figure 12-14). In general, zones of upwelling tend to be excellent fisheries because the high nutrient levels lead to abundant plankton, the primary food for fish (Box 12-1). When net water transport is onshore, water is piled up onshore and sinks, a process called **downwelling** that transports oxygen downward.

The frictional drag of wind on water creates the system of surface currents shown in Figure 12-15. Not surprisingly, this system of currents is in many ways similar to the system of winds described in Chapter 11. For example, the westward motion of the trade winds generates the westward-flowing equatorial currents. Likewise, the eastward-blowing westerlies steer the eastward-moving North Pacific and Gulf Stream currents. An obvious difference between the oceans and the atmosphere is that the oceans have boundaries (i.e., landmasses) that confine their circulation. As a result, surface circulation breaks up into large clockwise and counterclockwise rotating cells known as **gyres.**

blow southward along a coastline, the net water transport is to the west because this is the direction 90° to the right of the prevailing wind direction. Thus, if the land lies to the east of the ocean, the net water transport is offshore, and if the land lies to the west of the ocean,

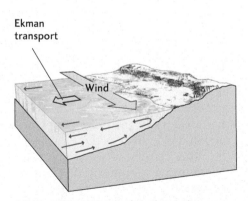

(a) Upwelling in the northern hemisphere

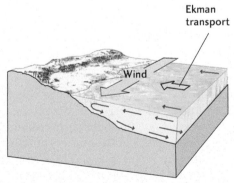

(b) Downwelling in the northern hemisphere

FIGURE 12–13 Coastal upwelling and downwelling are caused by winds blowing parallel to a coast and by Ekman transport. These diagrams show the patterns in the northern hemisphere. (a) Where land lies to the east and the wind is blowing southward, surface water moves offshore, enabling

deep currents to upwell in order to replace the water removed. (b) Where land lies to the west and the wind is blowing southward, surface water moves onshore, piling up and then sinking, or downwelling.

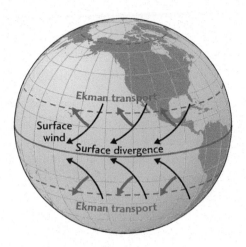

FIGURE 12–14 Equatorial upwelling is caused by Ekman transport of water away from the equator in response to the northeast and southeast trade winds.

These gyres and their associated currents play a strong part in transporting heat between the equator and high latitudes. In the northern hemisphere, the northward-flowing Gulf Stream (Atlantic Ocean) and Kuroshio (Pacific Ocean) currents carry heat to the high latitudes, and the southward-flowing California (Pacific Ocean) and Canary (Atlantic Ocean) currents transport cold water to the equator, where it is warmed (see Figure 12-15). Without the warm water of the Gulf Stream, temperatures over northern Europe would be 5°C to 10°C lower than they are at present. In the southern hemisphere, the Eastern Australia, Brazil, and Agulhas currents carry warm water poleward but are prevented from reaching Antarctica by the icy Antarctic Circumpolar Current produced by the southern hemisphere's westerly winds. Cut off from this source of heat, Antarctica is thermally isolated, causing its frigid temperatures.

Because the oceans have walls that block flow, winds can stack water up against them. The trade winds, for example, blow water westward, leading to a difference in ocean height between the western and eastern tropical Pacific of about 0.5 m.[26] Water flows downslope because of Earth's gravitational attraction, so the Equatorial Countercurrent flows from west to east, in opposition to the winds and the South and North Equatorial currents.

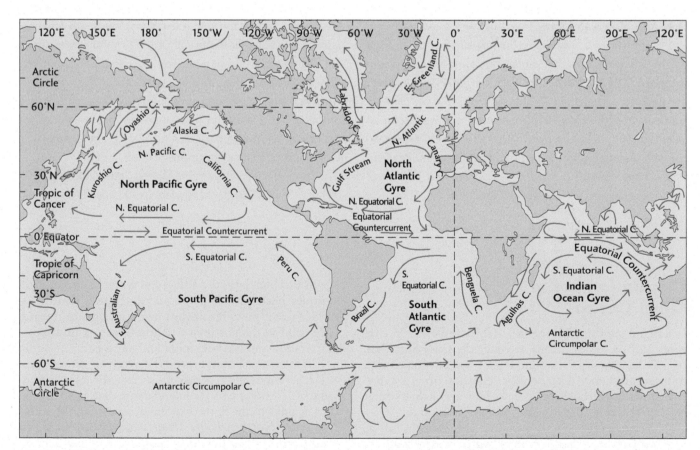

FIGURE 12–15 Surface currents in the world ocean circulate clockwise in the northern hemisphere and counterclockwise in the southern hemisphere. The flow of surface currents is similar to the pattern of atmospheric circulation.

BOX 12–1 GLOBAL AND ENVIRONMENTAL CHANGE

El Niño Climatic Events

Until the 1970s, Peru's anchovy fishery was the largest in the world, but the catch plummeted from some 13 million to less than 5 million tons over a period of 2 years in the early 1970s.[27] Although exacerbated by overfishing, the sudden decline in anchovy catch was due primarily to a major change in the coupled ocean–atmosphere system that sometimes appears around Christmas, and is called El Niño, Spanish for "the child." El Niño events are heralded by the formation of unusually warm water off the western coast of South America and occur irregularly with varying intensity, much the way floods do, at roughly 3– to 8–year intervals.[28]

During most years, the trade winds (see Chapter 11) blow from east to west across the Pacific Ocean in response to a pressure gradient between a high–pressure zone in the eastern tropical Pacific and a low–pressure zone near Indonesia. These winds pile up warm water in the western tropical Pacific, creating an area of high sea level (**Figure 1**). They also blow water away from the west coasts of North and South America, leading to the upwelling of deep, cold, nutrient–rich currents that replace the surface water blown westward. The sea surface temperature difference between the western and eastern Pacific reinforces the air pressure gradient, and the upwelling zone along the west coast of South America supports rich biological communities teeming with plankton and the fish that feed on them. Peruvian fishers have depended for decades on the vast schools of anchovies that thrive in the upwelling waters.

During El Niño events, the atmospheric and sea surface temperature conditions change dramatically. The eastern Pacific high–pressure and the Indonesian low–pressure zones weaken or even reverse, lessening the gradient and causing the trade winds to relax or change direction. No longer pushed westward, the warm pool of water around Indonesia sloshes eastward toward South America, cutting off the upwelling that sustains the fishery. The seesawing air pressure is known as the *Southern Oscillation,* and the entire phenomenon is often referred to as the **El Niño Southern Oscillation,** or ENSO for short. ENSO actually comprises three ocean–atmosphere states: El Niño; normal conditions, also called ENSO-neutral conditions; and La Niña, which is marked by anomalously cold temperatures in the eastern tropical Pacific and stronger–than–normal trade winds.[29]

The warm water that flows eastward during El Niños pumps heat into the atmosphere above the eastern Pacific, resulting in a low pressure zone and associated storms. Heavy rains drench the west coast of tropical South America and flood areas that are usually arid. El Niño events affect the rest of the world as well. For reasons that are not yet fully understood, they shift the positions of the winter jet streams, bringing elevated rainfall rates to the southern tier of the United States and drier–than–normal conditions to the Pacific Northwest, parts of Africa, India, and Australia.[30] La Niñas tend to have the opposite effect.

The record snowfalls in the United States' mid–Atlantic region in winter 2009–2010 were likely caused by a com-

Circulation in the Deep Ocean

Isolated from the atmosphere, deep-ocean currents owe their motion to variations in water density rather than to winds. Density is a function of water temperature and salinity, hence deep ocean circulation often is called **thermohaline circulation.** The colder and saltier water is, the greater its density and propensity to sink.

Thermohaline circulation can be thought of as a conveyor belt that transfers water warmed at the equator to high latitudes and cycles water chilled at high latitudes back toward the equator (Figure 12-16). As surface water flows northward from the equator toward Greenland in the northern Atlantic Ocean, it undergoes evaporation and becomes more and more saline. In addition, it loses much of its heat to the overlying atmosphere as Arctic winds blow across its surface. Eventually, this saline water becomes so cold and dense that, near Greenland, it sinks to the bottom of the ocean. From there, the cold, salty *North Atlantic Deep Water (NADW)* flows southward, crosses the equator, and eventually reaches

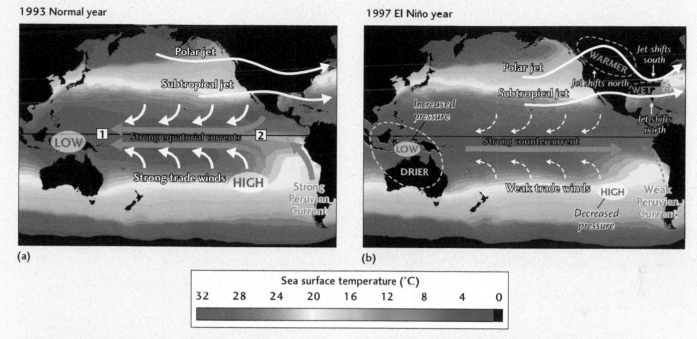

1993 Normal year

(a)

1997 El Niño year

(b)

Sea surface temperature (°C)

| 32 | 28 | 24 | 20 | 16 | 12 | 8 | 4 | 0 |

Figure 1 These satellite images illustrate the difference in sea surface temperature in 1993, a normal year, and 1997, an El Niño year. (a) During a normal year, the warmest waters are in the western Pacific (1), and a tongue of upwelling cool water, the Peruvian–Humboldt Current, spreads westward from South America (2). (b) During 1997, an El Niño year, water off the coast of South America was much warmer, and cool water of the Peruvian–Humboldt Current was absent at the surface.

bination of El Niño–supplied moisture and frigid air from the Arctic Oscillation (see Chapter 11). By mid–February, the Washington D.C. area already had received 185 cm (72.8 inches) of snow, shattering all previous records.[31] Baltimore and Philadelphia also set new records. At the same time, low snowfall rates and warmer–than–normal temperatures in British Columbia threatened to derail the 2010 Vancouver Winter Olympics, requiring snow to be brought in from the Canadian Rockies by truck during the early days of the competition.[32] Farther south along the West Coast, El Niño brought torrential rains, mudslides, and pounding surf to southern and central California.

The explanation of El Niño events is somewhat circular: It remains unknown whether they begin with an atmospheric

(continued)

the Southern Ocean, extending from 40° to 70° south latitude near Antarctica. From the South Atlantic, it rounds Africa and joins the circumpolar current that surrounds the frigid Antarctic continent.

A deep-water current also forms around Antarctica. Surface waters in the Southern Ocean are bitterly cold and highly salty because of the extensive sea ice formation around Antarctica. When seawater freezes, its dissolved salts are expelled into the surrounding seawater, producing highly saline brine at a rate of 20 to 50 million m^3/sec.[36] This cold, salty, and extremely dense Antarctic water sinks, slipping beneath the slightly warmer and less dense NADW that flows in from the north. Called *Antarctic Bottom Water (AABW)*, this salty and frigid current partially mixes with NADW to form *Common Water (CoW)* that flows northward into the Pacific and Indian oceans. The CoW wells up at the northern ends of these ocean basins, thereby delivering cold, nutrient-rich water to the surface. Surface currents then flow southward in the Pacific and Indian oceans to balance the northward-flowing bottom currents. These currents also flow westward, eventually returning to the southern

(continued)

Figure 2 Scientists use an array of buoys, such as this one, in the tropics (see the map) to gather data in order to monitor tropical ocean water conditions for fluctuations in temperature and wind, and to assess phenomena such as El Niño and La Niña events. *(NOAA)*

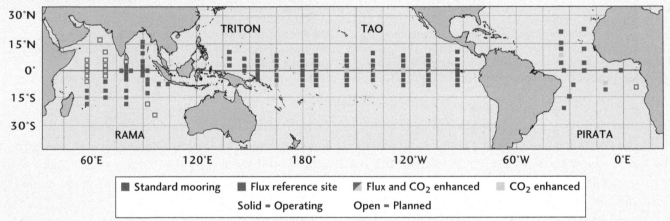

event or an oceanic event. Does the relaxation of the trade winds allow the Pacific warm pool to migrate eastward, or does the eastward migration of the warm pool cause a weakening of the trade winds as atmospheric pressures change in response to changing sea surface temperatures? Whatever the trigger, the reinforcing feedbacks between sea surface temperature and atmospheric pressure amplify the initial disturbance, allowing El Niño to develop. El Niño events appear to be increasing in frequency, with the most recent events in 1982–1983, 1986–1987, 1991–1992, 1994–1995, 1997–1998, 2002–2003, 2004–2005, 2006–2007, and 2009–2010.[33] Indeed, some scientists say the recurrence interval is every 2 to 5 years.[34]

Given the physical hazards and economic ramifications of El Niño events, many national governments and the United Nations are working to understand the mechanisms that cause El Niño and to develop long-range forecasts to better plan for the events. The Tropical Atmosphere Ocean project, funded primarily by the U.S. National Oceanic and Atmospheric Administration (NOAA) and Japan's Agency for Marine–Earth Science and Technology, uses 70 buoys moored in the tropical Pacific to measure water temperature and wind speed and direction to keep track of the evolving ocean over time (Figure 2).[35] As early as July 2009, these buoys sensed changes that predicted the winter 2009–2010 El Niño event.

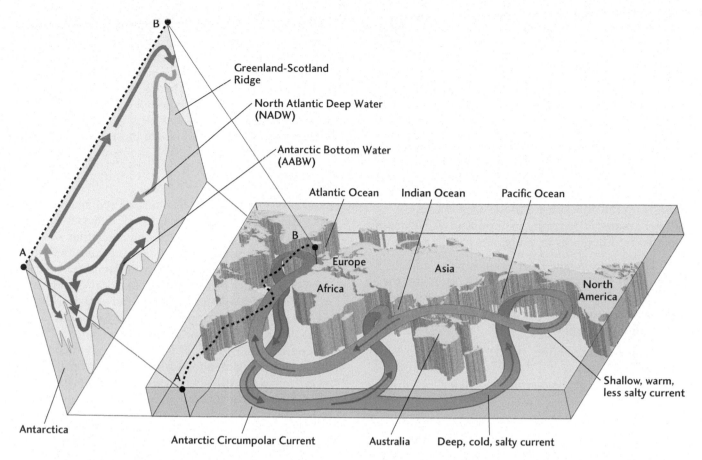

FIGURE 12-16 The model of an oceanic conveyor belt was developed by oceanographer Wallace Broecker of Columbia University's Lamont–Doherty Earth Observatory. The conveyor belt carries warm, salty water into the northern Atlantic Ocean, where it is cooled and becomes dense enough to sink into the deep ocean. From there, it flows southward along the seafloor at a rate of 10 million to 20 million m³/sec. Together with frigid bottom water produced around the margins of Antarctica, it travels northward into the Indian and Pacific ocean basins, only to upwell and flow back toward the south at the ocean surface.

A line of profile (left) drawn along the Atlantic Ocean from north to south (right) is used to plot seafloor bathymetry and the directions of flow of ocean currents. This profile illustrates the descent of surface water in the northern hemisphere along the Greenland–Scotland Ridge as it becomes the North Atlantic Deep Water (NADW) that then flows southward along the ocean bottom and returns to the surface near Antarctica. Antarctic Bottom Water (AABW) is the coldest, densest water in the ocean basins and is a by-product of sea ice formation, which expels salt and creates a very dense brine.

Atlantic, where they start their northward migration toward Greenland again (see Figure 12-16).

The conveyor belt model is consistent with much of the data about ocean behavior and has become a powerful tool for modeling climate change.[37] The conveyor belt might be able to start, stop, speed up, or slow down in response to changing amounts of freshwater added at the surface. Some scientists have linked rapid shifts from warm to cool global climate conditions in the past few tens of thousands of years to times when the conveyor belt might not have brought warm water to the northern Atlantic Ocean.[38] Although the ultimate cause of changes in the conveyor belt's pattern are as yet not fully

known, there is some concern about how global warming might affect the flow of this great ocean current and, as a consequence, affect future climate (see Chapter 15).

The surface part of the "conveyor belt" presently allows northern Europe to enjoy a fairly mild climate despite its high latitude. The flow of water in this system is immense—20 million m³ per second, about the size of 100 Amazon Rivers bundled together—and requires hundreds of years to complete a full cycle.[39] According to some climate models, if a melting Arctic sends a lot of freshwater into the North Atlantic, the salinity of the ocean may decline sufficiently that NADW formation slows or even stops.[40] Thus slowed, the great heat engine for northern Europe could diminish, cooling the region's climate, a *balancing feedback* caused by global warming. Other models suggest that this cooling would be counterbalanced by the warming from increased greenhouse gas concentrations in the atmosphere, so it is not possible at this time to say how northern Europe will be affected.[41]

■ Ocean circulation is divided into two primary types: surface mixed-layer currents and deep-ocean currents. In the surface mixed layer, the flow of water results from the movement of air masses—that is, wind—in the atmosphere. In the deep ocean, the flow of water is driven by variations in water density related to differences in temperature and salinity.

■ The surface mixed layer extends to a depth of about 150 m, varying based on the season and the geographic location. Because of wind mixing, the layer is fairly homogeneous in temperature and salinity throughout its depth.

■ Temperature, salinity, and density change dramatically between 150 and 1000 m, leading to the production of the thermocline, halocline, and pycnocline.

■ In comparison to the surface mixed layer, the deep ocean is far more homogeneous in temperature, salinity, and density, though it has variations in these properties that reflect the formation of deep-water masses in the North Atlantic and Antarctic oceans.

■ A conveyor belt model to describe ocean circulation has become a powerful tool for modeling climate change. Immense amounts of water warmed at the equator flow to high latitudes, becoming more saline because of evaporation and cooling as they lose heat. The water then flows back toward the equator as dense deep water that eventually joins the circumpolar current that surrounds Antarctica. Rapid shifts from warm to cool global climate conditions in the past few tens of thousands of years might have been caused by changes in the operation of this conveyor belt.

COASTAL PROCESSES AND THE HAZARDS OF LIVING BY THE SEA

Ocean circulation redistributes heat, thereby moderating Earth's climate and making our planet livable; and upwelling currents bring nutrients to ocean fisheries, thereby supporting the base of the food chain for marine ecosystems and providing our population with a source of protein. Not all oceanic processes and phenomena are as favorable to human activities, however. With population growth and development along coastlines quickening, the hazards of living at the edge of the sea are becoming increasingly acute. More than two-thirds of the human population lives within 400 km of a coast.[42] Forty percent of all people live within 100 km of a coast, a number that is expected to rise to 50 percent by 2015. Hurricanes (see Chapter 11), erosion, tsunamis, and rising sea level threaten those who live in this dynamic environment.

Beaches and Waves

Beaches are the narrow ribbons of sedimentary particles that lie at the interface of ocean water and the land surface. Their sediments originate from multiple processes: the erosive action of waves against the shoreline, which dislodges particles from land and smashes them into smaller pieces; stream transport of sediments to river mouths; and the pulverization of the shells and skeletons of marine organisms, such as mollusks and corals, that live in the nearshore environment. Due to the concentrated energy of waves at the shoreline zone, fine sediments (silts and clays) typically are winnowed away from coarser sediments and carried offshore, leaving the beach constructed of sand, pebble, or cobble-size particles.

Where sand is abundant and the seafloor slopes gently, **barrier islands**—popular destinations for tourists—commonly border the mainland (Figure 12-17). Separated from shore by water in estuaries, bays, and lagoons, these islands are elongate ridges of sand that parallel the coast. Hatteras Island in North Carolina, Miami Beach in Florida, and Padre Island in Texas are all barrier islands that formed 5000 to 7000 years ago, as rising sea level associated with the end of the last full glacial episode submerged elongate ridges and bars of sand on the previously exposed continental shelf. The seaward sides of barrier islands have beaches, whereas the landward sides often are covered with marshes, grasslands, dunes, and even forests. As sea level rises and falls over millennia, the positions of these barrier islands shift landward and seaward, respectively, across continental shelves.

The waves that are responsible for constructing the beaches on barrier islands and their adjacent mainlands are generated primarily by the winds, which exert a drag force on the water surface, stretching it in places and bunching it up in others. The wavelength (the distance between successive wave crests or troughs) of the disturbance depends on several factors: the wind speed; the length of the patch of ocean over which the wind blows, called the *fetch;* and the duration of time that the wind blows. An increase in any of these parameters raises the wavelength and speed at which the waves propagate.

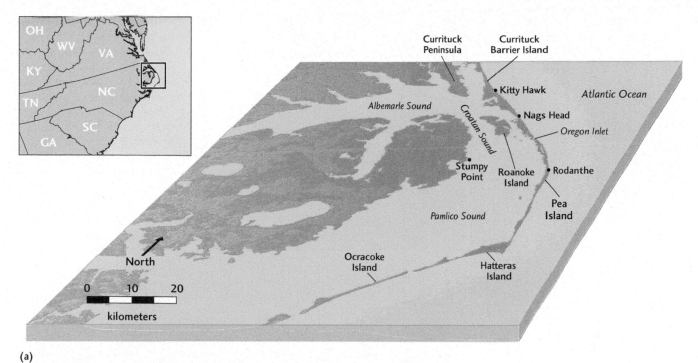

(a)

(b)

FIGURE 12–17 (a) Barrier islands off the coast of North Carolina form the chain of the Outer Banks, as shown here. (b) Pea Island National Wildlife refuge is located on the north end of Hatteras Island. *(b: Thomas R. Fletcher/Alamy)*

As a wave passes across the sea surface, it lifts and then lowers water molecules in such a manner that they trace out an orbital path (Figure 12-18). The energy imparted to the surface by the winds decreases with depth due to friction, so the orbits traced out by water molecules decrease in diameter with depth until they become negligible. The depth at which the influence of the winds is no longer felt, called **wave base,** lies at a depth below the surface approximately one-half the wavelength of the waves. The decay of wave energy with depth explains why a snorkeler floating on the sea surface can be tossed around by waves (sometimes resulting in seasickness!), while a scuba diver well below the surface feels nothing.

In the open ocean, wave base is far above the sea-floor, because the wavelength of wind-driven waves is typically only meters to hundreds of meters long while the ocean is thousands of meters deep. As waves approach shore, however, wave base intersects the continental shelf and generates frictional drag on the bottom of the water column (see Figure 12-18). This drag causes the bottom water to slow, but the surface water

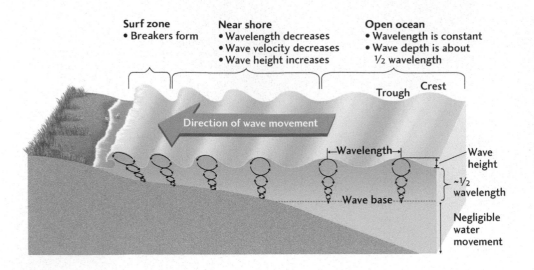

Surf zone
• Breakers form

Near shore
• Wavelength decreases
• Wave velocity decreases
• Wave height increases

Open ocean
• Wavelength is constant
• Wave depth is about ½ wavelength

Trough   Crest

Direction of wave movement

Wavelength

Wave height

~½ wavelength

Wave base

Negligible water movement

FIGURE 12–18 Orbital motion of water molecules in waves and wave base. Water molecules are lifted and lowered as a wave passes, tracing an orbital path. Wave base, the depth at which the influence of winds is negligible, lies at approximately one–half the wavelength.

continues to move forward at a higher speed, resulting in a pile-up of water that raises the wave height and reduces the wavelength. If the ratio of the wave height to the wavelength increases beyond a value of 1:7, the waves become oversteepened and break, generating the bubbly water that rushes up the beach, known as *surf*. The potential energy associated with wave elevation is converted to kinetic energy as the water falls, and this energy does work on the beach. That work consists primarily of moving particles up and down the beach, but wave energy also can compress air and water in rock fractures to further propagate fractures in rock and contribute to sea cliff erosion.

When waves strike the shore at an angle, the surf they produce moves up the beach at an angle, carrying particles with it. However, as the water recedes, it moves straight down (perpendicular to) the beach under the influence of gravity, which results in a zigzag travel path and the generation of a **longshore current** that travels parallel to shore (Figure 12-19). This current causes mainland beaches and barrier islands to migrate laterally over time.

Beach sands also move on and off shore in response to changing seasons. Strong storms during winter months produce large waves that erode the beach face and transport sand offshore. The remaining beach has a low slope and is backed by a steep scarp that marks the highest position of the eroding waves. The eroded sand remains stored in submerged bars until the summer season, when gentler waves push it back on shore, rebuilding the beach in the process and increasing its slope again.

Tides

Another process that shapes the coastal environment is the ebb and flow of tidal currents. Tides are twice-daily

changes in the elevation of the sea surface relative to the land surface in coastal zones. The English scientist Isaac Newton developed our modern understanding of tides in the late 1600s, reasoning that they were produced by the gravitational attraction of the Moon to Earth in conjunction with Earth's rotation.[43] This attraction causes the oceans to bulge outward toward the Moon on the side of Earth that faces the Moon (Figure 12-20). The solid Earth is also attracted toward the Moon, which causes it to move toward the Moon and away from the oceans on the side of Earth opposite the Moon. This creates a tidal bulge on the opposite side of the planet as well. As Earth undergoes its daily rotation, each location on its surface passes through the zone of gravitational influence of the Moon twice. The rising sea surface allows coastal areas to experience high tides at these times. In between high

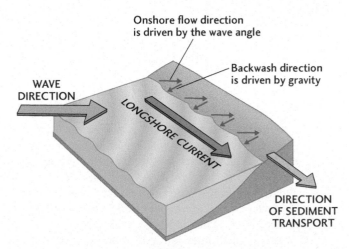

Onshore flow direction is driven by the wave angle

Backwash direction is driven by gravity

WAVE DIRECTION

LONGSHORE CURRENT

DIRECTION OF SEDIMENT TRANSPORT

FIGURE 12–19 Longshore sediment transport. Waves that strike the shore at an angle move back down perpendicular to the beach due to gravity, generating a longshore current that travels and carries sediment parallel to shore.

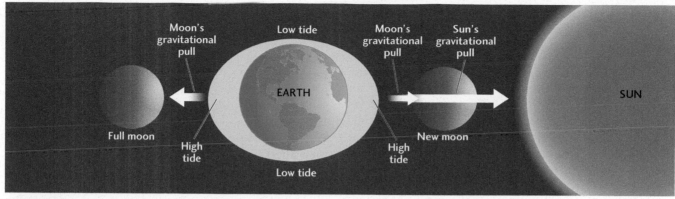

SPRING TIDE

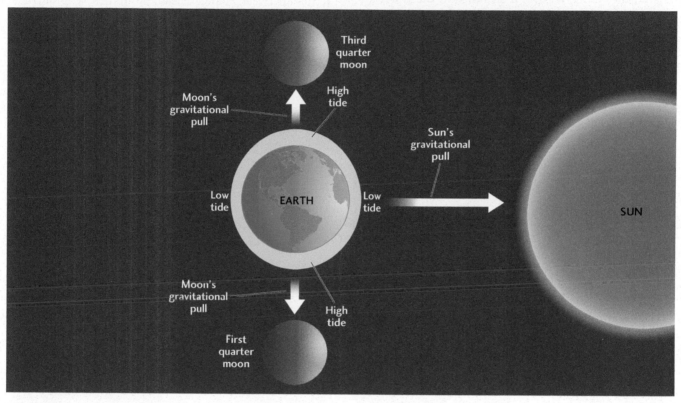

NEAP TIDE

FIGURE 12–20 Tides are produced by the gravitational pull of the moon and sun on the Earth combined with the Earth's daily rotation.

tides, the gravitational pull of the Moon diminishes, causing the ocean surface to subside such that the areas have low tides.

The Moon revolves around Earth once every 28 days. At those times when it lies in a straight line with Earth and the Sun, the tides are enhanced because of the combined gravitational pull of the Sun and Moon on the oceans. These enhanced tides are called *spring tides*. At times when the Moon lies at a right angle to the line connecting the Sun and the Earth, the tides are weaker and are referred to as *neap tides*.

No matter where the Moon is in its orbit around Earth, the total elevation difference between high tides and low tides varies with location and is strongly influenced by the geometry of coastlines. Bays and estuaries can serve as funnels for rising tides, elevating tides much higher than they would climb on straight coastlines. The Bay of Fundy in Nova Scotia is such a location and is renowned for its tidal range, which exceeds 15 m between high tide and low tide. The larger the tidal range, the faster and stronger the tidal currents and the more work they can do on the coastal environment.

Coastal Erosion and Attempts to Control It

The dynamic nature of the beach is the very thing that draws humans to live and visit there. Once beaches are developed, however, property owners are reluctant to let them change (Figure 12-21). While erosion is part of the natural geologic process that redistributes sediments along coasts, it is regarded as a hazard when it poses a threat to human-made structures and recreational beaches. Consider, for example, a tourist resort built near the upcurrent tip of a barrier island. Longshore currents moving sand will eventually move the island beyond the resort, leaving it stranded in the water and subject to wave attack. To address this problem, many schemes have been employed over the years, with varying costs and rates of success.

One of the most common strategies to slow or halt coastal erosion is to build **groins,** low barriers of rock or concrete that extend perpendicularly to the beach (Figure 12-22). The longshore current carries its load of sand until it encounters a groin. There, the slowing of the water causes deposition. Unfortunately, water moving around the groin is now relatively free of sand, with much potential to erode new material, so it erodes the beach on the downcurrent side of the groin. To prevent the beach from disappearing entirely, the downcurrent landowner must quickly build his or her own groin to trap sand. Typically, one groin leads to another, and then another, until the entire developed beach is covered with a groin field. The price of such engineered beach protection can be substantial.

Another strategy to stabilize a beach is to build jetties between barrier islands. **Jetties** are barriers

FIGURE 12–21 Resorts, high–rise buildings, roads, and tourist attractions are common on barrier islands, as in this example at Wrightsville Beach, North Carolina. Although barrier islands are ephemeral landforms consisting of unconsolidated sand, residents and officials make substantial efforts, typically at great cost, to stop sediment transport and preserve the islands in a fixed position. (*Jim Wark/Airphoto*)

FIGURE 12–22 Engineering strategies to protect narrow beaches. Groins protruding from the Miami Beach barrier island were used until the early 1990s in an attempt to protect its beaches from erosion. Although groins trap sand on their upcurrent side, the clear water flowing around them is more likely to erode sand on the downcurrent side, so beach protection is minimal. In the mid–1990s, after this photograph was taken, a major sand–replenishment project that cost millions of dollars replaced these groins with a beach that extends much farther seaward. (*Jim Wark/Airphoto*)

FIGURE 12–23 A seawall built parallel to the shore to deflect wave energy and minimize erosion. *(Jeffrey Willey/Alamy)*

placed to secure inlets and keep them from closing or migrating. When expensive marinas and waterfront developments are built adjacent to inlets, there is a strong incentive to keep them in place. Even before the U.S. construction boom of the 1970s to 1990s, inlets were stabilized with jetties to preserve lagoonal fisheries behind barrier islands. Seawater exchange through inlets brings marine water into the lagoons behind barrier islands and flushes out the part of the lagoon near the inlet. If the inlet closes or migrates, the salinity and water quality may change dramatically, disrupting commercial oyster, clam, and fishing grounds. For these reasons, a major task of the U.S. Army Corps of Engineers has been to stabilize inlets. Unfortunately, the same problem that affects groins affects jetties. The barrier impedes longshore drift, and sand builds up on the upstream side and erodes on the downstream side.

When waves threaten to reach shoreline structures, property owners sometimes construct bulkheads or **seawalls** parallel to the shore to deflect the erosive energy (Figure 12-23). Built from wood or metal pilings, concrete, or rock, these change the natural look and feel of the beach. They also change the natural flow of water and sediment up and down the beach, which can cause further problems. Sea walls deflect the full force of the incoming waves back outward. Thus deflected, the waves can scour the base of the wall, removing the sand and undermining the wall. The energy might also be deflected farther down the coast, affecting adjacent property owners who, in turn, build their own sea walls. Some coastal geologists refers to this as the

"Newjerseyization" of the shore, because that state's coasts have been heavily impacted by such engineering practices. They further point out that the walls, while momentarily protecting coastal property, also destroy the beach in front of them.

Another problem for beaches is declining sediment supply. Because rivers, which are the ultimate source of sand for beaches, have been dammed upstream on most continents, less sand reaches the coasts. Beach nourishment, or *replenishment,* is used widely to augment the amount of sand on a beach. Dredges or pumps bring sand from deeper waters offshore, from shipping channels that must be kept open, and from the upcurrent sides of groins and deposit it on recreational beaches, where the sand is bulldozed and distributed (Figure 12-24). Indeed, the U.S. Army Corps of Engineers is charged with keeping more than 40,000 km of coastline open to navigation, and dredges about 230 million m^3 of sediment from harbors and navigation channels annually. Some of this material is reused beneficially to replenish beaches.

Still, replenished beaches are notoriously short-lived and expensive. The sands dredged to resupply them are frequently finer grained than the original beach, making them more susceptible to erosion. In one particularly poignant example of the impermanence of nourishment activities, Ocean City, New Jersey, spent $5.2 million in 1982 to replenish a small section of beach.[44] Two-and-a-half months and several storms later, the beach was gone. Since that time, Ocean City has spent at least $93 million more federal, state, and

FIGURE 12–24 Dredging and beach replenishment at Wrightsville Beach, North Carolina. Sand is dredged from the continental shelf offshore, mixed with water, and pumped to the beach to replenish its supply of sand. *(Jack Dermid/Photo Researchers/Science Source)*

western coast of North America, for example, much expensive real estate is in peril of falling into the sea. In the Monterey Bay region of Central California, cliffs are retreating as much as 63 cm/yr.[48] Many of the cliffs bordering beaches there are composed of relatively soft sediment and rock that is easily eroded. The houses atop such cliffs are susceptible to collapse when coastal erosion undermines their foundations. Even coastlines in areas with little or no tectonic activity can be subject to cliff erosion, however. Such is the case along the shores of the Chesapeake Bay in the eastern United States, where weakly consolidated sediments form bluffs (Figure 12-25).

Cliff erosion is mainly caused by direct wave attack, which gradually erodes the rock. Waves also exploit joints and fractures in the rock as planes of weakness, tearing away large blocks at a time. Wind carries salt spray onto the cliffs, and as the water evaporates, salt crystals grow, slowly flaking off small rock fragments. The surface eventually crumbles. Earthquake-induced shaking may also bring down unstable portions of the cliff face. Even watering the lawn around a clifftop house might raise the pore pressure in the underlying rock, increasing the chance of failure (see the discussion of mass movement in Chapter 8).

Various engineering techniques have been tried to halt sea cliff erosion, but at best they can only slow the process, and at worst they might actually accelerate it. Concrete sheets poured over the cliffs erode as rapidly as the rock behind. At the base of the cliffs, protective structures (in the form of heaps of large boulders, rock-filled wire baskets called gabions, or pilings driven in front of the cliff face) are intended to break the force of

local tax dollars on beach replenishment.[45] Communities dependent on tourism contend that they have little choice but to continue to replenish the beach, but some geologists question the continuing taxpayer subsidies.[46] They also point out that beach nourishment impacts the sensitive shoreline ecosystem as bulldozers distributing sand crush and smother small creatures like mollusks and crustaceans that burrow in the sediment. It can take several months for populations to recover.[47] Similarly, dredging and bulldozing can increase water turbidity and negatively impact corals, which depend on photosynthesis for food. Sedimentary particles in the water column increase turbidity, which limits sunlight penetration of the water column.

Controlling Cliff Erosion

Cliff retreat as a result of the natural processes of coastal erosion is especially common along tectonically active coastlines throughout the world. Along the

FIGURE 12–25 Coastal erosion. Erosion has eaten away at the cliff beneath these waterfront homes in Maryland, along the western shore of the Chesapeake Bay. Trees, tires, and other debris are used in an attempt to protect the shore. *(Lowell Georgia/Science Source)*

oncoming waves. However, unless the protection covers the height of the cliff and extends for its entire length, the sea will eventually erode behind it, rendering it useless and increasing the rate of cliff erosion. Limiting or banning development along cliffs, or creating a buffer zone several hundred meters wide between structures and cliff edges, in conjunction with careful management practices, can slow the rate of erosion and limit the need for costly attempts to save properties in peril.

Tsunamis

While coastal erosion proceeds gradually and continuously and causes no human casualties, some coastal hazards are sudden and deadly. In Chapter 11 we discussed

the storm surge associated with hurricanes that causes coastal flooding and can lead to mass casualties when populations lack the information or resources to evacuate. Atmospheric circulation is not the only process that can generate massive coastal surge, however. When an offshore earthquake, volcanic eruption, or landslide suddenly displaces a large amount of seawater, it sets in motion a train of waves that radiates from the point of origin across the sea surface (Figure 12-26). The waves thus generated, called **tsunamis,** can have devastating effects upon reaching shore. One of the worst known earthquakes and subsequent tsunamis occurred on March 11, 2011 (see Chapter 3). Waves up to 40 m in height struck the Japanese coastline within tens of minutes after the Tōhoku megathrust earthquake, also known as the

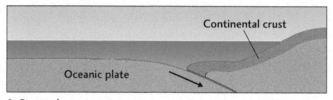

1. Pre-quake

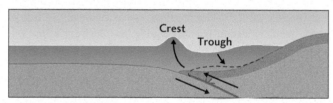

2. During an earthquake

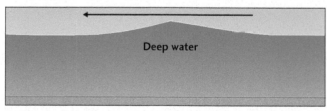

3. Wave movement

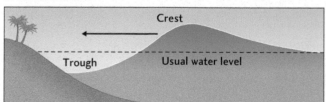

4. Coastal flooding

(a)

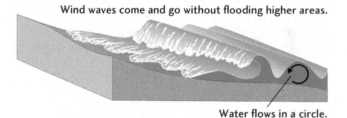

(b)

(c)

FIGURE 12–26 (a) A tsunami wave is generated when an offshore earthquake, volcanic eruption, or landslide displaces a large amount of seawater, setting in motion a train of waves that radiates from the point of origin across the sea surface. Tsunamis can cause great destruction upon reaching shore. (b) Unlike wind–driven waves, which move up and down

along beaches, tsunamis may advance far inland as a massive wall of water. (c) Tsunami waves up to 40 m in height struck the coast of Japan after the March 11, 2011, Great East Japan Earthquake, one of the worst earthquakes in history (photo in Miyako City, Iwate Prefecture). *(c: AFLO/Mainichi Newspaper/EPA/Newscom)*

Great East Japan Earthquake. With a magnitude of 9.0 (M_w), it was the largest earthquake known to have happened in Japan and one of the five largest earthquakes in the world since 1900. More than 100,000 buildings were damaged or destroyed, including several nuclear reactors, which spewed radioactivity into the environment for months, leading to the second-worst nuclear power disaster in history. The death toll from both the earthquake and tsunami was higher than 15,000, and at least 4000 people remain missing. Overall losses exceeded U.S. $300 billion.

The speed (C) of a tsunami is a function of the depth of the water the wave travels through (d) and Earth's gravitational acceleration (g, 9.8 m/s^2):

$$C = \sqrt{gd}$$

Given the great depth of the oceans, this relationship means that tsunamis can travel at speeds of 500 to 1000 km/hr or more in the open ocean, as fast as or faster than most jet airplanes. Surprisingly, the height of tsunami waves in the oceans rarely measures more than a few dozen centimeters, yet their wavelength can be as long as 200 km. As a result, a ship in the open ocean has little chance of noticing the waves.

Once a tsunami reaches shore, however, the increased frictional drag of the seafloor on the bottom of the wave slows the water's forward motion considerably, raising the amplitude of the wave and shortening the wavelength in the same way as it does for wind-driven waves. Because the wavelengths of tsunamis continue to be very long, however, the 1:7 ratio is not exceeded, so the onrushing wave does not break. Instead, it floods over the land surface like a very large, quickly moving tide. By the time a tsunami reaches shore, it can be tens of meters high, able to carry large ships, houses, trees, and other objects many kilometers inland. These objects act as battering rams, slamming into anything in their path and causing great destruction. Even people who are good swimmers may be crushed between pieces of debris or knocked unconscious and drowned.

Tsunamis can impact coastlines thousands of kilometers from the original displacing event, catching victims completely unaware of their approach. On December 26, 2004, for example, the magnitude 9.1 Sumatra-Andaman earthquake that struck Indonesia generated an immense train of tsunami waves. Produced by subduction of the Indian Plate beneath the Burma microplate, the earthquake ruptured a section of fault plane at least 500 km long and produced thrust displacements of up to 20 m.[49] The upward displacement of the water column above the fault zone generated tsunami waves that sped across the Indian Ocean, killing at least 225,000 people in 14 countries and leaving more than 1.5 million homeless (Figure 12-27). The heaviest casualties were in Indonesia itself, where at least 130,000 people were killed by a wall of water that measured more than 30 m deep.[50]

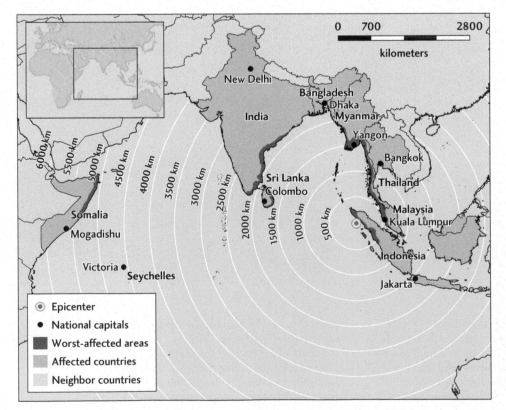

FIGURE 12–27 Earthquake-generated tsunami waves. Tsunami waves radiated outward from the hypocentral region of the December 2004 Sumatran-Andaman earthquake (magnitude 9.2), impacting coastlines up to thousands of kilometers away throughout the Indian Ocean. Hundreds of thousands of people died and more than 1.5 million were left homeless.

(a)

(b)

FIGURE 12-28 Tsunami destruction. The city of Banda Aceh, Indonesia, was obliterated by a tsunami wave produced by the Sumatra–Andaman earthquake, as shown in these before (a) and after (b) true–color satellite images. *(Digital Globe/ Getty Images)*

The Indonesian city of Banda Aceh was completely destroyed (Figure 12-28). Sri Lanka, India, and Thailand lost thousands of coastal residents, and the waves even killed 176 people in Somalia, more than 5000 km away.[51] Satellite imagery was helpful in showing the extent of the devastation and determining the locations of impacted populations.

Regrettably, at the time of the earthquake, the Indian Ocean region had little instrumentation to warn of tsunamis. Furthermore, many of the impoverished nations surrounding it had few resources to develop disaster management plans or to educate their populations on how to respond to tsunami threats. The number and types of active plate boundaries circling the Pacific Ocean make that region prone to earthquake-generated

tsunamis as well, but seismologists began installing a network of seismometers around the Pacific Rim as early as the 1940s. This network was followed in 2001 with the installation of tsunameters, equipment designed to directly detect tsunami waves in the open ocean (Figure 12-29). The Indian Ocean, however, did not have such equipment or early warning systems at the time of the Sumatran earthquake in 2004.

Each Deep-ocean Assessment and Reporting of Tsunamis (DART) station consists of a pressure transducer anchored to the ocean floor that can detect changes in the height of the overlying water column, and an adjacent moored buoy that communicates with overhead satellites. As tsunami waves cross the ocean, the increase in pressure is recorded by the DART

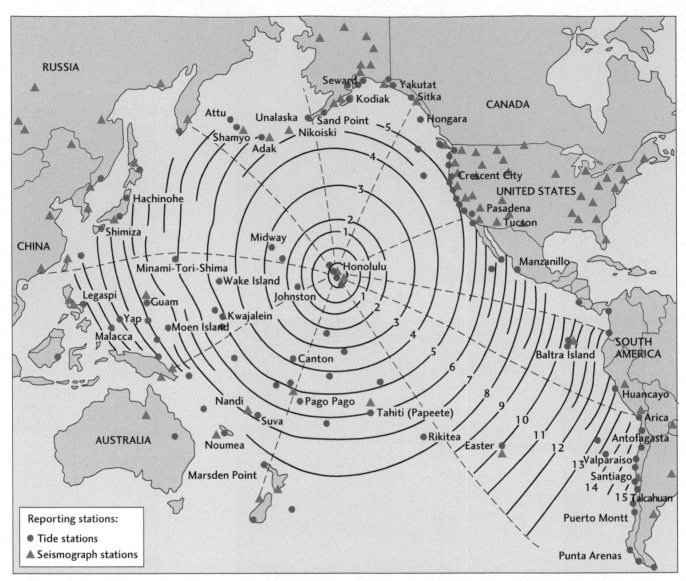

FIGURE 12–29 Tsunami warning system. A warning system established in the Pacific Ocean pinpoints earthquake sources of tsunamis immediately after the quake. Concentric lines indicate travel times: An earthquake–generated tsunami originating near Santiago, Chile, for example, would reach Honolulu in about 15 hours, whereas one coming from San Diego would reach the island of Hawaii in little more than 5 hours.

transducer, which relays the information to the buoy that sends it on to a satellite. The data then are relayed to the Pacific Tsunami Warning Center in Ewa Beach, Hawaii, and the West Coast and Alaska Tsunami Warning Center in Palmer, Alaska. At these stations, the data are used in conjunction with computer models to forecast the likely time of arrival of tsunami waves at coastlines around the Pacific and to estimate the height of those waves. Thirty-two DART stations are now in place around the Pacific Rim, with another 7 in the Caribbean and mid-Atlantic.[52]

The efficacy of this warning system was demonstrated during the February 27, 2010, magnitude 8.8 earthquake that struck the coast of Chile, devastating the city of Concepción and many smaller towns. Tsunami waves arrived at the closest DART station, off the coast of Chile, within 34 minutes of the earthquake. Scientists at the Pacific Tsunami Warning Center quickly used the information to forecast the likely time of arrival and wave amplitude in Hawaii and were able to issue evacuation orders. While the size of the tsunami turned out to be on the small end of the forecast by the time it hit Hawaii, the event proved an excellent test of the monitoring technology and of disaster management planning. Hawaii has instituted a network of sirens to warn coastal residents of impending danger and also has developed

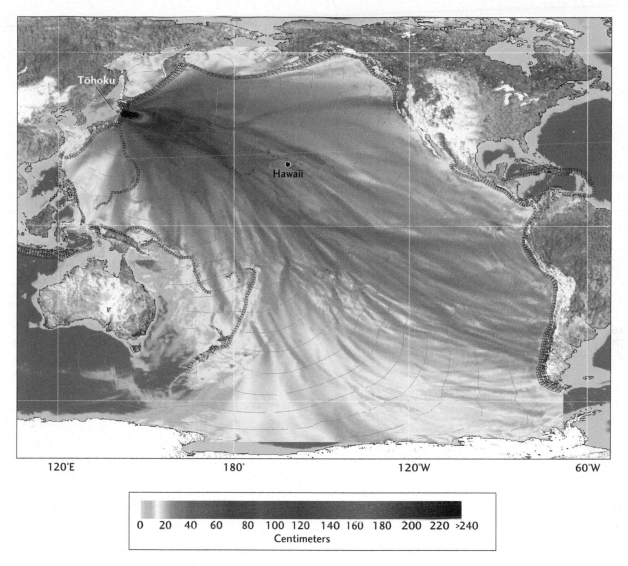

120°E 180° 120°W 60°W

0 20 40 60 80 100 120 140 160 180 200 220 >240
Centimeters

FIGURE 12–30 Predicting tsunami wave height.
Scientists use information about how tsunami waves interact with bathymetric features on the ocean floor and coastlines to model wave amplitude and travel time after an earthquake. This depiction, which uses color coding to show the tsunami amplitude, predicts the path, arrival time, and wave height for tsunami waves after the 2011 Tōhoku earthquake. The model predicts wave height of up to several tens of centimeters, for example, for the Pacific Northwest coast of the United States. *(NOAA Center for Tsunami Research)*

evacuation routes that are clearly marked. These steps, along with public education, will minimize loss of life in future, larger tsunamis, and were put in place after a 10.5-m tsunami devastated the city of Hilo in 1960 following another Chilean earthquake.[53]

Another test of this warning system was demonstrated a year later during the magnitude 9.0 Tōhoku earthquake in Japan on March 11, 2011. A DART buoy detected the earthquake 3 minutes after it happened, and the tsunami was recorded 25 minutes later. Scientists at the NOAA Center for Tsunami Research combined models of tsunami waves with information from DART buoys to predict the tsunami arrival time throughout the Pacific Ocean region (Figure 12-30). Their predictions for arrival times at Hawaii and the west coast of North America, for example, were accurate to within minutes.

Since the 2004 Sumatran earthquake and subsequent Indian Ocean tsunami, eight additional tsunameters have been installed: four in the Indian Ocean and four off the east coast of Australia. Scientists hope to eventually have an array of 22 stations to provide a level of warning similar to that of the Pacific Rim.[54] The United Nations has sponsored tsunami warning drills in which emergency management personnel from countries likely to be affected by tsunamis assess their preparedness.[55] While these measures will minimize loss of life at locations far from the source of future tsunamis, those living near the source will continue to be at risk. For this reason, it is imperative that coastal populations become educated about the relationship between earthquakes and tsunami hazards, and that they know to evacuate to higher ground as soon as shaking begins.[56]

■ Major hazards associated with oceans include hurricane storm surges, coastal erosion, and tsunamis.

■ On natural beaches, sand is removed by erosion and replenished by deposition from season to season and year to year, and over a period of hundreds to thousands of years, the rates of removal and replenishment are nearly equal. Recently, however, this long-term flow of sand has been interfered with by efforts to protect property from erosion and by damming streams, which limits sediment transport to the coast.

■ The water in waves moves in an orbital path. With increasing depth below the surface, orbits become smaller until a depth of no motion is reached. This depth is called wave base.

■ When wave base intersects the shore, the forward motion of waves is slowed by friction with the seafloor. Because waves farther offshore continue to move at high velocity, waves near shore steepen. If the ratio of wave height to wave length exceeds 1:7, waves break.

■ Many engineered structures, including groins, jetties, and seawalls, are used to stop the erosion of sand from beaches and the erosion of rock from cliffs, and to maintain inlets to harbors. Beach replenishment or nourishment involves replacing eroded sand, typically by using offshore dredging.

■ Tsunamis—sea waves generated by submarine earthquakes, volcanic eruptions, or landslides—displace a large amount of seawater and can reach heights of tens of meters as they approach shore.

■ A network of seafloor pressure transducers throughout the Pacific Ocean monitors tsunami wave heights to provide warnings. Construction of a similar array in the Indian Ocean was begun after a tsunami in 2004 killed more than 225,000 people.

HUMAN IMPACTS ON THE SEAS

The oceans are a great global commons that belongs to no single nation. The Dutch lawyer Hugo Grotius articulated this view in a 1609 publication entitled *Mare Liberum,* or *The Freedom of the Seas.*[57] During Grotius' lifetime, Spain and Portugal, then united under a single ruler, claimed the oceans for themselves, insisting that they had not only exclusive dominion over marine resources, but also the right to expel the ships of other nations. Grotius bristled at the idea that a single sovereign should control all maritime trade. Instead, he wrote, God had created all men equal, and as such, all men should enjoy access to maritime travel and the wealth of the seas (Box 12-2). Besides, Grotius wrote, the oceans and their resources were too vast to be altered by human activities; hence restrictions on their use were unnecessary.[58]

Since Grotius' time, population growth, industrialization, and the development of new technologies for resource exploitation render the notion of inexhaustible ocean resources obsolete. At the beginning of the 21st century, human population measured just over 6 billion. It reached 7 billion in 2012 and is projected to rise to nearly 9 billion by the middle of this century. As our numbers have grown, so too has our demand for protein resources from the oceans. With satellite navigation systems, radar tracking, deep-sea sounding devices, and giant nets that are many kilometers long and ensure no chance of escape, it is little wonder that we have depleted the stocks of some fish by as much as 90 percent relative to preindustrial levels.[59] We also have polluted the seas with our wastes. Our population currently produces 6 billion metric tons of solid waste and 10 billion tons of sewage sludge annually,[60] some fraction of which winds up in the seas. Whether intentionally or unintentionally introduced, the marine garbage patches and growing biological dead zones produced by these wastes testify to the need for better stewardship of marine resources.

Since Grotius' writings four centuries ago, we now know that the oceans are vast but not inexhaustible. As our human population grows, we are butting up against limits in Earth's natural systems. Both international cooperation and action from individuals will be needed to address these limits.

Waste Disposal and Polluted Runoff Along Continental Shelves

Coasts are the most threatened part of the ocean system because of their proximity to population centers. Already 40 percent of all people live within 100 km of the ocean, and this proportion is increasing as the global population becomes more urbanized and migrates to coastal cities. The greatest threat to the oceans is not the occasional oil spill from a tanker ship run aground or offshore drilling rig disasters (although such events can be catastrophic regionally), but the multitude of wastes from cities and other lands in human use worldwide. These include treated or partly treated sewage, runoff from streets and parking lots, chemical pollutants from industries and

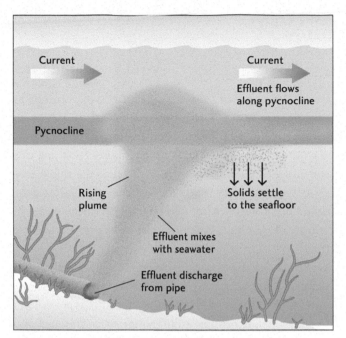

FIGURE 12–31 Effluent discharged from a pipe in a coastal zone forms a plume that rises to the surface and spreads downcurrent in seawater, typically along the pycnocline. Both industrial and sewage wastes have been discharged in this manner. Heavier solid particles settle to the bottom, forming a train of contaminated sediment that spreads away from the discharge pipe.

agriculture, and excess lawn fertilizer. Comprehensive coverage of these topics is beyond the scope of this book, but three topics—sewage waste, nutrient pollution, and plastics in the ocean—are discussed here. Oil spills from ruptured ship tanks and offshore drilling accidents, including the 2010 BP Horizon spill, are discussed in Chapter 13.

Sewage Waste *Sewage sludge,* a gummy mixture of organic and inorganic substances derived chiefly from human waste, is rich in nutrients as well as microbes such as bacteria and viruses. Globally, much sewage, both treated and untreated, is discharged directly into coastal zones through pipes (Figure 12-31), and some is dumped by ships. Other waste flows into septic systems, which are designed to separate liquid from solid waste prior to allowing the liquid to percolate through the soil. In theory, soil bacteria and plant roots take up nutrients in the waste and soil particles filter out any microbes.

A 1991 incident in South America illustrates the danger associated with untreated sewage discharges in coastal zones. An Asian freighter dumped its sewage in a harbor near Lima, Peru, where beds of shellfish provided villagers with a major source of food. Shellfish are filter feeders, so they rapidly ingested sewage bacteria and viruses, becoming contaminated. Unfortunately, the dumping incident coincided with a festival at which ceviche—raw shellfish marinated in lime juice—was served. Within days, villagers became violently ill from the Asiatic strain of the bacteria that cause cholera, a life-threatening disease characterized by severe intestinal distress and diarrhea.[67] Because of inadequate sanitation facilities, the disease spread rapidly as water supplies became contaminated and travelers carried the bacteria to other countries.[68] By 1993, more than 700,000 people in Central and South America were diagnosed with Asiatic cholera and more than 6000 died. As a result of this and similar tragedies elsewhere, many nations now prohibit raw and treated sewage disposal within set limits of their shores. Until 1991, for example, sewage sludge could be dumped at sea in the United States. The federal Ocean Dumping Ban Act of 1988 (an amendment to the Marine Protection, Research, and Sanctuaries Act of 1972) prohibited any further dumping of sewage sludge or industrial waste into the U.S. coastal waters after 1991.[69] Landfills have become a common repository for sewage sludge since the Ocean Dumping Ban was enacted.

Despite laws such as the Ocean Dumping Ban, problems with untreated sewage are not limited to the developing world. Indeed, poorly planned development has led to microbial contamination of beaches in many locations. In Florida, for example, septic systems often are constructed in areas of karst terrain, a limestone landscape known for its caverns and sinkholes as well as its rapid groundwater flow rates. Sewage introduced into this environment can flow rapidly toward the sea without being processed by soil bacteria or filtered through fine soil particles that can trap pathogens.[70] Development also involves the construction of roads, parking lots, and other impervious surfaces that speed water to storm drains. In many communities, especially older ones, storm sewers and sanitary sewers are combined, and when there is heavy rainfall, these sewers can overflow, spilling untreated human waste into stream channels that flow to the coast. Animal waste from dogs, cats, and wildlife also washes off of paved surfaces and into overflowing sewers.

In 2011, state, territorial, and tribal governments in the United States monitored 3650 beaches in 30 different states, five territories, and two tribal areas. Of these, 1575 (43 percent) were either closed to swimming because of excessive levels of bacteria or had issued public advisories warning of potential health hazards.[71] Swimming in these waters could introduce bacteria and viruses into the body through swallowed water or skin wounds. As in Peru, ingesting contaminated raw shellfish (for example, raw oysters on the half shell) also could prove deadly.

BOX 12-2 EARTH SCIENCE AND PUBLIC POLICY

Who Owns the Seas?

As the demand for marine resources increases, so too does the potential for international conflict, raising the questions: Who owns the seas? Who has the right to their fish, petroleum, minerals, and other resources? A century after Hugo Grotius, Dutch lawyer Cornelius van Bynkershoek had an answer. In a 1703 treatise entitled *Dissertatio de Dominio Maris,* he amended Grotius' *Freedom of the Seas* to assert that states with marine borders had the right to claim for themselves a narrow swath of ocean abutting their lands. The width of this so-called *territorial sea* should be the distance that could easily be defended from shore—for example, the distance of a cannon ball shot[61]—eventually accepted as 5 km (3 nautical miles).[62] Beyond this fringe, however, the seas should remain open to all. Van Bynkershoek's assertion held sway for the next 240 years, until president Harry Truman unilaterally declared in 1945 that the United States had exclusive claim to the entire continental shelf extending off of the country, a distance far wider than 5 km, particularly along the Atlantic and Gulf of Mexico coasts. Other nations immediately followed suit, claiming territorial expansions of various sizes.

To prevent total chaos and the outbreak of hostilities, beginning in 1958 the United Nations convened several conferences to divide the oceans into areas of national and international jurisdiction. By 1982, a formal treaty, entitled the UN Convention on the Law of the Sea (UNCLOS), recognized two zones around each coastal nation, a 22 km (12 nautical mile) territorial water zone that can be managed for national defense and customs purposes, and a larger 370 km (200 nautical mile) **Exclusive Economic Zone (EEZ).** Within their EEZs, nations were given exclusive rights to the ocean's natural resources, both those within the water column and those at the seabed, but they also were required to allow foreign ships involved in "innocent" activities to pass through. The remainder of the sea would stay open to all, though international agreements set quotas on fish catches in an attempt to limit overfishing.

Establishing the EEZs led to numerous campaigns to map the seafloor. In the United States, NOAA surveyed tens of thousands of square kilometers of seafloor along the Atlantic, Pacific, and Gulf coasts. To do this, NOAA used a new technique, multibeam sonar, in which a burst of sound is emitted every few seconds from a source mounted on a ship in a strip perpendicular to the direction the ship is traveling (the red area in **Figure 1**), and listening devices detect sound bouncing back from the seafloor (the green area). The velocity of sound in water is known, so the time it takes for sound waves to complete their travel path can be used to determine the depth and, therefore, the topography of the seafloor. Computers are used to analyze the data and produce detailed maps of the seafloor. These maps are valuable to petroleum companies exploring for oil and to geologists searching for faults that might cause earthquakes. Other countries have conducted similar mapping campaigns, and more than 120 nations have laid claim to their EEZs, removing the richest portions of the oceans—more than 40 percent of the seas—from the global commons (**Figure 2**).[63]

Though UNCLOS should have settled the question of maritime ownership and governance, problems remain.

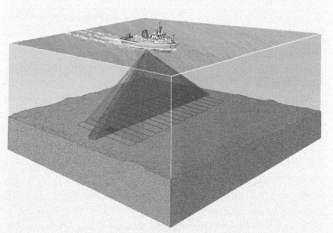

Figure 1 Mapping the seafloor. Multibeam sonar instruments on board a ship can map large areas of the seafloor in great detail, as shown in the Figure 12-4 map of the western coast of the United States. The ship's instruments send out sound waves, then record the waves' arrival as they reflect off the seafloor. *(Rapid City Journal)*

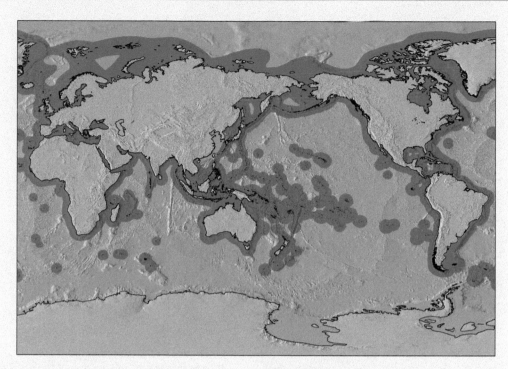

Figure 2 Exclusive Economic Zones. The UNCLOS treaty recognizes two zones around each coastal nation, a 22 km (12 nautical mile) territorial water zone that can be managed for national defense and customs purposes, and a larger 370 km (200 nautical mile) Exclusive Economic Zone (EEZ), shown here in green along continental margins and around islands. Nations have exclusive rights to marine resources within their EEZ, but must permit foreign ships passage.

In 1995, for example, the Canadian coast guard seized a Spanish fishing vessel and arrested its crew 45 km outside Canada's newly defined EEZ but within the Grand Banks fishing grounds on the country's continental shelf.[64] Claiming that the ship's operators violated internationally agreed-upon quota restrictions on turbot (also called Greenlandic halibut), the federal cabinet made use of a Canadian law enacted in 1994 allowing it to police all fishing activity on the Grand Banks, even outside the EEZ. By the argument of this law, Canada's jurisdiction extends to its entire continental landmass both above and below the sea, even where its undersea extent exceeds the limit of the EEZ. Indeed, the UNCLOS treaty allows nations to claim their entire continental shelves, with special permission.

In 2007, Canada got into another tussle, this time with Russia, the United States, Norway, the European Union, China, and Great Britain. Melting sea ice, most likely related to human-induced climate change, opened the Northwest Passage in the Arctic Ocean, making it possible for routine ship traffic that would greatly speed travel between Europe and Asia. In addition, the melting sea ice gave rise to dreams of oil and mineral exploration and previously unknown stocks of fish, and all of the nations bordering the Arctic wanted a share of its potential riches. Canada argues that much of the Northwest Passage lies within its internal waters inasmuch as ships have to navigate between islands that are part of its territory. It denies the claims of other nations that the passage is an international zone that they can use to access their own territorial claims. Russia upped the ante in August of 2007, using a submarine to plant its flag on the Arctic Ocean seafloor.[65] In response, Canada pledged to build an Arctic deep-sea port and to begin military patrols to defend its territory and the sensitive polar environment.[66] Time will tell how this situation plays itself out.

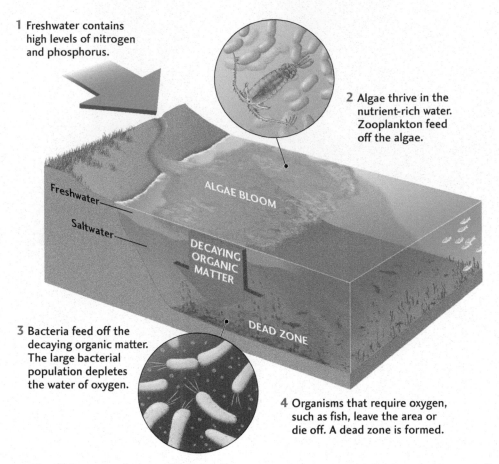

1 Freshwater contains high levels of nitrogen and phosphorus.

2 Algae thrive in the nutrient-rich water. Zooplankton feed off the algae.

Freshwater

Saltwater

ALGAE BLOOM

DECAYING ORGANIC MATTER

DEAD ZONE

3 Bacteria feed off the decaying organic matter. The large bacterial population depletes the water of oxygen.

4 Organisms that require oxygen, such as fish, leave the area or die off. A dead zone is formed.

FIGURE 12-32 Algal bloom and eutrophication.
Nutrient–rich water from rivers stimulates algal growth in nearshore waters. As algae and the zooplankton that feed on them die, bacteria feed on the detritus. This microbial activity consumes oxygen from the water, leading to the water being oxygen deficient, which is detrimental to marine life.

Nutrient Pollution Along with microbes, sewage contains nitrogen and phosphorus, as does fertilizer applied to lawns and farm fields. Atmospheric aerosols derived from fossil fuel combustion also contribute nitrogen to land and water bodies. While these elements are essential nutrients in seawater, there can be too much of a good thing (see Box 7-4). When excessive amounts are added to ocean and coastal ecosystems through sewer outfall pipes or river flow, the nutrients stimulate blooms of photosynthesizing algae in the photic zone near shore (Figure 12-32). Higher rates of algal growth cause more algae to die and decay, and the process of microbial decay consumes oxygen. As a result, oxygen is depleted from deep-ocean water and, if depleted enough, invertebrates (e.g., crabs, clams, and shrimp) and fish suffocate and die. Algal growth also causes water to become cloudy and blocks the transmission of sunlight to submerged aquatic vegetation, which then also dies and decays and further contributes to oxygen depletion. When dissolved oxygen levels fall below 2 mg per liter (the same as 2 ppm, or parts per million), water is said to be **hypoxic.**

A complete lack of oxygen in water is known as **anoxia,** and the water is said to be *anoxic.*

This process of nutrient enrichment followed by enhanced algal growth and oxygen depletion, known as **eutrophication,** is particularly severe in estuaries that are only partly open to the sea and have recirculating water (Box 12-3). Estuarine circulation patterns result in accumulation and retention of suspended sediments and nutrients, giving estuaries their high productivity, but the same circulation patterns can lead to an excess of nutrients as well as the trapping of toxic wastes, metals, and other pollutants. NOAA did two national assessments of estuarine eutrophication, the first in 1999 and the second in 2007.[72] These assessments determined that nutrient enrichment has resulted in negative impacts that range from moderate to high in 64 of the 99 estuaries that had adequate data for evaluation. Furthermore, conditions deteriorated in dozens of estuaries during the decade between the two assessments (Figure 12-33). The most severely impacted estuaries are those in more heavily populated

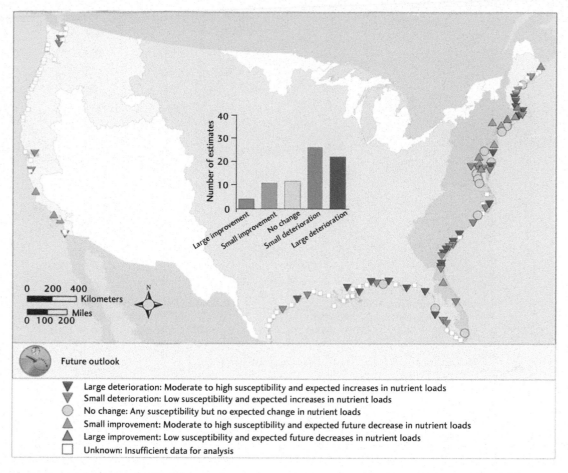

FIGURE 12–33 Eutrophication in U.S. estuaries.
A national estuarine eutrophication assessment indicates that of 99 estuaries that have enough data for analysis, 64 are moderately to severely impacted by eutrophication. There were small to large levels of deterioration (indicated by the downward–pointing triangles) in dozens of estuaries during the decade between the two assessments done in 1999 and 2007. The most severely impacted estuaries are those in more heavily populated regions, as in the mid–Atlantic region of the eastern United States. Conditions improved in some estuaries (upward–pointing arrows), however, and the management strategies used to reduce nutrient loading to those locations are role models for elsewhere. *(NOAA)*

regions, as in the mid-Atlantic region of the eastern United States.

Places where bottom water is depleted of oxygen are called **dead zones.** Eutrophication affects coastal zones throughout the world, and annually recurring dead zones are becoming increasingly numerous, particularly along heavily populated coastlines near the mouths of rivers (Figure 12-34). Rivers provide inputs of nutrients and sediment, both of which contribute to the process of eutrophication. First identified by scientists in the 1970s, there are now more than 400 known dead zones. Long-term observations indicate that some are growing in size.

One of the coastal areas most impacted by a growing dead zone is the Gulf of Mexico adjacent to the Mississippi delta. This seasonally recurring dead zone, the largest along U.S. coastal waters, typically waxes from spring to midsummer and disappears in the fall. The delta region receives the runoff of 41 percent of the conterminous United States[73] and drains areas of substantial agricultural significance (the bread basket in the Midwest and and the Corn Belt) where fertilizers are applied heavily, and hogs and chickens in large industrial farms generate enormous volumes of nutrient-rich manure. It also receives the sewage wastes of large metropolitan areas such as Minneapolis/St. Paul, St. Louis, and Cincinnati. For the past three decades, the U.S. Geological Survey (USGS) has measured the amount of nitrogen (N) and phosphorus (P) in the Mississippi River and its delta distributaries and found that more than 1,000,000 metric tons of N and 150,000 metric tons of P enter the Gulf each year.[74] Dr. Nancy Rabalais and her colleagues

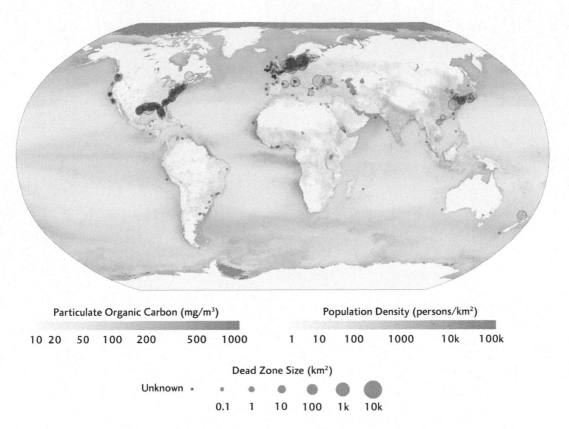

Particulate Organic Carbon (mg/m³)

10 20   50   100   200   500   1000

Population Density (persons/km²)

1   10   100   1000   10k   100k

Dead Zone Size (km²)

Unknown •   •   •   •   ●   ●   ●

0.1   1   10   100   1k   10k

FIGURE 12–34 Dead zones along coastlines. Dead zones are nearshore areas with oxygen–depleted, or anoxic, bottom water that are uninhabitable for invertebrates and fish that require oxygen dissolved in seawater. These zones are enriched in nutrients and carbon. In this figure, particulate organic carbon concentration in seawater is shown in shades of blue. There are more than 400 dead zones, and more are developing, particularly at the mouths of rivers that drain agricultural regions and in heavily populated areas (population density is shown in shades of brown), as along the Gulf of Mexico and Atlantic coasts of North America and the coasts of the British Isles in Europe. The size of some dead zones also is growing, with some now more than 10,000 km² in aerial extent (the circle size indicates the dead zone size). *(Robert Simmon & Jesse Allen; based on data from Robert Diaz, Virginia Institute of Marine Science [dead zones]; the GSFC Ocean Color team [particulate organic carbon]; and the Socioeconomic Data and Applications Center [population density])*

at the Louisiana Universities Marine Consortium have measured dissolved oxygen levels in the Gulf of Mexico and mapped a large zone of hypoxic water where fish and shellfish cannot survive (Figure 12-35). Between 2004 and 2008, the average size of this so-called Gulf dead zone measured 17,000 km², about the size of Lake Ontario.[75] In early 2011, scientists predicted that the Gulf dead zone might expand substantially, to as much as 24,000 km², as a result of flooding and nutrient run-off, but by August 2011, its size was only 17,521 km².[76] The smaller-than-predicted expansion was attributed to mixing of coastal water by strong winds and waves during a coastal storm.

Given the great economic and historic importance of Gulf Coast fisheries, such a large area with hypoxic conditions is cause for substantial concern, spurring efforts to reduce nutrient inputs. Since phosphorus typically binds to soil particles, changing farming practices

to minimize soil erosion can reduce the amount of phosphorus that reaches coastal areas. These practices include planting vegetative buffer strips between streams and farm fields to trap eroding soils and using contour plowing, in which farmers plow along lines of equal elevation to create miniature terraces that inhibit runoff. Minimizing overapplication of fertilizer and fertilizing only during months of active plant growth also can reduce nutrient runoff. Nutrient levels in sewage and industrial waters can be reduced through biological or chemical filtration, as well as by introducing effluents into constructed wetlands in which plants take up nitrogen and phosphorus.[77]

The Mississippi River Gulf of Mexico Watershed Nutrient Task Force, with representatives from the U.S. Environmental Protection Agency (EPA), the U.S. Department of Agriculture, NOAA, and the Department of the Interior, among others, hopes that efforts

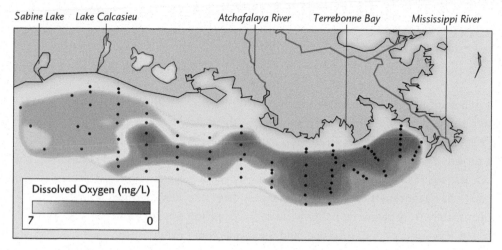

Sabine Lake   Lake Calcasieu   Atchafalaya River   Terrebonne Bay   Mississippi River

Dissolved Oxygen (mg/L)

7   0

FIGURE 12–35 Gulf of Mexico dead zone at the mouth of the Mississippi River in 2011. Since the mid–1980s, Dr. Nancy Rabalais and her colleagues at the Louisiana Universities Marine Consortium field station in Cocodrie, Louisiana, have measured and mapped dissolved oxygen contents along the Louisiana coast every summer (dots mark the locations of water samples). Concentrations lower than 2 mg/L are known as hypoxic and threaten marine life. Anoxic conditions are characterized by dissolved oxygen concentrations of 0 mg/L.

such as these will reduce the size of the Gulf hypoxic zone to 5000 km^2 by 2015.[78] The zone presently is three times that size, so this goal is unlikely to be met, but its ambitious nature speaks to the urgency of the problem.

Plastics in the Sea

Hypodermic needles and other medical wastes have washed up on New Jersey beaches since the 1980s, horrifying the public and prompting beach closures.[85] Determining the specific origin of the wastes has been unsuccessful, though some speculate that they were eroded from landfills or intentionally dumped. These waste accumulations, along with images of marine birds killed by entanglement in plastic, or of a whale that starved after swallowing a party balloon and ribbon that blocked its digestive tract, alerted many to the effects of human wastes on wildlife (Figure 12-36).

The intensity of human impact decreases away from the world's coastlines, and deep-ocean basins are still relatively pristine. Still, intermittent pollution occurs in most of the world ocean, especially along major transportation routes. Because of ocean mixing by currents, there are wastes even on remote islands. In 1997, scientists visiting the Scotia Arc, a chain of uninhabited islands off the coast of Antarctica, found numerous plastic bottles originating from Japan and Argentina, Styrofoam packing peanuts, plastic shipping bands, foam fishing-net floats, and fragments of metal oil drums washed up on beaches.[86]

Plastics are of particular concern because they generally are not biodegradable and therefore can persist in the environment for hundreds to thousands of years and perhaps longer if they sink to the deep sea.[87] Eighty percent or more of the waste found along shorelines and floating in the seas is plastic. There are two forms of plastics in the ocean: plastic pellets, which are the raw material from which plastic goods are manufactured, and plastic manufactured objects.

FIGURE 12–36 Marine debris threatens wildlife in many different ways. Here, a California sea lion has become entangled in fishing nets. Animals sometimes also ingest plastic particles, and exotic species hitch rides to new locations where they may become invasive and crowd out native species. *(© Paul A. Souders/Corbis)*

BOX 12–3 CASE STUDY

Chesapeake Bay and Population Growth

One of the most beautiful coastal wetland environments on the eastern shore of North America is Chesapeake Bay, a vast estuary that developed 6000 to 7000 years ago as rising sea level drowned the valleys of the Susquehanna River and its tributaries (see Figure 12–6). At that time, the estuary's upper reach was near present-day Annapolis, and sea level was approximately 9 m below its modern elevation.[79] Since then, sea level has continued to rise, and the bay has filled with the mud and sands of the rivers that flow into its quiet waters. At the mouth of each river, broad areas of silt, fine sand, and clay have been deposited and colonized by aquatic grasses, becoming marshes (grass- and sedge-dominated wetlands) and swamps (forested wetlands) that provide rich habitat for many species of nearshore estuarine plants and animals.

Some of the Chesapeake's environments are quite unusual. Adapted to life in wet, frequently flooded soil, bald cypress trees with knobby "knees" that protrude above wet mud are found at the mouths of Chesapeake rivers, farther north than anywhere else in the United States. Early English settlers described Chesapeake Bay as filled with thick beds of large oysters (see the chapter opening photo) and teeming masses of crabs, eel, and rockfish. Flocks of waterfowl, including ducks, geese, and osprey, covered its surface as far as the eye could see. During his exploration of the bay in 1608, English captain John Smith noted that it was a "faire bay encompassed but for the mouth with fruitful and delightsome land."[80] The sight was unlike anything in the settlers' experience in their more densely populated and developed homeland.

Since then, overfishing and rapid population growth along the shores and throughout the drainage area of the bay have placed stress on the bay's habitat (Figure 1). The bay is an important commercial shipping route and a regional center for the steel, leather, plastics, and chemical industries, which contribute numerous pollutants. Waters flowing into Chesapeake Bay drain some 166,000 km² (64,000 square miles) of New York, Pennsylvania, West Virginia, Virginia, Maryland, Delaware, and the District of Columbia, an area encompassing much of the industrial and urban eastern seaboard. Urban growth in the area has nearly doubled since 1950, and population has risen to 17.7 million people.[81]

Fifty rivers, with more than 100 sizable tributaries among them, flow into the Chesapeake, each carrying whatever has washed into it along its way seaward.

Between 1960 and 1994, the bay oyster catch fell by 98 percent (Figure 2),[82] and the spawning stock of female rockfish declined from nearly 82.7 million pounds to 5 million pounds from 1960 to 1984.[83] Their food supply declining, wild fowl populations also plummeted. Eelgrass beds once covered the bay shallows, forming part of the food web, providing habitat, and protecting bay mud from erosion. In the upper bay, the grasses now are almost completely gone, and even carefully planted beds in the middle estuary often don't survive, victims of eutrophication. In 2000, the Chesapeake Bay was declared an impaired water body under the federal Clean Water Act because of its high sediment and nutrient loads.

The causes of environmental stress in the bay are many and complex, and as a result are difficult to remedy. Enhanced sedimentation has sometimes been blamed on nonpoint sources that include ongoing deforestation, farming, and urban development, as well as erosion of coastal bluffs. Recently, however, scientists have discovered that much of the sediment entering the Chesapeake originated in colonial times, when land clearing, mining, and charcoaling (logging trees to make charcoal fuel) stripped hillsides of trees, promoting erosion. The sediments produced were trapped behind thousands of old mill dams that were built on bay tributaries for water power as early as the late 1600s. As these obsolete structures now breach or are purposely removed to promote fish migration, they release nonpoint source sediment from past land use practices to downstream reaches and the bay.

Other stresses on the bay are chemical in nature. Pesticides and herbicides that run off agricultural fields and chemically treated suburban lawns contain chlorinated hydrocarbons that have a high potential for human toxicity or are probable carcinogens. Many other compounds also enter the bay, and research suggests that almost all of them are either concentrated in the food chain or become entrained in the upper few centimeters of the bottom sediments where they can become deadly ground for burrowing organisms such as worms.

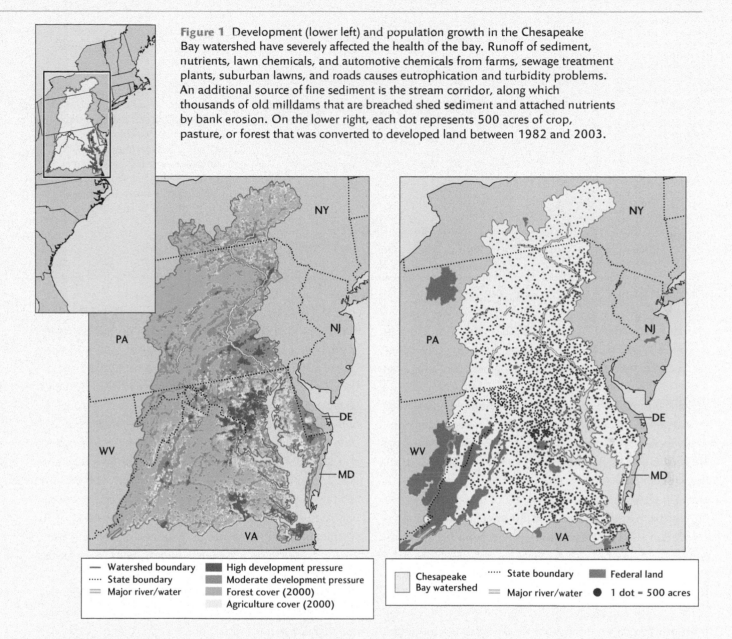

Figure 1 Development (lower left) and population growth in the Chesapeake Bay watershed have severely affected the health of the bay. Runoff of sediment, nutrients, lawn chemicals, and automotive chemicals from farms, sewage treatment plants, suburban lawns, and roads causes eutrophication and turbidity problems. An additional source of fine sediment is the stream corridor, along which thousands of old milldams that are breached shed sediment and attached nutrients by bank erosion. On the lower right, each dot represents 500 acres of crop, pasture, or forest that was converted to developed land between 1982 and 2003.

— Watershed boundary
···· State boundary
═ Major river/water

■ High development pressure
■ Moderate development pressure
■ Forest cover (2000)
□ Agriculture cover (2000)

□ Chesapeake Bay watershed
···· State boundary
═ Major river/water

■ Federal land
● 1 dot = 500 acres

Perhaps the most pressing threat to the bay's ecosystem, however, is a lack of oxygen in the brackish water as a result of nutrient pollution and eutrophication. Sewage and farm runoff contain high concentrations of nitrogen and phosphorus, and sediment from eroding sources upstream can contain substantial amounts of adsorbed phosphorus. These nutrients act as fertilizer for algae and cause algal blooms, the very high growth rates of green algae that lead to eutrophication.

The collapse of oyster populations and demise of many underwater grasses in the bay cause additional problems.

As filter feeders, oysters take in water and particles through their siphons and send clear water back out. At the time of Captain John Smith's visits to the bay, it is estimated that Chesapeake oysters filtered the entire estuary within a few days.[84] With their populations reduced to 2 percent of historic levels by a combination of overfishing, pollution, and more recently, disease, the bay's remaining oysters require nearly a year to do the same job. Without this vital ecological service, it is possible that the Chesapeake's water will remain cloudy, which will make it difficult for the important beds of underwater grasses to recover.

(continued)

(continued)

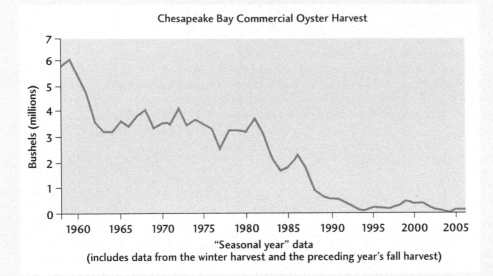

Chesapeake Bay Commercial Oyster Harvest

"Seasonal year" data
(includes data from the winter harvest and the preceding year's fall harvest)

Figure 2 Declining oyster harvest. As of 2005, the oyster harvest was only 3% of the 6 million bushels harvested in the 1950s. Losses are attributed to multiple causes that include pollution, loss of reef habitat, and disease. Overharvesting also has had an important and longer-term role, as commercial oyster fishing began in the 1700s and increased to a peak of 20 million bushels per year in the 1880s.

This submerged aquatic vegetation, or SAV, provides an essential habitat for many species, including blue crabs, eels, various fish, and waterfowl. The grassy meadows offer organisms not only food and shelter, but also produce oxygen through photosynthesis. A comprehensive study of Chesapeake Bay SAV by multiple agencies and institutions, including the Virginia Institute of Marine Science, the University of Maryland Center for Environmental Science, the USGS, NOAA, and the EPA, assessed the relations between water quality and the extent of underwater grasses during the years from 1984 to 2006. The study found a correlation between nitrogen levels and SAV: higher nitrogen levels resulted in reduced SAV coverage; lower levels of nitrogen resulted in expansion of the grasses. The responsiveness of SAV to nutrients in water indicates the importance of improving water quality—and therefore improving practices throughout the watershed—in order to restore ecosystem health.

States bordering the bay, in conjunction with the EPA and strong grassroots environmental groups, have mounted a "save the bay" campaign. The legislatures of two states with large areas within the bay watershed, Virginia and Maryland, created the Chesapeake Bay Commission (CBC) to coordinate their efforts to restore the estuary. The largest stream that drains to the bay is the Susquehanna River, for which most of the watershed lies in Pennsylvania; this state joined the CBC shortly after it was formed. Today the CBC consists of five legislators from each of the states of Maryland, Virginia, and Pennsylvania, one representative (a cabinet member) of the governor for each state, and three citizen members from each state, for a total of 21 members. The CBC works as an active partner with the EPA's Chesapeake Bay Program and other agencies and groups to understand the ecology of the bay, to identify federal and state policies needed to improve and sustain water quality and aquatic resources, and to persuade policy makers to take appropriate actions. In recent years, for example, the commission has advocated for federal funding to support upgrades to the Blue Plains Wastewater Treatment Plant in Washington, D.C. Upgrades to this massive treatment plant are anticipated to reduce annual nitrogen loads to the bay by as much as 4 million pounds.

Despite several decades of restoration efforts to improve water quality and aquatic resources in the Chesapeake Bay, less than 25 percent of the EPA's water quality goals have been met. In response in 2009, President Obama issued Executive Order 13508, which calls on the EPA to define a new generation of tools and to refine policies that will reduce sediment and nutrient loads to the Bay. In 2010, the EPA mandated that entities with waters that drain into the bay must not exceed pollution reduction allocations for nitrogen, phosphorus, and sediment, known as Total Maximum Daily Loads (TMDL). States are required to provide the EPA with "reasonable assurance" that water quality within their borders will improve by the 2025 implementation deadline.

Plastics affect the environment by contaminating beaches and killing marine animals that ingest plastic debris or become entangled in it. Impacts are difficult to quantify,[90] but a study from the 1980s concluded that 50,000 northern fur seals were killed through entanglement each year, and as many as 40,000 more by possible entanglement or other unknown factors, such as disease.[91] More recent studies examining the impact of plastics on wildlife have found that over 250 different species of marine organisms become entangled in or ingest plastic debris, including more than 100 species of sea birds, turtles, dolphins, fish, whales, seals, otters, crustaceans, and manatees.[92] The death of these organisms, when it is related to being entangled in abandoned fishing nets and lines, is known as *ghost fishing.*

Another hazard related to plastics is that they tend to attract toxic chemicals such as PCBs. These substances, previously used as insulating fluids in electrical transformers, are known carcinogens and have been banned in the United States and many other nations. Studies of PCBs in the ocean have found concentrations on plastic particles to be more than 100 times higher than in the surrounding seawater, and since PCBs are incorporated into fat cells and do not break down, they may be passed up the food chain in increasingly higher concentrations, a concern for wildlife and for people who consume fish and shellfish.[91]

A little-studied hazard of plastic pollution is that plastic debris can transport exotic organisms around the globe. While animals have been hitch-hiking rides on floating tree branches and other natural debris for millions of years, leading to the gradual dispersal of species, the sheer magnitude of the increase in floating debris since the dawn of the plastics era is cause for concern. Floating plastics appear to attract encrusting organisms, like barnacles, bryozoans, tube worms, and coralline algae. In turn, these organisms attract more mobile creatures that feed on or shelter in these species, turning each piece of floating plastic debris into a miniature ecosystem. While many of these organisms die out as the flotsam moves into waters that are less hospitable, it is possible that as the climate continues to warm, some of these species may find conditions in which they can reproduce, leading to species invasions that cause native species to decline.

Ocean Acidification

Another hazard of fossil fuel combustion is ocean acidification. Since the modern fossil fuel era began, levels of carbon dioxide in the atmosphere have increased from 280 to 400 ppm, leading to a concomitant increase of carbon dioxide in ocean water and a 30 percent increase in hydrogen ion concentration.[92] Carbon dioxide and other gases are transmitted from air to water across the ocean–atmosphere boundary, and about a third of the increased amount of carbon dioxide in the atmosphere is taken up by seawater. As can be seen in Equation 1 above, an increase in carbon dioxide in seawater drives the reaction to the right, producing hydrogen ions and thus lowering pH. This relative shift to lower pH values is known as **ocean acidification** (Figure 12-37a).

Many marine organisms manufacture their skeletons and protective shells from Ca^{2+} and CO_3^{2-} in the form of calcium carbonate ($CaCO_3$), but increased CO_2 in the oceans produces bicarbonate and lowers the availability of the carbonate ion required for building calcareous hard parts, as shown in the following reaction:

$$CO_2 + H_2O + CO_3^{2-} \rightarrow 2\ HCO^{3-} \quad \text{(Equation 3)}$$

Ocean scientists are concerned that acidification of the oceans could threaten the survival of calcifying organisms that form biogenic calcium carbonate.[93,94] Shells and skeletons could thin[95] and become more fragile, or simply dissolve away entirely (see Figure 12-37b).[96] Furthermore, all of the species affected are preyed upon by other organisms, so these losses might ripple upward through the oceanic food chain to affect the fish and shellfish upon which much of the human ...population is dependent for food. Disrupted food chains could also lead to changes in atmospheric chemistry and circulation, with profound implications for life on Earth. We discuss these issues further in Chapter 15.

Protecting the Oceans

Confusion over ownership of the seas during the 20th century led to uncertainty about whose responsibility it is to ensure their health. In the 1970s, the United Nations stepped in to fill some of the void, facilitating the creation of the International Convention for the Prevention of Pollution from Ships, better known as **MARPOL** (short for *marine pollution*).

Sovereign states that ratify the treaty (152 as of 2013) agree to be bound by regulations designed to reduce pollution from shipping, either through intentional dumping or accidental spills. For example, prior to new regulations, oil tankers filled their reservoirs with seawater after delivering their loads in order to ensure they had enough ballast to travel back across the ocean. Upon returning home, they flushed the oil-contaminated seawater out of their tanks to make them ready to accept the next load of oil. The MARPOL Convention altered this practice by requiring a commonsense and easy solution: Because oil floats on top of water, ballast water is to be decanted from the bottom of the tank, leaving the oily residue behind in the ship. MARPOL also required changes in design for new tankers to require separated oil and ballast water tanks

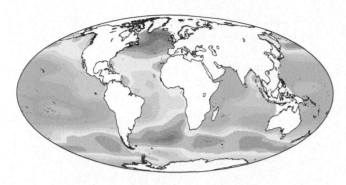

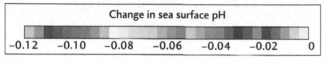

(a)

FIGURE 12-37 Ocean acidification. (a) Models of data for sea surface parameters that include dissolved inorganic carbon, alkalinity, temperature, and salinity indicate that the pH of surface seawater decreased in many parts of the oceans between the preindustrial 1700s and the 1990s. This ocean acidification, or relative shift to lower pH values, is the result of an increase in CO_2 in the atmosphere and, hence, in the oceans. Areas with missing data and landmasses are white. Color-coded values represent a change in annual mean sea surface pH. A value of −0.1, for example, indicates that mean annual sea surface pH at that location has decreased by 0.1. (b) Experiments have revealed that future CO_2-induced ocean acidification will make it harder for calcifying marine organisms to build their protective shells and skeletons. The pencil urchin (*Eucidaris tribuloides*) shown above was grown under normal levels of atmospheric CO_2 (400 ppm), and the one below was grown under levels predicted several centuries from now (2850 ppm). The magnesian calcite spines of the urchin grown under modern-day CO_2 levels are well-formed, while the spines of the urchin grown under future CO_2 levels are nearly completely dissolved away. *(b: Courtesy Justin Ries)*

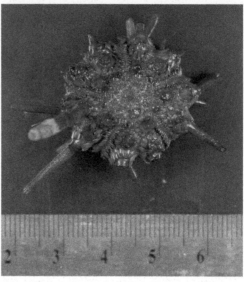

(b)

and to allow the ships to better withstand collisions or running aground.

Other MARPOL provisions address the practice of dumping sewage and solid waste in the ocean. Dumping of raw sewage is prohibited within 12 nautical miles of shore, and ports are required to have facilities for off-loading sewage wastes.[97] Furthermore, new ships must be outfitted either with sewage treatment equipment or a tank to store sewage for off-loading at a port. The MARPOL Convention also bans all intentional dumping of plastics at sea and severely restricts dumping of other garbage, such as food waste.[98] Furthermore, the Convention requires that all ships have a garbage management plan, written in the language spoken by the crew operating the vessel, and that records be kept of garbage disposal.

The United Nations also sponsored the London Dumping Convention Treaty (1972), which banned the dumping of hazardous substances such as mercury, cadmium, and high-level nuclear waste, and required special permits for dumping substances such as arsenic, lead, cyanides, pesticides, and fluorides. The Convention was updated in 1996, and now is called the London Protocol. It bans all dumping, with the exception of certain permitted substances such as dredged material, liquefied carbon dioxide from carbon dioxide capture and sequestration, fish by-products, natural organic materials, and inert geological materials.[99]

While the MARPOL Convention and London Protocol have put us on the right path toward cleaning up the seas, much marine pollution actually originates from land. Because of this, many nations instituted their own laws to deal with different contaminants introduced to their shores. In the United States, for example, the Ocean Dumping Ban Act of 1988 prohibited the dumping of sewage sludge and industrial waste into the ocean as of 1991.[100] Similarly, by the Ocean Dumping Regulations of 1989, Canada severely restricts the levels of toxic, carcinogenic, and radioactive substances that can be dumped along its coastal waterways.

Of course, laws banning intentional dumping can go only so far toward cleaning up the seas. Accidental pollution continues to occur, and the seas already contain substantial amounts of garbage. To deal with these issues, additional approaches are required. First, because the biggest challenge to the world ocean is the continued growth of the human population, the most pressing need is for improvements in recycling, reusing, and reducing wastes. More widespread plastics recycling, in particular, would significantly reduce that threat to the ocean. Second, fishing fleets can be engaged as partners in cleaning up the seas. In 2008, NOAA, in partnership with a private utility and nonprofit organizations, instituted a program to allow fishers to gather and dispose of derelict fishing nets, floats, and lines at no cost. The private utility burns the discarded debris and uses the energy of combustion to make electricity.[101] One ton of debris can generate enough electricity to power a single home for 25 days, and within its first 2 years, the program took in more than 539,000 pounds of marine debris in Massachusetts, New York, Rhode Island, New Jersey, Oregon, Maine, and Virginia. While this is a drop in the bucket compared to the amount of waste floating in the world ocean, the program points toward one part of a potential solution.

- Coastal nations have become increasingly territorial regarding their offshore continental shelf areas because of the wealth of resources—particularly fish, petroleum, and minerals—that are found there.

- According to the Law of the Sea, the area within 370 km (200 nautical miles) of a country's shore is defined as that country's Exclusive Economic Zone (EEZ), within which coastal nations have exclusive rights to the ocean's natural resources. The resources of the remainder of the sea are open to all nations.

- The greatest source of pollution to the oceans is runoff from continents in the form of treated or partly treated sewage, pollutants from streets and parking lots, chemical pollutants from industries, and excess fertilizer.

- Many human wastes, including sewage sludge and industrial waste, are disposed of on continental shelves. Sewage sludge adds nutrients to seawater, thus contributing to eutrophication, which depletes oxygen from the water. It also contaminates beaches and shellfish beds with pathogenic microbes.

- Nutrient enrichment has caused coastal eutrophication and problems with hypoxia and anoxia in many parts of the world. One of the most impacted sites is the Mississippi delta region (the Gulf dead zone).

- Untold numbers of marine species die every year through ingesting or becoming entangled in plastics. Additionallly, exotic species hitchhike on plastic debris to new locales, where they may become invasive.

- Just as carbon dioxide has accumulated in the atmosphere since the Industrial Revolution, carbon dioxide levels in seawater are increasing as a result of the burning of fossil fuels. The resulting ocean acidification threatens marine species that create shells and skeletons of calcium carbonate, as well as those species that depend upon them for food.

- The MARPOL Convention is an international agreement designed to reduce intentional and accidental releases of pollution in the world ocean.

CLOSING THOUGHTS

Ours is a watery world, yet for most of human history we have been confined to land. At the beginning of the 20th century, the depth of the oceans was known at only the few places where seafarers had dropped and measured the length of a weighted line. In both 20th century world wars, submarines traveled along the ocean floors, and in World War II, ships used a new technology—sonar—to determine water depth. Until then, the high seas were considered to begin a few kilometers from shore, and were open to any travelers. The oceans were this planet's last frontier.

Soon after World War II, continental shelves became precious for their resources, especially petroleum deposits and fisheries. Nations began laying claim to what had been a global commons, and years of conferences led to the Law of the Sea, an international code that delineates how the seafloor and overlying waters shall be apportioned. Clearly, we live in a time when limits to economic and population growth are being pushed to the extreme. Increasing exploitation of ocean and coastal resources puts pressure on the ocean environment.

But this is an opportune time as well. Until the 20th century, it was possible only to dream of how the seafloor might look. In 1943, the late ocean diver Jacques Cousteau and his compatriot Émile Gagnan, an engineer, invented and tested a device (later to become known as the Aqua-lung) that supplied air from tanks to an underwater diver. The Aqua-lung was the precursor to SCUBA, or Self-Contained Underwater-Breathing Apparatus. Today it is possible to go much deeper than a scuba diver can, if one is willing to enter a small submersible vehicle and sink many kilometers to the ocean floor. Scientists clamor for such opportunities, and those who have the chance to peer out a window and watch giant tube worms swaying on the edge of a hot, hydrothermal vent are considered the lucky ones.

One such oceanographer and explorer, Dr. Sylvia Earle, founded the Sylvia Earle Alliance (SEA) to generate international public support for "hope spots," protected marine areas with special environmental significance, as in the Ross Sea, which is partly covered by the Ross Ice Shelf in the Southern Ocean near Antarctica (Figure 12-38). With little pollution, this deep bay is one of the world's most ecologically intact nearshore ecosystems. Whereas 12 percent of land on Earth is protected in some fashion—for example, as national parks or monuments—less than 1 percent of the ocean is protected. SEA defines a marine protected area as a "space in the ocean where human activities are more strictly regulated than the surrounding waters." Clearly defined geographically, such spaces are to be recognized and managed through legal or other means so that they are conserved over the long term. During an interview in 2011, Dr. Earle described her inspiring vision for hope spots, with particular reference to the relatively unexplored Arctic Ocean:

"...if we take action while we still have some oysters and clams and the other creatures that make up the living systems, there's cause for hope that we can recover them. We can't go back to what they were 100 or 1,000 years ago. But they can be better than they are by taking proactive measures to restore their health."[102]

FIGURE 12–38 Orca whale in the Ross Sea. The Ross Sea is a "hope spot," a place identified by scientists and conservationists as being of exceptional biodiversity and ecological importance, one that is relatively unaltered by human activity or pollution. This deep bay off the southern coast of Antarctica has nutrient–rich seawater and is partly covered by the Ross Ice Shelf. *(AgeFotostock/SuperStock)*

SUMMARY

■ Ocean basins consist of continental shelves, slopes, and rises, as well as abyssal plains. They contain mid-ocean ridges along divergent plate boundaries and deep-sea trenches along subduction zones.

■ The action of plate tectonics is closing some oceans (notably, the Pacific) and expanding others (notably, the Atlantic). Coasts are defined by their tectonic setting. An active margin coast forms along a convergent boundary and therefore is mountainous and has a narrow shelf. Erosion is high because deep waters offer no resistance to the progress of high-energy waves. A passive margin coast lies far from a plate boundary and so has a broad continental shelf where shallow waters impede wave size.

■ Ocean water is saline because the rate of addition of some ions from rivers exceeds their rate of depletion. Ions such as calcium and bicarbonate, which are used by marine creatures to construct shells, have relatively short residence times in the ocean, while others, such as sodium and chloride, have very long residence times.

■ Nutrients essential to life are dissolved in ocean water and are used by phytoplankton to manufacture food. The phytoplankton are consumed by zooplankton, which in turn sustain other marine life.

■ Wind and gravity act on water, causing it to flow as ocean currents, which transport energy and matter from one part of Earth to another.

■ Surface currents are driven by winds and deflected by the Coriolis force. Each successively deeper layer in the water column is deflected relative to the layer above that is exerting the drag on it. Ekman transport is the overall water movement integrated over the entire depth experiencing the effect of the winds, and averages 90° to the prevailing winds. Ekman transport is important in distributing the ocean's nutrients because its interaction with land causes either upwelling or downwelling of nutrients along shores.

■ Deep-ocean currents occur because gravity pulls water along a density gradient created by vertical differences in temperature and salinity.

■ Deep-ocean currents move as if on a conveyor belt, circulating heat and matter around the globe.

■ Sea coasts are pleasant but hazardous places to live, liable to changes in sea level from global warming, and threatened by tsunamis and storm surges. Beach and cliff erosion are other hazards that stem from natural processes. Human attempts to mitigate these processes are costly and meet with little success.

■ Human activities and burgeoning coastal populations threaten coastal environments and the open ocean alike. Coastal development, sewage waste disposal, polluted runoff, the introduction of plastics, and oil spills all affect the coastal environment.

■ The desire to exploit the ocean's resources, combined with an understanding of the connection between those resources and ocean floor topography, has led to changes in territorial claims to the oceans. The United Nations Convention on the Law of the Sea gave each coastal nation an Exclusive Economic Zone (EEZ) 370 km wide adjacent to its shores.

KEY TERMS

ocean (p. 408)
sea (p. 408)
bathymetry (p. 409)
continental shelf (p. 409)
continental slope (p. 409)
continental rise (p. 409)
abyssal plain (p. 409)
active margin (p. 410)
passive margin (p. 411)
estuaries (p. 411)
phytoplankton (p. 414)

photic zone (or euphotic zone) (p. 414)
surface mixed layer (p. 416)
thermocline (p. 416)
halocline (p. 416)
pycnocline (p. 416)
deep ocean (p. 417)
Ekman transport (p. 417)
upwelling (p. 418)
downwelling (p. 418)

gyres (p. 418)
El Niño Southern Oscillation (p. 420)
thermohaline circulation (p. 420)
barrier islands (p. 424)
wave base (p. 425)
longshore current (p. 426)
groins (p. 428)
jetties (p. 428)

seawalls (p. 429)
tsunamis (p. 431)
Exclusive Economic Zone (EEZ) (p. 438)
hypoxic (p. 440)
anoxia (p. 440)
eutrophication (p. 440)
dead zone (p. 441)
ocean acidification (p. 447)
MARPOL (p. 447)

REVIEW QUESTIONS

1. Compare the shapes, sizes, locations, and underlying material of the following ocean-floor landforms: continental shelf, continental slope, continental rise, and abyssal plain.

2. How is plate tectonics changing the Pacific and Atlantic ocean basins?

3. What are the differences between leading-edge (active) and trailing-edge (passive) coastlines?

4. How do plate tectonics and climate change affect sea level, and what impacts do sea level changes have on coastlines?

5. What are the three most abundant major constituents (dissolved ions) in seawater? Why do they differ so much from the most abundant constituents in river water?

6. What is the source of calcium and bicarbonate ions in the oceans? How do these ions become bound in deep-sea sediments?

7. What are the two most important nutrients in seawater? Why is their amount in the surface ocean so different from that in the deep ocean?

8. What factors determine the pH of seawater?

9. What factors affect the flow of currents in the surface and deep ocean? How do ocean currents in these two zones differ?

10. Why is the net transport of water in the surface mixed layer at a right angle to the direction of the winds that cause its motion? What implications does this transport direction have for the vertical distribution of nutrients and oxygen in the water column?

11. What role does sea ice formation play in thermohaline circulation?

12. Why can a scuba diver swimming at 4 to 5 m depth encounter calm seas, while at the surface her boat is being tossed about on 0.5 to 1 m waves?

13. What processes add sand to barrier islands and beaches? What processes remove sand and transport it along the shore?

14. Describe two ways of attempting to control or halt coastal erosion. How has the damming of rivers contributed to increased coastal erosion on some coasts?

15. Why are efforts to control coastal erosion so costly and so rarely successful over long time periods (tens of years)?

16. What is a tsunami? What causes it? How does the tsunami warning system operate?

17. Briefly summarize the outcome of the United Nations conferences on the Law of the Sea regarding the rights of coastal nations to the ocean's natural resources. Define "Exclusive Economic Zone."

18. What are the major types of resources that nations wish to extract or harvest from the oceans?

19. List and briefly describe the major sources of pollution to the oceans.

20. What is eutrophication? How do nutrients contribute to its occurrence, and what locations are therefore particularly at risk?

21. Why is the status of oysters so critical to the health of Chesapeake Bay?

22. How is fossil fuel combustion changing the chemistry of Earth's oceans and what are the potential consequences for marine food chains?

23. What efforts have been made to clean up the seas?

THOUGHT QUESTIONS

1. Why is the extent of the continental margin important in resolving international disputes over rights to marine and seafloor resources? Give specific examples.

2. Which type of coastline—active or passive—is more likely to be associated with coastal erosion and cliff retreat? Why? Which type is more likely to be associated with large deltas? Why?

3. A tide gauge at a site in the state of Washington indicates that over the past few decades, sea level has been rising with respect to the landmass at a rate greater than would be expected from global measures of sea level. What are some possible explanations?

4. How might the Chesapeake Bay area have appeared about 20,000 years ago, when sea level was much lower than today? Was the coastline different then? Would an estuary have existed?

5. Why is the salinity of seawater in the Red Sea so much higher than that in the Atlantic Ocean?

EXERCISES

1. Make a sketch to answer the following questions. In the southern hemisphere, if wind is blowing northward along a coastline with a landmass to the west, is the direction of net water transport toward or away from the shore? Would you predict that there would be upwelling or downwelling along this coast?

2. Use an atlas to determine how much of a coastal city would be submerged if the West Antarctic ice sheet were to melt completely, raising sea level by about 5 m.

3. As mentioned in the section on cliff erosion, some sea cliffs in the Monterey Bay are retreating as fast as 63 cm/yr because of wave erosion. Using Google Earth, zoom in on W Cliff Dr., Santa Cruz, CA. Use the ruler tool to measure the distance between houses on West Cliff Drive and the ocean in meters. At a retreat rate of 63 cm/yr, how many years would it take before these houses plunge into the sea? What evidence can you find on the aerial photos that people are trying to control cliff erosion?

4. The Gulf Stream transports up to 150 million m^3 of water per second. How many Olympic-size swimming pools (50 m long × 25 m wide × 2 m deep) would it fill each second? Each year?

5. The average depth of the Indian Ocean is 3693 m. Given this depth, how fast would tsunami waves travel in kilometers per hour? Do an Internet search on the speed of the 2004 Indian Ocean tsunami. How well does your calculation compare?

SUGGESTED READINGS

Doney, S. C. "The Dangers of Ocean Acidification. *Scientific American* 294, no. 3 (2006): 58–65.

Garrison, T. S. *Oceanography: An Invitation to Marine Science,* 7th ed. Belmont, CA: Brooks/Cole, Cengage Learning, 2010.

Geist, E. L., V. V. Titov, and C. E. Synolakis. "Tsunami: Wave of Change." *Scientific American* 294, no. 1 (2006): 56–63.

Howarth, R., D. Anderson, J. Cloern, C. Elfring, C. Hopkinson, B. Lapointe, T. Malone, et al. "Nutrient Pollution of Coastal Rivers, Bays, and Seas." *Issues in Ecology,* no. 7 (2000): 1–15.

Jacobson, J. L., and A. Rieser. "The Evolution of Ocean Law." *Scientific American Presents: The Oceans* 9, no. 3 (1998): 100–105.

Jochum, M., and R. Murtugudde. *Physical Oceanography: Developments Since 1950.* New York: Springer Science+Business Media, 2006.

Kappel, C. V., B. S. Halpern, R. G. Martone, F. Micheli, and K. A. Selkoe. "In the Zone Comprehensive Ocean Protection." *Issues in Science and Technology.* (Spring 2009). Accessed February 22, 2010. http://www.issues.org/25.3/kappel.html.

Mallin, M. A. "Wading in Waste." *Scientific American* 294, no. 6 (2006): 52–59.

Mee, L. "Reviving Dead Zones." *Scientific American* 295, no. 5 (2006): 78–85.

National Research Council. *Tackling Marine Debris in the 21st Century.* Washington, DC: National Academies Press, 2009.

National Research Council. *Increasing Capacity for Stewardship of Oceans and Coasts: A Priority for the 21st Century.* Washington, DC: National Academies Press, 2008.

Pauly, D., and R. Watson. "Counting the Last Fish." *Scientific American* 289, no. 1 (2003): 42–47.

Pilkey, O., and K. Dixon. *From the Corps to Shores.* Washington, DC: Island Press, 1996.

Pilkey, O. H., and R. Young. *The Rising Sea.* Washington, DC: Island Press, 2009.

Pinet, P. R. *Invitation to Oceanography,* 5th ed. Sudbury, MA: Jones and Bartlett, 2009.

Sverdrup, K. A., and E. V. Armbrust. *An Introduction to the World's Oceans,* 10th ed. New York: McGraw-Hill, 2009.

Webster, P. J., and J. A. Curry. "The Oceans and Weather." *Scientific American Presents: The Oceans* 9, no. 3 (1998): 38–43.

Energy, Changing Earth, and Human-Earth Interactions

PART V

Energy

More than 300,000 years ago, perhaps even before one million years ago, our Stone Age ancestors warmed themselves by wood fires, the oldest human use of carbon-rich biomass as an energy resource for heat.[1] Since then, more concentrated sources of carbon have been discovered—ancient biomass buried and transformed into fossil fuels such as peat, coal, oil, and natural gas. Over a period of hundreds of millions of years, a small fraction of the flow of energy from the Sun has accumulated in aboveground biomass and an even smaller fraction in buried biomass and fossil fuels, giving Earth a very large energy stock.

Each year, we use about 17 cubic km (4 cubic mi) of that fossil biomass to power modern civilization, with 82 percent of world energy consumption from fossil fuels.[2] When burned, biomass and fossil fuels release not only energy but also carbon dioxide and other gaseous emissions that contribute to global warming. At the same time, burning these fuels depletes the fossil fuel supply. Because fossil fuels form over geologically long periods of time, they are considered nonrenewable resources; however, they are not the only energy sources available. For economic as well as environmental stability, the type of energy we use throughout the 21st century probably will depend more directly on greater energy efficiency and renewable resources, for example the Sun's radiation (including the ways in which it affects wind speed) and Earth's internal heat. In this chapter, we will explore the origin, distribution, and environmental impact of several important energy resources on Earth. We will:

✔ Trace flows of energy through Earth's systems.

✔ Investigate ways to recover, use, and conserve nonrenewable fossil fuels (oil, natural gas, and coal).

✔ Examine nuclear energy and waste disposal.

As in this peat bog in Ireland, plants use solar energy and carbon dioxide to make carbon–rich biomass, which can accumulate in the pedosphere and lithosphere as peat, coal, natural gas, and oil. When burned, these fossil fuels release their stored chemical energy in the form of heat, and release the carbon, hydrogen, and oxygen that comprised them as gases to the atmosphere. *(Ned Haines/Science Source)*

457

✔ Examine solar energy, the only perpetual source of energy available worldwide.

✔ Consider the environmental impacts of using different energy sources.

✔ Reflect on the "new energy era" foreseen by some analysts, one marked by greater efficiency and use of renewable energy resources and little or no use of fossil fuels.

ENERGY AND HUMANS

Humans use immense amounts of energy, much of it for electricity and illumination (Figure 13-1). Each discovery and development of new sources of energy has resulted in major technological, societal, and environmental changes.

Prior to the mid-1800s, much of the world's energy consumption came from burning wood; the rest came from water, animals, and wind. These would all be considered *renewable resources* by today's standards. By 1900, however, coal had become the dominant source of energy (Figure 13-2). Used in steam engines that pumped water from mines and powered mills, coal was pivotal to the early industrialization of Europe and North America. Subsequent development of the internal combustion engine, which could burn liquid fuel, set the stage for major technological changes. The search for underground oil, from which gasoline is produced, intensified, and discoveries outpaced consumption throughout the 20th century. By World War II, oil was nearly as important as coal to world energy demands, and the importance of oil has increased since then. With these shifts in use of energy resources have come new forms of environmental pollution and the need for environmental regulation (see Chapters 11 and 15).

By the 1970s, oil provided the world with more than 40 percent of its energy. Then OPEC (Organization of the Petroleum Exporting Countries, a cartel formed in 1960) created a series of energy crises in oil-importing nations by repeatedly raising its oil prices. It suddenly became clear that much of the world's oil, perhaps one-half to two-thirds, was located in the Middle East and in what generally had been considered, until that time, to be a resource-deficient desert.

Industrialized nations responded in several ways to become less dependent on foreign oil. They increased their reliance on natural gas, sought new sources of domestic oil (e.g., in the Gulf of Alaska, the Alaska North Slope, and the North Sea), increased efforts to conserve energy, and made some attempts to develop alternative sources such as solar energy. Consequently, oil prices dropped, and by the 1980s OPEC was contributing a

Figure 13–1 Cloud–free images of Earth at night reveal the glow of human activity powered largely by fossil fuels. Most of the light is from cities and reveals highway networks, but wildfires show up outside cities, as well as the lights from fishing boats, gas flares, and oil drilling or mining operations. As noted by NOAA scientist Christopher Elvidge, "Nothing tells us more about the spread of humans across the Earth than city lights." This image was assembled from satellite data acquired in 2012 to create a composite global view. *(NASA Earth Observatory image by Robert Simmon, using Suomi NPP VIIRS data provided courtesy of Chris Elvidge [NOAA National Geophysical Data Center]. Suomi NPP is the result of a partnership between NASA, NOAA, and the Department of Defense.)*

U.S. Primary Energy Consumption Estimates by Source, 1775–2010

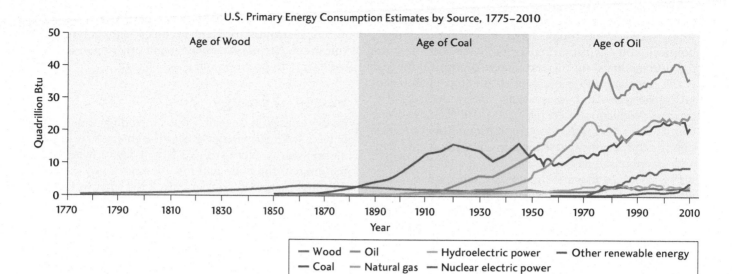

Age of Wood Age of Coal Age of Oil

— Wood — Oil — Hydroelectric power — Other renewable energy
— Coal — Natural gas — Nuclear electric power

Figure 13–2 Since 1850, world energy consumption has increased markedly and shifted from burning biomass (mostly wood) to fossil fuels, with oil surpassing coal as the most used energy source in the mid–20th century.

smaller share of the world's energy consumption. Nevertheless, when Iraq invaded Kuwait in 1990 and took over its oil fields, many nations, including the United States, sent troops to defend the Persian Gulf and their interests in its fossil fuels. As summed up in a report by energy analysts Christopher Flavin and Nicholas Lenssen of the Worldwatch Institute soon after, "Not only is the world addicted to cheap oil, but the largest liquor store is in a very dangerous neighborhood."[3] That situation has changed little during the two decades since, and geopolitical issues in the Middle East continue to provoke military responses from oil-consuming nations.[4]

Today we use enormous amounts of energy resources, mostly from the fossil fuels oil, natural gas, and coal, and mostly for the purposes of transportation and electricity (Figure 13-3).[5] The extraction and processing

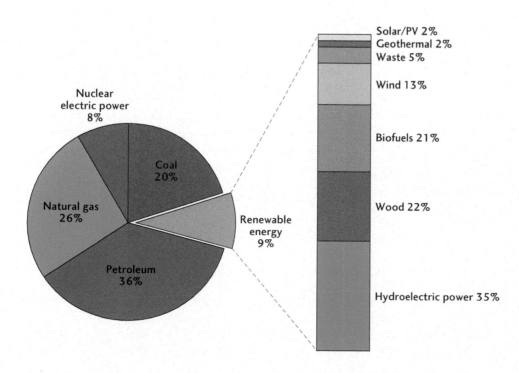

Figure 13–3 The world's primary energy consumption in 2011 was comprised mostly (82%) of the nonrenewable, carbon–rich fossil fuels oil, natural gas, and coal. The remainder was about evenly split between nuclear electric power and renewable energy sources.

of these resources to produce and deliver energy services such as electricity have significant economic, political, and environmental costs. Some fossil fuel extraction methods, such as mountaintop removal for coal in the Appalachian region, have substantial environmental impacts on land, ecosystems, and waterways. The use and production of energy from fossil fuels are major contributors to atmospheric pollution. Environmental impacts to the atmosphere include haze, smog, acid rain, and greenhouse gas emissions (see Chapter 11). Freshwater resources also are impacted by energy use. Nearly 40 percent of all freshwater withdrawals are for thermal power plants in the United States, and half a gallon of water is consumed for each kilowatt hour of energy services provided by power plants.

Some experts think that reliance on oil has reached its peak and that consumption of natural gas will peak between 2030 and 2050. Although coal could provide us with energy for centuries, its use has much greater environmental impact than other fossil fuels. Many scientists predict that increased efficiency and nonfossil energy resources that are renewable—especially solar and wind energy—will become increasingly important throughout the 21st century and into the future. We will discuss each of these energy resources, but first, we'll examine Earth's overall energy system.

EARTH'S ENERGY SYSTEM

Energy on Earth makes the planet a *dynamic system* (Figure 13-4; see Chapter 1). All processes and changes occur because of energy, the ability to do work. Earth generates energy in its interior and receives energy from the Sun. Energy from these two sources, plus small amounts from gravitational attraction among the Moon, the Sun, and Earth, flows through all Earth systems and powers the various cycles of rock, water, nutrients, and other substances.

Our Moon, in contrast, is a *static system,* considered geologically and environmentally dead because its interior is cold and lacking in energy and because it has no atmosphere to capture and transfer solar energy. As a consequence, there is no volcanism on the Moon, no mountain-building, no plate formation at spreading ridges or subduction at convergent boundaries, no soil formation, and no flowing water in rivers or moving masses of atmospheric moisture in storms. The Moon's surface is relatively calm, so craters and pits formed during the period of meteor bombardment more than 4 billion years ago appear fresh, as if formed yesterday. Earth was not spared this heavy bombardment, yet few craters appear on its surface. Most of those that do are subtle remnants of their original form, and are much younger than the craters on the Moon. On Earth, most evidence of the planet's early history has been nearly or completely destroyed by plate tectonics (e.g., subduction), volcanism, erosion, weathering, and other processes that use energy to do work and enact change.

States of Energy

Energy exists in several states, including kinetic and potential. **Kinetic energy** (KE) is the energy of a body in motion and is expressed as $KE = 0.5 \, mv^2$, where m is the mass of the body and v is the velocity of the body. **Potential energy** (PE), in contrast, is the energy of a body that results from its position within a system. Potential energy is related to the gravitational attraction between bodies and is expressed as $PE = mgh$, where m is mass of the body, g is acceleration of the body due to the gravitational force, and h is the relative height of the object.

Boulders perched on a hillside contain potential energy as long as they are not moving. If a process—perhaps an earthquake—causes the ground to shake or give way, the boulders will move downhill under the influence of gravity. Before the rockfall begins, the rocks contain potential energy proportional to their distance from Earth's center. As they roll and bounce down the slope, their potential energy is converted into kinetic energy and **thermal energy** (due to the motion of atoms). By the time the boulders reach the bottom of the hill, all their potential energy has been converted into thermal energy as a result of frictional resistance.

This example demonstrates a fundamental principle of energy: *energy is completely conserved when work is done.* While the energy has undergone a change in state (from potential to kinetic and thermal), the amount of potential energy before the work was done equals the amount of kinetic and thermal energy expended to do the work.

Energy can be transferred between bodies in the universe, as in the transfer of thermal energy. All bodies contain internal thermal energy, which arises from the random motions of their atoms. The term **heat** describes the transfer of thermal energy from one body to another. In a body receiving thermal energy, the added energy causes the atoms in the body to speed up. The temperature of a body is a measure of the average speed at which its atoms move. When heat is transferred to a body, its temperature rises because its atoms are moving faster.

Energy, Work, and Power

Energy is the ability to do work (e.g., lifting or hammering an object or raising its temperature), and this work is measured in many different units, including British thermal units and electron volts, but the most common convention among Earth scientists is to measure energy in either joules or calories. One *joule* (J) is defined as $1 \, kg*m^2/sec^2$. The *calorie* (cal) is defined as the amount

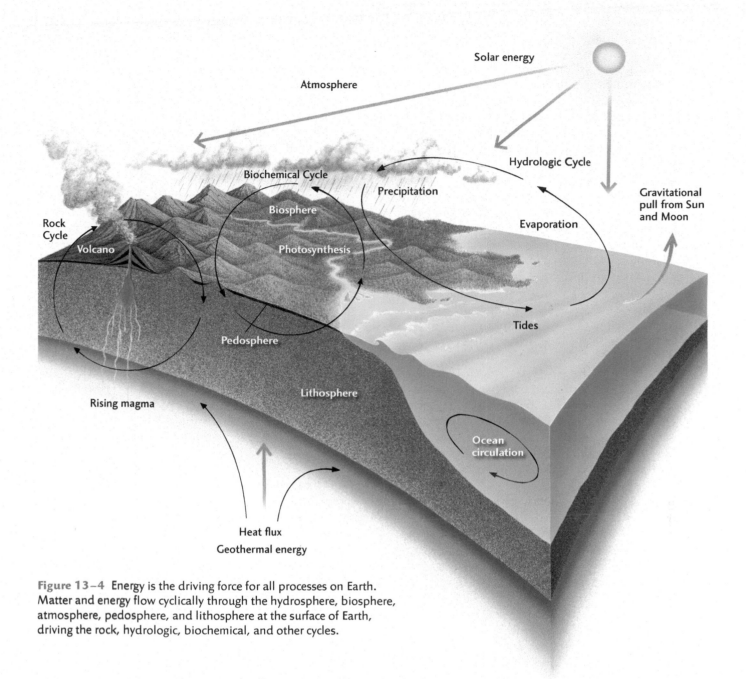

Solar energy

Atmosphere

Biochemical Cycle

Biosphere

Photosynthesis

Precipitation

Hydrologic Cycle

Evaporation

Gravitational pull from Sun and Moon

Rock Cycle

Volcano

Pedosphere

Tides

Rising magma

Lithosphere

Ocean circulation

Heat flux
Geothermal energy

Figure 13–4 Energy is the driving force for all processes on Earth. Matter and energy flow cyclically through the hydrosphere, biosphere, atmosphere, pedosphere, and lithosphere at the surface of Earth, driving the rock, hydrologic, biochemical, and other cycles.

of energy needed to raise the temperature of 1 g of liquid water by 1°C. One calorie is equivalent to 4.184 J. When contemplating how many calories our food contains, we refer to Calories (upper case C), each of which is equivalent to 1000 cal, or 4184 J. A candy bar, for example, contains about 400 Calories, or 400,000 calories, enough to raise the temperature of 400 kg of water by 1°C. The human body (typically 50 to 80 kg) needs to maintain a temperature of 37°C, so it uses more calories when the surrounding temperature is lower than the body's temperature. Because about 70 percent of the human body is water, much of its caloric energy is used to maintain the temperature of water at 37°C.

The British thermal unit, abbreviated as Btu (or sometimes BTU), is equal to about 1055 J. It is the amount of energy needed to heat one pound of liquid water by one degree Fahrenheit at one atmosphere of pressure. A wooden match produces about one Btu when burned. Humans use enormous amounts of energy, so the Btu are given in units of **quads**, which are one quadrillion (10^{15}) Btu (or 1.055×10^{18} J). The U.S. Department of Energy uses quads for national and international energy budgets. In 2011, for example, the United States used about 98 quads of energy from all sources.

Doing work over a period of time is expressed as **power**, the rate at which energy is used, transferred, or transformed. A common unit for power is the **watt**, named for James Watt, the Scottish engineer who developed the steam engine that was so vital to the Industrial Revolution in the mid-18th century. A watt (W) is 1 J per second. If a steam engine generates 14,914 J per second, for example, it is using 14,914 W.

Watt himself developed the unit called the horsepower (hp) in order to compare the output of steam engines to that of the work horses they could replace. The earliest steam engines, typically fueled with coal, were able to replace hundreds of horses in doing work such as pumping water from wells, mines, and canals, all imperative tasks for industrialization. One hp is equivalent to about 746 W. The steam engine, able to generate 14,914 W, for example, would be rated as a 20 hp engine (14,914 W × 1 hp/746 W = 20 hp).

Sources of Energy

The flow of energy through the lithosphere, pedosphere, hydrosphere, atmosphere, and biosphere drives the movement and transformation of matter that make Earth a dynamic planet capable of sustaining life. Just as matter flows from one reservoir to another, as in the rock and hydrologic cycles, so does energy, through the transfer of electricity or heat.

Three sources of energy, two internal and one external, power all the processes on Earth (see Figure 13-3). The foremost source, supplying 99.98 percent of Earth's energy, is the Sun. Smaller amounts of energy are created within Earth through spontaneous radioactive decay of some elements in rocks, and some residual heat from early Earth formation still exists. An even smaller amount of energy is supplied by the gravitational attraction within the Earth–Sun–Moon system and drives the movement of ocean water as tides.

The inflow and outflow of energy at Earth's surface can be quantified for the three sources of energy to the whole Earth system: solar radiation, internal thermal energy, and tidal energy. Earth receives 173,000 × 10^{12} W of power via solar radiation from the Sun annually. In comparison, the heat energy produced within Earth itself yields only 32 × 10^{12} W of power, and the energy related to tides, a mere 3 × 10^{12} W.

Solar Energy The greatest source of energy at Earth's surface is electromagnetic radiation from the Sun. This external energy source provides an inflow of 173,000 × 10^{12} W, or 173,000 terawatts (tW), per year and makes up 99.98 percent of the energy flow on Earth. Solar energy is the ultimate origin of nearly all other energy sources on Earth, including the energy stored in wood, coal, oil, natural gas, and wind. In the Sun, hydrogen atoms at extremely high temperatures and pressures combine to form helium atoms in a process called fusion. The fusion reaction releases radiant energy that is transmitted through space in all directions, and Earth intercepts some of this energy.

Some incoming solar energy (less than 1 percent) is converted by plants to a form of chemical potential energy. During photosynthesis, plants combine solar energy with water and carbon dioxide to produce the carbohydrates on which the whole biosphere depends for food and fuel. When we burn wood in a campfire, we release its stored chemical energy, which we can then use to warm ourselves or to cook food. Under the proper geological conditions, some biomass may be converted to oil, coal, or natural gas. These are called **fossil fuels** because they come from plants and animals that lived many tens to hundreds of millions of years ago. When we drive cars that run on gasoline, we are using solar energy that might be many tens to hundreds of millions of years old!

The amount of energy stored within fossil fuels is extremely small compared with the amount of solar energy received by Earth. In any 20-day period, Earth receives as much energy from the Sun as is stored in all the existing fossil fuel reserves. For this reason, solar energy could become an important energy source in the 21st century.

Earth's Internal Energy Earth makes its own supply of internal energy, of which about 32 tW a year reaches the planet's surface, through the spontaneous decay of certain elements in minerals. In this process, known as *radioactive decay*, parent atoms convert to other atoms by losing or gaining subatomic particles in their nuclei (discussed more fully in Chapter 6). Radioactive decay also releases relatively small amounts of thermal energy that are transferred through the surrounding rock.

Although much smaller than the influx of solar energy, it is Earth's internal heat energy, along with some residual heat from early Earth formation, that drives lithospheric plate motions, which in turn result in earthquakes and volcanism (discussed in Chapters 2, 3, and 5). Earth's heat energy also can turn groundwater to steam, producing geysers and geothermal fields.

As plates move, potential energy can be stored in rocks, much as you store energy in an elastic band by stretching it (discussed in Chapter 3). When such rocks finally snap, as a band stretched too far would, the amount of energy released can be quite large. One of the largest relatively recent earthquakes in California, the moment magnitude 6.7 Northridge earthquake that occurred just north of Los Angeles in 1994, released about 7.1 × 10^{14} J of energy. This amount is equivalent to about 1000 nuclear bombs the size of those dropped during World War II. A much larger earthquake, the moment magnitude 9.0 Tōhoku earthquake that caused the great tsunami in Japan in 2011, released 1.41×10^{18} J of energy, about 10,000 times more than the Northridge earthquake. For comparison, annual energy use in the United States is about 94×10^{18} J.

Gravitational Attraction Earth derives energy from gravitational attraction between itself and the Moon, the Sun, and the other bodies in the solar system. Although minor compared even with Earth's internal energy, the energy derived from gravitation causes both Earth's crust and ocean water to change shape with time in a regular fashion, resulting in rock tides as well as ocean tides. The kinetic energy of winds and waves—like all kinetic energy—ultimately is converted to heat by friction, then dissipates in the atmosphere.

Earth's Energy Cycle and Budget

The energy system of Earth is in an approximately *steady-state condition;* the amount of energy received by the whole Earth system is approximately equal to the amount of energy flowing out of the system. An energy budget, as with a monetary budget, is an accounting of inputs and outputs. A steady-state condition is like a balanced financial budget. Because the Sun supplies all but 0.02 percent of Earth's energy, the inflow and outflow of solar energy are a close approximation of Earth's energy budget.

As shown in Figure 13-5, more than half of all incoming solar radiation is returned to space before doing any work at Earth's surface. About 25 percent is reflected to space by particles or clouds in the atmosphere and another 5 percent is reflected by Earth's surface, while 25 percent is absorbed by clouds and the atmosphere. This leaves about 45 percent of incoming solar radiation to be absorbed by Earth's surface. About two-thirds of this absorbed energy (29 percent of the original incoming radiation) is cycled through Earth's environmental systems and reradiated, while one-third (16 percent of the original incoming radiation) is converted to heat (thermal energy).

The wavelengths of reradiated energy are longer than those of the original (or reflected) solar radiation. Because clouds and greenhouse gases such as carbon dioxide trap energy at longer wavelengths relatively easily, reradiated energy has a strong influence on Earth's climate. If you add the percentages in Figure 13-5, you can see that inflow of solar energy is equal to outflow, with 30 percent of the inflow leaving Earth as reflected energy and 70 percent leaving as reradiated energy. Earth's energy budget is balanced.

Energy Transfers and Photosynthesis

The flow of energy through the biosphere, pedosphere, and lithosphere is of special human concern because these systems are the sources of food and fuel energy. The biosphere depends on energy stored during **photosynthesis,** the process by which chlorophyll in green plants converts solar energy into chemical energy in the form of organic (carbon-based) molecules (see Chapter 7). The simplest photosynthetic reaction produces a carbohydrate called glucose ($C_6H_{12}O_6$). The energy stored in glucose and other organic molecules can be released if the carbon bonds are broken in a reaction with oxygen. *Respiration* is the process by which animals oxidize organic molecules to convert their chemical energy to heat, releasing carbon

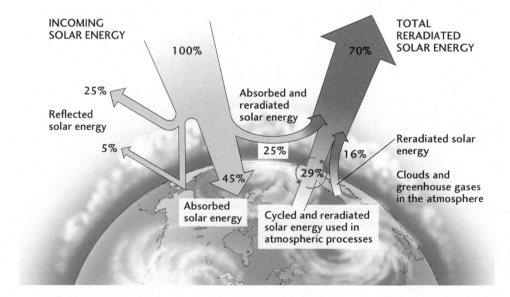

Figure 13–5 Some solar energy is reflected and some is absorbed and reradiated to outer space. Inflow of energy is equal to outflow, keeping Earth's energy budget balanced.

dioxide and water as by-products. This process is vital to animals, for they are not able to use solar energy directly to synthesize their own carbohydrates.

Decomposition and combustion are also processes that release heat energy by oxidizing organic molecules. *Decomposition* is the relatively slow oxidation of dead organic matter by microbes that take in oxygen from the atmosphere. This process is essential to soil and peat formation (see Chapter 8). *Combustion* is the much more rapid oxidation that occurs when organic matter ignites and burns, generating both heat and light. This process is of importance to understanding how wood, charcoal, and fossil fuels are used to extract the energy stored in biomass for a useful energy resource:

		Chemical energy released		
$6 O_2$ +	$C_6 H_{12} O_6$	$\longrightarrow$	$6 H_2O$ +	$6 CO_2$
(Oxygen removed from atmosphere, hydrosphere, biosphere, and pedosphere)	(Glucose removed from biosphere)	(Heat released to organism by respiration or to environment by decomposition or combustion)	(Water returned to hydrosphere)	(Carbon dioxide returned to atmosphere, pedosphere, and biosphere)

The energy released by this reaction is substantial. Oxidizing 1 kg of carbon—whether by respiration, decomposition, or combustion—releases about 33 million joules of stored energy, enough to drive an efficient subcompact car several tens of kilometers or more.

Because of the reciprocal reactions of photosynthesis and oxidation of organic matter, the energy inflow into the biosphere is nearly equivalent to the outflow. Of the 173,000 tW of solar energy that reaches Earth's surface annually, only 100 tW is stored in the biosphere through photosynthesis of organic molecules, contributing to the formation of soils and peat on land and organic oozes on the ocean floor. Most of the carbon in the biosphere, about 99.9 percent, is returned to the atmosphere as carbon dioxide after organisms die and decay as a result of microbial activity requiring oxygen.

A tiny remainder, 0.1 percent of the flux of carbon through the biosphere, is buried in sediments and protected from oxidation, becoming part of the lithosphere. There, organic matter may be transformed by pressure and heat into fossil fuels, including petroleum, coal, oil shale, and tar sand. Although the rate at which solar energy becomes stored in fossil fuels is small, it has accumulated in the lithosphere over millions to hundreds of millions of years, and so the total stock of fossil fuels is large. Examples of these stocks include the vast oil fields of the Middle East, the tar sands of Canada, and the oil shales of North Dakota. Extracting and burning fossil fuels releases stored energy that originated as sunlight and was used by organisms for photosynthesis long ago.

Resources, Energy Currency, and End-Use Services

Primary energy, such as that stored in coal or available from the flow of water and wind, must be converted to some form of *energy currency* in order to provide an end-use service. Crude oil, for example, is converted to motor gasoline at an oil refinery. It must be distributed for use, typically by pipelines, or in tanks carried on ships and in trucks. Once pumped into the tank of an automobile, gasoline can be combusted to release mechanical energy that renders a service, in this case the movement of a car from one point to another. The distance traveled is the work done with that energy source. As another example, coal is burned at a power station to generate electricity, another form of energy currency, and that electricity is distributed via transmission lines in order to provide end-use services such as lighting a lamp by which one could read a book.

The modern world currently relies on nonrenewable fossil fuels for about 82 percent of its primary energy needs (for heating, transportation, and electricity; see Figure 13-3). Nuclear power, renewable biomass, hydroelectricity, wind, and solar energy supply most of the rest. Nuclear electric power accounts for about 8 percent of world energy use. Although the uranium oxides from which nuclear power is derived contain the largest stock of nonrenewable energy on Earth, the risks of radiation exposure and challenges of radioactive waste disposal have impacted full-scale development of this energy resource. Geothermal energy from Earth's interior heat flow is essentially a perpetual resource at a few locations but is insignificant in the world energy tally. Direct solar energy and wind energy are perpetually available and environmentally sustainable, and their use is increasing worldwide. The origin, economic potential, and environmental effects of each of these energy resources, as well as others of local importance, are described in the following sections.

■ Three sources of energy are responsible for all processes on Earth: solar energy, Earth's internal thermal energy, and gravitational energy.

■ Solar energy contributes 99.98 percent of the energy that drives processes on Earth, but Earth's internal thermal energy drives plate tectonics, while the gravitational attraction among Earth, the Moon, the Sun, and other bodies in the solar system drives the tides.

■ Earth's energy system is in a steady state; the amount of energy received by the whole Earth system is approximately equal to the amount of energy flowing out of the system.

■ External energy from the Sun and Moon and internal energy from Earth's interior flow through Earth's systems and sometimes are stored temporarily before being lost to space as waste heat.

■ Energy flows into the biosphere through photosynthesis and is temporarily stored in biomass and fossil fuels in the form of organic carbon compounds. Respiration, decomposition, and combustion all release stored energy by oxidizing organic molecules.

■ About 82 percent of the world's energy supply is from fossil fuels, although the amount from renewable resources (e.g., biofuels, solar energy, and wind energy) has been increasing in recent years.

PETROLEUM: CRUDE OIL AND NATURAL GAS

The time period since the mid-20th century has been called the Petroleum Age, with remarkable technological advances made possible largely by the energy stored in petroleum (see Figure 13-2). **Petroleum** is a flammable liquid or gaseous form of organic matter composed of **hydrocarbons**—compounds consisting of carbon and hydrogen atoms (Table 13-1). The origin of the word is Greek, with *petro* for rock and *oleum* for oil. Commonly, *petroleum* refers to any hydrocarbon mixture that can be extracted through a drill pipe, and includes crude oil and natural gas. **Crude oil** is the unrefined liquid part of petroleum (in its early days of use sometimes called "black gold"), and **natural gas** is the gaseous form of petroleum. Both forms of petroleum—crude oil and natural gas—are easier to transport than coal, a solid hydrocarbon fossil fuel.

At Earth's surface, hydrocarbon molecules that have four or fewer carbon atoms occur as gases, whereas those with five or more occur as a liquid. Natural gas molecules contain less carbon and more hydrogen than oil molecules do, making it less dense, but both oil and natural gas are less dense than water and therefore both are able to float on it's surface. Natural gas is a mixture of several hydrocarbon gases, but consists mostly of methane (70 percent to 90 percent). In the methane molecule (CH_4), four hydrogen atoms surround one carbon atom. Ethane (C_2H_6), propane (C_3H_8), and butane (C_4H_{10}) are also part of the natural gas mixture, each with greater numbers of carbon atoms relative to hydrogen atoms. Crude oil (unrefined) consists of multiple chains, branches, and rings of hydrocarbon molecules,

	Natural gas	Crude Oil	Coal
Carbon (C)	76	84.5	81
Hydrogen (H)	24	13	5
Sulfur (S)	trace–0.2	1.5	3
Nitrogen (N)	0	0.5	1
Oxygen (O)	0	0.5	10

TABLE 13–1 Average Elemental Composition of Fossil Fuels (% by weight)

*Source: From John M. Hunt, *Petroleum Geochemistry and Geology*, New York: W. H. Freeman and Company, 1995.

with up to 60 carbon atoms in a given molecule. As discussed later, crude oil is refined through thermal and chemical processes to generate different fuel products, such as gasoline, jet fuel, kerosene, and diesel fuel. These products can be separated from the crude oil because of their different boiling points.

A few terms are used to refer to different types of oil and natural gas production. Natural gas occurs alone in subsurface reservoirs, but it also can be found mixed with oil. When mixed, the natural gas is dissolved in crude oil and can be trapped above the liquid (see discussion of geologic traps on following pages). Natural gas that is found with oil during production is referred to as *associated natural gas*. In the United States, however, most natural gas (about 80 percent) that is produced is *nonassociated*, that is, not associated with oil production.

To date, most oil and natural gas production have come from high-permeability reservoirs and are referred to as **conventional oil and natural gas**. Increasingly, more oil and natural gas are extracted from low-permeability deposits and are referred to as **unconventional oil and natural gas**. Examples of unconventional natural hydrocarbons discussed later include tight oil and shale gas.

Origin of Petroleum

Scientists who examined the oil produced from early wells were puzzled by its origin. Its composition indicated an organic origin, but the hydrocarbons in petroleum are chemically somewhat different from carbohydrates, proteins, and other organic molecules. The key difference is the abundance of oxygen in other organic compounds and its nearly complete absence in petroleum (see Table 13-1). Nevertheless, many studies have shown that hydrocarbons are produced by the transformation of different kinds of organic matter. During transformation, oxygen is lost from the organic molecules, commonly in the form of water vapor.

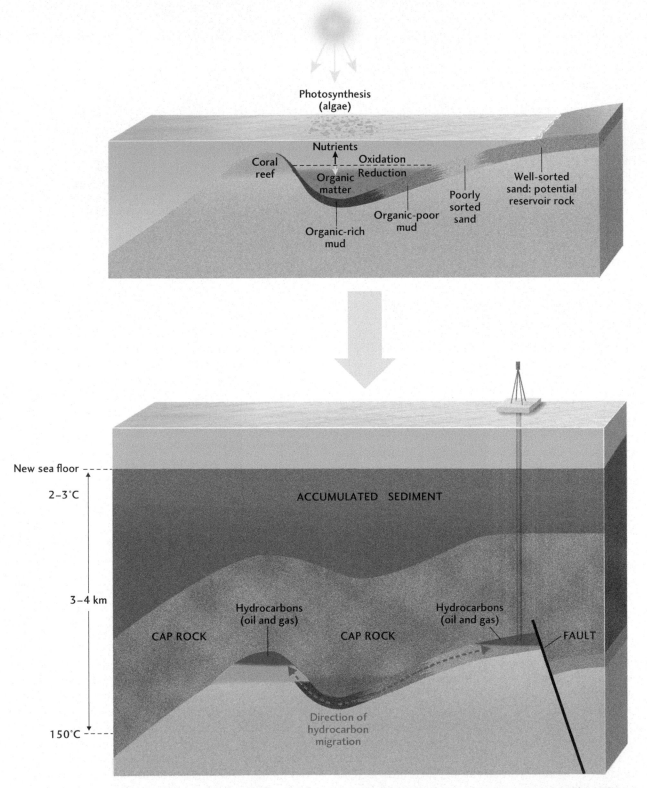

Figure 13–6 Marine environments where potential petroleum source and reservoir rocks might occur are shown. In a depression with restricted circulation landward of a limestone reef, mud and unoxidized organic matter are trapped and, as a consequence, accumulate with time. Eventually, the reef and depression are buried by sediment and, in this example, the strata are folded and faulted. The reef, fold, and fault form traps for petroleum as it migrates (red arrows) from source to reservoir rocks after maturation.

The most common source of organic material for petroleum is planktonic algae, or phytoplankton, the single-celled photosynthesizing marine organisms that are abundant along continental shelves (see Chapter 12).[6] When plankton and other marine organisms sink to the ocean floor after death, most are eaten or decomposed, but in restricted water that is depleted of oxygen, scavengers seldom venture and decay is unlikely.[7] Such environments exist in deep basins along continental shelves, where fine-grained sediments (silt and clay-rich mud) and organic matter accumulate and are buried over prolonged periods of time (Figure 13-6). The petroleum-rich Monterey formation on the California coast, for example (Figure 13-7), is the uplifted remnant of such a sedimentary basin. The thick sediments at river deltas also favor hydrocarbon formation. Some 90 metric tons of plant matter is needed to produce about a gallon of gasoline (a product of crude oil refining).

Coastal swamps, bayous, river plains, and deltas also are associated with the production and trapping of large amounts of organic matter that might avoid oxidation if buried in wetlands. Terrestrial plants, including trees, might be converted during burial to coal, but oil and gas also can be produced by their decomposition. The Mississippi delta's offshore extension into the Gulf of Mexico has trapped sediment and organic matter from the North American continent for 150 million years, and today yields abundant supplies of both oil and natural gas.

How are substances such as microscopic aquatic plants transformed to become the oil and natural gas that eventually are used in our home furnaces, automobiles, or power plants? As mud and organic matter are deposited in sedimentary basins, the deeper and older layers are subjected to increasing pressure from overlying sediment and increasing temperature from Earth's internal heat (see Figure 13-6). A number of changes occur in the sediments as a result. First, the pressure expels water from the pore spaces—just as squeezing forces water from a sponge—and compresses the sedimentary particles into an increasingly firm mass. Below about 1000 m, the sediments are compacted into mudstone and shale. Some clay minerals contain water, but below about 3000 m, water is squeezed from clay particles and into adjacent pore spaces, generating very high pressure in the sediments. If a well is drilled to these depths, the localized release of pressure can push the water and any oil and gas more than 1000 m above the ground surface. Such *blowouts,* discussed later, are extremely dangerous to workers on oil rigs, because the hydrocarbons are highly combustible.

While water is squeezed out of pore spaces and migrates upward through cracks in the compacted sediments, organic compounds remain behind, stuck to clay grains. They, too, are heated and squeezed during burial. The chemical bonds between their atoms begin to break at depths greater than about 1000 m, depending on the local rate of temperature increase with depth below Earth's surface. As the bonds break, lighter organic molecules are formed that become the components of hydrocarbons. At temperatures between about 60°C and 120°C, oil is generated, and at temperatures greater than about 100°C, natural gas forms. For typical geothermal gradients (between 1.8°C and 5.5°C per 100 m), petroleum is most likely to form at depths between 1000 and 8000 m.

Oil and gas are less dense than the *source rock,* the parent rock where the hydrocarbons were generated, so the buoyant fluids will rise and seep out at Earth's surface unless trapped by a relatively impermeable layer, called *cap rock.* Typical cap rocks are fine-grained or crystalline, and include shale, mudstone, and salt. Any geologic structure that confines the flow of petroleum, causing it to pool beneath cap rock, is a **petroleum trap.** Petroleum migrates upward from the source rocks where it formed, generally dark marine shales and mudstones. These same rocks can act as traps to hydrocarbons that migrate from other source rocks.

The most common petroleum trap, the **anticlinal trap,** forms where sedimentary layers of variable permeability are folded into a dome or arch called an *anticline*

Figure 13–7 The Monterey formation, a sequence of banded sedimentary rocks named for the town of Monterey, occurs throughout much of western California and is an important source of fossil fuels. Geoscientists have calculated that 1 m of the banded rock represents nearly 5000 years of sedimentation. The average thickness of the formation is 300 m. Layers in this photograph represent approximately 30,000 years of geologic time. *(Courtesy Richard J. Behl, California State University Long Beach)*

(see arched rocks on the bottom of Figure 13-6). Petroleum moving up through an anticline pools in the structure's more porous and permeable rock layers. These *reservoir rocks* generally are sandstones and fractured limestones. A **fault trap** occurs where the movement of rock along a fault juxtaposes permeable reservoir rocks with impermeable cap rocks (see Figure 13-6). A **salt dome trap** forms when halite (NaCl), which accumulates in sedimentary deposits via evaporation of ion-rich water, flows upward through surrounding rock and forms large, impermeable, dome-shaped masses. Halite flows upward for the same reason that water and hydrocarbons do—it is less dense than surrounding rocks. It is also somewhat plastic because layers of sodium chloride molecules glide past one another under pressure. When subjected to tectonic forces and the weight of overlying sediments, the salt flows, pushes against the denser and more rigid surrounding rock, and causes it to fold into broad, dome-shaped structures with impermeable salt cores. Petroleum pools between the outside walls of the salt core and the inner walls of the surrounding rock.

Finding, Extracting, and Refining Petroleum

At the start of the 20th century, wells were drilled where petroleum was found seeping out at the surface. Seeps occur where the Monterey formation crops out along the California coast, for example, and early in the 20th cen-

tury oil wells dotted the landscape in southern California (see Figure 13-7). However, oil seeps are shallow and their yield typically is small. For deeper deposits, prospectors would try drilling at the top of the highest hill, because petroleum tends to collect beneath arched rocks. High hills, however, do not necessarily have arched rocks located beneath them.

A few geologists began to suspect that the location of oil hundreds of meters below the ground could be inferred from mapping the types and orientations of rocks at Earth's surface, a method called *structural mapping*. They combined their maps with information about subsurface rocks from existing drill holes. Structural mapping provided a more reliable means of finding arched rocks than the occurrence of hills. By 1916, oil companies had hired teams of geologists to map areas with oil potential, commonly associated with folded or faulted rocks. Geologists have remained in charge of oil exploration decisions ever since, predicting and finding oil both onshore and offshore throughout the world, in Saudi Arabia, Alaska, Texas, Venezuela, the North Sea, Indonesia, and the region of the Gulf of Mexico and the Caribbean.

Today, geologists use a variety of technologies to find potential petroleum reservoirs. One such exploratory method is the use of *seismic reflection* experiments (Figure 13-8). In this technique, energy is produced at Earth's surface, usually by explosives on land or a sound-generating device offshore. The sound waves produced from the energy source radiate outward through the

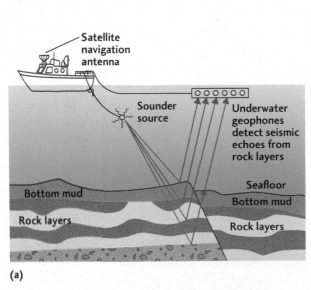

(a)

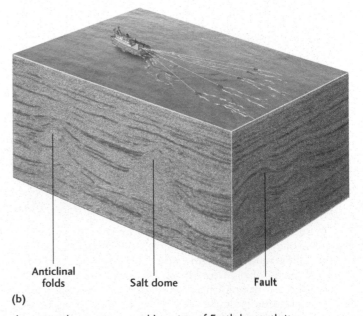

Anticlinal folds   Salt dome   Fault

(b)

Figure 13–8 (a) A ship towing a sound source and underwater geophones acquires data that are used to produce three-dimensional seismic reflection diagrams depicting the structure and layering of Earth beneath its surface. (b) Geologists use such data to find potential petroleum traps.

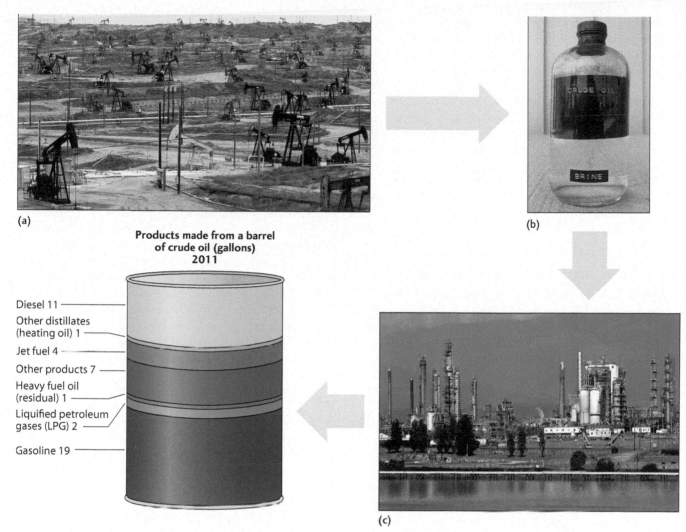

(a)

(b)

Products made from a barrel of crude oil (gallons) 2011

Diesel 11
Other distillates (heating oil) 1
Jet fuel 4
Other products 7
Heavy fuel oil (residual) 1
Liquified petroleum gases (LPG) 2
Gasoline 19

(c)

Figure 13–9 Drilling rigs (as here in Bakersfield, California) (a), pump crude oil (b) that is sent to refineries (c) where it is separated into usable products that include gasoline, diesel fuel, and jet fuel. A U.S. barrel (bbl) of crude oil is 42 gallons, but yields 45 gallons of petroleum products. *(a: Ken James/Bloomberg via Getty Images; b: Frank Relle; c: Thinkstock)*

underlying material, reflecting off different layers and returning to the surface at different times. The arrival times of the reflected seismic waves are recorded by special receivers called geophones. Combining the data from an array of geophones, geologists can construct a three-dimensional picture of the material beneath Earth's surface, helping them to locate potential petroleum traps and reservoirs.

Seismic reflection data and computer technology have been used to find petroleum in many types of sedimentary basins. For example, three-dimensional models generated by seismic reflection experiments can reveal the location of ancient deltaic distributary channels made of sand. Each of these might be a potential petroleum reservoir to be discovered during drilling. With this information, exploration geologists recommend where and how deep to drill, then wait anxiously as drillers approach that level to see if their reasoning leads to another source of oil or if they have a dry hole. While some holes bear oil, many are dry. Even with greater understanding of sedimentary basins and advanced computer technology, finding oil still requires a bit of luck.

Oil and natural gas can be found on land and at sea (Figure 13-9), but ocean drilling is more expensive, requiring the use of large platforms. The use of highly sophisticated technology has led to extraction of petroleum from undersea areas at increasingly greater water depths (now exceeding 2000 m) and total well depths (now more than 6000 m) over the past few decades. However, extracting and transporting petroleum in the ocean can cause substantial harm to the marine environment if an accident occurs, as discussed below.

Once extracted, crude oil is transported—typically by ocean tanker, pipeline, rail, or truck—to refineries (see Figure 13-9). Oil refineries process the crude oil into a number of compounds, boiling it and condensing its vapors in a series of stages so that progressively lighter and smaller hydrocarbon molecules can be collected and separated. From heavier to lighter carbon compounds, petroleum products include lubricating oils, diesel fuel, kerosene, and—most importantly at present—gasoline for automobiles. Natural gas is transported almost exclusively by high-pressure pipeline because of its low density and gaseous state. About 305,000 miles of pipeline exist in the United States today to transport natural gas.

Peak Oil and Gas: How Long Will Crude Oil and Natural Gas Last?

Geologist M. King Hubbert is said to have commented that he found it hard to decide which is more remarkable, that it took 600 million years for the Earth to make its oil, or that it took 300 years to use up much of it.[8] Oil is a nonrenewable resource: the amount that exists today is finite, and the rate of its replenishment is very slow relative to the scale of human interest. As a result, someday the world could run out of oil. The timing of that "someday" is the source of much debate among scholars. Predicting when we might run out of oil requires that we know how much exists and how much we might use each year until all of it is consumed. If we were to stop using oil immediately, for example, then the amount that remains would not be consumed. If we were to start using it at higher rates, then it would take less time to consume what remains.

How much of this precious resource exists, how much of it already has been used, and most importantly, how much remains for future use? Analysts estimated in 2011 that 1474 billion barrels of crude oil could be recovered from known fields.[9] This amount is called the oil **reserve**, because its existence is known and current technologies make its recovery economically viable under existing economic and operating conditions. In other words, reserves are the best current estimate of the amount of a resource that can be produced given current prices and technology. In comparison, total global use of crude oil from 1870 to 2009 was at least 944 billion barrels.[10]

If the estimated amount of recoverable oil is only about 56 percent more than has been used in the past approximately 140 years, and given that the rate at which we use oil has increased during that time, how much longer will supplies last? The world consumes 32 billion barrels of oil each year. At that rate, it would take only about 46 years to use up the estimated 1474 billion barrels of oil remaining. However, the rate at which the world uses oil has increased steadily since the first wells were drilled, because the global population is increasing and developing countries are striving to raise their standard of living. If the rate of consumption continues to increase as much as it has in recent years, the amount of recoverable oil that is left might be used up in even less time.

On the other hand, balancing the alarm over oil depletion are new and "enhanced recovery" technologies, such as extraction of oil from deeper water locations or horizontal drilling that enables greater recovery from a reservoir. Substitutions for oil are growing, as with the increased use of the lighter component of petroleum, natural gas (see Figure 13-2). Furthermore, undiscovered resources might exist in places where the geologic conditions are right, but where little or no exploration has yet been done. Some U.S. federal lands that might have technically recoverable petroleum are currently protected and not available for exploration and drilling. The possibility of extracting from them has led to contention and debate between those seeking to preserve exceptional value ecosystems and those advocating the production of resources at those locations (Box 13-1).

The *ratio of reserves to production* (known as the RPR or R/P ratio) is a key indicator in the energy industry and can be used for any nonrenewable resource. Because the RPR is a ratio of a quantity (reserves) and the production of that quantity for a unit of time (e.g., quantity/year), the units of the ratio are in years. For oil, the world RPR estimated in 2011 was 54 years, and for natural gas it was 64 years.[11] (The world estimate for coal, a solid hydrocarbon, was 112 years.) The RPR is not synonymous with how many years of a resource's use remain, because both the estimate of reserves and the rate of production change over time.

Many analysts discuss when the amount of global oil production might reach its peak. Most predictions vary only by several decades, ranging from about 2025 to 2050.[12] This concept of "**peak oil**" has garnered substantial attention from the media and public. Hubbert himself predicted in a now famous paper published in 1956 that the U.S. production of oil in the lower 48 states would increase exponentially until about 1965, when it would "peak," and after that time the production would diminish.[13] In essence the peak is the point at which demand outpaces production, or the "beginning of the end," when the end of production is in sight.[12]

Hubbert made a graph to illustrate the rate of depletion with respect to the amount of the resource available (i.e., its reserves). Now known as "Hubbert's curve," this graph shows time on the horizontal (x) axis and petroleum production on the vertical (y) axis (Figure 13-10a). Hubbert's curve can be modeled mathematically with a *logistic formula* in which only three parameters are

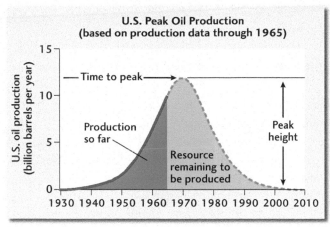

(a)

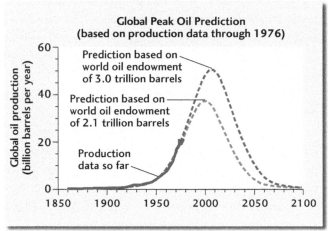

(b)

Figure 13–10 (a) Hubbert's curve shows U.S. oil production, using data through the mid–20th century. (b) Global oil production, using data through 1976, showing Hubbert's prediction of global oil depletion.

required to make the curve: the slope, the height of the peak, and the timing of the peak. Note that the mathematical curve is symmetrical; the rising and falling limbs are mirror images of one another. Hubbert assumed that oil production (rising limb) and depletion (falling limb) follow a similar trend.

Historical production data for a given resource, in this case oil, are used to determine the slope (i.e., the rate of increase in production with time). When Hubbert developed his curve in the mid-20th century, production data prior to that time determined the slope of his curve (see Figure 13-10a). The area under the curve represents the total amount of the resource that is estimated to be discoverable and technically extractable (i.e., the resource endowment). If all the oil available

for use was produced and added up, it would be equal to the resource endowment, or area under the logistic curve. This amount determines the height and timing of the peak if the curve is assumed to be logistic and symmetrical about a peak in production. In his 1956 paper, Hubbert estimated the resource endowment for oil in the conterminous United States to be 150 to 200 billion barrels. The higher possible resource endowment adds only a few years difference in the timing of the peak, however, from 1965 to 1971. Production of crude oil in the United States did indeed peak in 1970, within the time period predicted by Hubbert.

Hubbert evaluated world peak oil data in 1976, using estimates of 1.35 to 2.1 trillion barrels for the resource endowment, and made a curve similar to that he had made earlier for the United States (Figure 13-10b). He predicted the time of peak oil production for the world as 1995 to 2000, with the exact year depending upon the size of the resource endowment. A quarter-century later, however, the estimate of world oil resource endowment had increased nearly 50 percent, from 2.1 to 3.0 trillion barrels. Nevertheless, it only increased the prediction for time of peak by a few years, to the year 2005. By 2012, however, the world oil endowment was estimated to be even larger and oil production still had not peaked.

Some economists and scientists argue that regardless of a peak in oil production, we will never run out of oil because we will have switched to cheaper, cleaner, more reliable substitutes for oil before that happens. In essence, the production rate might diminish rapidly, perhaps so quickly that the rate of decline would not be a mirror image to that of the increase in production with time on a graph. One argument is that new technologies are likely to make the use of substitute power sources increasingly feasible. In other words, the resource that is being depleted might become obsolete at the same time it becomes scarcer. In the case of fossil fuels, it's also possible that increasing concern about the environmental effects of their use, such as global warming from greenhouse gas emissions, might also play an important role.

Looking back on the history of fossil fuel use, Christopher Flavin and Nicholas Lenssen of the Worldwatch Institute reflected on what might happen in our energy future:

History suggests that major energy transitions—from wood to coal or coal to oil—take time to gather momentum. But once economic and political resistance is overcome and the new technologies prove themselves, things can unfold rapidly. If the past is any guide, unexpected events, new scientific developments, and technologies not yet on the drawing board could push the pace of change even faster than we have suggested.[14]

BOX 13–1 EARTH SCIENCE AND PUBLIC POLICY

The Essential Trade–off between Fossil Fuels and the Environment

Described as an American Serengeti that pulses "with million–year–old ecological rhythms,"[15] the Arctic National Wildlife Refuge (ANWR) teems with caribou, muskoxen, wolves, polar bears, snow geese, fish, and other wildlife (**Figure 1**).[16] For decades, Congress has debated whether or not to open ANWR's coastal plain for petroleum exploration and production. Because of ANWR's status as a protected area, however, oil companies must get congressional approval in order to explore and drill for petroleum.

First established as a National Refuge through the Alaska National Interest Lands Conservation Act (ANILCA) in 1980, ANWR is now the largest National Wildlife Refuge, with 78,000 km^2 of wetlands, coastal plain, and river deltas (**Figure 2**). U.S. National Wildlife Refuges are protected areas that are managed by the U.S. Fish and Wildlife Service. Begun in 1903, this system of public lands and waters has been set aside for the purposes of conservation and management of fish, wildlife, and plant resources, with protection also provided for overall biological integrity and biodiversity. About 40 percent of the ANWR coastal tundra is protected further as a national Wilderness Area, a special

designation given to places that are pristine and untrammeled by human activity.

In addition to a plethora of ecological resources, the continental shelf along ANWR has the conditions favorable to formation and preservation of petroleum. In Section 1002 of the ANILCA, Congress deferred decisions for future management of one part of ANWR, an area of 6100 km^2 referred to as the 1002 area. The possibility for future drilling remained open for this area because of its potential for holding significant amounts of petroleum. The area also is environmentally significant, because at least 200 different species, including caribou, muskox, and many water birds, migrate through and breed in it.

Oil companies argue that tapping this field could reduce American reliance on imported oil and provide jobs and revenue for Alaskans. Environmentalists counter that opening this area of irreplaceable biodiversity to exploration and production would result in construction of numerous roads, pipelines, drilling platforms, and related extraction and transportation infrastructure that would lead to inevitable environmental degradation. Furthermore, they argue, the oil that might be produced would only meet domestic demand

Figure 1 At ANWR (Arctic National Wildlife Refuge), caribou migrate north along Kongakut River to calving ground in early summer. *(John Schwieder/Alamy)*

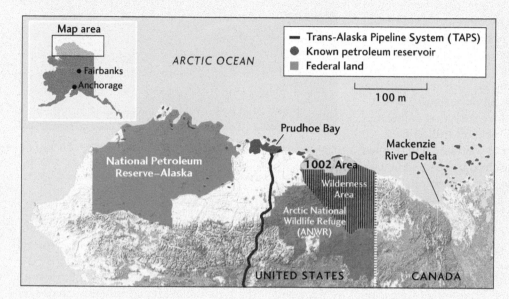

Figure 2 Map of northern coast of Alaska showing the National Petroleum Reserve, ANWR (with its encompassed wilderness area), and the Trans–Alaska Pipeline. ANWR is the largest National Wildlife Refuge in the United States. Future management and resource extraction in the area designated as Section 1002 are controversial, with drilling for oil still a possibility.

for several months to years. A comparable amount could be saved by increasing the miles–per–gallon fuel requirements for new automobiles, or by making power plants and buildings more energy efficient.

Estimating how much as yet undiscovered petroleum might be recovered from ANWR is central to this debate. In a 1998 report and later updates, the U. S. Geological Survey used regional geologic information to estimate the amount of crude oil that might be technically recoverable (i.e., *prospective resource*) from the ANWR coastal plain.[17] Their estimate of 5.7 to 16.0 billion barrels is about 5 – 13 percent of that for undiscovered, technically recoverable oil that is estimated to exist in the rest of the United States (approximately 120 billion barrels). The 1002 area of ANWR is estimated to have only 4.3 to 11.8 billion barrels of technically recoverable undiscovered oil. The low and high estimates represent the 5% and 95% probability range. *Proved reserves* of crude oil and natural gas liquids for the United States, in comparison, are 29 billion barrels.

As with other environmental topics raised in this book, scientific information and analysis are essential to informed decision–making. Choices about whether or not to use a resource in an area of environmental value, on the other hand, boil down to evaluating what the National Academy of Sciences has referred to as the "essential trade-off". In 2003, at the request of Congress, the National Academies Press released a report from the Committee on Cumulative Environmental Effects of Oil and Gas Activities on the Alaska North Slope. Congress requested that the committee review information about oil and gas activities on the North Slope, and assess possible cumulative impacts of these activities on the region's environment and resources. A key finding in the report was that industrial petroleum development results in undesirable, inevitable effects on the physical and biotic environments, as well as on human societies in the region. Furthermore, these effects accumulate over time due to numerous small decisions, typically made independently of one another. Cumulative effects create patterns of environmental perturbation that can be spread over large areas and persist over long periods of time. Referring to the essential trade–off for the Alaska North Slope in particular, the report concludes: "Whether the benefits derived from oil and gas activities justify acceptance of the inevitable accumulated undesirable effects that have accompanied and will accompany them is an issue for society as a whole to debate and judge."[18]

Unconventional Fluid Hydrocarbons: Oil Sands, Shale Oil, and Gas

Recall that unconventional hydrocarbons are those that are extracted from low-porosity, low-permeability geologic reservoirs or acquired by other unconventional means. Oil sands typically are unconsolidated or partly consolidated sandstones that contain a dense form of viscous hydrocarbon called bitumen. Tight sands are hard, highly compressed sandstones with relatively low permeability that might contain oil or natural gas. Shale is a fine-grained, dense sedimentary rock that sometimes has sufficient organic matter to yield oil (called *tight oil*) and natural gas (called *shale gas*). Methane gas trapped in coal seams is referred to as coalbed methane. Here we discuss oil sands and shale gas in more detail, as production of these unconventional sources has increased substantially in recent decades.

Figure 13–11 A sample of the tar–like substance from oil sands at the Syncrude Canada Ltd. mine in Alberta, before it is processed into crude oil. *(Lara Solt/Dallas Morning News/Corbis)*

Bituminous (Oil) Sands A type of petroleum deposit that occurs as dense, extremely viscous **bitumen** (commonly called tar) in some unconsolidated sands and partly consolidated sandstones is called **bituminous sands, oil sands** or **tar sands**. These different names for hydrocarbons that resemble molasses are generally synonymous. Bitumen seeps out at Earth's surface along riverbanks and other low places, and has been used by humans for millennia. In ancient Egypt and Mesopotamia, for example, it was used to seal the seams on boats so that they would not leak.

Early geologists in Canada explored seeps of bitumen from Cretaceous sandstone along riverbanks, and they estimated that enormous amounts of oil might be stored in this rock formation (Figure 13-11). Today, proven reserves of bituminous sands are estimated to be about 250 billion barrels worldwide, with nearly 71 percent in Canada.[19] Remarkably, 98 percent of Canada's oil reserves occur in oil sands. Oil sands became economically and technically recoverable in 2003, increasing Canada's proven reserves of oil by an order of magnitude.[20] In 2007, about 44 percent of Canadian oil production was from oil sands. Large deposits of oil sands also occur in other countries, with large reserves in Kazakhstan and Russia, but even these are small in comparison to those in Canada. At present, only Canada produces oil from oil sands at a large commercial scale.

Too viscous to flow and be pumped from a well, the semisolid to solid state hydrocarbon in bituminous sands is an *unconventional petroleum deposit*. Instead, it must be extracted either by surface mining, much the same way that coal is strip-mined, or by various in situ (underground) techniques that inject steam, hot air, or solvents into the sand in order to reduce its viscosity so that it flows to wells (Figure 13-12). Surface mining is viable where oil sands are shallow (up to several tens of

meters deep), and in situ techniques are used if the oil sands occur at greater depths.

Extracting oil from oil sands generally requires more energy and water than conventional oil extraction. This is the reason that the process only became economically viable in the past decade, as the price of oil rose on the world market. In Canada's largest known deposit, the Athabasca oil sands, about two tons of sand are mined for each barrel of oil produced and more than 90 percent of the bitumen is recovered from the sand. As with the tailings generated in other types of surface mining, the sand and bitumen that remain are returned to the mine site and reclaimed. Dilutants are added to the bitumen so it can be sent via pipeline to a refining facility for further upgrading.

Shale Oil and Gas Shale is a fine-grained, dense rock that sometimes has sufficient organic matter to yield oil and natural gas, but getting wells that are economically viable from such low-permeability rock requires new technologies that include horizontal drilling and rock fracturing. With deregulation of the natural gas industry in the 1990s, improved extraction technology, and construction of cleaner burning gas-fired power plants, drilling in unconventional gas reservoirs has increased in recent decades. Supporters of oil and gas from shale in the United States also point to the value of relying more heavily on domestic than foreign fuel sources, especially those foreign sources where governments are politically unstable. The Energy Information Administration projects an increase for unconventional gas projects from 44 percent to 50 percent of U.S. supplies from 2008 to 2018, respectively.

In conventional oil and gas reservoirs, liquid hydrocarbons migrated from organic-rich source rocks to permeable reservoir rocks and were trapped there by an

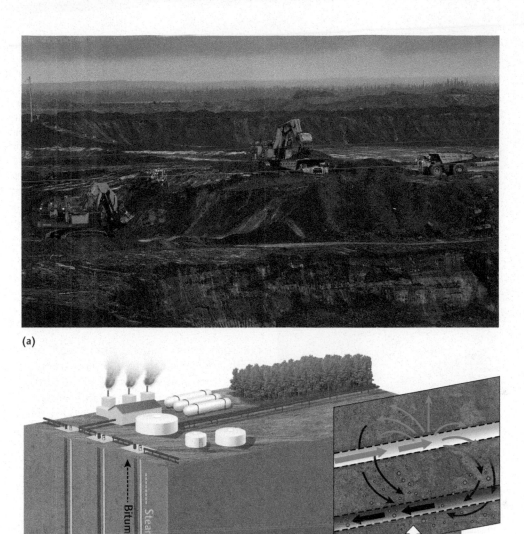

(a)

(b)

Figure 13-12 (a) The Shell Oil Jackpine open-pit mine in Canada uses trucks that are three stories tall, weigh 1 million pounds and cost $7 million each. (b) Steam-assisted gravity drainage (SAGD) is a thermal production technology that extracts bitumen from subsurface oil sands deposits. SAGD uses a horizontal well pair configuration—an upper injection well and a production well drilled, in parallel, 5 m below. Steam is injected under high temperature and pressure into the oil sands reservoir, heating the reservoir, reducing the bitumen's viscosity and enabling it to flow into the production well. Water to generate the steam is typically sourced from deep reservoirs in the area. SAGD development plans involve recycling more than 90% of the water required for steam generation. SAGD production technology has a significantly smaller surface disturbance footprint than oil sands mining operations, and doesn't require tailings ponds (see Box 13-2). *(a: Michael S. Williamson/The Washington Post via Getty Images)*

adjacent relatively impermeable rock. The organic-rich source rocks themselves can contain crude oil and natural gas if some portion of these fluid hydrocarbons were inhibited from migration by the low permeability of the rock in which the organic matter originated. An example is the occurrence of oil and natural gas in shale, a fine-grained sedimentary rock.

The hydrocarbons in shale are distributed through-out the rock, mostly in small pore spaces that include natural fractures, but some natural gas is attached to

(a)

Figure 13-13 (a) The Devonian age Marcellus formation is a brittle shale with multiple thin fractures, some oriented vertically. The combination of horizontal drilling and hydrofracking enables more oil and gas to be pumped from greater volumes of the rock to production wells. (b) The Marcellus shale was deposited as mud in an inland sea that was located west of an ancient mountain belt. Large amounts of organic matter that accumulated in this sedimentary basin were altered at high temperature and pressure to become natural gas. Contours show that the shale thickens from west to east, and is thickest in eastern Pennsylvania. *(a: ©Lvlock/Graeme Teague Photography; b: U.S. Energy Information Administration)*

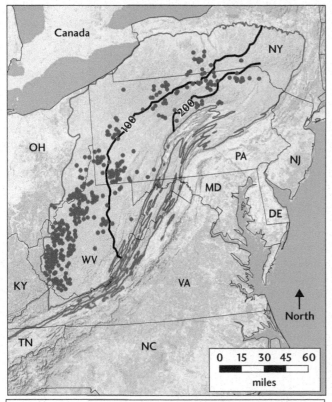

• Wells producing from the Marcellus Shale (some also commingle production from other formations)

— Marcellus Shale gross thickness, ft (Wrightstone)

— Extent of Marcellus Shale (USGS, Wrightstone)

(b)

(adsorbed on) mineral grains and organic matter. As with conventional fluid hydrocarbons, those found in unconventional sources were buried deeply by geologic processes such that heat and pressure converted the organic material into hydrocarbons. The pore spaces in shale are small and poorly connected, so fluid hydrocarbons do not drain out of them easily or quickly. Shales are brittle, however, and can be highly fractured as a result of tectonic processes over long periods of geologic time. Fractures provide a conduit for hydrocarbons to flow through shale.

An example is the Marcellus shale, a black Devonian age rock formation in the Appalachian Mountains that spans states from Tennessee and Kentucky in the south to Pennsylvania and New York in the north (Figure 13-13). During the Devonian period, a shallow inland sea existed in this region. As a result of plate tectonic movement that occurred over millions of years, mountains rose in the east, shedding muddy sediment westward that gradually filled the sea. Today this organic-rich muddy sediment forms the Marcellus formation, shale that typically lies at least a mile deep in Earth's crust and varies from a few tens of feet to nearly 900 feet in thickness. Thin fractures in the Marcellus shale oriented at different angles to one another produce a network of fractures from which natural gas could drain from a relatively large volume of rock and flow to wells. However, most of the fractures are vertical, so a borehole that is drilled vertically intersects relatively few of them.

Drilling a well horizontally (after going to a certain depth with a vertical well) results in many more intersections with vertical fractures and hence a greater yield of fluid hydrocarbons (Figure 13-14). Although horizontal drilling was done as early as the 1930s, investment in research and development by the federal government led to substantial improvements in directional and

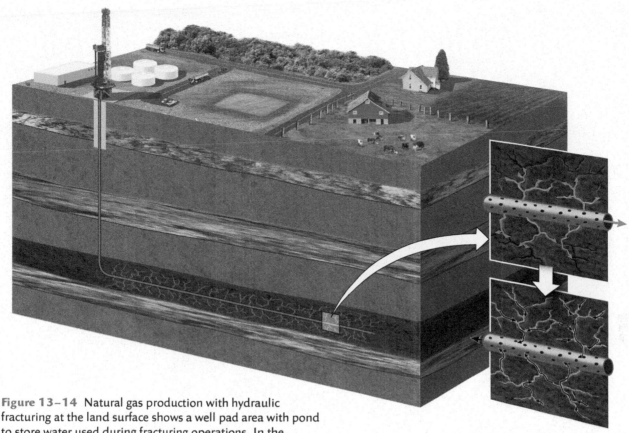

Figure 13–14 Natural gas production with hydraulic fracturing at the land surface shows a well pad area with pond to store water used during fracturing operations. In the subsurface, the well can be seen to bend to become horizontal. From this horizontal pipe, the rock formation is fractured and a slurry of water with chemicals and proppants is forced into the rock under high pressure. As a result, fractures are propped open to release natural gas.

horizontal drilling techniques since the 1970s. Today, horizontal drilling can extend laterally up to 2 miles within shale.

Combined with hydraulic fracturing, it now is possible to extract large amounts of oil and natural gas from great volumes of shale. Also known as fracking or hydrofracking, hydraulic fracturing is a technique that pumps millions of gallons of water with chemical additives and sand or ceramic beads (called *proppants*) down a well and into the surrounding shale after microexplosions have been used to create multiple small, artificial fractures in the rock. Sand or ceramic beads are forced into the microfractures and used to prop, or hold open the fractures, so that more oil and gas can be drawn up the well to the surface. Today, 90 percent of natural gas wells in the United States are developed through hydraulic fracturing.

This combination of horizontal drilling and hydraulic fracturing has led to a boom in oil and natural gas production from U.S. shale formations in Texas (the Barnett and Eagle Ford shales), North Dakota (the Bakken shale), and the eastern United States (the Marcellus shale) (Figure 13-15). In 2000, 1 percent of natural gas was from shale gas; as of 2011, it was 30 percent, and the U.S. Energy Information Administration predicts that shale gas might account for nearly half of the U.S. supply of natural gas by 2035.[21]

The state of North Dakota now has one of the lowest rates of unemployment in the United States, thanks to thousands of new jobs from the oil and gas industry associated with horizontal drilling, hydrofracking, water transport, and transportation of produced oil and natural gas, among other related activities. The state is installing thousands of kilometers of new pipeline in order to transport the growing supply of oil produced each year. In March of 2013, after its oil production doubled in a mere two years, North Dakota surpassed Alaska to become the nation's second highest producer of oil, after

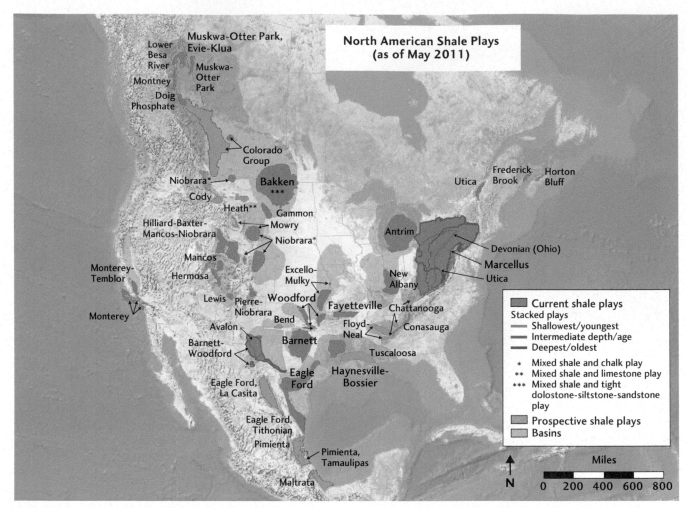

Figure 13–15 Numerous shale formations in North America contain oil and natural gas that can be extracted with a combination of horizontal drilling and hydraulic fracturing. These rocks are known as "plays." Those that are both thickest and shallowest usually are the most highly desirable for drillers and producers. *(U.S. Energy Information Administration)*

Texas, the top-ranking state. A map of North Dakota with more than 3,000 active and inactive oil wells and their one- to two-mile-long lateral extents in the subsurface reveals the tentacles of hydraulic fracturing underground (Figure 13-16). Some estimates are that North Dakota eventually will have many tens of thousands of oil wells.

Immense amounts of water are needed to develop a well with hydraulic fracturing. A single well might use between 1.2 and 5 million gallons of water, and more is needed if the well is refractured during its lifetime.[22] About 10 to 30 percent of the water that is used to "frack" a well is returned to the surface with the hydrocarbons. The rest remains underground. The water that returns, known as **flowback,** has a different composition than the water used initially. For the Marcellus shale, for example, this flowback contains salt and other naturally occurring elements. Investigations of this flowback have shown that it has the chemical signature of ancient brine, a result of its marine origin when the muddy sediment was deposited during the Devonian period.[23]

Environmental Impacts Associated with Petroleum Use

Environmental impacts occur during extraction, transportation, refining, and use of petroleum products. Aside from greenhouse gases that fossil fuel combustion releases (discussed in Chapter 11), the environmental concerns about petroleum use center on land disturbance, spills, leaks, accidents, water use and disposal, and the impacts of enhanced recovery technologies (especially hydraulic fracturing) on seismic risk and water supplies (Table 13-2).

Oil Pollution and Spills Most of the oil pollution in oceans is derived from ships that flush their oil tanks at

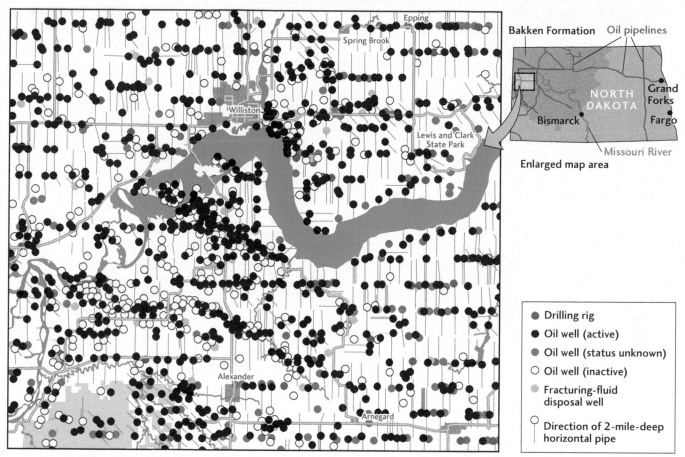

Figure 13-16 Thousands of wells are tapping oil from the Bakken formation, a Devonian shale. Each dot on the map is a well where a vertical borehole first was drilled, and the lines coming out from each well indicate the position of horizontal wells that extend laterally underground from these vertical positions.

sea and from industrial and urban runoff from continents. In addition to this persistent source of pollution, accidental oil spills can cause severe damage to local marine or coastal ecosystems. Most oil spills occur when tankers run aground and their storage compartments are pierced, as happened when the *Exxon Valdez* tanker ship crashed into a rocky reef in Alaska's Prince William Sound. About 1500 miles of shoreline was coated in oil, and several hundred thousand birds and marine mammals died. More than 40,000 tons of crude oil was spilled.

Some spills occur as a result of well blowouts at offshore rigs. Well blowouts occur because petroleum exists under conditions of extremely high fluid pressure, which can be released suddenly during drilling. The largest oil spill in U.S. history occurred in 2010 when a blowout preventer at the *Deepwater Horizon* offshore oil rig failed to fully engage during drill pipe collapse (Box 13-2). Water depth at the well was approximately 1,500 m (5,000 feet). The great pressure under such deep water made it difficult to cap the well and stop the gushing of oil into the Gulf of Mexico. It is estimated that

210 million gallons (4.9 million barrels; 780,000 m^3) of oil were spilled into the Gulf over the course of nearly 3 months before the blowout was stopped.

When oil accumulates on an animal's skin, breathing and movement become difficult for the animal. For birds, oil can limit or completely inhibit their ability to fly, dive, and float. Some drown as a result. Feathers lose their ability to repel water, so birds can also succumb to hypothermia. Some animals that ingest oil die immediately; others die at later times as a result of lung, liver, and kidney damage.

The three stages of cleanup that occur after an oil spill in the marine environment are called response, recovery, and restoration. Depending on the spill location, multiple agencies might be involved during each of these stages. During the initial response stage, the immediate goal is to prevent as much oil as possible from escaping. Less dense than water, oil floats and forms an *oil slick* that is several millimeters in thickness. In a response referred to as *mechanical containment,* booms, barriers, skimmers, and large, absorbent buffers are placed on the ocean water surface to contain

TABLE 13–2 Environmental Impacts Associated with Fossil Fuel Use

Fuel	Extraction	Transportation/Refining	Delivery and Use
Oil	• Spills during drilling • Hydraulic fracturing • Abandoned wells • Water use/disposal • Wildlife habitat encroachment	• Tanker accidents • Spills along pipelines • Refinery accidents • Water use	• Emissions and leaks from gas stations • Fire hazards • Emission of gases to atmosphere (carbon dioxide, nitrous oxides, sulfur oxides) • Emission of particulate matter to atmosphere
Natural gas	• Spills during drilling • Hydraulic fracturing • Abandoned wells • Water use/disposal • Wildlife habitat encroachment • Seismic risk from hydraulic fracturing and wastewater injection wells	• Leaks at compressor stations • Accidents and leaks along pipelines	• Leaks along pipelines • Fire hazards • Emission of gases to atmosphere (carbon dioxide, nitrous oxides)
Coal	• Strip mining, land disturbance, habitat loss • Mountaintop removal, valley filling, and habitat loss • Wildlife habitat encroachment • Water pollution	• Railroads and trucks • Coal preparation • Water use	• Coal ash • Emission of gases to atmosphere (carbon dioxide, nitrous oxides, sulfur oxides) • Emission of particulate matter to atmosphere • Acid rain • Release of mercury, heavy metals, and organic matter to environment, soils, waterways • Water use
Oil (tar) sands	• Strip mining, habitat loss • Wildlife habitat encroachment	• Energy- and carbon-intensive refining processes • Water use	• Similar to oil

and capture the spreading oil slick.[32] Oil already on the water must be isolated and kept from washing up on the coastline. Chemical dispersants are sprayed on the oil in order to speed up its dispersion and biodegradation. These chemicals reduce oil's surface tension, causing it to form smaller droplets that disperse more rapidly and, as a result, are more likely to evaporate and biodegrade. Microbes and other biological agents can be applied to speed up degradation of the oil. When oil biodegrades,

its hydrocarbon molecules are broken apart and become fatty acids, other organic molecules, and carbon dioxide.[33] With time, gases escape from the remaining oil, leaving a dense, viscous residue that sinks to the bottom of the ocean and may smother organisms on the ocean floor. The residue can persist for many years, hampering recovery of the ecosystem.

If spilled oil does reach shore, the cleanup is more difficult because waves can distribute the oil across tidal zones. Natural processes, including evaporation, oxidation, and biodegradation, might provide some cleaning, but these processes are slow and generally inadequate during a spill. Physical methods are used to speed up the cleaning, and these include pressure washing, raking, and bulldozing material that has been soaked and coated with oil.

Highly publicized disasters, such as the *Exxon Valdez* spill in 1989 and the *Deepwater Horizon* accident and spill in 2010, have made the public wary of efforts to increase oil drilling in coastal regions. The demand for energy resources is in direct conflict with the desire to preserve habitat and protect wilderness resources. On the other hand, a variety of factors cause nations to consider ways to increase use of their domestic energy resources. Prominent among these factors is the desire to eliminate or reduce dependence on foreign imports of energy resources. An area of much contention in the United States is the Arctic National Wildlife Refuge (ANWR) along the Alaska North Slope (see Box 13-1). National debate about whether or not to explore for and produce petroleum from this protected area spans more than three decades.

Oil Sands Extracting, processing, and transportation of oil sands have a number of environmental impacts (see Table 13-1). Extraction and processing methods for oil sands are generally more energy- and water-intensive than for lighter types of crude oils. Cold-region environments such as boreal forest and muskeg (peat-rich wetland) are impacted by open-pit mining and tailings ponds. Oil spills from pipelines are a concern, as they are with conventional oil. However, for analyses that include data from source through to distribution and use (that is, from wells to wheels), greenhouse gas emissions per unit of energy produced from oil sands are estimated to be as much as 20 percent higher than for conventional crude oil that is imported, refined, and consumed in the United States.[34] The reasons for higher greenhouse gas emissions are not only the energy- and resource-intensive extraction methods, but also the fact that oil sands typically have higher amounts of carbon and sulfur than lighter crude oils.

Shale Gas and Oil from Hydraulic Fracturing The environmental impacts associated with hydraulic fracturing are most often related to water use and wastewater disposal. In regions of relative water scarcity, obtaining the millions of gallons of water needed for each well can be challenging. Although some of the water is reused, the flowback from hydraulic fracturing can contain hazardous chemicals, both added and natural, from the rock formation. This highly salty fluid sometimes is treated before disposal or reuse, and in other instances it is injected into deep wastewater wells. For the Bakken shale of North Dakota, the majority of wastewater is injected into deep wastewater wells. Spills, leaks, and other accidental releases of the wastewater can result in contamination of water bodies, aquifers, and surrounding areas. Wastewater injection wells have been associated with increased occurrences of relatively small earthquakes, another concern related to hydraulic fracturing and its procedures. Some earthquakes linked to wastewater injection wells in the state of Ohio have been as large as magnitude 4.0.[35]

One of the most widely cited concerns associated with hydraulic fracturing is the contamination of groundwater that is used for drinking water supplies. The fact that drillers need not disclose the composition of chemicals that are added to the slurry of fracturing water and sand also has added to the public's sense of wariness and concern. Although hydraulic fracturing generally is done thousands of feet below the surface and most groundwater aquifers are shallow (one to two hundred feet, typically), residents are concerned at the possibility of methane or other substances seeping upward from fractured rocks into shallow groundwater systems.

Multiple studies and tests by the federal and different state governments have been done to determine whether known instances of contaminated water are the result of hydraulic fracturing. The most careful studies show no evidence for contamination from hydraulic fracturing, yet troubling incidents have occurred in association with well casing and wastewater disposal.[36] In some cases, substandard well casing has been the cause of shallow groundwater contamination. Accordingly, it is thought that higher standards, better practices, and effective regulation could avoid contamination due to casing problems in the future.[37] Substandard casing isn't an issue only for hydraulic fracturing processes; using low-quality materials could cause pollution in other types of oil and gas drilling activities, too.

Many support shale gas as a better option than coal for generating electricity; natural gas produces about half the amount of carbon dioxide per kilowatt-hour of electricity that is generated. Some recent studies indicate, however, that methane leaks at the ground surface of shale gas wells might offset this benefit, though the results are contested by some. Scientists are investigating sites at different locations in the United States in order to determine the magnitude of emissions due to such leaks. Leaks can be controlled with improved gaskets, better maintenance, and monitoring. Furthermore, producers have economic incentives to minimize leakage of the product that they intend to sell.

■ Petroleum, found in pore spaces in sediments and sedimentary rocks, occurs in two forms: crude oil and natural gas. The latter is less dense and typically occurs above crude oil in a petroleum reservoir.

BOX 13–2 EARTH SCIENCE AND PUBLIC POLICY

Deepwater Drilling and Well Blowouts

On April 20, 2010, the *Deepwater Horizon* drilling rig in the Gulf of Mexico—one of 600 deepwater wells in the Gulf—was destroyed as the result of an uncontrolled underwater well blowout and subsequent explosion and fire (Figure 1). The accident was devastating for numerous reasons, but new legislation, safety regulations, and funds for conservation, restoration, and scientific research might lead to long-term positive outcomes for the Gulf region, and possibly reduce the chances of future spills occurring.[24] Companies responsible for the accident are paying billions of dollars in settlement costs and fines that must be used to better understand and reduce the risks that oil production poses to human health and the environment. These efforts will include coastal barrier island restoration projects that might improve the ecological vitality of some parts of the Gulf coast.

Human, environmental, and economic impacts of the oil spill were considerable. Eleven workers were killed and 17 were injured. Spreading outward from its seafloor source

at rates as high as 62,000 barrels a day (9,900 m³/d),[25] the oil impacted marine habitat, coastal wetlands, estuaries, and wildlife, causing extensive damage to the fishing and tourism industries in the Gulf (Figure 2). For months, scientists could track a continuous plume of oil at least 35 km long.[26] The ocean area impacted was about the size of the state of Oklahoma, some 180,000 km². The first oil reached shore within several weeks, but continued to wash ashore along the coasts and barrier islands of Louisiana, Mississippi, Florida, and Alabama for over a year. Ultimately, possibly as much as 790 km (491 miles) of coastline was contaminated by oil. Economic interests of thousands of fishermen and businesses were harmed. Potential economic impacts from losses in revenue, profits, wages, and jobs were predicted to be as high as $9 billion during the first 7 years after the spill.[27]

Thousands of species were affected by the oil spill, including fish, birds, mollusks, crustaceans, sea turtles, and marine mammals (e.g., dolphins, whales, and manatees). Scientists found that oil entered the ocean food chain

(a)

(b)

Figure 1 (a) The *Deepwater Horizon* drilling rig in the Gulf of Mexico (66 km off the coast of Louisiana) had a well blowout on April 20, 2010, that led to an explosion and fire. Oil gushed uncontrollably from the well for nearly 3 months.

(b) The growing oil slick could be seen in coastal water off the Mississippi Delta in satellite images, as in this image of oil illuminated by sunlight that was acquired on May 24, 2010. *(a: AP Photo/U.S. Coast Guard; b: NASA Goddard Space Flight Center)*

through zooplankton, which are single-celled animals that drift in ocean water.[28] The primary food source for young fish, shrimp, and other animals, zooplankton provided the entry point for oil to work its way up through the food web to higher trophic levels (see Chapter 7). In addition, the oil served as a catalyst for plankton and other material in the ocean to form clumps that fell to the seafloor at rates ten times those of normal sedimentation. Scientists dubbed this a "dirty blizzard" that might account for some of the missing oil that didn't reach shore.[29]

Of the marine species impacted in the Gulf of Mexico, 14 are *endangered* and protected by the U.S. Endangered Species Act, the Marine Mammal Protection Act, and the Migratory Bird Treaty Act. Another 40 marine species in the Gulf that are not protected by any U.S. federal laws are identified as *threatened* on the International Union for Conservation of Nature's (IUCN) Red List (see Chapter 7 for discussion of species protection). Red List species garner special attention for international inventories and conservation actions.

Impending environmental and economic losses led to immediate U.S. federal action. President Obama put a 6-month moratorium on drilling new deepwater wells and appointed a presidential commission to examine the incident and safety regulations. The U.S. government committed to making all responsible parties accountable for the costs of cleanup and retribution for damage. Those parties include British Petroleum (BP), the oil company that was leasing the *Deepwater Horizon* drilling rig during the accident, and Transocean Ltd., the company that owned the rig. After an investigation, Assistant Attorney General Lanny A. Breuer of the U.S. Justice Department's Criminal Division stated that the explosion of the rig resulted from "BP's culture of privileging profit over prudence."[30]

In November 2012, BP admitted to its criminal conduct and agreed to plead guilty to felony manslaughter, envi-

Figure 2 Workers responded to minimize the amount of oil gushing from the well and to contain it at the ocean surface. As oil reached shore, other workers rushed to clean contaminated beaches and wildlife. *(NOAA and Georgia Department of Natural Resources)*

ronmental crimes, and obstruction of Congress. Among the 14 counts filed in court were violations of the Clean Water Act and Migratory Bird Treaty Act. The company agreed to pay $4 billion in criminal fines and penalties, making it the largest criminal resolution in U.S. history. From a second settlement of federal charges in January, 2013, Transocean Deepwater Inc. agreed to pay $1.4 billion in civil and criminal fines and penalties for violations of the Clean Water Act.[31] As the result of new legislation signed by President Obama in 2012, the RESTORE Act (Resources and Ecosystems Sustainability, Tourist Opportunities, and Revived Economies of the Gulf Coast States Act of 2011), 80% of the settlement funds are to be used for economic and ecological restoration projects along the Gulf Coast. Furthermore, the Transocean defendants must improve operational safety and emergency response capabilities at all of their drilling rigs in U.S. waters.

■ Petroleum forms largely from planktonic algae, single-celled floating marine plants that sink to the ocean floor upon death, accumulate in sediments, and become hydrocarbons at the high pressures and temperatures associated with deep burial and tectonic deformation.

■ The environmental impacts of producing, transporting, and burning petroleum for energy are many and include the emission of carbon dioxide—a greenhouse gas—into the atmosphere, ocean pollution, and habitat disruption.

■ National debate on whether or not to open protected land on the North Slope of Alaska to petroleum exploration and production results from the essential trade-offs that policy makers and the public face when deciding between increased production of domestic energy resources and protection of environmental resources.

COAL

Coal is far more widely distributed and abundant than conventional petroleum reservoirs.[38] Similarly to petroleum, coal is composed primarily of carbon and hydrogen compounds (see Table 13-1), and most coal is contained within sedimentary rocks. Unlike petroleum, however, coal is a solid rather than a liquid or gas and usually contains elements besides hydrogen and carbon. These elements include sulfur and nitrogen, which combine with oxygen and water in the atmosphere and hydrosphere to produce the sulfuric and nitric acids that contribute to acid rain and acid mine drainage. Because of coal's sulfur and nitrogen content, the drainage waters from coal mines and gaseous emissions from coal-burning power plants, if untreated, contain more pollutants than the drainage and emissions associated with petroleum extraction and combustion.

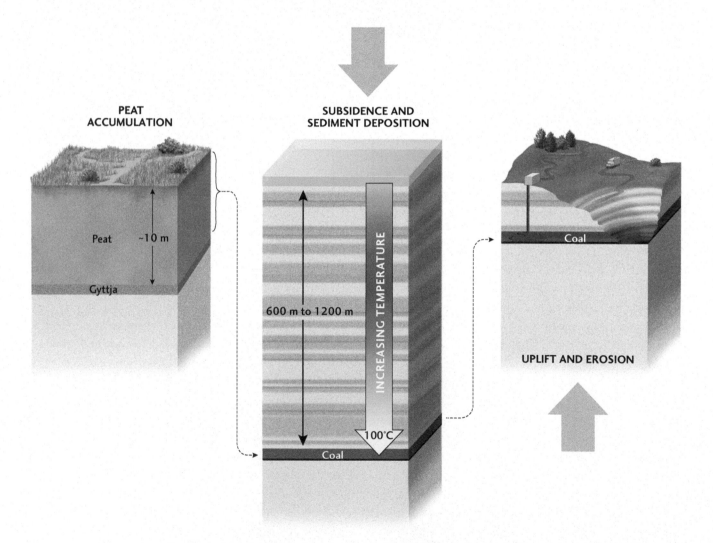

Figure 13–17 Partial decay of peat in a wetland environment forms black, gel–like gyttja, a precursor to coal. With deep burial (up to thousands of feet) under layers of younger sediment and organic matter, and with higher temperatures at depth, the gyttja is thermally altered to become coal.

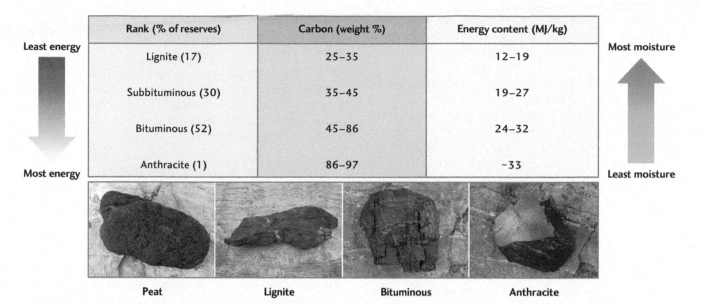

	Rank (% of reserves)	Carbon (weight %)	Energy content (MJ/kg)	
Least energy	Lignite (17)	25–35	12–19	**Most moisture**
	Subbituminous (30)	35–45	19–27	
	Bituminous (52)	45–86	24–32	
Most energy	Anthracite (1)	86–97	~33	**Least moisture**

Peat	Lignite	Bituminous	Anthracite

Figure 13–18 During coal formation, the percentage of carbon increases with deeper burial and greater thermal alteration. The highest ranking (most altered) coals consist of more than 80% carbon by weight, and have the least moisture and highest energy content. *(Donna Beaver Pizzarelli/ USGS)*

Coal is formed from land plants that once lived in swamps, marshes, and coastal plains, whereas most petroleum originated from marine phytoplankton that lived along and near continental shelves. Because petroleum migrates after it is formed and collects in permeable rocks adjacent to traps, the occurrence of conventional petroleum tends to be more restricted than the occurrence of coal, which is found in the broadly distributed areas where it formed. In this regard, coal deposits are more similar to shale oil and gas resources.

Origin of Coal

Layers of coal buried in Earth's crust are the tombs of ancient wetland ecosystems. Evidence of a wetland origin for coal is abundant in the variety of plant fossils that geologists and miners find in coal beds. Common fossils in coal seams from the Mississippian and Pennsylvania periods (together called the Carboniferous), a time of widespread coal formation, include primitive ferns and spore-bearing trees that once grew in coastal swamps, estuaries, marshes, and river floodplains. These wetland environments were similar to the modern Great Dismal Swamp in coastal Virginia and North Carolina, or the Okeefenokee Swamp that spans southern Georgia and northern Florida.

Under certain geologic conditions, organic matter from wetlands can accumulate to become beds of solid fossil fuels (hydrocarbons) over long periods of time (Figure 13-17). Depending on the depth of burial of organic matter in wetland environments, peat or different types of coal are formed. The general process, described in more detail below, is as follows: With increasing decay and thermal alteration of organic matter during progressive burial, the amount of water and other volatile gases (e.g., carbon dioxide) decreases. Simultaneously, the percent carbon, and hence heating value, of the coal increase. This entire process of transformation, from organic decay to coal formation with burial and increasing temperature, is called **coalification**. With increasing coalification, first lignite, then bituminous, and finally anthracite coals are produced (Figure 13-18).

The process of coalification begins as plants die in a wetland. Aerobic bacteria (those that require oxygen) and fungi decompose the organic matter, partly breaking down cellulose and other substances to produce tea-colored humic matter (see Chapter 6) and gases that include methane (CH_4), carbon dioxide (CO_2), and ammonia (NH_3). This loose decomposed material, called **peat,** is compacted easily. Peat has a high moisture content, contains less than 50 percent carbon, and burns freely when dried (see Table 13-2). Any mineral matter contained in the peat will not burn, but instead it forms small gritty particles called *ash*.

With time and continued accumulation of partly decayed organic material, the aerobic bacteria consume all available oxygen that was trapped in the original peat. Subsequent decomposition is continued at depth (typically greater than half a meter) by anaerobic bacteria (those that do not require oxygen). These bacteria generate organic acids as a waste product, so the pH of water within the deposit gets lower. Even the anaerobic bacteria then die. As a result, organic decay ceases. During this stage, **gyttja,** a black organic substance with the consistency of a gel, is formed, typically at a depth of 6 m to 10 m within the accumulating deposit. At the surface, fresh organic matter added to the growing deposit continues to decay, first becoming peat and then gyttja, so the gyttja layer at the base of the peat increases in thickness over time.

If buried deeply enough by overlying sediment to reach certain temperatures, this organic-rich gyttja begins to transform to coal through low-grade metamorphic processes. Sediment that buries the organic-rich gel might be from sea level rise, or it might be associated with the evolving floodplain and delta of a large river. For deep burial to occur, however, the stack of sediment needs to subside over time. At depths of about 600 m to 1200 m or so, depending on the local geothermal gradient, the temperature reaches 100°C. At this critical temperature, water is driven off and the molecules of organic matter can be broken, or thermally cracked, to form carbon-rich coal.

Depending on the pressure and heat exerted on peat, it may be transformed into different grades, or *ranks*, of coal (see Figures 13-17 and 13-18). The softest is **lignite**, a brown sedimentary rock that is about 20 to 25 percent carbon and gives off much more heat than peat when burned. If deep burial further compresses lignite, thus removing more water from it, it becomes harder, brown-black sedimentary rocks known as **subbituminous coal** or **bituminous coal**, which are about 35 to 45 percent and 45 to 86 percent carbon, respectively.

During coalification, if bituminous coal is heated enough to recrystallize, it becomes a shiny, dense, hard, black, metamorphic rock known as **anthracite coal**. Anthracite coal is more than about 86 percent carbon. If anthracite were to metamorphose even further, it would ultimately become pure carbon, in the form of the mineral graphite. As with peat, mineral matter in any of these three types of coal is much less combustible than is carbon and becomes ash when burned.

Coals commonly occur as numerous layers, which coal miners call seams, ranging from one to several meters thick and separated by beds of sandstone, siltstone, shale, and limestone. The layers of clastic and chemical sedimentary rocks separating the coal seams formed when the ancient swamps where coal is found were periodically submerged.

Global Distribution of Coal Deposits

Consider the following facts about the global distribution of coal:

■ The largest known coal deposits occur in sedimentary and metamorphic rocks of about the same age—359 million to 299 million years old—in the United States, Europe, the former Soviet Union, and China. For this reason, that time is named the Carboniferous period (the Pennsylvanian and Mississippian periods in North America). Fossils in these coal deposits indicate they formed in tropical and subtropical coastal swamps through which large rivers flowed seaward.

■ All these deposits contain numerous coal seams, sometimes as many as 100, typically separated by beds of marine shale and limestone. The sequences of coal, shale, and limestone are cyclic, repeated over and over in the thick deposits.

What do these clues tell us about the origin and current distribution of much of the world's coal? How does a geologist interpret these facts? If the areas that now contain Carboniferous coals once were tropical or subtropical swamps along the ancient coastlines of North America, Europe, and Asia, these landmasses must have been much farther south, closer to the equator than they are now. Continental drift is the explanation: these continents during the Carboniferous period were much closer to one another and straddled the equator (Figure 13-19; see also Chapter 2).

As for the number of sequences of coal, shale, and limestone, these indicate that sea level rose and fell many times during the Carboniferous period, inundating continental lowlands and enabling wetlands to cover vast areas. With each rise in sea level, coastal swamps would have been drowned and buried by marine mud that later became shale and limestone. With each fall in sea level, the swamps and river floodplains would have extended out from the continents, forming new swamps in which peat accumulated and later became coal. Although some of the Carboniferous coal deposits now are located in mountain belts, such as the Appalachians, they mark the locations of ancient lowlands periodically inundated by shallow seas.

Why would the Carboniferous sea level have risen and fallen so many times? The most likely cause is an ice age in which ice masses grew and then melted over and over again. Geologic evidence indicates that when the northern continents were coming together along the equator, the southern continents (South America, Australia, India, Africa, and Antarctica) already were assembled as the supercontinent Gondwana, which was positioned over the South Pole. Glacial deposits and rocks scoured by glacial ice provide a picture of an ice sheet much like the modern Antarctic ice sheet. Just as ice sheets and smaller glaciers repeatedly have advanced and retreated for the past few million years, causing sea level to rise and fall more than 100 m, the Gondwana ice sheet varied in size and affected the Carboniferous sea level.

Few coal deposits are older than Carboniferous because vascular land plants only evolved about 470 million years ago, but coals that are younger do exist. Some of these young coals are located along the low interiors of continents that were inundated by high sea levels since Carboniferous time. Global sea level was especially high

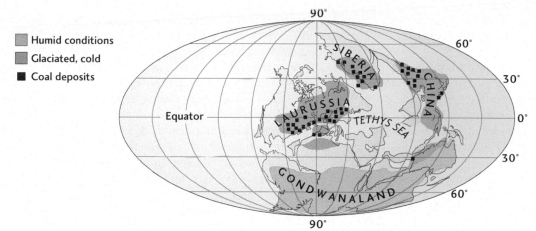

Figure 13–19 When continents are arranged in their Carboniferous positions of about 359 million to 299 million years ago, the modern coal beds are located along or near the equator, surrounding a large body of water known as the Tethys Sea. A polar ice cap over the Gondwana supercontinent (Gondwanaland) repeatedly grew and shrank throughout Carboniferous time, providing the fluctuating sea level that enabled numerous swamps and coal seams to develop one atop the other around the coastal margins of the Tethys Sea.

during the Cretaceous period 145 million to 66 million years ago, when global temperature reached some of the highest values in Earth's history. In North America, a large Cretaceous sea spread throughout the lowland west of what are now the Rocky Mountains. Just as in Carboniferous time, fluctuating sea level favored the development of many layers of coal that now are mined throughout the region.

Coal Production and Reserves

Because there are much greater reserves of coal than any other fossil fuel, and new mining methods can recover greater amounts than in the past, the use of coal is likely to continue to grow in the future. At present, it is the second most important global primary energy source, providing 30 percent of all energy consumed, and is used primarily to generate electricity at power plants. For electricity, coal is the primary source of energy (41 percent for the world, and 45 percent for the United States). China and India, both rapidly growing countries, are particularly dependent on coal for their energy needs and probably will continue to increase their use of this resource in the future.

In the past ten years, coal production increased by 48 percent, the most rapid growth in use of any fossil fuel. At the same time, global energy demand increased 28 percent. China's coal consumption during this decade more than doubled (170 percent increase), growing at a rate of 10.4 percent per year. For the rest of the world, coal consumption increased 16 percent from 2000 to 2010, at a growth rate of 1.5 percent per year. In 2010, the United States and China accounted for about 65 percent of the world's production of coal.

Because much coal is exposed at Earth's surface, geologists can more easily estimate the recoverable reserves of coal than of petroleum. By measuring the thickness of different coal seams and the area they cover, geologists have estimated that world reserves are about 861 billion metric tonnes, sufficient to last for at least 112 years at current rates of use (i.e., the R/P ratio). For the United States, the R/P ratio is 228 years.

The largest coal reserves occur in the United States, the Russian Federation (the former Soviet Union), and China.[39] As with petroleum, advances in mining technologies may raise the estimate of how long we can rely on coal. For instance, deep *underground mining* in the eastern United States once provided the nation with most of its coal, but now nearly 70 percent of it is produced by surface mining in western states.

Coal Mining

Worldwide, 40 percent of coal is mined at the surface and the remainder is mined from underground (see Figure 13-17). In the United States, 71 percent is mined at the surface and the rest is mined underground. Surface mining is done when coal is within about 60 m of the ground surface. Where coal seams are relatively thick, close to horizontal, and less than about 60 to 100 m below ground, surface mining is cheaper and safer than underground mining. It also is more efficient at removing most (90 percent or more) of the coal. With time, the more accessible seams of coal have been mined and mining companies have developed new strategies to mine coal profitably.

Underground Mining In underground mining, as much as three-quarters of the available coal must be left as roof support to keep mine openings from collapsing. One common method for deep mining where coal seams are roughly horizontal is called "room and pillar." Mining progresses along a seam, removing "rooms" of coal, and pillars of rock are left standing to support the mine roof.

Surface Mining The three most common forms of surface mining are open-pit, strip, and mountaintop removal. All begin with the removal of vegetation cover and topsoil. During **open-pit mining,** large terrace pits are dug by a combination of drilling and blasting the coal and overburden (unwanted rock that overlies the coal). Heavy machinery such as draglines, some as tall as 40-story buildings, is used to remove the blasted rock to expose the coal seams. **Strip mining** is a similar process to open-pit mining except that the overburden is removed in strips.

Mountaintop Removal and Valley Fill A relatively new mining technique in steep terrain is **mountaintop removal mining.** In order to expose seams of coal up to about 300 m deep, the tops of mountains or summit ridges are removed and leveled with explosives and heavy machinery (Figure 13-20). In several months, an entire geologic record of hundreds of thousands to millions of years of subsiding wetland environments can be blasted, crushed, and redistributed in order to extract its hydrocarbon fuel source.

In mountainous terrain, disposal options for the overburden (the non-coal rock) are limited. The immense amount of debris sometimes is placed back on the ridge and graded, but more often it is packed into the bottoms of adjacent headwater valleys in a practice known as **valley fill.** The headwater refers to the beginning, or head, of a stream. Proponents of mountaintop removal view it as one of the most efficient and cost-effective ways to mine coal. About one ton of coal is recovered for approximately every 16 tons of terrain displaced.

Mountaintop removal for coal mining is especially common in Kentucky, West Virginia, Tennessee, and Virginia, in the Appalachian Mountains of the eastern United States. The practice is highly controversial, however, because of the sheer magnitude of physical destruction and impact on waterways, biodiversity, and local human communities.[40] It is the largest earthmoving activity in the United States.

Environmental Impacts of Mining and Burning Coal

Coal is arguably the most environmentally damaging fossil fuel energy source. All stages of the coal energy cycle, from mining to processing and combustion, affect the environment. Surface mining is safer than underground mining for mine workers, but can lead to slope instability, unsightly and dangerous open pits, and soil and sediment erosion. Reclamation of operating coal mines is currently required in many nations, including the United States, but old, unreclaimed mines still present a signifi-

cant hazard to surrounding communities. Furthermore, it is costly and laborious to reclaim surface-mined land in arid and semiarid regions where vegetation and water are scarce. Both surface and underground mines can contribute to acid mine drainage.

Coal combustion results in numerous atmospheric emissions that are linked to a variety of environmental problems, including acid rain, smog, and human health issues. A number of laws in the United States, including the Surface Mining Control and Reclamation Act, the Clean Water Act, and the Clean Air Act are used to minimize and regulate the environmental impacts of coal mining and combustion.

Mountaintop Removal Mining With mountaintop removal mining, a particularly common practice in the Appalachian Mountains, the combination of cutting down the original summit and filling in the adjacent valley causes extreme topographic, hydrologic, and ecological change (Box 13-3). An ancient belt of tectonically uplifted crust in the eastern United States, the Appalachian Mountains are heavily forested and have unique biological diversity and freshwater streams. Headwater streams comprise 70 percent or more of watersheds in Appalachia and are particularly important to ecosystem functions.[41] The U.S. Environmental Protection Agency (EPA) estimated, however, that about 11.5 percent of forests in Kentucky, West Virginia, Tennessee, and Virginia would be destroyed or degraded as a result of mountaintop removal by the year 2012.[42] This is equivalent to an area that is larger than the state of Delaware. More than 1,000 miles of streams would be buried with rubble from the former mountaintops. Over time, mines have extended along ridges and the size of individual mines has grown, so the resulting impacts are cumulative.

In a 2008 report, the EPA evaluated scientific publications and environmental impact statements on mountaintop mining in Appalachia in order to determine the mining practice's impact on the environment. Based on a review of studies done on more than 1200 stream segments, the EPA concluded that springs and small perennial streams are permanently lost, certain ions and heavy metals are higher in downstream waterways, water quality can become so degraded as to be lethal for standard test organisms, and aquatic organism communities are significantly degraded and less diverse.[43] Finally, the EPA also determined that water runoff from mountaintop removal areas has high loads of silt, iron, sulfur compounds, and heavy metals that include zinc and selenium.

These findings were further supported by a comprehensive analysis of numerous scientific studies and water quality data that was reported in 2010.[44] In this study, the scientists concluded that the environmental impacts of mountaintop removal and valley filling are so serious and irreversible that the situation can be considered a failure of policy development and enforcement. The

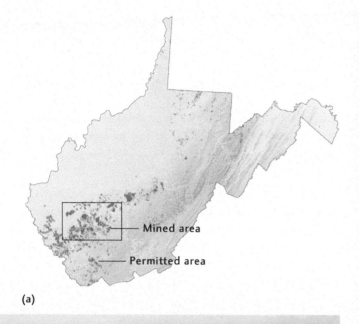

(a)

September 17, 1984

July 30, 1995

(b)

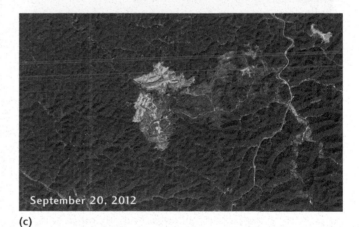

September 20, 2012

(c)

Figure 13–20 (a) Large areas of West Virginia have been affected by mountaintop removal mining (light red), and additional areas have been permitted but not yet mined (blue). The outlined area contains some of the state's largest mines, including the Hobet–21 Mine and the Kayford Mine, each of which is more than 10,000 acres. (b) An aerial view of a large mountaintop removal mine reveals exposed coal seams. (c) Based on data from the *Landsat 5* and *Landsat 7* satellites, these natural–color (photolike) images document the growth of the Hobet Mine as it moves from ridge to ridge between 1984 to 2012. *(a: NASA map by Robert Simmon and Jesse Allen, based on topographic data from the Shuttle Radar Topography Mapping (SRTM) mission and mine permit data from the W.V. Department of Environmental Protection; NASA image by Jesse Allen, based on data from the MODIS Rapid Response Team, Goddard Space Flight Center; b: ZUMA Press, Inc./Alamy; c: NASA)*

authors noted that in addition to the loss of headwater streams, contaminants persist in streams and wetlands below valley fills, forests are destroyed, and biodiversity is reduced. Clearly, they wrote, reclamation and the mandatory mitigation strategies intended to compensate for lost habitat and ecosystem functions were insufficient and ineffective.

You might wonder how it is possible to dispose of so much waste rock from mining into the headwater areas of streams, as these are ecologically and

BOX 13–3 GEOLOGIST'S TOOLBOX

Visualizing Mountaintop Removal in 3–D with Remote Sensing and Lidar

It is possible to visualize the environmental impact of mountaintop removal mining in three dimensions over time using two different remote sensing technologies to construct topography. They are called *stereophotogrammetry* and *lidar*. With stereophotogrammetry, the three-dimensional (3–D) coordinates of points on an object, such as Earth's surface, are determined with photographs taken from different positions. Common points on two or more aerial photos are identified and the angles between them and known points (with coordinate data) are calculated to determine the coordinates of each unknown point by a process of triangulation. An example of 3–D coordinates for a position on Earth's surface would be longitude, latitude, and elevation, also referred to as the x, y, and z coordinates.

The word *lidar* began as a combination of the words "light" and "radar," but more recently has been viewed as an acronym for *light detection and ranging*. Radar, an acronym for radio detection and ranging, is used to calculate distance, or range, to an object by measuring the time it takes for pulses of radio waves to travel from a source, for example, a radio dish, to objects in their pathway and then, after bouncing off the objects, back to the location of the transmitter. Using a laser instrument that sends light in the form of laser pulses, lidar is used in a similar way in order to measure distances from the laser source to Earth's surface. The cloud of returning data is used to portray topography in 3–D. Lidar instruments can be mounted in airplanes (airborne lidar) to generate high-resolution 3–D topographic data for large areas of Earth's surface. These data include returns from vegetation, buildings, and any other object at the surface, but the data can be processed so as to create a "bald Earth" view. A bald Earth view removes vegetation, buildings, and other features to reveal the bare ground surface topography.

Digital elevation models, or DEMs, are models of Earth's surface that are generated with coordinate data from technology such as stereophotogrammetry and lidar. From DEMs produced at different times, it is possible to quantify and evaluate landscape change. In our example, we will look at how the terrain was altered by mountaintop removal and valley filling during coal mining at the Spruce No. 1 Mine in West Virginia, a controversial mine that is one of the largest in the Appalachian Mountains (**Figure 1**).[48]

The "before mining" 3–D terrain for the Spruce No. 1 Mine was obtained from topographic data acquired with aerial stereophotogrammetry in 2003. A DEM based on these data shows the topography of mountaintops and valleys at the mine location in 2003 (see Figure 1). The "after mining" 3–D terrain was obtained from airborne lidar data collected in 2010. With the pre- and post-mining terrain models, the difference in elevation between the two DEMs can be determined. Topographic profiles, which are two dimensional plots of elevation versus distance, can be generated from the data. Before and after topographic profiles reveal how the tops of mountains are gone, and the valleys are filled in with the crushed material from the former knobs. It also is possible to calculate how much volume change has occurred. In this case, about 7.4 million cubic meters of rock was removed from the mountaintop and placed as fill in the adjacent valley headwater area.

---

hydrologically important parts of watersheds. You might also recall from Chapter 9 that the federal Clean Water Act generally prohibits discharge of pollutants into U.S. waterways and requires a permit from the EPA for those discharges that are allowed. However, the Clean Water Act has a provision by which the U.S. Army Corps of Engineers can give permits for the discharge of "dredged or fill material" into water bodies. This practice has been common for construction projects. In 2002, the George W. Bush administration adopted a new regulation that defined mine spoil from surface mines as "fill material" and gave the Corps the lead role in permitting mountaintop removal and valley fill operations. In effect, the practice of dumping mine waste into waterways from mountaintop removal was legalized.

Even so, the Clean Water Act states that dredged or fill material should not be discharged into the aquatic ecosystem if it might cause significant degradation.

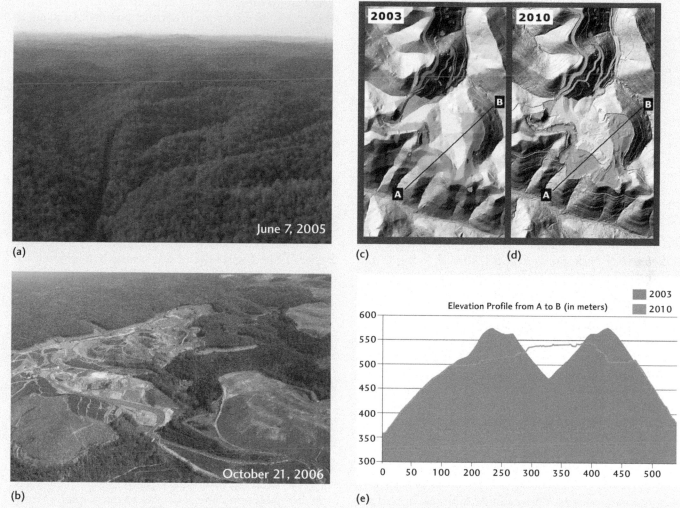

Figure 1 Photos of hills in West Virginia before (a) and during (b) mountaintop removal and valley filling illustrate the impact of this type of mining for coal. The differences can be visualized with topographic data from high-resolution mapping of Earth's surface. These topographic data, from airborne lidar, were used to generate hillshade images in order to illustrate the shape of the mountaintop before (c) and after (d) mountaintop removal. The black line represents the location of a vertical profile shown in (e), illustrating the removal of two peaks and filling of the former low valley between them.
(a, b: Vivian Stockman, flyover courtesy SouthWings; courtesy Ohio Valley Environment. c, d, e: Yolandita Rivera/SkyTruth)

According to the EPA, no fill material will be permitted if a less damaging (to the aquatic environment) alternative exists, or if the nation's waters would be significantly degraded.[45] Another relevant law that was altered by the G. W. Bush administration is the stream buffer zone protection provision of the Surface Mining Control and Reclamation Act of 1977. This is the primary federal law for regulating the environmental effects of coal mining. The Stream Buffer Zone Rule prohibits surface coal mining activities and disturbances within 100 feet of streams. In December 2008, at the end of the G. W. Bush presidency, this provision was revoked for cases of disposal of mining waste rock in headwater waterways.

One of the most controversial and publicized cases involving environmental impacts of mountaintop removal regards the Spruce No. 1 Mine.[46] A proposed expansion would have made it one of the largest mountaintop removal mines in the Appalachian Mountains. The dredge-and-fill permit required by the Clean Water Act (Section 404) for this expansion was denied by the EPA because of its potential environmental impact on waterways, but this ruling has been contested in a series of court cases that were appealed and overturned. As of 2013, the permit veto was upheld by a federal court. The EPA decision contains statements that are relevant to understanding environmental issues and Earth systems:

> [The EPA is] acting under the law and using the best science to protect water quality, wildlife and Appalachian communities, who rely on clean waters for drinking, fishing and swimming. These streams contain vibrant wildlife communities and play a critical role in sustaining the quality of downstream waters. These six miles of streams are among the last remaining high-quality streams within a watershed that has been fundamentally altered by prior and ongoing surface coal mining activities.[47]

Undergound Coal Mine Fires Coal fires in abandoned mines, particularly underground mines, are a problem because so much coal is left in place. Today more than 100 underground fires are burning throughout the United States, and many of these cannot be extinguished. Thousands more burn in mines in at least 22 other countries.[49] The mazes of abandoned tunnels and shafts below ground provide ample air flow and oxygen to keep fires burning for years. Underground mining also results in surface subsidence above collapsed tunnels, thereby damaging buildings, roads, and railways.

Acid Mine Drainage When sulfide minerals in coal are exposed to the atmosphere, as in tailings, they react with oxygen and water to form iron oxides and hydroxides as well as sulfuric acid (see Chapter 11). These substances can drain from mines into surface water. This **acid mine drainage** increases the acidity of local streams, thus increasing their metal content and affecting the biota. Aquatic plants and animals are sensitive to the acidity of their habitat, and organisms begin to die as the pH of water falls below their tolerance levels. Furthermore, iron oxides and hydroxides precipitate out of the water as a yellow-orange mass called yellowboy that coats the stream bottom and smothers aquatic life. The problem is prevalent in the eastern United States and northern Europe, where precipitation is high and coals and coal-bearing strata contain more sulfides. Although acid mine drainage has been treated in many areas of the United States, thousands of kilometers of waterways and tens of thousands of square kilometers of land still are affected.

Atmospheric Emissions and Other Waste from Coal Combustion Coal combustion results in atmospheric emissions that include carbon dioxide, sulfur dioxide, nitrogen oxide, and heavy metals. During 2011, coal consumption in the United States contributed 34 percent, approximately 5472 million metric tons, of the energy-related emissions of carbon dioxide.[50] Coal emits more carbon dioxide per unit of energy than any other fossil fuel.

Sulfur released to the atmosphere by human activities is the main contributor and precursor to acid deposition in the form of **acid rain**. In the early 1800s, global anthropogenic sulfur dioxide emissions were nearly zero, but they increased with industrialization and peaked in the 1980s (Figure 13-21).

Although North America and Europe currently are the largest contributors to global sulfur dioxide emissions, they also have achieved the greatest emission reductions in recent decades. Scientists used NASA satellite data to determine that sulfur dioxide in the atmosphere near major coal-burning power plants in the eastern United States declined by nearly half between the measurement intervals of 2005–2007 and 2008–2010.[51] These satellite-based estimates are confirmed by measurements of the concentration of sulfur dioxide in the atmosphere. At the same time, however, other rapidly growing Asian countries that rely heavily on coal for energy are producing greater emissions of sulfur dioxide to the atmosphere. These countries contribute an increasing proportion of sulfur dioxide emissions each year (40 percent by 2005).

The decline in sulfur dioxide in the United States is attributed to the Clean Air Interstate Rule, a law passed in 2005.[52] This law called for deep reductions in sulfur dioxide emissions over time and a cap on emissions that will decrease with time.[53] Many power plants have installed pollution control devices to remove sulfur from their fuel sources and to limit the release of sulfur dioxide in their emissions.

Sulfur emissions also can be reduced by burning low-sulfur coals. Coals that formed in coastal swamps, typical of the eastern coal deposits of the United States, have higher sulfur content than do coals formed in freshwater (inland sea) swamps typical of western deposits because the amount of sulfate ions is much higher in salty coastal waters than in continental freshwater. The sulfate ions combine with other elements, such as iron, to form sulfide minerals, such as pyrite, in the coals. The average sulfur content of eastern coal is 2.5 percent, whereas for western coal it is only 0.3 percent.

Finally, combustion of coal produces other wastes, including ash from mineral matter and small amounts of a variety of metals. Ash is collected in power plants and usually disposed of as solid waste, though some forms of ash are reused, for example as a component of cement. Some metals that in large amounts can be toxic to organisms are emitted to the atmosphere during combustion. These include selenium (Se), mercury (Hg), and

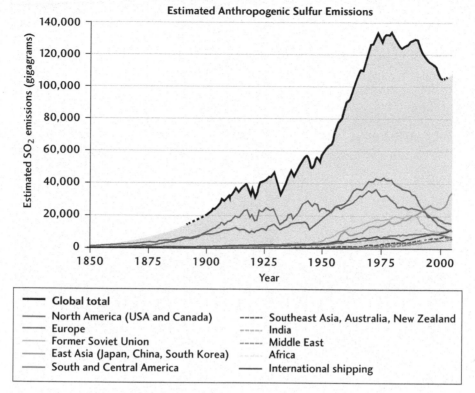

Figure 13–21 Coal combustion is the largest source of sulfur emitted to the atmosphere and biosphere by human activities. As countries industrialize, their per capita demand for energy increases, and so does their emission of sulfur if they use high–sulfur fossil fuels. Offsetting this increase is a more recent trend of reductions in sulfur dioxide emissions from industrialized countries that have established pollution controls, as has happened in the United States in recent decades.

arsenic (As). Those coals with relatively large amounts of these substances are less desirable for burning.

Carbon Capture and Storage

With increasing concern regarding global warming and ocean acidification, reduction of carbon dioxide emissions to the atmosphere from human activities has become a central issue among policy makers. Long-term geologic storage of some of the CO_2 waste via carbon capture and storage (CCS, also known as carbon capture and sequestration) is one of a range of solutions to mitigate greenhouse gas emissions (Figure 13-22).[54] Large sources of CO_2 emissions, such as power plants that use fossil fuels, are good candidates for applying this relatively new waste storage technology. The U.S. Department of Energy, through the National Energy Technology Laboratory (NETL), funds most CCS-related work currently, though there are a few other companies currently seeking to build a commercial-scale CCS industry.[55]

Carbon dioxide storage from natural CO_2 sources has been utilized to enhance oil recovery from petroleum-bearing reservoirs since the 1970s. During enhanced oil recovery, carbon dioxide (in gas, liquid, or supercritical state) is injected into a geologic formation to push out residual petroleum. In a similar way, but with different technical details, deep geologic formations—typically greater than about 800 to 1000 m—can be used to sequester CO_2 waste.[56] For sequestration, however, the project scales and potential volumes of CO_2 storage are much greater.

For a power plant, the gas must be separated and captured from the power plant's effluent stream, then compressed to a supercritical state. It is transported via trucks or more efficiently through pipelines to a sequestration site. The CO_2 is injected into the subsurface through a Class VI injection well for CCS. This new class of wells was created by the U.S. Environmental Protection Agency, under its authority through the Safe Drinking Water Act (see Chapter 9) to protect groundwater sources of drinking water and to ensure strict regulation of CCS projects.[57]

Sedimentary rocks are the best repositories for underground geological storage of carbon dioxide. These rocks typically have higher porosity and permeability than metamorphic or igneous rocks. Porous, permeable rocks are best suited for injecting large quantities of liquids and gases into a formation. To sequester the carbon dioxide, some sort of trap is needed, just as with the

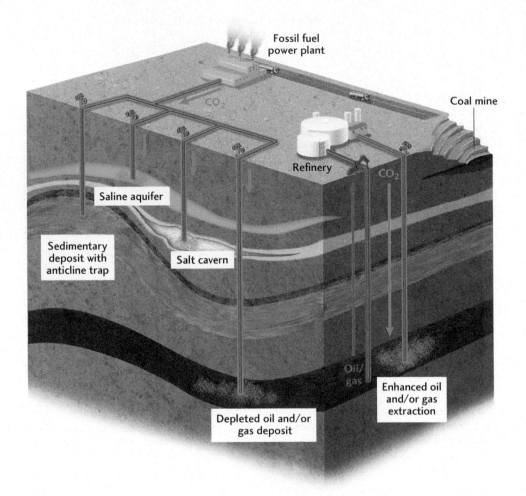

Figure 13–22 Ideas for ways to sequester, or store, carbon in the subsurface include pumping carbon dioxide into saline aquifers, salt caverns, and folded sedimentary rocks with cap rocks that would trap the gas. Other possibilities are to pump carbon dioxide into rocks from which oil or gas has been depleted. Finally, carbon dioxide could be used to enhance recovery of oil and gas.

trapping of petroleum in a reservoir rock. In addition, areas with active volcanism and faults are not well suited for long-term geologic storage because of the potential for leakage to the surface. Recall, for example, the case of Cameroon, where a natural carbon dioxide buildup was released from a lake in a volcanically active area, killing 1700 people by suffocation.

Anticlines are the ideal structural trap for CO_2 sequestration. The CO_2 is injected into porous and permeable formations (e.g., sandstones and carbonates) that are capped by impermeable formations (e.g., shales). Because CO_2 is more buoyant than the surrounding reservoir brine, it rises slowly to the top of the geologic structure and is trapped vertically by the cap rock and horizontally by the anticline structure. This ensures four-way capture of the CO_2 so that it does not migrate and so that it can be monitored indefinitely.

The U.S. government, through the NETL, is working to develop a portfolio of commercial-scale CCS technologies for deployment in 2020. One challenge is something called an "energy penalty"; there is a substantial amount of energy needed to capture CO_2, thus adding to the waste that is produced and needs to be stored. Through research and development with associated pilot projects, the agency aims to lower the costs of CCS, to reduce the energy penalty associated with CO_2 capture at large sources, and to better understand long-term safety issues associated with storage. A long-term aspiration is for new power plants and other facilities to have CCS technologies and for existing facilities to be retrofitted for such technology.

■ Coal forms from land plants that are deeply buried and biochemically and thermally altered. The greater the depth and higher the temperature of burial, the higher the grade of coal and carbon content, ranging from low in peat and lignite to high in bituminous and anthracite coal.

■ Major coal deposits are associated with convergent plate boundaries where conditions were favorable to the formation of peat in extensive coastal wetlands, the burial of layers of peat under sediments associated with changing sea level, and the deformation of layers of peat and coal during plate collision. All these conditions existed during and after the Carboniferous period some 300 million to 360 million years ago, when the continents that composed Pangaea were drifting together and sea level fluctuated in response to polar glaciation.

■ Large reserves of coal occur on many continents, particularly North America and Asia, which are currently estimated to be sufficient to last for up to 241 years or so, and even more coal might exist as undiscovered resources.

■ The environmental impacts of mining, transporting, and burning coal for energy are many; two of the greatest are the emission of carbon dioxide and sulfur dioxide into the atmosphere, the latter contributing to acid rain. In addition, sulfur dioxide that reaches the hydrosphere from mine tailings and abandoned mine tunnels produces acid mine drainage. To mitigate greenhouse gas emissions, some of the CO_2 waste from fossil fuel use can be stored in geologic deposits such as sedimentary rocks with high porosity and permeability. This technology is known as carbon capture and storage or carbon capture and sequestration (CCS).

GEOTHERMAL ENERGY

The energy available from Earth's interior is called **geothermal energy**. The steam and hot water produced by Earth's heat flow can be used to heat buildings or to drive turbines and generate electricity at geothermal power plants (Figure 13-23). Geothermal energy is considered a renewable resource because Earth has such high heat content relative to the rate at which it could be used as an energy resource. It also is regarded as a secure, sustainable, and economically and environmentally viable energy source, producing little in the way of pollutants. It provides 0.3 percent of world electricity and 0.4 percent of U.S. electricity, ranking it fourth in this list of renewable energy resources in the United States (after hydroelectricity, biofuels, and wind). Since 2004, worldwide geothermal capacity has grown at a rate of about 5 percent per year.

Geothermal energy has been made use of where heat flow is high and very hot or where molten rock occurs within several kilometers of the surface, conditions primarily found along plate margins. The concentration of volcanoes around the Pacific Ocean known as the "ring of fire" marks convergent plate boundaries, which yield substantial geothermal energy for Japan and New Zealand. Spreading centers along the Mid-Atlantic Ridge supply 90 percent of Iceland's electricity, with some Icelandic cities currently heated entirely by geothermal

Figure 13–23 Geothermal energy tapped from steam or water heated underground by Earth's internal heat is a valuable source of energy at a number of places on Earth, primarily along active plate boundaries. This geothermal power plant is located in Iceland, near the Krafla volcano. *(Frantisek Staud/Alamy)*

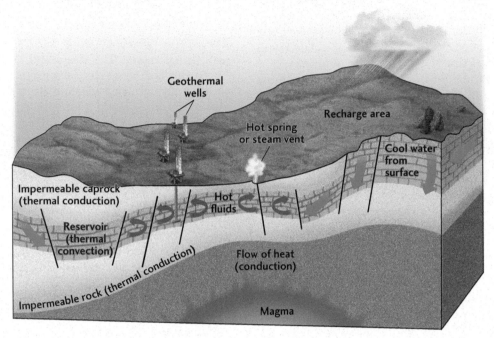

Figure 13–24 Four main elements comprise a geothermal system: a heat source; a fractured rock reservoir that stores and enables the flow of heated water; a water convection and recharging system that transfers the heat; and sometimes a rock structure, or cap, that serves as a heat insulator to maintain high water temperatures.

energy. Geothermal energy also has the potential to be harnessed along the East African Rift. Hot spots are ideal locations for geothermal energy resources, and the Hawaiian Islands have much geothermal potential.

Four main elements comprise a geothermal system: a heat source such as a shallow magmatic intrusion or area with high geothermal gradient; a fractured rock reservoir that stores and enables the flow of heated water; a water convection and recharging system that transfers the heat; and sometimes a rock structure, or cap, that serves as a heat insulator to maintain high water temperatures (Figure 13-24). Where cap rocks are fractured or faulted, fluid can leak from the reservoir, forming hot springs, geysers, and fumeroles.

Geothermal energy can reach Earth's surface through natural convection processes within a geothermal system. Groundwater that percolates downward through porous and permeable rocks until it touches hot rock is heated, expands, and circulates back toward the surface as hot water or steam. Just as wells are drilled to extract groundwater, geothermal wells are drilled into rock to extract hot water and steam from its pore spaces.

In the United States, areas with high geothermal gradients and heat flows are concentrated in the western states, where volcanism and tectonic activity occur (Figure 13-25). Geothermal plants operate in Alaska, California, Hawaii, Idaho, New Mexico, Nevada,

Oregon, Wyoming, and Utah. In 2010, the United States used about 30 percent of the geothermal electricity generated worldwide (3093 of 10,310 megawatts [MW] installed capacity). The Geysers, a geothermal field in northern California with 22 geothermal power stations, is the largest supply of natural steam in the world and provides the northern California coastal region with about 60 percent of its electricity demand. Other top-ranking countries for geothermal energy use are the Philippines, Indonesia, Mexico, Italy, New Zealand, Iceland, and Japan (in rank order from top down).[58]

A 2008 assessment of the nation's geothermal resources by the U.S. Geological Survey, mandated by the Energy Policy Act of 2005, determined that the potential electric power generation from identified geothermal systems is 9047 MW. In addition, the potential electric power generation from undiscovered geothermal resources is about 30,033 MW. The assessment identified 241 geothermal systems with moderate (90°C to 150°C) to high temperature (over 150°C) (see Figure 13-25). It also estimated that an additional 517,800 MW of electricity could be generated by technologies that fracture reservoirs with low permeability but otherwise favorable conditions for geothermal systems. As a result, the study concluded that geothermal resources could have a much more significant role in the U.S. energy mix in the future.

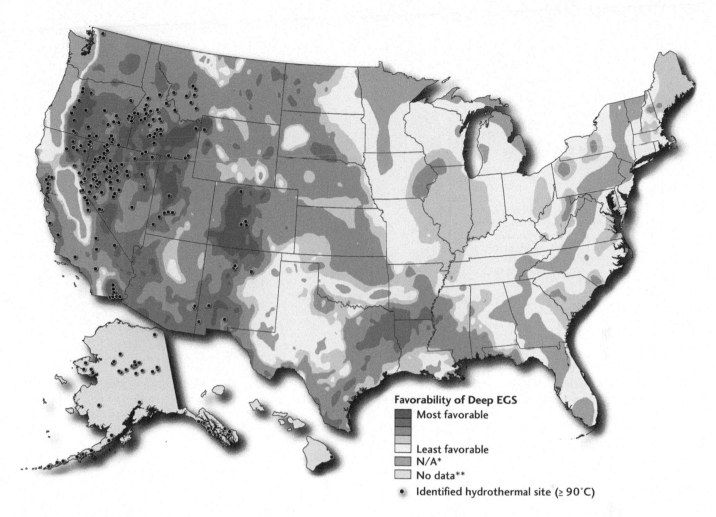

Favorability of Deep EGS
- Most favorable
- Least favorable
- N/A*
- No data**
- • Identified hydrothermal site (≥ 90°C)

Figure 13–25 The National Renewable Energy Laboratory produced a map of identified geothermal resources for the United States, including locations of known hydrothermal sites and areas that are favorable for deep enhanced geothermal systems. Note that the highest favorability sites are mostly located in western states, within areas of active tectonic and volcanic activity (e.g., the Cascade Mountains of Oregon and Washington states). *(National Renewable Energy Laboratory, U.S. Department of Energy)*

■ Geothermal energy is a renewable resource commonly available along active plate margins where heat flow is high and especially where steam or hot water flowing through a fractured reservoir is abundant.

■ The United States is the world's leading producer of electricity from geothermal energy.

■ The environmental costs of geothermal energy production are minimal.

NUCLEAR ENERGY

Nuclear energy provides about 8 percent of total world commercial energy use. A nuclear power plant consists of reactors in which nuclear reactions release heat that can run turbines or drive other mechanical processes that, in turn, generate electricity (Figure 13-26). With fossil fuels, combustion of hydrocarbons releases thermal energy and waste products such as carbon dioxide gas to the atmosphere. With nuclear power, the waste product is radioactive substances. Some of these will remain radioactive for thousands of years. The challenges of finding suitable disposal sites for radioactive waste have limited uranium from replacing fossil fuels as the world's major source of electricity.

The Nuclear Fuel Cycle

The nuclear fuel cycle begins with uranium mining and processing that supply fuel to reactors and ends with

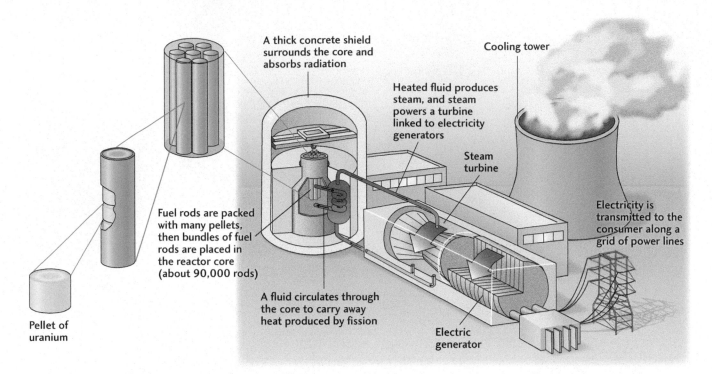

Figure 13-26 A nuclear power plant has a reactor core that produces energy from a bundle of metal fuel rods. Each fuel rod contains many small pellets of uranium oxide enriched in uranium-235. Some rods contain boron, which controls the reaction rate by absorbing neutrons. Nuclear fission produces heat, which is used to turn a turbine and generate electricity. If insufficient coolant is present to remove heat, a meltdown will destroy the reactor container and release radiation to the environment.

storage of radioactive waste products (Figure 13-27). Uranium-235 is used in nuclear reactors because it is **fissile,** which means that it can sustain a chain reaction of nuclear fission (discussed below). It is the only fissile isotope found in significant quantities in nature. About 0.7 percent of natural uranium is the isotope uranium-235. Whether through spontaneous decay or decay induced in a **nuclear reactor,** the isotope uranium-235 has the potential to release 2,500,000 times the amount of energy released by the combustion of an equal amount of carbon.

Volcanic processes bring uranium minerals to Earth's crust, where they remain in igneous rocks or become incorporated into sedimentary rocks through geochemical processes. Thirty percent of the world's uranium ore bodies are found at unconformities between metamorphosed sedimentary rocks and younger, overlying sandstones (Figure 13-28). Mineralization occurs after deposition of the overlying sandstone by hydrothermal alteration, as hot (110°C–140°C) salty groundwater in Earth's crust is moved through the rocks, dissolving uranium in some places and concentrating it as uranium oxide minerals in others.

Worldwide known recoverable uranium resources are estimated to be 5.4 million tonnes (in 2009), with 31 percent in Australia, 12 percent in Kazakhstan, 9 percent each in Canada and Russia, and the remainder in multiple other countries. About 1 percent of that amount was produced for nuclear fuel in 2010. The amount of uranium ore on Earth is estimated to be sufficient to last at least a century at present rates of production. In addition, about a third of global uranium reactor fuel now is derived from secondary sources, such as decommissioned nuclear warheads, reenrichment of uranium mining waste, and reprocessing of uranium and mixed oxide fuel.

About a third of the world's uranium is mined with an in situ leaching process that extracts uranium in solution; most other uranium mined for nuclear power comes from open-pit or underground mining. The mined or leached uranium is processed to make a product called yellowcake, a powder that consists of about 80 percent uranium oxide (see Figure 13-28). Yellowcake is used, typically with further processing, to make fuel pellets that go into long fuel rods used in nuclear power plants. Uranium fuel pellets are about the size of a fingertip, and each pellet contains the energy equivalent of 17,000 cubic feet of natural gas, or 1780 pounds of coal, or 149 gallons of oil.

THE NUCLEAR FUEL CYCLE

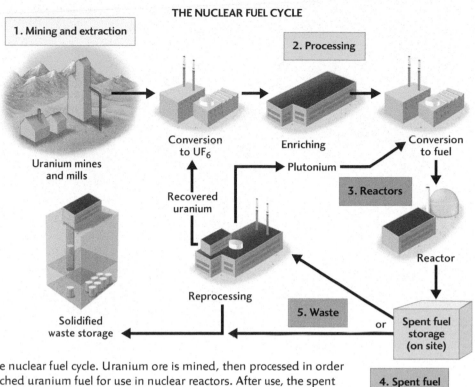

1. Mining and extraction

Uranium mines and mills

2. Processing

Conversion to UF$_6$

Enriching

Plutonium

Recovered uranium

Conversion to fuel

3. Reactors

Reactor

Solidified waste storage

Reprocessing

5. Waste

or

Spent fuel storage (on site)

4. Spent fuel

Figure 13–27 The nuclear fuel cycle. Uranium ore is mined, then processed in order to produce an enriched uranium fuel for use in nuclear reactors. After use, the spent fuel is stored, commonly on site in the United States, or reprocessed to a different waste product for storage.

(a)

(b)

Figure 13–28 (a) In this example at the Ranger Mine in Australia's Northern Territory, the ore body is the uranium-mineralized part of the Lower Proterozoic Cahill formation, a metamorphosed sedimentary rock sequence exposed in the wall of the open pit. The overlying rock is the Middle Proterozoic Kombolgie sandstone, forming the mountains in the background. (b) Uranium minerals are mined from rock, crushed, and chemically processed to make yellowcake, a powder that consists of about 80% uranium oxide. *(a: Independent Picture Service/Alamy; b: ©AREVA, Jean–Marie Taillat)*

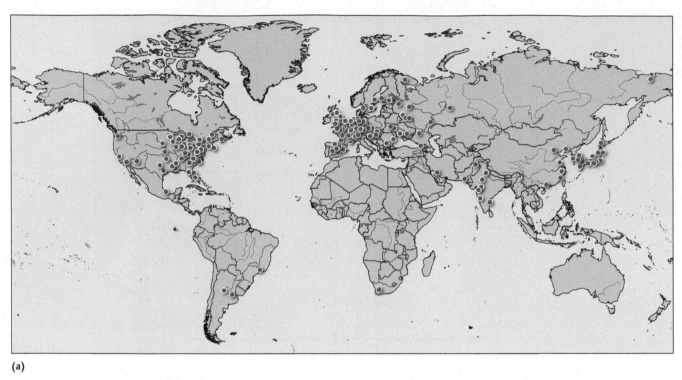

(a)

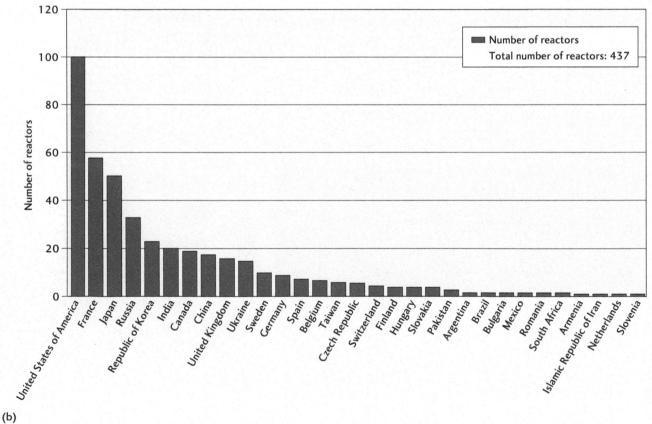

(b)

Figure 13–29 (a) Map of nuclear power plants in the world and (b) number of reactors per country.

There were more than 430 nuclear reactors in the world in 2013 (Figure 13-29). In a typical commercial nuclear power plant, the induced decay of uranium-235 releases heat, which boils water into steam, which turns a turbine to generate electricity (see Figure 13-26). A nuclear power plant is similar to other electric power plants run by steam except that the heat comes from nuclear reactions. Keeping the heat under control is critical to keeping the plant intact and preventing radiation leaks.

Fission and Fusion

Two types of nuclear reactions can release energy: *fission,* a type of nuclear decay, and *fusion,* the combining of atomic nuclei. Both reactions release energy, but so far only fission has proved practical for generating substantial amounts of electricity.

Nuclear fission is the splitting of a large nucleus into smaller nuclei by bombarding it with neutrons. Nuclear fission energy has its origin in the formation of Earth and the solar system. During the explosions of massive stars called supernovae, high temperatures and pressures fused smaller nuclei into larger nuclei of

the atomic mass of iron or greater, including uranium. Large atomic nuclei were preserved in Earth, but some are unstable and spontaneously decay to smaller nuclei, providing natural clocks by which to date minerals (see Chapter 6). The nuclei of other heavy isotopes, such as uranium-235, are stable under natural conditions, but can be induced to break apart and release energy when bombarded by neutrons.

In a fission reaction, some mass is converted to energy, and the amount lost is proportional to the amount of energy released. Inside a nuclear reactor, uranium-235 is bombarded with neutrons to split the isotope, starting the fission reaction. Neutrons ejected from the uranium nucleus can cause further fissions in other uranium-235 atoms in a *chain reaction* (Figure 13-30). In addition, some fission products themselves are naturally radioactive and will continue to decay from one radioactive isotope to another. The kinetic energy of these fission fragments is converted to heat, which is used to make the steam that drives a turbine to generate electricity.

Nuclear fusion is the welding together of atomic nuclei, a reaction that occurs naturally only in stars, where temperatures exceed 100 million degrees Celsius

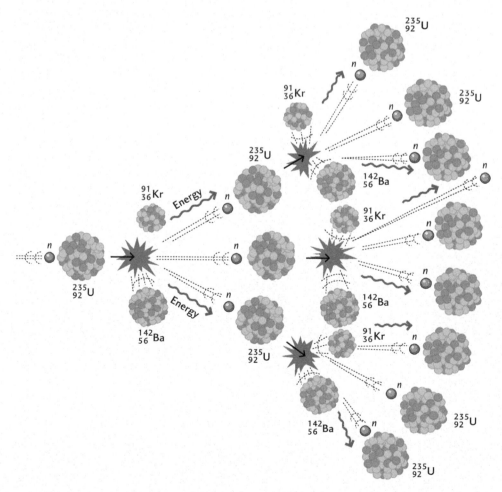

Figure 13-30 When a uranium-235 nucleus is bombarded by a neutron, it splits into two smaller atoms (krypton and barium), and several neutrons are ejected from the nucleus. These neutrons then bombard other uranium-235 atoms, triggering a chain reaction that will continue until most atoms are split or the neutrons are absorbed by a nonfissionable substance. At nuclear power plants, the chain reaction can be kept under control with boron solution, a substance that absorbs neutrons. (The lower number shown for each nucleus is the number of protons.)

and pressures are extremely large. In our Sun and other stars, hydrogen atoms fuse to form helium atoms and release large amounts of energy in the process. Hydrogen can be dissociated from water with an electric current and could be a potential energy source on Earth. However, achieving the temperatures and pressure necessary to initiate fusion requires more energy than the reaction tends to generate. No safe and economical commercial fusion power plant has yet been built.

Nuclear Power Plant Safety

The nuclear power industry has been affected by changes in public opinion—partly in response to several nuclear power plant accidents—and its growth has waxed and waned as a result of political decisions. The first nuclear power plant in the United States was built in 1957 in Shippingport, Pennsylvania. By the late 1970s, the United States had 54 operating nuclear power plants and the U.S. Nuclear Regulatory Commission, which regulates the nuclear industry in the United States, had issued construction permits for 174 new reactors. Three accidents involving nuclear power plants since then, however, have affected public opinion regarding nuclear energy, as well as the permitting and building costs of new power plants.

In 1979, after an accident at the Three Mile Island nuclear power plant near Harrisburg, Pennsylvania, no construction permits for new plants were issued until 2012. Loss of cooling fluid in the reactor core resulted in its partial meltdown. A **meltdown** occurs when the temperature in the reactor becomes too high, melting the uranium-laden fuel rods in the core and destroying the reactor container. Once the protective container is breached, radioactive material is released to the environment. As engineers at Three Mile Island struggled to control the temperature in the core and prevent a complete meltdown, the governor called out the National Guard and ordered the evacuation of surrounding areas. Fortunately, the engineers were able to gain control of the reactor temperature in time to avoid catastrophic failure. Some radioactive gases were released at the time of the accident, but subsequent studies indicate that no statistically significant increases in mortality occurred in the local population.

An accident at a nuclear power plant in Chernobyl, Ukraine, in 1987 was even more serious than that at Three Mile Island, releasing radioactive waste to the atmosphere that spread throughout the northern hemisphere. Telltale signs of radiation in the atmosphere alerted workers at a Swedish nuclear power plant that an accident had occurred upwind of their own facility. Not until Sweden confronted the Soviet government was news of the meltdown released to the world. At Chernobyl, the core reached temperatures as high as 3000°C,

completely destroying the reactor container and releasing as much as 80 percent of the radioactive fuel in a giant explosion. The reactor was on standby status at the time, yet somehow it surged to 50 percent of its capacity in a short period, resulting in an uncontrollable nuclear chain reaction. The accident caused 31 immediate deaths from massive radiation exposure. Within a week, radioactive gases and particles drifted downwind and spread over Scandinavia and Western Europe, contaminating the landscape with radioactive fallout. To this day soils in the immediately surrounding area are contaminated. Most of the several billion people in the northern hemisphere were exposed to small amounts of radiation from Chernobyl over the next week, and those living within several tens of kilometers of the reactor were exposed to extremely high doses. Long-term exposure to radiation in the areas of greatest fallout could have resulted in as many as 30,000 deaths over the next several decades.

In March 2011, the Tōhoku, Japan, earthquake (magnitude 9.0) generated a tsunami up to 6 m in height along the Japan coastline (see Chapter 3). Eleven nuclear reactors were shut down automatically. At three of those, meltdowns occurred. Explosions occurred at the three reactors due to the build up of hydrogen gas after cooling systems failed. In addition to removing the heat that builds up from radioactive decay processes in a reactor, the cooling system maintains the temperature of pools in which spent fuel rods are stored. Backup diesel generators for emergency situations were destroyed at some power plants because the tsunami overtopped seawalls designed to protect them from tsunami waves. Hundreds of thousands of nearby residents were evacuated. External power lines were destroyed as well. Some radioactive water was released at one plant. Later, radioactive substances were detected in tap water and soil in multiple districts surrounding the reactors. Scientists continue to investigate the effects of this disaster in terms of the release of radiation to air, water, and soil, and its incorporation into the food web.

Before the Tōhoku earthquake and tsunami, nuclear power plants provided 30 percent of Japan's electricity. After the earthquake, 11 nuclear reactors in Japan were shut down. As of May 2013, eight of these were operating again. The earthquake and tsunami have altered Japan's, and perhaps other countries', long-term plans for relying on nuclear energy for electricity, and particularly those countries in seismically active zones (see Figure 13-29).

Nuclear Waste

Concerns about nuclear power focus on three issues: extraction of nuclear fuel at mine sites and nuclear power plant safety, both discussed above, nuclear weapons proliferation, a topic for another book, and nuclear waste disposal, discussed below.

Radioactive Mine Wastes Like most mining, uranium mining produces residual material that is disposed of as tailings. The volume of material is not large compared to that of other mining operations, but it is far more dangerous because the tailings are radioactive. Although most of the uranium is extracted from the ore deposit, its radioactive "daughter" products are not, and the radioactivity levels still are many times higher than the background levels to which we are exposed in everyday life. About 200 million metric tons of waste is stored in licensed tailings piles in the United States, but more than 26 million metric tons of radioactive tailings also exist at 24 abandoned sites.[59] Of the abandoned uranium tailings sites in the United States, 17 were remediated as of 1999.[60] In a typical remediation method, the tailings are moved to a multibarrier repository designed to reduce emissions of radioactivity, prevent infiltration of water, and prevent erosion for up to 1000 years. Unfortunately, at least 10,000 years may be needed for the tailings to dissipate enough radioactivity to be at levels considered safe.

Disposal of Nuclear Waste One of the most challenging problems associated with nuclear energy is the disposal of its radioactive waste products. Nuclear waste classed as *low-level* includes the machinery, equipment, cleanup residues, and protective clothing used at nuclear power plants, hospitals, and research laboratories. Nuclear waste classed as *high-level* includes the highly radioactive liquid and solid residue from spent nuclear fuel rods and weapons production. After several years, fuel rods in a reactor core must be replaced, because although still radioactive, they contain insufficient uranium-235 for fission reactions. The rods, with their load of radioactive decay products, must be removed and replaced. Some daughter products decay into stable atoms within a few years, but others remain radioactive for many thousands of years.

The U.S. government requires states to provide for the timely disposal of *low-level* radioactive waste generated within their borders. Several sites accept waste from other states.[61] But finding a site that has the appropriate conditions for disposal can be challenging. After 20 years of scientific investigation, a site near Carlsbad, New Mexico, was approved as the first long-term underground repository for low-level military radioactive waste. The Carlsbad site began accepting military waste—gloves, aprons, and lab equipment—in 1999.[62] The site is underlain by thick beds of salt, some of which have flowed upward and formed salt domes. Salt is a good repository rock because it is ductile and does not break when deformed, so it has few or no fractures along which wastes might leak. As salt flows upward, it will

entomb the waste, sealing it off from the surroundings. In addition, salt is highly soluble in groundwater, so its very existence as a solid indicates that water—which can corrode radiation-proof canisters—is not likely to exist in the repository.

At year's end in 2011, about three quarters of high-level radioactive waste from spent fuel was in temporary storage under water at more than 77 nuclear power plant locations across the United States, and the rest was stored in dry casks. Another 2000 metric tons are generated each year (Figure 13-31).[63] Spent fuel rods and wastewater from the 300 or so nuclear power plants in other nations also require permanent disposal. Disposal

Figure 13-31 Spent fuel rods stored temporarily under water at a nuclear power plant. Many nuclear power plants around the world rely on temporary storage while nations decide how and where to store their high-level radioactive waste permanently. Although the fuel rods no longer have enough uranium-235 to be useful in a reactor core (unless processed), they still are hot enough to glow under water and will remain radioactive for thousands of years. *(Timothy Fadek/ Bloomberg via Getty Images)*

must isolate the waste for approximately 10,000 years, by which time the amount of radioactivity will be comparable to that of a similar size uranium ore deposit. Isolating something from the biosphere, atmosphere, and hydrosphere for that long is difficult on a dynamic planet. Furthermore, disposal sites must be in remote locations that can be protected from sabotage.

During the approximately 40 years of nuclear power generation, no permanent repository for high-level nuclear waste has yet begun operating in the United States. In 2010, the U.S. Department of Energy withdrew the license for a waste repository at Yucca Mountain, Nevada, that had been under scientific investigation for permanent disposal of waste for decades. The U.S. government continues to investigate options for long-term permanent storage of high-level nuclear waste.

■ Nuclear energy plants use neutrons to split the nuclei of uranium-235 atoms. In this process, called fission, energy and more neutrons are released, and these neutrons split other atoms to generate more energy in a chain reaction.

■ Uranium-235 is obtained from uranium minerals, typically oxides, and is used to produce fuel pellets that are inserted into the core of a nuclear reactor. After several years, the fuel rods are spent and must be replaced. Because they are still highly radioactive, long-term storage is a scientific challenge and a political problem.

■ Accidents at nuclear power plants occur if the reactor is not cooled sufficiently to prevent a meltdown in the reactor's core and the radiation-proof casing around the core is damaged as a consequence.

RENEWABLE ENERGY

A number of energy resources are considered to be naturally replenishing or perpetually replenishable if managed with care, and as a consequence are referred to as **renewable**. In contrast, nonrenewable energy resources are depletable, in that they are not created or restored at the same rates at which they are consumed (see Figure 1-23). Some renewable resources are available in such large amounts that they are essentially inexhaustible. These include wind and solar energy. Others must be managed carefully in order to avoid depletion. These include hydropower, geothermal sources, and biomass. Geothermal energy was discussed in an earlier section of this chapter. Here, we focus on solar, wind, hydropower, and biofuel energy resources.

Recall from Figure 13-3 that only 9 percent of the world's primary energy consumption in 2011 was obtained from renewable energy resources. Of the world's electricity, however, about 19 percent was obtained from renewable energy resources. Worldwide, solar photovoltaics, wind, and biofuels are the fastest growing in terms of installed capacity. Furthermore, during the past 40 years the costs of generating electricity with solar photovoltaics and wind have declined in concert with technological improvements. As a result, *levelized costs* for renewable energy resources are becoming competitive with those of conventional energy resources (gas, coal, and nuclear).[64] Levelized costs measure the overall competitiveness of an electricity-generating technology by including the costs per kilowatt-hour to build and operate a power plant, obtain fuel, and maintain the plant for the operating life of the plant.

Solar Energy

The Sun's energy can be used in numerous ways to provide light, heat, and electricity, and it is, from a human perspective, essentially an undepletable resource.[65] The obvious advantages of solar energy are its overall amount and accessibility. It is the most abundant energy source on Earth, with 20 days of sunshine equivalent to the energy contained in the world's recoverable fossil fuels. Another positive aspect of solar energy, in contrast to fossil fuels and nuclear fission, is that solar recovery and power-generating systems produce essentially no pollutants and disturb little land.[66] Furthermore, once a solar energy system is installed and has operated long enough to pay back the initial cost of construction, the supply of energy is free, and the only costs are those of maintenance.

On the other hand, even though a large amount of solar energy reaches Earth's surface, its concentration at any one spot is relatively small (Figure 13-32). The global annual average on land is 175 watts per square meter.[1] For comparison, a laptop computer might use 15 to 60 watts. In a residence, a typical incandescent lightbulb uses 25 to 100 watts, a compact fluorescent bulb uses 5 to 30 watts, and an advanced LED bulb uses 5 to 20 watts.

Solar radiation is an energy flow and must be intercepted in order to be useful. The amount of incoming direct solar radiation is concentrated in the world's middle latitudes, the southwestern United States, Australia, and the Sahara and Kalahari deserts in Africa (see Figure 13-32). In the United States, for example, the hot, dry southwestern states receive more than twice as much sunlight as the chilly, damp northeastern states. Furthermore, receipt of solar energy in cool, moist areas frequently is hampered by heavy cloud cover. A cloud-covered area intercepts relatively little solar energy. The primary technological challenge of using solar energy is to devise solar collectors that can intercept enough of the Sun's rays to meet energy demands. The second is to store enough solar energy to provide a sufficient and uninterrupted supply when little solar energy is intercepted.

Solar energy can be used to provide heat and hot water through passive and active solar heating systems.

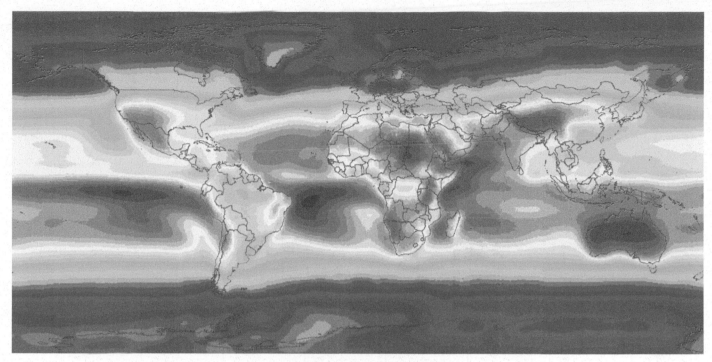

Yearly Mean of Irradiance in W/m²

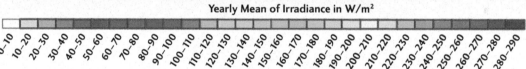

0-10 10-20 20-30 30-40 40-50 50-60 60-70 70-80 80-90 90-100 100-110 110-120 120-130 130-140 140-150 150-160 160-170 170-180 180-190 190-200 200-210 210-220 220-230 230-240 240-250 250-260 260-270 270-280 280-290

Figure 13-32 The global distribution of average annual solar radiation that reaches a horizontal surface on the ground. *(Mines Paris Tech/Armines 2008)*

A *passive solar heating system* intercepts solar energy and allows it to pass through to indoor surfaces that convert the energy to low-temperature heat, but it does not allow the heat energy to pass back out readily. Any ordinary window can function as a passive solar heating system, but a deliberately designed passive system generally is structured with energy-efficient windows strategically placed and insulation to keep heated air from escaping. Buildings are oriented to maximize the amount of solar radiation that can be collected and are built with thick concrete or adobe walls and floors and other materials that can store a lot of heat and release it slowly (Figure 13-33). Salt has a high heat capacity and sometimes is impregnated in timbers used for construction to enhance their heat-storage capacity. Windows, shades, and sometimes shutters are opened and closed throughout the day to regulate the flow of heat into and throughout a solar-heated house. In most modern designs, components of a passive solar system are wired to a computer that monitors and controls the actions of each part of the system.

In *active solar heating,* solar collectors typically are mounted on rooftops and connected to pipes and pumps that circulate heated air or water throughout a building. Excess heat can be stored in tanks of hot water or in other materials, such as rocks. As new technologies are developed and tested, active solar heating is becoming more common because of its ability to store excess heat. From 2010 to 2011, for example, the use of solar water and space heating increased 25 percent, reaching a total global capacity of 232 gigawatts (thermal, not electric).

One of the primary uses for energy in industrialized nations is the generation of electricity to run appliances and provide light. Solar energy can be converted directly to electricity in devices called photovoltaic (PV) cells. **Photovoltaic cells** are made of a *semiconductor* material, so called because at high temperatures it conducts electricity almost as well as metals do, but at lower temperatures its conductivity practically disappears. One of the most popular semiconductors is silicon, because it is abundant, inexpensive, and easy to manipulate. Molten silicon is poured into sheets thinner than paper, so they heat rapidly in sunlight. The cooled and solidified sheets are cut to a variety of shapes and sizes and coated in glass or plastic for protection. Sunlight shining on a photovoltaic cell excites electrons in the semiconductor

(a)

Figure 13–33 (a) This structure in Snowmass, Colorado, houses the Rocky Mountain Institute, an energy and conservation organization founded by Hunter and Amory Lovins. The southern part of the facility is partly built into the ground for additional insulation and summer cooling. Nearly all heat, hot water, and lighting are provided by the Sun.
(b) A common design for a passive solar house uses a sunspace with insulated windows facing the direction of greatest incoming solar radiation to collect solar energy that heats inside air. Vents enable warm air to circulate, return to the sunspace when cooled, and be rewarmed. The angle of the Sun is higher in summer and lower in winter, so trees that have summer foliage can be used to block solar radiation during warmer months, and additional vents can be opened for greater cooling. *(a: Printed with permission from Rocky Mountain Institute)*

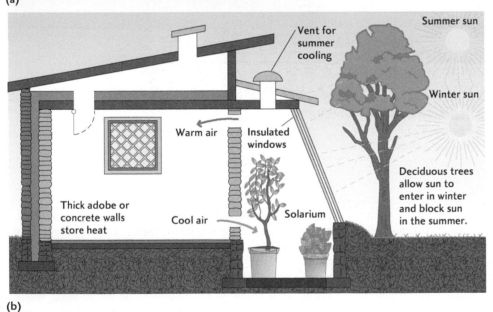

(b)

material, which then flow and create an electric current. Each cell produces just a tiny amount of energy, so many of them are wired together to form a *solar panel*. A number of panels are mounted either on a rooftop or on a structure that can follow the Sun throughout the day (Figure 13-34). Current technology relies on banks of batteries to store electricity and supply it to a building as needed.

Alternatively, electricity can be generated in a **solar-thermal power plant**. One design, called a solar power tower, uses reflective solar collectors that track the Sun and concentrate heat and light on a central receiver mounted atop a tower (see Figure 13-34). Fluid in the receiver is heated to the boiling point, producing steam that is used to generate electricity by turning a turbine in much the same way that steam-driven turbines are used in other types of power plants. The largest such plant in the world is the Ivanpah Solar Electric Generating System in the southern California desert, with 173,500 heliostat (sun-tracking) mirrors spread across 3500 acres of public land focusing solar energy onto the top of each of three solar power towers (see Figure 13-34).[67] Near completion, it is anticipated to provide 392 megawatts of electricity, sufficient to provide power for 140,000 homes. The carbon dioxide offset is estimated to be equivalent to that of the emissions from 70,000 cars. Dozens of

(a)

Figure 13-34 (a) The flat parts of the roof of the Minneapolis Convention Center host the largest solar array in the region as of 2010, with 600 kW of solar photovoltaic panels (2613 panel array) that provide 5 percent of the center's power and offset carbon emissions that would be generated if fossil fuels were used instead. (b) The Ivanpah solar-thermal power plant will be the largest in the world when completed and operating (anticipated by 2014). Heliostats spread across 3500 acres of public land in the desert of southeastern California focus solar energy on three solar power towers that will generate up to 370 megawatts of electricity, enough for 140,000 houses. *(a: Courtesy Meet Minneapolis Convention and Visitors Association; b: Jim West/Alamy)*

(b)

solar-thermal power plants now exist in the world, with most located in the United States, Spain, and China, and a total generating capacity of about 12 gigawatts among the three countries.

Environmental impacts from solar energy use are relatively minimal.[68] The scale of the system correlates with level of impact on land use. Solar electric systems range in size from small rooftop photovoltaic arrays to utility-scale photovoltaic systems and large-scale solar-thermal power plants. Utility-scale photovoltaic systems and concentrating solar power plants require about 1 km^2 of land for every 15 to 60 million watts of electricity generated.[68] Like fossil fuel and nuclear power plants, concentrating solar-thermal power plants use some water for cooling, although the amount used varies widely with different types of plants.

Wind Energy

Uneven solar heating of Earth's surface generates areas of high and low pressure that lead to moving air, or wind (Figure 13-35). Wind moves at average speeds that range from about 2 to 11 m/s, with weather patterns, irregularities in Earth's surface, the planet's rotation, water bodies, and vegetation causing variations in wind speed. This moving air can be used as an energy

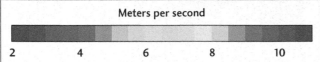

Meters per second

2   4   6   8   10

Figure 13–35 World average of wind speed (m/s) at 100 m above sea level, the height of modern wind turbine hubs. Values are derived from a global climate computer model.

resource (Figure 13-36; see Chapter 11). For centuries, people have harnessed the kinetic energy of wind to do mechanical work, using windmills to lift water from wells and grind grain in mills, and using sails to propel sailboats and sailing ships.

Today, wind power is the most rapidly growing of all energy resources used for utilities, and wind-generated electricity is becoming economically competitive with conventional sources of electricity (4 to 8 cents per kilowatt-hour). By late 2011, worldwide capacity from wind power generation was 237,000 megawatts. China, with the highest growth rates for wind power, is expanding its capacity by 60 percent over the next few years.[69] The hubs of modern wind turbines reach heights of 100 m, and at this height about 80 terrawatts of cost-competitive wind power (i.e., wind speed greater than 7 m/s) is available over Earth's land surface and near-shore coastal areas worldwide.[70] This is equivalent to about five times the total amount of energy used worldwide in 2008.

Like direct solar radiation, wind is an energy flow and needs to be intercepted in order to be converted into a usable form of energy. Places with consistently high wind speeds are ideally suited for utilizing wind energy. One such area is the Great Plains of the central United States, sometimes called the Saudi Arabia of Wind. The Great Plains have great wind energy potential in much the same way that Saudi Arabia has immense stores of oil. Examining Figure 13-35, one can see that the Great Plains have average annual wind speeds greater than

7 m/s, faster than in most other parts of the United States. This world map of wind speeds also reveals that coastal areas have some of the highest wind speeds, and for this reason many wind turbines are located in near-shore coastal areas, with some even located offshore. Another area with potential for wind energy utilization is the high-speed wind corridors along ridgelines of mountain belts, such as along the Appalachian and Rocky mountains in the United States.

At present, most wind energy is converted into electricity by wind turbines, a modern adaptation of ancient windmill technology (see Figure 13-36). More than 200,000 wind turbines existed worldwide as of 2012. China, the United States, Germany, and Spain lead the world in operating wind power capacity. The energy in wind turns the blades of a turbine, typically two or three blades that are shaped like propellers and mounted about a rotor. The rotor connects to a main shaft that spins a generator, creating electricity that is sent to an electrical grid, a network that uses transmission and distribution lines to supply customers with electricity.

Wind farms pose fewer environmental impacts than most other energy sources. The turbines have small physical footprints relative to the amount of electricity produced, and those located offshore have even less impact with respect to competing land use. Furthermore, wind turbines are compatible with other land use, including grazing, farming, and forestry. No fuel source is mined or transported, and no substances are emitted to air or

(a)

Figure 13–36 (a) Modern wind turbines are located in areas where wind speed is at least 7 m/s. (b) Facing into the wind, the turbine blades spin and turn a shaft that generates electricity that is distributed to an electrical substation and wider grid for use. *(a: Iberdrola Renewables, Inc./Courtesy NREL, U.S. Department of Energy)*

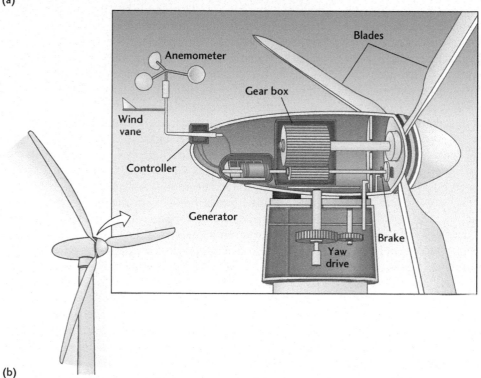

(b)

water. Like solar power, using wind power offsets emissions from other fuel sources, and as a result is a way to reduce overall carbon dioxide emissions worldwide. Finally, no water is required for cooling, a common concern with fossil fuel and nuclear power plants, especially in dry regions.

Some negative impacts of using wind power do exist, including visual disturbance, as along the crest of a scenic mountain or historic village. Wind farms, the roads used to access them, and electricity distribution lines can alter habitat and displace species. The last of these is minimal, though, in comparison to habitat disruption from activities such as aboveground coal mining. Of most concern is the impact of wind turbines on animals, and particularly on bats and birds.[69] The air-pressure changes caused by spinning blades can cause bats to die from internal hemorrhaging if they fly through the wake of the blades. The blades on turbines can kill birds by direct

collision, and some wind farms are located along migratory bird routes or near the habitat of hawks, eagles, and other birds of prey. Blades move at speeds up to hundreds of kilometers per hour. Studies show, however, that wind farms account for relatively few avian deaths, and that other human-associated factors are larger contributors (e.g., domestic pets, autos, window glass). Nevertheless, some critics point out that migratory birds of prey are unlikely to die from these other causes and are more likely to be impacted by wind turbines because the birds use high-speed wind corridors at the altitudes of turbine blades. Furthermore, some of these large birds are already endangered or few in number. Wind power operators are using a variety of strategies to avoid avian deaths, including better siting, use of fewer and larger turbines, use of radar to detect incoming flocks, and even temporary shut downs when birds approach during migration times.

Hydropower

Energy from flowing water, like energy in wind, comes ultimately from the Sun, as solar energy drives the hydrologic cycle. Flowing water has been used for thousands of years to turn waterwheels and more recently to turn turbines that generate electricity in hydroelectric power plants. Because of rainfall variability and the need to control the rate of energy production, dams are built to store water in reservoirs and release it over waterfalls, generating electricity as needed (see Chapter 9, particularly Figures 9-17 and 9-20). Most hydroelectric plants are part of multiuse projects, with flood control, recreation, irrigation, or other associated services.

Hydropower generates more electricity than any other renewable energy resource, and has an advantage over most other renewable energy resources in that its energy potential can be stored (e.g., in a reservoir). About 16 percent of the world's electricity and 6 percent of the world's total energy are produced from hydropower. The estimated world resource base is 9.2×10^{19} J per year, of which only about 14 percent has been developed. Asia and Africa have the largest resource bases, with 30 percent and 19 percent of the world total, respectively. North America and Europe have developed the largest proportions of their resource bases (48 percent and 27 percent, respectively).

In addition to energy production, the benefits of dams to humans include flood control, water storage, and recreational facilities, but these benefits do not always outweigh the environmental costs. Creating new large hydroelectric dams is expensive, can have substantial social impacts, and causes profound damage to river systems. Dams impede the flow of water and change flow rates, disrupt aquatic life, trap sediment upstream, cause erosion downstream, drown river valleys, prevent nutrients from reaching adjacent floodplains, provide breeding grounds for mosquitoes, and block fish migration patterns. Furthermore, large reservoirs that flood vegetated valley bottoms can release large amounts of methane and carbon dioxide as the plants decompose.

Recall from Chapter 9 (see Figure 9-18) that in order to build the Three Gorges Dam, the largest and most complex hydroelectric dam in the world, China submerged hundreds of kilometers of riverside ecosystems. The elevated pool of water upstream of the 181-meter-high dam extends 600 km upstream, submerging 13 cities and 466 towns. More than 1.3 million people were displaced, the greatest relocation in human history. The benefits, however, include staving off a severe energy shortage and slowing the expansion of more coal-burning power plants, which already supply about 65 percent of China's electricity.[71] In contrast, hydropower provides about one-fifth of China's electricity.

New dam construction in the United States and most of the developed world is likely to be minimal in the future. Dam construction probably will continue in developing nations, despite the environmental consequences, because of the ever-pressing need for energy. Asia and South America, in particular, are currently expanding their hydropower capacity.

Biomass

Plant matter and animal waste, called **biomass,** can be burned directly for its energy, or it can be converted to gaseous or liquid **biofuels** to be used in the same ways as fossil fuels (Figure 13-37). Overall, the proportion of energy provided from biomass has declined over the past few centuries, but wood, animal dung, and other biological materials are still used for heating and cooking by more than 2 billion people worldwide. Globally, biomass supplies 14 percent of total energy demand. In countries such as Canada and the United States that rely heavily on fossil fuels, biomass accounts for only 4 to 5 percent of energy use, but in developing nations, it accounts on average for 36 percent of energy use.

An example of a relatively common biofuel is the use of fermented corn or sugarcane to make an alcohol-gasoline fuel mixture. Ethyl alcohol (also called ethanol, with chemical formula CH_3CH_2OH) can be mixed with gasoline for use in automobile fuel tanks. Ethanol is used in the United States both as an octane booster and to reduce dependence on foreign oil. It is added to gasoline in blends of up to 85 percent ethanol. Cellulosic biomass can be used as a feedstock for ethanol production, and this newly developing technology can alleviate the disadvantage of using cropland and a food product (e.g., feed corn for animals) for fuel. When corn prices rise, however, its advantage for use as a fuel source diminishes. Ethanol production for fuel has grown rapidly in the United States, increasing in production capacity by 373 percent from 2001 to 2008.

In the United States, biomass energy supplied 2 percent of energy demand in 2011 (see Figure 13-3). Of this amount, 2.87×10^{18} J was generated from wood and wood waste, and the remainder from agricultural and industrial waste and municipal solid waste. The use of municipal waste as an energy source could be expanded dramatically. About 144 million tons of municipal wastes are generated in the United States every year. Of this amount, 10 percent is incinerated, 10 percent is recycled, and the remainder is placed in a landfill. If all nonrecyclable municipal waste were incinerated to generate electricity, this resource could supply 8 percent of U.S. electricity consumption. The major drawback of using this resource is that burning organic waste generates carbon dioxide and other potentially harmful emissions. Without stringent emission controls, a large increase in the use of biomass energy would have increasingly severe environmental penalties that would at some point surmount the environmentally positive aspects of this energy source.

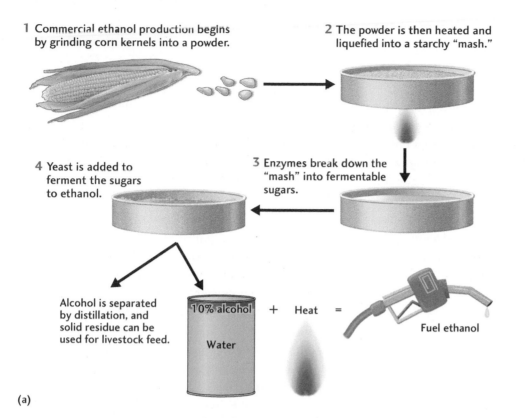

1 Commercial ethanol production begins by grinding corn kernels into a powder.

2 The powder is then heated and liquefied into a starchy "mash."

4 Yeast is added to ferment the sugars to ethanol.

3 Enzymes break down the "mash" into fermentable sugars.

Alcohol is separated by distillation, and solid residue can be used for livestock feed.

10% alcohol Water + Heat = Fuel ethanol

(a)

(b)

Figure 13–37 (a) Ethanol (ethyl alcohol) is produced by milling corn (or sugar cane or cellulose), then heating and adding yeast to the liquefied mash to ferment the sugar and convert it to ethanol. (b) This product is added to gasoline and can be used in automobiles, as in the blend known as E–85 (85% ethanol, 15% gasoline) shown at this pump in Washington, DC. *(b: David R. Frazier Photolibrary, Inc./Alamy)*

Biomass is a renewable energy stock as long as its supply is replenished at the same rate that it is consumed. This requires at least one tree to be planted for every tree cut down, one hectare of crops to be planted for every hectare harvested, and no net loss of soil organic matter. In practice, maintaining equal rates of replenishment and depletion is difficult, especially in those parts of the world where fuel wood is the primary energy resource and its supplies are scarce.

Energy Efficiency and Conservation

The greatest hope for the world's energy future is in improving the efficiency of energy use, and some energy analysts anticipate that greater emphasis on energy efficiency and conservation will lead to a "new energy era" (Box 13-4).[72] Because of physical laws governing energy conversions, some energy waste is unavoidable.

BOX 13-4 CASE STUDY

The "New Energy Era": Green Buildings and The Road Taken

Forecasting what might happen with energy use in the future is difficult, but an important lesson was learned from past efforts to do so. In the journal *Foreign Policy* in 1976, Amory Lovins, a physcist who studies energy issues, described two paths for energy futures.[73] Referring to these possible energy futures as "soft" and "hard" paths, Lovins began the article—possibly the journal's most frequently reprinted—by quoting from the last stanza of a well-known poem by Robert Frost, "The Road not Taken":

> Two roads diverged in a wood, and I—
> I took the one less traveled by,
> And that has made all the difference.
> – Robert Frost

Asking what road America was following with its energy policies, Lovins then wondered which path the country might follow instead, cautioning that the choices were mutually exclusive. The path America was following at the time, the "hard" path, seemed likely to be more of the same. It portended continued expansion of centralized technologies such as massive coal-burning power plants (dubbed "future technologies whose time has passed" by Lovins) and increasingly greater energy consumption, particularly in the use of electricity. This path was inefficient: it was associated with substantial mismatches in scale and losses of energy during the conversion from resource to service, as in the process of burning coal at high temperature to generate electricity to illuminate a lightbulb. Large economic and sociopolitical problems, such as the use of military intervention to maintain stability in the Persian Gulf, loomed along this path, albeit a familiar one.

The soft path, on the other hand, would combine efficiency, conservation, and a shift to more renewable energy resources such as solar and wind, all while doing more with less. Adding home insulation, turning off lights not in use, carpooling, recycling, bicycling, and using other energy saving practices were listed for this energy future. This path would better match scales between resources and end-use tasks. Lovins wrote that he chose the phrase "soft path" to describe certain energy technologies because they have highly attractive technical, economic, and political attributes. In particular, he described them as "flexible, resilient, sustainable and benign." If chosen, Lovins predicted, this path might offer an end to nuclear proliferation as well as other socioeconomic and geopolitical advantages.

For both paths, Lovins made schematic graphs of energy futures for the next half century from 1975 to 2025. Meant simply to be illustrative examples, the hard path showed much greater use of nuclear energy and coal and continued use of large amounts of oil and gas. Overall energy use tripled. The soft path, in marked contrast, showed a long-term decrease in total energy use; diminished use of coal, oil, and gas to insignificant amounts; and a substantial expansion in new soft energy technologies (e.g., solar energy).

In 1982, Amory Lovins cofounded the Rocky Mountain Institute (RMI), a nonprofit "think-and-do tank" based in Colorado that promotes sustainability, especially with respect to energy and resource efficiency (see Figure 13-33a). In 2008, RMI became a partner in an innovative deep retrofit of the Empire State Building (built 1929–1931) in New York City, which has led to substantial energy savings of more than $4 million annually (**Figure 1**). The Empire State Building's green renovation reduced its energy use by about 38%, earning it the prestigious LEED Gold certification in 2011.[74] The building's numerous energy efficient upgrades, including an overhaul of all 6514 of its operating windows, reduced its carbon emissions by an estimated 105,000 metric tons over 15 years. Furthermore, the office building uses only renewable energy from its electricity supplier, so it is now carbon neutral, meaning that it has a net zero carbon footprint (net zero carbon emissions after measuring carbon released and balancing with equivalent amount sequestered, offset, or purchased with carbon credits).

Ever visionary and forward thinking, Lovins wrote a new call to action in the pages of *Foreign Policy* in 2012 and, with his team at RMI, published a book that expanded on the original article's ideas.[75] Titled *Reinventing Fire: Bold Business Solutions for the New Energy Era,* the book provides a blueprint for how private enterprise, civil society, and military innovation can lead to a growing U.S. economy that uses no oil, coal, or nuclear energy and less natural gas by 2050.[72] Even better, these "bold business solutions" would save trillions of dollars in comparison to a business-as-usual approach. The power of the market—smart decisions motivated by lower costs—would drive this transition away

Figure 1 The Empire State Building is the tallest building in the western hemisphere to have been awarded LEED certification for its massive "green" overhaul, energy efficiency, use of renewable energy, and net zero carbon footprint. Renovations to the nearly centennial age building took place during 2008–2010 and included overhauled windows, installation of insulation, and an energy management system that tenants can control via the web. The approach is being copied in other large commercial buildings worldwide. *(Mark Kelly/Alamy)*

from fossil fuels and toward even greater efficiency and use of more renewable energy resources.

As one example, buildings are an enduring capital investment, lasting many decades in most cases, and 75% of electricity use in the United States is to power buildings. Their energy use is the major contributor to greenhouse gas emissions from urban areas. Retrofitting existing commercial buildings in the United States has the potential to recoup savings of more than a trillion dollars in the form of energy that would have been lost otherwise over the next 40 years. The benefits are even greater, in that such retrofitting also would reduce greenhouse gas emissions.

It turns out that Lovins was remarkably prescient in his 1976 article. Forecasting future energy consumption on the soft path that could be taken, he projected a much smaller increase in total primary energy consumption than did those in government and industry at the time. Looking back, we now know that he was right, but most of the energy savings resulted from improvements in efficiency rather than increased use of renewable energy sources. Renewable energy resources still have much room for improvement. If Lovins is right this time, their potential will be more fully realized in the next few decades, and fears of running out of fossil fuels will be as historically quaint as those of 19th-century citizens who worried about running out of whale oil for their lights. Lovins' calls the business solutions in *Reinventing Fire* "spherically sensible," in that choosing the soft path makes sense whether one is interested in saving money, reducing carbon emissions, or enhancing global security.

The first law of thermodynamics, or the law of conservation of energy, states that in all physical and chemical changes, energy is neither created nor destroyed, but it may be converted from one form to another. By the second law of thermodynamics, when energy is converted from one form to another, some of the energy always is degraded to lower-quality, more dispersed energy that has less ability to do work.

A simplified example of the proportion of energy wasted in order for a coal-burning power plant to provide electricity illustrates the nature of these two energy laws (Figure 13-38). Chemical energy stored in coal is converted to heat (thermal energy) when it is burned, but only 35 percent of the heat energy is captured and used to make the steam that turns a turbine in a generator to produce electricity. The rest is *waste heat*, heat that is given off to the surroundings and dispersed by the random motion of air and water molecules. The conversion from chemical to electrical energy, therefore, is 35 percent efficient.

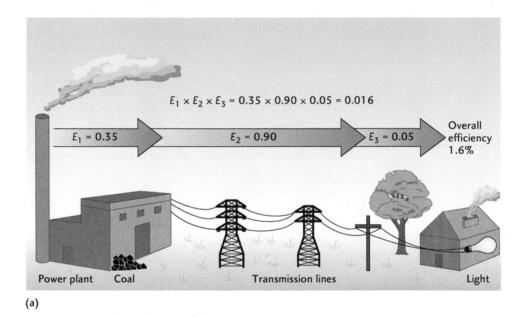

$$E_1 \times E_2 \times E_3 = 0.35 \times 0.90 \times 0.05 = 0.016$$

$E_1 = 0.35$ $E_2 = 0.90$ $E_3 = 0.05$ Overall efficiency 1.6%

Power plant   Coal   Transmission lines   Light

(a)

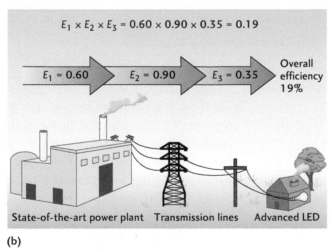

$$E_1 \times E_2 \times E_3 = 0.60 \times 0.90 \times 0.35 = 0.19$$

$E_1 = 0.60$ $E_2 = 0.90$ $E_3 = 0.35$ Overall efficiency 19%

State-of-the-art power plant   Transmission lines   Advanced LED

(b)

Figure 13–38 (a) Fuel that burns at a high temperature—as much as 600°C—is used to generate electricity to burn incandescent lightbulbs at temperatures of less than 100°C. In the process, much energy is lost as waste heat. To find the total efficiency for the conversion process shown, the efficiencies (*E*) of each step are multiplied.

(b) State-of-the-art gas–burning power plant has higher efficiency than the coal–burning power plant in the example in (a), and the advanced LED bulb has higher efficiency than the incandescent bulb in (a), resulting in an increase in overall efficiency from 1.6% to 19%.

The **energy efficiency** of an energy conversion process is the percentage of energy not lost as waste heat. When electricity is transmitted along power lines, 10 percent of the original energy captured is converted to heat because of frictional resistance as the moving electrons rub against other materials in the lines. Transmission, therefore, is 90 percent efficient. When the transmitted electricity is converted into light in an incandescent bulb, 95 percent of the electrical energy that made it to the bulb is converted to waste heat. The conversion from electricity to incandescent light, therefore, is only 5 percent efficient. From the first conversion to the last, 98.4 percent of the energy that went into this system was not used to do work, but instead was converted to waste heat. Only 1.6 percent of the energy input did work. Therefore, 0.016 is the measure of the efficiency of this whole energy conversion system. A system's energy efficiency is equal to the percentage of total energy input that does work, and can be calculated by multiplying the efficiencies of each step in the energy conversion process. Changing to a state-of-the-art gas-burning power plant and an LED lightbulb, both of which have higher efficiencies, results in a gain in overall efficiency of 17.4 percent.

All power plants waste energy in producing electricity. The wasted energy is heat. Most power plants are only 30 to 40 percent efficient, thus converting 60 to 70 percent of their initial energy input to waste heat that is dissipated into the environment. Roughly half the waste heat from fossil fuel power plants is released directly into the atmosphere through tall stacks. The remainder is released to cooling waters, which then transport the waste heat to the environment, either into the atmosphere through cooling towers or into a lake or river.

Of the 84 percent of commercial energy that is wasted in the United States, 41 percent is the unavoidable result of the second law of thermodynamics. The other 43 percent of wasted energy is avoidable. It dissipates from poorly insulated buildings that lose energy, from "petropig" vehicles that swill gasoline because they get few miles per gallon, and from careless habits such as leaving windows open when air conditioning or heating is in use. Photographs of houses taken with infrared-sensitive film can reveal "hot spots," places where heat is lost to the surroundings. In many cases such total losses are astoundingly high, equivalent to losses that would occur from large holes in the walls or from windows left open year-round.

What might motivate more people to invest in better insulation, drive more efficient cars, turn off lights when rooms are empty, and practice other energy-saving habits? The United States uses more energy per unit of gross domestic product than any other nation, nearly twice as much as Japan and most western European nations. If Americans were to reduce the 43 percent of avoidable energy waste, their energy costs would be reduced by about $300 billion each year, and the amount of air pollution and nuclear waste generated by power plants would drop significantly. Perhaps the cause of such waste in the United States is that energy is relatively cheap. Gasoline, for example, costs less now than it did in 1920 when price is adjusted for inflation, and is much cheaper than in Europe and Japan. Fossil fuels and nuclear energy would cost more if the full cost of their use, in particular the "external costs" of pollution and environmental degradation were accounted for in market prices.

■ Solar energy is the largest source of energy available, but its widespread use is limited somewhat by its relatively low concentration in any given area, especially at high latitudes, and by its unreliable availability because of cloud cover. New technologies are emerging rapidly, however, to optimize use of solar energy, and all solar technologies have relatively little to no environmental impact.

■ Solar energy can be used to heat the air and water in buildings, and it can be converted directly to electricity with photovoltaic cells.

■ With current technologies, concentrating solar electric power plants have the highest costs for producing electricity but have become economically viable in the past decade.

■ Solar energy is the ultimate source of wind and water power to generate electricity at wind farms and hydroelectric dams. The environmental impacts of power production at such facilities are less than those of fossil fuels and nuclear power, although hydroelectric dams significantly affect local stream systems and their habitat.

■ Wind power is the most rapidly growing of all energy resources for utilities, and wind-generated electricity is becoming economically competitive with conventional sources of electricity. Environmental impacts are relatively minor, with the exception of avian and bat mortality, but modern turbines and better siting are mitigating this impact.

■ New dam construction probably is likely to continue in developing nations, especially throughout Asia and Africa, because of their economic growth and their ever-pressing need for energy. Dams and their reservoirs can have substantial environmental and social consequences.

■ Solar energy is the ultimate source of chemical energy stored in biomass, which provides nearly a third of the world's people, primarily in developing countries, with energy in the form of plant matter (primarily wood) and animal waste. Biofuels are used with gasoline in some countries to reduce use of fossil fuels. The problems in using biomass include sustaining it as a renewable resource and limiting carbon dioxide emissions.

■ Substantial reductions in energy use can and have been achieved with greater attention to energy efficiency and conservation, such as the use of LED instead of incandescent or compact fluorescent lightbulbs, the replacement of old windows with new-technology ones, carpooling, and bicycling for transportation.

CLOSING THOUGHTS

People devote enormous amounts of time and ingenuity to acquiring and using energy resources, and those sources of energy in turn increase our technological capacity to find and use additional energy resources. Think of the first producing oil well in the United States drilled to a depth of about 20 m in Titusville, Pennsylvania, in 1857, and now the use of horizontal drilling, GPS technology, and hydrofracking to extract droplets of oil from fine-grained shale at depths and lateral distances of many thousands of meters. Think of mountaintop removal and valley filling to get at coal seems in the Appalachian Mountains, or the building of thousands of heliostats in the California desert to reflect solar radiation to a "power tower" that generates electricity for hundreds of thousands of people who need air conditioning and other services in order to live in the desert. The energy industry is one of constant exploration, innovation, and technological advancement.

Each discovery and development of new sources of energy has resulted in major technological, societal, and environmental changes. With shifts in dominant sources of energy during the past two centuries—from biomass (primarily wood) to coal to petroleum—have come new forms of environmental pollution and the need for environmental regulations. Societies and governments grapple with the impacts of using different forms of energy, for example by adding laws to protect water and air, or by providing incentives to use more renewable sources and energy-efficient technologies.

Meanwhile, scientists investigate how much of a given energy resource exists, such as estimating petroleum reserves. They track the ways in which greenhouse gases, such as carbon dioxide, emitted during fossil fuel burning flow through and accumulate in different parts of the Earth system, including the atmosphere, oceans, and biosphere. They monitor the impacts of valley filling on headwater streams and ecosystems, and they examine the effects of dams on rivers, sediment transport, and riparian plant and animal communities. Scientists also are involved in restoration, a process that requires understanding of how systems operate naturally, how they have been altered, and how best they can be restored so that lost ecosystem services can be regained.

At the beginning of this chapter we noted that some experts think oil consumption has reached its peak and that natural gas use will peak between 2030 and 2050. Meanwhile, however, rapidly industrializing countries in Asia, particularly China, are using more and more coal as their economies expand. Nevertheless, some energy analysts anticipate that greater energy efficiency and use of renewable energy sources will surpass fossil fuels such as oil and coal as the dominant drivers in world energy use during this century. Renewable technologies generally have little to no environmental impact, so if these analysts' predictions are correct, the world could be quite different by the next century.

SUMMARY

■ Earth, a dynamic system, receives solar energy from the Sun, generates its own internal thermal energy, and derives gravitational energy from itself, the Sun, the Moon, and other bodies in the solar system.

■ Earth's processes and changes are powered through a cyclical flow of matter and energy among the hydrosphere, biosphere, atmosphere, pedosphere, and lithosphere.

■ Petroleum (including crude oil and natural gas) is a nonrenewable energy resource originating as organic matter, primarily planktonic algae, which was buried under thousands of meters of sediments and sedimentary rocks and subjected to relatively high temperature and pressure over tens to hundreds of millions of years.

■ Coal is a nonrenewable energy resource, which commonly occurs as seams within sedimentary and metamorphic rocks. Coal originates from the partial decay and burial of land plants in wetland environments during times of changing sea level and/or relative lowering of land (i.e., subsidence).

■ Estimates of coal reserves are greater than those of any other fossil fuel. The recoverable reserves of coal are more accurately estimated than those of petroleum due to exposure of coal seams at Earth's surface.

■ Extracting, refining, transporting, and using fossil fuels such as crude oil, natural gas, and coal can result in environmental impacts that include greenhouse gas emissions, wildlife habitat disruption, and pollution of fresh water resources, oceans, and soils.

■ Geothermal energy, a renewable energy resource, utilizes steam and water that are heated underground to generate electricity in locations where heat flow is high and steam or hot water flow through a fractured reservoir, usually along the margins of tectonic plates.

■ Nuclear energy, which uses uranium-235 for nuclear fission, supplies about 8 percent of total world commercial energy use despite scientific and political concerns associated with nuclear plant safety, weapons proliferation, and waste disposal.

■ Solar and wind energy, both renewable energy resources with minimal environmental impacts, are becoming more economically viable and efficient due to new technologies.

■ New dam construction is likely to continue in developing nations due to their economic growth and high energy demand despite the environmental and social concerns associated with hydropower.

■ While the use of energy stored in biomass is most prominent in developing nations, some countries are adding ethanol to gasoline to reduce use of fossil fuels.

KEY TERMS

kinetic energy (p. 460)
potential energy (p. 460)
thermal energy (p. 460)
heat (p. 460)
quads (p. 461)
power (p. 462)
watt (p. 462)
fossil fuels (p. 462)
photosynthesis (p. 463)
petroleum (p. 465)
hydrocarbons (p. 465)
crude oil (p. 465)
natural gas (p. 465)
conventional oil and
    natural gas (p. 465)

unconventional oil and
    natural gas (p. 465)
petroleum trap (p. 467)
anticlinal trap (p. 467)
fault trap (p. 468)
salt dome trap (p. 468)
reserves (p. 470)
peak oil (p. 470)
bitumen (p. 474)
bituminous sands
    (p. 474)
oil sands (p. 474)
tar sands (p. 474)
flowback (p. 478)
coalification (p. 485)

peat (p. 485)
gyttja (p. 485)
lignite (p. 486)
subbituminous coal
    (p. 486)
bituminous coal (p. 486)
anthracite coal (p. 486)
open-pit mining (p. 488)
strip mining (p. 488)
mountaintop removal
    mining (p. 488)
valley fill (p. 488)
digital elevation models
    (DEMs) (p. 490)
acid mine drainage (p. 492)

acid rain (p. 492)
geothermal energy (p. 495)
fissile (p. 498)
nuclear reactor (p. 498)
nuclear fission (p. 501)
nuclear fusion (p. 501)
meltdown (p. 502)
renewable (p. 504)
photovoltaic cells (p. 505)
solar-thermal power plant
    (p. 506)
biomass, (p. 510)
biofuels (p. 510)
energy efficiency (p. 515)

REVIEW QUESTIONS

1. When did the dominant energy source for human endeavors change from wood to coal to petroleum? Why?

2. What are the three main sources of energy on Earth? Of these, which is the largest? Which is responsible for earthquakes? Which is responsible for the energy released when fossil fuels are burned?

3. What compounds are used by plants during photosynthesis? What compounds are released?

4. What are the similarities and differences between respiration, decomposition, and combustion?

5. What is the difference between carbohydrates and hydrocarbons?

6. In what two main forms does petroleum occur? Why?

7. What type(s) of organism is thought to be the major source of petroleum? What particular geologic conditions are conducive to its transformation to oil?

8. Name and describe three types of petroleum traps.

9. Give examples of typical source rocks and cap rocks for hydrocarbons.

10. List three environmental drawbacks associated with extracting, transporting, or using petroleum and coal.

11. Explain how oil is extracted from depths of many kilometers (both onshore and offshore).

12. What is hydrofracking, and why is it becoming more widespread and important to the energy industry? Why wasn't hydrofracking more common 40 years ago?

13. Compare and contrast rock types from which petroleum is extracted via hydrofracking with those from which it is extracted via typical well drilling and pumping.

14. Why is natural gas considered to be the cleanest fossil fuel?

15. Why is it unlikely that the hydrocarbons extracted from bituminous (oil) sands will supplant petroleum as a major source of energy in the near future?

16. From what does coal form? What geologic conditions are most conducive to its formation?

17. Why did the Carboniferous period result in thick deposits of coal on several continents?

18. What is acid mine drainage? How can it be controlled?

19. Compare and contrast the different ways in which coal is mined and give examples of the environmental impacts associated with each type.

20. How are emissions of sulfur dioxide gas from burning coal reduced at power plants?

21. What is carbon capture, why is it practiced, and what might be its benefits to the environment?

22. Why is geothermal energy practical only within limited areas?

23. What is the nature and source of the fuel used in the core of a nuclear reactor?

24. Nuclear reactors are being built in many countries: What are the reasons for growth of the nuclear industry? What events have impacted this industry's growth (e.g., earthquakes or accidents), and how have the events affected the use of nuclear power?

25. List environmental drawbacks associated with using nuclear energy.

26. What are some of the technological challenges to utilizing the full potential of solar energy?

27. List major environmental advantages associated with using solar energy rather than fossil fuels.

28. List the advantages and disadvantages of using water energy and wind energy to generate electricity.

29. What are the first and second laws of thermodynamics? What is energy efficiency?

30. How have changes in energy efficiency affected the growth of energy consumption during the past few decades?

31. What percentage of the energy used to generate electricity at a coal-burning power plant is lost as waste heat? Why is the amount so large?

32. Give examples of different ways to use energy more efficiently.

THOUGHT QUESTIONS

1. How are the processes of photosynthesis and respiration by plants examples of transfers of solar energy from one Earth reservoir to another?

2. If the annual amount of solar energy stored in plants that become buried to form fossil fuels is very tiny, why do such large accumulations of fossil fuels exist on Earth?

3. What types of changes in our lifestyles might occur if we were to switch from oil to natural gas as a primary energy source?

4. Predictions of the configuration of tectonic plates and continents for the next few hundred million years indicate that a new landmass will form as the Pacific Ocean closes and the Americas collide with Asia. Explain whether this configuration is or is not conducive to the formation of thick, extensive seams of coal.

5. If a solar power plant loses 15 percent of its energy in waste heat while boiling water, 30 percent in running a turbine, and 10 percent in transmitting the electricity through power lines, what is the energy efficiency of the system?

EXERCISES

1. Make a list of all the energy transfers that occur from the time that solar energy enters Earth's atmosphere, to when it is stored in water vapor that falls on a continent as rainwater, to when it becomes part of a river along which hydroelectric plants use the falling water to generate electricity, and, finally, to when it is used to light a home.

2. If you were a political advisor to a developing nation in an arid region with few surface water resources and a largely rural population, what arguments would you present to encourage development of solar energy resources?

SUGGESTED READINGS

Craig, J. R., D. J. Vaughan, and B. J. Skinner. 2010. *Earth Resources and the Environment,* 4th ed. Upper Saddle River, NJ: Prentice Hall.

Freese, B. 2004. *Coal: A Human History.* New York: Penguin Books.

Hyne, N. J. 2001. *Nontechnical Guide to Petroleum Geology, Exploration, Drilling, and Production,* 2nd ed. Tulsa, OK: Pennwell Books.

Lenssen, N., and C. Flavin. 1996. "Sustainable Energy for Tomorrow's World: The Case for an Optimistic View of the Future." *Energy Policy* 24 (9) (September): 769–781. doi:10.1016/0301-4215(96)00060-2.

Lovins, A. B. 2011. *Reinventing Fire: Bold Business Solutions for the New Energy Era.* White River Junction, VT: Chelsea Green Publishing.

National Research Council. 2009. "America's Energy Future: Technology and Transformation: Summary Edition." Washington, DC: The National Academies Press. http://www.nap.edu/catalog.php?record_id=12710.

Shepherd, M. 2009. *Oil Field Production Geology,* 1st ed. American Association of Petroleum Geologists.

Smil, V. 2006. *Energy: A Beginner's Guide.* Oxford, UK: Oneworld.

Smil, V. 2008. *Oil: A Beginner's Guide.* Oxford, UK: Oneworld.

Yergin, D. 2008. *The Prize: The Epic Quest for Oil, Money and Power.* New York: Free Press.

Understanding Climatic and Environmental Change

Mount St. Helens, in Washington State, erupted on May 18, 1980, sending an enormous avalanche of rock and volcanic ash into the landscape that destroyed vegetation and dammed streams. Coldwater Creek, for example, turned into Coldwater Lake, and was surrounded by a blanket of ash meters thick and studded with charred trees. Over the next several years, ash carried downslope by runoff washed into the lake.

To collect the ash as it entered, geologists Roger Anderson, Edward Nuhfer, and Walter Dean installed funnel-shaped sediment traps in the lake.[1] Every 15 days, the traps automatically released plastic beads, which landed on top of whatever sediments had fallen in during that time period; in this way, they kept track of deposition history. Comparison of the collected sediments with meteorological records revealed that each sedimentary layer in the traps had been deposited by runoff from a specific storm. Furthermore, each layer's thickness corresponded directly to the intensity of the storm, the thicker layers indicating heavier storms.

Today, we have rain gauges to keep track of duration and intensity of precipitation events, but during most of Earth's history such instruments did not exist. Fortunately, the observations made at Coldwater Lake confirm that sedimentary layers are good **proxies**—reliable indicators of environmental conditions. To reconstruct Earth's environmental history and predict future change, geologists study ancient sedimentary layers and other proxies for environmental events. In this chapter, we:

✔ Examine the origin of Earth's climate.

✔ Analyze some of the causes of climatic and environmental change.

✔ Discover how proxies such as sediments, tree rings, pollen grains, cave deposits, and glacial ice are used to unravel Earth's history of environmental change.

✔ Discuss how numerical models are used to understand past climates.

The Cowlitz River in Washington State quickly became choked by uprooted trees and mudflows of volcanic ash and soil in the wake of the disastrous 1980 eruption of Mount St. Helens. (*Diane Cook and Len Jenshel/Getty Images*)

INTRODUCTION

Between 1987 and 1992, California experienced one of the worst droughts in its recorded history.[2] Six years of below-average rainfall might have devastated the state's enormous agricultural and industrial economy were it not for water available from vast reservoirs constructed in the Sierra Nevada mountain range in the late 1800s and early 1900s to catch melting snow. With a population of nearly 38 million, California's economy ranks eighth largest in the world.[3] It is built on high value agricultural crops such as oranges, strawberries, wine grapes, and almonds, as well as high technology, aerospace engineering, tourism, and entertainment. All of these industries rely on imported surface water for their continued existence.

Coastal California, where San Diego, Los Angeles, and San Francisco are located, is home to 87 percent of the state's population[4] but receives only 15 percent of its precipitation (Figure 14-1a). Most of the rain and snow fall instead on the Sierra Nevada and northern coastal ranges. As a result, coastal cities must import their water hundreds of kilometers from its source via aqueducts (Figure 14-1b), and during the six-year drought, residents watched in dismay as water levels in mountain reservoirs dwindled. By the third year of the drought, water volumes had fallen to 40 percent of storage capacity. Not knowing when the situation would improve, in 1991 the State Water Project cut off the supply of water to farmers and dropped the amount given to municipalities to 30 percent of what they had requested.

To cope with the loss of water resources, farmers were forced to rely on limited groundwater supplies, and coastal cities imposed water rationing. Residents were prohibited from watering their lawns and washing their cars, restaurants served water only upon request, hotels and developers of new homes were required to install low-flow showerheads and toilets, and water-wasters were fined. With no end to the drought in sight, some cities made plans to construct desalination plants, such as those used in Saudi Arabia and other arid Middle Eastern countries, to extract drinking water from the sea. Then, as abruptly as it began, the drought ended, and many of these plans were left on hold.

The 1987–1992 drought was not an isolated incident, however. Indeed, California went into another multi-year drought in 2007, with annual precipitation values around 70 percent of average through the end of 2009. A strong El Niño (see Chapter 12) in winter 2009–2010 provided some relief, but unfortunately, geologic evidence shows that droughts in California can last much longer than a few years. Tree stumps discovered at the bottom of modern marshes, lakes, and streams in the Sierra Nevada (Figure 14-2) indicate

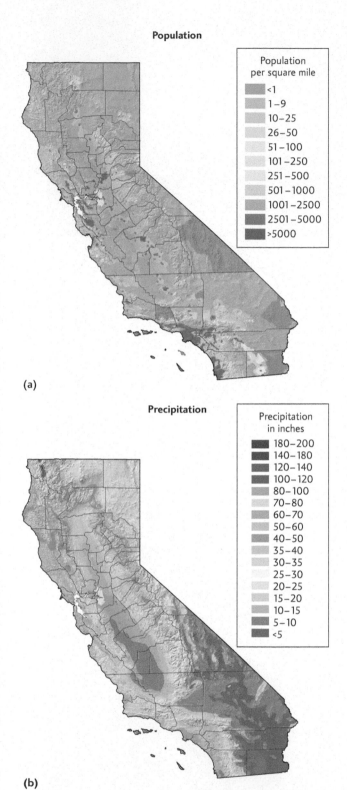

FIGURE 14–1 (a) The majority of California's population lies along the coast and in the state's Central Valley, where many of the fruits and vegetables consumed by the United States are produced. (b) Precipitation is generally low in these areas, which are therefore supplied with water via aqueducts from reservoirs in the Sierra Nevada and the Cascade mountains as well as with water taken from the Colorado River.

FIGURE 14–2 A drowned forest in the West Walker River is evidence of a drought in California within the past 1000 years that lasted more than a century, long enough for trees to grow in the dry riverbed. As conditions became wetter, ending the drought, water began flowing down the river channel again, submerging the trees and killing them. *(Garry Hayes/Geotripper Images)*

that twice in the geologically recent past these areas were dry long enough for trees to grow before they subsequently refilled with water.[5] Radiocarbon dating of these stumps, coupled with counting of tree rings, indicates that one drought lasted longer than 200 years, ending in 1112 CE, and the other, 140 years, ending in 1350 CE. Rising water levels drowned these forests as the droughts ended. Were such droughts to recur today, it could prove catastrophic.

Climatic changes have proven devastating to human societies in the past. A centuries-long drought was likely responsible for the downfall of the Mayan civilization in Mexico in the period 750–900 CE.[6] Sediments deposited in Lake Chichancanab on the Yucatán peninsula revealed oxygen isotopes and organisms that indicate an episode of extreme drying from 800 to 1000 CE. This extended drought could have caused famine and persuaded the Maya to migrate to more hospitable environments. Similarly, tree-ring studies reveal that the Anasazi people, native Americans of the four corners region on the Colorado plateau, abandoned their cliff dwellings during protracted droughts of the 11th to 13th centuries,[7] and that the early American settlements of Jamestown and Roanoke likely succumbed to droughts in the late 1500s and early 1600s.[8] Nor were changes in rainfall the only factors causing societal distress. Colder conditions and expansion of glaciers during the 14th century caused crop failures and famine that played a role in driving the Vikings out of Greenland.[9]

As these examples illustrate, human societies stand to benefit from studying the climatic and environmental history of their local regions, as well as of the whole Earth. In particular, we need to understand how past changes in climate affected such environmental factors as water resources. If climate changes are found to be cyclical, their recurrence may be predictable, a circumstance that would be of great help in water resource planning and land management. Even if these events are random and therefore unpredictable, understanding them can help us develop contingency plans. In the same way that people in earthquake-prone areas prepare for the next big tremor, people in drought-prone areas may prepare for the next severe dry spell. The city of Santa Barbara, California has already acted on the lessons of 1987–1992 by building a desalination plant. Although presently not in use, the plant stands ready for the next protracted drought.[10]

CLIMATE ON TERRESTRIAL PLANETS

"Civilization exists by geological consent, subject to change without notice." This quote from historian Will Durant (1885–1981) is as true of the climate system as of geological hazards such as earthquakes, volcanic eruptions, and floods. Still, for most of Earth's history, climate has been sufficiently favorable to allow life to originate and evolve into millions of species. Why do environmental conditions suitable for life exist on Earth but not on Mercury, Venus, or Mars, the other terrestrial (rocky) planets in our solar system? For many years conventional wisdom held that Earth's temperature and climate were hospitable to life because our planet happened to form at the "right" distance from the Sun. By this reasoning, the frigid −55°C average surface temperature of Mars occurs because that planet formed too far from the Sun, while the blistering 460°C average temperature of Venus[11] and 179°C temperature of Mercury[12] are attributable to their closeness to the Sun.

This "Goldilocks" hypothesis is flawed, however. For the first 2.5 billion years of Earth's history, the Sun gave off so little radiation that our planet would have been frozen had there not been some other temperature-regulating mechanism in operation.[13] Likewise, satellite imagery of Mars shows that it was not always a frozen wasteland. The most likely explanation for the deep canyons covering much of the red planet's surface is that liquid water—and, therefore, temperatures greater than 0°C—must have existed at one time (Figure 14-3). Another flaw in the hypothesis is that although Venus is farther from the Sun, it is actually warmer than Mercury. Clearly, then, distance from the Sun is not the sole factor determining a planet's temperature. Comparison of Earth, Mars, and Venus reveals that a habitable planet probably requires a carbon cycle, plate tectonics, a hydrologic cycle, and possibly a magnetic field. Paradoxically, a habitable planet may also require life!

Earth's Temperature and the Faint Young Sun Problem

According to well-accepted models of stellar evolution—the way in which stars form, develop, and die—the amount of light given off by the Sun has increased through time. At the time Earth formed, solar luminosity was likely 20 to 30 percent lower than the present value. If early Earth had an atmosphere of today's composition, a 30 percent reduction in solar output would have produced an average surface temperature of about −20°C, and as the Sun's luminosity increased, Earth's temperature would not have climbed above freezing until roughly 2 billion years ago (Figure 14-4).[14] Geologic evidence suggests that early Earth was quite warm, however. Sedimentary rocks reveal that our planet had liquid oceans by 3.8 billion years ago, and the oldest fossils discovered to date are 3.5 billion years old.[15]

What enabled Earth to maintain warm surface temperatures during a time of low solar energy output? Geologists refer to this puzzle as the **faint young Sun problem.**

FIGURE 14–3 A false-color image of the surface of Mars. Deeply incised gully channels, showing meandering and braiding patterns, indicate that water once flowed on this now frozen planet. *(NASA/JPL/University of Arizona)*

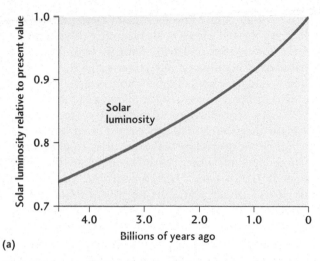

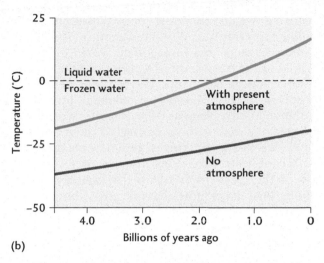

(a)

(b)

FIGURE 14-4 (a) The hypothetical increase in solar luminosity over time and (b) the corresponding increase in Earth's surface temperature, assuming an atmosphere of today's composition. With our current atmosphere, Earth would not have become warm enough to support liquid water until about 2 billion years ago. If there had been no atmosphere at all, Earth would still be frozen today.

The solution lies in the evolution of Earth's atmosphere. If the early atmosphere had a higher concentration of greenhouse gases than are present today, the heat retained could have kept Earth's oceans liquid. Later, as solar luminosity increased, greenhouse gas levels would have had to decline to prevent Earth from overheating. While such a scenario might seem unlikely, it turns out that methane and carbon dioxide are capable of meeting these requirements.

Methane Recall from Chapter 11 that early Earth contained little or no free oxygen in its atmosphere. Anoxic conditions are conducive to the growth of a class of microbes known as **methanogens,** which consume carbon dioxide and make methane as a waste product. Methanogenic microbes belong to the domain Archaea and are among the earliest forms of life.[16] Today they are found in only a few select locations where anoxic conditions prevail, such as in the digestive tracts of animals and in rice paddies, peat bogs, and landfills where organic matter is decaying. On an anoxic early Earth, however, they were likely quite abundant, boosting levels of methane in the atmosphere as much as 1000 times over modern levels[17] and accounting for as much as half of the greenhouse warming needed to keep Earth warm under lowered solar luminosity.

Another domain of microbes, the bacteria, was also present on early Earth and gradually spelled the downfall of the methanogens. Among these organisms were the **cyanobacteria,** the first cells to use oxygenic photosynthesis (see Chapter 7). Oxygen is toxic to methanogens, so as oxygen levels in the atmosphere climbed, the influence of methanogens decreased and methane levels dropped.[18] Thus it seems that life played an important role in buffering Earth's temperature as the Sun's radiation output increased.

Carbon Dioxide In Chapter 7 we introduced the inorganic carbon cycle. Recall that carbon dioxide and water combine to make carbonic acid (H_2CO_3), which dissociates into hydrogen (H^+) and bicarbonate ions (HCO_3^-). Chemically reactive hydrogen displaces calcium and other metals from silicate minerals during chemical weathering, and these ions travel in river water to the oceans. There, foraminifera, mollusks, corals, and other organisms use the calcium and bicarbonate to make calcium carbonate shells and skeletons. As these organisms die, their hard parts litter the ocean floor, from which they are subducted into Earth's interior. Carbonates can also form by inorganic precipitation, and these sediments likewise are carried into the mantle by the slow and steady movement of Earth's tectonic plates. Within subduction zones, metamorphic reactions between calcium carbonate and quartz create new silicate minerals and release carbon dioxide back to the atmosphere through volcanic eruptions, completing the cycle.

Now consider the role of silicate weathering in an enhanced greenhouse world early in Earth's history. As the Sun's luminosity increased, the increased energy would have promoted evaporation, driving more water out of the oceans and into the atmosphere. The water vapor would have combined with the more abundant carbon dioxide, fallen to Earth as carbonic acid rain, and caused an increase in silicate weathering.

Weathering would also have been enhanced by the warming temperatures. With greater rates of weathering, more calcium and bicarbonate would have been delivered to the oceans, thereby increasing the production of carbonate sediments. The resulting decrease in atmospheric carbon dioxide would have limited the greenhouse effect as long as the carbonates were not all subducted and their carbon returned to the atmosphere as CO_2. We know that this in fact happened because Earth's continents have grown in size over time, and many of the rocks from which they are constructed are limestones made of marine calcium carbonate sediments that did not subduct. Thus, weathering seems to be a viable mechanism for keeping Earth at a fairly stable temperature.

Mars: A Frozen Planet

The evidence for liquid water on Mars' surface (see Figure 14-3) indicates that it also started out with higher greenhouse gas levels than at present. What might have caused the red planet to cool? Based on numerical modeling results and data from Mars missions, two scenarios are presently favored: 1) Mars lost most of its greenhouse gases to outer space, or 2) most of the carbon dioxide in the early Martian atmosphere is tied up in carbonate minerals. We'll examine each in turn.

Collisions of high-energy particles from outer space (called the solar wind) with the Martian atmosphere may have driven off atmospheric gases, a phenomenon known as **sputtering**.[19] On Earth, the magnetic field generated by the sloshing of the liquid outer core around the solid inner core deflects such particles, shielding our atmosphere from bombardment. Mars, on the other hand, has a very weak magnetic field. It also has about one-ninth the mass of Earth, causing it to have a much lower gravitational attraction than our planet. The combination of the weak magnetic field and low gravitational attraction may have allowed Mars' greenhouse gases to escape the bounds of the planet when bombarded, thereby triggering the deep freeze.

In an alternative scenario, silicate weathering on Mars may have produced carbonates in an early Martian ocean or in soils. In the case of oceanic deposition, carbonates were likely produced without the aid of biological activity as concentrations of calcium and bicarbonate gradually increased and reached chemical saturation.[20] Soil carbonates could have been produced much the same way as on Earth, through the interaction of acidic groundwater with soil minerals. The carbon thereby locked into ocean sediments and soils would have been permanently isolated from the Martian atmosphere due to the lack of plate tectonics on that planet that could recycle the carbonates back into atmospheric carbon

dioxide. Why does Mars lack plate tectonics? Its small size gives the red planet a surface area to volume ratio nearly double that of Earth. With its internal heat dissipated over a comparatively larger surface area, ancient Mars would have cooled much more quickly than Earth, losing the energy necessary for plate tectonics.

Whatever scenario turns out to be correct, scientists estimate that the initial concentration of atmospheric carbon dioxide on Mars was equivalent to about 2 bars of pressure and subsequently dropped to the present value of 0.006 bars.[21] In comparison, Earth's atmospheric CO_2 concentration is equivalent to 0.0003 bars. Although the Martian atmosphere contains more CO_2 than Earth's, the amount is insufficient to keep Mars warm because that planet is much farther from the Sun and receives less solar radiation than Earth.

Venus: A Runaway Greenhouse Effect

If Mars lost carbon dioxide from its atmosphere, Venus experienced the opposite problem. At present, 99.9 percent of the carbon on Earth is tied up in rocks, whereas on Venus, nearly the same proportion is found in the atmosphere. In fact, Venus' atmosphere is 96 percent carbon dioxide and is about 100 times denser than Earth's (Figure 14-5).[22] The reason for the superabundance of carbon dioxide—and the resulting runaway greenhouse effect—is the planet's lack of water. Without liquid water, silicate weathering reactions are impossible. Thus, CO_2 belched out by Venusian volcanoes accumulated in the atmosphere and generated an intense greenhouse effect.

Venus may initially have had oceans, but today it is nearly dehydrated. How did the planet lose its water? A process active on Earth may indicate the answer. Strong ultraviolet radiation coming in from the Sun causes water vapor in the upper levels of Earth's atmosphere to break apart into hydrogen and oxygen gases, a process called **photodissociation** (see Figure 14-5a). The oxygen remains in the atmosphere, but the hydrogen drifts to outer space because its small mass allows it to escape Earth's gravitational pull. Because very little water ever reaches the upper atmosphere of Earth, our planet's loss of water by photodissociation is minimal. Recall from Chapter 11 that Earth's water vapor is largely restricted to the troposphere (0–10 km elevation) because the decline in temperature with elevation causes vapor to condense close to Earth's surface.

Its nearness to the Sun gave Venus a higher starting surface temperature than Earth. Therefore surface water on Venus would have evaporated early in the planet's history,[23] and the resulting heat and humidity probably caused Venus' atmospheric temperature to drop very

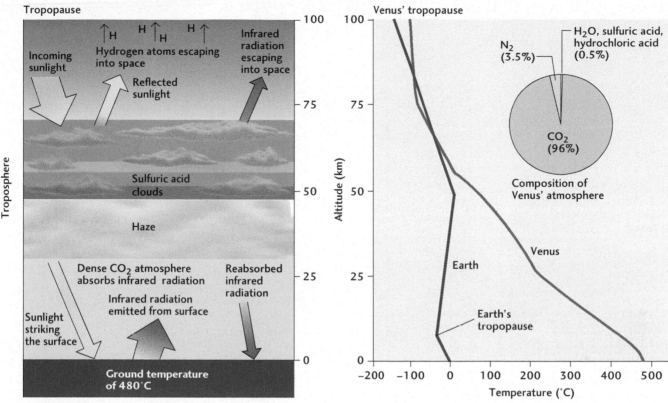

(a) Structure of Venus' atmosphere

(b) Atmospheric temperatures of Venus and Earth

FIGURE 14–5 The greenhouse effect on Venus.
(a) Thick sulfuric acid clouds block much of the incoming solar radiation, but dense carbon dioxide clouds underneath them trap most of the outgoing infrared radiation emitted by Venus' surface, keeping the planet hot. Water is lost through the photodissociation of vapor in the upper atmosphere and the escape of hydrogen to outer space. (b) As a consequence of Venus' exaggerated greenhouse effect compared with Earth's, the average surface temperature of Venus is about 30 times that of Earth.

slowly with altitude compared with the environmental lapse rate on Earth (see Figure 14-5b). Therefore, the tropopause on Venus would have been found at about 150 km elevation.[24] At so high an elevation, water vapor would have photodissociated easily to oxygen and hydrogen, and the hydrogen would have escaped the planet's gravitational pull, ending the hydrologic cycle necessary to strip carbon dioxide from the Venusian atmosphere.

■ Earth's surface temperature has remained within a range suitable for sustaining life because of the interaction of three cycles: plate tectonics, the hydrologic cycle, and the carbon cycle. In addition, Earth's magnetic field and the presence of life itself have played a role in maintaining our planet's habitable temperature.

■ Water at the surface of Mars is presently frozen because the planet has lost most of its greenhouse gases. The loss is thought to have occurred either because carbon dioxide was removed from the atmosphere and stored in oceanic or soil carbonates or because of loss of Mars' atmosphere to outer space through sputtering.

■ Venus has a runaway greenhouse effect because there is insufficient water vapor in the planet's atmosphere to remove the massive amounts of carbon dioxide responsible for warming.

CAUSES OF CLIMATIC CHANGE

As the examples of Earth, Venus, and Mars show, climate is determined by complex interactions among interior processes, the pedosphere, biosphere, hydrosphere, and atmosphere, as well as distance from the Sun and variations in solar radiation intensity over time (Figure 14-6). In contrast to its terrestrial neighbors, Earth has been hospitable to life for at least 3.5 billion years, but that does not mean that our planet's temperature has

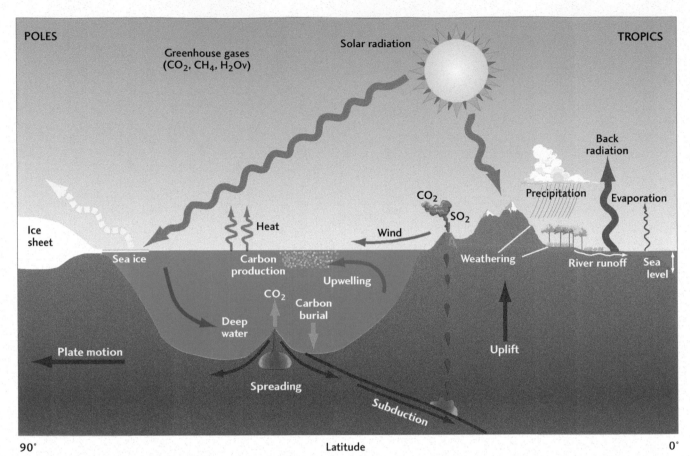

FIGURE 14–6 Earth's climate reflects complex interactions between the lithosphere, pedosphere, hydrosphere, atmosphere, biosphere, and incoming solar radiation and reflects an integration of numerous processes acting over different timescales. *(From W. F. Ruddiman,* Earth's Climate: Past and Future, *2nd edition, © 2008 by W. H. Freeman and Company)*

remained static. Instead Earth has experienced warm intervals alternating with glaciations (Figure 14-7). These climatic changes have come about because of variations in the configuration of the continents and changes in atmospheric composition, among other factors, and are responsible for the conditions we experience today. Furthermore, climate changes have occurred on a variety of timescales. Volcanic eruptions can cool our planet for a few years, whereas plate tectonic movements lead to climatic changes that play out over millions of years. Let's explore some of the factors that may have led Earth's temperature to change over time.

Climate and the Evolution of Life

As discussed above, life can have a profound impact on Earth's climate because of its influence on atmospheric composition. Methanogenic microbes likely contributed to Earth's early warmth, only to die out in most places as the planet warmed and photosynthetic cyanobacteria produced oxygen. Life may also have been responsible for some of Earth's ice ages. The first occurred about 2.4 billion years ago as oxygen levels in Earth's atmosphere began to climb dramatically (see Figure 14-7).[25] Recall from Chapter 4 that prior to this time, rocks known as banded iron formations were deposited in the oceans. Because iron is insoluble in its oxidized state, it is difficult to dissolve and transport today. In a world with low or nonexistent oxygen, however, iron would have been in its soluble reduced state, capable of being transported to the oceans and then deposited. The fact that the glaciation period of 2.4 billion years ago occurred simultaneously with the end of banded iron formation production suggests that both events were caused by a rise in oxygen levels related to the proliferation of cyanobacteria. Since cyanobacteria use carbon dioxide for photosynthesis and produce oxygen that kills methanogens, the resulting change in greenhouse gas levels may have plunged Earth into an early ice age.[26]

Another ice age that may have been caused by the existence of life on Earth lasted for about 80 million years during the Carboniferous and Permian periods of the Paleozoic era (see Figure 14-7).[27] Land plants had evolved a bit earlier, during Silurian time, increasing

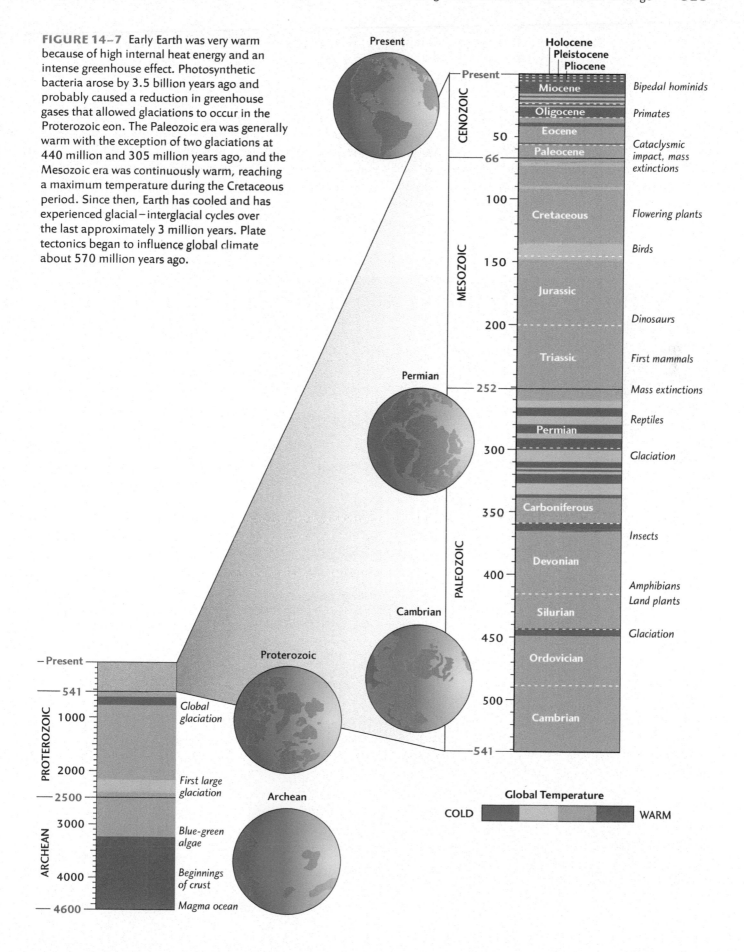

FIGURE 14–7 Early Earth was very warm because of high internal heat energy and an intense greenhouse effect. Photosynthetic bacteria arose by 3.5 billion years ago and probably caused a reduction in greenhouse gases that allowed glaciations to occur in the Proterozoic eon. The Paleozoic era was generally warm with the exception of two glaciations at 440 million and 305 million years ago, and the Mesozoic era was continuously warm, reaching a maximum temperature during the Cretaceous period. Since then, Earth has cooled and has experienced glacial–interglacial cycles over the last approximately 3 million years. Plate tectonics began to influence global climate about 570 million years ago.

the percentage of Earth's surface covered by organisms using oxygenic photosynthesis. The vast coal swamps for which the Carboniferous period is named suggest that the ice age that occurred during this time was related to terrestrial plant drawdown of atmospheric carbon dioxide. In fact, ice sheets grew and shrank repeatedly, causing sea level to fall and rise. These fluctuations are recorded in repeated cycles of sediments that are alternately marine (beach sands and offshore mud) and non-marine (river channel, floodplain, and delta deposits). Called **cyclothems,** these sequences contain much of the coal found in the United States today (Figure 14-8).

Influence of Plate Tectonics on Climate

Plate tectonics influences climate by changing the distribution of continents between the equator and poles, by uplift of land to higher altitudes, and by seafloor spreading and volcanic eruptions. Recall from Box 8-1 that the collision of India and Asia, which began about 50 to 55 million years ago[28] and continues today, uplifted land to elevations that permitted glaciation, which in turn promoted physical and chemical weathering that may have removed CO_2 from the atmosphere and caused

(a)

(b)

FIGURE 14–8 (a) Plants flourished in the swamps of the Carboniferous period, as seen in this reconstruction. Much of the coal we use today originates from this ancient vegetation. (b) A cyclothem sequence. The dark layers are the remains of coastal coal swamps. Note the person, for scale. *(a: Sheila Terry/ Science Source; b: Raymond Coveney)*

global cooling (see the long-term cooling trend since the Eocene in Figure 14-7). In addition, the uplifted lands act as a barrier to the flow of moisture-laden air from the Indian and Pacific oceans, which over time has caused large parts of interior China to become desert.[29]

Plate tectonics has also influenced the circulation of the oceans over geologic time by affecting the configuration of continents and ocean basins. As an example, the Antarctic ice sheet developed 34 million years ago when Antarctica and South America began to split apart, establishing a flow of water around Antarctica called the Antarctic Circumpolar Current (see Chapter 12, Figure 12-15). Because this current deflects heat-bearing equatorial waters, it may have contributed to the freezing over of the southern continent.[30]

Tectonic Influences on Earth's Reflectivity In addition to creating topographic barriers and opening or closing oceanic gateways, tectonic changes in the distribution of the continents and oceans affect climate by determining how much solar radiation can be absorbed at Earth's surface at different latitudes. The ability of Earth's surface materials to absorb incoming radiation is related to their color. Dark soils, deep blue seawater, and dark green leaves absorb energy much more readily than do snow and ice, which tend to reflect most energy. Reflectivity, also known as **albedo,** is valued from 0 to 1, with 0 denoting a perfectly absorptive surface and 1 denoting a perfectly reflective surface. Earth's average albedo value is 0.3, which means that, on average, our planet reflects 30 percent of incoming energy. Recall from Chapter 11 that most of this albedo comes from

Earth's clouds. Still, local albedo values differ depending on the degree of cloud cover and the type of material making up Earth's surface.

Latitude affects oceanic albedo because of the angle of incidence of the Sun's rays. Have you ever been in a boat or on a beach at sunrise or sundown and seen glare from the Sun coming off the water, although around noon you noticed no such effect? When solar radiation strikes water from a high angle, water absorbs most of it and gets warm, but when the angle is low, water absorbs little of the incoming radiation, which instead glances off the water into your eyes. Similarly, at tropical latitudes where sunlight strikes from a high angle, close to 90°, ocean water absorbs the light almost entirely, reflecting only 3 to 10 percent of the energy, for an albedo value of about 0.03 to 0.1.[31] The same ocean water at high latitudes reflects 10 to 100 percent of incoming solar radiation. Meanwhile, the albedo of land typically ranges from 0.15 to 0.45, comparable to the albedo of water at higher latitudes. Snow covered surfaces can be even more reflective, ranging from 0.4 to 0.95. Thus, land areas concentrated at both poles with a large ocean circling the globe at low latitudes would lead to a lower albedo and higher average temperature for Earth than would the reverse situation (Figure 14-9).

In both cases, the intensity of radiation received from the Sun is greatest in the low latitudes (where solar radiation strikes the planet's surface at nearly right angles; see Chapter 11) and is much lower at high latitudes (where incoming radiation is inclined relative to the planet's surface). If the ocean encircles the equator, the water absorbs nearly all the incoming high-intensity

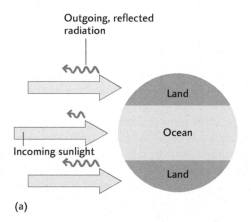
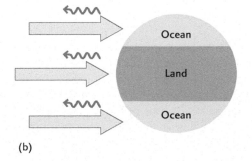

FIGURE 14–9 The distribution of land and ocean affects global temperature and climate. (a) On average, lands at the poles with an ocean circulating between them would warm the planet because, compared with land, the ocean absorbs more light at low latitudes, where the greatest amount of solar radiation is received by the planet. (b) Oceans at the poles with land between them would, therefore, have the opposite effect. Arrows striking the planet represent incoming solar radiation; arrows leaving the planet represent reflected radiation. Shorter outgoing arrows denote that more energy has been absorbed by the planet's surface than longer outgoing arrows, which represent more reflection.

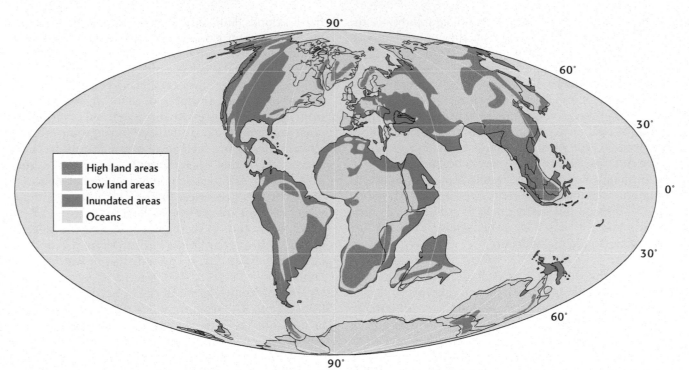

FIGURE 14-10 The distribution of land and ocean during the late Cretaceous period, about 95 million years ago. Extensive shallow seas inundated much of the continental landmass. Sea level was very high during the Cretaceous partly because of accelerated rates of seafloor spreading, which created young, buoyant oceanic crust that displaced ocean water onto the continents. In addition, an enhanced greenhouse effect, brought about by increased CO_2 outgassing to the atmosphere, led to melting of ice sheets, contributing to sea level rise.

energy, while the higher latitude landmasses reflect a substantial amount of the low-intensity radiation. In the reverse situation, however, with land encircling the equator and oceans at the poles, much more of the high-intensity radiation hitting the equator would be reflected back to space, leading to lower average temperatures on Earth. Just this scenario may have contributed to several episodes of extreme glaciation during the late Protero-zoic (see Figure 14-7; Box 14-1).

Seafloor Spreading Another impact of plate tectonics on climate involves changes in rates of seafloor spread-ing, which influence both greenhouse gas levels and the depths of the ocean basins. Geologists have found that the Cretaceous period was a time of more rapid seafloor spreading than today.[35] The increase in spreading rate is thought to have led to an enormous volume of carbon dioxide in the atmosphere as upwelling magmas along oceanic ridges released dissolved CO_2. Scientists believe that Earth's atmosphere during the Cretaceous period contained 4 to 10 times the CO_2 that it contains today, leading to an enhanced greenhouse effect and glob-ally warm temperatures.[36] No evidence of glaciation exists in any rocks from this time, suggesting that all ice sheets and glaciers must have melted. Furthermore,

the rapid spreading rates caused a decrease in the depth of the ocean basins because young, hot crust is more buoyant than old, cold crust (see Chapter 4). Ocean water therefore flooded continents and covered much of the midwestern United States, Africa, South America, and Europe with shallow seas, further contributing to warmth (Figure 14-10).

Volcanic Eruptions While increased volcanic activity ultimately acts to increase the concentration of green-house gases in the atmosphere and cause warming, in the short term (months to years), volcanic eruptions may cause a cooling of Earth by injecting ash, sulfuric acid droplets, and sulfate particles high into the troposphere and stratosphere where they reflect incoming solar radia-tion (see the discussion of Tambora and Mount Pinatubo in Chapter 5).

Influence of the Oceans on Climate

The ocean, through its large volume, low albedo, and efficient circulation, has a huge capacity to store and redistribute heat absorbed from the Sun. For this reason Earth's climate, as "buffered" by the ocean, is gener-ally stable, hospitable to life, and suffers few extremes in temperature except in places remote from the coasts where climate is described as continental.

As mentioned in Chapter 12, Earth scientists now conceive of oceanic circulation as comparable to a large conveyor belt (see Figure 12-16). In this model, warm water from low latitudes presently flows northward in the Atlantic as a surface current known as the Gulf Stream and its extension, the North Atlantic Current. Through evaporation, the salinity of this surface current increases, and through northward migration, the temperature of the current decreases. At high latitudes, this cold, salty water sinks, forming a mass of water called North Atlantic Deep Water that flows at depth back toward the equator.

This circulation is thought to be responsible for maintaining relatively mild winter temperatures in Western Europe through sea–air heat exchange as the midlatitude westerly winds blow over the Gulf Stream.

In the past, this oceanic conveyor belt appears to have slowed or stopped, shutting off the transport of heat to high latitudes. At the height of the last ice age about 20,000 years ago, for example, meltwater from a massive ice sheet that covered northern North America flowed down the ancestral Mississippi River. As Earth warmed and the ice sheet retreated, meltwater eventually flowed eastward through the Hudson and St. Lawrence rivers and into the North Atlantic (Figure 14-11).

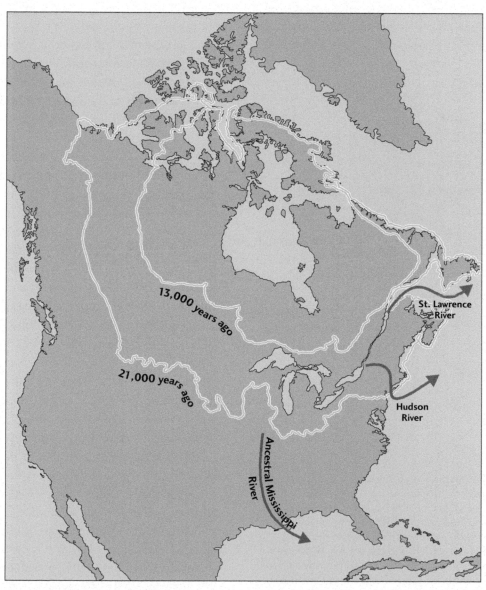

FIGURE 14–11 Routes for glacial meltwater during deglaciation of North America. At the peak of glaciation, meltwater flowed down the ancestral Mississippi River, but as ice sheet retreat commenced, meltwater began flowing down the Hudson and St. Lawrence rivers into the North Atlantic.

BOX 14–1 CASE STUDY

Snowball Earth

Between 750 and 580 million years ago, Earth appears to have repeatedly frozen over for tens of millions of years at a time. These episodes, termed **snowball Earth events,** were marked by average surface temperatures nearly as cold as modern Mars (−50°C) and sea ice that stretched from the poles all the way to the equator.[32] Evidence for the events is found on many continents and consists of oceanic sediments that were uplifted by tectonic processes and are now found on land. Three distinct forms of sedimentation encapsulate a freezing episode (**Figure 1**): layers of coarse sand, gravel, and boulder debris; a layer of iron–rich sediment sandwiched between the clastic sediments; and a layer of carbonate that abruptly overlies the upper clastic debris layer.[33]

Geologists interpret these layers as follows: As ice sheets expanded, their seaward margins spawned icebergs that rafted clastic sediments into the open ocean, creating the layers of sand, gravel, and boulders that lie atop typical fine-grained oceanic sediments. As temperatures continued to plunge, sea ice expanded to cover the entire ocean surface, cutting off iceberg production, isolating the world ocean from the atmosphere, and killing off most photosynthetic organisms. The resulting decline in oxygen allowed reduced iron to accumulate in the seas. With the whole Earth frozen, silicate weathering ceased, yet plate tectonics continued, allowing volcanoes to spew carbon dioxide into the atmosphere. After several million years, greenhouse gas levels grew high enough to begin melting the ice, perhaps 1000 times the pre–snowball level. As oxygen re–entered the oceans, the dissolved iron precipitated into layers analogous to the banded iron formations earlier in Earth's history.[34] In addition, as ocean water reappeared, waves began to lap at the edges of continental glaciers, reinitiating iceberg formation and the deposition of *ice–rafted debris.* As the ice melted, the high levels of carbon dioxide in the atmosphere drove the planet into an extreme greenhouse climate that promoted enhanced silicate weathering and the rapid deposition of carbonate layers on top of the ice–rafted debris.

From the geologic evidence, this icehouse/hothouse oscillation appears to have occurred at least twice, and perhaps as many as four times in the late Proterozoic. What initiated these monstrous swings in temperature? Earth scientists aren't yet completely certain, but we know from paleomagnetic evidence (see Chapter 2) that the continents straddled the equator at the time it began (**Figure 2**), a configuration that promotes cooler planetary temperatures from albedo effects (see Figure 14–9). In addition, the warm wet conditions of the tropics likely enhanced chemical weathering of the continents, drawing down carbon dioxide levels in the atmosphere. With this continental configuration, anything that initiated ice growth likely started a runaway process because of a climatic feedback known as the

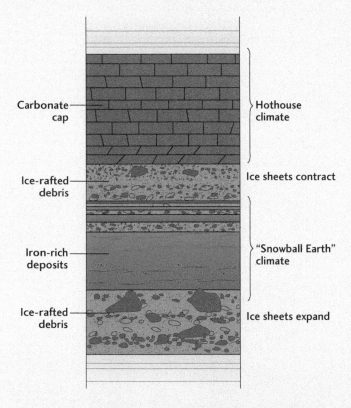

Carbonate cap — Hothouse climate

Ice-rafted debris — Ice sheets contract

Iron-rich deposits — "Snowball Earth" climate

Ice-rafted debris — Ice sheets expand

Figure 1 Evidence from marine sediments for snowball Earth episodes in the late Precambrian. As ice sheets expanded around the globe, sand, gravel, and boulders associated with iceberg flotillas deposited on top of finer grained deep–sea muds. When all of Earth's surface froze, oxygen depletion allowed transport of reduced iron to the oceans, which later oxidized when greenhouse gas levels grew sufficiently high to induce melting and reintroduction of oxygen into the oceans. High levels of carbon dioxide in the atmosphere enhanced weathering of the continents and provided the ions necessary for precipitation of carbonates.

(continued on page 536)

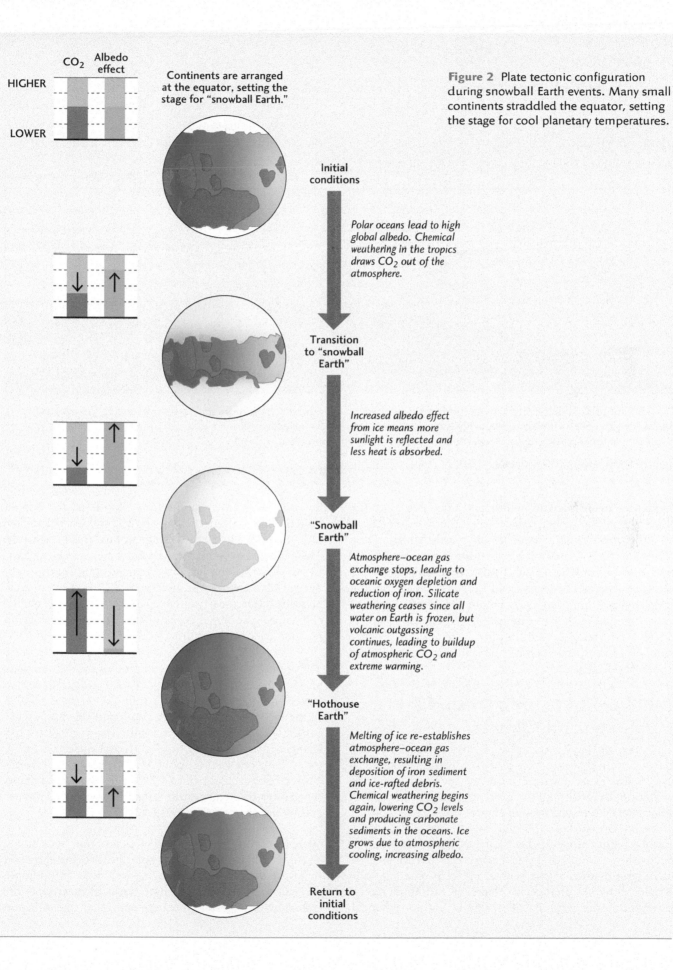

Figure 2 Plate tectonic configuration during snowball Earth events. Many small continents straddled the equator, setting the stage for cool planetary temperatures.

CO₂   Albedo effect

HIGHER

LOWER

Continents are arranged at the equator, setting the stage for "snowball Earth."

Initial conditions

Polar oceans lead to high global albedo. Chemical weathering in the tropics draws CO₂ out of the atmosphere.

Transition to "snowball Earth"

Increased albedo effect from ice means more sunlight is reflected and less heat is absorbed.

"Snowball Earth"

Atmosphere–ocean gas exchange stops, leading to oceanic oxygen depletion and reduction of iron. Silicate weathering ceases since all water on Earth is frozen, but volcanic outgassing continues, leading to buildup of atmospheric CO₂ and extreme warming.

"Hothouse Earth"

Melting of ice re-establishes atmosphere–ocean gas exchange, resulting in deposition of iron sediment and ice-rafted debris. Chemical weathering begins again, lowering CO₂ levels and producing carbonate sediments in the oceans. Ice grows due to atmospheric cooling, increasing albedo.

Return to initial conditions

(continued)

ice–albedo feedback (**Figure 3**): As ice begins to grow, its highly reflective surface sends more incoming solar energy back to space, thereby cooling the climate. This promotes additional ice growth that reflects still more solar energy, causing yet more cooling, and so on. The ice–albedo feedback also works in reverse: As ice begins to melt, it exposes ocean water and soils that absorb more solar energy than ice, thereby causing climate to warm. The increased warming leads to further ice melting and exposure of additional low albedo material.

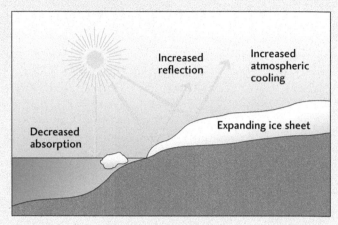

Figure 3 The ice–albedo feedback is a reinforcing feedback. As ice grows, more incoming solar energy is reflected back to outer space, thereby cooling Earth's surface and atmosphere.

As Earth cools, more liquid water freezes to ice, increasing the reflectivity of the surface and producing still colder temperatures.

Evidence from marine sediments indicates that the fresh glacial meltwater reduced the salinity of the North Atlantic, prompting a reduction in North Atlantic Deep Water formation in six different episodes between 15,800 and 12,600 years ago. Each freshening event triggered a return to colder conditions and ice sheet expansion that lasted several hundred years before conditions began to warm again. On shorter timescales, we noted in Chapters 11 and 12 that ocean–atmosphere interactions produce climatic phenomena such as hurricanes and the El Niño Southern Oscillation.

Influence of Earth's Orbital Parameters on Climate

Thus far we have examined processes internal to the Earth system that influence global climate. Astronomical cycles also exert a strong influence on Earth's climatic evolution by affecting the amount and distribution of solar energy received by our planet.

Earth's Elliptical Orbit The shape of Earth's orbit influences the amount of solar energy our planet receives at different times of the year. A perfectly circular orbit would show no change in planetary radiation receipt throughout the year. Earth's orbit is slightly elliptical,

however, and Earth is currently closest to the Sun in January and farthest away in July. Our planet therefore receives more energy on average in January than in July. The **eccentricity**—the deviation from perfect circularity—of Earth's orbit has not remained constant throughout geologic history but undergoes a 100,000-year cycle between 0.005 (nearly circular) and 0.06 (slightly elliptical), causing subtle changes in the amount of radiation received over time (Figure 14-12a).[37]

Tilt of the Spin Axis The orientation of Earth's spin axis also influences the distribution of solar radiation. As discussed in Chapter 11, Earth's spin axis is tilted relative to the incoming rays from the Sun, and this **tilt** is what causes Earth to experience seasons. The angle of tilt has changed over geologic time and continues to change today. In a cycle lasting about 41,000 years, the tilt varies between about 22.2° and 24.5°.[38] The greater the angle, the greater the temperature range between the summer and winter seasons (Figure 14-12b).

Wobble of the Spin Axis Like a top that is winding down, Earth wobbles on its spin axis, so the direction of tilt of the spin axis changes over time. In addition, the entire elliptical path that Earth takes around the sun slowly migrates through space, a phenomenon known as *precession of the ellipse.* Together, these two cycles change the timing of the seasons, a phenomenon known as **precession of the equinoxes**.[39] Recall that winter occurs when a hemisphere is tilted away from the Sun and summer when the same hemisphere is tilted toward the Sun. Currently, therefore, winter occurs from December to March in the northern hemisphere and from June to September in the southern hemisphere (Figure 14-12c). However, 11,500 years from now, when Earth's spin axis is midway through its wobble cycle and its elliptical orbit has moved about half way through its full precession, winter will occur from June to September in the northern hemisphere and from December to March in the southern hemisphere. At the same time, the projection of Earth's spin axis into space will shift away from Polaris, the North Pole star, toward other stars in the sky. In 23,000 years, Earth will have completed a full precessional cycle, bringing the spin axis back into line with Polaris. Winter and summer will then fall in the same calendar months as they do today.

The importance of precession in determining the amount of solar energy received by any location on Earth depends on the degree of ellipticity of Earth's orbit. Winter in the northern hemisphere currently occurs when Earth is closest to the Sun in its orbit, and summer occurs when Earth is farthest from the Sun (see Figure 14-12c). The result is that winters and summers in the northern hemisphere are relatively mild today because the seasons are offset somewhat by Earth's proximity to the Sun. Eleven thousand years in the future, northern hemisphere winter will occur when Earth is farthest from the Sun, and summer when Earth is closest to the Sun. In this instance, because Earth's proximity to the Sun and hemispheric tilt will amplify each other's effects, summers will be hotter than they are today and winters colder. The southern hemisphere will experience just the opposite effects.

Milankovitch Cycles The variations in Earth's orbital parameters have come to be known among *paleoclimatologists*—scientists who study past climate and the history of climate change—as **Milankovitch cycles**, after Serbian scientist Milutin Milankovitch (1879–1958), who built on the theory of Scottish janitor and scientist James Croll (1821–1890) that orbital variations could lead to varying receipts of solar radiation over time

FIGURE 14–12 Changes in Earth's orbital parameters cause changes in the amount of solar radiation the planet receives. (a) Orbital eccentricity determines Earth's distance from the Sun at any given time of the year. (b) As the tilt angle increases, the temperature range between summer and winter increases at any given location. (c) Precession, or wobble, of the spin axis together with orbital eccentricity determines whether the seasons will be extreme or mild.

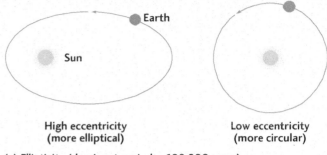

(a) Ellipticity (dominant period = 100,000 years)

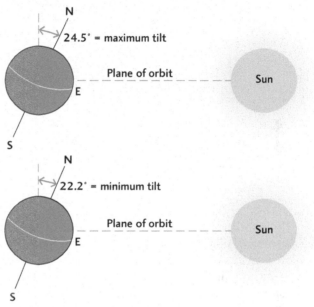

(b) Tilt of the axis (period = 41,000 years)

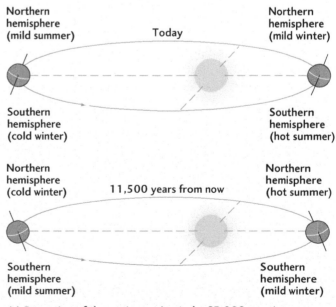

(c) Precession of the equinoxes (period = 23,000 years)

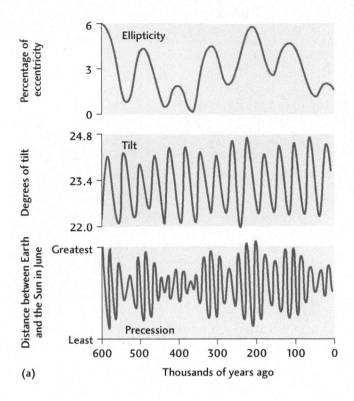

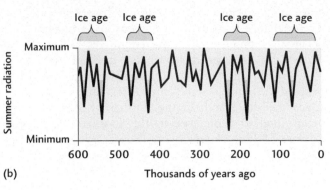

(b)

FIGURE 14–13 (a) The eccentricity, tilt, and precession cycles for the last 600,000 years. Note that the eccentricity cycle has a 100,000–year period, the tilt cycle a 40,000–year period, and the precession cycle a 23,000–year period. (b) Solar radiation in June at 65°N latitude over the last 600,000 years, calculated by Milutin Milankovitch from the eccentricity, tilt, and precession cycles shown in (a). The shaded areas represent minima in incoming solar radiation— times when Milankovitch believed Earth was glaciated.

that would affect climate.[40] Combining the effects of the eccentricity, tilt, and precession cycles, Milankovitch calculated the *insolation*—the amount of incoming solar energy—hitting all latitudes of the surface of Earth over the last 600,000 years (Figure 14-13). He surmised that episodes of minimum insolation would have led to cold climates, and perhaps glaciation, and episodes of maximum insolation would be characterized by very warm climates.

Milankovitch considered the amount of energy reaching 65°N latitude in June to be of primary importance in determining whether glaciation will occur in the northern hemisphere, because this is the latitude at which ice sheets begin to grow.[41] He theorized that glaciation would occur during times of relatively mild winters and cool summers. Warmer air in winter can transport more moisture as snow to the ice sheets, and cooler air in summer can minimize melting. By identifying periods of particularly low values of summer solar radiation over the last 600,000 years, Milankovitch suggested dates for periods of glaciations evident in the Alps; his correlations are shown as the shaded areas in Figure 14-13b. Milankovitch thereby became the first person to suggest absolute dates for these glaciations. These dates have since been confirmed through radiocarbon dating and other techniques that had not yet been developed when Milankovitch did his work. As a result, most climate scientists accept that Milankovitch cycles were important drivers of climate change on Earth in the preindustrial era.

Influence of Humans on Climate

Just as the evolution of microbes and plants has influenced climate over geologic time, so too has the evolution of humans. While climate scientists are in overwhelming agreement that our species has fundamentally altered atmospheric composition and Earth's temperature through our combustion of fossil fuels since the onset of the industrial revolution (a topic we examine in detail in Chapter 15), human influences on the climate system may have begun much earlier. In 2003, paleoclimatologist William Ruddiman published a provocative paper in which he proposed that the Holocene interglaciation we are living in has been anomalously long due to the advent of agriculture thousands of years ago.[42] Given the Milankovitch cycle theory, summers should have begun cooling after 11,000 years ago, the time of the most recent peak in solar radiation. Instead, climate has remained quite warm.

Ruddiman suggested two possible mechanisms for the long-lasting warmth—deforestation to create farmland and the start of rice paddy agriculture. Ice core records indicate that carbon dioxide levels in Earth's atmosphere began rising about 8000 years ago, the very time that archeological evidence shows early farmers living in southeastern Europe began to clear forests to plant crops first cultivated in the Middle East (Figure 14-14). Using the slash and burn method, these farmers released carbon stored in trees back to the atmosphere, increasing the level of the greenhouse gas carbon dioxide and

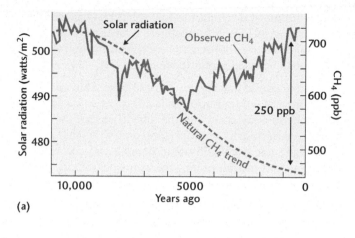

(a)

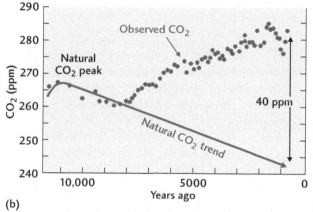

(b)

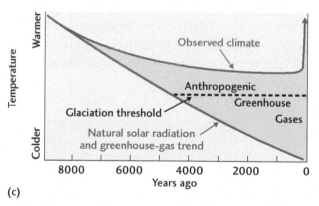

(c)

FIGURE 14-14 Greenhouse gas concentrations measured in bubbles in glacial ice reveal that carbon dioxide and methane concentrations rose during the early to middle Holocene at the same time as solar radiation was declining. In all prior interglacial episodes studied, greenhouse gas concentrations varied in concert with solar radiation changes. *(From W. F. Ruddiman, "The Early Anthropogenic Hypothesis: Challenges and Responses,"* Reviews of Geophysics *45 [2007].)*

effectively expanding the acreage of wetlands. Inasmuch as wetlands contain abundant decaying vegetation in a low-oxygen environment, the start of rice paddy agriculture increased the amount of methane, a potent greenhouse gas, released to Earth's atmosphere, which may also have kept Earth warm.

Ruddiman's ideas have been met both with intrigue and skepticism over the last decade. Some scientists doubt that the small numbers of humans on Earth in the early Holocene could have had much of an impact on greenhouse gas concentrations, particularly in view of the fact that the farmers were not contributing fossil carbon in the form of oil, coal, or natural gas to the atmosphere. These scientists call for natural variability in the climate system and its feedbacks to explain the long episode of Holocene warmth. The problem with this argument is that all previous interglaciations that have been studied show that greenhouse gas concentrations in the atmosphere varied in sync with changes in solar radiation. As radiation increased, greenhouse gas levels rose, and as radiation decreased, greenhouse gas levels fell. Only the Holocene interglaciation has been different. Furthermore, declines in carbon dioxide content seen in the last 2000 years coincide with major wars and plagues that caused human population declines. In an agrarian world, Ruddiman argues, population die-offs likely led to reforestation of abandoned farm fields, with photosynthesis sequestering carbon dioxide in trees and causing atmospheric concentrations to decline. Time will tell if Ruddiman's hypothesis about the cause of Holocene warmth is correct. In the meantime, it has spurred much fascinating new research at the intersection of geology, archeology, atmospheric science, oceanography, and glaciology.

Climatic Feedbacks

While changes in solar energy inputs or in any of Earth's systems can cause changes in climate, reinforcing feedbacks within the climate system can amplify and balancing feedbacks can reduce climatic variability. As mentioned in Box 14-1, the tendency of ice to perpetuate itself is a reinforcing feedback mechanism. As ice caps grow, they reflect increasing amounts of energy back to space. As a consequence, less incoming solar radiation is absorbed by Earth's surface, resulting in declining temperatures and more ice growth, a continuously self-amplifying cycle. Another reinforcing feedback in the climate system is the water vapor–temperature feedback.[43] Water vapor is a strong greenhouse gas, responsible for most of the surface warming Earth's atmosphere provides. Recall also from Chapter 11 that warm air can hold more water vapor than can cold air. Because of this fact, anything that causes Earth to begin warming will allow more vapor to be held by the atmosphere, which in turn will increase the warming. The fact that Earth

buffering some of the temperature decline produced by Milankovitch cycle cooling. Deforestation gradually spread to Western Europe, India, and China as agriculture took hold, increasing the carbon dioxide concentration of the atmosphere further. Then, around 5000 years ago, farmers in Asia began creating irrigated rice paddies,

has never experienced a runaway greenhouse climate like Venus has indicates that balancing feedbacks must exist to stabilize climate.

Silicate weathering and the production of carbonates is one such balancing feedback that keeps Earth's temperature in a habitable range. Another such mechanism arises from phytoplankton (microscopic aquatic plants) photosynthesis. On sunny days, the surface mixed layer of the ocean warms markedly, leading to density stratification that prevents nutrients from the deeper ocean from mixing into surface waters. At the same time, high levels of ultraviolet (UV) light penetrate the water. The combination of the reduction of nutrients and increased UV cause physiologic stress for phytoplankton, which respond by producing dimethyl sulfide gas (DMS) that leaks out of the surface ocean and into the atmosphere (Figure 14-15).[44] There the DMS undergoes a number of reactions that convert it to sulfate aerosols, which act as cloud nucleation sites (see Chapter 11). As clouds form, they block incoming solar radiation, cooling the atmosphere and surface mixed layer of the ocean, and reducing the stress on the phytoplankton. As a result, less DMS is produced, and the clouds begin to dissipate, allowing Earth to warm again. Phytoplankton physiology, then, may help to maintain Earth's temperature at a fairly uniform value and reduce climatic variability.

How do we know that these processes were going on in the past? How can we learn what climate and environment were like at different times in Earth's history? The answers lie in clues contained in the geologic record and in numerical model simulations of the climate system.

- Evolution of different life-forms has changed the composition of Earth's atmosphere over time, leading to changes in greenhouse gas concentrations that occasionally caused glaciations.

- Plate tectonic impacts on climate arise from changes in continental configuration that affect oceanic and atmospheric circulation and the albedo of Earth's surface; variations in seafloor spreading rates that affect greenhouse gas levels and ocean depths; and volcanic eruptions that introduce into the atmosphere both greenhouse gases and aerosols that block solar energy.

- Oceanic circulation has changed over time in response to glacial meltwater inputs, and changes in circulation have led to climatic fluctuations. The coupled ocean–atmosphere system is also responsible for climatic events such as El Niño and for the production of hurricanes.

- Processes that cause climate change operate on distinctive timescales: Volcanic eruptions can cause climate to change for a few years, whereas rearrangements of continents and ocean basins by plate tectonics cause climate to change over millions of years. Evolution of life has caused changes over tens of millions to billions of years.

- Recent glaciations on Earth have been driven by cycles in Earth's orbital parameters, such as tilt and wobble of the spin axis and eccentricity of the orbit, which combine to produce insolation minima.

- Reinforcing feedbacks such as the ice-albedo and water vapor–temperature feedbacks amplify climatic

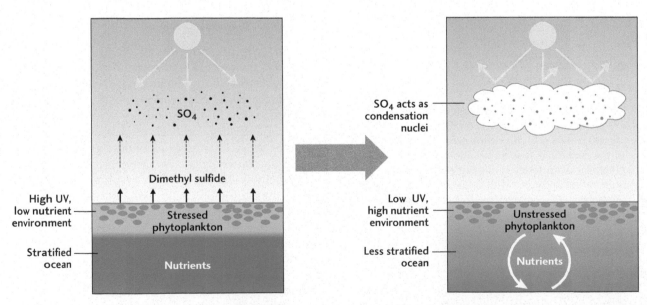

FIGURE 14–15 The phytoplankton–DMS climate balancing feedback. Physiological stress on sunny days causes phytoplankton to produce dimethyl sulfide gas, which acts as a cloud nucleator, reducing solar energy receipts at the surface ocean and reducing stress.

variability, while balancing feedbacks such as the silicate weathering cycle and phytoplankton–DMS cycle tend to stabilize climate.

INDICATORS OF ENVIRONMENTAL CHANGE

Scientists have identified many climatic and environmental indicators in the geologic record that provide information about temperature, precipitation, and vegetation distribution at different times in the past. These indicators fall generally into one of two categories, geologic or biologic. The sediment layers deposited in Coldwater Lake, described at the beginning of this chapter, are

geologic indicators of rainfall intensity. Tree rings, used to infer the length of droughts in California, are an example of biological indicators.

Geologic Records of Climate and Environment

Earth's environmental history is written in desert landforms, lake shorelines, glacial landforms, ice, stalagmites, and ocean and lake sediments, among other proxies.

Arid Environments Dunes are a common feature of today's deserts. Sculpted by winds that blow sand grains up one side of the dune and down the other (Figure 14-16), they move forward over time in the

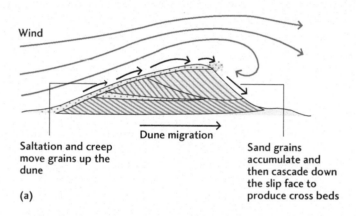

FIGURE 14–16 The formation of cross beds in dune sands. (a) Grains are blown up the "stoss" (upwind) side of the dune and are redeposited on the "lee" (downwind) side of the dune, leading to distinctive layering that indicates the direction of the prevailing wind. (b) The shape of this dune reveals that the prevailing wind direction is from left to right as this photo is oriented. The upwind side of the dune (left) has a low slope, up which grains are blown. At the crest of the dune, they avalanche down the steep downwind side. *(b: Kirsten Menking)*

direction of the prevailing winds. If you were to take a shovel and dig into a dune, you would see the results of the forward motion as a series of inclined layers of sand, known as **cross beds,** which point in the direction of the prevailing wind.

If you found 350-million-year-old rocks made of cross-bedded sands that had become cemented together, you might assume that these sands indicate a past desert environment. However, dunes, like many other sedimentary deposits, can form in a variety of environments. All they require is a supply of sand and active winds. Therefore, while frequently found in deserts, they are also common along coasts. To establish whether the dunes formed in a desert or a coastal environment requires additional information and a good understanding of modern depositional environments. If the dunes are deposited in a sedimentary environment of sands containing shells of clams and oysters, muds containing marine fossils, and limestones produced by the accumulation of shells of dead marine organisms, the dunes indicate coastal rather than arid conditions (Figure 14-17). If, however, the dunes occur with sequences of debris flow deposits, muds devoid of marine fossils, and evaporite salts, they probably indicate an arid environment. The muds and salts are deposited in ephemeral lakes that fill during desert downpours and dry up within a few days to weeks. Further evidence of aridity could be found in polygonal **mud cracks** that form when lakes dry up and the drying mud contracts. Only after we identify the depositional environment in which the dunes formed can we make climatic interpretations.

Desert or coast, cross beds that develop in dunes contain an accurate record of past prevailing wind direction. In addition, the shapes of the dunes themselves often give directional information.

Glacial Landforms Evidence of past glaciation, which indicates colder temperatures than prevail today, is much less ambiguous than that presented by dunes. Glaciers form from an ongoing buildup of snow deposited at high altitudes or high latitudes where summer temperatures do not rise high enough to melt the previous winter's accumulation, hence they are indicators of cold climates. As snow continues to fall, the weight of new snow compresses that which fell several years earlier, causing recrystallization of the snowflakes into solid masses of ice. If the ice accumulates on sloping topography, it eventually becomes thick enough that its lower portion will flow under the influence of gravity, and that flow generates a variety of distinctive landforms that are not reproduced in any other environment.

Moraines are ridges of unsorted sediment left behind as a glacier melts. They form in a variety of ways at the front and sides of a glacier and underneath the ice. Most commonly, moraines form when debris-laden ice reaches the end of the glacier and melts, dropping its sedimentary load (Figure 14-18). Front-end moraines can also form when a glacier pushes rocks and soil ahead of it in the manner of a snowplow. If the climate becomes warmer, the ice may melt, causing the glacier to retreat and leave the moraine behind.

Glaciers also drag rocks embedded in the ice over bedrock. Sharp points on the moving rocks scratch into the underlying bedrock, creating a series of grooves known as **striations** that indicate the flow direction of the ice (Figure 14-19). The continual grinding of rock against rock is a very effective erosional process. As a result, mountain glaciers quickly deepen and widen their valleys, forming steep-sided U-shaped valleys in contrast to the V-shaped valleys produced by streams (see Figure 14-18). *Plucking,* in which glacial ice tears away large chunks

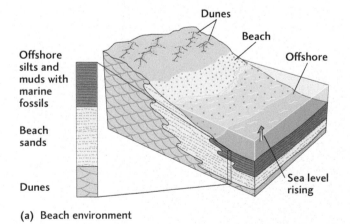

(a) Beach environment

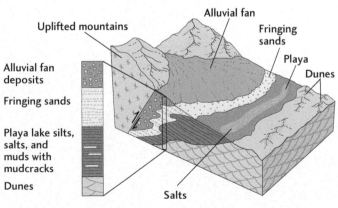

(b) Desert environment

FIGURE 14–17 Dune formation in (a) a beach environment and (b) a desert environment.

PRE-GLACIAL TERRAIN

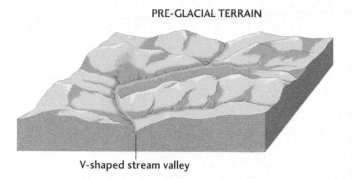

V-shaped stream valley

GLACIAL PERIOD

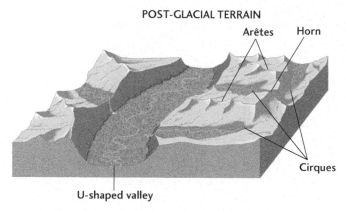

Glacier   Moraines   Arêtes   Horn

Cirques

ICE

ICE

Erosion through abrasion   Erosion through plucking

POST-GLACIAL TERRAIN

Arêtes   Horn

Cirques

U-shaped valley

FIGURE 14–18 A typical glaciated landscape. The advance and retreat of a glacier within its valley organizes debris into ridges called moraines, which consist of a variety of grain sizes from small clay particles to large boulders, and carve the valley into characteristic forms.

FIGURE 14–19 Glacial striations in Banff National Park, Alberta, Canada, were created when an angular rock embedded in the glacial ice was dragged over the underlying bedrock. Such striations can be used to infer the direction of past ice flow. *(Gilbert S. Grant/Science Source)*

of underlying bedrock, also contributes to the erosion. The thin spine of rock left between two glacial valleys, known as an **arête,** and the spires (or **horns**) found near the headwalls (or **cirques**) of several glaciers are also clear evidence that a landscape was once glaciated.

Ice sheets leave behind somewhat different landforms than do mountain valley glaciers because they are unconfined (Figure 14-20a). Vast deposits of **till,** a mixture of particles ranging in size from boulder to clay, reflect the processes of plucking and grinding that occur beneath glaciers. Chunks of ice left behind in the till sheet as the ice retreats melt to form **kettle lakes** (Figure 14-20b). Minnesota, the "land of 10,000 lakes," owes its geography largely to this process. In contrast, pockets of glacial debris contained within

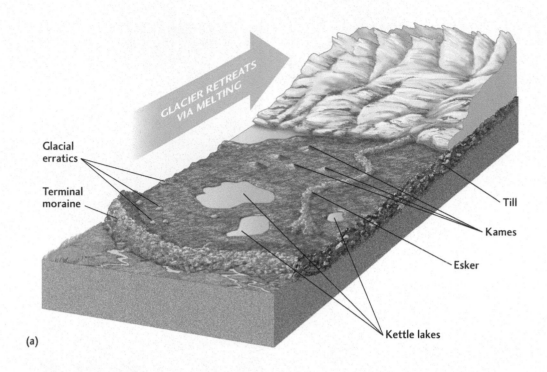

(a)

(b)

FIGURE 14–20 (a) Landforms diagnostic of continental ice sheets. (b) This kettle lake in Minnesota, so called because of its round shape, formed when a chunk of glacial ice melted on a till sheet. *(b: Carlyn Iverson/Science Source)*

the ice deposit during melting to form conical mounds called **kames.** Debris-laden meltwater streams flowing at high pressures underneath glacial ice deposit their loads of sediment in long, winding ridges called **eskers.** Another common feature of continental ice sheets is the **glacial erratic,** a boulder that has traveled hundreds and sometimes thousands of kilometers from its source region, frequently over hills. Glacial ice is the only medium on Earth dense and viscous enough to carry these large stones.

Glacial Ice A wealth of information regarding past climate can be found in glacial ice itself. The ice forms in readily distinguishable annual layers that vary in thickness in response to temperature variations in the atmosphere. Cold air cannot hold nearly as much moisture as can warm air, so during cold years, very little snow falls on ice sheets, producing thin layers of glacial ice. In contrast, warmer years are marked by increased precipitation as moisture-laden air masses travel poleward, producing thicker layers. The **oxygen isotopic composition** of ice also reveals past atmospheric temperature. There are three isotopes of oxygen: ^{16}O (8 protons and 8 neutrons), ^{17}O (8 protons and 9 neutrons), and ^{18}O (8 protons and 10 neutrons). Of these, ^{16}O, the isotope with lowest mass, is by far the most abundant, constituting 99.76 percent of all oxygen.[45] The heaviest isotope, ^{18}O, is second most abundant at nearly 0.2 percent, and ^{17}O is quite rare. Water molecules may contain any of these oxygen isotopes.

When water evaporates from ocean basins, molecules containing ^{16}O evaporate preferentially because less energy is required to evaporate molecules of lower mass. Atmospheric water vapor is therefore enriched in ^{16}O and depleted in ^{18}O compared to the ocean water from which it was evaporated. When vapor condenses to form clouds, molecules containing ^{18}O condense first and rain out of the air more readily than molecules containing ^{16}O. The colder the atmosphere, the more the heavy ^{18}O water molecules precipitate out, leaving the resulting cloud mass made of lighter isotopes. Consequently, storm systems may have lost much of their ^{18}O by the time they reach high latitudes, and the colder the atmosphere, the less ^{18}O in the precipitation (Figure 14-21). Warmer temperatures allow more ^{18}O to travel to high latitudes.

Paleoclimatologists use an instrument called a *mass spectrometer* to measure the ratio of ^{16}O to ^{18}O atoms in glacial ice, expressing the ratio with the following equation:

$$\delta^{18}O = \left[\frac{\left(\frac{^{18}O}{^{16}O}\right)_{sample}}{\left(\frac{^{18}O}{^{16}O}\right)_{standard}} - 1 \right] * 1000‰$$

where $(^{18}O/^{16}O)_{sample}$ is the ratio of ^{18}O to ^{16}O in a sample of glacial ice, and $(^{18}O/^{16}O)_{standard}$ is the isotopic ratio in a standard material of known isotopic composition.

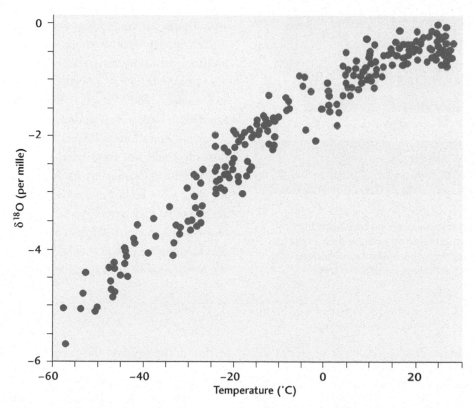

FIGURE 14–21 Relationship between atmospheric temperature and oxygen isotopic composition of precipitation. The warmer the air, the greater the amount of water vapor containing the heavy isotope ^{18}O, and the more positive the $\delta^{18}O$ value.

BOX 14–2 EMERGING RESEARCH

Earth's Flickering Temperature and Climatic Seesaw

In the last several decades, numerous expeditions to Greenland and Antarctica have used hollow drills to recover tens of kilometers of glacial ice extending deep into the ice sheets and hundreds of thousands of years backward in time (Figure 1). The work is not for the faint of heart, requiring weeks or months of camping in tents on ice sheets where temperatures during the short summer drilling season can measure as low as −35°C.[46] Nor are the working spaces a reprieve from the frigid conditions, given that temperatures must be kept below freezing to ensure the stability of the extracted ice. Drillers and scientists work in parkas, mit-

tens, and thick–soled boots to avoid the danger of frostbite. Though the working conditions challenge even the hardiest of souls, the information the ice contains has revolutionized how we think about climate variability on Earth.

By studying layer thickness, dust content, gas bubble composition, and oxygen and hydrogen isotopes, scientists have made remarkable discoveries. Prior to ice core research, it was thought that Earth's temperature changed gradually in response to the Milankovitch cycles. The difference between glacial and interglacial temperatures was about 5°C to 8°C and the transition from one climate state to the other unfolded over thousands or even tens of thousands of years. The seeming stability of climate throughout the Holocene fit with this gradualist view. Ice cores taken from Greenland painted a very different picture, however. For example, during the last ice age, abrupt swings of temperature measuring as great as 8°C to 16°C appear to have occurred over time spans as short as a couple of decades or less;[47] in some instances, temperatures changed more than 1°C per year (Figure 2)! Cores taken from Antarctica have revealed similar, though smaller, climate pulses. Interestingly, every transition to colder temperatures in Antarctica appears to have been matched by a transition to warmer temperatures in Greenland, and vice versa, as though the climate system acted like a huge seesaw during glacial times.[48]

What caused these frequent and abrupt changes? They certainly could not have been produced by variations in Earth's orbital parameters because those changes occur much too slowly. The ending of the events about 10,000 years ago suggests that they were in some way related to the presence of ice sheets in North America and Europe, which retreated almost completely by that time. Sediments from the floor of the North Atlantic tend to confirm this observation because

(a)

(b)

Figure 1 (a) Scientists removing an ice core from the drill string. (b) Note the annual layering and the dust in the ice. (a: Courtesy Douglas Hardy, University of Massachusetts Geosciences; b: Courtesy National Ice Core Laboratory, NSF, and USGS)

The lower the proportion of ^{18}O in a layer of glacial ice, the colder the atmosphere must have been when the ice was deposited as snow and the more negative the $\delta^{18}O$ value (Box 14-2).

In addition to temperature information, glacial ice contains a record of past winds. Atmospheric dust

settles on the ice, and the grain size of the dust particles is a direct indicator of wind strength: Faster winds are capable of transporting larger grains, whereas gentler winds transport finer particles. Glacial ice also contains a record of past volcanic eruptions as acid aerosols and ash particles injected high into the atmosphere and

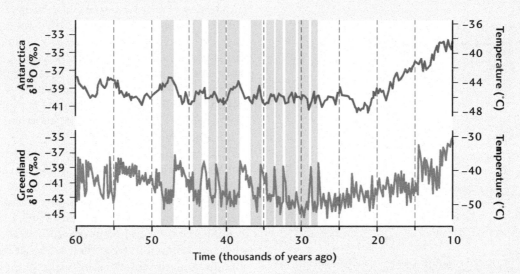

Figure 2 Greenland (NGRIP) and Antarctic (Byrd) ice core records show rapid changes in oxygen isotopic composition and, therefore, atmospheric temperature during the late Pleistocene. These temperature swings occurred too rapidly to have been caused by Milankovitch cycle forcing and have been nicknamed the "flickering switch" of climate. Note also that the Greenland and Antarctic records are frequently inverse of each other, a pattern that has been dubbed the climatic seesaw.

they show layers of debris occasionally interrupting the ordinary pattern of sedimentation. These layers suggest that the North American ice sheet sometimes launched armadas of icebergs into the North Atlantic, which carried loads of debris far into the ocean as the icebergs slowly melted.

If the iceberg armadas were sufficiently large, they may have significantly freshened the surface waters of the North Atlantic (see Chapter 12), thereby shutting off the formation of North Atlantic Deep Water. As a consequence, the global ocean conveyor system would have ground to a halt, discontinuing the transport of warm water to high northern latitudes and allowing temperatures around Antarctica to warm. Several of the climate shifts recorded in the Greenland ice cores are synchronous with the periods of ice–rafted debris deposition in the North Atlantic.[49] However, because some others do not correlate with changes in marine sedimentation, the climatic shifts remain somewhat of a mystery. In addition, if the shifts were caused by releases of icebergs to the North Atlantic, scientists must find the mechanism that produced the iceberg flotillas and answer such questions as why the North American ice sheet only occasionally released large quantities of ice to the ocean. Some recent numerical modeling work suggests that cycles in sunspots, which affect the amount of radiation given off by the Sun, may be responsible,[50] but definitive proof remains to be gathered.

stratosphere deposit onto ice thousands of kilometers away from the eruptive source (Figure 14-22).

Perhaps the most amazing feature of glacial ice is its ability to record past atmospheric composition. As snow recrystallizes to form ice, most of the air between adjacent snowflakes is squeezed out of the accumulating pile, but a small fraction remains and forms bubbles. Scientists sample these bubbles and measure the concentrations of gases like carbon dioxide and methane. By comparing gas levels to the oxygen isotopic composition of ice within individual layers, they have been able to show that glacial periods in

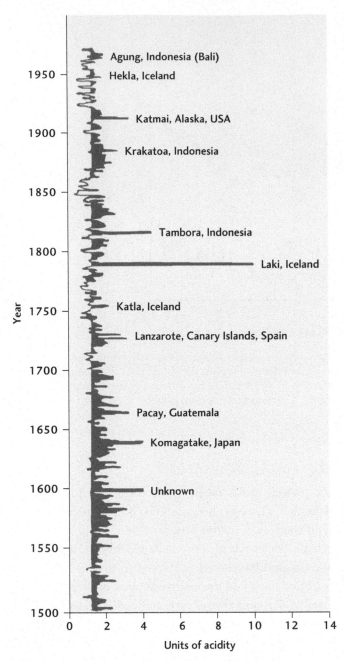

FIGURE 14–22 The acidity of glacial ice. High values correspond to periods of volcanic eruptions.

the last several hundred thousand years were marked by low greenhouse gas concentrations (Figure 14-23). Conversely, interglacial periods, like the warm interval we are presently living in, had higher levels. The annual layering of glacial ice also allows us to know when colder and warmer intervals occurred.

Lake Shorelines and Sediments In the western United States, basins that are presently arid show evidence of once having contained large lakes. This evidence consists of two types of shoreline development, constructional and destructional (Figure 14-24), as well as of sediments deposited within the lakes. **Destructional shorelines** are produced when waves strike the same location over and over again, cutting away at the land and forming a wave-cut bench. These benches are commonly covered with a beach deposit of rounded gravels. **Constructional shorelines** are usually identified by an algally produced $CaCO_3$ deposit known as *tufa,* which is made when springs of freshwater empty into the margins of a saline lake. In either case, both types of shoreline are evidence of a previously wetter climate.

Lakes also contain a variety of sediments that yield climatic information. For example, recall from the chapter opener that scientists linked the number and thicknesses of sediment layers in Coldwater Lake to the number and intensity of rainstorms. Lakes also collect pollen grains and plant macrofossils (twigs, leaves, seeds, etc.) blown in from their adjacent drainages and contain aquatic fossils that yield climatic information (see biological proxies below). In so-called closed basin lakes—lakes that lack outflow streams—water from the surrounding drainage basin delivers dissolved solids to the lake, which gradually increase in concentration over time as lake water evaporates. Depending on the types of dissolved solids present, a variety of evaporite minerals precipitate in the water column and deposit on the lake floor. If inflow ceases, evaporation leads to a succession of mineral deposits, with the least soluble (such as calcium carbonate) depositing first and the most soluble (such as halite) precipitating last. Variations in the types of salts present in a sediment core taken from the lake therefore record variations in the precipitation–evaporation balance of the lake over time.

Calcium carbonate precipitation in lakes provides additional information on hydrologic balance because it contains a record of the oxygen isotopic composition of lake water that also can be used to infer lake level. Evaporation preferentially removes water with ^{16}O from the lake because it is lighter than water with ^{18}O. Therefore, the water in lakes that are subject to high rates of evaporation tends to become heavier with time. When a lake receives runoff from its surrounding drainage basin or precipitation on its surface, that runoff and precipitation are usually very light isotopically.[51] If there is more inflow to the lake than outflow through evaporation, the lake will grow bigger and contain lighter water. If, however, there is more outflow from the lake than inflow, the lake will shrink and contain heavier water. Calcium carbonate crystals in the lakebed record these isotopic variations and, therefore, also record fluctuations in the size of the lake.

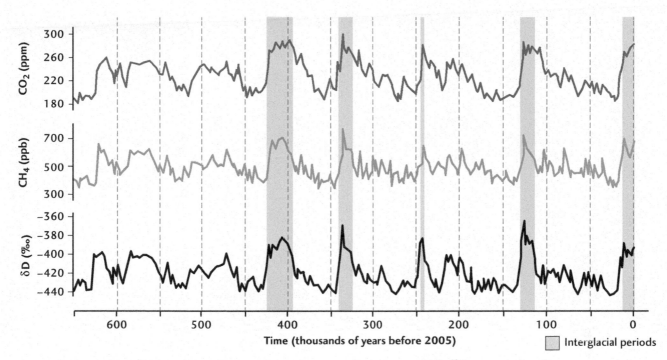

FIGURE 14–23 The Antarctic EPICA/Dome C ice core reveals a striking correspondence between the deuterium isotopic composition (δD) and greenhouse gas levels (CO_2 and CH_4) in bubbles trapped in the ice. Deuterium is a heavy isotope of hydrogen, and like ^{18}O is more abundant in atmospheric water vapor when temperatures are warm and less abundant when temperatures are cold.

Cave Deposits Another geologic deposit that holds oxygen isotopic information is the **speleothem,** a calcium carbonate cave formation (Figure 14-25). Speleothems form when groundwater carrying dissolved calcium carbonate drips into a cave and precipitates. The groundwater originates as rainfall at Earth's surface, and as the rain infiltrates through soils, it combines with carbon dioxide produced by root respiration and decay of

(a)

(b)

FIGURE 14–24 Destructional and constructional shorelines at Pyramid Lake, northwestern Nevada. (a) Destructional wave–cut benches along hillslope (in distance). (b) Constructional tufa towers formed at the margins of the lake. *(Kirsten Menking)*

(a)

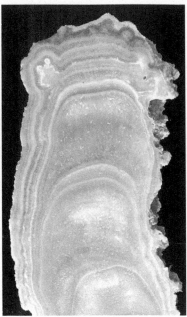

(b)

FIGURE 14–25 (a) Speleothems forming in a cave. (b) Cross section through a speleothem showing growth banding. *(a: Auscape/UIG/Getty Images; b: Courtesy Victor Polyak)*

organic matter to make carbonic acid. In areas of limestone, the carbonic acid dissolves the mineral calcite, putting calcium into solution:

$$H_2O \ + \ CO_2 \ \rightarrow \ H_2CO_3$$
water + carbon dioxide → carbonic acid

$$H_2CO_3 \ + \ CaCO_3 \ \rightarrow \ Ca^{2+} \ + \ 2 \ HCO_3^-$$
carbonic acid + calcite → calcium + bicarbonate

Because the air in caves typically contains lower amounts of carbon dioxide than the air in soils, when groundwater drips into a cave, it is suddenly oversaturated with dissolved carbon dioxide in bicarbonate form. As a consequence, the chemical reactions shown above reverse direction, allowing calcite to precipitate and carbon dioxide to bubble out of the groundwater into the cave atmosphere. The precipitating calcite creates stalactites and stalagmites—cave formations that hang from the cave ceiling or build up from the cave floor as water droplets fall to the ground. Eventually, as they continue to grow by chemical precipitation, these formations may fuse to become a continuous column connecting the cave ceiling and floor.

The fact that speleothems originate in rainfall at Earth's surface makes them useful climate archives. As in the case of ice cores, changes in air temperature over time may cause changes in the oxygen isotopic composition of rainfall that are preserved as the rainfall becomes groundwater and then eventually precipitates calcium carbonate. These variations may be related to changes in global temperature, to changes in the season when most precipitation falls (e.g., winter precipitation would have a more negative $\delta^{18}O$ value than summer precipitation), or to changes in the source area of the precipitation (e.g., air masses originating over the tropics would have a more positive $\delta^{18}O$ value than those originating over the poles). Variations in speleothem $\delta^{18}O$ values also arise from changes in the temperature at which the calcite precipitates, in other words from changes in the air temperature within the cave, itself directly related to mean annual air temperature at Earth's surface. It is the job of the paleoclimatologist to interpret the isotopic fluctuations over time in light of what is known about the present climate and about past changes derived from other proxy records and numerical modeling simulations.

Because they form underground, speleothems are among the most challenging climatic proxies to obtain. Geologists travel deep into Earth, often slithering on their bellies through tight passageways that connect cave rooms. At the same time, they carry rock-climbing equipment to navigate vertical passageways, and hammers, chisels, and rock drills to collect their samples. All of this equipment and the heavy samples themselves must then be brought back to the surface, and the total distances traversed can be hundreds of meters to a kilometer or more. Once at the surface, however, speleothems give up their climatic information readily. Geologists split them in half lengthwise to reveal their growth bands, use dental drills to extract tiny samples of calcite, and then introduce those samples into mass spectrometers to measure their oxygen isotopic compositions. Along with radiometric dating, these samples can provide a continuous record of climatic variation over tens or even hundreds of thousands of years.

Using these techniques, scientists in New Mexico have found oxygen isotopic variations in speleothems that mimic the isotopic variations measured on the Greenland ice sheet (see Box 14-2). In New Mexico's Fort Stanton Cave, isotopic fluctuations in speleothem calcite have been traced to shifts in the position of the polar jet stream (see Chapter 11) that affect the amounts of summer and winter precipitation striking that region.[52] During times when the Greenland ice sheet showed cooling episodes, the jet stream was apparently deflected southward, bringing greater amounts of winter precipitation to Fort Stanton Cave that resulted in more negative $\delta^{18}O$ values. Episodes of abrupt warming over Greenland brought a reduction in winter moisture to New Mexico and a more positive $\delta^{18}O$ value as the jet stream shifted northward and warmer summer moisture increased in importance.

Biological Records of Climate and Environment

Organisms grow larger and reproduce more when environmental conditions are favorable to them. An analysis of fossils and living organisms thus contributes to an understanding of past environments.

Fossils The presence of specific fossils indicates when organisms flourished, suffered, or became extinct. Crocodile bones dating to the Eocene epoch (57 million to 34 million years ago) have been found buried in sediments in Utah, Colorado, Wyoming, and the Dakotas.[53] Crocodiles are reptiles—cold-blooded animals that require temperatures warmer than freezing in order to survive. Currently they are found in the parts of Africa, Australia, Asia, and the Americas that are warm year round. In the United States, for example, they are found only in states along the Gulf Coast, but the states in which *fossil* crocodiles have been found presently experience harsh winters with very cold temperatures. Therefore, paleoclimatologists believe that temperatures in these northern states must have been substantially warmer in the Eocene, perhaps more akin to temperatures observed today in Florida and Louisiana. Furthermore, crocodiles live in swampy habitats quite unlike modern conditions in Utah, Colorado, Wyoming,

and the Dakotas. Not only were these states warmer in the Eocene, they must have been wetter.

In addition to qualitative information about climate, some fossils can provide quantitative estimates of temperature. Coral reefs grow only at temperatures warmer than 16°C to 17°C,[54] and their growth is optimized at temperatures between 23°C and 25°C; today, these conditions are found only in the tropics, between 30°N and 30°S latitude. Therefore, fossil corals indicate past warm water temperature. Furthermore, coral growth bands vary in width with changes in temperature, as does coral oxygen isotopic composition, thus providing a sensitive record of temperature change in tropical waters.[55]

The fossils of some organisms provide information not only about temperature but also about the salinity of the environment they lived in. Ostracodes and diatoms (microscopic animals and plants), as well as mollusks and fish, are present in many streams and lakes. Some of these organisms prefer very fresh water, whereas others thrive only in more saline environments. Often the species found in a lake can be used to infer changes in lake level in closed basin lakes: Water becomes salty when there is more evaporation than inflow, just as it becomes isotopically heavy. Thus, changes in salinity, like changes in isotopic composition, reflect variations in lake size.

Packrat Middens Packrats are small rodents that construct nests, called **middens,** out of twigs, leaves, and bark from their immediate surroundings. Because they are made of biological materials, middens contain carbon, and therefore can be dated by radiocarbon. Furthermore, identification of the twigs and other plant debris can be used to determine what climate existed in the region at the time the midden was constructed, since different plants require different climatic conditions for their existence. Cacti, for example, are found primarily in deserts.

Pollen Pollen grains from different plant species have very distinctive shapes and can be used to infer what plants were present at different times in the past. *Palynologists,* scientists who study pollen, typically extract a core of sediments from the floor of a lake into which pollen has been deposited. They then use radiocarbon methods to determine the age of the sediment at different depth horizons in the core and identify the pollen grains to assess what the climate was at each depth or age horizon. Using pollen from many different locations in the eastern United States, scientists have shown how forests responded to the retreat of the North American ice sheet from 18,000 years ago (the time of the last glacial maximum) to the present (Figure 14-26). Cold-loving species were initially found in southern states, but migrated northward as the ice sheet retreated.[56] Today,

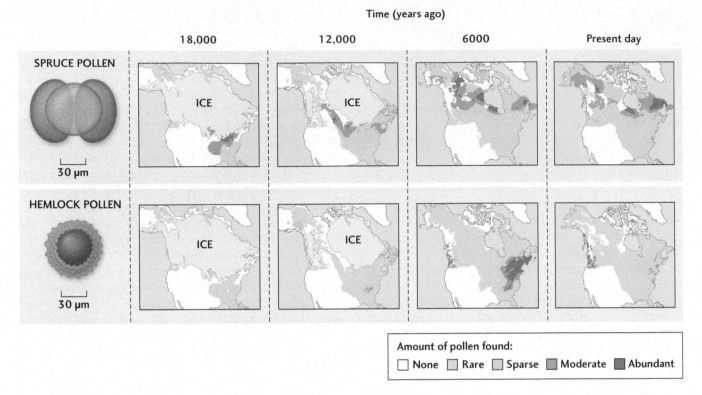

Time (years ago)

18,000   12,000   6000   Present day

SPRUCE POLLEN

30 μm

HEMLOCK POLLEN

30 μm

Amount of pollen found:

☐ None   ☐ Rare   ☐ Sparse   ■ Moderate   ■ Abundant

FIGURE 14–26 Spruce and hemlock pollen have unique shapes that allow them to be readily identified. Maps of pollen distribution show the progress of these species as climate warmed and the large ice sheet that blanketed much of North America during the last glacial maximum retreated. Spruce migrated northward, seeking a colder climate, and hemlock developed in areas previously too cold for its survival.

species such as spruce and fir are found in abundance only in Canada and the northern United States.

Tree Rings Trees growing in a climate that varies seasonally go through a growth phase and a dormancy phase each year. During the growth phase, new wood is added to the outer layer of the tree (Figure 14-27). This **early wood** is composed of large, spongy cells and is very low in density. As fall approaches, the tree starts to become dormant, but before it does, it lays down a layer of **late wood**, which consists of very densely packed, smaller cells.[57]

The widths of the early and late wood rings are directly tied to climatic conditions. If growing conditions are optimal, the tree will show a fat growth ring, but if growing conditions are poor, a narrow ring will indicate that very little growth occurred in that year. Many factors influence growth, including cloudiness, availability of nutrients in the soil, temperature, precipitation, and the amount of food stored within the tree. Of these, the two most important factors are usually temperature and precipitation.

For trees living in a continuously warm climate such as experienced in the American Southwest, the limiting

factor on growth rate tends to be precipitation, with arid years yielding very little growth and wet years yielding more growth. In a rainy environment, temperature might be the limiting factor. Thus, tree-ring widths indicate how the limiting climatic factor has changed over time. This kind of analysis has yielded climatic records that go back several thousand years.

Tree rings are particularly useful climatic indicators because each band represents one year of growth. Radiocarbon dating can be used to determine the approximate age of one of the bands, and the total amount of time spanned by the tree can be found simply by counting the rings.

Marine Organisms and Oxygen Isotopes During glacial periods, water containing ^{16}O is preferentially extracted from the ocean basins to form ice sheets. As a result, ocean water becomes heavier, that is, more enriched in $H_2^{18}O$, during glacial periods. Tiny single-celled organisms living in the oceans known as **foraminifera** (Figure 14-28) secrete shells of calcium carbonate ($CaCO_3$) in equilibrium with the ocean water surrounding them. Thus, their shells contain a record of the past isotopic composition of the ocean, a record that

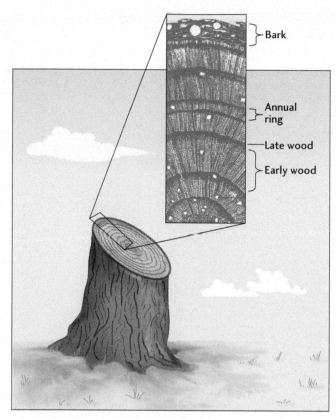

FIGURE 14-27 A cross section of a conifer tree trunk showing early wood and late wood rings. Each early wood–late wood couplet represents one year of growth.

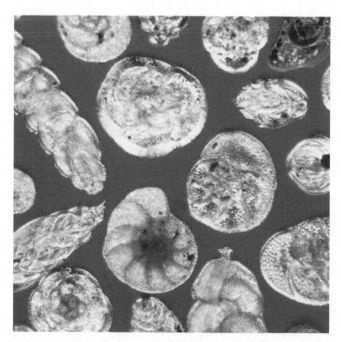

FIGURE 14-28 Geologists use the shells of foraminifera such as these to determine the oxygen isotopic composition of the ocean, from which they can infer global ice extent. *(Alfred Pasieka/Science Photo Library/Science Source)*

can be used to determine when glaciations occurred on Earth.

Using devices called piston corers (essentially, hollow drills), geologists have sampled ocean sediments all over the world. The resulting cores usually contain a continuous record of sedimentation for thousands to hundreds of thousands of years. As with the oxygen isotopic composition of glacial ice, scientists use stable isotope mass spectrometers to measure the abundance of ^{18}O and ^{16}O atoms in foraminifera extracted from the sediment cores, expressing their results with the $\delta^{18}O$ equation shown above. Higher proportions of ^{18}O suggest the existence of ice sheets, whereas lower proportions indicate little ice. The record of isotopic fluctuations now extends backward in time 65 million years[58] and reveals such events as the Paleocene–Eocene thermal maximum, a time when Earth was much warmer than at present; the onset of glaciation in Antarctica; and the growth of ice sheets in the northern hemisphere (Figure 14-29a). The high proportion of ^{18}O at about 20,000 years ago corresponds to the last glacial maximum, when an ice sheet covered all of Canada and stretched south into parts of the United States (Figure 14-29b).

■ Indicators of environmental change can be broadly classified as either geologic or biologic.

■ Arid environments may be recognized from the presence of sand dunes and mud cracks. Cross-bedding in dunes indicates the prevailing wind direction.

■ Erosional and depositional landforms produced by glaciation are distinctive and allow geologists to infer the past extent of glacial ice. Glacial ice itself contains a wide variety of climatic information: Oxygen isotopes in the ice allow past air temperature to be determined, bubbles contain gases that record past atmospheric composition, and dust particles may record volcanic eruptions and wind strength.

■ Lake shorelines in the arid western United States testify to previously wetter climates.

■ Minerals precipitated into closed basin lakes reflect the degree of evaporative concentration of lake water and are therefore correlated to lake volume and the precipitation–evaporation balance.

■ Cave deposits, called speleothems, contain oxygen isotopic records that reflect variations in temperature at Earth's surface, changes in seasonality of moisture, or changes in moisture sources.

■ Fossils of plants and animals provide ecological information about temperature, amounts of precipitation, and lake salinity.

■ Plants are used in a variety of ways in paleoenvironmental studies: The width of tree rings can indicate temperature or precipitation variations from

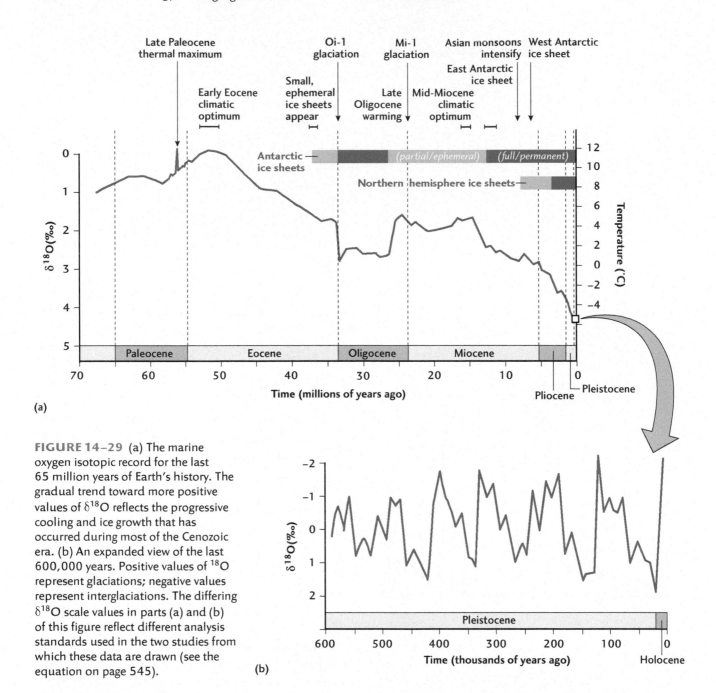

FIGURE 14–29 (a) The marine oxygen isotopic record for the last 65 million years of Earth's history. The gradual trend toward more positive values of $\delta^{18}O$ reflects the progressive cooling and ice growth that has occurred during most of the Cenozoic era. (b) An expanded view of the last 600,000 years. Positive values of ^{18}O represent glaciations; negative values represent interglaciations. The differing $\delta^{18}O$ scale values in parts (a) and (b) of this figure reflect different analysis standards used in the two studies from which these data are drawn (see the equation on page 545).

year to year, and pollen and packrat middens record vegetational distribution at different times in the past.

■ Oxygen isotopes in foraminifera provide information on the amount of ice on Earth over time.

GLOBAL CLIMATE MODELS

In the last few decades as computing power has increased, scientists have turned increasingly toward numerical models of the climate system to interpret the proxy record. These models, called **global climate models** or

general circulation models (both **GCMs**), use mathematics to simulate atmospheric and/or oceanic processes and conditions under different land cover or greenhouse gas scenarios. The basic construction of GCMs is a three-dimensional grid of boxes that represents Earth's surface as well as different levels in the atmosphere or depths in the ocean (Figure 14-30). Atmospheric GCMs, for example, represent Earth's surface as a series of boxes containing either land or water, and have 10 to 20 overlying layers of atmosphere broken into boxes defined by latitude and longitude lines. Each of these boxes is populated with a mixture of gases (nitrogen,

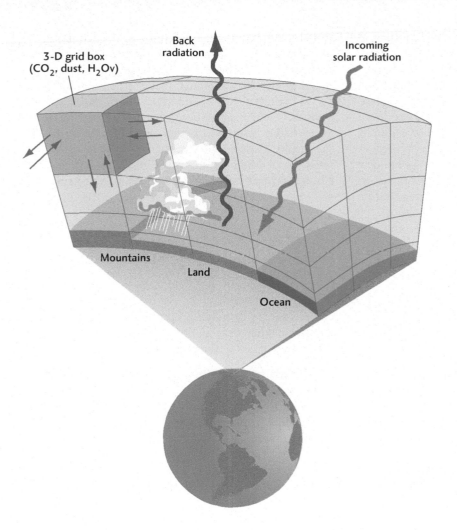

FIGURE 14–30 Global climate models incorporate physical laws of conservation of mass, energy, and momentum to simulate atmospheric and oceanic conditions under different climatic forcing scenarios. *(From W. F. Ruddiman,* Earth's Climate: Past and Future, *2nd ed., © 2008 by W. H. Freeman and Company)*

oxygen, water vapor, carbon dioxide, etc.) and aerosols and allowed to interact with incoming solar radiation and outgoing radiation from Earth's surface. The latter includes both solar energy reflected by Earth's surface and infrared energy emitted by the surface after solar radiation absorption.

Within each atmosphere box, fundamental laws of physics apply. For example, the ideal gas law relates gas pressure, volume, and temperature to each other, and conservation of mass requires that if solar energy heats the air within one grid box and causes its density to decline through thermal expansion, that the mass associated with the density drop be transferred to adjacent grid boxes rather than simply disappearing. In this way, the models create winds that move gases and energy from one box to another. Ocean GCMs operate in a very similar fashion, the differences being that the oceans have walls, and the fluid they simulate is denser and more viscous than the atmosphere. Just as conservation of mass and energy leads to winds in atmospheric GCMs, these same fundamental physical laws generate currents in the simulated oceans. The most sophisticated GCMs include both oceanic and atmospheric circulation

and may also simulate changes in vegetation and/or ice cover over time.

Climate modelers use GCMs to conduct two different types of experiments. In **sensitivity experiments,** they test the response of the models to a change in a specific parameter. For example, they might compare a simulation of Earth's climate with the present greenhouse gas composition to an Earth with much lower carbon dioxide and methane, such as we know occurred during the last glacial maximum (recall the ice core record of atmospheric greenhouse gas concentrations). By changing only the greenhouse gas composition, they can determine whether such a change is sufficient to explain the colder conditions that existed at that time. Modelers might also explore the impact of tectonics on climate by experimenting with opening and closing different oceanic gateways such as Drake's Passage between Antarctica and South America.[59]

Another type of model experiment is the **climate reconstruction.** In these experiments, modelers use the best available information to specify the topography and makeup (ice, desert, forest, etc.) of the land surface, the depth of the oceans and extent of coastlines, greenhouse gas concentration of the atmosphere, and solar radiation inputs for a particular period of time they wish to simulate. After allowing the model to reach equilibrium, they generate maps of surface temperature, wind speed and direction, and other climatic variables that they can compare to data derived from proxies. The maps produced can also drive further proxy investigations. For example, a climate reconstruction might show a change in precipitation or temperature occurring in a location previously not investigated. Seeing these maps spurs researchers to go on a hunt for proxy data to either confirm or refute the results of the model.

Planning for future changes in climate and the environment is increasingly reliant on climate modeling simulations that forecast changes in temperature and precipitation patterns associated with increasing greenhouse gas levels. To determine whether models are suitable for such predictions, scientists test their abilities to simulate present and past climates. The better the fit between model outputs and proxy data in the past, the more likely the model will correctly respond to perturbations designed to simulate the future. Our knowledge and measurement of past events will always be incomplete and spottier in some time intervals and regions than in others, however.[60] Likewise, our understanding of some aspects of the climate system (e.g., clouds, discussed in Chapter 15) remains poor.

Another difficulty is that much of the physics of the climate system takes place at the molecular level (e.g., the absorption of solar radiation by individual carbon dioxide, methane, or ozone molecules) and at the timescale of nanoseconds, and computing power will never be great enough to simulate such small-scale processes on a planetary scale and over long geological time periods.[61] These limitations mean that models require us to make assumptions that introduce uncertainty to the results. Despite these problems, models are valuable tools for helping us study the basic operation of the climate system, but a degree of reasonable skepticism is warranted when using models to predict the future.

■ Global climate models (general circulation models) use the laws of physics to mathematically simulate atmospheric and/or oceanic conditions for different periods in Earth's history.

■ Sensitivity experiments test the response of the climate system to a change in the configuration of continents and ocean basins or to different amounts of greenhouse gases in the atmosphere.

■ Climate reconstruction experiments attempt to simulate a particular time period in Earth's history and require detailed information on past topography, atmospheric composition, and the amount of solar radiation striking at the time of interest. Reconstructions are compared to proxy data from the same time interval to determine how well the simulation matches the known conditions.

CLOSING THOUGHTS

Earth's surface environments are in a constant state of flux because of the actions of many processes, some of which cause linear changes in climate and environment, others of which are more cyclic. These processes operate on a variety of timescales, some occurring so slowly that they are nearly unrecognizable on the timescale of a human life. As we go about our daily business, we do not notice that the continents beneath our feet are migrating to new positions, nor do we recognize the tiny changes in solar radiation intensity that accompany annual variations in Earth's orbit about the Sun. Nevertheless, the climate at any given point in time is dependent on these and many other factors, some understood by modern science, some probably yet to be discovered.

The continuous cycling of carbon between the atmosphere, hydrosphere, biosphere, and geosphere has worked to make Earth uniquely habitable among the planets in our solar system. However, this does not mean that all Earth environments are hospitable or that those that presently are will always remain so. Climates have changed dramatically throughout geologic time. The tools that we have at our disposal to assess these past changes are of varying kinds. Some give quantitative information about changes in temperature, precipitation, and vegetation patterns. Others are more qualitative, telling us only if climate was colder or warmer, wetter or drier. Still, any information we can gather about past climate and environment is critical to our understanding of how Earth works and what we might expect for the near future.

As the human population grows, our ability to adjust to changes in climate becomes less certain. Modern societies cannot, as our early ancestors did, simply pack their bags and move to a new location if the current environment becomes inhospitable. We must be able to develop contingency plans to deal with climatic and environmental change in each community. The first step toward this goal is a full understanding of how Earth's climate system operates and our planet's history of global environmental change.

SUMMARY

■ Were it not for the interaction of the hydrologic and carbon cycles with plate tectonics, Earth's surface temperature would probably be too warm or too cold to support liquid water and, therefore, life. Venus and Mars are each missing one of these three vital cycles and therefore neither planet can support life.

■ Climate is the by-product of interactions among the lithosphere, pedosphere, atmosphere, hydrosphere, biosphere, and solar radiation.

■ Early Earth must have had a substantially larger concentration of greenhouse gases than at present to counteract the much lower luminosity of the young Sun. That this is true is evident from sedimentary rocks, which indicate that Earth had liquid water as early as 3.8 billion years ago.

■ Processes that change climate operate on a variety of timescales. Volcanic eruptions can cool the planet for a matter of years. Variations in Earth's orbital parameters drive climate changes that occur over tens of thousands of years, and plate tectonics alters climate over time spans of millions to tens of millions of years. Climate at any point in Earth's history represents the aggregate of these different processes and timescales.

■ Earth has been repeatedly glaciated in the last few million years because of cyclical variations in its orbital parameters that together produced periods of low insolation. Once ice began to grow, the reinforcing feedback mechanism based on the reflectivity of ice probably enhanced that growth. Earlier glaciations were likely related to evolution of different photosynthetic life-forms or to changes in the configuration of continents and ocean basins on Earth's surface.

■ Climate is recorded by a number of processes acting at Earth's surface, allowing us to reconstruct Earth's history of environmental change.

■ Evidence of previously arid conditions includes sand dunes, mud cracks, and salts deposited in saline lakes.

■ Abandoned shorelines found high above valley floors are evidence that climate was once much wetter in those regions than it is today, whereas tree stumps rooted at the bottom of lakes suggest a previously drier climate.

■ Glaciers and ice sheets leave behind both erosional and depositional landforms. Some of these landforms indicate the past extent of ice, and some show direction of flow of the ice.

■ Oxygen isotopes can be used to determine past atmospheric temperature, extent of ice on land, and relative lake level.

■ Vegetational distribution, as determined from pack-rat middens and pollen grains, can give information about past rainfall amounts and temperatures.

■ In the last few decades, numerical models have become widely used tools to help understand Earth's history of climatic change and to help predict future changes.

KEY TERMS

proxies (p. 521)
faint young sun problem (p. 524)
methanogens (p. 525)
cyanobacteria (p. 525)
sputtering (p. 526)
photodissociation (p. 526)
cyclothems (p. 530)
albedo (p. 531)
snowball Earth events (p. 534)
ice-albedo feedback (p. 536)

eccentricity (p. 536)
tilt (p. 536)
precession of the equinoxes (p. 537)
Milankovitch cycles (p. 537)
cross beds (p. 542)
mud cracks (p. 542)
moraines (p. 542)
striations (p. 542)
arête (p. 543)
horns (p. 543)
cirques (p. 543)

till (p. 543)
kettle lakes (p. 543)
kames (p. 544)
eskers (p. 544)
glacial erratic (p. 544)
oxygen isotopic composition ($\delta^{18}O$) (p. 545)
destructional shorelines (p. 548)
constructional shorelines (p. 548)
speleothem (p. 549)

middens (p. 551)
early wood (p. 552)
late wood (p. 552)
foraminifera (p. 552)
global climate models/ general circulation models (GCMs) (p. 554)
sensitivity experiments (p. 555)
climate reconstruction (p. 555)

REVIEW QUESTIONS

1. What cycles on Earth presently keep its surface temperature within a livable range?

2. Which of these cycles is Mars missing? Venus?

3. What Earth system changes allowed our planet to maintain a habitable climate despite steadily increasing solar luminosity?

4. What evidence is there for snowball Earth events, and what allowed Earth to emerge from icehouse conditions?

5. What role does life play in setting Earth's temperature?

6. How, and on what timescales, does plate tectonics influence climate change?

7. What role does the ability of a surface to absorb or reflect radiation play in setting global temperature?

8. How does oceanic circulation influence the climate of northern Europe?

9. What are Milankovitch cycles?

10. What conditions are favorable to the growth of ice sheets in the northern hemisphere?

11. What are climatic balancing feedbacks? How do they work?

12. If summer solar radiation were to increase, would the ice-albedo feedback lead to an increase or a decrease in the amount of ice on land? Explain your answer.

13. What kinds of evidence can we examine to discover how climate has changed throughout Earth's history?

14. In what ways are oxygen isotopes used to unravel Earth's history of climate change?

15. How are plants used in studying climate change?

16. What indicators supply information about the relative wetness of past environments?

17. What indicators supply information about past wind speed and wind direction?

18. What are general circulation models and for what purpose are they used?

THOUGHT QUESTIONS

1. Will Earth's climate always remain suitable for life, or will our planet someday become like Venus or Mars?

2. Suppose that snow and ice were black rather than white. Would the reinforcing feedback mechanism responsible for the formation of glacial ice continue to operate on Earth? Why or why not? What would happen to global temperature if Antarctica, the Arctic Ocean, and Greenland, presently covered by white snow and ice, were instead covered by black snow and ice?

3. Paleontologists have removed hundreds of bones from the La Brea tar pits, a naturally occurring sticky swamp of tar in Los Angeles. Many of these bones have been identified as belonging to wooly mammoths and have been radiocarbon dated at 20,000 years old. What does the presence of these bones imply about the past climate of Los Angeles?

4. Suppose that global warming caused the Greenland ice sheet to melt. What might the consequences for oceanic circulation be? What kind of experiment could you devise to test your hypothesis?

EXERCISES

1. Pick any place on Earth. What conditions would be necessary to create the hottest possible summer at that location? The coldest possible winter? Use what you know about Earth's orbit and spin axis, reflectivity or absorption of energy, and temperature as a function of elevation to explain your answer.

2. Take a walk in a wooded area and note how many different species of plants you see. You don't have to be able to identify the plants to do this; just carry a sketchbook and draw a picture of the leaves each kind of plant has. Then, pick an area 10 ft by 10 ft and record how many of each different kind of plant you find there. Now examine the leaf litter on the ground. Are all the plants in the area represented by leaves in the litter, or are leaves of some plants missing? Why might some plants not be represented in the litter? Are there leaves in the litter that come from plants outside your defined area? How did they get there? Are the relative proportions of leaves in the litter the same or different from the relative proportions of plants in the area? Why? If you were to try to infer what vegetation once flourished in an area based solely on a deposit of leaves, what difficulties might you have? These are the difficulties encountered by palynologists when they make environmental interpretations based on pollen grains.

SUGGESTED READINGS

Alley, R. B. *The Two-Mile Time Machine: Ice Cores, Abrupt Climate Change, and Our Future.* Princeton: Princeton University Press, 2002.

Bell, J. "The Red Planet's Watery Past." *Scientific American* (December, 2006): 62–69.

Catling, D. C., and K. J. Zahnle. "The Planetary Air Leak." *Scientific American* (May, 2009): 36–43.

Frakes, L. A. *Climates Throughout Geologic Time.* New York: Elsevier, 1979.

Hodell, D. A., J. H. Curtis, and M. Brenner. "Possible Role of Climate in the Collapse of Classic Maya Civilization." *Nature* 375 (1995): 391–394.

Hoffman, P. F., and D. P. Schrag. "Snowball Earth." *Scientific American* (January, 2000): 68–75.

Imbrie, J., and K. P. Imbrie. *Ice Ages.* Hillside, NJ: Enslow, 1979.

Kasting, J. F. "When Methane Made Climate." *Scientific American* (July, 2004): 78–85.

Kasting, J. *How to Find a Habitable Planet.* Princeton: Princeton University Press, 2010.

Linden, E. *The Winds of Change: Climate, Weather, and the Destruction of Civilizations.* New York: Simon and Schuster, 2006.

Ruddiman, W. F. *Earth's Climate: Past and Future.* 3rd Edition. New York: W. H. Freeman and Company, 2013.

Ruddiman, W. F. *Plows, Plagues, and Petroleum: How Humans Took Control of Climate.* Princeton: Princeton University Press, 2005.

Stine, S. "Extreme and Persistent Drought in California and Patagonia During Mediaeval Time." *Nature* 369 (1994): 546–549.

Humans and the Whole Earth System: Living in the Anthropocene

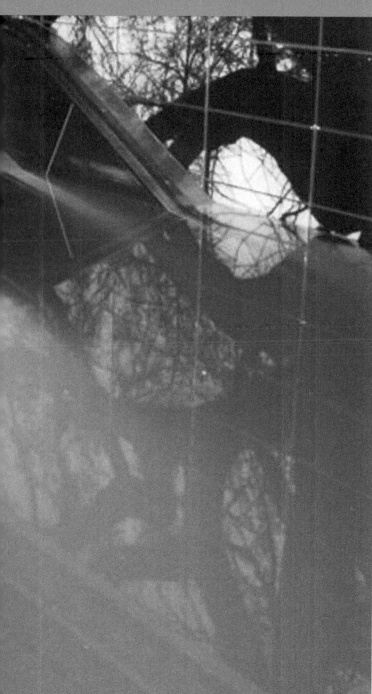

In late October of 2012, a Category 1 hurricane churning up the eastern coast of the United States merged with a winter storm coming from the west to deliver a devastating blow to communities in the Northeast. A ridge of high pressure over Greenland prevented the storm from moving out to sea, steering it directly into northern New Jersey and New York City. As if these conditions weren't bad enough, the storm coincided with a full moon that amplified the tides and raised surging waters still higher. Dubbed Superstorm Sandy, or Frankenstorm because of its close association with Halloween, the cyclone overran many coastal cities, floating houses off of their foundations, breaking in windows and doors, and leaving layer upon layer of mud and sand in streets, yards, and living rooms.

In a situation reminiscent of the flooding of New Orleans by Hurricane Katrina, lower Manhattan was submerged by more than 4 m of seawater.[1] The ocean poured down ventilation grates, filling several of the city's subway tunnels and bringing mass transit to a halt. For the first time since 1888, the New York Stock Exchange suspended trading for two days as millions of East Coast residents dealt with power outages. Damage to property was estimated at more than $62 billion and at least 125 people lost their lives throughout the Atlantic states.[2] Climate scientists blamed sea level rise associated with a century of global warming for some of the devastation, predicting a grim future for coastal cities around the world.

In this chapter we:

✔ Explore the impact of human activities on Earth systems.

✔ Discuss what is presently known about human modification of climate.

Superstorm Sandy brought more than 4 m of water to lower Manhattan, flooding many miles of the New York City subway system. *(AP Photo/Metropolitan Transportation Authority)*

✔ Describe what the future might hold as anthropogenic emissions of greenhouse gases continue.

✔ Discuss strategies to prevent catastrophic greenhouse warming and to adapt to climatic changes that are now inevitable.

INTRODUCTION

"Knee high by the 4th of July" goes the old adage regarding the growth of corn in the midwestern United States. But while the U.S. Department of Agriculture predicted a record-breaking crop in the spring of 2012,[3] by the time the nation celebrated its Independence Day, extreme drought had settled into the interior (Figure 15-1). Corn stalks withered as day after day of unrelenting heat baked soils, provoking comparisons to the Dust Bowl years of the 1930s. Along with corn, fields of soybeans,

grazing pastures, and watering holes all dried up, and livestock owners were faced with difficult decisions: Try to hang on and hope for rain, or slaughter herds it had taken years or even decades to build.[4]

While the drought proved catastrophic for many farmers, its disruptions were not limited to the agricultural sector. Tinderbox conditions throughout the western states led to massive forest fires in Idaho, Montana, and Utah. In New Mexico, lightning sparked the largest fire in state history, burning more than 1126 square km of land in the Gila National Forest.[5] Firefighters waged numerous campaigns to put out the flames, but it seemed that just as one blaze was extinguished another popped up somewhere else. More than a billion dollars[6] was spent to fight the fires as over 2.8 million hectares[7] went up in smoke. The drought and fires destroyed food supplies for wildlife, leading hungry bears to break into suburban trash cans, homes, and shops.[8] One black bear in Estes Park, Colorado, was caught on the video security

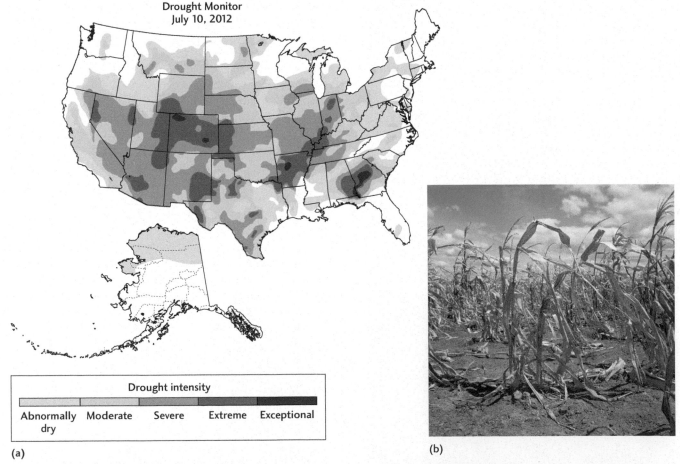

Drought Monitor
July 10, 2012

Drought intensity

| Abnormally dry | Moderate | Severe | Extreme | Exceptional |

(a)

(b)

FIGURE 15-1 (a) Summer of 2012 saw widespread drought across the United States. (b) Crop losses associated with the drought totaled more than $20 billion dollars. *(a: The U.S. Drought Monitor is produced in partnership by the National Drought Mitigation Center at the University of Nebraska-Lincoln, the U.S. Department of Agriculture, and the National Oceanic and Atmospheric Administration. Map courtesy NDMC-UNL; b: Nigel Cattlin/Alamy)*

system of the Rocky Mountain Chocolate Factory breaking in seven separate times to haul away chocolates and other delicious treats.

On the East Coast, severe weather associated with sultry atmospheric conditions killed 22 people and knocked out power to 3 million.[9] While the power was out, trains stopped running, leaving commuters stranded, and food rotted in grocery store refrigerators as temperatures soared past 38°C. County emergency management offices in Virginia went into action, opening shelters in which those vulnerable to heat stroke could cool themselves. The heat and associated drought were so extreme that the Mississippi River stood at its lowest level in decades, exposing sand bars and making navigation treacherous. After a barge ran aground, the U.S. Coast Guard closed 18 km of the river to ship traffic in order to allow the U.S. Army Corps of Engineers to conduct dredging operations to deepen the channel.[10]

On the other side of the globe, the warm summer of 2012 brought torrential rains to China.[11] Nearly 18 cm fell in only 10 hours in Beijing, with suburbs deluged by nearly 46 cm.[12] Massive flooding disrupted the lives of more than 6 million people and caused at least 95 deaths

across 17 different provinces. Chinese farm animals were even less fortunate: some 170,000 perished in the onrushing waters that left farmers with nearly $1 billion in damages.[13] In Europe, record-breaking heat in August set off wildfires in Spain, Italy, France, and Croatia, and officials feared a repeat of 2003 when more than 20,000 people died during a heat wave.[14] Farther north, satellite data revealed that 97 percent of the surface of the Greenland ice sheet melted in July, breaking all previously observed records (Figure 15-2).[15] In normal years, only about half the surface melts, and most of the water quickly refreezes. In 2012, meltwater releases exceeded all previously studied years[16] as warmer than normal temperatures over the Arctic persisted for weeks. These same warm temperatures melted vast expanses of the Arctic ice cap, which showed its lowest summertime extent since satellite observations began in 1979.[17]

In January of 2013, it was Australia's turn to roast under the hot summer Sun. The continent experienced its hottest summer in recorded history as temperatures soared as high as 54°C.[18] Temperatures in fact climbed so high that meteorologists had to add additional colors to their weather maps to depict the new conditions. Fires swept across the Australian states of New South

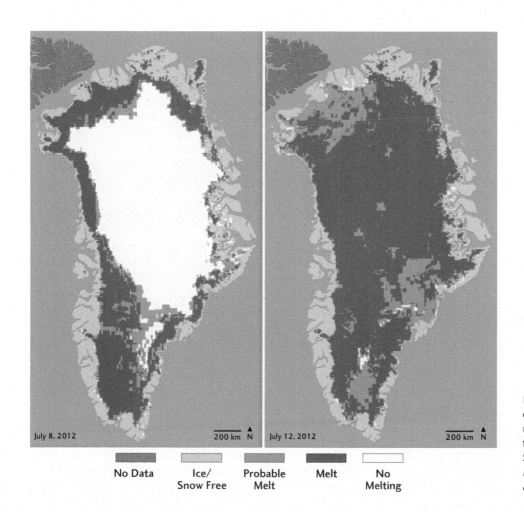

July 8, 2012   200 km N   July 12, 2012   200 km N

No Data   Ice/   Probable   Melt   No
           Snow Free   Melt   Melting

Figure 15–2 Abnormally warm conditions in the Arctic led to melting on nearly all parts of the surface of the Greenland Ice Sheet in summer of 2012. *(Nicolo E. DiGirolamo, SSAI/MASA GSFC, and Jesse Allen, NASA)*

Wales and Tasmania, burning 1280 km^2 of land. These catastrophes followed on the heals of massive flooding in March of 2012 during which 70 percent of New South Wales, along with parts of the states of Queensland and Victoria,[19] an area equivalent to France and Germany combined, went under water.

The extreme heat and associated hydrologic disruptions seen around the globe led many to ponder whether human-induced climatic change might be to blame and whether these conditions might represent a "new normal" for the planet. Combustion of fossil fuels, which accelerated with the advent of the industrial revolution in the late 18th century, has pumped about half a trillion metric tons[20] of the greenhouse gas carbon dioxide into Earth's atmosphere. Along with increased concentrations of methane from rice paddy agriculture, livestock production, and industrial processes, as well as releases of heat-trapping chlorofluorocarbons, these increases appear to be warming Earth's climate and generating greater hydrologic variability. Globally, July 2012 was the fourth warmest July on record, while for the northern hemisphere land surface, it was warmer than any prior July since record keeping began in 1880.[21] Warming, in fact, appears to be accelerating, with nine of the ten warmest years on record having occurred since the year 2000 and the tenth in 1998.[22]

As global human population has grown and our standard of living has risen, our species has become the dominant influence on many Earth systems. From increases in climate-warming greenhouse gases to changes in biogeochemical cycles to wholesale reorganizations of landscapes and water supplies, our species is leaving its mark on the surface of planet Earth, leading many to propose that we have entered an entirely new geologic epoch, coined the **Anthropocene**. The long-term consequences of human activities remain unclear, though an ever-increasing body of scientific knowledge indicates we have cause for concern. Decisions made by those of us alive today will determine whether future generations inherit a degraded planet quite different from our own or a planet capable of providing for human and ecosystem needs.

PLANETARY BOUNDARIES

In 2009, sustainability scientist Johan Rockström and his colleagues published a paper titled "Planetary Boundaries: Exploring the Safe Operating Space for Humanity."[23] In it, they attempted to identify Earth system processes critical to human well-being and to assess how those processes have been altered by human hands. They noted that the Holocene epoch has been a time of relative stability during which people developed agriculture, built permanent settlements, and made phenomenal

technological advances. They also cautioned that this stability might be transitory and sought to recognize thresholds within critical processes capable of tipping Earth into entirely new environmental states that could be disruptive to society (Box 15-1). While not exhaustive, Rockström and his colleagues developed a list of nine factors in which exponential growth of the human population and its associated consumption of resources threatens planetary stability (Figure 15-3):

- Atmospheric aerosol loading
- Biodiversity loss
- Changes in biogeochemical cycles
- Changes in land use
- Chemical pollution
- Climate change
- Global freshwater use
- Ocean acidification
- Stratospheric ozone depletion

In developing this list, the Planetary Boundaries team recognized that feedbacks within the Earth system might amplify responses to human-induced changes and noted that some systems might display nonlinear behavior. They also realized that some processes are interdependent, meaning that some boundaries might shift as others are crossed. Climate change, for example, affects the distribution and amount of global freshwater supplies, such that what might be a sustainable water use at the present may far exceed what's sustainable in a world beset by drought. Rockström and his colleagues therefore did not develop absolute thresholds beyond which environmental systems would inalterably change. Rather, they noted the uncertainty inherent in such determinations and called for more research.

While the absolute values of planetary thresholds remain fuzzy, the idea that Earth systems are interconnected and that humans are having a marked impact on these systems could not be more clear and serves as a useful framework for this concluding chapter of our book. We have addressed many of the **planetary boundaries** identified by Rockström and his colleagues in previous chapters. We provide summaries of those boundaries here, reserving the bulk of our discussion for human-induced climate change. In carrying out our analysis, we make use of the deep time perspective of Earth science, which provides us with myriad examples of former planetary states quite different from the present. As humans continue to alter the environment, these former worlds serve as cautionary examples of what we may have in store in centuries to come. We conclude with a discussion of strategies to avoid the worst possible outcomes.

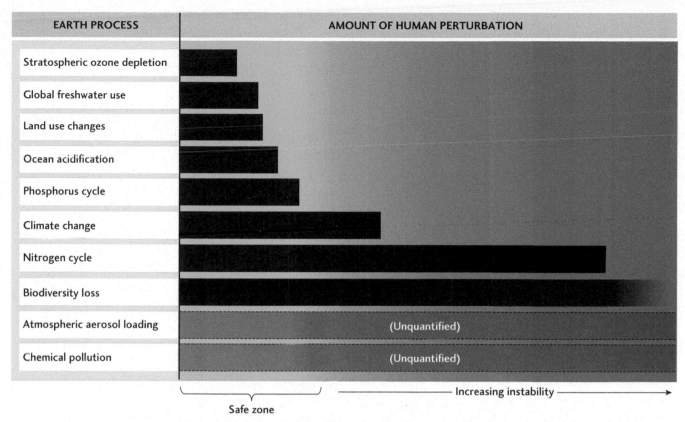

Figure 15–3 Relative impacts of human activities (black bars) on the Earth systems that govern planetary stability according to the Planetary Boundaries team. Impacts confined to the green zone are considered "safe" whereas impacts extending into the orange and red zones may cause planetary instability and substantial changes from current conditions.

Atmospheric Aerosol Loading

In Chapter 11, we discussed several examples of human-introduced pollutants in the atmosphere and their consequences for human and ecosystem health. We noted that the impacts of these pollutants depend on their chemical reactivities and residence times. Some compounds, such as nitrogen and sulfur oxides, react quickly with cloud droplets and fall to earth as acid rain where they cause ecological damage, while others are more chemically inert and remain in the atmosphere for longer time periods. The shorter-lived species have local effects whereas the longer-lived species may be distributed farther, having regional or possibly even global impacts. Given the myriad behaviors of different pollutants, the Planetary Boundaries team declined to set a threshold value for atmospheric aerosol loading (see Figure 15-3), recognizing that much more research remains to be done. As a result, it is not yet possible to say whether current aerosol levels are within a "safe operating space for humanity" or have exceeded a threshold to permanent ecological change.

Biodiversity Loss

Chapter 7 explained that as the human population has grown, landscape modification for the purpose of agriculture, settlement construction, and resource extraction has led to habitat fragmentation and loss. Land transformation, combined with overhunting and introduction of invasive exotic species, has led to a rise in the number of species classified as threatened or critically endangered. Anthropogenic climate change will push still others into the threatened and endangered category. Should these organisms go extinct in coming decades, Earth will experience species loss rates comparable to those during the five major mass extinction episodes earlier in its history (Chapter 6). Inasmuch as each species occupies a particular ecological niche and contributes its own set of ecosystem services to the environment (Chapter 7), species loss may lead to radical changes in ecosystem functioning. While it is difficult to determine a biodiversity loss rate that will preserve the major features of the biosphere into which humans evolved, the Planetary Boundaries team points to the background geologic record of

BOX 15–1 GLOBAL AND ENVIRONMENTAL CHANGE

Tipping Points

In the past decade, scientists have given increasing attention to the idea that Earth's environmental systems contain **tipping points** that separate widely varying stable states. The idea of tipping points is often illustrated by a rocking canoe (**Figure 1**). The canoe has two stable states, either right–side up or upside down. Paddlers can rock the canoe back and forth to a certain extent, but if they rock too vigorously, the canoe will flip over, tipping them and their gear into the water. The exact point separating the two states is known as the tipping point, and the slightest of nudges at that point can provoke a catastrophic change from one state to the other. The problem for the paddlers is in knowing where the tipping point lies!

Earth's environmental systems contain many tipping points that lead to local and/or global impacts. On the local scale, rising inputs of the nutrients nitrogen and phosphorus to streams can trigger eutrophication in lakes (Chapter 7), pushing clear, well–oxygenated waterbodies into turbid, anoxic states incapable of supporting invertebrates and fish. On the global scale, the ice core record from Greenland (Chapter 14, Box 14–2) reveals that Earth lurched back and forth between warmer and colder states during the last ice age. The cause of the transitions between states remains unclear but must have involved some sort of tipping point. Another global state transition occurred with the evolution of cyanobacteria, whose photosynthetic production of oxygen paved the way for multicellular life (Chapter 11).

As humanity continues its uncontrolled experiment in fossil fuel combustion, Earth's environmental systems may encounter a variety of tipping points. The Arctic ice cap has

Figure 1 Paddlers accidentally push their canoe past its tipping point, dropping themselves into the water as the canoe heads toward its stable upside down state. *(Courtesy Lynn University)*

extinctions of 1.8 per million species per year as a lower bound. They note that humans are presently treading in dangerous territory as modern extinction rates have the potential to exceed those in the background record by 100 to 1000 times, and they suggest a maximum rate of 10 extinctions per million species per year for planetary stability (see Figure 15-3).

Changes in Biogeochemical Cycles

In addition to biodiversity loss, Chapter 7 addressed human impacts on biogeochemical cycles. Humans have modified the global nitrogen cycle through the cultivation of nitrogen-fixing crops and by application of synthetic fertilizers to fields and lawns. We have further changed the cycle through combustion of nitrogen-rich fossil fuels and by discharging untreated sewage into the ground and waterways. The consequences of nitrogen cycle disruptions include eutrophication of lakes and coastal zones (Chapter 12) and contamination of drinking water supplies (Chapter 10). Reactive nitrogen in fertilizer also contributes to the production of the greenhouse gas nitrous oxide,[26] meaning that changes in biogeochemical cycles impact global climate change.

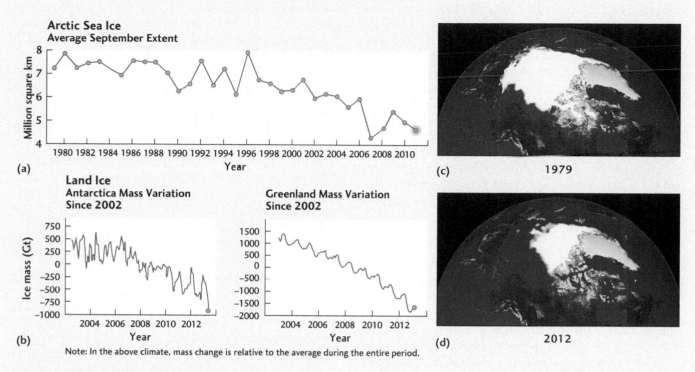

Note: In the above climate, mass change is relative to the average during the entire period.

Figure 2 Arctic sea ice has both thinned and shrunk in extent over the past three decades as climate has warmed. Sea ice in summer may entirely disappear within a matter of decades.

(a: NSIDC; b: NASA/University of California, Irvine; c, d:NASA/ Goddard Scientific Visualization Studio)

gotten significantly thinner and less extensive over the last three decades as climate has warmed, and it appears that we're headed for a future in which Arctic summers may be entirely ice free (**Figure 2**).[24] The conversion of highly reflective sea ice to highly absorptive ocean water means that the Arctic will warm further, making recovery of summer sea ice nearly impossible. While loss of Arctic sea ice has no impact on sea level, the warming pole may cause the Greenland ice sheet to melt, with the ice–albedo feedback driving continued warming and more ice melt. Inasmuch as the ice sheet contains enough ice to raise sea level by 6 to 7 m,[25] such a scenario would be highly undesirable for coastal communities.

Humans have made similar changes to the global phosphorus cycle, and the Planetary Boundaries team notes that past mass extinctions caused by large-scale anoxic events in the oceans may have originated due to increased influx of phosphorus to the ocean basins. Based on the potential harm to many of Earth's systems, they recommend no more than a tenfold increase above background weathering rates for phosphorus (no more than 10 megatons per year) and a fourfold reduction in the introduction of reactive nitrogen to Earth's surface (from 150 to 35 megatons per year). By these estimations, our species is coming close to the threshold for phosphorus but has already vastly exceeded the threshold for nitrogen (see Figure 15-3).

Changes in Land Use

As mentioned under biodiversity loss, the growth of the human population has led to large-scale land transformation for the purpose of agriculture, settlement, and extraction of mineral, energy, and timber resources. By United Nations Food and Agriculture Organization estimates, 13 million hectares of forest are felled each year,[27] and the Planetary Boundaries team notes that fully 12 percent of Earth's ice-free surface is used to grow crops for humanity. They suggest that no more than 15 percent (see Figure 15-3) be devoted to this purpose in order to preserve habitat for other organisms. They also suggest that the most productive agricultural lands be protected from residential and commercial development in order to avoid pushing agriculture into more marginal locations, and they call for changes in agricultural practices to conserve soils and their fertility. These may include such practices as crop rotations, conservation tillage, and development of new crop varieties that do not require soil disturbance (Chapter 8).

While the Planetary Boundaries team does not discuss a threshold for urbanization, it is clear that just as a growing human population will need more food, so too will it require additional living space. This need for additional space calls for careful management to minimize future impacts on the biosphere, hydrosphere, and pedosphere because, in addition to causing habitat fragmentation and loss, changes in land use trigger changes in the hydrologic cycle. Cutting down forests to construct housing developments and shopping malls decreases the canopy storage capacity of trees and covers porous soils with pavement, both of which speed runoff toward stream channels, exacerbating flooding. Pavement also reduces the infiltration needed to supply baseflow to streams during dry periods, threatening the health of aquatic ecosystems (Chapters 9 and 10), and poor management of construction sites leads to soil loss (Chapter 8).

Chemical Pollution

Tens of thousands of synthetic chemicals have been developed since the advent of the Industrial Age to serve a variety of purposes, including killing agricultural pests, preventing children's pajamas from catching on fire, and creating a seemingly infinite supply of plastic products. These chemicals enter the environment either purposefully, as in the case of pesticides, or unintentionally, as in the breakdown of plastic bags along roadsides. For the vast majority, their impacts on the environment are unknown: testing has never been carried out. However, even for those chemicals that have been tested, information remains scant; only a limited number of organisms have been subjected to experiments, and these organisms have shown widely varying tolerances to exposure. The ecological and Earth system impacts of chemical pollution are therefore presently impossible to quantify, and it is not possible to determine whether we have passed a planetary threshold for chemical pollution.

As the Planetary Boundaries team notes, some chemical pollution thresholds have most definitely been exceeded. The toxic impacts of polychlorinated biphenyls (PCBs), for example, have rendered fish caught in the Hudson River of New York State unfit for consumption by children and women who may become pregnant. The health effects for the fish themselves include reproductive failure, tumors, immune system problems, and death. Inasmuch as the fish are part of a larger ecosystem, these problems may have broader ecological and Earth system impacts that we have not yet begun to comprehend.[28]

Global Freshwater Use

As discussed in Chapters 9 and 10, our species uses nearly 4000 km^3 of water annually for drinking, growing crops, sanitation, and industrial purposes, among other uses.[29] Where settlements are located in water-scarce regions, we construct aqueducts hundreds of kilometers long to draw water from wetter areas, drill hundreds of meters into the subsurface to extract water from ancient aquifers, and sometimes reroute entire rivers. Withdrawals and diversions have led rivers and lakes to dry up, with devastating consequences for aquatic life. The once mighty Colorado River, for example, dwindles to a trickle in northern Mexico, victim of water abstractions to municipalities such as Los Angeles and Las Vegas and to farms in the deserts of Arizona, California, and northern Mexico. The river stops short of its delta in the Gulf of California most of the time, cutting off freshwater flows that once supported a variety of fish and shellfish. Regrettably, the Colorado is only one of several rivers that frequently run out of water before reaching the sea. Others include Asia's Indus and Yellow rivers, Australia's Murray River, and North America's Rio Grande.[30] Diverting the Amu Dar'ya and Syr Dar'ya rivers led Asia's Aral Sea to desiccate, causing the loss of 60,000 fishing jobs[31] and subjecting the region to choking dust storms (Figure 15-4).

While these examples are dramatic and devastating, the Planetary Boundaries team notes that perhaps an even more serious problem involves the disruption of climate through human tinkering with the hydrologic cycle. Desiccation of the Aral Sea, for example, reduced the supply of water vapor to the atmosphere in central Asia, leading to higher rates of evaporation and a further drying of the region.[32] Likewise, clear-cutting of tropical rain forests diminishes the amount of water transpired by trees, thereby reducing the amount of water vapor in the atmosphere over the forests and causing a concomitant decline in the amount of rain falling on the forest. As rain forests such as that in the Amazon River Basin are felled to produce cattle pasture and to grow soybeans and oil palms, the region risks conversion to grassland, with insufficient moisture recycling within the system to support trees.[33] Since tropical rain forests are biodiversity hot spots, such transformations also risk species extinctions.

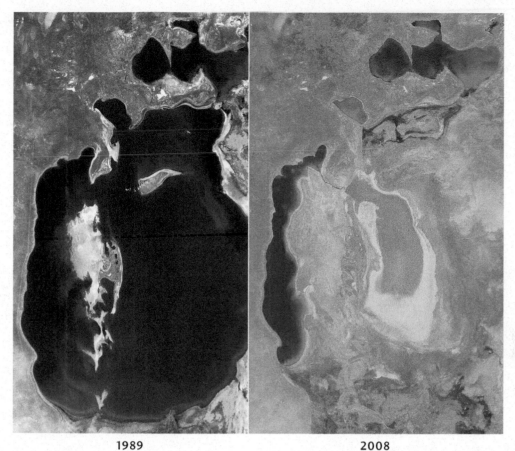

1989	2008

Figure 15–4 Central Asia's Aral Sea has desiccated over the last several decades, a victim of water diversions from the Amu Dar'ya and Syr Dar'ya rivers that once fed it. Once a source of moisture to the atmosphere, the dried–up lake bottom now produces toxic dust storms that cause respiratory problems for populations downwind. *(NASA)*

It is challenging to place definite limits on human freshwater use, but the Planetary Boundaries team suggests a value of 4000 to 6000 km³/yr as an upper bound to maintain ecosystem stability. Our current consumption falls slightly below this level, but as population growth continues and agricultural demands increase, we may exceed this value. Furthermore, it is important to consider that for many local environments (e.g., the Aral Sea), we have already far surpassed a sustainable use threshold.

Stratospheric Ozone Depletion

As mentioned in Chapter 11, the concentration of ozone in Earth's stratosphere declined dramatically with the invention and release of chlorofluorocarbon gases used as refrigerants and propellants in aerosol spray cans and for Styrofoam manufacturing. Thinning of the ozone layer has been most severe over Antarctica due to chemical processes that are enhanced by extremely cold temperatures, though thinning of ozone has occurred worldwide. Because ozone serves as a protective shield against ultraviolet radiation given off by the Sun, thinning has resulted in increased exposure to UV rays. Higher exposure, in turn, is causing greater frequency of sunburn and skin cancer-causing genetic damage in places like Australia. Ecosystem impacts include stunted growth in plants and reproductive and developmental impairments in aquatic algae, invertebrates, and fish.[34]

As the Planetary Boundaries team notes, humanity clearly exceeded the threshold for safety for stratospheric ozone depletion over Antarctica. The tremendous success of the Montreal Protocol, a United Nations–led effort to ban ozone-depleting substances (Chapter 11), has meant that global ozone layer destruction has been stopped, however, and should reverse in coming decades. We have therefore not exceeded the planetary threshold for stratospheric ozone depletion, thanks to the willingness of atmospheric scientists to speak out about the problem and of national governments to listen and take action.

■ Human population growth and consumption of resources are altering all parts of the Earth system. These alterations have the potential to move the planet toward a new environmental state that could be disruptive to society.

■ Tipping points are thresholds in Earth's systems that separate stable environmental states that may be quite different from one another. The loss of Arctic sea ice represents one such threshold and may result in extensive melting of the Greenland ice sheet with associated sea level rise.

■ Sustainability scientists are attempting to determine the nature and values of different Earth system tipping points in order to guide policy changes designed to minimize the possibility of catastrophic ecosystem change.

■ Many planetary thresholds are difficult or impossible to quantify at the present. In addition, some planetary thresholds have the potential to affect others as their values are exceeded. Climate change, for example, will impact global freshwater use and biodiversity loss.

ANTHROPOGENIC CLIMATE CHANGE

One of the most vexing problems of the late 20th and early 21st centuries has been **anthropogenic climate change,** which is challenging not only because of the incredible complexity of Earth's climate system but also

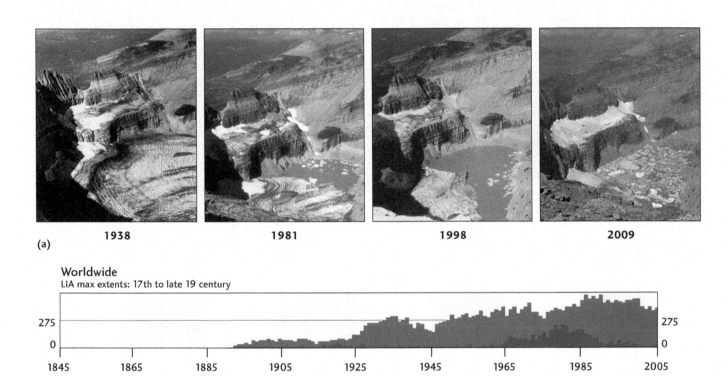

(a)   1938   1981   1998   2009

Worldwide
LIA max extents: 17th to late 19 century

(b)

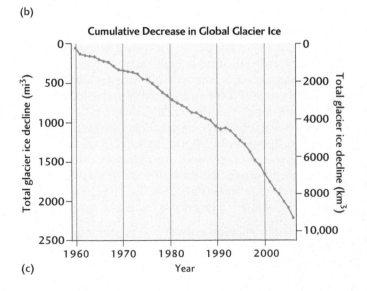

(c)

Figure 15–5 (a) Since the Industrial Revolution began, retreating glaciers have outnumbered advancers as globally warming temperatures melt ice. The iconic glaciers of Glacier National Park in Montana may disappear by 2030 as warming continues. (b) This chart shows advancing (blue) and retreating (red) glaciers worldwide from 1845 through 2005. (c) Loss of glacial ice since 1960. *(a: 1938: Hileman Photo, courtesy GNP Archives; 1981: Carl Key photo, USGS; 1998: Dan Fagre photo, USGS; 2009: Lindsey Bengtson photo, USGS; b: Zemp, M., I. Roer, A. Kaab, M. Hoelzle, F. Paul, and W. Haeberli, eds. Global Glacier Changes: Facts and Figures. Zurich: UNEP, World Glacier Monitoring Service, 2008. http://www. grid.nep.ch/glaciers/)*

because climate science has become politicized, much to the consternation of climate scientists. Well-funded disinformation campaigns promoted by a variety of special interest groups, along with sloppy journalistic practices that give equal time and space to opposing minority viewpoints, have muddied the waters for the general public and created a great deal of confusion about the nature of the problem and the urgency with which we need to address it (Box 15-2). In this section we delve into the current state of the science and what it can and cannot tell us about the future. What we conclude is that there is incontrovertible evidence that Earth is warming and that the primary cause is human activities such as fossil fuel combustion and deforestation. This warming poses a threat to many species, including our own, but the exact magnitude of the problem is difficult to determine because it depends on future human behaviors that are challenging to predict. Whether the globe will pull together to address the problem, as it did with the Montreal Protocol, remains to be seen.

Evidence for a Warming Earth

While global warming skeptics still occasionally claim that there is no evidence that Earth's surface is warming, disappearing glaciers and rising sea levels prove otherwise. Photographic and survey evidence compiled by the World Glacier Monitoring Service, an agency of the United Nations Environment Program, shows that while a few glaciers have been advancing over the last century, an overwhelming majority have been in retreat[40] (Figure 15-5). Inasmuch as glaciers require temperatures

below freezing to accumulate and persist, widespread retreat proves that Earth's surface has been warming during this time. Indeed, the iconic glaciers of Glacier National Park in Montana have fallen prey to a warming climate and may entirely vanish by 2030.[41]

More than a century of observations demonstrate that at the same time that glaciers have been retreating, sea level has been rising (Figure 15-6). While some of this rise is related to thermal expansion of seawater that has occurred as Earth's surface has warmed and another fraction is related to groundwater extraction and use, a large fraction (about 35 percent in data examined from 1961–2008)[42] has resulted from the melting ice itself, and sea level rise appears to be accelerating as time goes on. Accompanying the physical evidence of glacial retreat and sea level rise, instrumental measurements of temperature change made on the ground, and more recently by satellites, confirm that Earth's surface is warming (Figure 15-7).[43] When these records are extended through the use of paleoclimatic proxies such as tree rings (Chapter 14), the recent century and a half of warming appears even more striking, superimposed as it is on a record of gradual temperature decline (Figure 15-8).

Causes of Climatic Change

In Chapter 14, we took a look at the fundamental components of Earth's climate system. We learned that our planet is kept warm by a blanket of water vapor, carbon dioxide, methane, and other greenhouse gases and that this blanket has thinned over time as the Sun gradually grew hotter. We discussed the fact that Earth's

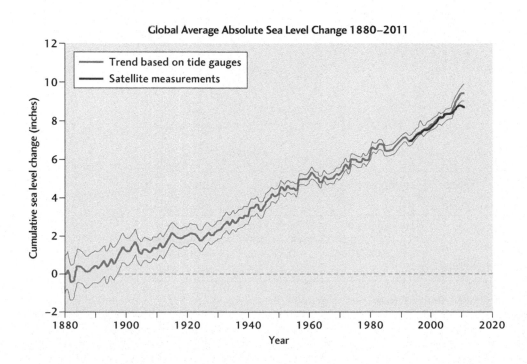

Figure 15-6 Sea level rise over the last century measures more than 25 cm and appears to be accelerating.

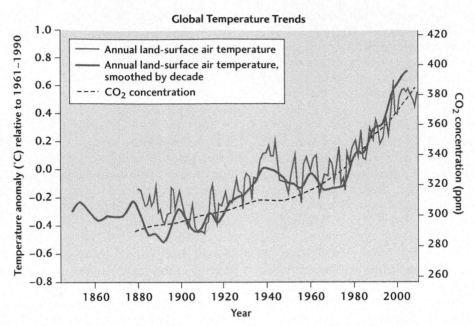

Figure 15–7 Measurements made on the ground and more recently by satellites show that Earth's surface temperature has risen by about 0.9°C since 1850, with warming accelerating since the 1970s.

temperature is maintained at a level conducive to life by the interaction of plate tectonic, hydrologic, and biogeochemical cycles that determine greenhouse gas levels in the atmosphere on long geologic timescales.

We further noted that climate has changed in the past due to factors such as plate tectonic reorganizations of continents and ocean basins, changes in the amount of solar radiation striking the planet, and massive volcanic

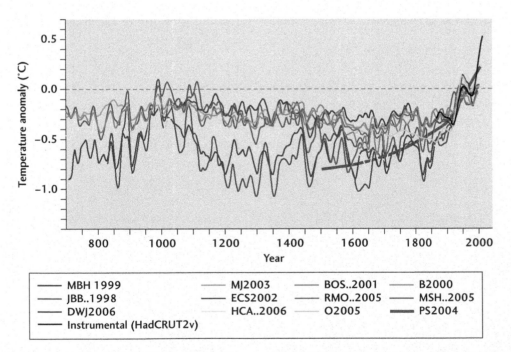

Figure 15–8 The past century and a half of warming contrast with stability or gradual cooling seen in tree ring and other climatic proxy records of the last millennium. *(From Jansen, E., J. Overpeck, K. R. Briffa, J.–C. Duplessy, F. Joos, V. Masson–Delmotte, D. Olago, et al. "Chapter 6: Palaeoclimate," in* Climate Change 2007: The Physical Science Basis. Contribution of Working Group I to the Fourth Assessment Report of the Intergovernmental Panel on Climate Change, *edited by S. Solomon, D. Qin, M. Manning, Z. Chen, M. Marquis, K. B. Averyt, M. Tignor, and H. L. Miller. Cambridge, UK: Cambridge University Press, 2007.)*

BOX 15–2 EARTH SCIENCE AND PUBLIC POLICY

In 2007, the Union of Concerned Scientists (UCS), a non-partisan organization dedicated to the promotion of scientific approaches to societal problems, published a report titled "Smoke, Mirrors & Hot Air: How ExxonMobil Uses Big Tobacco's Tactics to Manufacture Uncertainty on Climate Science."[35] In it, the organization documents how, in the late 1980s and in conjunction with other energy and automobile producers, the world's largest oil and gas company helped found the Global Climate Coalition (GCC), an industry group created for the purpose of casting doubt on climate change research. At the time, information on Earth's climate system was more limited than at present, and legitimate questions arose within the scientific community over the role of humans versus other causal agents in global warming. According to UCS, the GCC exploited the debate, promulgating the view that global warming science was rife with uncertainty and suggesting that more research should be done before any sort of actions were taken to curb greenhouse gas emissions.

As scientific studies of the climate system accelerated in the 1990s and 2000s, it became clear that Earth's surface was warming at an increasing rate and that fossil fuel combustion was largely to blame, leading many members of the GCC to withdraw from the group and to begin investing in alternative energy research. Rather than join them, ExxonMobil helped constitute the Global Climate Science Team, an organization dedicated to furthering uncertainty. Throughout the late 1990s and 2000s, reports the UCS, the company spent tens of millions of dollars on lobbyists and political campaign contributions to persuade congressional leaders not to take action on curbing greenhouse gas emissions in the United States. It also funded nonprofit organizations and public policy think tanks that took skeptical positions on the science of global warming and strategies to mitigate it. One of these, the Heartland Institute, went so far as to publish its own response to work conducted by the United Nations–sponsored **Intergovernmental Panel on Climate Change (IPCC)**.

Manufacturing a Controversy

The IPCC contains representatives from 194 nations and consists of thousands of scientists who author and review reports on the current state of climate change research for the benefit of the general public, educators, researchers, and policy makers.[36] Scouring the peer-reviewed climate science literature, these scientists discuss both those aspects of the climate system about which much is known and those areas in which considerable uncertainty remains. Its careful, transparent, and consensus-based approach to summarizing the state of knowledge about Earth's climate system makes the IPCC the ultimate authority on climate change research.

The Heartland Institute's report plays on the name of the IPCC as it seeks to downplay the role of human activities in climatic change. With the title "Climate Change Reconsidered: 2011 Interim Report of the Nongovernmental International Panel on Climate Change"[37] and with its reference to being a joint effort of the Science and Environmental Policy Project and of the Center for the Study of Carbon Dioxide and Global Change, it might appear to a casual reader to have the same legitimacy as the IPCC reports. Rather than the thousands of scientists contributing to the IPCC's climate assessments, however, the Heartland Institute publication was written by three scientists, with contributions from seven others. One of these scientists is a coauthor of a work titled "The Many Benefits of Atmospheric CO_2 Enrichment: How Humanity and the Rest of the Biosphere Will Prosper from This Amazing Trace Gas that So Many Have Wrongfully Characterized as a *Dangerous Air Pollutant!*"[38] With concerted disinformation campaigns such as these, it is no wonder that members of the general public express confusion over the state of climate change science. The fact that the mainstream media often lend the same weight to the pronouncements of global warming skeptics as they do to the overwhelming majority of climate scientists (more than 97 percent in a 2009 survey),[39] who acknowledge global warming as real and caused primarily by humans, does nothing to help clarify the matter.

---

eruptions that have spewed Sun-blocking ash and aerosols into the stratosphere. Given the complexity of the climate system and the multitude of factors that can influence temperature, how can we be certain that the warming that Earth is currently experiencing isn't part of a natural cycle?

The answer arises from the vast array of paleoclimatologic data that have been compiled over the last several decades combined with information garnered from numerical modeling experiments. The rate of warming that has occurred over the past century is simply inconsistent with natural geologic processes that bring about climatic change. At about 0.74°C per century,[44] it far surpasses temperature changes expected from gradual plate tectonic reorganizations of Earth's continents and ocean basins (0.00002°C per century) (Figure 15-9).[45] Indeed, Earth has been cooling since the Paleocene–Eocene boundary,[46] around 55 million years ago, so not only is the temperature change of the wrong order of magnitude, it is also of the wrong sign to have been caused by tectonic changes. Similarly, the expected temperature change over the past century due to variations in Earth's orbital parameters should be about a 0.02°C cooling, again a much smaller magnitude of change and one of opposite sign to the observed warming that has occurred.

In Chapter 14, we noted that millennial scale oscillations in temperature of unknown origin occurred repeatedly during the late Pleistocene and are recorded in ice cores from Greenland and Antarctica. Similar, though much smaller in amplitude, oscillations have occurred throughout the Holocene,[47] and yield rates of climatic change less than 0.02°C per century (see Figure 15-9), so these too are incapable of explaining the warming that has occurred since the Industrial Revolution began. The only physical process that even comes close to being able to cause temperature changes in the tenths of a degree per century is variation in solar energy output.[48] Solar scientists have demonstrated a correspondence between the number of sunspots on the Sun's surface (dark-colored patches caused by magnetic activity) and the amount of energy given off by our star, and attribute a particularly cold interval in the last millennium in part to a marked reduction in the number of sunspots (Chapter 11). Still, changes in solar radiation appear to be insufficient to cause the rapid warming Earth has experienced over the past century.[49]

The Case for Human Sources

Given that natural processes appear incapable of generating the warming of the last century, what then is to blame? Measurements of greenhouse gas concentrations made at the Mauna Loa Observatory in Hawaii since the late 1950s reveal that carbon dioxide levels in the atmosphere have been rising (Figure 15-10). While these measurements are not in and of themselves proof of anything, when combined with information garnered from ice cores, they present a compelling culprit. As mentioned in Chapter 14, gas bubbles encased in the ice contain records of ancient atmospheric conditions. Ice core scientists use fine needles to extract air from these bubbles, passing it through instruments that measure its gas concentrations. Ice layers also reveal past atmospheric temperatures

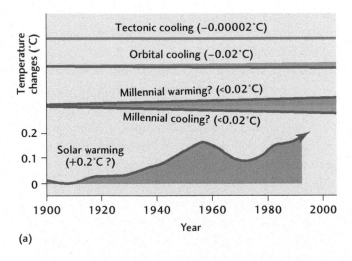

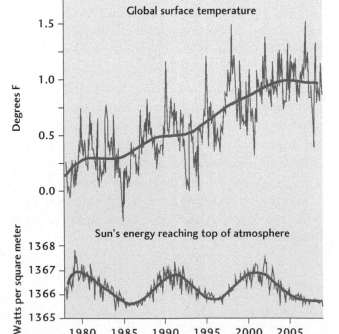

Figure 15–9 (a) The magnitude of warming seen over the past 150 years cannot be explained by natural processes such as plate tectonic reorganizations of continents and ocean basins, variations in Earth's orbital parameters, or changes in solar energy inputs. (b) These factors are too small and sometimes of the wrong sign to explain observed temperature changes. *(a: Ruddiman, W. F. Earth's Climate: Past and Future, 2nd ed. [p. 407]. New York: W. H. Freeman, 2007; b: From Karl, T. R., J. M. Melillo, and T. C. Peterson, eds. Global Climate Change Impacts in the United States. Cambridge, UK: Cambridge University Press, 2009.)*

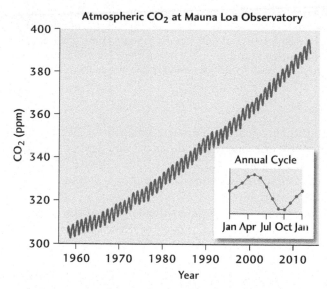

Atmospheric CO₂ at Mauna Loa Observatory

Figure 15–10 Carbon dioxide concentrations measured at the Mauna Loa Observatory in Hawaii since 1958 have risen in response to human combustion of fossil fuels and deforestation. The annual cycle in values arises from the annual growth and decay of vegetation in the northern hemisphere. During the summer growing season, carbon dioxide levels fall in response to photosynthesis. During the fall and winter, decay of vegetation returns some of this carbon dioxide to the atmosphere. The great expanse of ice (Antarctica) and desert (Australia) in the southern hemisphere minimizes that hemisphere's impact on the annual cycle.

from isotopes of hydrogen and oxygen contained in their water molecules. Over the last 650,000 years, carbon dioxide concentrations and atmospheric temperatures derived from isotopic measurements varied together (Chapter 14, Figure 14-22), with low concentrations of CO_2 found during glacial intervals and higher concentrations found during interglacials. Methane concentrations have shown similar variations with temperature. Given past relationships between greenhouse gas concentrations and temperature, there is no reason to doubt that the increases in greenhouse gases that have occurred since the advent of the Industrial Revolution should result in climatic warming (Box 15-3). This idea is not a novel one; Swedish scientist Svante Arrhenius proposed a relationship between greenhouse gas concentrations and temperature as early as 1896.

The concentrations of carbon dioxide in preindustrial ice core bubbles measure between 180 and 300 parts per million volume (ppmv). Our present concentration of 400 ppmv thus vastly exceeds that of any time in the past 650,000 years. Similarly, methane concentrations have climbed from preindustrial values of 400 (glacials) to 700 (interglacials) parts per billion volume (ppbv) to the present value of 1840 ppbv,[50] and nitrous oxide concentrations have risen from a preindustrial level of 270 ppbv

to a present value of 326 ppbv.[51] Not only have levels of these and other greenhouse gases increased over the last several decades, their rate of increase has accelerated over time (Figure 15-11).

While all of this is true, a global warming skeptic might still ask how we know that changes in greenhouse gas levels in the atmosphere are associated with human activities and not caused by heretofore-unrecognized natural processes. The answer arises from the fact that at the same time greenhouse gas concentrations have been rising, atmospheric oxygen levels have been falling, consistent with the combustion of fossil fuels (Figure 15-12).[52] Changing carbon isotopic values of the atmosphere over time also reveal humans as the source of increasing greenhouse gas concentrations. Fossil fuels and present-day vegetation contain relatively less of the stable isotope ^{13}C than do the oceans or volcanic eruption gases. Gradual declines in the ratio of ^{13}C to ^{12}C in the atmosphere over time (see Figure 15-12) therefore reflect our combustion of coal, oil, and natural gas as well as land clearance for the purpose of agriculture and settlement. Perhaps the most obvious human culpability in the changing composition of the atmosphere lies in the concentration of chlorofluorocarbon gases. These gases have no natural source and did not exist prior to about 1950, proving unequivocally that increasing emissions are related to human activities.[52]

The warming potential of different greenhouse gases is large compared to that associated with solar variability (Figure 15-13) and is measured relative to the average amount of solar energy striking the top of Earth's

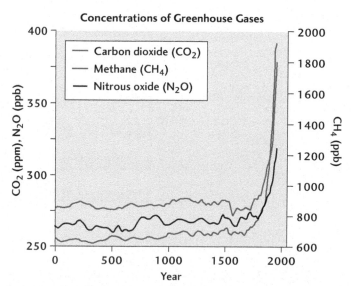

Concentrations of Greenhouse Gases

Figure 15–11 The ice core record of greenhouse gases shows that concentrations have increased exponentially since the advent of the Industrial Revolution and industrial–scale agriculture. Concentrations are now higher than at any time in the past 650,000 years.

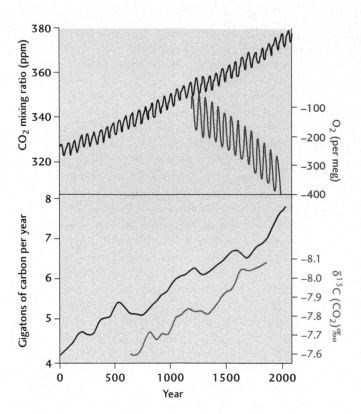

Figure 15–12 As carbon dioxide concentrations have risen, oxygen levels have declined, implicating combustion of fossil fuels as the source of CO_2 rise. At the same time, the isotopic composition of carbon in the atmosphere has shifted as more and more fossil carbon is added to the atmosphere.

atmosphere (called the solar constant). The so-called **radiative forcing (RF)** related to sunspot cycles measures only about 0.12 W/m², whereas the RF associated with heightened postindustrial carbon dioxide concentrations measures 1.66 W/m².[53] While this value might appear small compared to the solar constant value of 1370 W/m², numerical modeling experiments show that it is large enough to cause temperatures to rise by the amount observed over the past 150 years.

Additional evidence that Earth's surface is warming as a result of human-produced greenhouse gases arises from measurements of stratospheric temperatures since the late 1950s. At the same time that Earth's surface and troposphere have warmed, the stratosphere has cooled (Figure 15-14).[54] While this might initially seem

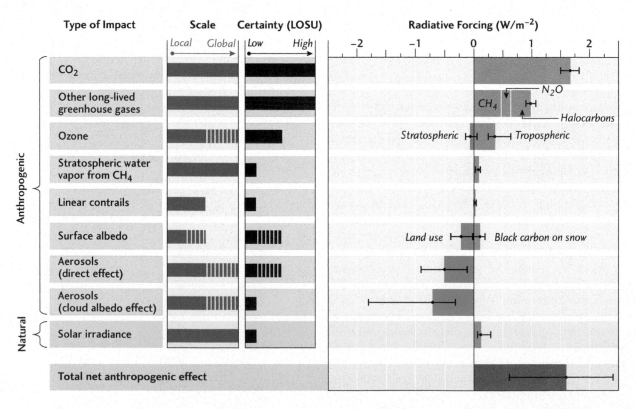

Figure 15–13 The IPCC *Assessment Report 4,* released in 2007, attempted to quantify the warming or cooling potential of different human impacts along with natural solar variability. Positive forcings are colored orange and lead to warming whereas negative forcings are blue and lead to cooling. Range bars reflect the potential maximum and minimum values for each impact. The greater the span of the bars, the more poorly understood the impact. Note that when all of the human impacts are summed, they surpass the forcing expected from changes in solar energy inputs. LOSU refers to the level of scientific understanding. Clouds are poorly understood (Low) whereas greenhouse gases are very well understood (High).

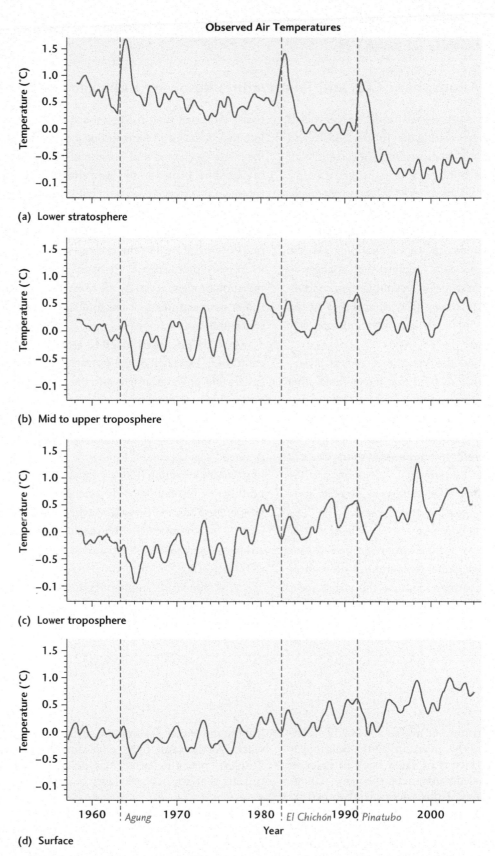

Observed Air Temperatures

(a) Lower stratosphere

(b) Mid to upper troposphere

(c) Lower troposphere

(d) Surface

Agung   El Chichón   Pinatubo

Figure 15–14 Temperature measurements reveal that over the same time that Earth's surface and troposphere have been warming, the stratosphere has been cooling, consistent with increasing greenhouse gas concentrations in the atmosphere. Dashed lines represent major volcanic eruptions that threw ash particles and sulfuric acid droplets into the stratosphere. These aerosols cooled the surface and troposphere while warming the stratosphere.

BOX 15–3 EMERGING RESEARCH

Atmospheric CO_2 and Temperature: Resolving a Chicken and Egg Problem

As ice core records were collected from the Greenland and Antarctic ice sheets during the 1980s, 1990s, and 2000s, a climatic conundrum arose from observations of greenhouse gas levels in bubbles and associated isotopic proxies for atmospheric temperature.[55] While variations in carbon dioxide concentrations appeared to precede changes in temperature recorded in layers of ice from Greenland, the opposite was true for Antarctica. Global warming skeptics seized upon the Antarctic data, arguing that changes in Earth's temperature drive changes in greenhouse gas concentrations rather than the other way around, and they sought to use this as justification for lack of action on anthropogenic greenhouse gas emissions.

As scientists examined the ice core record of greenhouse gas fluctuations, they realized that it generated other interesting questions. For example, where did the excess carbon dioxide present in the atmosphere during interglacial time periods go during glaciations? More expansive ice sheets and shifts from forests to tundra vegetation ruled out uptake of carbon dioxide by terrestrial vegetation, and the surface ocean seemed an unlikely reservoir given its constant exchange of gases with the atmosphere.[56] The deep ocean appeared to be the only reservoir capable of hosting the excess carbon, but how did the carbon get into it? One possibility is that declining global temperatures drove more carbon dioxide into the oceans because the gas is more soluble in cold water than it is in warm.[57] Another possibility is that photosynthetic organisms living in the surface ocean increased their biological activity, incorporating CO_2 into their tissues and transporting it to the deep ocean when they were consumed and incorporated into the fecal pellets of organisms higher on the food chain. The exposure of the continental shelves as sea level fell, combined with stronger wind speeds associated with an increasing pole to equator temperature gradient, may have blown greater amounts of nutrient-rich dust into the oceans, stimulating the increase in primary production.[58] Perhaps, then, changes in solar radiation brought about by the Milankovitch cycles triggered colder temperatures that led to changes in carbon dioxide concentrations.

A study published in 2012 attempts to reconcile the contrasting behavior of the northern and southern hemisphere and provides insight into the transfer of atmospheric carbon between the deep ocean and atmosphere.[59] Climate scientist Jeremy Shakun and his colleagues compiled temperature records from 80 sites situated equatorward of the Antarctic and Greenland ice sheets, finding that during the most recent transition from glacial to interglacial conditions, northern hemisphere records generally track the changes in temperature seen in Greenland while southern hemisphere records track Antarctica. When records for the whole globe are averaged together, temperature rises after increases in CO_2 (Figure 1).

Why should the southern hemisphere warm before the northern? Shakun and his colleagues suggest that increasing solar radiation striking the high northern latitudes between

to contradict global warming, it makes sense when considered carefully. The energy given off by the Sun that is absorbed by Earth's surface is radiated back to space in the infrared part of the electromagnetic spectrum (Chapter 11). On its way, some of this energy is absorbed by greenhouse gases in the atmosphere, which radiate it back to the surface as well as to outer space. The higher the concentration of greenhouse gases in the atmosphere, the more times infrared energy will bounce back and forth between Earth's surface and the atmosphere before

leaving the planet, leading the surface and troposphere to warm. At the same time, since outgoing energy is being "caught" near the surface, less reaches the stratosphere, causing that atmospheric layer to cool.

■ Instrumental measurements combined with climatic proxy data reveal that Earth's surface temperature has been warming since the Industrial Revolution began. This warming is a result of increasing concentrations of greenhouse gases in Earth's atmosphere resulting from

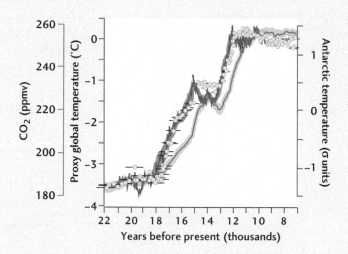

Figure 1 Proxies for temperature in Antarctica and the southern hemisphere (red line) lead changes in atmospheric carbon dioxide content (yellow circles) while global temperature (blue line) lags carbon dioxide. *(Shakun, J. D., P. U. Clark, F. He, S. A. Marcott, A. C. Mix, Z. Liu, B. Otto-Bliesner, A. Schmittner, and E. Bard. "Global Warming Preceded by Increasing Carbon Dioxide Concentrations during the Last Deglaciation." Nature 484 (April 2012): 49–54. ©2012 Rights managed by Nature Publishing Group.)*

21,500 and 19,000 years ago initiated ice sheet melting that led to significant influxes of freshwater into the North Atlantic. This freshening slowed or halted the thermohaline circulation that brings warm waters from the tropical south Atlantic northward (Chapter 12). Since it was no longer losing heat to the northern hemisphere, the southern hemisphere began to warm, melting sea ice around Antarctica and allowing carbon dioxide stored in ocean waters to escape to the atmosphere. Increasing carbon dioxide concentrations in the atmosphere in turn drove global warming that led to the decline of northern hemisphere ice sheets. If this scenario turns out to be correct, carbon dioxide would have acted both as a climatic chicken and as an egg, but of particular relevance to the anthropogenic climate change question, global warming followed increases in greenhouse gas concentrations.

fossil fuel combustion, deforestation, and other human activities.

■ Natural causes of climatic variability, including plate tectonic reorganizations of Earth's surface, orbital cycles, and changes in solar energy outputs are too small and/or have the wrong sign of temperature change to be responsible for the warming.

■ Carbon isotopes along with changes in atmospheric oxygen content and chlorofluorocarbon gases demonstrate incontrovertibly that human activities are the source of greenhouse gas changes.

■ As Earth's surface and troposphere have warmed, the stratosphere has cooled, consistent with higher greenhouse gas levels in the atmosphere.

■ Ice core and other paleoclimatological proxy evidence reveal that at the end of the last glaciation, global temperatures rose in response to rising carbon dioxide levels.

IMPACTS OF RISING GREENHOUSE GAS CONCENTRATIONS

Given that emissions of greenhouse gases from human activities are causing Earth to warm, the next concerns we need to address are how much warming we might expect for particular concentrations of greenhouse gases and whether or not the rising temperatures are problematic. The former question has been answered primarily through climate modeling experiments (Chapter 14). The United Nations Intergovernmental Panel on Climate Change *Assessment Report 4,* published in 2007, lists 23 different climate models used to predict the impact of increasing carbon dioxide concentrations relative to the value in 1850.[60] These models have been tested extensively and found to successfully reproduce the

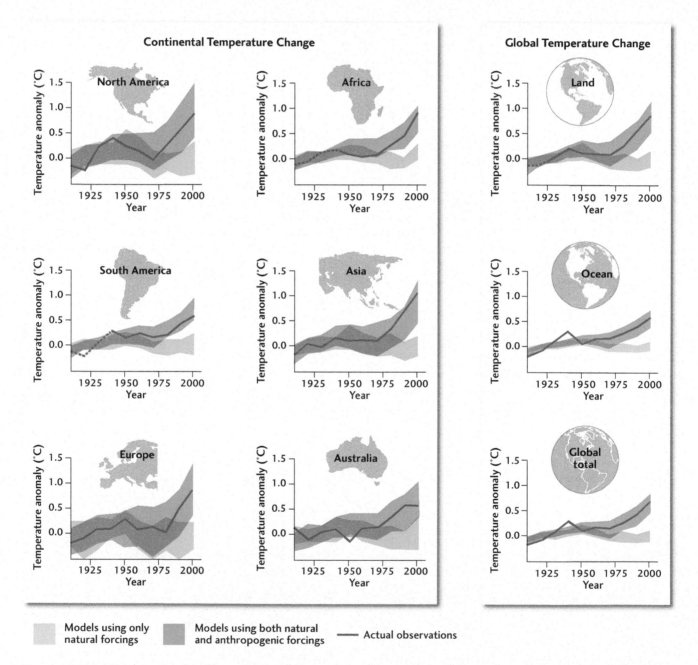

Figure 15–15 Observations of Earth's average surface temperature over the past century are successfully reproduced by a suite of 14 different climate models that incorporate observed variability in solar radiation, atmospheric aerosols from volcanic eruptions, and known changes in greenhouse gas concentrations.

major features of modern climate (for example, seasonal changes in temperature and precipitation, monsoon circulation, and oscillations in surface pressure—see Chapter 11). Most have also been used to simulate conditions at the last glacial maximum, 21,000 years ago, when ice cover was far more extensive than it is today and carbon dioxide levels were a third lower than preindustrial values, and at 6000 years ago, when orbital parameters caused greater than present solar radiation values. For all three time periods, the pattern of oceanic and land surface temperatures, precipitation, wind speed and direction, and other climatic variables created in the simulations has been compared to modern climatic and proxy paleoclimatic data, showing good agreement (Figure 15-15). This agreement lends confidence to the idea that we can use these models to forecast how climate will evolve as greenhouse gas levels change.

Greenhouse Gas Scenarios

Predicting exactly what the climate of Earth will be at the end of the 21st century and beyond is a challenging task, in part due to the limitations of climate models, but more importantly because future conditions depend on human behaviors we presently have no ability to predict. The climate science community has addressed the first problem by using many different models with varying parameterizations of the natural processes governing energy absorption, radiation, cloud formation, and other factors, along with different spatial resolutions (e.g., size and number of grid boxes representing the oceans and atmosphere) to come up with a range of estimates for climatic changes associated with different greenhouse gas concentrations in the atmosphere. They have addressed the latter problem by devising a number of greenhouse

gas emission scenarios reflective of potential socioeconomic and sociopolitical changes in the future (Table 15-1).[61] These scenarios include continually rising greenhouse gas emissions, declining emissions associated with a transition to alternative energy sources, and emissions associated with a mix of alternative and fossil energy sources. The scenarios also take into consideration different population growth rates and varying levels of international cooperation. IPCC reports present the outcomes of these different scenarios for the decades spanning 2020–2029 and 2090–2099. Potential impacts of global warming are summarized in Table 15-2, with impacts that are expected to be deleterious to ecosystems and human societies in red, and impacts that may prove beneficial in green. We address many of these topics below.

Temperature

If current carbon dioxide emission rates (about 7 gigatons of carbon per year)[62] continue, we will reach atmospheric concentrations double those of preindustrial values later this century. Climate modeling experiments suggest that doubling CO_2 will lead to a global temperature rise of between 1.5°C and 4.5°C[49] (Figure 15-16). The range of possible values arises from differing mathematical representations of climatic processes that lead to varying sensitivities of the modeled climate system to CO_2. It also depends on our ability to interpret paleoclimatic data to derive quantitative measures of past global temperature. Earth's average temperature in 1850, when carbon dioxide levels were 280 ppmv, measured approximately 15°C. If Earth was 4°C colder at the last glacial maximum when carbon dioxide concentration measured 180 ppmv, then a relatively large change in CO_2 led to a

TABLE 15–1 Examples of Climate Change Scenarios				
Scenario Name	Population	Economic Growth	Energy Use and sources	International Exchange
A1B	Peaks at 9 billion in 2050, then declines	Rapid, toward higher efficiency	Mix of fossil and alternative energies	Extensive, lower levels of income inequality globally
B1	Peaks at 9 billion in 2050, then declines	Rapid, toward service and information economy	Increased efficiency and alternative energy; less material intensity.	Extensive, emphasis on economic, social, and environmental sustainability
A2	Continuously increasing	Slow, regionally oriented economic development	Slow technological change	Low, world of self-reliant nations

TABLE 15–2 Potential Climate Change Impacts in the 21st Century Identified by the IPCC

Phenomenon and direction of trend	Likelihood based on IPCC model scenarios	Agriculture, forestry, and ecosystems	Water resources	Human health	Industry, settlement, and society
Warmer conditions over most land areas, fewer cold days and nights, more frequent hot days and nights	Virtually certain (>99% chance)	Increased crop yields in colder environments; decreased yields in warmer environments; increased insect outbreaks	Reduced water resources for populations relying on snowmelt; increased variability of rainfall events	Reduced human mortality from decreased cold exposure	Reduced energy demand for heating; increased energy demand for cooling; declining air quality in cities; reduced disruption to transport due to snow, ice; effects on winter tourism/ski resorts
Increasing frequency of warm spells/heat waves over most land areas	Very likely (>90% chance)	Reduced crop yields in warmer regions due to heat stress; increased danger of wildfire	Increased water demand; increased water quality problems, e.g., algal blooms	Increased risk of heat-related mortality, especially for the elderly, chronically sick, very young, and socially isolated	Reduction in quality of life for people in warm areas without appropriate housing
Increased frequency of heavy precipitation events in most areas	Very likely (>90% chance)	Damage to crops; soil erosion; inability to cultivate land due to waterlogging of soils	Adverse effects on quality of surface and groundwater; contamination of water supply; water scarcity may be relieved in some areas	Increased risk of deaths, injuries, and infectious disease outbreaks	Disruption of settlements, commerce, transport, and societies due to flooding; pressures on urban and rural infrastructures; loss of property
Increased area affected by drought	Likely (>66% chance)	Land degradation; lower yields/crop damage and failure; increased livestock deaths; increased risk of wildfire	More widespread water stress	Increased risk of food and water shortage; increased risk of malnutrition; increased risk of water- and food-borne diseases	Water shortage for settlements, industry and societies; reduced hydropower generation potentials; potential for population migration
Intense tropical cyclone activity increases	Likely (>66% chance)	Damage to crops; uprooting of trees; damage to coral reefs	Power outages causing disruption of public water supply	Increased risk of deaths, injuries, water- and food-borne diseases; posttraumatic stress disorders	Disruption by flood and high winds; withdrawal of risk coverage in vulnerable areas by private insurers; potential for population migrations; loss of property
Increased incidence of extreme high sea level (excludes tsunamis)	Likely (>66% chance)	Salinization of irrigation water, estuaries, and freshwater systems	Decreased freshwater availability due to saltwater intrusion	Increased risk of deaths and injuries by drowning in floods; migration-related health effects	Costs of coastal protection versus costs of land use relocation; potential for movement of populations and infrastructure

Source: Modified from IPCC, AR4—Synthesis for Policymakers, p. 53.

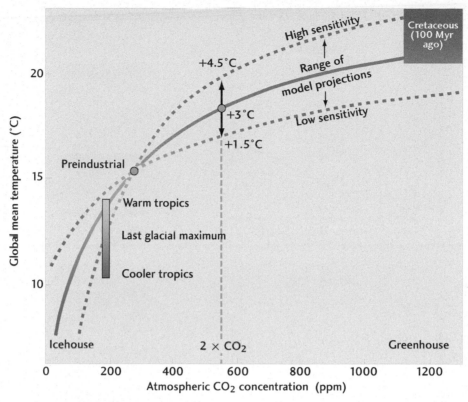

Figure 15–16 The expected amount of warming associated with a doubling of atmospheric carbon dioxide concentration by later this century depends on the sensitivity of the climate system to changes in greenhouse gases. A highly sensitive response would lead to greater amounts of warming than a lower sensitivity. The degree of sensitivity is not yet fully known. *(Ruddiman, W. F. Earth's Climate: Past and Future, 2nd ed. [p. 339]. New York: W. H. Freeman, 2007.)*

relatively small change in global temperature. If, on the other hand, as many data suggest, Earth was 7°C colder at the last glacial maximum, then the same change in CO_2 led to a much larger temperature response. Moving toward a doubled CO_2 world, a higher climatic sensitivity would lead to a greater temperature rise than would a lower sensitivity. More paleoclimatic proxy data need to be collected before we can determine precise sensitivity of climate to greenhouse gas concentrations. Furthermore, sensitivity in a warmer world may be different from that during icehouse conditions.

Whichever sensitivity scenario turns out to be correct, all model experiments show a warming of Earth's surface with increasing greenhouse gas concentrations (Figure 15-17). Note that all of the IPCC scenarios described in Table 15-1 show a warming by 2020–2029 of 1–1.5°C relative to a reference period spanning 1980–1999, with the greatest rise at the polar latitudes. By 2090–2099, when CO_2 levels have surpassed twice their preindustrial values, temperatures are predicted to rise to conditions the planet has not witnessed in tens of millions of years. If we conceive of the temperature distribution of any point on Earth's surface as a bell curve, with the center of the curve situated at the average temperature and the tails representing the extremes of hot and cold conditions

presently experienced, then a warming of the climate will generate a shift in temperature distribution (Figure 15-18). This shift will make what had previously been rare heat waves much more common and will make cold conditions far less common.[63]

Another way to conceive of the climatic shift that is underway is to think of it as a geographic change in the location of a particular area. For example, by 2090–2099, the climate of New Hampshire will be approximately equivalent to the present climate of North Carolina or Virginia, depending on which IPCC emission scenario prevails (Figure 15-19).

Sea Level Rise

As mentioned earlier in this chapter, more than a century of observations have demonstrated that sea level has been rising as Earth's surface temperature has warmed. From 1961 to 2003, the amount of ice lost from the Greenland and Antarctic ice sheets was enough to raise sea level 0.5 +/– 0.18 mm/yr.[64] When looking at the interval 1991–2003, the loss was sufficient to raise sea level 0.77 +/– 0.22 mm/yr, suggesting that **sea level rise** is accelerating along with warming. While fractions of a millimeter per year may seem inconsequential,

Projected Global Surface Warming Under Alternate Emission Scenarios

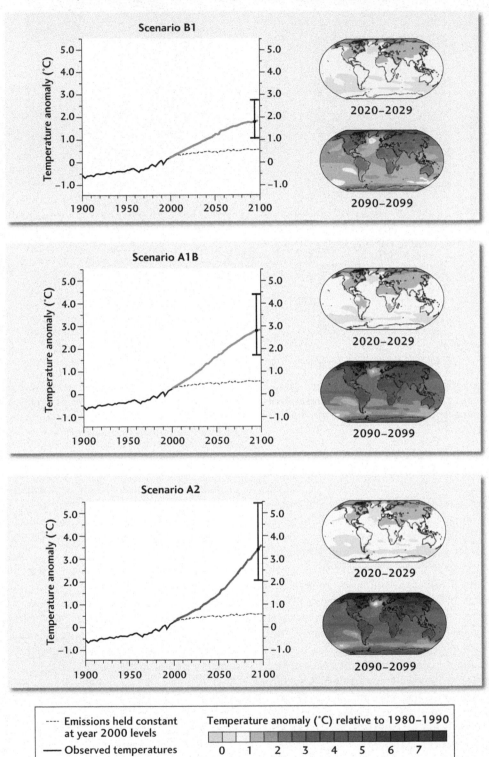

Figure 15–17 Expected warming for different IPCC emissions scenarios, which are based on different assumptions for population growth, energy use, and global information exchange. In all cases, warming is greatest at high latitudes due to loss of ice through the ice–albedo feedback.

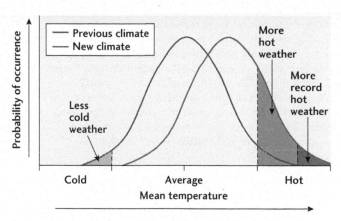

Figure 15–18 Global warming shifts the distribution of temperatures in any given location toward higher values. This shift increases the number of days experiencing record heat waves and decreases the frequency of cold events with consequences for ecosystems, water resources, and human health.

over many decades they add up to substantial changes. Furthermore, additional sea level rise is forecast to occur with thermal expansion of seawater as Earth warms. For the IPCC scenarios shown in Table 15-1, sea level rise by 2090–2099 is projected to measure from 0.18–0.38 m for the B1 scenario to 0.23–0.51 m for the A2 scenario (Figure 15-20a),[65] though more recent work suggests sea level is likely to rise by 0.8 m by century's end.[66] Due to the slow response of ice sheets to climatic changes, continued melting over the next several centuries will raise sea levels higher still. How high depends on which scenario prevails, but if temperature changes of the sort projected for the doubled CO_2 world continue for many centuries, the entire Greenland ice sheet will melt, leading to a sea level rise of 7 m.[67] Additional sea level rise associated with the melting of Antarctica is also possible.

Earth actually experienced a significant melting of the Greenland ice sheet some 125,000 years ago. At that time, the Milankovitch cycles (Chapter 14) increased the amount of solar radiation striking the high northern latitudes. This, combined with feedbacks in the climate system such as the ice-albedo feedback, caused the atmosphere over Greenland to warm by up to 3.5°C in summer.[24] This warming led to sufficient melting to drive 1.9–3.0 m of sea level rise as the ice sheet shrank to 30–60 percent of its current volume.[68] Total sea level rise at that time was on the order of 5–7 m, with the remainder caused by thermal expansion of seawater and

Geographic Equivalents to Projected Future Summer Temperatures and Precipitation

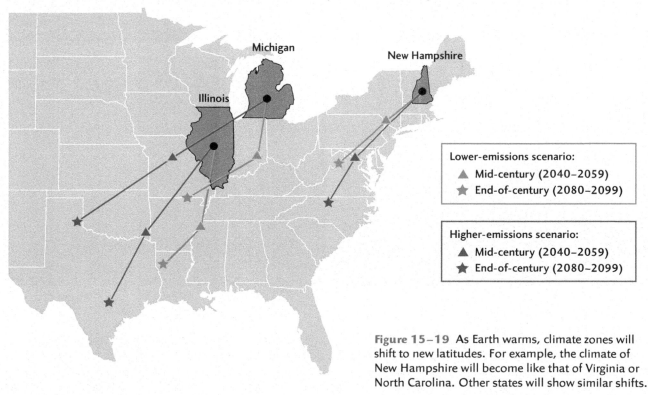

Figure 15–19 As Earth warms, climate zones will shift to new latitudes. For example, the climate of New Hampshire will become like that of Virginia or North Carolina. Other states will show similar shifts.

Global Average Sea Level Rise (1990–2100)

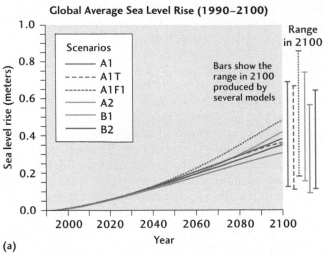

Figure 15–20 (a) Expected sea level rise by the end of the 21st century for different IPCC emissions scenarios. Range bars reflect the suite of models run for each scenario, each of which parameterizes climate slightly differently. (b) Maps of coastal inundation associated with different amounts of sea level rise that may occur over the next several hundred to thousand years if greenhouse gas emissions are not brought under control. (b: *NOAA Geophysical Fluid Dynamics Laboratory*)

Sea Level Rise

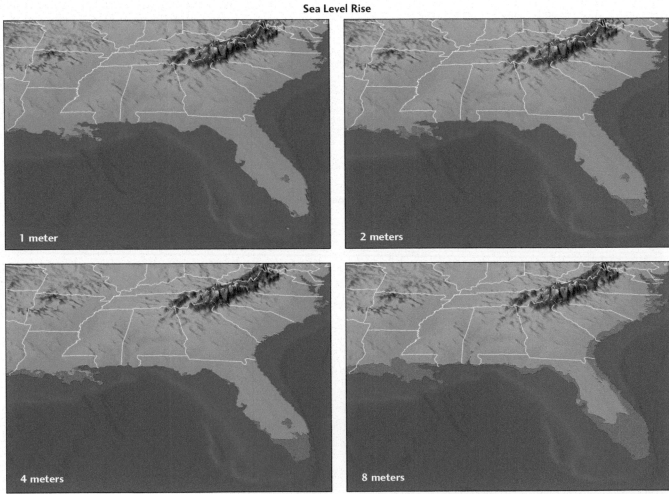

possible melting of the Antarctic ice sheet. Such a rise, were it to happen in the future, is sufficient to drown southern Louisiana, southern Florida, and half of the nation of Bangladesh, among other locations (see Figure 15-20b). Every millimeter of sea level rise that occurs over the coming centuries also makes coastal zones more vulnerable to catastrophic events such as hurricanes and tsunamis and increases the risk of saltwater intrusion into coastal aquifers. Clearly this is a grave concern for all nations that border Earth's ocean basins.

Changes to the Hydrologic Cycle

As Earth's surface warms and the pole to equator temperature gradient changes, we might expect changes in atmospheric circulation that could lead to changes in precipitation patterns. One of the largest areas of uncertainty in climate modeling is the formation and dissipation of clouds, which makes forecasting precipitation responses to a warmer world challenging.[69] Figure 15-21 illustrates precipitation patterns predicted for the winter and summer seasons in the decade 2090–2099 for the IPCC emission scenario A1B. The results reflect the aggregate output of the 23 climate models used in the IPCC analysis. Blue colors represent increases in precipitation whereas reds and oranges represent declines. The pattern of dots superimposed on the colors shows those locations where more than 90 percent of the climate models yield the same sign of precipitation change whereas the white regions are locations where less than 66 percent of models agree. The model results suggest that areas presently experiencing dry climates (southwestern United States, the Sahara) will experience further drying, whereas some currently wet areas (Bangladesh) will experience increasing precipitation. Of great concern for agriculture, the models cannot predict with certainty how precipitation will change in the summer growing season for the world's breadbasket regions, including the plains of North America and central Asia.

A warmer atmosphere can hold greater amounts of water vapor (Chapter 11), so when precipitation does fall, it is expected to be heavier.[70] In addition, the intensity of tropical storms and hurricanes is projected to increase, with stronger winds and greater rainfall amounts. As a result of these changes, flooding and landslides may become more common. At the same time, warming of the Arctic is expected to cause the atmospheric Hadley cell to expand poleward. Since the descending limb of this cell is associated with arid conditions, its northward migration is expected to expand areas in the subtropics that experience frequent drought.[71] Model experiments suggest that the southwestern United States and northern Mexico may experience permanent changes in precipitation toward Dust Bowl–type conditions. Inasmuch as the southwestern United States already experiences frequent fires and water resource challenges, global warming is likely to make these problems worse.

Additional problems for water resources lie in changes in the form of precipitation that is falling. A shift from winter snow to rainfall reduces the amount of water stored on the landscape. Similarly, reductions in snowpack lower groundwater infiltration and make landscapes more susceptible to drought. Loss of glacial ice on the Tibetan Plateau threatens the water resources of half a billion people in South Asia who depend on the Indus, Ganges, and Brahmaputra rivers for drinking and irrigation water supplies.[72] Melting ice contributes as much as 45 percent of the total flow of these rivers. Other glaciated mountain regions (e.g., the Andes, Alps, Sierra Nevada, and Rocky mountains) will be likewise affected, with implications for water resources and hydroelectric power generation potential.

Impacts on Human Health

In the summer of 2003, Europe was struck by a heat wave that led to the deaths of over 20,000 people, over half of which occurred in France.[73] Northern Europe is accustomed to relatively cool summers, and few homes have air conditioning. At a time when many families were hitting the beaches on their summer vacations, elderly people in overheated apartments succumbed to

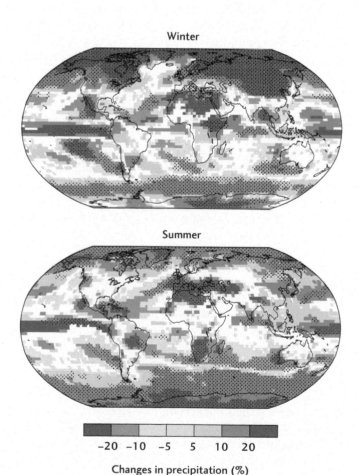

Winter

Summer

-20 -10 -5   5   10   20

Changes in precipitation (%)

Figure 15–21 Climate model predictions for precipitation changes suggest that already dry areas may experience greater aridity in the future, whereas wet regions may become wetter. Regrettably, models do not show sufficient agreement in many locations (white areas) to be able to predict the sign of precipitation changes. This is particularly problematic in the world's grain belts in summer.

heat stroke. As global warming continues, the frequency of heat waves such as this is expected to increase and with it, heat-related mortality.

In addition to the direct impacts increasing heat may have on populations, global warming has the potential to alter disease transmission. At present, the microorganisms that cause malaria are restricted to environments in which the temperature is 20°C or higher, confining them to the tropics. As global warming progresses, however, warmer temperatures will spread poleward, bringing mosquitoes infected with malaria to higher latitudes. Other diseases will also migrate as warming continues. Organisms causing Lyme disease and encephalitis have recently expanded into higher altitude regions of the Czech Republic, most likely in response to warming,[75] and West Nile virus is spreading in the United States as warmer and wetter conditions lead to additional mosquito breeding habitat.[76] Elevated carbon dioxide levels in the atmosphere may also affect human health through their impacts on plants. A study conducted in an experimental forest owned by Duke University in which carbon dioxide levels can be manipulated from specially designed emission towers found that higher concentrations of the greenhouse gas led to higher growth rates for poison ivy vines.[77] Not only did the vines grow more vigorously, the toxic oil they produced was more allergenic than usual!

Additional impacts of global warming on human health may include malnutrition and food insecurity in developing nations beset by droughts or severe weather (e.g., hail) that damages crops. Water scarcity may also lead to outbreaks of diseases such as cholera as people are forced to drink water contaminated with sewage.[75] Warmer temperatures exacerbate air quality problems as well. For example, tropospheric ozone formed from compounds in car exhaust develops more readily as temperatures rise, which may increase respiratory distress for people with asthma and other lung problems.

Ecosystem Impacts of a Warming Planet

The impacts of rising greenhouse gas concentrations affect not only humans but also the other species with which we share Earth. As temperatures warm, plant communities adapted to cooler climates can no longer survive in their current locations, triggering ecosystem shifts. Similarly, species favoring warmth are able to expand their range. A study conducted in the Ural Mountains of Russia found that the Siberian larch tree has expanded its range up mountainsides by 20 to 60 m in the past century in response to local warming of 0.9°C and a twofold increase in precipitation.[78] This expansion has come at the expense of tundra vegetation that requires cooler conditions. Animals are likewise affected.

The loss of Arctic sea ice has made polar bears "poster children" for global warming, and other species are also affected. These include cold-water fish species such as bull trout,[79] which are being driven out of streams as water temperatures rise, and many species of migratory birds, which may use day-length cues to know when to migrate and which may find their food supply disrupted by changes in temperature.[80]

When climatic change intersects with habitat fragmentation, additional ecosystem impacts arise.[27] Plants that might otherwise migrate upward or poleward in search of cooler climates may be prevented from doing so as human land uses hinder their dispersal. Seeds landing on parking lots and suburban lawns will not find suitable conditions in which to take root or will be prevented from growing by lawn maintenance. Furthermore, the intersection of habitat fragmentation and climate change leads to isolation and genetic inbreeding that inhibits populations from developing evolutionary responses to warming. The success of invasive species may also be promoted by climatic change, not only because of warming but also because of changes to the hydrologic cycle that bring about more extreme weather events that generate disturbance.[27]

Just as more humans may succumb to diseases formerly restricted to the tropics, warming will exacerbate disease transmission for plants and animals. While little is known of the impact of warming on native plant diseases, studies of some agricultural plants indicate that various pathogens may stunt growth under higher CO_2 concentrations, outweighing any gains in growth associated with increased temperatures and photosynthesis enhanced by carbon dioxide fertilization.[81] Amphibians appear also to be suffering declines and outright extinctions as a result of diseases whose transmission is being enhanced by global warming. Frogs and toads in the mountains of Costa Rica have been sickened by a fungus that has become more widespread as temperatures have risen.[82] Sixty-seven percent of the genus *Atelopus* (Monteverde harlequin frogs), which contains 110 different species, has gone extinct in recent decades as a result of this pathogen.

One of the most concerning aspects of greenhouse gas increases is their impacts on marine ecosystems. Corals have a symbiotic relationship with algae that live in their cells and carry out photosynthesis. As the oceans warm, corals expel these zooxanthelle, losing a large part of their food supply and increasing their mortality. Corals that have undergone such a change appear ghostly white and are said to have been "bleached." **Coral bleaching** affects not only the corals themselves, but also the fish and other organisms that live off of them. Large areas of reef have been stressed in this way in recent years. For example, almost 60 percent of Australia's Great Barrier Reef succumbed to bleaching in

2002,[83] and Asia is expected to lose up to 30 percent of its reefs in the next three decades, in part due to ocean warming.[84] Inasmuch as coral reefs are like the tropical rain forests of the oceanic realm, these losses represent a major threat to biodiversity in the future.

Temperature is not the only factor presenting a problem for marine life. The increasing concentration of carbon dioxide in the atmosphere also increases its absorption by the oceans. There it combines with seawater to form carbonic acid, leading to a gradual decline in oceanic pH (Figure 15-22). Indeed, pH has fallen by 0.1 unit since 1750.[85] This **ocean acidification** may affect organisms that secrete calcium carbonate shells and skeletons, because carbonic acid reacts with calcium carbonate to liberate calcium ions and carbon dioxide gas. Many of these organisms are at or near the base of the oceanic food chain (e.g., foraminifera and coccolithophorids), and so this is cause for concern; however, much more research needs to be done to determine the scope of the problem.

Lessons from Former Worlds

Earth is a very different planet now than at any point in its long history, but some conditions in the past may prove relevant as we try to understand the potential impacts of anthropogenic climate change. As mentioned earlier, higher solar radiation receipts 125,000 years ago led to melting of the Greenland ice sheet that drove sea levels 4 to 6 m higher than present. At that time, humanity existed in vastly smaller numbers and lived a nomadic subsistence lifestyle. As we contemplate a similar scenario brought about by rising greenhouse gas levels, it is instructive to recall the catastrophic flooding of New Orleans in 2005 during Hurricane Katrina, of Myanmar in 2008 by Typhoon Nargis, of northern Japan in 2011 by the Tōhoku earthquake tsunami, and of coastal New Jersey and New York City in 2012 by Superstorm Sandy. The gradual melting of the Greenland ice sheet over the next several centuries has the potential to levy similar damage to shorelines all across the globe at a time when our population will likely exceed nine billion and live increasingly in urbanized coastal areas.

If nothing is done to curtail the growth of greenhouse gas emissions, our planet will achieve carbon dioxide concentrations equivalent to those 34 million years ago by the end of this century.[86] In the absence of carbon sequestration, continuing to combust fossil fuels until every accessible drop of oil, piece of coal, and volume of gas are used up would put enough carbon into the atmosphere to raise concentrations to 2000 ppmv, a value not seen in more than 100 million years.[87] If these scenarios seem extreme, consider the fact that carbon dioxide concentrations are presently growing at a rate of 2 ppmv/yr[88] and that a boom in the discovery and exploration of natural gas has set back the development of non-fossil energy sources in the United States, one of the world's biggest emitters of greenhouse gases (Chapter 13). Consider also that China presently builds a new coal-burning power plant every month.[89]

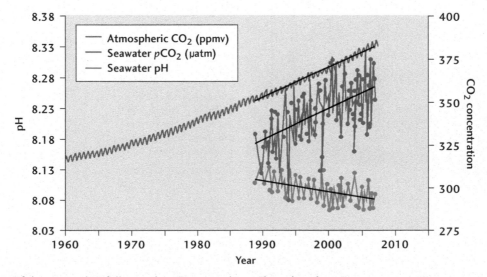

Figure 15-22 pH of the oceans has fallen at the same time that carbon dioxide concentration has risen. pH decline threatens the health of organisms that secrete calcium carbonate shells and skeletons.

BOX 15–4 EMERGING RESEARCH

Frozen Methane: A Climatic Bomb Waiting to Go Off?

As Earth's surface warms, Arctic permafrost is thawing. This perennially frozen ground contains as much as 1700 billion tons of carbon locked up in organic matter that accumulated over many millennia.[91] Along with the organic matter, the permafrost contains mineral materials and frozen groundwater. The latter is commonly found in lenses in the soil as surface and pore water is drawn toward ground that is freezing, segregating the water in the soil from the other solids. When temperatures warm, these lenses melt, leading the soil above them to collapse and forming a depression that fills with water, a feature known as a **thermokarst lake** (Figure 1).[92]

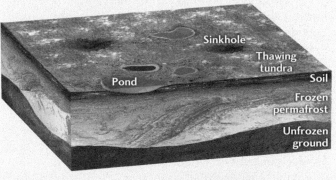

1 Ice in the frozen soil melts and the ground subsides, forming sinkholes that fill with water, becoming ponds.

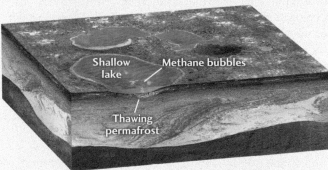

2 Ponds merge into lakes. The water thaws the soil below and microbes decompose the organic matter anaerobically, producing methane.

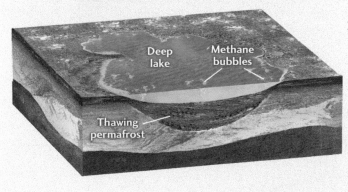

3 Deepening lakes thaw permafrost, frozen earth that is far richer in organic matter. It decomposes as well, generating numerous methane bubbles that rise to the lake surface and burst into the atmosphere.

Figure 1 Melting ice in permafrost produces depressions in which lakes form. Organic matter in lake bottom sediments is subject to anaerobic decomposition, which produces methane, a greenhouse gas 25 times more potent in its warming potential than carbon dioxide. *(Kevin Hand)*

At the warm temperatures brought about by even a doubling of CO_2, feedbacks in the climate system have the potential to bring about still more warming. The ice-albedo feedback has been mentioned already, but another, and potentially quite dangerous, feedback involves the thawing of Arctic permafrost and the release of presently frozen carbon (Box 15-4). Indeed, Earth is thought to have experienced such a thawing 55 million years ago, a time known as the **Paleocene–Eocene thermal maximum**.[86] At this time, deep ocean tempera-

Figure 2 Dr. Katey Walter Anthony samples methane from a seep in a frozen lake. Tossing a match on a hole cut into the frozen lake surface causes the methane to catch fire. *(Mark Thiessen/National Geographic Society/Corbis)*

The warming of permafrost also activates bacteria that consume the soil organic matter. When growing in sediments at the bottom of lakes where oxygen levels are low, these bacteria use anaerobic decomposition to convert the organic matter to methane.[92] Bubbles of the gas rise to the surface of thermokarst lakes and enter the atmosphere where they contribute to the global warming problem. Methane is a far more potent greenhouse gas than is carbon dioxide, producing 25 times as much warming per volume of gas. This climatic feedback is presently not incorporated into the general circulation models scientists use to predict the future impact of greenhouse gas increases on Earth's temperature. Small wonder then that many scientists are worried. Could melting permafrost unleash an unstoppable and catastrophically large warming?

Aquatic ecologist Katey Walter Anthony and her colleagues have been working to quantify the rate at which Arctic permafrost is thawing and releasing methane.

Trudging across the Siberian and Alaskan tundra, they have deployed bubble traps to capture and measure the volume of methane released from thermokarst lakes. Bubbles in the ice covering some of these lakes reveal that the bulk of the methane rises from restricted sites called seeps. Dr. Anthony discovered that the seeps produced methane when she broke through the ice bubbles and lit a match over the resulting hole (**Figure 2**). The flames produced by the expelled gas shot more than 5 m high, an incongruous sight in a land of ice and water![93] Through her measurements and maps of thermokarst lakes, Dr. Anthony estimates that somewhere between 14 and 35 million metric tons of methane are being released by permafrost thawing throughout the Arctic every year. This rate of release is sufficient to raise Earth's temperature 0.32°C by the end of the 21st century, on top of the warming expected from increasing carbon dioxide concentrations, which is clearly a cause for concern.

tures rose by 5°C to 6°C in less than 10,000 years, sea surface temperatures at high latitudes rose by as much as 8°C, and acidification of ocean water from increased carbon dioxide concentrations led to widespread dissolution of carbonate sediments deposited on the seafloor.[90] The acidification of the oceans drove mass extinctions of sea bottom dwelling foraminifera, and the changing temperature and precipitation conditions around the globe led to major migrations of mammals in the northern hemisphere. It took nearly 200,000 years for the increased carbon in the atmosphere to be removed by carbon cycle processes.[90] It is scenarios such as the Paleocene–Eocene thermal maximum that have led some scientists to suggest that atmospheric carbon dioxide levels should be brought down to a value no higher than 300 to 350 ppmv. Numerical modeling experiments show that such a level would likely restrict warming to 2°C relative to preindustrial temperatures, a temperature rise that, while still deleterious, is unlikely to surpass the threshold for tipping Earth into an entirely new environmental state.

Climate Change Benefits

As Table 15-2 illustrates, not all impacts of global warming are likely to be negative. Warmer temperatures will reduce human mortality due to cold exposure in higher latitudes and will lower heating costs in winter. Longer summers in these areas may lead to increased crop yields. A shift of precipitation from snow to rain will lead to fewer transportation disruptions and to declining use of road salt and other deicing chemicals that have contaminated streams, lakes, and groundwater supplies. Furthermore, melting of the Arctic Ocean's sea ice may make global ship traffic much more efficient by opening up routes connecting Europe and the east coast of North America to Asia.[94] Melting sea ice also makes additional oil and gas exploration possible. Undoubtedly there will be climatic "winners" as well as "losers," though for the globe as a whole, it appears that the negative impacts of climate change far outweigh the positives.

■ Climate modeling experiments predict a range of temperature increases in response to rising greenhouse gas concentrations. This range results from differences in the ways in which models represent the physics of heat absorption and radiation, the generation of clouds, and other aspects of the climate system.

■ The range in temperature predictions also arises from different assumptions about future human behavior and how it may influence greenhouse gas emissions. Regardless of the model physics or socioeconomic scenario assumed, all models show a warming over the next century. Warming will be greatest at the polar latitudes as reflective ice and snow melt and are replaced by energy-absorbing soil and ocean water.

■ Rising temperatures will continue to melt glaciers and ice sheets, raising sea level by tens of centimeters by century's end, which will threaten coastal areas.

■ Changes to Earth's hydrologic cycle may cause currently dry areas to experience further drying and presently wet locations to become even wetter. These changes are expected to increase fires associated with drought as well as flooding. Hurricane intensity may also increase.

■ Changes in precipitation patterns and loss of glaciers threaten water resources for people across the globe.

■ Rising temperatures can influence human health directly through increased exposure to excessive heat as well as indirectly through impacts on disease transmission.

■ Rising temperatures combined with habitat fragmentation and loss threaten individual species as well as entire ecosystems. Both terrestrial and marine ecosystems are at risk, with coral reef communities among the most vulnerable to temperature changes. Corals and other calcium carbonate secreting organisms are also threatened by oceanic acidification as increased absorption of carbon dioxide into the oceans reduces seawater pH.

■ As the Arctic warms, thawing permafrost is releasing methane gas to the atmosphere. Methane is more potent as a greenhouse gas than carbon dioxide is, and it will cause further global warming.

■ Not all impacts of global warming are expected to be negative. Reduced mortality due to cold exposure, a longer growing season at higher latitudes, and reduced disruption of transportation by ice and snow are potential benefits of climatic change.

COMING TO GRIPS WITH THE ANTHROPOCENE

The fact that Earth is warming as a result of human activities requires some sort of response if we wish to prevent the most negative of outcomes for humanity and the rest of the biosphere. Numerous actions have been proposed, ranging from eliminating emissions of greenhouse gases to armoring coastlines to withstand future sea level rises to using geoengineering strategies to increase the amount of incoming solar energy reflected to outer space. We address these topics briefly here, noting that while science can point the way toward different strategies to deal with the problem of global warming, only changes in human behavior and political will can lead to their implementation.

Greenhouse Gas Reduction Strategies

Earlier this century, Princeton scientists Stephen Pacala and Robert Socolow described a way to stabilize and ultimately reduce global greenhouse gas emissions. They identified several sectors of the economy that could be altered through a mix of technological and social innovations to prevent further growth of carbon dioxide levels in the atmosphere (Figure 15-23). These include

ANNUAL EMISSIONS
In between the two emissions paths is the "stabilization triangle." It represents the total emissions cut that climate-friendly technologies must achieve in the next 50 years.

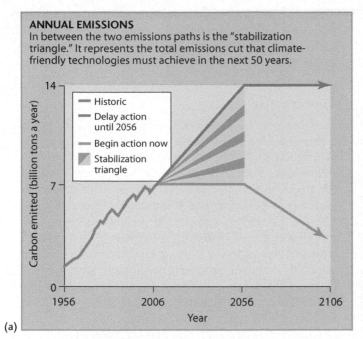

Figure 15–23 (a) To stabilize carbon dioxide levels in the atmosphere and eventually reduce them, we must eliminate future growth in emissions that would occur as a result of global population increase and rising standards of living. (b) Scientists Stephen Pacala and Robert Socolow have identified a number of ways in which we might do so, with each wedge of the CO_2 reduction pie representing an elimination of 1 billion tons per year of the expected future growth in emissions. *(a: Illustrated by Jen Christiansen, from Socolow, R. H., and S. W. Pacala. "A Plan to Keep Carbon in Check." Scientific American (September, 2006): 50–57; b: Illustration by Janet Chao)*

THE WEDGE CONCEPT
The stabilization triangle can be divided into seven "wedges," each a reduction of 25 billion tons of carbon emissions over 50 years. The wedge has proved to be a useful unit because its sizes and time frame match what specific technologies can achieve. Many combinations of technologies can fill the seven wedges.

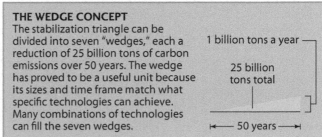

(a)

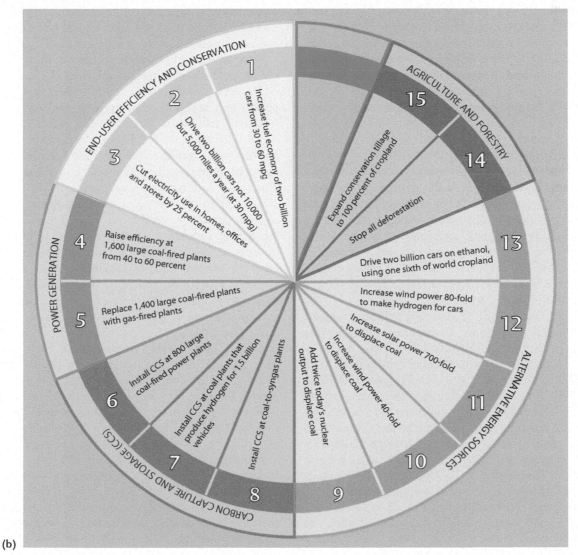

(b)

increasing fuel efficiency standards for motor vehicles, replacing inefficient lighting and appliances in buildings, preventing the construction of additional coal-burning power plants, developing technology to capture carbon dioxide released at existing power plants and sequester it in the ground, generating more power through alternative energy sources such as wind and solar in addition to nuclear energy, expanding opportunities for telecommuting, eliminating deforestation, implementing no-till agricultural practices, and reducing birth rates to slow the expected growth in population that will occur over the next 50 years. Each innovation they proposed was designed to cut 1 billion tons per year of carbon dioxide from the expected growth in emissions that would occur in the absence of any action.

At the time Pacala and Socolow wrote their paper, they estimated that it would be necessary for the human population to eliminate seven billion tons of carbon emissions per year in order to stabilize atmospheric carbon dioxide levels at year 2004 levels by 2054, with further reductions necessary thereafter to bring levels down to the "safe" value of 350 ppmv. Regrettably, political inaction has not yet brought their suggestions to fruition, though some progress has been made. In 2011, U.S. President Barack Obama negotiated an agreement with 13 different automobile manufacturers to raise the fuel efficiency of cars and pickup trucks to 35.5 miles per gallon from 2012 to 2016 and then to 54.5 miles per gallon by 2025.[95] Not only will this be helpful toward reducing greenhouse gas emissions, it will save drivers nearly $1.7 trillion at the pump. Additional energy savings will come from the phase out of incandescent lightbulbs, which generate more waste heat than they do light. Much more efficient compact fluorescent lightbulbs and light emitting diodes (LEDs) slash energy consumption by 75 percent or more, and are replacing incandescents as a result of the Energy Independence and Security Act passed by the U.S. Congress in 2007.[96] This legislation also mandates new efficiency standards for appliances such as walk-in freezers and refrigerators.

Additional work will be required to "decarbonize" Earth's economy, and other countries are leading the way. In Germany, for example, solar energy now powers eight million homes,[97] an amazing achievement for a nation of 82 million people and a climate that includes a lot of cloudy skies and rain. Another alternative energy standout is Denmark, which generates 25 percent of its electricity through wind power. The nation hopes to achieve 50 percent wind energy by 2020.[98] China also hopes to increase its reliance on renewables to 20 percent by 2020, and is spending $473 billion to meet its goals over the next few years.[99] These are all hopeful trends, but they may still be insufficient to avoid a climatic tipping point in the near future if population and living standards continue to rise. Each additional person added to the globe increases the total impact of humanity, and if that person is born into an industrialized society, his or her personal impact reflects the accoutrements of that society, including purchases of energy-intensive consumer goods like automobiles and appliances.

Geoengineering

In the absence of sufficient action to reduce greenhouse gas emissions, some scientists have proposed **geoengineering** strategies to keep Earth's surface from overheating. Such strategies can be as simple as changing the albedo of Earth's surface by painting rooftops, roads, and sidewalks white and changing the types of crops that are planted in agricultural areas to ones that are more reflective of incoming solar energy, though these strategies are expected to have limited impact by themselves.[100] Another proposed "radiation management" technique is using high-flying aircraft to spray sulfate aerosols into the stratosphere in an attempt to block incoming solar radiation and cool the climate. As noted in Chapter 5, such a strategy has the potential to cause other problems because sulfate aerosols are precursors of acid rain and because altering atmospheric aerosol levels has the potential to change precipitation patterns that may cause droughts in some nations.[101] Furthermore, blocking incoming solar radiation does nothing to address the problem of carbon dioxide diffusion into the oceans and declining seawater pH. Another suggestion, to launch trillions of light-blocking pieces of plastic or metal into orbit around Earth,[102] also does nothing to address ocean acidification, and the logistical problems involved in getting these materials into space makes this potential solution highly improbable.

A second class of geoengineering strategies involves carbon dioxide removal from the atmosphere. Preventing further deforestation and allowing trees to regrow in areas previously logged for agriculture or timber can take up carbon dioxide through photosynthesis and sequester it in roots, trunks, and tree limbs. Another way to remove carbon dioxide from the atmosphere is to increase the rate at which it reacts with silicate minerals during chemical weathering. This can be done by mining rocks rich in minerals like olivine and then pulverizing them and spreading the powder onto fields, because the increase in surface area brought about by pulverization greatly increases the rate of chemical weathering. The problem with this strategy is that it requires approximately 7 km^3 per year of rock to be mined and ground up annually to offset the amount of carbon dioxide humans presently produce. Inasmuch as this is twice the rate of global coal production, such a solution would lead to major environmental damage of its own.[100]

Fertilizing the surface ocean with nitrogen, phosphorus, or iron is another geoengineering strategy that has been proposed. Here the idea is to increase the biomass of phytoplankton at the base of the food chain in an attempt to sequester carbon in organic tissues that sink to the ocean floor, much as may have happened during glaciations. Several experiments have been carried out over the past two decades to test this idea, but they have shown only limited success, since some other factor becomes limiting as the fertilization experiment continues. Furthermore, some scientists are concerned about implementing such a strategy more broadly because an increase in the amount of biomass decaying in the deep oceans has the potential to deplete oxygen levels. Using phosphorus to fertilize the oceans for climate mitigation also reduces the amount of this vital nutrient available for agriculture. And, as is the case with many of the geoengineering proposals described here, the costs of carrying out large-scale ocean fertilization experiments are likely to be prohibitive.

Chapter 13 described another strategy to remove the products of fossil fuel combustion from the atmosphere: carbon sequestration at power plants. So-called carbon capture and storage would take carbon dioxide from flue gases, compress it and pump it into the ground in places like old oil fields.[103] Carbon dioxide is already used to enhance oil recovery in some locations, so this is a technology that is feasible, though it would have to be scaled upward substantially, with pipelines constructed to move the captured CO_2 to deep disposal sites.

Adaptations to a Warmer World

Even if we were able to stabilize atmospheric carbon dioxide concentrations at current levels tomorrow, Earth's surface would continue to warm for decades to a century or more. The continued warming is due to the fact that the melting of the polar ice caps imparts inertia to the climate system. The ice-albedo feedback (Chapter 14) changes reflective ice and snow into absorptive ocean water and sediments. The more energy the poles absorb, the warmer they become, melting additional ice and snow and making the high latitudes even more absorptive. Since Earth will continue to warm even if we stabilize greenhouse gas emissions, we need to prepare for the resulting sea level rise and disruptions to the hydrologic cycle we can expect in the future. These preparations include raising the elevation of structures in coastal areas to deal with an increased frequency of flooding. The sewage treatment plant for the city of Boston, for example, sits on a peninsula in the Massachusetts Bay and was constructed with sea level rise in mind. The facility was elevated 1.9 ft higher than it might otherwise have been, in order to ensure that treated sewage could flow into the bay via gravity throughout the plant's expected 50-year lifespan.[104]

In another example, New Jersey Governor Chris Christie recently approved revised flood maps created by the Federal Emergency Management Agency designed to make communities more resilient as sea level rise continues.[105] These maps require people living in areas likely to be flooded by events like Superstorm Sandy to elevate their homes or face exorbitant flood insurance rates. Elevating existing housing has become a multimillion dollar industry in southern Louisiana in the aftermath of Hurricane Katrina, and this industry may expand north to New Jersey under the new rules. Companies excavate trenches beneath load-bearing beams in order to install house jacks, then raise the entire structure, and build a new foundation beneath it (Figure 15-24). Staircases, or in some cases, elevators, are added to the design to allow people to access their newly elevated homes.

Figure 15–24 Elevating homes to make them more flood–resistant has become a multimillion dollar industry in Louisiana. *(AP Photo/Bill Haber)*

Other strategies to deal with sea level rise include moving entire communities to higher ground, building higher levees to keep the ocean out, and abandoning coastal development altogether. In Louisiana, the extensive development of New Orleans and its surrounding suburbs makes community relocation impossible. As a result, the U.S. Army Corps of Engineers has undertaken several projects in the aftermath of Hurricane Katrina to raise levees, construct floodwalls and closeable gates across waterways, and build pumping stations to remove water that makes its way behind the defensive line. For example, a 2-mile-long and 26-foot-high surge barrier recently completed at the western edge of a coastal embayment called Lake Borgne will prevent storms like Hurricane Katrina from being able to push water into the city from the east (Figure 15-25).[106] This barrier

includes gates that remain open during fair weather to allow for ship traffic but that can be closed in the event of storms. England has had a similar barrier across the Thames River since the early 1980s to prevent once-in-a-thousand-year storm surges and high tides from Europe's North Sea from flooding London (Figure 15-26a).[107] Venice, Italy, is also preparing for sea level rise with Project Moses, 78 gates that can be raised to keep water from entering the Venice Lagoon from the Adriatic Sea.[108] The lagoon is separated from the sea by a series of barrier islands, and the gates are being constructed across the tidal inlets between them to prevent storm surges from destroying the city (Figure 15-26b).

Of course, coastal areas are not the only locations that have to contend with increased flooding in a warmer world. People living adjacent to streams in

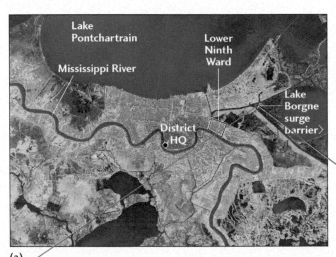

(a)

Figure 15–25 (a) The Lake Borgne surge barrier is designed to prevent storm surges from the Gulf of Mexico from entering the canal system that both provides navigation through and drainage of water from New Orleans and its suburbs. Breaching of canal walls during Hurricane Katrina led to the catastrophic flooding of the city's lower 9th ward neighborhood. "District HQ" is the U.S. Army Corps of Engineers District Headquarters. (b) Aerial view of the Lake Borgne surge barrier. *(a: NASA Landsat image composed by Angela King, © 2005 Geology.com; b: U.S. Army Corps of Engineers, New Orleans District)*

(b)

Figure 15–26 (a) The Thames barrier consists of a series of massive gates that can be raised to block a surge of water from flowing up the river into the heart of London. As sea level rises, the use of this barrier is expected to increase. (b) Project Moses consists of a series of gates set into tidal inlets between barrier islands separating the Venice Lagoon from the Adriatic Sea. The gates will be raised during extreme high tides and storm events. *(a: Skyscan Photolibrary/Alamy)*

(a)

Historic Venice

Venice Lagoon

Site of Project MOSES barrier gates

Adriatic Sea

Venice

ITALY

Venice Lagoon

Adriatic Sea

SEA BED

AIR

The gates are filled with seawater when closed.

Air is pumped into the gate to raise a barrier against extreme high tide or storm surge.

(b)

regions experiencing higher intensity storm events also need to prepare for increased flooding. Toward this end, the state of New Jersey has expanded its Blue Acres program, in which taxpayer funds are used to buy up flood-prone properties and convert them to open space, though the program is currently severely underfunded.[109] New Jersey and several other states are taking a hard look at restoring floodplains and redesigning bridges to cope with higher flood volumes. They're also investigating how building materials might need to change as warmer temperatures lead to increased buckling of pavements.[110]

In the southwestern United States, adaptations involve preparing for a future with far less water. The city of Chandler, Arizona, for example, offers rebates to citizens who convert their lawns to xeriscapes, landscapes that include drought tolerant native vegetation such as yucca, cactus, and desert willow.[111] Rebates are also offered for irrigation controllers to prevent accidental overwatering. The city of Peoria, Arizona, offers kits to convert high-volume showerheads in older homes to low-flow fixtures.[112] These kits also contain bags that can be filled with water and placed inside of toilet tanks. When the tank fills after flushing, less water enters it due to the displaced volume in the bag, leading to less water being used on the next flush. Strategies such as these have made southwestern cities such as Tucson, Arizona, and Santa Fe, New Mexico, among the most water efficient cities in the United States (Chapter 9 and Chapter 10).

International Agreements

In Chapter 11, we noted that the global community came together to solve the problem of stratospheric ozone depletion through an international agreement known as the Montreal Protocol. This agreement phased out the production of chlorofluorocarbon gases and has arrested the loss of Earth's protective shield against the Sun's ultraviolet rays. As with ozone, several United Nations–led efforts have attempted to address the problem of global warming. In 1992, the United Nations Conference on Environment and Development, commonly nicknamed the "Earth Summit," was held in Rio de Janeiro, Brazil, and involved 172 national governments.[113] This conference produced the United Nations Framework Convention on Climate Change, an international treaty designed to slow the growth in greenhouse gas emissions and to promote sustainable development. This was followed, in 1995, by a stronger agreement known as the Kyoto Protocol. This treaty required the world's major industrialized states to commit to binding emission targets by 2008–2012 relative to values in the year 1990 (Table 15-3).

The Kyoto Protocol set up market-based programs to assist nations in meeting their greenhouse gas targets. These include the creation of a carbon market in which countries with higher emissions pay to exceed their quotas, with the funds going to support alternative energy, reforestation, or increased efficiency programs in other nations.[100] To date, 190 nations have ratified the Kyoto Protocol, although the United States of America has not. One of the reasons the United States has balked has to do with the fact that the treaty does not include some of the most rapidly growing economies in the world, specifically China and India. At the time it was first negotiated, these nations had small greenhouse gas emissions and

TABLE 15–3 The Kyoto Protocol Commitments by Nation

Nation	Emissions target (% of 1990 emissions)
Australia	108
Austria	91
Belgium	92
Bulgaria	92
Canada	94
Croatia	92
Czech Republic	92
Denmark	92
Estonia	92
Finland	92
France	92
Germany	92
Greece	92
Hungary	94
Iceland	110
Ireland	92
Italy	92
Japan	94
Latvia	92
Liechtenstein	92
Lithuania	92
Luxembourg	92
Monaco	92
Netherlands	92
New Zealand	100
Norway	101
Poland	94
Portugal	92
Romania	92
Russian Federation	100
Slovakia	92
Slovenia	92
Spain	92
Sweden	92
Switzerland	92
Ukraine	100
United Kingdom	92
United States of America	93

Source: Modified from: http://unfccc.int/resource/docs/convkp/kpeng.pdf.

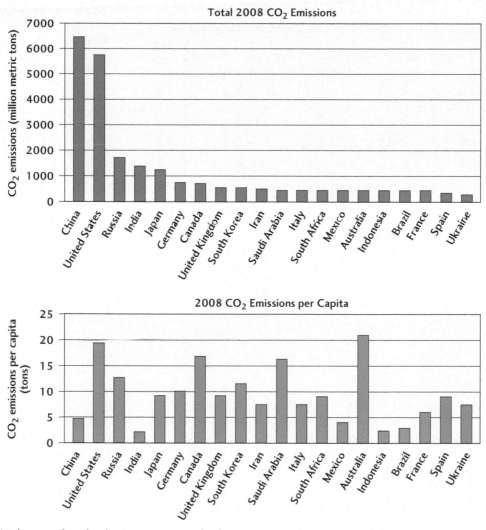

Figure 15–27 In the past decade, the increasing standard of living in China has led that nation to become the world's largest producer of carbon dioxide emissions. On a per capita basis, however, Americans and Australians far surpass the emissions of the average citizen of China, India, Brazil, and other rapidly developing economies.

hundreds of millions of citizens living in poverty. When we wrote the first edition of this book, for example, the average Chinese citizen used a bicycle for transportation. Economic growth has transformed the country, and rising affluent and middle classes have meant a demand for automobiles and electrification that have now increased the nation's greenhouse gas emissions. When total emissions are considered, China is the world's biggest producer of carbon dioxide. On a per capita basis, however, the United States still far surpasses China in emissions (Figure 15-27) and historically has contributed more to the greenhouse gas problem than any other nation. Indeed, if every person in the world lived the way Americans do, it would require more than four Earths to supply all the resources needed (Figure 15-28). The U.S. refusal to commit to binding greenhouse gas reductions therefore makes it appear selfish and obstructionist

during climate negotiations.[114] Still, the U.S. representatives to the climate conventions do have a point. Emissions must be brought down by all nations if we are to avoid extreme warming. Unfortunately, the United States has resisted attempts to require wealthy nations to financially assist poorer countries in developing alternative energy supplies to support rising living standards. Only time will tell whether politics and diplomacy will change this situation.

■ Greenhouse gas emissions can be reduced through a mixture of technological and social innovations that include reducing human population growth; ending deforestation; promoting the use of alternative energy as well as greater efficiency of transportation, lighting, appliances, and buildings; and developing carbon sequestration technology. Nations such as Germany

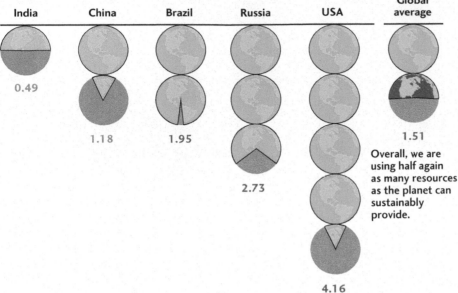

Resource Use by Society
How many Earths would we need if everyone lived like a resident of these countries?

India — 0.49
China — 1.18
Brazil — 1.95
Russia — 2.73
USA — 4.16
Global average — 1.51

Overall, we are using half again as many resources as the planet can sustainably provide.

Figure 15–28 Resource use by different societies varies markedly. If every person on Earth lived like the average American, we would need 4.16 Earths to provide all the necessary resources. If, on the other hand, every person lived like a citizen of India, we would only need half the current planet to provide for our needs.

and Denmark are leading the way on alternative energy production.

■ Geoengineering strategies seek to decrease the amount of solar energy absorbed by Earth's surface or to remove carbon dioxide from the atmosphere. Many of these proposed solutions are logistically or financially impractical or have environmental consequences of their own.

■ Adaptations to a warmer world include elevating homes in coastal areas to deal with sea level rise, constructing higher levees and flood walls, buying out property owners in locations that are frequently flooded, and increasing water use efficiency in arid areas.

■ The Kyoto Protocol is an international treaty that seeks to reduce greenhouse gas emissions of industrialized nations through the establishment of market-based incentives. The United States has refused to ratify the treaty because it does not place binding emissions reduction targets on China and India, the two most populous nations in the world and countries experiencing rapid economic growth, both of which are producing an increase of greenhouse gases.

CLOSING THOUGHTS

When the earliest hominids first began walking across Africa (Chapter 6), Earth's destiny changed. While populations were initially low enough to permit our ancestors to survive as subsistence foragers and hunters, continuing growth eventually necessitated the development of agriculture and permanent settlements. Technological advances kept pace with population growth, allowing us to produce more food per acre of cultivated land and to support increasing numbers of people. As the means of production became mechanized, fewer people were required to till the land, allowing human ingenuity to develop other goods and services that led to the modern consumer culture we presently live in. Along with this consumer culture have come vast changes to the soils, air, and water of our planet.

Currently, humanity moves more soil than all geologic processes combined,[115] fixes more nitrogen than all natural processes combined, and emits more greenhouse gases to the atmosphere than all natural processes combined. These factors, along with the wastes that we produce, are leaving an indelible mark upon the landscape. We are creating a new stratigraphic horizon that will be visible millions of years hence, much like the dinosaurs of the Mesozoic. Like them and most other species that have at one time inhabited Earth, we will probably disappear in the future. Whether that disappearance happens as a result of natural processes or due to our own hands, at present we live in an extraordinary time in Earth history and one that is worth attempting to extend.

Evolution of the human mind has led to incredible levels of scientific understanding, technological achievements, and artistic expressions. We have the capability to reduce our personal and collective impacts on the environment and to shift toward a more sustainable future. We remain hopeful that our species will accelerate its movement to live more lightly upon Earth. With your enhanced understanding of the grandeur of our planet, you can play a role in helping us to achieve this goal.

SUMMARY

■ As a result of population growth and increasing standards of living, human activities now outpace many natural processes both in spatial scale and in rates of change. For example, humans move more soil than the combined total of that transported by wind, rivers, glacial ice, and mass movements, and humans fix more nitrogen than does the rest of the biosphere.

■ Impacts of human activities have the potential to tip Earth into an environmental state different from the one in which civilization began, prompting some scientists to try to quantify thresholds within Earth's systems and to determine "safe" levels of human disturbance.

■ Many planetary thresholds are as yet unknown due to limited scientific study. Lags built into some systems mean that we may have already passed a threshold for sustainability, the full consequences of which will unfold in the future.

■ One of the most urgent problems facing Earth's species is the combination of climatic change and habitat fragmentation. Instrumental records combined with climatic proxy data indicate that Earth's surface is warming rapidly, and seed dispersal may not be able to keep up with coming temperature and precipitation changes, particularly in locations where natural ecosystems have been converted to agricultural lands or human settlements.

■ Natural sources of climatic variability such as plate tectonic movements, Earth's orbital cycles, and volcanic eruptions cannot explain the rapid warming of Earth's surface that has occurred since the Industrial Revolution began. Many lines of evidence point instead toward human combustion of fossil fuels, deforestation, and increases in agricultural and industrial processes that produce greenhouse gases.

■ Numerical modeling experiments show that increasing greenhouse gas emissions will lead to further warming of Earth's surface, melting of glaciers and ice sheets, sea level rise, and alterations of the hydrologic cycle. These changes have implications for human well-being in addition to ecosystem integrity.

■ The absolute magnitude of warming and its consequences for precipitation and evaporation depend on human behavior and are therefore difficult to predict. If humans continue to increase their consumption of fossil fuels, carbon dioxide levels in the atmosphere could climb to levels that have not been seen in over 30 million years. If, on the other hand, our species gets serious about investing in alternative forms of energy, carbon dioxide concentrations could stabilize.

■ Even if greenhouse gas emissions were to stabilize tomorrow, slow-acting feedbacks in the climate system, such as the ice albedo feedback, will lead to additional warming over the next few centuries. As a result, populations need to prepare for higher sea levels, greater frequencies of flooding and drought, and socioeconomic disruptions caused by severe weather, fires, and tropical disease transmission.

■ Catastrophic climate change remains avoidable if all nations work to reduce greenhouse gas emissions through a combination of technological and social changes that include reducing birth rates; increasing the efficiency of transportation, housing, and appliances; eliminating and reversing deforestation; and reducing consumption.

KEY TERMS

Anthropocene (p. 564)
planetary boundaries (p. 564)
tipping points (p. 566)
anthropogenic climate change (p. 570)

Intergovernmental Panel on Climate Change (p. 573)
radiative forcing (p. 576)
sea level rise (p. 583)

coral bleaching (p. 588)
ocean acidification (p. 589)
thermokarst lake (p. 590)
Paleocene–Eocene thermal maximum (p. 590)

geoengineering (p. 594)
xeriscapes (p. 598)
Kyoto Protocol (p. 598)
carbon market (p. 598)

REVIEW QUESTIONS

1. What factors have led many scientists to suggest that we are living in a new geologic era known as the Anthropocene?

2. What are the "planetary boundaries" identified by Johan Röckstrom and his colleagues and why are these boundaries important to humanity?

3. Why is it difficult to place absolute values on some planetary boundaries?

4. About which boundaries are we most uncertain and why?

5. Explain the concept of a tipping point.

6. What lines of evidence prove that Earth's surface has been warming over the last century and a half?

7. How do we know that this warming is the result of human activities rather than due to natural processes?

8. What might explain the fact that at the end of the last glaciation, Antarctic temperatures rose prior to increases in greenhouse gas levels whereas northern hemisphere and average global temperatures rose in response to rising greenhouse gas concentrations?

9. Why is it so difficult to predict what Earth's temperature will be in the future?

10. What are the likely consequences of rising surface temperatures for humanity?

11. What impacts will rising surface temperatures have on other species, and what compounding factors may exacerbate these impacts?

12. Why are climate scientists concerned about the impact of rising temperatures on Arctic permafrost?

13. What are some potential benefits of anthropogenic climate change?

14. What existing technologies can we use to reduce greenhouse gas emissions?

15. Describe a geoengineering strategy designed to reduce the amount of solar radiation striking Earth's surface. How feasible is this strategy? Does it have any unintended environmental consequences?

16. What kinds of adaptations will be required as climate continues to warm?

17. Why has the Kyoto Protocol thus far failed to garner the same success as the Montreal Protocol to limit stratospheric ozone destruction?

THOUGHT QUESTIONS

1. In Chapter 7, we learned that energy is lost from one trophic level to the next and as a result there are declining numbers of organisms up the food chain. One owl, for example, may need to consume 300 small rodents annually to survive, and those rodents in turn may require tens of thousands of seeds and nuts. Given trophic energy losses, what impact might human dietary choice have on Earth's planetary boundaries? If all people eschewed vegetables and became more carnivorous, how might greenhouse gas emissions, fertilizer use, and freshwater demand change? What if all people became vegans? How might you alter your own diet to minimize your impact on Earth's systems?

2. What responsibilities (if any) do the world's industrialized nations have to assist poorer countries in acquiring a higher standard of living while at the same time limiting their greenhouse gas emissions?

3. Take stock of your personal energy consumption. Do you leave your computer running all night even though you're asleep? Are the lights burning in your dorm room or home even when you're not there? Do you drive your car across campus to visit the library or do you walk? What lifestyle choices could you make to reduce your production of greenhouse gases?

EXERCISES

1. Presume that your nation has not yet ratified the Kyoto Protocol. Write a letter to a government official (e.g., congressional representative or senator in the United States) in which you outline the case for ratifying it. Include information describing the evidence for anthropogenic climate change and its likely impacts. Discuss the possible outcomes resulting from inaction and propose both technological and sociological solutions that will help your nation reduce its greenhouse gas emissions.

2. If you live in the United States, visit http://downloads.globalchange.gov/usimpacts/pdfs/climate-impacts-report.pdf to download the 2nd National Climate Assessment of the United States Global Change Research Program. Go to the Regional Climate Change Impacts section and read about the region where you live. What are the predicted temperature and precipitation changes coming to your area as greenhouse gases continue to rise? How are food production and tourism expected to change? How will human health be affected? What recommendations would you make to a regional planning board that is beginning to consider how to prepare for the future?

3. Because anthropogenic climate change is not only a scientific and technological problem but is also affected by socioeconomic and political factors, all people have a role to play in coming up with solutions. What methods and knowledge associated with your educational major or potential major could you use to increase awareness of the issue and/or to bring about change? Write down your ideas and consider how you might implement them. A music major, for example, might choose to write a short opera about global warming, whereas a chemistry major might work on designing a better battery to store solar and wind energy. What talents can you contribute to promoting a more sustainable lifestyle in our society?

4. Calculate the amount of carbon dioxide you produce each year by driving. To carry out this analysis, first estimate the number of miles you drive, then use the fuel efficiency of your car to determine the number of gallons of gasoline that you burn (e.g., 5000 miles divided by 25 miles per gallon = 200 gallons). Assuming that a gallon of gasoline weighs about 6 lbs and is about 84 percent carbon by weight, how many pounds of carbon do you burn each year? The atomic mass of carbon is 12 and of oxygen is 16. Take the number of pounds of carbon you burn each year and multiply by the ratio of the molecular mass of carbon dioxide to the atomic mass of carbon to determine the total number of pounds of carbon dioxide you produce by driving annually. How might you reduce your carbon dioxide emissions?

SUGGESTED READINGS

Anthony, K. W. "Methane: A Menace Surfaces." *Scientific American* (December, 2009): 68–75.

Bernstein, L., P. Bosch, O. Canziani, Z. Chen, R. Christ, O. Davidson, W. Hare, et al. *Climate Change 2007: Synthesis Report, Intergovernmental Panel on Climate Change*. Intergovernmental Panel on Climate Change, 2007.

Calderia, K. "The Great Climate Experiment: How Far Can We Push the Planet?" *Scientific American* (September, 2012): 78–83.

Greene, C. H. "The Winters of Our Discontent." *Scientific American* (December, 2012): 51–55.

Karl, T. R., J. M. Melillo, and T. C. Peterson, eds. *Global Climate Change Impacts in the United States*. Cambridge, UK: Cambridge University Press, 2009.

Kolbert, E. *Field Notes from a Catastrophe: Man, Nature, and Climate Change*. New York, NY: Bloomsbury USA, 2006.

Mann, M. E. *The Hockey Stick and the Climate Wars: Dispatches from the Front Lines*. New York, NY: Columbia University Press, 2012.

Musser, G. "The Climax of Humanity." *Scientific American* (September, 2005): 44–47.

Oreskes, N., and E. M. Conway. *Merchants of Doubt: How a Handful of Scientists Obscured the Truth on Issues from Tobacco Smoke to Global Warming.* New York, NY: Bloomsbury Press, 2011.

Rockström, J., W. Steffen, K. Noone, A. Persson, F. S. Chapin III, E. F. Lambin, T. M. Lenton, et al. "A Safe Operating Space for Humanity." *Nature* 461 (2009): 472–475.

Socolow, R. H., and S. W. Pacala. "A Plan to Keep Carbon in Check." *Scientific American* (September, 2006): 50–57.

Weart, S. R. *The Discovery of Global Warming,* 2nd ed. Cambridge, MA: Harvard University Press, 2008.

Zalasiewicz, J., M. Williams, W. Steffen, and P. Crutzen. "The New World of the Anthropocene." *Environmental Science and Technology* 44 (2010): 2228–2231.

Classification of Biological Organisms

To facilitate communication among scientists throughout the world, all living organisms are classified according to a system developed in the 18th century by Swedish botanist Carl von Linne, or Carolus Linnaeus, (1707–1778) and modified as biological knowledge increased. In the Linnaean classification, organisms are grouped into a hierarchy of sets and subsets, with species the narrowest of the subsets and the kingdom and domain the overarching set. Today, many of the species classifications that Linnaeus described have been revised.

A *species* is a population whose members are able to interbreed freely under natural conditions. Above species in the Linnaean hierarchy is the *genus* (plural *genera*), a group of species that are very similar and of more or less immediate common ancestry. Related genera make up *families,* related families make up *orders,* related orders make up *classes,* related classes make up *phyla,* and related phyla make up *kingdoms.* In general, the life-forms on Earth are divided into five kingdoms: animals, plants, fungi (molds, yeasts, and mushrooms), protists (amoebas, paramecia, and algae), and monerans (bacteria and related organisms). All of the kingdoms are classified under the higher order of *domain,* which are Bacteria, Archaea (nonbacterial prokaryotes that survive in some of Earth's most extreme climates), and Eukarya.

Life forms are generally referred to by genus and species names, which by convention are written in italics; the genus name begins with a capital letter. The genus and species designation for modern human beings is *Homo sapiens,* meaning "thinking man." Using ourselves as an example, we can trace our classification through increasing levels of detail.

Domain: Eukarya (organisms comprised of Eukaryotic cells, which are defined by a membrane-surrounded nucleus containing DNA)
Kingdom: Animals (Eukaryotic organisms with many cells that live by taking in complex organic molecules)
Phylum: Chordates (animals with a notochord, a rudimentary spinal column)
Class: Mammals (animals that secrete milk to nourish their young)
Order: Primates (apes, monkeys, and related animals)
Family: Hominids (primates that walk upright)
Genus: *Homo* (large-brained hominids)
Species: *sapiens* (modern humans)

Appendix 2

Periodic Table of Elements

Characteristics associated with different electron shell patterns are indicated on the table.

Noble gasses: outer shells filled; no tendency to gain or lose electrons

Strong tendency for outermost electrons to be lost to uncover full outer shell

Tendency to fill outer electron shell by electron sharing and gain or loss of electrons

Strong tendency to gain electrons to make full outer shell

Tendency to lose electrons from inner shells

1 H Hydrogen																	2 He Helium
3 Li Lithium	4 Be Beryllium											5 B Boron	6 C Carbon	7 N Nitrogen	8 O Oxygen	9 F Fluorine	10 Ne Neon
11 Na Sodium	12 Mg Magnesium											13 Al Aluminum	14 Si Silicon	15 P Phosphorus	16 S Sulfur	17 Cl Chlorine	18 Ar Argon
19 K Potassium	20 Ca Calcium	21 Sc Scandium	22 Ti Titanium	23 V Vanadium	24 Cr Chromium	25 Mn Manganese	26 Fe Iron	27 Co Cobalt	28 Ni Nickel	29 Cu Copper	30 Zn Zinc	31 Ga Gallium	32 Ge Germanium	33 As Arsenic	34 Se Selenium	35 Br Bromine	36 Kr Krypton
37 Rb Rubidium	38 Sr Strontium	39 Y Yttrium	40 Zr Zirconium	41 Nb Niobium	42 Mo Molybdenum	43 Tc Technetium	44 Ru Ruthenium	45 Rh Rhodium	46 Pd Palladium	47 Ag Silver	48 Cd Cadmium	49 In Indium	50 Sn Tin	51 Sb Antimony	52 Te Tellurium	53 I Iodine	54 Xe Xenon
55 Cs Cesium	56 Ba Barium	57 La Lanthanum	72 Hf Hafnium	73 Ta Tantalum	74 W Tungsten	75 Re Rhenium	76 Os Osmium	77 Ir Iridium	78 Pt Platinum	79 Au Gold	80 Hg Mercury	81 Tl Thallium	82 Pb Lead	83 Bi Bismuth	84 Po Polonium	85 At Astatine	86 Rn Radon
87 Fr Francium	88 Ra Radium	89 Ac Actinium	104 Rf Rutherfordium	105 Ha Hahnium	106 Sg Seaborgium	107 Bh Bohrium	108 Hs Hassium	109 Mt Meitnerium									

Rare elements

58 Ce Cerium	59 Pr Praseodymium	60 Nd Neodymium	61 Pm Promethium	62 Sm Samarium	63 Eu Europium	64 Gd Gadolinium	65 Tb Terbium	66 Dy Dysprosium	67 Ho Holmium	68 Er Erbium	69 Tm Thulium	70 Yb Ytterbium	71 Lu Lutetium
90 Th Thorium	91 Pa Protactinium	92 U Uranium	93 Np Neptunium	94 Pu Plutonium	95 Am Americium	96 Cm Curium	97 Bk Berkelium	98 Cf Californium	99 Es Einsteinium	100 Fm Fermium	101 Md Mendelevium	102 No Nobelium	103 Lr Lawrencium

Elements of major abundance in Earth's crust

Elements of lesser abundance but of importance to Earth's systems and environmental geosciences

1 ← Atomic number
H ← Chemical symbol
Hydrogen ← Element name

Units and Conversions

In the English system of measurement, which is used by most Americans, length is measured in inches, feet, and miles, and mass is measured in units of ounces, pounds, and tons. In the metric system, which is used by many other countries and the scientific community world-wide, length is measured in centimeters, meters, and kilometers, and mass is measured in grams, kilograms, and metric tons. Both systems use the same units for time: seconds, minutes, hours, and years.

The metric system is based on powers of 10. A meter is equal to 10 decimeters, or 100 centimeters, or 1000 millimeters; 10 meters is called a dekameter, 100 meters a hectometer, and 1000 meters a kilometer. Each of these numbers can be expressed as a power of 10: 10 is 10 to the first power, or 10^1; 100 is 10 to the second power, or 10^2 (10×10); and 1000 is 10 to the third power, or 10^3 ($10 \times 10 \times 10$). Numbers smaller than 1 are expressed as negative powers: one-hundredth of a meter, or 0.01m, is the same as 10^{-2}, or $1 \div (10 \times 10)$. The table below lists some metric numbers with the most commonly used prefixes and their abbreviations.

In this book, we generally use the metric system, because it is the standard for scientific work. However, a few special cases warrant the use of different measures. Particularly in the United States, climatic and hydrologic data typically are recorded in English units. Rainfall, for example, commonly is given in inches per year, and stream discharge in cubic feet per second. Rather than convert standard data recorded by government agencies, we have retained the English system in these instances. In addition, navigators commonly measure ocean depth in fathoms, and governments used this unit in establishing offshore territorial rights.

Metric Numbers

$10^{-9} = 0.000000001$	nano (n)	$10^2 = 100$	
$10^{-8} = 0.00000001$		$10^3 = 1000$	
$10^{-7} = 0.0000001$		$10^4 = 10,000$	
$10^{-6} = 0.000001$	micro (μ)	$10^5 = 100,000$	
$10^{-5} = 0.00001$		$10^6 = 1,000,000$	mega (M)
$10^{-4} = 0.0001$		$10^7 = 10,000,000$	
$10^{-3} = 0.001$	milli (m)	$10^8 = 100,000,000$	
$10^{-2} = 0.01$	centi (c)	$10^9 = 1,000,000,000$	giga (G)
$10^{-1} = 0.1$	deci (d)	$10^{10} = 10,000,000,000$	
$10^0 = 1$		$10^{11} = 100,000,000,000$	
$10^1 = 10$	deka (da)	$10^{12} = 1,000,000,000,000$	tera (T)

The charts below provide the factors needed to convert from one system of units to another in measuring length, area, volume, mass, pressure, energy, power, and temperature. Begin at the left: To convert a value expressed in inches to the equivalent in centimeters, for example, multiply by 2.54.

Conversion Charts

from	to	multiply by
Length		
centimeters	inches	0.3937
fathoms (1 fathom = 6 feet)	meters	1.8288
feet	meters	0.3048
inches	centimeters	2.5400
kilometers	miles (statute)	0.6214
kilometers	feet	3280.84
meters	feet	3.2808
meters	yards	1.0936
meters	inches	39.37
miles (statute)	kilometers	1.6093
miles (nautical)	kilometers	1.8531
yards	meters	0.9144
Area		
acres (U.S.; 1 acre = 43,560 square feet)	hectares	0.4047
hectares (1 hectare = 10,000 square meters)	acres	2.471
square centimeters	square inches	0.1550
square inches	square centimeters	6.4516
square feet	square meters	0.0929
square meters	square feet	10.764
square meters	square yards	1.1960
square kilometers	square miles	0.3861
square miles	square kilometers	2.590
Volume		
acre feet	cubic meters	1234
barrels of petroleum (1 barrel = 42 U.S. gallons)	cubic meters	0.159
cubic centimeters	cubic inches	0.06102
cubic feet	cubic meters	0.02832
cubic inches	cubic centimeters	16.3871
cubic meters (1 cubic meter = 1000 liters)	cubic feet	35.314
cubic meters	cubic yards	1.3079
cubic yards	cubic meters	0.7646
gallons (U.S.)	liters	3.7853
liters (1000 cubic centimeters)	quarts (U.S.)	1.0567
liters	gallons	0.2642
quarts	liters	0.9463

from	to	multiply by
Mass		
grams	ounces	0.03527
kilograms	pounds	2.20462
tonnes (metric tons) (1 metric ton = 1000 kilograms)	short tons	1.1023
ounces	grams	28.34952
pounds	kilograms	0.45359
tons (1 short ton = 2000 pounds)	kilograms	907.1848
tons (short)	tonnes (metric tons)	0.90718
Pressure		
bars (1 bar = 0.98692 atmosphere)	pascals	10^5
bars	newtons/square meter	10^5
kilograms/square centimeter	atmospheres	0.96784
kilograms/square centimeter	bars	0.98067
kilograms/square centimeter	pounds/square inch	14.2233
pascals (1 newton/square meter)	bars	10^{-5}
pounds/square inch	kilograms/square centimeter	0.70307
Energy (see Figure A-1)		
Btu (British thermal units)	ergs	1.054×10^{10}
Btu	joules	1.054×10^3
Btu	quads	10^{-5}
foot pounds	joules	1.356
ergs	calories (gram)	2.39006×10^{-8}
ergs	Btu	9.48451×10^{-11}
ergs	joules	10^{-7}
joules	ergs	10^7
joules	calories (gram)	0.2390
joules	Btu	9.484×10^{-4}
quads	Btu	10^{15}
quads	joules	1.05×10^{18}
Power		
ergs/second	watts	10^{-7}
horsepower (U.S.)	watts	7.4571×10^2
Btu/minute	watts	1.758×10^1
watts	ergs/second	10^7
watts	horsepower (U.S.)	0.001341
watts	Btu/minute	0.05688

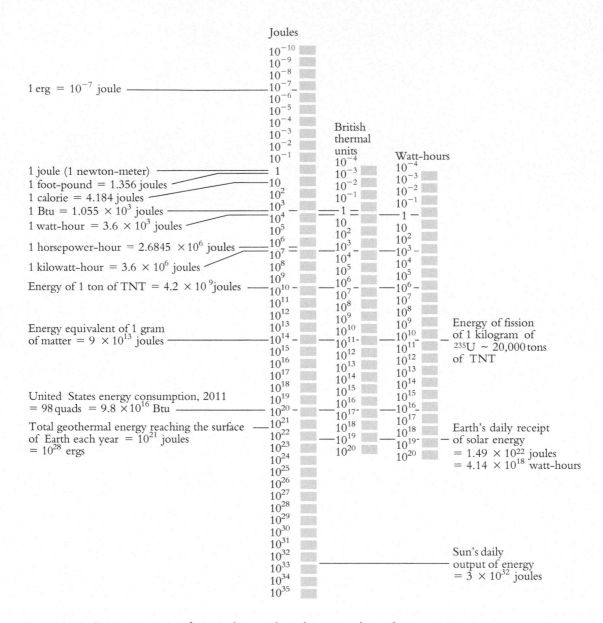

Figure A-1 Energy conversion from joules to selected, commonly used energy units.

Conversion Chart

from	to	compute
Temperature (see Figure A-2)		
Fahrenheit (°F)	Celsius (°C)	(°F − 32)/1.8
Celsius (°C)	Fahrenheit (°F)	(°C × 1.8) + 32

Measuring Energy

Energy is measured in many different units; the most common are joules, calories, British thermal units, and watts.

One joule (J) is defined as 1 kg-m²/sec². For a sense of what a joule is, consider how many joules of energy can be produced by burning 1 kg of carbon. Complete combustion releases 33 million J of energy, enough to drive an efficient subcompact car for at least 15 km.

Because global energy budgets and energy demands involve large amounts of energy, preferred units for these uses are the exajoule (10^{18} J), petajoule (10^{15} J), terajoule (10^{12} J), and gigajoule (10^{9} J). The world's annual energy consumption is currently more than 300×10^{18} J (300 exajoules). In comparison, a human being needs 3×10^{9} J of energy each year, supplied by food, to breathe, move, and function. People generally use much more energy than this, however, because they also drive cars, fly in airplanes, and use goods manufactured with the aid of energy resources. In the United States, per capita energy consumption based on all types of activities is 320×10^{9} J per year. The calorie is defined as the amount of energy necessary to raise the temperature of 1 g of liquid water 1°C. One calorie is equivalent to 4.184 J. The energy content of our food is measured in Calories (upper case C), each of which is equivalent to 1000 calories or 4184 J. A single candy bar contains about 400 Calories, or 400,000 calories, enough to raise the temperature of 400 kg of water by 1°C. The human body typically weighs 50–80 kg, and more than 70 percent of this weight is water. Normal body temperature is 37°C, so a great proportion of food intake is used to maintain the temperature of water at 37°C. To keep itself at a fairly constant temperature, the body burns an increased number of calories when the ambient temperature is lower than the body's temperature.

One of the most common units for measuring energy resources in the English-speaking world is the British thermal unit, or Btu. A Btu is the amount of energy required to raise the temperature of 1 pound of water 1°F from 39.2°F to 40.2°F. One barrel of oil, which contains 42 U.S. gallons, is equivalent to 5.8 million Btu. When discussing a large amount of an energy resource, such as that obtained from millions or billions of barrels of oil, the preferred unit of measure is the quad, which represents one quadrillion (10^{15}) Btu.

The amount of energy used over a given period of time is a measure of power, and a common unit for power is the watt. A watt is 1 J per second, and a terawatt is 10^{12} watts. Earth's annual energy inflow is more than 173,000 terawatts. In comparison, energy requirements for all human activities in a single year are a mere 10 terawatts. It is evident, therefore, that although supplies of nonrenewable energy resources might become exhausted, the world will not run out of energy.

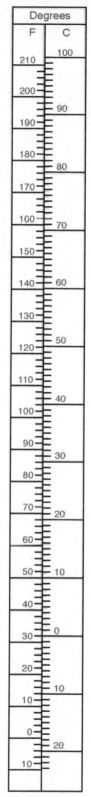

Figure A–2 Comparison of temperature equivalents of Fahrenheit and Celsius degrees.

Appendix 4

Properties of Common Minerals

Mineral	Composition	Chemical Classification	Hardness	Cleavage	Common Color(s)	Streak Color	Other Properties
Minerals with metallic luster (in general order of decreasing hardness)							
Pyrite	FeS_2	sulfide	6-6.5	none	pale brass yellow	greenish black, brownish black	cubic crystals common
Magnetite	Fe_3O_4	oxide	5.5-6.5	none	black	black	strongly attracted to a magnet
Hematite	Fe_2O_3	oxide	5.5-6.5	none	silver to dark gray	reddish brown	reddish rust very common
Uraninite	UO_2 to U_3O_8	oxide	5-6	none	black to dark brown	black to dark brown	dense; radioactive
Limonite	$Fe_2O_3 \cdot H_2O$	oxide (hydrous)	1-5.5	none	silver to golden-brown	yellowish brown	yellowish rust very common
Sphalerite	ZnS	sulfide	3.5-4	perfect in six directions	yellowish brown	brownish to light yellow	resinous luster; streak smells like rotten eggs
Chalcopyrite	$CuFeS_2$	sulfide	3.5-4	none	brass yellow	greenish black, brownish black	iridescent tarnish
Copper	Cu	native element	2.5-3	none	copper, commonly stained green	reddish copper	can be cut with a knife; dense
Galena	PbS	sulfide	2.5	perfect in three directions at right angles	silver gray	gray to black	cubic crystals common; dense
Gold	Au	native element	2.5	none	gold	golden yellow	malleable; can be flattened without breaking; dense
Graphite	C	native element	1	one perfect	pencil-lead gray	black	forms slippery flakes; feels greasy
Minerals with nonmetallic luster (in general order of decreasing hardness)							
Garnet	$Al_2(SiO_4)_3$ + other metallic elements	silicate	6.5-7.5	none	dark red to black	none	commonly 12-sided crystals; conchoidal fracture
Quartz	SiO_2	silicate	7	none	clear, milky, purple, rose, smoky	none	conchoidal fracture; crystal faces common
Olivine	$(Fe,Mg)_2SiO_4$	silicate	6.5-7	none	green	none	conchoidal fracture
Potassium feldspar (orthoclase)	$KAlSi_3O_8$	silicate	6-6.5	two perfect, at right angles	white to pink	none	prism-shaped crystals common
Plagioclase feldspar	$NaAlSi_3O_8$	silicate	6-6.5	two perfect, nearly at right angles	white to dark gray	none	fine, straight striations
Amphibole group (hornblende is most common member)	complex Ca, Na, Mg, Fe, Al silicates	silicate	6	two, intersecting at 56° and 124°	dark green to black or brown	white to pale green	long, six-sided crystals, irregular grains, and fibers
Pyroxene group	$XYSi_2O_6$ (X,Y = Ca, Mg, and Fe)	silicate	6	two perfect, nearly at right angles	dark green to black	white to pale green	eight-sided stubby crystals and granular masses
Fluorite	CaF_2	halide	4	perfect in four directions	highly variable	none or white	glassy; cubic crystals
Malachite	$CuCO_3 \cdot Cu(OH)_2$	carbonate	3.5-4	one perfect, but rarely seen	green	green	copper ore mineral

Mineral	Composition	Chemical Classification	Hardness	Cleavage	Common Color(s)	Streak Color	Other Properties
Minerals with metallic luster (in general order of decreasing hardness)							
Azurite	$2\,CuCO_3 \bullet Cu(OH)_2$	carbonate	3.5-4	perfect in two directions	blue	blue	copper ore mineral
Dolomite	$CaMg(CO_3)_2$	carbonate	3.5-4	perfect in three directions at oblique angles	clear, yellow, gray, or pink	none	glassy; crystals with rhomb-shaped faces
Serpentine	$H_4Mg_3Si_2O_9$	silicate	2.5-4	one perfect	light to blackish green	white	splintered or layered appearance; greasy feel
Aragonite	$CaCO_3$	carbonate	3.5	poor in two directions	clear or white	none or white	massive or slender, needlelike crystals; effervesces with dilute hydrochloric acid
Calcite	$CaCO_3$	carbonate	3	perfect in three directions at oblique angles	clear, yellow, white	none or white	glassy; crystals with rhomb-shaped faces; effervesces with dilute hydrochloric acid
Biotite	$K(Mg,Fe)_3AlSi_3O_{10}(OH)_2$	silicate	2.5-3	perfect in one direction	brown to black	black	transparent in thin sheets; flakes are elastic
Muscovite	$KAl_2(AlSi_3O_{10})(OH)_2$	silicate	2-3	perfect in one direction	clear to yellow-brown	white	transparent in thin sheets; flakes are elastic
Bauxite	$Al_2O_3 \bullet 2\,H_2O$	oxide (hydrous)	2.5	none	white, tan, red, brown, black	tan to brownish red	pebblelike character; earthy appearance
Halite	$NaCl$	chloride	2.5	three perfect, at right angles	clear to dark gray	none or white	glassy; cubic crystals; salty taste
Sulfur	S	native element	1.5-2.5	none	yellow to red	pale yellow	conchoidal fracture; brittle
Kaolinite	$Al_2Si_2O_5(OH)_4$	silicate	2-2.5	one perfect	white, yellow, pink, reddish	white	chalky; soft, earthy masses
Gypsum	$CaSO_4 \bullet 2\,H_2O$	sulfate	2	one perfect	clear to yellow	white	glassy, transparent plates or satin-white rods
Talc	$Mg_3Si_4O_{10}(OH)_2$	silicate	1-1.5	one perfect	silvery to greenish white	white	layered appearance; greasy feel

A'a The Hawaiian name for a fast-moving, high viscosity basaltic lava flow that develops a rough surface texture characterized by broken fragments of chilled material, called clinkers.

Absolute age The numerical age of a rock, fossil, or geologic feature. Compare *Relative age.*

Abyssal plain A flat or nearly flat plain on the deep-ocean floor.

Acid mine drainage Acidic runoff from a mine or mine dump.

Acid rain Precipitation with a pH less than 5.0, usually caused by human activities that introduce significant amounts of acids into the local hydrologic cycle.

Active margin A coastline associated with a convergent margin where an oceanic tectonic plate is subducting beneath an oceanic or a continental plate.

Adiabatic cooling The decline in temperature of parcel of air when it is expands in volume as a result of a decrease in confining pressure.

Aerosols Extremely fine particles or droplets carried in atmospheric suspension.

Aftershock One of the smaller, residual shocks that frequently occur soon after the main earthquake.

Air mass A parcel of air of homogeneous temperature, humidity, and density that moves as an entity.

Albedo A measure of the amount of incoming solar energy reflected by a substance. A perfectly reflective substance has an albedo value of 1, whereas a substance that absorbs all solar energy incident upon it has an albedo of 0.

Alluvium Sediment deposited by a stream.

Angle of repose The steepest angle of slope that a sediment can maintain without becoming unstable.

Anoxia A complete lack of oxygen in water.

Anthracite coal The highest-grade, metamorphic?form of coal; it is black, harder and shinier than the sedimentary forms, and contains more than 90 percent carbon.

Anthropocene A term coined in the last few decades to suggest that Earth has entered a new geological epoch dominated by human activities.

Anthropogenic climate change Climate change brought about by the activities of humans, including fossil fuel combustion, deforestation, and a variety of agricultural practices.

Anticlinal trap A petroleum trap formed when layers of rock of varying porosity are folded into an arch and petroleum accumulates within the arch.

Apparent polar wander The apparent movement of the poles. The magnetization directions of volcanic rocks of different ages erupted on the same continent point toward different magnetic poles, making it appear that the north pole has wandered over time when in fact the continents have migrated across Earth's surface while the pole has remained largely fixed in position.

Aquifer A permeable body of rock or sediment that stores and transmits enough water to supply wells or springs.

Arête A thin spine of rock separating two glaciated valleys.

Artesian spring A natural flow of groundwater at Earth's surface that occurs when water under high pressure in a confined aquifer is able to reach the surface, for example along a fault.

Artesian well A well that taps water under high pressure in a confined aquifer and needs little or no pumping.

Ashfall deposits Layers of volcanic ash deposited downwind of an erupting volcano.

Asthenosphere The part of Earth that extends for about 250 km below the lithosphere and which deforms in a plastic manner.

Atmosphere The envelope of gases that surrounds Earth's surface.

Autotrophs (or primary producers) Organisms (commonly plants) capable of producing their own food from inorganic substances through the absorption of energy from sunlight (photosynthesis) or chemical reactions (chemosynthesis).

Balancing feedback Change to a system that leads to further events that reverse the direction of the original change.

Barrier island A long, narrow island nearby and parallel to the shore, separated from the mainland by either open water or salt marsh.

Base flow The inflow from groundwater that determines the average water level in a stream channel or river.

Base level The level below which a stream cannot continue to erode a channel; usually sea level.

Bathymetry The measurement of ocean depth, for the purpose of mapping seafloor topography.

Best management practices Practices designed to prevent soil erosion, reduce runoff, and limit the use of pesticides and fertilizers.

Biochar A product of pyrolysis that is more durable than, and contains more than double the carbon of, ordinary biomass.

Biodiversity The multiplicity of life-forms on Earth.

Biofuel A liquid or gaseous alternative to fossil fuels that is produced from biomass.

Biogeochemical cycle The movement of an element (commonly a nutrient element such as phosphorus, carbon, or nitrogen) through Earth's different systems via biological, chemical, and physical processes.

Biological amplification The accumulation and concentration of chemical compounds at each succeeding level in the food chain; also referred to as biomagnification.

Biological sediments Sedimentary matter derived from carcasses of organisms.

Biomass The quantity of living matter in a certain area; also, plant and animal matter and their wastes (e.g., dung), which can be used as a fuel source.

Biomes Major biological communities determined by climate, geology, and topography. Primary terrestrial biomes include forests, grasslands, deserts, and tundra.

Bioremediation The use of microbes to clean up oil spills and contaminated soils and aquifers.

Biosphere The part of Earth's exterior shell that includes living and nonliving organic matter.

Bioturbation The disintegration and mixing of organic and mineral matter and aeration of soil, which often intensifies physical weathering.

Bitumen A type of viscous, unconventional petroleum deposit that cannot be recovered through commercial drilling; commonly referred to as tar.

Bituminous coal A black sedimentary rock that is the intermediate grade of coal and contains 60 to 90 percent carbon.

Bituminous sands A form of unconventional petroleum located within geologic reservoirs of sand, water, and clay that is recovered through open pit and in situ mining techniques.

Body wave A seismic wave that travels through Earth's interior, e.g., P- and S-waves.

Bombs Fragments of lava greater than 64 mm in diameter ejected during a volcanic eruption that cool to form rocks that rain out around the volcanic vent. Travel through the air often results in streamlined forms.

Bowen's reaction series Formation of a series of minerals during magma cooling as a result of fractional crystallization; named for Norman Bowen, the scientist who first identified the phenomenon.

Braided stream A high-energy stream carrying such a large load of sediment that the water forms many small interweaving channels.

Caldera A large crater (more than 1.5 km in diameter) resulting from the collapse of a volcano into the emptied magma chamber.

Carbon market A financial market designed to reduce greenhouse gas emissions over time by providing incentives to nations or polluting industries to increase energy efficiency, develop alternative energy resources, and plant trees on deforested land, among other activities.

Carbon sequestration A geoengineering strategy in which carbon is deposited in a reservoir to reduce the amount of carbon dioxide in the atmosphere.

Carbonation The chemical weathering process that occurs when carbonic acid dissolves minerals and rocks, particularly calcite and limestone.

Carrying capacity The population of a species that can be supported with food, water, and other necessities by a given area of land.

Catalytic converter A system in the engine of a car that turns smog-producing waste into relatively benign gases.

Channelization The replacement of natural streambanks with artificial ditches, usually in order to straighten and deepen the natural channel and protect the surrounding land against floods.

Chemical sediments Sediments that originate from chemical weathering processes that partially or wholly dissolve minerals.

Chemical weathering The dissolving or altering of minerals to other forms as a result of chemical reactions that take place in rocks exposed to water and the atmosphere. Compare *Physical weathering*.

Chlorofluorocarbons (CFCs) A group of organic molecules that are able to release chlorine atoms, which then destroy ozone molecules in the stratosphere.

Cinder (or scoria) cones Small volcanoes typically made of basaltic tephra. Slopes are typically steep, reflecting a high angle of repose for the angular rock fragments that make up the cinders.

Cirques Bowl-shaped depressions found at the tops of valley glaciers and carved by glacial erosion.

Clastic sedimentary rocks Particles deposited physically, by wind, running water, or ice.

Clay minerals Hydrous aluminum silicate minerals with a layered structure, such as kaolinite.

Clean Air Act A federal law regulating the emission of pollutants into both indoor and outdoor air.

Clean Water Act (CWA) A federal law regulating pollutant levels and discharge of pollutants within waters of the United States.

Cleavage The property by which minerals break at certain regular planes of weakness along the crystal lattice.

Climate The long-term atmospheric and surface conditions that characterize a particular region. Compare *Weather*.

Climate reconstruction An experiment carried out with a global climate model/general circulation model in which past topography, makeup of the land surface, ocean bathymetry, greenhouse gas levels, and incoming solar radiation receipts are prescribed. The model is run to equilibrium, and maps of climatic variables such as surface temperature, wind speed and direction, and precipitation are made for comparison to data derived from proxies.

Closed system A system that can exchange energy but not matter across its boundaries. Compare *Isolated system; Open system*.

Coalification The process by which decay transforms organic matter into coal during burial under conditions of increasing temperature and pressure.

Cone of depression The cone-shaped dip in the water table that forms around a well when the amount of water being removed from the well exceeds the flow of water into the well.

Confined aquifer An aquifer in which the water is held under pressure between strata that are impermeable or have very low permeability.

Conservation tillage The practice of leaving some crop residue on the soil surface with the purpose of minimizing soil erosion and retaining soil nutrients.

Constructional shoreline A shoreline made up of tufa, sedimentary rock formed from freshwater deposits where springs empty into the margins of a saline lake.

Continental drift The horizontal movements of the continents relative to each other across Earth's surface.

Continental rise A sediment-covered region of the seafloor that rises from the abyssal plain to meet the continental slope.

Continental shelf The submerged area of a continent that lies between the coast and the upper continental slope.

Continental slope The portion of the seafloor that slopes steeply away from the continental shelf and ends in the continental rise.

Continent–continent collision The collision of two tectonic plates consisting of continental lithosphere. The low density of each plate relative to the underlying asthenosphere prevents either plate from subducting. Consequently, continent–continent collisions result in folding and reverse faulting of the crust and produce high mountain ranges.

Convection The mechanism of heat distribution in a liquid or plastic solid that is heated from the bottom; hot material from the bottom rises and cooler surface material sinks, producing circulation.

Conventional oil and natural gas Forms of petroleum extracted from high-permeability reservoirs through a drill pipe, occurring as a flammable liquid or in the gaseous state. Compare *Unconventional oil and natural gas*.

Convergent boundary A boundary where lithospheric plates collide, resulting in either subduction or continental collision with crustal thickening.

Coral bleaching A phenomenon in which corals expel symbiotic algae (called *zooxanthelle*) from their tissues, which reduces their food supply and leads to increased mortality. The remaining coral skeleton has a white color, making the coral appear to have been bleached.

Core Earth's center, composed mainly of iron and nickel and differentiated into two concentric regions: an outer layer kept liquid by extreme temperatures and pressures and an inner core that even higher pressures make solid.

Coriolis effect The force resulting from Earth's rotation about its axis that causes moving objects in the northern hemisphere to be deflected to the right of their direction of motion and moving objects in the southern hemisphere to be deflected to the left.

Crater A steep-sided depression located at the summit of a volcano and produced by a combination of explosive removal of the mountaintop during eruption and collapse due to downward draining of magma into the emptied magma chamber at the end of an eruption.

Creep The slow movement of soil and rock downhill under the influence of gravity.

Critical minerals Minerals assessed as having unique properties for performing essential functions for human activity, of limited availability in space and time, and having no satisfactory substitute.

Critical zone The outermost zone at Earth's surface, where rocks, soil, water, gases, and living organisms interact in complex ways to regulate habitat and determine the availability of life-sustaining resources.

Critically endangered A designation for species that occupy less than 100 km^2 of space on Earth, have a population less than 250 individuals, and/or a probability of going extinct in the wild of at least 50 percent within 10 years or three generations.

Cross beds The inclined planes of deposition in a sediment or sedimentary rock formed by the action of wind or water.

Crude oil The liquid form of petroleum.

Crust Earth's outermost solid layer, made up mostly of basalt in the oceanic crust and granite in the continental crust.

Crystallization The formation of crystals from a gas or liquid.

Curie point The temperature below which magnetic minerals lock in a permanent magnetization direction. For example, as a basaltic lava flow cools, the mineral magnetite it contains acquires a magnetization parallel to Earth's magnetic field at the site of the eruption once the lava flow cools below its Curie point temperature of 570°C.

Cyanobacteria Microbes that use oxygenic photosynthesis to generate their own food; among the earliest forms of life on Earth.

Cyclothems Repeated cycles of sedimentary rock layers that alternate between marine and nonmarine deposits and typically include shale, limestone, and coal.

Darcy's law A formula used to quantify the flow of fluids through a porous medium, such as an aquifer; according to this law, the rate of groundwater flow is proportional to the hydraulic gradient of the fluid and the hydraulic conductivity of the porous medium.

Daughter product An atom formed as the result of the radioactive decay of an isotope that is the parent element.

Dead zone A part of the ocean with hypoxic or anoxic conditions, resulting in the death of sessile organisms and out-migration of mobile organisms. Dead zones are typically the byproducts of eutrophication.

Deep ocean The parts of oceans that reach depths of at least 1 km below the ocean surface.

Demographic uncertainty Uncertainty regarding the ability of a population to reproduce itself based on reproductive characteristics including age and sex ratios. Compare *Environmental uncertainty; Genetic uncertainty.*

Depositional environment The environment in which sediments accumulate and often become sedimentary rocks.

Desertification The transformation of land to deserts by a number of processes, including loss of vegetation, reduced soil fertility, soil erosion, and the trampling of soil by animals.

Destructional shoreline A shoreline produced when waves strike the same place repeatedly, producing a wave-cut bench.

Detritivores Animals that consume dead organic matter such as fallen leaves and twigs.

Digital elevation models (DEMs) Models of Earth's surface that are generated with coordinate data from technology such as stereophotogrammetry and lidar, making it possible to quantify and evaluate landscape change over time.

Discharge (groundwater) The movement of groundwater to water bodies at the surface.

Discharge (stream) The rate of water flow through a stream, measured in volume per unit time.

Divergent boundary A boundary between two lithospheric plates that are moving apart, allowing magma to rise up between them and form new crust, usually along a mid-ocean ridge.

Downwelling The downward movement of water from the surface toward the deep ocean in response to Ekman transport of surface water. This process delivers dissolved oxygen to the deep ocean. Compare *Upwelling.*

Drainage basin An area of land that contributes water to a stream or lake.

Drainage divide The boundary between two drainage basins that have separate stream systems.

Drainage network The network of tributaries, large and small, belonging to a stream system.

Dynamic system A system in which energy is used to do work that causes the condition, or state, of the system to change with time.

Early wood Low density, large, spongy cells added to a tree's trunk and limbs during the growing season. Compare *Late wood.*

Earth system science The scientific study of the interconnected whole-Earth systems, such as the lithosphere, hydrosphere, and atmosphere, the cycling of matter and energy through them, and the changes that occur in them with time.

Earthquake cycle The repeated buildup and release of elastic energy along a fault due to gradual movements of Earth's crust on opposite sides of the fault.

Eccentricity The degree to which a planet's orbit deviates from a circle.

Ecological niche The way an organism lives in its environment, including its response to climatic conditions, the type of food it eats and how it acquires that food, its reproductive behavior, and its interactions with other organisms.

Ecological restoration The process of returning a degraded ecosystem to a close approximation of its condition prior to disturbance and impairment.

Ecological succession Replacement of one community of organisms with another over time due to changes in light levels and microclimate brought about by the organisms themselves.

Economic concentration factor The concentration factor necessary to make recovery of a mineral economically feasible.

Ecosystem An ecological unit comprising the resident organisms and the environment they inhabit.

Ecosystem services Ecological processes and resources that provide benefits to humanity.

Ekman transport The movement of water to the left of the wind direction in the southern hemisphere and to the right in the northern hemisphere as a result of the Coriolis effect.

El Niño Southern Oscillation A coupled oceanic–atmospheric phenomenon characterized by a reduction in the atmospheric pressure contrast between the western and eastern tropical Pacific Ocean, a slowing of the trade winds, and an eastward migration of warm water toward the west coast of South America (El Niño) or the reverse of all of these conditions (La Niña), with global consequences for precipitation patterns and more local impacts on South American fisheries.

Elastic rebound model The scientific theory that sudden movements of rock along a fault and the earthquakes associated with them result from the release of elastic energy along the fault.

Endangered A designation for species that occupy less than 5000 km^2 of space on Earth, have a population less than 2500 individuals, and/or a probability of going extinct in the wild of at least 20 percent within 20 years or five generations.

Endemic Restricted to a particular geographic location.

Endemic species Species that are restricted to a particular geographic location.

Endocrine disruptor Synthetic chemicals that can mimic or block hormones and disrupt normal functions when absorbed into the body.

Energy The ability to do work.

Energy efficiency The percentage of energy not lost as waste heat.

Environmental lapse rate The decline in temperature with elevation in the troposphere caused by adiabatic cooling.

Environmental uncertainty Uncertainty regarding the ability of a population to reproduce itself based on external factors such as unpredictable weather events, natural catastrophes or disease. Compare *Demographic uncertainty; Genetic uncertainty.*

Eon The largest unit of geologic time, of which three are formally recognized: Archean, Proterozoic, and Phanerozoic.

Epicenter The point on Earth's surface directly above the hypocenter, or focus, of an earthquake.

Epoch A subdivision of a geologic period, which can be chosen to correspond to a stratigraphic sequence.

Era A division of geologic time intermediate in length between a period and an eon.

Erosion The transport of rock and sediment over Earth's surface, largely by gravity, water, wind, and ice.

Eskers Long, winding ridges of sediment deposited in sub-glacial meltwater tunnels.

Estuary A partially enclosed body of water along the coast, where freshwater and saltwater mix.

Eutrophication The process that begins when an extreme influx of nutrients (often from sewage and agricultural runoff) into a body of water encourages unusually large algae blooms, which prevent light from penetrating the water and thereby inhibit subsurface photosynthesis and deplete the water of oxygen.

Evaporite A sedimentary rock that originates as a precipitate from water as it evaporates from a closed basin.

Exclusive Economic Zone (EEZ) The area extending 200 nautical miles (322 km) seaward from a country's shoreline, within which that country has exclusive mineral and fishing rights.

Exfoliation joints Fractures in rock that are parallel or sub-parallel to Earth's surface and divide rock into multiple thin slabs, or sheets. Also called sheeting joints.

Exponential growth The repeated multiplication of a quantity by a given exponent; this type of growth is approximated in populations, in which doublings of the population occur in such quick succession that the total number of organisms competing for resources increases rapidly.

Extinct The condition of a species when every last member of that species has died out, thereby ending that species' existence on Earth.

Extirpated The condition of a species when it is lost from a particular geographic location where it formerly existed.

Extrusive igneous rock A fine-grained igneous rock that has cooled from lava ejected onto Earth's surface.

Faint young sun problem A geological paradox resulting from the realizations that our sun put out much less solar radiation early in its history, which should have led to Earth's being frozen, and that there is evidence for liquid water on Earth at the same time. Resolution of the paradox requires that Earth must have had much higher levels of greenhouse gases early in its history that could have kept it warm during the period of a weaker sun.

Fault trap A petroleum trap in which a fault has juxtaposed permeable and impermeable rocks so that petroleum accumulates in the permeable rocks.

Faunal and floral succession, principle of The assumption that fossil assemblages of species of plants and animals succeed each other over time; the basis for relative age dating.

Federal Land Policy Management Act A federal law regulating the management of public lands, particularly in regard to the requirement of a permit and an environmental assessment for any activity that will disturb the ground surface, as well as the reclamation of disturbed land after mining.

Felsic magmas A term describing a light-colored igneous rock, lava, or magma with high silica content and low iron and magnesium content.

Ferrel cell A zone of transition between the warm, tropical Hadley cell and the cold, polar cell in which surface winds generally flow from west to east (the westerlies) but are also influenced by seasonal differences in land and ocean water temperatures.

Fissile The ability to sustain a chain reaction of nuclear fission, such as with the isotope uranium 235.

Fissure eruption Eruptions of basaltic lava along a line of weakness in Earth's crust several hundred meters to kilometers in length.

Flood basalt Large-scale fissure eruptions thought to be associated with hot spot volcanism. Aerial extents exceed tens of thousands of square kilometers, with individual flows up to 50 m thick and total thickness of accumulated flows in the thousands of meters.

Flood frequency analysis Prediction of the frequency with which floods of a particular magnitude tend to occur.

Floodplain A nearly planar landform composed of sediment that lies on either side of a stream and is commonly underwater during floods.

Flow A basic mechanism of mass movement in which debris moves similarly to a fluid and no clear plane of failure occurs within or below the moving mass.

Flowback Fluid containing water, salt, and other naturally occurring elements that have returned to the surface after water is injected into the ground during hydraulic fracturing.

Flux The movement of material and energy from one reservoir to another.

Foliation A set of planes, produced by deformation, in a metamorphic rock; the planes can be either flat or wavy.

Food chains A collection of organisms related to one another through food consumption. All food chains begin with autotrophs, which are consumed by herbivores. These herbivores in turn may be consumed by omnivores or carnivores, which may themselves be consumed by other omnivores or carnivores.

Foraminifera Surface and deep-sea protozoa with calcium carbonate shells.

Forecast An imprecise statement about the timing, location, and extent of a geological hazard event such as an earthquake or volcanic eruption. For example, based on past events, geologists can say with certainty that there will be future eruptions of volcanoes in the Cascade Range of the United States, but cannot say which volcano will erupt next, nor when, nor how extensive the eruption is likely to be.

Foreshock A smaller earthquake that precedes a larger tremor and that often changes the state of stress along the fault plane, thereby setting the conditions for the larger event. When they occur, foreshocks are located in the same general area and happen minutes to weeks before the main event.

Formation A set of sedimentary rocks with similar characteristics that covers a considerable area.

Fossil The naturally preserved remains, imprint, or cast of a plant or an animal.

Fossil fuels Combustible deposits of altered organic material, such as coal, crude oil, and natural gas.

Front The boundary between two air masses.

General Mining Act of 1872 A federal law providing regulations for acquiring and protecting mining claims on public lands.

Genes Segments of DNA holding genetic instructions for organisms.

Genetic diversity The genetic variability among individuals in a population of a single species.

Genetic uncertainty Uncertainty regarding the ability of a population to reproduce itself based on low levels of genetic diversity, which can promote the propagation of life-threatening genetic defects. *Compare Demographic uncertainty; Environmental uncertainty.*

Geoengineering Attempts to mitigate the impacts of rising greenhouse gas levels on Earth's climate through large-scale projects such as injecting sulfate aerosols into the atmosphere to block incoming solar radiation.

Geologic column A composite stratigraphic section that shows the subdivisions of all or part of geologic time.

Geologic hazard A natural phenomenon, process, or event with the potential for negative effects on human life, property, or the environment.

Geologic timescale The subdivision of geologic time into eras, periods, and epochs, determined by means of stratigraphy, paleontology, and radiometric dating.

Geosphere The solid part of the Earth system, including Earth's rocky crust, mantle, and metallic core.

Geothermal energy Energy obtained from the heat contained in the volcanic regions of Earth's crust.

Glacial erratic Sediment transported by ice as much as hundreds, and sometimes thousands, of kilometers from its source region.

Glass A noncrystalline solid lacking a regularly repeating array of elements.

Global climate models/general circulation models (GCMs) Numerical models of Earth's climate system designed to run on supercomputers and incorporating fundamental physical principles such as the conservation of mass and energy. These models are used to study the sensitivity of Earth's climate to variations in greenhouse gas levels, changes in plate tectonic configurations, and other parameters as well as to simulate past and future conditions.

Greenhouse effect The warming of Earth's surface by heat reradiated from carbon dioxide and other heat-absorbing gases in the atmosphere.

Groin A structure built perpendicular to a beach to catch sand transported alongshore.

Groundwater Underground water below the level at which all open spaces in rock and sediment are saturated. See also *Water table.*

Gully A small stream channel several to tens of meters deep, tens to hundreds of meters wide, and up to several kilometers long; typically formed by rapid incision into sediments associated with deforestation or the erosion of soil.

Gyres Clockwise (northern hemisphere) and counter-clockwise (southern hemisphere) circulation cells within the surface ocean produced by the drag of the winds on the ocean surface and the continental boundaries of the ocean basins.

Gyttja A black, gel-like, organic substance formed during early stages of the coalification process, typically at depths of 6–10 m.

Habitat The place where a particular organism lives, as determined by its food, water, and shelter requirements.

Hadley cell A circulation loop in Earth's atmosphere caused by the Coriolis effect.

Half-life The time needed for half the number of atoms of a given radioactive isotope to decay.

Halocline The change in salinity with depth between the surface mixed layer and deep ocean.

Harmonic tremor Seismic activity associated with the movement of magma that often occurs prior to volcanic eruption.

Heat A term describing the transfer of thermal energy from one body to another.

Heave A type of mass movement in which material on a hillslope is alternately raised and lowered in response to alternating expansion and contraction, typically caused by repeated wetting and drying or freezing and thawing.

Heterotrophs (or **consumers**) Animals that must eat plants or other animals to acquire the energy they need to maintain themselves, grow, and reproduce.

Horns Spires of rock in glaciated terrain produced by cirque erosion from valley glaciers that radiate outward from a mountain peak.

Hot spot A spot on a lithospheric plate directly over a mantle plume, characterized by volcanic activity.

Hothouse climates Warm climate conditions globally relative to average climatic conditions over geologic time. Compare *Icehouse climates.*

Humus Decayed organic matter found in a soil.

Hurricane A large storm with winds that blow at speeds greater than 120 km per hour.

Hydraulic conductivity The volume of groundwater that moves through a unit of area perpendicular to the direction of flow in a unit of time; also referred to as hydraulic permeability.

Hydraulic gradient The change in water pressure over horizontal distance in an aquifer.

Hydrocarbon An organic compound containing only hydrogen and carbon; hydrocarbons may be solid, liquid, or gas, and are major components of coal, crude oil, and natural gas.

Hydrograph A graph that displays the amount of water flowing past a particular location in a drainage network as a function of time, with the amount of water expressed as either stage (relative height with respect to a datum) or volume (i.e., discharge).

Hydrologic budget equation The statement that inflow of water into a reservoir or system is equal to outflow plus or minus changes in stock; for the surface water system, inflow is precipitation and outflow is evapotranspiration and runoff. The hydrologic budget is the balance between rates of precipitation, evapotranspiration, surface runoff, and groundwater recharge.

Hydrologic cycle The path traveled by water as it moves from the ocean to the atmosphere, back to Earth's surface as precipitation, and returns to the sea through streams.

Hydrolysis A chemical weathering reaction between a silicate mineral and water, usually producing clay minerals.

Hydrosphere The shell of water around Earth's surface, made up of oceans, surface water, groundwater, water vapor, and glaciers.

Hydrothermal ore deposits Minerals deposited or altered by the circulation of hot groundwater.

Hypocenter (or **focus**) The point within Earth's crust where an earthquake occurs.

Hypothesis An explanation of observations based on physical principles.

Hypoxic Having a dissolved oxygen concentration in water below 2 mg/L.

Ice-albedo feedback A reinforcing feedback in which ice growth leads to cooler atmospheric temperatures as less solar energy is absorbed, which drives still more ice growth and even higher albedo values that yield yet cooler atmospheric temperatures, and so on.

Icehouse climates Cool climate conditions globally relative to average climatic conditions over geologic time. Compare *Hothouse climates.*

Igneous ore deposits Deposits of ore minerals within igneous rocks; typically originate along volcanic arcs in zones of plate convergence.

Igneous rock Rock formed from cooling magma or lava.

Immiscible contaminants In groundwater, contaminants that are insoluble.

Inbreeding The production of offspring by closely related individuals.

Infiltration capacity The ability of the ground to absorb water, usually expressed in terms of rate of infiltration.

Injection wells Wells through which water is pumped into the ground to maintain the pressure of the reservoir or to dispose of chemical wastes.

Inner core See *Core*.

Insolation The amount of solar radiation that reaches Earth's surface.

Intergovernmental Panel on Climate Change A United Nations–sponsored organization of scientists charged with reviewing and assessing peer-reviewed climate science literature and summarizing its findings for the benefit of policy makers.

Intertropical Convergence Zone (ITCZ) The location where the northeast and southeast trade winds converge and rise. The position of the ITCZ changes seasonally, but is generally near the equator.

Intrusive igneous rock The new rock formed when magma that has intruded into the rock cools.

Inversion An increase in temperature with elevation in the troposphere.

Island arc See *Volcanic arc*.

Isolated system A system that does not allow either energy or matter to cross its boundaries. Compare *Closed system; Open system*.

Isotope One of the possible forms of the same element, differing in the number of neutrons in the nucleus but having a constant number of protons.

Jet streams Narrow, high-speed currents of air in the upper troposphere that flow from west to east and are associated with the large temperature and pressure gradients at the boundaries of the atmospheric circulation cells.

Jetty A structure built out from the shore to prevent accumulation of sediment in a harbor.

Kame A conical mound of sediment deposited by the melting of debris-laden glacial ice.

Kettle lakes Lakes produced by the melting of ice blocks embedded in glacial till left behind by retreating continental ice sheets.

Kinetic energy (KE) The energy of a moving body.

Kyoto Protocol A 1995 United Nations–sponsored treaty requiring the world's major industrialized states to commit to binding greenhouse gas emission targets by 2008–2012 relative to values in the year 1990.

Lahar A mudflow made of volcanic ash mixed with water.

Land subsidence The slow sinking of an area of crust without major deformation.

Late wood Densely packed small cells added to a tree's trunk and limbs prior to seasonal dormancy. Compare *Early wood*.

Lava tube A tunnel constructed when a lava flow develops a chilled crust at its surface that insulates the molten material inside, allowing the flow to continue moving downslope and eventually drain the tunnel.

Leachate A liquid mixture of water, chemicals, and metals produced in water that has infiltrated a landfill, come into contact with waste, and moved into the zone of saturated groundwater.

Levee A ridge of fine-grained sediment deposited along the banks of a river during floods.

Levee setback Levees rebuilt in a location farther back from the channel than their original placement, allowing a portion of the floodplain to flood in order to reduce the level of flooding downstream and to minimize potential flood losses.

Lignite A brown sedimentary rock that is the lowest grade of coal, softest and with the lowest carbon content (less than 60 percent). Compare *Anthracite; Bituminous coal*.

Limiting nutrients Inorganic substances that, when present in small quantities, limit biological growth. Phosphorus and nitrogen are the most common limiting nutrients.

Linear growth The change (i.e., growth) in a variable that is directly proportional to the amount of time that has elapsed. For example, the amount of water in a reservoir might increase the same amount each day, so the volume of water increases by that incremental amount times the number of days that have elapsed.

Liquefaction The phenomenon in which solids behave more like liquids as a result of shaking during events such as earthquakes.

Lithification The process of converting sediment into rock by compaction, cementation, and crystallization.

Lithosphere Earth's outermost solid layer, about 100 km thick, made up of the crust and upper mantle. See also *Pedosphere*.

Longshore current An ocean current that moves parallel to the coastline and is produced by waves striking the shoreline at an angle.

Lower mantle Dense, rigid rock of the mantle that lies below a depth of about 670 km and extends to a depth of about 2890 km.

Mafic A term describing rocks or magmas rich in iron and magnesium.

Magma chamber A reservoir of molten rock found below Earth's surface that typically lies beneath and feeds an active volcano.

Magnetic polarity timescale A record of reversals in Earth's magnetic field derived from radiometrically dated lava flows.

Mantle The layer of Earth, composed of silicate minerals rich in iron and magnesium, that lies between the core and the crust.

Mantle mesosphere Dense, rigid rock of the mantle that lies below a depth of about 670 km and extends to a depth of about 2890 km.

MARPOL The nickname of a United Nations–sponsored treaty (the International Convention for the Prevention of Pollution from Ships) banning intentional dumping of petroleum products and plastics in the ocean and restricting the disposal of sewage, food waste, and other garbage at sea.

Mass balance equation An equation that expresses the relative difference between inflow and outflow of water in a reservoir and from which changes in the stock of water within a reservoir (e.g., an aquifer) can be calculated.

Mass extinction The loss of numerous species over a limited span of geologic time.

Mass movement The movement of rock or soil downhill under the influence of gravity; can be either slow, as in creep, or very fast, as in landslides or debris flows.

Mass wasting The erosion of Earth materials that occurs when soil and rock move downslope as a result of gravity.

Maximum Contaminant Level An enforceable regulation placing a maximum level at which a contaminant can be present in drinking water.

Maximum Contaminant Level Goal The determination of a level at which a given contaminant in drinking water is likely to result in no adverse health effects.

Meandering stream A stream characterized by a channel that winds freely on a broad floodplain in prominent, sinuous curves.

Meltdown The melting of the structure within a nuclear power plant that contains atomic reactor fuel; caused by uncontrolled fission.

Mesocyclone A 2–10-km-diameter vortex of rising air rotating about a semi-vertical axis in a storm system.

Metamorphic rock Rock that has been altered by heat or pressure so that its texture or composition has been transformed.

Meteorological drought A drought that is the result of a precipitation deficit that occurs in a region over an extended period of time, typically at least one season in duration.

Methanogens Microbes that consume carbon dioxide and make methane as a waste product. These organisms may have provided some of the elevated greenhouse gas levels necessary for early Earth to be warm during a period of lower solar radiation (see *Faint young sun problem*).

Midden An accumulation of twigs, leaves, and bark collected by packrats, used for carbon dating and reconstructions of past climates.

Mid-ocean ridge A long, underwater mountain chain found at a divergent plate boundary, where new ocean floor is being created.

Milankovitch cycles Variations in Earth's orbital parameters—eccentricity, tilt, and precession—that lead to varying receipts of solar radiation over time, and hence to changing climatic conditions on Earth.

Mineral A natural, inorganic solid with a definite chemical composition and crystal structure.

Miscible contaminants In groundwater, contaminants capable of dissolving in water.

Modified Mercalli intensity scale (MMI) A method of rating the intensity of an earthquake, I to XII, based on the nature of the damage caused and the subjective perception of the event rather than on the actual amount of energy. Formerly known as the Mercalli scale.

Mohs hardness scale A scale on which minerals are rated in respect to hardness from 1 to 10 by comparison with other minerals.

Moment magnitude scale A scale of earthquake magnitude based on the product of the area of slippage along a fault, the amount of slip, and rock rigidity.

Monsoons Winds created by the presence of a high plateau in the interior of a continent.

Moraine A deposit of poorly sorted glacial till left behind by a glacier as it retreats; varieties include lateral, medial, and terminal.

Mountaintop removal mining A form of surface mining common in steep terrain, where the tops of mountains or summit ridges are removed and leveled with explosives and heavy machinery to expose seams of coal.

Mudcrack An indicator of dessication in fine-grained sedimentary deposits.

National ambient air quality standards (NAAQS) Standards set by the Environmental Protection Agency for the maximum levels of six atmospheric pollutants that have proved harmful to human and environmental health.

Natural catastrophe A natural event, such as an earthquake, volcanic eruption, or hurricane, that generates ecological disturbance.

Natural disaster A sudden and destructive environmental change as a result of long-term geologic processes that appears to occur without notice.

Natural gas The gaseous form of petroleum.

Natural selection A process that favors reproductive success to those individuals best adapted for survival, and that winnows individuals least fit for a given environment.

Nitrogen cycle The continuous flow, or cycling, of nitrogen from one Earth system to another.

Non–point source pollutants Pollutants that occur in runoff from the land surface and cannot be traced to a specific source.

Nonrenewable resources Resources such as fossil fuels that have a finite stock and are exhaustible.

Normal fault A type of fault in which a hanging wall block (rock above an inclined fault plane) slides downward relative to a footwall block (rock below the fault plane). This form of faulting results in horizontal lengthening of Earth's crust and is associated with extensional forces such as occur along mid-ocean ridges.

Nuclear fission The breaking of the nucleus of an atom into other, lighter elements, a processes that releases energy

Nuclear fusion The combination of two atomic nuclei to form one larger nucleus, a process that releases energy.

Nuclear reactor An instrument with the ability to start and sustain a nuclear fission reaction

Ocean The reservoir of salt water that covers 71 percent of Earth's surface. Compare *Sea*.

Ocean acidification Declining pH of the oceans in response to absorption of increased atmospheric carbon dioxide levels; threatens marine organisms that secrete calcium carbonate shells and skeletons.

Ocean trench A long depression on the ocean floor along the boundary between two lithospheric plates where one is being subducted beneath the other.

Oil sands Refer to *Bituminous sands*.

Open system A system that allows both matter and energy to flow across its boundaries. Compare *Closed system; Isolated system*.

Open-pit mining Mining done by excavating a large area of the surface, usually to obtain large quantities of low-grade ore.

Ore deposit A rock containing ore minerals in sufficient concentrations and quantity to be mined economically.

Ore mineral A mineral that is commercially valuable and that occurs in concentrations large enough to be mined economically.

Original horizontality, principle of The assumption that all sedimentary rocks were deposited originally in horizontal layers.

Outer core See *Core*.

Overbank deposition The deposition of fine sediment on the floodplain of a river during a flood.

Overconsumption Describes the situation in which a population consumes resources at rates that result in resource depletion and/or environmental degradation.

Overland flow Movement of water at Earth's surface that occurs when soil is fully saturated or the land surface (e.g., pavement) is impermeable.

Overpopulation Describes the situation in which a population exceeds the available resources needed to sustain its size, often perpetuating environmental degradation.

Oxidation A process of chemical weathering in which an element combines with oxygen to form oxide or hydroxide minerals.

Oxygen isotopic composition ($\delta^{18}O$) A measure of the relative abundance of light (^{16}O) and heavy (^{18}O) isotopes of oxygen in a variety of materials (for example, glacial ice, precipitation, and carbonate lake sediments), which can serve as a proxy for atmospheric temperature, global ice volume, and other climatic variables.

P (primary) wave The faster subsurface seismic wave generated by an earthquake, traveling through solid rock as a compressional wave. Compare *S (secondary) wave*.

Pāhoehoe The Hawaiian name for a slow-moving, low-viscosity basaltic lava flow that develops a ropey or billowy surface texture as it cools and solidifies.

Paleocene-Eocene thermal maximum A time interval when Earth underwent a warming of 5°C–6°C in less than 10,000 years and ocean acidification occurred as a result of increased greenhouse gas levels in the atmosphere. These changes led to mass extinctions of a variety of marine organisms as well as major migrations of North American mammals.

Paleoclimate The history of past climate changes on Earth.

Paleoseismology The study of past earthquakes from the records they leave behind in geologic and archeological materials.

Parent element An unstable element that through radioactive decay forms daughter products.

Passive margin A coastline located far from plate boundaries. Compare *Active margin*.

Peak oil A concept related to the timing of maximum global oil production, after which production would diminish.

Peat A deposit of partially decayed plant remains that accumulated in a wetland environment and still contains substantial amounts of moisture; the basis for coal formation.

Pedosphere The layer of decomposed rock particles and organic matter covering Earth's surface.

Period A unit of geologic time that subdivides an era.

Permeability The ability of a rock to allow the flow of groundwater through its pores. See also *Hydraulic conductivity*.

Perpetual resources Resources that are inexhaustible on the human timescale.

Petroleum A liquid or gaseous form of organic matter composed of hydrocarbons, compounds consisting of atoms of carbon and hydrogen. See also *Crude oil* and *Natural gas*.

Petroleum trap Any geologic structure that confines the flow of petroleum.

Photic zone (or **euphotic zone**) The uppermost layer of the ocean, where enough light penetrates to allow photosynthesis.

Photodissociation The decomposition of complex molecules into simpler molecules due to the absorption of solar radiation (e.g., conversion of water vapor into oxygen and hydrogen gases).

Photosynthesis The metabolic process by which plants use chlorophyll and solar energy to convert water and carbon dioxide to larger organic molecules.

Photovoltaic cell A component of solar panels, made of a semiconductor material, that transforms solar energy into electrical energy.

Physical weathering The mechanical processes by which a rock is broken into smaller pieces. Compare *Chemical weathering*.

Phytoplankton Microscopic plants that float at or near the surface of water columns and lie at the base of aquatic food chains.

Placer An ore deposit formed when relatively dense ore minerals, such as gold, become mechanically segregated from other minerals, typically in streambeds along the quiet-water zones on the outsides of meander bends or in deep pools at the base of a waterfall.

Planetary boundaries A proposed framework of thresholds that define a safe operating space for humanity within certain limits.

Plate tectonics The scientific theory that describes the origin and interaction of lithospheric plates, explaining the occurrence of continental drift, earthquakes, volcanoes, and mountains.

Point source pollutant A pollutant that is released from an identifiable source.

Polar cell Air circulating between the poles, where it descends, and a roughly 60° latitude belt, where it rises; part of the tricellular model of atmospheric circulation, including the Ferrel and Hadley cells.

Polar easterlies Surface winds associated with the atmospheric polar cells that blow from an easterly direction, under the influence of the Coriolis force, from the poles toward the 60° latitude belt.

Pollution The contamination of a substance with another, undesirable, material.

Pollution per unit of resource used The amount of pollution produced per unit resource used, a component of environmental degradation. See also *Overconsumption* and *Overpopulation*.

Pool An area of deep water in a stream.

Population viability analysis A quantitative determination, generally based in statistics, of the probability of extinction of a given population.

Porosity The ratio of pore volume in a rock to its total volume, expressed as a percentage.

Potential energy (PE) The energy of a body that results from its position within a system.

Potentially renewable resources Resources that can be depleted in the short term by rapid consumption but can be replaced in the long term by natural rates of replenishment.

Potentiometric surface A hypothetical plane located above the top of a confined aquifer to which water would rise if a well were drilled into the aquifer.

Power The rate at which energy is used, transferred, or transformed. It describes the action of work over a period of time.

Precession of the equinoxes The phenomenon in which the tilt of Earth's axis changes with time, resulting in a gradual shift of the onset of the different seasons (and hence the equinoxes) over a period of thousands of years.

Predator–prey cycles Oscillations in the population sizes of predators and their prey based upon food availability. When prey are abundant, the predator population increases, but this increase leads to a decline in prey availability, which in turn causes the predator population to decrease. With decreasing predation, prey populations again expand and the cycle continues.

Prediction A precise statement about the timing and location of a geological hazard event. For example, geologists used seismic data and measurements of volatile gas releases to predict the eruption of Mount Pinatubo within a matter of days after the volcano began to show signs of activity.

Proxies Geological and biological indicators of past environmental conditions.

Pycnocline The change in density with depth between the surface mixed layer and deep ocean, determined by changes in temperature and salinity with depth.

Pyroclastic debris Fragmented materials such as boulders, cinders, and ash that are ejected from a volcano. Also called tephra.

Pyroclastic flow Avalanches of ash, rock fragments, and hot gases erupted from a volcano that move down the volcano's flanks under the influence of gravity.

Pyrolysis The process of heating biomass at a low temperature (400°C–500°C) in the absence of oxygen.

Quad A measurement equivalent to one quadrillion (10^{15}) British thermal units (Btu), or (1.055×10^{18} J).

Radiative forcing A measure of the warming potential of different climatic factors (e.g., increased greenhouse gas levels, changing surface albedo) expressed in terms of equivalent increases in solar radiation output.

Radioactive (isotope) The emission of energy and/or subatomic particles from the nuclei of unstable isotopes; radioactive elements include radon and uranium.

Radiocarbon dating A procedure for dating a sample of organic matter by determining the amount of carbon-14 within the sample and comparing it to the amount of carbon-14 thought to have existed in the atmosphere at the time of the organism's death.

Radiometric dating The dating of rocks by measuring the abundance of radioactive isotopes and their stable products.

Rainsplash A process of the erosion of soil in which raindrops hit exposed (unvegetated) soil during intense rainstorms and lift fine particles above the soil surface, causing some of the loosened soil to move downslope.

Recharge basins Basins that have been constructed to hold water in order to resupply groundwater by allowing the water to infiltrate over time.

Recurrence interval The average length of time separating repetitions of a geologic event such as a flood or earthquake.

Refraction The bending of seismic rays in response to changes in density and material properties within Earth.

Reinforcing feedback Change to a system that promotes further change in the same direction the system is moving.

Relative age The age of one geologic feature relative to another without reference to absolute age.

Renewable Describes the ability of a resource to be naturally replenished, either with or without managed care.

Reserves Deposits of minerals or fossil fuels that are recoverable but not yet excavated. Compare *Resources*.

Reservoir A place of residence for a store of a particular material, e.g., the pedosphere in relation to carbon and nitrogen.

Resilience The ability to recover from ecological disturbances such as fire, hurricanes, volcanic eruptions, or disease.

Resources Deposits of minerals, fossil fuels, or other valuable material, discovered or undiscovered, that may be used at some time in the future. Compare *Reserves*.

Respiration The conversion of organic molecules in the presence of oxygen into carbon dioxide, water, and energy, which are used by organisms to carry out metabolic processes.

Restoration ecology A scientific discipline devoted to restoring species distributions and ecological functioning to landscapes degraded by human activities.

Reverse faulting A style of faulting in which a hanging wall block (rock above an inclined fault plane) slides upward relative to a footwall block (rock below the fault plane). This form of faulting results in horizontal shortening and is associated with compressional forces such as occur in continent–continent collisions.

Riffle An area of shallow water in a stream.

Rill A small stream channel several centimeters in width and depth that typically forms by rapid incision into sediments as a result of deforestation or the erosion of soil.

Riparian buffer A strip of vegetation along a stream channel intended to provide environmental benefits by reducing the impacts of human land use.

Risk The magnitude of potential deaths, injuries, or loss of property associated with a specific hazard.

Rock cycle The geologic cycle governing the production, alteration, and destruction of rocks resulting from processes such as volcanism, weathering, erosion, and lithification.

S (secondary) wave The slower subsurface seismic wave generated by an earthquake that arrives after the P (primary) wave; it travels as a shearing wave but cannot travel far through water. Compare *P (primary) wave*.

Safe Drinking Water Act (1974) A federal law ensuring the safety of U.S. drinking water.

Saline intrusion The infiltration of saltwater into the freshwater supply of coastal regions.

Salt dome trap A petroleum trap in which salt forms large, impermeable, dome-shaped masses that trap petroleum along their walls.

Saturation vapor pressure The maximum amount of water vapor that a parcel of air can contain given its temperature.

Scarp A break in slope that results from movement along a fault and offsets physical and structural features such as rock layers and roads.

Scientific method A method of investigation that involves the collection of data, the formation of hypotheses, and the testing of those hypotheses.

Sea A body of salt water largely surrounded by land but connected to the world ocean.

Sea level rise An increase in the elevation of the sea surface brought about by melting glaciers and ice sheets as well as thermal expansion of seawater as Earth's temperature warms.

Seafloor spreading The process by which new seafloor is produced at mid-ocean ridges, causing neighboring lithospheric plates to move apart.

Seamount A submerged island.

Seawall A structure made of rock or concrete built to protect land along the shore from waves.

Sedimentary rock Rock lithified from deposits of sediment.

Sedimentary structures Evidence of sediment transport processes left in the rock record, such as ripple marks and cross beds.

Seismic wave Any of the elastic vibrations produced by earthquakes, such as primary waves, secondary waves, and surface waves.

Seismogram A graphic representation of the passage of seismic waves from an earthquake.

Seismometer An instrument that detects and records the passage of seismic waves from an earthquake.

Sensitivity experiments Experiments performed on global climate models/general circulation models designed to assess the response of Earth's climate system to a change in greenhouse gas levels, plate tectonic configuration, or other parameters.

Sheet wash erosion The removal of successive layers of rock or soil from a gentle slope by thin sheets of running water.

Shield volcano A gently sloping volcanic landform with a very broad base, formed by successive eruptions of basaltic lava that spreads out in thin sheets.

Silicate minerals Minerals based on the silicate ion (SiO_4); the most abundant mineral group in Earth's crust.

Silicates See *Silicate minerals*.

Sinkhole A topographic feature caused by the collapse of underground limestone caverns.

Slash-and-char An alternative form of the slash-and-burn technique in which organic matter is partially, rather than fully, burned.

Slide A type of mass movement in which relatively cohesive blocks of material move, or fail, along a well-defined plane.

Smog A complex mixture of ozone, nitrogen oxides, and hydrocarbons in the lower atmosphere that results from industrial activities and automobile exhausts.

Snowball Earth events Intervals lasting tens of millions of years and existing between 750 and 580 million years ago when Earth apparently froze over completely.

Soil Loose material that accumulates on Earth's surface and is composed of clay, sand, and humus.

Soil horizons A unique layer of soil that differs from others because of its color, texture or composition.

Soil profile A section of soil through all its soil horizons.

Solar-thermal power plant A power plant that generates electricity by means of solar energy and heat energy.

Special Flood Hazard Areas (SFHAs) Areas classified by the Federal Emergency Management Agency as being within a flood zone that would be inundated by flood waters from events with a given probability of exceedance.

Species diversity The number and type of species in a particular location.

Species richness The total number of species in a region or ecological community.

Specific retention The ratio of groundwater retained by surface tension to total volume of water.

Specific yield The ratio of the volume of water that is drained from an aquifer by gravity to its total volume.

Speleothem A calcium carbonate deposit formed by the outgassing of carbon dioxide in dripping groundwater in a cave.

Sputtering The loss of atmospheric gases due to collision with high-energy particles from outer space.

Stage The relative elevation of water in a stream with respect to a datum plane.

Static system A system in which no work is done and no change in state occurs.

Steady state A system in which inflows balance outflows.

Stick-slip Displacement along a fault plane characterized by long intervals of no movement punctuated by brief episodes of rapid movement. Such behavior results in earthquakes.

Stock The content of a reservoir at any given time.

Strata Layers of rock.

Stratigraphy The study of the origin and significance of strata of sedimentary rocks.

Stratosphere The layer of the atmosphere above the troposphere; a major site of ozone production.

Stratovolcano A cone-shaped volcanic landform made up of alternating layers of viscous lava and pyroclastic material.

Stress The amount of force acting on a body per unit area.

Striations Grooves found in rocks that have been ground against one another during transport by glaciers.

Strike-slip faulting A style of faulting characterized by horizontal movement of fault blocks past one another along a vertical or nearly vertical fault plane.

Strip mining A form of surface mining in which the overburden is removed in strips.

Subbituminous coal A hard, brown-black, sedimentary rock that is the intermediate grade of coal between lignite and bituminous coal and contains 35–45% carbon.

Subduction The sinking of one plate beneath another that occurs at convergent plate boundaries.

Superfund The short name for the Comprehensive Environmental Response, Compensation and Liability Act of 1980 and its amendments.

Superfund National Priorities List (or Superfund list) The U.S. Environmental Protection Agency's list of sites where a release or threatened release of hazardous substances may endanger public health, welfare, or the environment.

Superposition, principle of The assumption that a stratum that overlies another stratum is the younger of the two; the basis for stratigraphy.

Surface mixed layer The layer of the ocean, up to about 200 m deep, that is mixed by the wind.

Surface tension The attraction between molecules at the surface of a liquid.

Surface wave A seismic wave, slower than a secondary wave, that follows the surface of Earth.

Sustainability Long-term management of resources so as not to degrade or impair environmental quality or to reduce the stock of a resource below its replenishment ability.

Sustainable agriculture Agricultural production that meets current needs without compromising resources for future generations, while promoting environmental health, economic profit, and social and economic equity.

Tailings Waste rock discarded after mining.

Tar sands Refer to *Bituminous sands.*

Tephra Fragmented materials such as boulders, cinders, and ash that are ejected from a volcano. Also called pyroclastic debris.

Theory A hypothesis that has survived repeated testing.

Thermal energy Heat-related energy produced by the motion of atoms.

Thermocline The drop-off of temperature with depth between the surface mixed layer and deep ocean.

Thermohaline circulation The density-driven circulation of the whole ocean in response to temperature and salinity variations at the surface.

Thermokarst lake A lake produced when ice lenses contained in perennially frozen ground (permafrost) melt, causing the land surface to collapse and creating a depression that fills with water.

Till A glacial deposit consisting of a mixture of particles ranging from clay- to boulder-size.

Tilt The angle between Earth's spin axis and a line drawn perpendicular to the plane of Earth's orbit about the sun. Higher angles of tilt correspond to greater seasonal variability, whereas no tilt would lead to a lack of seasonal variations in solar radiation receipts.

Tipping points Thresholds between different stable states of a system.

Tolerable soil loss The amount of erosion of soil, or rate of soil loss, below which that soil can sustain a high level of crop productivity indefinitely.

Topsoil The organic-rich, upper layer of a soil that is produced by weathering processes and the accumulation of plant and animal matter.

Tornado A violently rotating wind that forms a vertical, elongated vortex; common in the midwestern United States.

Total Maximum Daily Loads (TMDLs) An estimate of the upper limit of a pollutant that a water body can receive and yet still meet water quality standards.

Trade winds Tropical surface winds associated with the atmospheric Hadley cells that blow toward the equator and respond to Coriolis deflection. In the northern hemisphere, these winds blow from the northeast toward the southwest and are called the northeast trade winds. In the southern hemisphere, they blow from the southeast toward the northwest and are known as the southeast trade winds.

Transform plate boundary A plate boundary where two lithospheric plates slide past each other.

Triangulation The use of P- and S-wave arrival times at three or more seismographic stations to determine the location of earthquake epicenters.

Tropical storm A large, rotating storm over the ocean; less powerful than a hurricane.

Troposphere The layer of the atmosphere, closest to Earth's surface, where weather occurs.

Tsunami A large water wave caused by undersea earthquakes or volcanic eruptions.

Turbidity A measure of water clarity that typically refers to the amount of suspended fine sediment in the water.

Unconfined aquifer An aquifer that is not held under pressure by an impermeable layer.

Unconformity A plane between two layers of rock that represents a time when no deposition occurred and/or erosion removed some of the lower surface before deposition resumed.

Unconventional oil and natural gas Forms of petroleum extracted from low-permeability reservoirs, including examples such as tight oil and shale gas. Compare *Conventional oil and natural gas.*

Unifying theory A scientific theory that has undergone extensive testing and survived all challenges such that it has become accepted as fact; for example, Newton's law of gravitation, which states that there is a physical attraction between any two bodies in the universe dependent on their masses and the distance between them. Also called a universal law.

Universal law A scientific theory that has undergone extensive testing and survived all challenges such that it has become accepted as fact; for example, Newton's law of gravitation, which states that there is a physical attraction between any two bodies in the universe dependent on their masses and the distance between them. Also called a unifying theory.

Universal soil loss equation A mathematical tool for predicting rates of the erosion of soil under a constellation of varying conditions, including variations in soil character, climate, land cover, and topography.

Upwelling The upward movement of water from the deep ocean toward the surface in response to Ekman transport of surface water. This process delivers dissolved nutrients to the surface ocean. Compare *Downwelling*.

Valley fill The practice of placing the debris and overburden that has accumulated from mountaintop removal mining into adjacent headwater valleys.

Virtual water The amount of water theoretically "embedded" within a product, based on the amount of water that was used to produce and transport the product.

Viscosity A measure of a liquid's resistance to flow.

Volatiles Dissolved gases contained within magma.

Volcanic arc A chain of volcanic islands formed at a subduction zone by melted material from the subducted plate that rises upward through the overriding plate.

Water footprint An estimate of the total amount of freshwater a person uses, either directly or via the production of goods and services that he or she consumes.

Water hardness The concentration of calcium and magnesium ions in water.

Water spreading The practice of spreading runoff water over a large land area to infiltrate permeable regions where the water table is not close to the surface.

Water table The two-dimensional surface beneath Earth's surface that divides the saturated zone, in which pore spaces in the rock are filled with groundwater, from the overlying unsaturated zone.

Watershed Land area defined by drainage divides that encompasses the surface water runoff of a stream network with its main stem and tributary channels.

Watt A common unit for power, equivalent to 1 J per second.

Wave base The depth in the ocean below which the overlying winds have no influence on water movement.

Weather Short-term (e.g., daily) fluctuations in temperature, wind speed, and precipitation in a particular region. Compare *Climate*.

Westerlies Surface winds associated with the atmospheric Ferrel cells. These winds blow from the 30° latitude belt where air is descending toward the 60° latitude belt where air is rising and are acted upon by the Coriolis force, giving them a generally west to east flow direction.

Wetlands Areas characterized by extremely moist soils, e.g., swamps and marshes.

Windshear The changing direction and velocity of winds with elevation above Earth's surface.

Xeriscapes Landscapes surrounding homes that use drought-tolerant native vegetation that requires little or no watering in place of water-guzzling turf grasses.

Chapter 1

1. Cullen, N. J., Sirguey, P., Mölg, T., Kaser, G., Winkler, M., and Fitzsimons, S. J. 2013. "A Century of Ice Retreat on Kilimanjaro: The Mapping Reloaded." *The Cryosphere* 7: 419–431. doi:10.5194/tc-7-419-2013.

2. Thompson, L. G., Mosley-Thompson, E., Davis, M. E., Henderson, K. A., Brecher, H. H., Zagorodnov, V. S., Mashiotta, T. A., et al. 2002. "Kilimanjaro Ice Core Records: Evidence of Holocene Climate Change in Tropical Africa." *Science* 298: 589–593. doi:10.1126/science.1073198.

Moulton-Howe, L. 2001. "Disappearing Glaciers Evidence of a Rapidly Warming Earth." Accessed July 22, 2013. http://dwij.org/pathfinders/linda_moulton_howe/linda_mh1.htm.

3. Pepin, N.C., Duane, W. J., and Hardy, D. R. 2010. "The Montane Circulation on Kilimanjaro, Tanzania and Its Relevance for the Summit Ice Fields: Comparison of Surface Mountain Climate with Equivalent Reanalysis Parameters." *Global and Planetary Change* 74: 61–75. doi:10.1016/j.gloplacha.2010.08.001.

4. Mölg, T., Großhauser, M., Hemp, A., Hofer, M., and Marzeion, B. 2012. "Limited Forcing of Glacier Loss through Land-Cover Change on Kilimanjaro." *Nature Climate Change* 2: 254–258. doi:10.1038/nclimate1390.

5. Barr, J., and Chander, A. 2013. "Africa without Ice and Snow." *Environmental Development* 5: 146–155. doi:10.1016/j.envdev.2012.10.003.

6. United Nations. 1987. "Report of the World Commission on Environment and Development." General Assembly Resolution 42/187, December 11.

7. National Research Council. 2001. *Basic Research Opportunities in Earth Science*. Washington, D.C.: The National Academies Press.

8. Mora, C., Tittensor, D. P., Adl, S., Simpson, A. G. B., Worm, B. 2011. "How Many Species Are There on Earth and in the Ocean?" *PLoS Biology* 9: e1001127. doi:10.1371/journal.pbio.1001127.

9. Population Reference Bureau's DataFinder, http://www.prb.org/DataFinder.aspx. Accessed July 31, 2013.

10. Zalasiewicz, J., et al. 2008. "Are We Now Living in the Anthropocene?" *GSA Today* 18: 4–8. doi:10.1130/GSAT01802A.1.

11. Rockström, J., Steffen, W., Noone, K., Persson, Å., Chapin, F. S. III, Lambin, E., Lenton, T. M., et al. 2009. "Planetary Boundaries: Exploring the Safe Operating Space for Humanity." *Ecology and Society* 14: 32.

Chapter 2

1. Schwarzbach, M. 1986. *Alfred Wegener, the Father of Continental Drift*, p. 241. Madison, WI: Science Tech.

2. Oreskes, N. 1999. *The Rejection of Continental Drift: Theory and Method in American Earth Science*. New York: Oxford University Press.

3. Monroe, J. S., and Wicander, R. 2005. *The Changing Earth: Exploring Geology and Evolution*, 4th ed, p. 243. Belmont, CA: Thomson Brooks/Cole.

4. Wertenbaker, W. 1974. *The Floor of the Sea: Maurice Ewing and the Search to Understand the Earth*, p. 81. Boston: Little, Brown and Co.

5. Dineley, D. L. 2000. "Harry Hammond Hess (1906–69)." In *The Oxford Companion to the Earth*, edited by P. L. Hancock and B. J. Skinner, 494–509. New York: Oxford University Press.

6. Ryan, J. W., and Clark, T. A. 1988. "NASA/Crustal Dynamics Project Results: Tectonic Plate Motion Measurements with Mark-III VLBI." In *The Impact of VLBI on Astrophysics and Geophysics: Proceedings of the 129th IAU Symposium, Cambridge, MA, May 10–15, 1987*, edited by M. J. Reid and J. M. Moran, 339. Dordrecht: Kluwer Academic Publishers.

7. Flynn, J. J., and Wyss, A. R. 1998. "Recent Advances in South American Mammalian Paleontology." *Trends in Ecology and Evolution* 13 (11):449–454.

8. Stanley, S. M. 2009. *Earth System History*, 3rd ed., pp. 178–183. New York: W. H. Freeman and Company.

9. Oreskes, N. 1999. *The Rejection of Continental Drift: Theory and Method in American Earth Science*, p. 10. New York: Oxford University Press.

10. Oreskes, N. 2003. "From Continental Drift to Plate Tectonics." In *Plate Tectonics: An Insider's History of the Modern Theory of the Earth*, edited by N. Oreskes, 3–27. Boulder, CO: Westview Press.

11. Condie, K. 2000. "Crustal Composition and Recycling." In *The Oxford Companion to the Earth*, edited by P. L. Hancock and B. J. Skinner, 192. New York: Oxford University Press.

12. Aviso. "Jul. 2002: Doris Measures Plate Motion." Accessed July 28, 2010. http://www.aviso.oceanobs.com/en/news/idm/2002/jul-2002-doris-measures-plate-motion/index.html.

13. Parker, D. E., Wilson, H., Jones, P. D., Christy, J. R. and Folland, C. K. 1995. "The Impact of Mount Pinatubo on World-Wide Temperatures." *International Journal of Climatology* 16 (5): 487–497.

14. Self, S., Gertisser, R., Thordarson, T., Rampino, M. R., and Wolff, J. A. 2004. "Magma Volume, Volatile Emissions, and Stratospheric Aerosols from the 1815 Eruption of Tambora." *Geophysical Research Letters* 31 (20):L20608.

15. Nikonov, A. A. 1989. "The Rate of Uplift in the Alpine Mobile Best." *Tectonophysics* 163 (3–4): 267–276.

16. Chang, K. 2011. "Quake Moves Japan Closer to U.S. and Alters Earth's Spin." *New York Times*, March 13. http://www.nytimes.com/2011/03/14/world/asia/14seismic.html.

17. National Police Agency of Japan. 2011. "Damage Situation and Police Countermeasures Associated with 2011 Tohoku District – Off the Pacific Ocean Earthquake." June 6, http://www.npa.go.jp/archive/keibi/biki/higaijokyo_e.pdf.

18. Raff, A. D., and Mason, R. G. 1961. "Magnetic Survey Off the West Coast of North America, 40° N. Latitude to 52° N. Latitude." *Geological Society of America Bulletin* 72: 1267–1270.

Mason, R. G. and Raff, A. D. 1961. "Magnetic Survey Off the West Coast of North America, 32° N. Latitude to 42° N. Latitude." *Geological Society of America Bulletin* 72:1259–1265.

19. Vine, F. J. 2003. "Reversals of Fortune." In Oreskes, *Plate Tectonics*, 46–66.

20. Morley, L. W. 2003. "The Zebra Pattern." In Oreskes, *Plate Tectonics*, 67–85.

21. Atwater, T. 2003. "When the Plate Tectonic Revolution Met Western North America." In Oreskes, *Plate Tectonics*, 243–263.

22. Oliver, J. 2003. "Earthquake Seismology in the Plate Tectonics Revolution." In Oreskes, *Plate Tectonics*, 155–166.

Molnar, P. "From Plate Tectonics to Continental Tectonics: An Evolving Perspective of Important Research, from a Graduate Student to an Established Curmudgeon." In Oreskes, *Plate Tectonics*, 288–328.

23. GLGArcs. 2010. "Introduction to the Landforms and Geology of Japan." Accessed August 6. http://www.glgarcs.net/intro/subduction.html.

24. Anderson, D. L. 2006. "Plate Tectonics; the General Theory: Complex Earth Is Simpler Than You Think." In "Earth and Mind: How Geologists Think and Learn About the Earth," *Geological Society of America Special Paper* 413, edited by C. A. Manduc and D. W. Mogk, 29–38.

25. Anderson, D. L., and Natland, J. H. 2005. "A Brief History of the Plume Hypothesis and Its Competitors: Concept and Controversy." In "Plates, Plumes, and Paradigms," *Geological Society of America Special Paper* 388, edited by G. R. Foulger, J. H. Natland, D. C. Presnall, and D. L. Anderson, 119–145.

26. Wolfe, C. J., Solomon, S. C., Laske, G., Collins, J. A., Detrick, R. S., Orcutt, J. A., Bercovici, D., and Hauri, E. H. 2009. "Mantle Shear-Wave Velocity Structure Beneath the Hawaiian Hot Spot." *Science* 326 (5958):1388–1390. doi:10.1126/science.1180165.

27. Wilson, J. T. 1963. "A Possible Origin of the Hawaiian Islands." *Canadian Journal of Physics* 41 (6):863–870. doi:10.1139/p63-094.

28. Clague, D. A., and Dalrymple, G. B. 1987. "The Hawaiian-Emperor Volcanic Chain, Part I: Geologic Evolution." *U.S. Geological Survey Professional Paper* 1350.

29. Tarduno, J. A., Duncan, R. A., Scholl, D. W., Cottrell, R. D., Steinberger, B., Thordarson, T., Kerr, B. C., et al. 2003. "The Emperor Seamounts: Southward Motion of the Hawaiian Hotspot Plume in Earth's Mantle. *Science* 301(5636): 1064–1069.

Chapter 3

1. Israel Ministry of Foreign Affairs. 2001. "Beit She-An—A Biblical City and Scythopolis—A Roman-Byzantine City." Accessed January 14, 2010. http://www.mfa.gov.il/MFA/History/Early%20History%20-%20Archaeology/Beit%20She-an%20-%20A%20Biblical%20City%20and%20Scythopolis%20-%20A.

2. Satake, K., Shimazaki, K., Tsumi, Y., and Ueda, K. 1996. "Time and Size of a Giant Earthquake in Cascadia Inferred from Japanese Tsunami Records of January 1700." *Nature* 379:246–249.

3. Nelson, A. R., Asquith, A. C., and Grant, W. C. 2004. "Great Earthquakes and Tsunamis of the Past 2000 Years at the Salmon River Estuary, Central Oregon Coast, USA." *Bulletin of the Seismological Society of America* 94:1276–1292.

Atwater, B. F., Tuttle, M. P., Schweig, E. S., Rubin, C. M., Yamaguchi, D. K., and Hemphill-Haley, E. 2003. "Earthquake Recurrence Inferred from Paleoseismology." In *Developments in Quaternary Science 1: The Quaternary Period in the United States*, edited by A. R. Gillespie, S. C. Porter, and B. F. Atwater. New York: Elsevier, p. 331–350.

4. Mahoney, M. 2011. "The Japan Earthquake & Tsunami and What They Mean for the U.S." Accessed June 7, 2011. http://nthmp.tsunami.gov/taw/downloads/japan-earthquake-3-17-2011.pdf.

5. National Police Agency of Japan. "Damage Situation and Police Countermeasures Associated with 2011 Tohoku District—Off the Pacific Ocean Earthquake." June 6, 2011. http://www.npa.go.jp/archive/keibi/biki/higaijokyo_e.pdf.

6. Kubota, Y. "Update 1—Japan Makes New Nuclear Safety Vows After Quake Disaster." Reuters, June 7, 2011. http://www.reuters.com/article/2011/06/07/japan-nuclear-idUSL3E7H70X520110607.

7. National Research Council. 2003. *Living on an Active Earth: Perspectives on Earthquake Science*, pp. 23–24. Washington, D.C.: National Academies Press.

8. Scholz, C. H. 2002. *The Mechanics of Earthquakes and Faulting*, 2nd ed, p. 180. Cambridge: Cambridge University Press.

9. **U.S. Geological Survey.** 2009. "The Northern California Earthquake, April 18, 1906: Rupture Length and Slip." Accessed January 14, 2010. http://earthquake.usgs.gov/regional/nca/virtualtour/earthquake.php.

10. **Harden, D. R., Stenner, H., and Blatz, I.** 2001. "The Calaveras and San Andreas Faults in and Around Hollister." In *U.S. Geological Survey Bulletin 2188: Geology and Natural History of the San Francisco Bay Area: A 2001 NAGT Field-Trip Guidebook*, edited by P. W. Stoffer and L. C. Gordon. Reston, VA: U.S. Geological Survey.

11. **Lienkaemper, J. J., Galehouse, J. S., and Simpson, R. W.** 2001. "Long-Term Monitoring of Creep Rate Along the Hayward Fault and Evidence for a Lasting Creep Response to 1989 Loma Prieta Earthquake," *Geophysical Research Letters* 28:2265–2268.

12. **Scholz, C. H.** 2002. *The Mechanics of Earthquakes and Faulting*, 2nd ed., p. 167. Cambridge: Cambridge University Press.

13. **Reid, H. F.** 1910. "The Mechanics of the Earthquake, The California Earthquake of April 18, 1906." *Report of the State Investigation Commission*, vol. 2, Washington, D.C.: Carnegie Institution of Washington.

14. **Scholz, C. H.** 2002. *The Mechanics of Earthquakes and Faulting*, 2nd ed, p. 179. Cambridge: Cambridge University Press.

15. **Ide, S., Beroza, G. C., Shelly, D. R., and Uchide, T.** 2007. "A Scaling Law for Slow Earthquakes." *Nature* 447:76–79.

Forrest, M. R. 2007. "Slow Earthquakes." Southern California Earthquake Center. Accessed August 8, 2010. http://www.scec.org/news/00news/feature000401.html.

16. **Grotzinger, J., and Jordan, T.** 2010. *Understanding Earth*, 6th ed., p. 340. New York: W. H. Freeman.

17. **Winthrop, J.** 1755. *A Lecture on Earthquakes; Read in the Chapel of Harvard-College in Cambridge, N.E. November 26th 1755: On Occasion of the Great Earthquake Which Shook New-England the Week Before.* Boston: Edes & Gill.

18. **Stein, R. S., and Bucknam, R. C.** 1986. "Quake Replay in the Great Basin." *Natural History* 95:28–35.

19. **Brumbaugh, D. S.** 1999. *Earthquakes: Science and Society*, p. 27. Upper Saddle River, NJ: Prentice Hall.

20. **Brumbaugh, D. S.** 1999. *Earthquakes: Science and Society*, p. 35. Upper Saddle River, NJ: Prentice Hall.

21. **Lillie, R. J.** 1999. *Whole Earth Geophysics*, p. 213. Upper Saddle River, NJ: Prentice Hall.

22. **Bolt, B. A.** 2010. "Inge Lehmann." Berkeley, CA: University of California, Berkeley, Seismographic Station. Accessed August 6, 2010. http://www.physics.ucla.edu/~cwp/articles/bolt.html.

23. **IRIS (Incorporated Research Institutions for Seismology).** 2011. "Transportable Array." Accessed June 7, 2011. http://www.usarray.org/researchers/obs/transportable.

24. **U.S. Geological Survey.** 2011. "Magnitude 9.0—Near The East Coast of Honshu, Japan." Accessed June 7, 2011. http://earthquake.usgs.gov/earthquakes/recenteqsww/Quakes/usc0001xgp.php#summary.

25. **IRIS (Incorporated Research Institutions for Seismology).** 2011. "Aftershocks of Japan M9.0 2011/03/11 05:46:23." Accessed June 7, 2011. http://www.iris.edu/dms/products/experimental/aftershockmovies/Japan.aftershocks.MP4.

26. **Stein, R. S.** 2003. "Earthquake Conversations." *Scientific American* 288:72–79.

27. **Shahid, A.** 2011. "Aftershocks Rock Japan Following Earthquake and Tsunami, Could Hamper Recovery Efforts." *New York Daily News*, March 12, 2011. Accessed June 7, 2011. http://articles.nydailynews.com/2011-03-12/news/29139712_1_earthquake-aftershock-main-shock-magnitude.

28. **Hergert, T., and Heidbach, O.** 2010. "Slip-Rate Variability and Distributed Deformation in the Marmara Sea Fault System," *Nature Geoscience* 3:132–135.

29. **U.S. Geological Survey.** 2000. "Implications for Earthquake Risk Reduction in the United States from the Kocaeli, Turkey, Earthquake of August 17, 1999." *U.S. Geological Survey Circular* 1193.

30. **Brumbaugh, D. S.** 1999. *Earthquakes: Science and Society*, p. 81. Upper Saddle River, NJ: Prentice Hall.

31. **U.S. Geological Survey.** 2010. "Earthquake Facts and Statistics." Accessed January 14, 2010. http://earthquake.usgs.gov/earthquakes/eqarchives/year/eqstats.php.

32. **U.S. Geological Survey.** 2009. "Magnitude 9.1—Off the West Coast of Northern Sumatra." Accessed January 14, 2010. http://earthquake.usgs.gov/earthquakes/eqinthenews/2004/usslav/#details.

33. **Holzer, T. L.** 1998. "The Loma Prieta, California, Earthquake of October 17, 1989—Liquefaction." *U.S. Geological Survey Professional Paper* 1551-B.

34. **U.S. Geological Survey.** 2009. "Historic Earthquakes: Northridge, California." Accessed January 14, 2010. http://earthquake.usgs.gov/earthquakes/states/events/1994_01_17.php.

35. **U.S. Geological Survey.** 2010. "Magnitude 7.0—Haiti Region." Accessed August 6, 2010. http://earthquake.usgs.gov/earthquakes/recenteqsww/Quakes/us2010rja6.php#details.

36. **U.S. Geological Survey.** 2010. "Magnitude 8.8—Offshore Maule, Chile." Accessed August 6, 2010. http://earthquake.usgs.gov/earthquakes/eqinthenews/2010/us2010tfan/.

37. **Ken-Tor, R., Agnon, A., Enzel, Y., Stein, M., Marco, S., and Negendank, J. F. W.** 2001. "High-Resolution Geological Record of Historic Earthquakes in the Dead Sea Basin." *Journal of Geophysical Research B: Solid Earth* 106:2221–2234.

Bookman, R., Stein, M, and Enzel, Y. 2009. "Station 5: The Ze'elim ("East") Gully: Deep Window into the Holocene Dead Sea." In *GLOCOPH Israel 2009, 25th October–3rd November Field Trip Guidebook,* edited by Y. Enzel, N. Greenbaum, and M. Laskow, 81–85. Jerusalem: Hebrew University Press.

38. Sieh, K. E., and Jahns, R. H. 1984. "Holocene Activity of the San Andreas Fault at Wallace Creek, California." *Geological Society of America Bulletin* 95:883–896.

39. Lienkaemper, J. J., Dawson, T. E., Personius, S. F., Seitz, G. G., Reidy, L. M., and Schwartz, D. P. 2002. "A Record of Large Earthquakes on the Southern Hayward Fault for the Past 500 Years." *Bulletin of the Seismological Society of America* 92:2637–2658.

40. Machette, M. N., Personius, S. F., Nelson, A. R., Schwartz, D. P., and Lund, W. R. 1991. "Wasatch Fault Zone, Utah—Segmentation and History of Holocene Earthquakes." *Journal of Structural Geology* 13:151–164.

Eldredge, S. N., and Clarke, V. 1996. "The Wasatch Fault." *Utah Geological Survey Public Information Series* 40.

41. Csontos, R., and Van Arsdale, R. 2008. "New Madrid Seismic Zone Fault Geometry." *Geosphere* 4:802–813.

42. National Research Council. 2003. *Living on an Active Earth: Perspectives on Earthquake Science,* p. 132. Washington, D.C.: National Academies Press.

43. Calais, E., Free, A. M., Van Arsdale, R., and Stein, S. 2010. "Triggering of New Madrid Seismicity by Late-Pleistocene Erosion." *Nature* 466:608–611.

44. Schweig, E., Gomberg, J., and Hendley II, J. W. 1995. "The Mississippi Valley—'Whole Lotta Shakin' Goin' On.'" *U.S. Geological Survey Fact Sheet* 168-95.

Hough, S. 2010. "On the New Madrid Strain Rate/Release Discrepancy: Reexamining the Observational Underpinnings of Sacred Exotic Cows." *Seismological Research Letters* 81:359.

45. Field, E. H., Dawson, T. E., Felzer, K. R., Frankel, A. D., Gupta, V., Jordan, T. H., Parsons, T., et al. 2008. "The Uniform California Earthquake Rupture Forecast, Version 2 (UCERF 2)." USGS Open File Report 2007-1437, CGS Special Report 203, SCEC Contribution #1138, Version 1.1.

46. U.S. Geological Survey. 2009. "The Parkfield, California, Earthquake Experiment." Accessed January 14, 2010. http://earthquake.usgs.gov/research/parkfield/index.php.

47. Bakun, W. H., Aagaard, B., Dost, B., Ellsworth, W. L., Hardebeck, J. L., Harris, R. A., Ji, C., et al. 2005. "Implications for Prediction and Hazard Assessment from the 2004 Parkfield Earthquake." *Nature* 437:969–974.

48. Zoback, M., Hickman, S., and Ellsworth, W. 2010. "Scientific Drilling into the San Andreas Fault Zone." *EOS* 91:197–204.

49. Schleicher, A. M., van der Pluijm, B. A., and Warr, L. N. 2010. "Nanocoatings of Clay and Creep of the San Andreas Fault at Parkfield, California." *Geology* 38:667–670.

50. Roger Bilham. Interview by Amy Goodman on *Democracy Now!* March 1, 2010. Accessed August 8, 2010. http://www.democracynow.org/2010/3/1/seismologist_roger_bilham_in_recent_earthquakes.

51. Yeats, R. S. 2001. *Living with Earthquakes in California: A Survivor's Guide,* p. 333. Corvallis, Oregon: Oregon State University Press.

52. Yeats, R. S. 2001. *Living with Earthquakes in California: A Survivor's Guide,* p. 237. Corvallis, Oregon: Oregon State University Press.

53. California Seismic Safety Commission. 2005. *Homeowner's Guide to Earthquake Safety.* Sacramento, CA: State of California.

54. Ghafory-Ashtiany, M., and Hosseini, M. 2008. "Post-Bam Earthquake: Recovery and Reconstruction." *Natural Hazards* 44:229–241.

55. Bradford, M. 2009. "Shoddy Construction Faulted in Deadly Quake." *Business Insurance.* April 27, 2009. Accessed January 15, 2010. http://www.businessinsurance.com/article/20090426/ISSUE01/100027535.

56. BBC News. April 6, 2009. "Power Italian Quake Kills Many." Accessed January 14, 2010. http://news.bbc.co.uk/2/hi/7984867.stm.

57. *New York Times.* May 6, 2009. "Sichuan Earthquake." Accessed January 15, 2010. http://topics.nytimes.com/topics/news/science/topics/earthquakes/sichuan_province_china/index.html.

58. Southern California Earthquake Center. 2010. "The Great California Shakeout." Accessed June 6, 2011. http://www.shakeout.org/.

59. Pittman E. 2010. "Earthquake Early Warning System Coming to California." Accessed August 4, 2010. http://www.emergencymgmt.com/disaster/Earthquake-Early-Warning-System-California.html.

Pounders, E., and Gordon, L. 2009. "Earthquake Early Warning System Possible." USGS Newsroom. Accessed August 4, 2010. http://www.usgs.gov/newsroom/article.asp?ID=2366&from=rss_home.

Gorwyn, A. 2010. "Richard Allen: 'Earthquake Early Warning System Possible for California.'" Accessed August 4, 2010. http://earthsky.org/earth/richard-allen-earthquake-early-warning-system-possible-for-california.

60. Birmingham, L. 2011. "Japan's Earthquake Warning System Explained." *TimeWorld.* Accessed June 7, 2011. http://www.time.com/time/world/article/0,8599,2059780,00.html.

61. Simpson Strong-Tie Company. 2010. "World's Largest Earthquake Shake Table Test in Japan." Accessed June 7, 2011. http://www.strongtie.com/about/research/capstone.html.

62. U.S. Geological Survey. 2009. "Historic Earthquakes: San Fernando, California." Accessed January 15, 2010. http://earthquake.usgs.gov/earthquakes/states/events/1971_02_09.php.

63. **State of California Department of Conservation.** 2007. "California Geological Survey—Alquist-Priolo Earthquake Fault Zones." Accessed January 15, 2010. http://www.consrv.ca.gov/CGS/RGHM/AP/Pages/Index.aspx.

64. **State of California.** 2007. "Alfred E. Alquist Seismic Safety Commission." Accessed January 14, 2010. http://www.seismic.ca.gov/about.html.

Chapter 4

1. **Shea, N.** 2008. "Cavern of Crystal Giants." *National Geographic,* November.

2. **García-Ruiz, J. M., Villasuso, R., Ayora, C., Canals, A., and Otálora, F.** 2007. "Formation of Natural Gypsum Megacrystals in Naica, Mexico." *Geology* 35:327–330.

3. **Van Driessche, A. E. S., García-Ruiz, J. M., Tsukamoto, K., Patino-Lopez, L. D., and Satoh, H.** 2011. "Ultraslow Growth Rates of Giant Gypsum Crystals." *Proceedings of the National Academy of Sciences* 108:15721–15726.

4. **Sanna, L., Saez, F., Simonsen, S., Constantin, S., Calaforra, J.-M., Forti, P., and Lauritzen, S.-E.** 2010. "Uranium-Series Dating of Gypsum Speleothems: Methodology and Examples." *International Journal of Speleology* 39:35–46.

5. **Mineral Information Institute.** 2011. "Mineral Usage per Capita." http://www.mii.org/pdfs/percapita.pdf.

6. **Kristof, N. D., and WuDunn, S.** 1994. *China Wakes: The Struggle for the Soul of a Rising Power.* New York: Vintage.

7. **van Oss, H. G.** 2011. *2009 Minerals Yearbook: Cement [Advanced Release].* U.S. Geological Survey. http://minerals.usgs.gov/minerals/pubs/commodity/cement/myb1-2009-cemen.pdf.

8. **Menzie, D., Tse, P.-K., Fenton, M., Jorgenson, J., and van Oss, H.** 2004. *China's Growing Appetite for Minerals.* U.S. Geological Survey Open-File Report 2004-1374.

9. **International Copper Study Group.** 2010. *The World Copper Factbook 2010.* Available at http://www.icsg.org.

10. **Doebrich, J.** 2009. "Facts About Copper: Copper Uses, Resources, Supply, Demand and Production Information." U.S. Geological Survey. http://pubs.usgs.gov/fs/2009/3031/FS2009-3031.pdf.

11. **National Research Council of the National Academies.** 2008. *Minerals, Critical Minerals, and the U.S. Economy.* Washington, DC: National Academies Press.

12. **U.S. Geological Survey.** 2010. *Mineral Commodity Summaries 2010,* p. 6. Washington, DC: U.S. Government Printing Office.

13. **Long, K. R., Van Gosen, B. S., Foley, N. K., and Cordier, D.** 2010. "The Principal Rare Earth Elements Deposits of the United States—A Summary of Domestic Deposits and a Global Perspective." *Scientific Investigations Report 2010–5220.* U.S. Geological Survey. http://pubs.usgs.gov/sir/2010/5220/pdf/SIR2010-5220.pdf#page=10.

14. **U.S. Geological Survey.** "Rare Earths Statistics and Information." Accessed June 28, 2012. Available at http://minerals.usgs.gov/minerals/pubs/commodity/rare_earths/.

15. **Haxel, G. B., Hedrick, J. B., and Orris, G. J.** "Rare Earth Elements—Critical Resources for High Technology." *U.S. Geological Survey Fact Sheet* 087-02. Accessed January 31, 2012. http://pubs.usgs.gov/fs/2002/fs087-02/.

16. **Zielinski, S.** 2010. "Rare Earth Elements Not Rare, Just Playing Hard to Get." *Surprising Science* (blog), *Smithsonian,* November 18. http://blogs.smithsonianmag.com/science/2010/11/rare-earth-elements-not-rare-just-playing-hard-to-get/.

17. **U.S. Department of Energy.** 2012. *Critical Materials Strategy: December 2011.* http://energy.gov/sites/prod/files/DOE_CMS2011_FINAL_Full.pdf.

18. **U.S. Geological Survey.** 2012. "Chromium Statistics and Information." In *Mineral Commodity Summaries 2012.* http://minerals.usgs.gov/minerals/pubs/commodity/chromium/mcs-2012-chrom.pdf.

19. **International Copper Study Group.** 2010. *The World Copper Factbook 2010.* Available at http://www.icsg.org.

20. **U.S. Geological Survey.** "Lead Statistics and Information." Accessed January 31, 2012. http://minerals.usgs.gov/minerals/pubs/commodity/lead.

21. **National Mining Association.** "Fast Facts About Coal." Accessed February 4, 2012. http://www.nma.org/statistics/fast_facts.asp.

22. **U.S. Environmental Protection Agency.** "Summitville Mine." Accessed February 5, 2012. http://www.epa.gov/superfund/accomp/factsheets04/summit.htm.

23. **HDICuris.** "The Florence (AZ) Copper Project: Building a Next Generation Copper Producer." Accessed February 6, 2012. http://www.curisresources.com/s/Home.asp.

HDICuris. 2012. "Florence Copper: Building a Next Generation Copper Producer." http://www.curisresources.com/i/pdf/CUV_FactSheet.pdf.

24. **U.S. Department of the Interior.** *Bureau of Land Management Mining Laws.* Accessed July 3, 2012. http://www.blm.gov/wo/st/en/info/regulations/mining_claims.html.

25. **World Bank Group.** *Environmental, Health, and Safety Guidelines.* 2007. Accessed July 3, 2012. http://www1.ifc.org/wps/wcm/connect/Topics_Ext_Content/IFC_External_Corporate_Site/IFC+Sustainability/Sustainability+Framework/Environmental,+Health,+and+Safety+Guidelines.

26. **Newman, C. J., Collins, D. N., and Weddick, A. J.** 2000. *Recent Operation and Environmental Control in the Kennecott Smelter Corporation.* Magna, UT: Kennecott Utah Copper Corporation.

27. **Dobb, E.** 1996. "Pennies from Hell: In Montana, the Bill for America's Copper Comes Due." *Harper's,* November, 39–54.

Dobb, E. 2000. "New Life in a Death Trap." *Discover Magazine,* December 1. http://discovermagazine.com/2000/dec/featnewlife.

28. Berkeley Pit Public Education Committee. "Summer 2009: Comprehensive Berkeley Pit Information." http://www.pitwatch.org/2009.htm.

29. U.S. Environmental Protection Agency. "National Priorities List." Accessed February 4, 2012. http://www.epa.gov/superfund/sites/npl.

U.S. Environmental Protection Agency. "Superfund Program: Silver Bow Creek/Butte Area." Accessed February 4, 2012. http://www.epa.gov/region8/superfund/mt/sbcbutte/index.html.

Chapter 5

1. Foxworthy, B. L., and Hill, M. 1982. "Volcanic Eruptions of 1980 at Mount St. Helens: The First 100 Days." *U.S. Geological Survey Professional Paper* 1249.

2. Decker, R., and Decker, B. 2006. *Volcanoes,* 4th ed., p. 51. New York: W. H. Freeman.

3. Decker, R., and Decker, B. 2006. *Volcanoes,* 4th ed., pp. 55–58. New York: W. H. Freeman.

4. Foxworthy, B. L., and Hill, M. 1982. "Volcanic Eruptions of 1980 at Mount St. Helens: The First 100 Days." *U.S. Geological Survey Professional Paper* 1249.

5. Decker, R., and Decker, B. 2006. *Volcanoes,* 4th ed., p. 46. New York: W. H. Freeman.

6. Sherrod, D. R., Scott, W. E., and Stauffer, P. H. 2008. "A Volcano Rekindled: The Renewed Eruption of Mount St. Helens, 2004–2006." *U.S. Geological Survey Professional Paper* 1750.

7. Simkin, T., and Siebert, L. 2000. "Earth's Volcanoes and Eruptions: An Overview." In *Encyclopedia of Volcanoes,* edited by H. Sigurdsson, p. 254. New York: Academic Press.

8. Sigurdsson, H. 2000. "Volcanic Episodes and Rates of Volcanism." In *Encyclopedia of Volcanoes,* edited by H. Sigurdsson, p. 273. New York: Academic Press.

9. Decker, R., and Decker, B. 2006. *Volcanoes,* 4th ed., pp. 134–137. New York: W. H. Freeman.

10. Schmincke, H.-U. 2004. *Volcanism,* p. 42. New York: Springer Verlag.

11. Delmelle, P., and Stix, J. 2000. "Volcanic Gases." In *Encyclopedia of Volcanoes,* edited by H. Sigurdsson, p. 805. New York: Academic Press.

12. Raymond, L. A. 2002. *Petrology: The Study of Igneous, Sedimentary, and Metamorphic Rocks,* 2nd ed., p. 34. Boston: McGraw Hill.

13. Chester, D. 1993. *Volcanoes and Society,* p. 74. London: Edward Arnold.

14. Newhall, C., Hendley II, J. W., and Stauffer, P. H. 1997. "The Cataclysmic 1991 Eruption of Mount Pinatubo, Philippines." *U.S. Geological Survey Fact Sheet* 113-97. http://pubs.usgs.gov/fs/1997/fs113-97/.

15. Raymond, L. A. 2002. *Petrology: The Study of Igneous, Sedimentary, and Metamorphic Rocks,* 2nd ed., p. 140. Boston: McGraw Hill.

16. Walker, G. P. L. 2000. "Basaltic Volcanoes and Volcanic Systems." In *Encyclopedia of Volcanoes,* edited by H. Sigurdsson, p. 284. New York: Academic Press.

17. U.S. Geological Survey Hawaiian Volcano Observatory. "Summary of Puʻu ʻOʻo–Kupaianaha Eruption, Kilauea Volcano, Hawaiʻi." Accessed August 15, 2012. http://hvo.wr.usgs.gov/kilauea/summary/main.html.

18. Walker, G. P. L. 2000. "Basaltic Volcanoes and Volcanic Systems." In *Encyclopedia of . Volcanoes,* edited by H. Sigurdsson, p. 286. New York: Academic Press.

19. Decker, R., and Decker, B. 2006. *Volcanoes,* 4th ed., p. 266. New York: W. H. Freeman.

20. Parfitt, E. A., and Wilson, L. 2008. *Fundaments of Physical Volcanology,* p. 4. Malden, MA: Blackwell Publishing.

21. Lipman, P. W. 2000. "Calderas." In *Encyclopedia of Volcanoes,* edited by H. Sigurdsson, p. 643. New York: Academic Press.

22. U.S. Geological Survey. 2009. "Yellowstone Caldera, Wyoming." Accessed June 9, 2011. http://vulcan.wr.usgs.gov/Volcanoes/Yellowstone/description_yellowstone.html.

23. U.S. Geological Survey. 2008. "CVO Website—Crater Lake, Oregon." Accessed June 9, 2011. http://vulcan.wr.usgs.gov/Volcanoes/CraterLake/description_crater_lake.html.

24. U.S. Geological Survey Volcano Hazards Program. 2009. "USGS: Volcano Hazards Program—Long Valley Caldera Geology and History." Accessed June 9, 2011. http://volcanoes.usgs.gov/lvo/about/history.php.

25. Decker, R., and Decker, B. 2006. *Volcanoes,* 4th ed., p. 177. New York: W. H. Freeman.

26. Hill, D. P., Bailey, R. A., Sorey, M. L., Hendley II, J. W., and Stauffer, P. H. 2000. "Living with a Restless Caldera—Long Valley, California." *U.S. Geological Survey Fact Sheet* 108-96, online ver. 2.1. Accessed January 20, 2010. http://pubs.usgs.gov/fs/fs108-96/.

27. Klimasauskas, E., Bacon, C., and Alexander, J. 2005. "Mount Mazama and Crater Lake: Growth and Destruction of a Cascade Volcano." *U.S. Geological Survey Fact Sheet* 092-02, online ver. 1.0. Accessed January 20, 2010. http://pubs.usgs.gov/fs/2002/fs092-02/.

28. Simkin, T., and Siebert, L. 2000. "Earth's Volcanoes and Eruptions: An Overview." In *Encyclopedia of Volcanoes,* edited by H. Sigurdsson, 259. New York: Academic Press.

29. **U.S. Geological Survey Volcano Hazards Program.** 2009. "Questions About Super Volcanoes." Accessed June 9, 2011. http://volcanoes.usgs.gov/yvo/about/faq/faqsupervolcano.php.

30. **Foxworthy, B. L., and Hill, M.** 1982. "Volcanic Eruptions of 1980 at Mount St. Helens: The First 100 Days." *U.S. Geological Survey Professional Paper* 1249.

31. **Schmincke, H.-U.** 2004. *Volcanism*, p. 260. New York: Springer Verlag.

32. **Chenowith, K.** 2001. "Two Major Volcanic Cooling Episodes Derived from Global Marine Air Temperature, AD 1807–1827." *Geophysical Research Letters* 28:2963–2966.

33. **Mills, M. J.** 2000. "Volcanic Aerosol and Global Atmospheric Effects." In *Encyclopedia of Volcanoes,* edited by H. Sigurdsson, p. 933. New York: Academic Press.

34. **Zerefos, C. S., Gerogiannis, V. T., Balis, D., Zerefos, S. C., and Kazantzidis, A.** 2007. "Atmospheric Effects of Volcanic Eruptions as Seen by Famous Artists and Depicted in Their Paintings." *Atmospheric Chemistry and Physics* 7:4027–4042.

35. **Parker, D. E., Wilson, H., Jones, P. D., Christy, J. R., and Folland, C. K.** 1995. "The Impact of Mount Pinatubo on World-Wide Temperatures." *International Journal of Climatology,* 16:487–497.

36. **Kunzig, R.** 2008. "A Sunshade for Planet Earth." *Scientific American* 299:46–55.

37. **Trenberth, K., and Dai, A.** 2007. "Effects of Mount Pinatubo Volcanic Eruption on the Hydrological Cycle as an Analog of Geoengineering." *Geophysical Research Letters* 34. doi:10.1029/2007GL030524.

38. **Tilmes, S., Müller, R., and Salawitch, R.** 2008. "The Sensitivity of Polar Ozone Depletion to Proposed Geoengineering Schemes." *Science* 30:1201–1204.

39. **Alaska Volcano Observatory.** "About Alaska's Volcanoes." Accessed August 15, 2012. http://avo.alaska.edu/volcanoes/about.php.

40. **Alaska Volcano Observatory.** "Katmai Reported Activity." Accessed August 15, 2012. http://www.avo.alaska.edu/volcanoes/activity.php?volcname=Katmai&eruptionid=502&page=basic.

41. **Guffanti, M., and Miller, E. K.** 2002. "Reducing the Threat to Aviation from Airborne Volcanic Ash." Paper presented at the 55th Annual International Air Safety Seminar, Dublin, Ireland, November 4–7. Available at http://volcanoes.usgs.gov/ash/trans/aviation_threat.html.

42. **Hendry, E. R.** 2010. "What We Know From the Icelandic Volcano." *Smithsonian Magazine,* April 22. http://www.smithsonianmag.com/science-nature/What-We-Know-From-the-Icelandic-Volcano.html.

43. **BBC News.** 2010. "Flight Disruptions Cost Airlines $1.7bn, Says IATA." Last updated April 21, 2010. Accessed June 9, 2011. http://news.bbc.co.uk/2/hi/business/8634147.stm.

44. **Miller, T. P., and Casadevall, T. J.** 2000. "Volcanic Ash Hazards to Aviation." In *Encyclopedia of Volcanoes,* edited by H. Sigurdsson, 915–930. New York: Academic Press.

45. **Sorey, M. L., Farrar, C. D., Gerlach, T. M., McGee, K. A., Evans, W. C., Colvard, E. M., Hills, D. P., et al.** 2007. "Invisible CO_2 Gas Killing Trees at Mammoth Mountain, California." *U.S. Geological Survey Fact Sheet* 172-96, online ver. 2.0. Accessed January 20, 2010. http://pubs.usgs.gov/fs/fs172-96/.

46. **Parfitt, E. A., and Wilson, L.** 2008. *Fundamentals of Physical Volcanology,* p. 171. Malden, MA: Blackwell Publishing.

47. **Schmincke, H.-U.** 2004. *Volcanism,* p. 259. New York: Springer Verlag.

48. **Smithsonian Global Volcanism Program.** 2011. "Lonquimay." Accessed June 8, 2011. http://www.volcano.si.edu/world/volcano.cfm?vnum=1507-10=&volpage=photos.

49. **Schmincke, H.-U.** 2004. *Volcanism,* p. 238. New York: Springer Verlag.

50. **U.S. Geological Survey Volcano Hazards Program.** 2010. "U.S. Volcanoes and Current Activity Alerts." Accessed January 20, 2010. http://volcanoes.usgs.gov/.

51. **Newhall, C. G., and Self, S.** 1982. "The Volcanic Explosivity Index (VEI): An Estimate of Explosive Magnitude for Historical Volcanism." *Journal of Geophysical Research* 87(C2): 1231–1238. doi:10.1029/JC087iC02p01231.

52. **Smithsonian Global Volcanism Program.** "USGS: Volcano Hazards Program—Long Valley Caldera Geology and History." Accessed August 20, 2012. http://www.volcano.si.edu/index.cfm.

53. **Helz, R. L.** 2005. "Monitoring Ground Deformation from Space." *U.S. Geological Survey Fact Sheet* 2005-3025. Accessed June 9, 2011. http://volcanoes.usgs.gov/activity/methods/insar//public_files/InSAR_Fact_Sheet/2005-3025.pdf.

54. **U.S. Geological Survey Volcano Hazards Program.** 2010. "Recent Ups and Downs of the Yellowstone Caldera—2007 Article." Accessed June 9, 2011. http://volcanoes.usgs.gov/yvo/publications/2007/upsanddowns.php.

55. **Crandell, D. R., and Mullineaux, D. R.** 1978. "Potential Hazards from Future Eruptions of Mount St. Helens Volcano, Washington." *U.S. Geological Survey Bulletin* 1383-C.

56. **U.S. Geological Survey Volcano Hazards Program.** 1999. "USGS: Volcano Hazards Program—Long Valley Caldera Geology and History." Accessed January 20, 2010. http://volcanoes.usgs.gov/volcanoes/long_valley/long_valley_geo_hist_13.html.

57. **Monastersky, R.** 1991. "Perils of Prediction." *Science News* 139:376–379.

58. **BBC.** 2002. "Volcano Hell." January 17. Last updated September 2005. Accessed January 20, 2010. http://www.bbc.co.uk/science/horizon/2001/volcanohelltrans.shtml.

59. Barberi, F., Martini, M., and Rosi, M. 1990. "Nevado del Ruiz Volcano (Colombia): Pre-Eruption Observations and the November 13, 1985 Catastrophic Event." *Journal of Volcanology and Geothermal Research* 42:1–12.

60. Decker, R., and Decker, B. 2006. *Volcanoes*, 4th ed., pp. 145–146. New York: W. H. Freeman.

61. ICAO News Centre. 2010. "ICAO Assesses Situation of Air Transport Following Eruption of Eyjafjallajökull Volcano in Iceland." April 20. http://icaopressroom.wordpress.com/2010/04/20/icao-assesses-situation-of-air-transport-following-eruption-of-eyjafjallajokull-volcano-in-iceland/.

62. Driedger, C. L., and Scott, W. E. 2008. "Mount Rainier—Living Safely with a Volcano in Your Backyard." *U.S. Geological Survey Fact Sheet* 2008-3062. http://pubs.usgs.gov/fs/2008/3062/fs2008-3062.pdf.

63. Sorenson, D. G. 2010. "Summary of Land-Cover Trends—Puget Lowland Ecoregion." U.S. Geological Survey Land Cover Trends Project. Accessed January 20, 2010. http://landcovertrends.usgs.gov/west/eco2Report.html.

64. Pierce County Department of Emergency Management. 2008. "Mount Rainier Volcanic Hazards Plan." Working Draft October 2008. http://www.co.pierce.wa.us/xml/abtus/ourorg/dem/emdiv/vhrp/rainerplan10-2008.pdf.

65. McClintock, J. 1999. "Under the Volcano." *Discover*, November 1. http://discovermagazine.com/1999/nov/underthevolcano1713.

66. Washington State Department of Transportation. 2009. "Orting Bridge for Kids Fact Sheet." Accessed January 20, 2010. http://www.wsdot.wa.gov/NR/rdonlyres/A451CD37-242E-490D-A11E-B608006C2A10/60482/BridgeForKidsFactSheet.pdf.

67. Chester, D. 1993. *Volcanoes and Society*, p. 205. London: Edward Arnold.

68. Smithsonian Global Volcanism Program. 2010. "Etna: Eruptive History." Accessed January 21, 2010. http://www.volcano.si.edu/world/volcano.cfm?vnum=0101-06=&volpage=erupt.

69. Barberi, F., Carapezza, M. L., Valenza, M., and Villari, L. 1993. "The Control of Lava Flow During the 1991–1992 Eruption of Mt. Etna." *Journal of Volcanology and Geothermal Research* 56:1–34.

70. Cowell, A. 1992. Zafferana Etnea Journal; It's Plug Up Mt. Etna or Go the Way of Pompeii, *New York Times,* April 25 edition.

Chapter 6

1. Klein, R. G. 2009. "Darwin and the Recent African Origin of Modern Humans." *Proceedings of the National Academy of Sciences* 106:16007–16009.

2. Gould, S. J. 1987. *Time's Arrow, Time's Cycle: Myth and Metaphor in the Discovery of Geological Time*. Cambridge, MA: Harvard University Press.

3. Beus, S. S., and Morales, M. 2003. *Grand Canyon Geology*. New York: Oxford University Press.

4. Rudwick, M. J. S. 1976. *The Meaning of Fossils: Episodes in the History of Palaeontology*. Chicago: University of Chicago Press.

Baucon, A. 2010. "Leonardo Da Vinci, the Founding Father of Ichnology." *PALAIOS* 25:361–367.

5. Holland, H. D., and Turekian, K. K. 2010. *Radioactive Geochronometry: A Derivative of the Treatise on Geochemistry*. London: Academic Press.

Berry, W. B. N. 2003. "Chronometry of Sediments and Sedimentary Rocks." *Treatise on Geochemistry* 7:325–349. Amsterdam: Elsevier.

6. Darwin, C. 1888. *The Origin of Species by Means of Natural Selection*. New York: TY Crowell.

7. Watson, J. D., and Crick, F. H. C. 1953. "Molecular Structure of Nucleic Acids." *Nature* 171:737–738.

8. Mould, R. F. 1995. "Röntgen and the Discovery of X-Rays." *British Journal of Radiology* 68:1145–1176.

9. Mould, R. F. 1998. "The Discovery of Radium in 1898 by Maria Sklodowska-Curie (1867–1934) and Pierre Curie (1859–1906) with Commentary on Their Life and Times." *British Journal of Radiology* 71:1229–1254.

10. Wager, L. R. 1964. "The History of Attempts to Establish a Quantitative Time-Scale." *Geological Society, London, Special Publications* 1:13–28.

11. Bowring, S. A. 1999. "Priscoan (4.00–4.03 Ga) Orthogneisses from Northwestern Canada." *Contributions to Mineralogy and Petrology* 134:3.

12. Wilde, S. A. 2001. "Evidence from Detrital Zircons for the Existence of Continental Crust and Oceans on the Earth 4.4 Gyr Ago." *Nature* 409:175.

13. U.S. Geological Survey. "Age of the Earth." Accessed July 23, 2012. http://pubs.usgs.gov/gip/geotime/age.html.

14. Walter, R. C. 1994. "Age of Lucy and the First Family: Single-Crystal $^{40}Ar/^{39}Ar$ Dating of the Denen Dora and Lower Kada Hadar Members of the Hadar Formation, Ethiopia." *Geology* 22:6–10.

15. White, T. D. et al. 2009. "Ardipithecus Ramidus and the Paleobiology of Early Hominids." *Science* 326:64–86.

16. *Science*. 2008. "*Ardipithecus ramidus*." Accessed July 23, 2012. http://www.sciencemag.org/site/feature/misc/webfeat/ardipithecus/.

17. Keeling, C. D. et al. 1976. "Atmospheric Carbon Dioxide Variations at Mauna Loa Observatory, Hawaii." *Tellus* 28:538–551.

18. Petit, J. R. et al. 1999. "Climate and Atmospheric History of the Past 420,000 Years from the Vostok Ice Core, Antarctica." *Nature* 399:429–436.

19. Steffensen, J. P. et al. 2008. "High-Resolution Greenland Ice Core Data Show Abrupt Climate Change Happens in Few Years." *Science* 321:680–684.

20. Alley, R. B. 2002. *The Two-Mile Time Machine: Ice Cores, Abrupt Climate Change, and Our Future.* Princeton, NJ: Princeton University Press.

Chapter 7

1. Gende, S. M., and Quinn, T. P. 2006. "The Fish and the Forest." *Scientific American* 295:84–89.

2. Strobel, B. 2004. "Ecosystem Responses to a Large-Scale Salmon Carcass Enrichment Project." Accessed July 22, 2011. http://www.sandyriverpartners.org/reports/CarcassMonitoringInterim2004.pdf.

Biby, R. E., Fransen, B. R., Bisson, P. A., and Walter, J. K. 1998. "Response of Juvenile Coho Salmon (*Oncorhynchus kisutch*) and Steelhead (*Oncorhynchus mykiss*) to the Addition of Salmon Carcasses to Two Streams in Southwestern Washington, U.S.A." *Canadian Journal of Fisheries and Aquatic Sciences* 55:1909–1918.

3. Dempster, W. F. 1999. "Biosphere 2 Engineering Design." *Ecological Engineering* 13:31–42.

4. Severinghaus, J. P., Broecker, W. S., Dempster, W. F., MacCallum, T., and Wahlen, M. 1994. "Oxygen Loss in Biosphere 2." *EOS, Transactions of the American Geophysical Union* 75:33, 35–37.

5. Alling, A., Nelson, M., and Silverstone, S. 1993. "Life Under Glass: The Inside Story of Biosphere 2." Oracle, AZ: The Biosphere Press.

6. Cohen, J. E., and Tilman, D. 1996. "Biosphere 2 and Biodiversity: The Lessons So Far." *Science* 274:1150–1151.

7. Naeem, S., Chapin III, F. S., Costanza, R., Ehrlich, P. R., Golley, F. B., Hooper, D. U., Lawton, H. J., et al. 1999. "Biodiversity and Ecosystem Functioning: Maintaining Natural Life Support Processes." *Issues in Ecology* 4:2–11.

8. Kaufmann, R. K., and Cleveland, C. J. 2008. *Environmental Science*, ch. 5. Boston: McGraw Hill Higher Education.

9. Marino, B. D. V., Mahato, T. R., Druitt, J. W., Leigh, L., Lin, G., Russell, R. M., and Tubiello, F. N. 1999. "The Agricultural Biome of Biosphere 2: Structure, Composition and Function." *Ecological Engineering* 13:199–234.

10. Silverstone, S. E., Harwood, R. R., Fraco-Vizcaino, E., Allen, J., and Nelson, M. 1999. "Soil in the Agricultural Area of Biosphere 2 (1991–1993)." *Ecological Engineering* 13:179–188.

11. de Secada, C. A. G. 1985. "Arms, Guano, and Shipping: The W. R. Grace Interests in Peru, 1865–1885." *Business History Review* 59:597–621.

12. Leonard, A. 2008. "When Guano Imperialists Ruled the Earth." Accessed July 18, 2011. http://www.salon.com/technology/how_the_world_works/2008/02/29/guano_imperialism.

13. Kaufmann, R. K., and Cleveland, C. J. 2008. *Environmental Science*, ch. 9. Boston: McGraw Hill Higher Education.

14. Kaufmann, R. K., and Cleveland, C. J. 2008. *Environmental Science*, p. 185. Boston: McGraw Hill Higher Education.

15. Karter, A. J., and Dieterich, R. A. 1989. "Population Dynamics: An Introduction for Alaskan Reindeer Herders." Agriculture and Forestry Experiment Station, University of Alaska, Fairbanks, Bulletin 81. Accessed July 5, 2012. http://www.uaf.edu/files/snras/B81.pdf.

Talbot, S. S., Talbot, S. L., Thomson, J. W., and Schofield, W. B. 2001. "Lichens from St. Matthew and St. Paul Islands, Bering Sea, Alaska." *Bryologist* 104:47–58.

Klein, D. R., and Shulski, M. 2009. "Lichen Recovery Following Heavy Grazing by Reindeer Delayed by Climate Warming." *Ambio* 38:11–16.

16. Ricklefs, R. E. 2008. *The Economy of Nature*, 6th ed., ch. 15. New York: W.H. Freeman.

17. Sheriff, M. J., Krebs, C. J., and Boonstra, R. 2011. "From Process to Pattern: How Fluctuating Predation Risk Impacts the Stress Axis of Snowshoe Hares During the 10-Year Cycle." *Oecologia* 166:593–605.

18. Tyson, R., Haines, S., and Hodges, K. E. 2010. "Modelling the Canada Lynx and Snowshoe Hare Population Cycle: The Role of Specialist Predators." *Theoretical Ecology* 3:97–111.

19. Gauthier, S., Bergeron, Y., and Simon, J-P. 1996. "Effects of Fire Regime on the Serotiny Level of Jack Pine." *Journal of Ecology* 84:539–548.

20. Ricklefs, R. E. 2008. *The Economy of Nature*, 6th ed., p. 380. New York: W.H. Freeman.

21. Ellis, E. C., and Ramankutty, N. 2008. "Putting People in the Map: Anthropogenic Biomes of the World." *Frontiers in Ecology and Environment* 6:439–447.

22. MacDonald, G. 2003. *Biogeography: Introduction to Space, Time and Life*, pp. 168–171. New York: John Wiley and Sons.

23. Pickett, S. T. A., Cadenasso, M. L., Grove, J. M., Boone, C. G., Groffman, P. M., Irwin, E., Kaushal, et al. 2011. "Urban Ecological Systems: Scientific Foundations and a Decade of Progress." *Journal of Environmental Management* 92:331–362.

24. Fritts, S. H., Bangs, E. E., Fontaine, J. A., Johnson, M. R., Phillips, M. K., Koch, E. D., Gunson, J. R. 1997. "Planning and Implementing a Reintroduction of Wolves to Yellowstone National Park and Central Idaho." *Restoration Ecology* 5:7–27.

25. Fortin, D., Beyer, H. L., Boyce, M. S., Smith, D. W., Duchesne, T., and Mao, J. S. 2005. "Wolves Influence

Elk Movements: Behavior Shapes a Trophic Cascade in Yellowstone National Park." *Ecology* 86:1320–1330.

26. **Ripple, W. J., and Beschta, R. L.** 2003. "Wolf Reintroduction, Predation Risk, and Cottonwood Recovery in Yellowstone National Park." *Forest Ecology and Management* 184:299–313.

27. **Clifford, F.** 2009. "Wolves and the Balance of Nature in the Rockies." *Smithsonian*, February 2009. February. Accessed July 7, 2012. http://www.smithsonianmag.com/science-nature/Howling-Success.html?c=y&page=1.

28. **Primack, R. B.** 2002. *Essentials of Conservation Biology.* Sunderland, MA: Sinauer.

29. **Australian Museum.** 2011. "What Is Biodiversity?" Accessed July 11, 2011. http://australianmuseum.net.au/What-is-biodiversity.

30. **Chapin, F. S. I., Zavaleta, E. S., Eviners, V. T., Naylor, R. L., Vitousek, P. M., Reynolds, H. L., Hooper, D. U., et al.** 2000. "Consequences of Changing Biodiversity." *Nature* 405:234–242.

31. **Purvis, A., and Hector, A.** 2000. "Getting the Measure of Biodiversity." *Nature* 405:212–219.

32. **Tilman, D.** 2000. "Causes, Consequences and Ethics of Biodiversity." *Nature* 405:208–211.

33. **Cork, S.** 2001. "How Ecosystem Services Relate to One Another." In Ecosystem Services Project, *Natural Assets: An Inventory of Ecosystem Goods and Services in the Goulburn Broken Catchment.* Accessed July 12, 2011. http://www.ecosystemservicesproject.org/html/publications/docs/Natural_Assets_LR.pdf.

34. **Costanza, R., d'Arge, R., de Groot, R., Farber, S., Grasso, M., Hannon, B., Limburg, K., et al.** 1997. "The Value of the World's Ecosystem Services and Natural Capital." *Nature* 387:253–260.

Daily, G. C., Alexander, S., Ehrlich, P. R., Goulder, L., Lubchenco, J., Matson, P. A., Mooney, H. A., et al. 1997. "Ecosystem Services: Benefits Supplied to Human Societies by Natural Ecosystems. *Issues in Ecology* 2:1–16.

35. **UN Food and Agriculture Organization.** 2011. "State of the World's Forests 2011." Accessed July 12, 2011. http://www.fao.org/docrep/013/i2000e/i2000e.pdf.

36. **FAO Fisheries and Aquaculture Department.** 2010. *The State of World Fisheries and Aquaculture 2010*, p. 3. Accessed July 12, 2011. http://www.fao.org/docrep/013/i1820e/i1820e.pdf.

37. **Daily, G. C., Alexander, S., Ehrlich, P. R., Goulder, L., Lubchenco, J., Matson, P. A., Mooney, H. A., et al.** 1997. "Ecosystem Services: Benefits Supplied to Human Societies by Natural Ecosystems. *Issues in Ecology* 2:1–16.

38. **Sumner, D. A., and Borris, H.** "Bee-conomics and the Leap in Pollination Fees." Giannini Foundation of Agricultural Economics. Accessed July 15, 2011. http://aic.ucdavis.edu/research/bee-conomics-1.pdf.

Apis Hive Company. 2011. "Pollination Recap." Accessed July 15, 2011. http://www.apishive.com/pollenation-serives/pollination-services-fees.

39. **Communicating Ecosystem Services Project.** 2011. "Ecosystem Services Fact Sheets: Water Purification." Accessed July 19, 2011. http://www.esa.org/ecoservices/comm/body.comm.fact.wate.html.

40. **New York City Environmental Protection Department.** 2010. "New York City 2010 Drinking Water Supply and Quality Report." Accessed July 19, 2011. http://www.nyc.gov/html/dep/pdf/wsstate10.pdf.

41. **Goucher, B. A.** 2011. "Facts About the Trees in the Catskill Mountains." eHow.com. Accessed July 19, 2011. http://www.ehow.com/facts_5491619_trees-catskill-mountains.html.

42. **Schneiderman, J. S.** 2000. "From the Catskills to Canal Street: New York City's Water Supply." In *The Earth Around Us: Maintaining a Livable Planet,* edited by J. S. Schneiderman, 166–180. New York: W.H. Freeman.

43. **U.S. Environmental Protection Agency.** 2011. "History of the New York City Watershed Filtration Avoidance Determination." Accessed July 19, 2011. http://www.epa.gov/Region2/water/nycshed/FADHistory.pdf.

44. **Watershed Agricultural Council.** 2010. "Whole Farm Planning." Accessed July 19, 2011. http://www.nycwatershed.org/ag_planning.html.

45. **Smithsonian Institution Human Origins Program.** 2011. "Human Evolution Timeline Interactive." Accessed July 14, 2011. http://humanorigins.si.edu/evidence/human-evolution-timeline-interactive.

46. **Cohen, J. E.** 1995. *How Many People Can the Earth Support,* pp. 77–78. New York: W.W. Norton.

47. **Marean, C. W.** 2010. "When the Sea Saved Humanity. *Scientific American* 303:54–61.

48. **Erb, K-H., Krausmann, F., Gaube, V., Gingrich, S., Bondeau, A., Fischer-Kowalski, M., and Haberl, H.** 2009. "Analyzing the Global Human Appropriation of Net Primary Production—Processes, Trajectories, Implications. An introduction." *Ecological Economics* 69: 250–259.

49. **Foley, J. A., DeFries, R., Asner, G. P., Barford, C., Bonan, G., Carpenter, S. R., Chapin, F. S., et al.** 2005. "Global Consequences of Land Use." *Science* 309:570–574.

50. **Vitousek, P. M., Aber, J., Howarth, R. W., Likens, G. E., Matson, P. A., Schindler, D. W., Schlesinger, W. H., et al.** 1997. "Human Alteration of the Global Nitrogen Cycle: Causes and Consequences." *Issues in Ecology* 1:1–16.

51. **Hammond, A. L.** 1971. Phosphate Replacements: Problems with the Washday Miracle. *Science* 172:361–363.

52. **Laws, E. A.** 2000. *Aquatic Pollution: An Introductory Text*, 3rd ed. New York: John Wiley and Sons.

53. **Wetzel, R. G.** 1983. *Limnology,* 2nd ed. Philadelphia, PA: Saunders College Publishing.

54. **Litke, D. W.** 1999. "Review of Phosphorus Control Measures in the United States and Their Effects on Water Quality." *U.S. Geological Survey Water-Resources Investigations Report* 99-4007.

55. **Koch, W.** 2010, June 30. "16 States Ban Phosphate-Laden Dishwasher Soap." *USA Today*. Accessed April 13, 2013. http://content.usatoday.com/communities/greenhouse/post/2010/06/16-states-ban-phosphate-laden-dishwasher-soap/1#.UWm2KnCeS0w.

56. **U.S. Environmental Protection Agency.** 2008. *The Great Lakes: An Environmental Atlas and Resource Book*. Accessed July 22, 2011. http://epa.gov/greatlakes/atlas/glatch4.html#Eutrophication.

57. **National Park Service.** 2011. "Tallgrass Prairie National Preserve." Accessed July 15, 2011. http://www.nps.gov/tapr/index.htm.

58. **Fahrig, L.** 2003. "Effects of Habitat Fragmentation on Biodiversity." *Annual Reviews in Ecology and Evolutionary Systematics* 34:487–515.

59. **Chalfoun, A. D., Thompson III, F. R., and Ratnaswamy, M. J.** 2002. "Nest Predators and Fragmentation: A Review and Meta-Analysis." *Conservation Biology* 16:306–318.

60. **National Invasive Species Information Center.** 2011. "Plants—Giant Hogweed." Accessed July 18, 2011. http://www.invasivespeciesinfo.gov/plants/hogweed.shtml.

61. **Cornell University Cooperative Extension.** 2011. "Giant Hogweed." Accessed July 18, 2011. http://nyis.info/plants/GiantHogweed.aspx#HumanImpact.

62. **Cornell University Cooperative Extension.** 2011. "Zebra Mussel and Quagga Mussel." Accessed July 18, 2011. http://nyis.info/animals/ZebraAndQuaggaMussel.aspx.

63. **Mack, R. N., Simberloff, D., Lonsdale, W. M., Evans, H., Clout, M., and Bazzaz, F.** 2000. "Biotic Invasions: Causes, Epidemiology, Global Consequences and Control." *Issues in Ecology* 5:1–20.

64. **Bohlen, P. J., Scheu, S., Hale, C. M., McLean, M. A., Migge, S., Groffman, P. M., and Parkinson, D.** 2004. "Non-Native Invasive Earthworms as Agents of Change in Northern Temperate Forests." *Frontiers in Ecology and the Environment* 2:427–435.

Frelich, L. E., Hale, C. M., Scheu, S., Holdsworth, A. R., Henegha, L., Bohlen, P. J., and Reich, P. B. 2006. "Earthworm Invasion into Previously Earthworm-Free Temperate and Boreal Forests." *Biological Invasions* 8:1235–1245.

65. **Groffman, P. M., Bohlen, P. J., Fisk, M. C., and Fahey, T. J.** 2004. "Earthworm Invasion and Microbial Biomass in Temperate Forest Soils." *Ecosystems* 7:45–54.

66. **Costello, D. M., and Lamberti, G. A.** 2008. "Non-Native Earthworms in Riparian Soils Increase Nitrogen Flux into Adjacent Aquatic Ecosystems." *Oecologia* 158:499–510.

67. **Oates, J. F., Abedi-Lartey, M., McGraw, W. S., Struhsaker, T. T., and Whitesides, G. H.** 2000. "Extinction of a West African Red Colobus Monkey." *Conservation Biology* 14:1526–1532.

68. **McGraw, W. S., and Oates, J. F.** 2007. "Miss Waldron's Red Colobus." In *Primates in Peril: The World's 25 Most Endangered Primates 2006–2008*, R. A. Mittermeier et al. (compilers), p. 10. Unpublished report, IUCN/SSC Primate Specialist Group, International Primatological Society, and Conservation International, Arlington, VA.

69. **Conservation International.** 2011. "Extinction Threat Growing for Mankind's Closest Relatives." Accessed July 11, 2011. http://www.conservation.org/newsroom/pressreleases/Pages/Extinction-primates-red-list-IUCN.aspx.

70. **Barnosky, A. D., Matzke, N., Tomiya, S., Wogan, G. O. U., Swartz, B., Quental, T. B., Marshall, C., et al.** 2011. "Has the Earth's Sixth Mass Extinction Already Arrived?" *Nature* 471:51–57.

71. **Morris, W., Doak, D., Groom, M., Kareiva, P., Fieberg, J., Gerber, L., Murphy, P., et al.** 1999. *A Practical Handbook for Population Viability Analysis*. Washington, DC: The Nature Conservancy.

72. **Shaffer, M. L.** 1981. "Minimum Population Sizes for Species Conservation." *BioScience* 31:131–134.

73. **Meffe, G. K., and Carroll, C. R.** 1994. *Principles of Conservation Biology*. Sunderland, MA: Sinauer.

74. **BBC News.** 2006. "The Story of the Stephens Island Wren." Accessed July 22, 2011. http://www.bbc.co.uk/dna/h2g2/A11484902.

75. **Diamond, J. M.** 1975. "The Island Dilemma: Lessons of Modern Biogeographic Studies for the Design of Nature Preserves." *Biological Conservation* 7:129–146.

Preston, F. W. 1960. "Time and Space and the Variation of Species." *Ecology* 41:611–627.

Terborgh, J. 1974. "Preservation of Natural Diversity: The Problem of Extinction Prone Species." *BioScience* 24:715–722.

76. **Darlington, P. J.** 1957. *Zoogeography: The Geographical Distribution of Animals*. New York: John Wiley and Sons.

77. **Jackson, S. T., and Hobbs, R. J.** 2009. "Ecological Restoration in the Light of Ecological History. *Science* 31:567–569. doi:10.1126/science.1172977.

Chapter 8

1. **Mann, C. C.** 2002. "The Real Dirt on Rainforest Fertility." *Science* 297:920–923. doi:10.1126/science.297.5583.920.

2. **Mann, C. C.** 2000. "Earthmovers of the Amazon." *Science* 287:786–789. doi:10.1126/science.287.5454.786.

Mann, C. C. "1491." 2002. *Atlantic Monthly*, March, 41–53.

3. **Glaser, B.** 2007. "Prehistorically Modified Soils of Central Amazonia: A Model for Sustainable Agriculture in the Twenty-First Century." *Philosophical Transactions of the Royal Society B: Biological Sciences* 362:187–196. doi:10.1098/rstb.2006.1978.

4. **Lal, R.** 2009. "Soils and Sustainable Agriculture: A Review." In *Sustainable Agriculture* 1., edited by E. Lichtfouse et al., 15–33. Dijon, France: Springer Science.

5. **Marris, M.** 2006. "Putting the Carbon Back: Black Is the New Green." *Nature* 442:624–626, doi:10.1038/442624a.

6. **Woods, W. I., and McCann, J. M.** 1999. "The Anthropogenic Origin and Persistence of Amazonian Dark Earths." *Yearbook. Conference of Latin Americanist Geographers* 25:7–14.

7. **Marris, E.** 2006. "Putting the Carbon Back: Black Is the New Green." *Nature* 442:624–626. doi:10.1038/442624a.

8. **Lehmann, J.** 2007. "A Handful of Carbon." *Nature* 447:143–144. doi:10.1038/447143a.

Sombroek, W., Ruivo, M. L., Fearnside, P. M., Glaser, B., and Lehmann, J. 2003. "Amazonian Dark Earths as Carbon Stores and Sinks." In *Amazonian Dark Earths*, edited by J. Lehmann et al., 125–139. Dordrecht: Kluwer Academic Publishers.

9. **Pessenda, L. C. R., Gouveia, S. E. M., and Aravena, R.** 2001. "Radiocarbon Dating of Total Soil Organic Matter and Humin Fraction and Its Comparison with ^{14}C Ages of Fossil Charcoal." *Radiocarbon* 43:595.

10. **National Research Council of the National Academies, Committee on Challenges and Opportunities in Earth Surface Processes.** 2010. *Landscapes on the Edge: New Horizons for Research on Earth's Surface.* Washington DC: National Academies Press.

11. **Martel, S. J.** 2011. "Mechanics of Curved Surfaces, with Application to Surface-Parallel Cracks." *Geophysical Research Letters* 38:L20303. doi:10.1029/2011GL049354.

12. **Hall, K.** 1999. "The Role of Thermal Stress Fatigue in the Breakdown of Rock in Cold Regions." *Geomorphology* 31:47–63. doi:10.1016/S0169-555X(99)00072-0.

13. **Gómez-Heras, M.** 2006. "Surface Temperature Differences Between Minerals in Crystalline Rocks: Implications for Granular Disaggregation of Granites Through Thermal Fatigue." *Geomorphology* 78:236. doi:10.1016/j.geomorph.2005.12.013.

Hall, K., Guglielmin, M., and Strini, A. 2008. "Weathering of Granite in Antarctica: II. Thermal Stress at the Grain Scale." *Earth Surface Processes and Landforms* 33:475–493. doi:10.1002/esp.1617.

14. **Walder, J. S., and Hallet, B.** 1986. "The Physical Basis of Frost Weathering: Toward a More Fundamental and Unified Perspective." *Arctic and Alpine Research* 18:27–32. doi:10.2307/1551211.

Hallet, B., Walder, J. S., and Stubbs, C. W. 1991. "Weathering by Segregation Ice Growth in Microcracks at Sustained Subzero Temperatures: Verification from an Experimental Study Using Acoustic Emissions." *Permafrost and Periglacial Processes* 2:283–300. doi:10.1002/ppp.3430020404.

15. **Raymo, M. E., and Ruddiman, W. F.** 1992. "Tectonic Forcing of Late Cenozoic Climate." *Nature* 359:117–122. doi:10.1038/359117a0.

Kump, L. R., Brantley, S. L., and Arthur, M. A. 2000. "Chemical Weathering, Atmospheric CO_2, and Climate." *Annual Review of Earth and Planetary Sciences* 28:611–667. doi:10.1146/annurev.earth.28.1.611.

16. **Molnar, P., and England, P.** 1990. "Late Cenozoic Uplift of Mountain Ranges and Global Climate Change: Chicken or Egg?" *Nature* 346:29–34. doi:10.1038/346029a0.

17. **Carter, D. L., Mortland, M. M., and Kemper, W. D.** 1986. "Specific Surface." In *Methods of Soil Analysis, Part I*, 2nd ed., edited by A. Klute, 413–422. Madison, WI: American Society of Agronomy and Soil Science Society of America.

18. **Darwin, C.** 1892. *The Formation of Vegetable Mould, Through the Action of Worms, with Observations on Their Habits*, pp. 308–316. London: J. Murray.

19. **Calvaruso, C., Turpault, M.-P., and Frey-Klett, P.** 2006. "Root-Associated Bacteria Contribute to Mineral Weathering and to Mineral Nutrition in Trees: A Budgeting Analysis." *Applied and Environmental Microbiology* 72:1258–1266. doi:10.1128/AEM.72.2.1258-1266.2006.

Uroz, S., et al. 2009. "Mineral Weathering by Bacteria: Ecology, Actors and Mechanisms." *Trends in Microbiology* 17:378–387. doi:10.1016/j.tim.2009.05.004.

20. **Bonneville, S., Smits, M. M., Brown, A., Harrington, J., Leake, J. R., Brydson, R., and Benning, L. G.** 2009. "Plant-Driven Fungal Weathering: Early Stages of Mineral Alteration at the Nanometer Scale." *Geology* 37:615–618. doi:10.1130/G25699A.1.

21. **Schnitzer, M. and Khan, S. U.** 1978. *Soil Organic Matter.* Amsterdam: Elsevier.

22. **Quinton, J. N., Govers, G., Van Oost, K., and Bardgett, R. D.** 2010. "The Impact of Agricultural Soil Erosion on Biogeochemical Cycling." *Nature Geoscience* 3:311–314. doi:10.1038/ngeo838.

23. **Food and Agriculture Organization of the United Nations.** "Digital Soil Map of the World." Accessed February 7, 2013. http://www.fao.org/nr/land/soils/digital-soil-map-of-the-world/en/.

24. **Eswaran, H., Beinroth, F., and Reich, P.** 1999. "Global Land Resources and Population Supporting Capacity." *American Journal of Alternative Agriculture* 14:129–136.

25. **Pimental, D., and Hall, C. W.** 1989. *Food and Natural Resources.* San Diego, CA: Academic Press.

26. **World Food Programme.** "Hunger: Who Are the Hungry?" Accessed June 20, 2013. http://www.wfp.org/hunger/who-are.

27. **Malthus, T. R.** 1999. In *An Essay on the Principle of Population,* edited by G. Gilbert. New York: Oxford University Press.

28. **United Nations Food and Agriculture Organization,** based on the work of Bot, A., J., Nachtergaele, F. O., and Young, A. 2000. "Land Resource Potential and Constraints at Regional and Country Levels." *World Soil Resources Report* 90:114.

29. **Eswaran, H., Lal, R., and Reich, P. F.** 2001. "Land Degradation: An Overview." In *Responses to Land Degradation,* (Proceedings of the Second International Conference on Land Degradation and Desertification, Khon Kaen, Thailand), edited by E. M. Bridges, I. D. Hannam, L. R. Oldeman, F. W. T. Pening de Vries, S. J. Scherr, and S. Sompatpanit, 20–35. New Delhi: Oxford Press.

30. **Syvitski, J. P. M., Vörösmarty, C. J., Kettner, A. J., and Green, P.** 2005. "Impact of Humans on the Flux of Terrestrial Sediment to the Global Coastal Ocean." *Science* 308:376–380. doi:10.1126/science.1109454.

Syvitski, J. P. M., and Kettner, A. 2011. "Sediment Flux and the Anthropocene." *Philosophical Transactions of the Royal Society A: Mathematical, Physical and Engineering Sciences* 369:957–975. doi:10.1098/rsta.2010.0329.

31. **Walter, R. C., and Merritts, D. J.** 2008. "Natural Streams and the Legacy of Water-Powered Mills." *Science* 319:299–304. doi:10.1126/science.1151716.

32. **Hillel, D.** 1992. *Out of the Earth: Civilization and the Life of the Soil.* Berkeley: University of California Press.

33. **Schubert, S. D., Suarez, M. J., Pegion, P. J., Koster, R. D., and Bacmeister, J. T.** 2004. "On the Cause of the 1930s Dust Bowl." *Science* 303:1855–1859. doi:10.1126/science.1095048.

34. **Cook, B. I., Miller, R. L., and Seager R.** 2009. "Amplification of the North American 'Dust Bowl' Drought Through Human-Induced Land Degradation." *Proceedings of the National Academy of Sciences* 106:4997–5001. doi:10.1073/pnas.0810200106.

35. **Lowdermilk, W. C.** 1975. *Conquest of the Land Through Seven Thousand Years.* Washington, DC: U.S. Dept. of Agriculture, Soil Conservation Service.

36. **Montgomery, D. R.** 2007. "Soil Erosion and Agricultural Sustainability." *Proceedings of the National Academy of Sciences of the United States of America* 104:13268–13272. doi:10.1073/pnas.0611508104.

37. **FAO Media Center.** 2011, September 7. "Global Soil Partnership For Food Security Launched at FAO: New Effort to Assure Soils Future Generations." Accessed May 5, 2012. http://www.fao.org/news/story/en/item/89277/icode/.

38. **Omuto, C. T., Balint, Z., and Alim, M. S.** 2011. "A Framework for National Assessment of Land Degradation in the Drylands: A Case Study of Somalia." In *Land Degradation & Development,* edited by A. Cerdà. Published electronically October 20, 2011. doi:10.1002/ldr.1151.

39. **Millennium Ecosystem Assessment.** 2005. *Ecosystems and Human Well-Being: Desertification Synthesis.* Washington, DC: World Resources Institute. Accessed October 22, 2012. http://www.maweb.org/documents/document.355.aspx.pdf.

40. **Dregne, H. E., and Chou, N-T.** 1992. "Global Desertification Dimensions and Costs." In *Degradation and Restoration of Arid Lands,* edited by H. E. Dregne, 249–282. Lubbock: Texas Tech University.

41. **Schertz, D. L.** 1983. "The Basis for Soil Loss Tolerances." *Journal of Soil and Water Conservation* 38:10–14.

42. **Montgomery, D. R.** 2007. *Dirt: The Erosion of Civilizations.* Berkeley: University of California Press.

43. **Committee on Challenges and Opportunities in Earth Surface Processes, National Research Council.** 2010. *Landscapes on the Edge: New Horizons for Research on Earth's Surface.* Washington, DC: National Academies Press.

44. **Conservation Technology Information Center.** *Conservation Tillage Facts.* Accessed December 2, 2012. http://www.ctic.org/resourcedisplay/281/.

45. **U.S. Geologic Survey.** "Landslides Hazard Program." Accessed October 15, 2012. http://landslides.usgs.gov/.

46. **Spignesi, S. J.** 2004. *Catastrophe!: The 100 Greatest Disasters of All Time,* p. 60. New York: Citadel Press.

Chapter 9

1. **Syed, T. H., et al.** 2010. "Satellite-Based Global-Ocean Mass Balance Estimates of Interannual Variability and Emerging Trends in Continental Freshwater Discharge." *Proceedings of the National Academy of Sciences.* doi:10.1073/pnas.1003292107.

2. **Battersby, S., and Le Page, M.** 2001. "Flood of Floods: Here Comes the Rain." *New Scientist* 2804: 44–47.

3. **Trenberth, K. E.** 2011. "Changes in Precipitation with Climate Change." *Climate Research* 47: 123. http://nldr.library.ucar.edu/repository/assets/osgc/OSGC-000-000-000-596.pdf.

4. **Intergovernmental Panel on Climate Change.** 2007. *Climate Change 2007: The Physical Science Basis: Contribution of Working Group I to the Fourth Assessment Report of the Intergovernmental Panel on Climate Change.* New York: Cambridge University Press.

5. **Shiklomanov, I. A.** 2000. "Appraisal and Assessment of World Water Resources." *Water International* 25: 11–32. http://www.tandfonline.com/doi/abs/10.1080/02508060008686794#.UdHnxZyOmtw.

6. **Southern Nevada Water Authority.** "Water Smart Landscapes Rebate." Accessed June 23, 2013. http://www.snwa.com/rebates/wsl.html.

7. **Meko, D. M. et al.** 2007. "Medieval Drought in the Upper Colorado River Basin." *Geophysical Research Letters* 34: 10705; Woodhouse, C. A., Gray, S. T., and Meko, D. M.

2006. "Updated Streamflow Reconstructions for the Upper Colorado River Basin." *Water Resources Research* 42: W05415. doi:10.1029/2005WR004455.

8. **Pyper, J., and ClimateWire.** 2011. "Colorado River Faces Flood and Drought—Becoming Less Reliable?" *Scientific American*. http://www.scientificamerican.com/article .cfm?id=colorado-river-faces-flood-and-drought.

9. **U.S. Bureau of Reclamation.** "Colorado River Basin Water Supply and Demand Study." Accessed July 13, 2012. http://www.usbr.gov/lc/region/programs/crbstudy.html.

10. **Barnett, T. P., and Pierce, D. W.** 2009. "Sustainable Water Deliveries from the Colorado River in a Changing Climate." *Proceedings of the National Academy of Sciences* 106: 7334–7338. doi:10.1073/pnas.0812762106; Committee on the Scientific Bases of Colorado River Basin Water, National Research Council. 2007. *Colorado River Basin Water Management: Evaluating and Adjusting to Hydroclimatic Variability*. National Academies Press.

11. **Blum, M. D., et al.** 2000. "Late Pleistocene Evolution of the Lower Mississippi River Valley, Southern Missouri to Arkansas." *Geological Society of America Bulletin* 112: 221–235. doi:10.1130/0016-7606(2000)112<221:LPEOTL >2.0.CO;2.

Rittenour, T. M., Blum, M. D., and Goble, R. J. 2007. "Fluvial Evolution of the Lower Mississippi River Valley During the Last 100 K.y. Glacial Cycle: Response to Glaciation and Sea-level Change." *Geological Society of America Bulletin* 119: 586–608. doi:10.1130/B25934.1.

12. **Ashley, S. T., and Ashley, W. S.** 2008. "Flood Fatalities in the United States." *Journal of Applied Meteorology and Climatology* 47: 805–818. doi:10.1175/2007JAMC1611.1.

13. **National Weather Service, Office of Climate, Water, and Weather Services.** "Weather Fatalities." Accessed August 10, 2012. http://www.weather.gov/os/hazstats.shtml.

14. **Brody, S. D., Highfield, W. E., and Kang, J. E.** 2011. *Rising Waters: The Causes and Consequences of Flooding in the United States*. Cambridge, UK: Cambridge University Press.

15. **Spatial Hazard Events and Losses Database for the United States.** Accessed August 9, 2012. http://webra.cas .sc.edu/hvri/products/sheldus.aspx.

16. **Mutel, C. F.** 2010. *A Watershed Year: Anatomy of the Iowa Floods of 2008*. Iowa City: University of Iowa Press.

17. **King, R. O.** 2011. "National Flood Insurance Program: Background, Challenges, and Financial Status." Congressional Research Service, Report for Congress.

18. **Michel-Kerjan, E., and Kunreuther, H.** 2011. "Redesigning Flood Insurance," *Science* 333: 408–409. doi:10.1126/science.1202616.

19. **Federal Emergency Management Agency.** "Flooding and Flood Risks: Understanding Flood Maps." Accessed August 23, 2012. http://www.floodsmart.gov/floodsmart/pages/ flooding_flood_risks/understanding_flood_maps.jsp.

20. **Pinter, P.** 2005. "One Step Forward, Two Steps Back on U.S. Floodplains," *Science* 308: 207–208. doi:10.1126/ science.1108411.

21. **Lieb, D. A., and Slater, J.** 2011. "APNewsBreak: FEMA Flood Buyouts Top $2B Since 1993." Associated Press. Accessed June 23, 2013. http://cnsnews.com/news/article/ apnewsbreak-fema-flood-buyouts-top-2b-1993.

22. **Carter, J. M., Williamson, J. E., and Teller, R. W.** 2002. *The 1972 Black Hills-Rapid City Flood Revisited*. Fact Sheet, USGS Fact Sheet. http://pubs.usgs.gov/fs/fs-037-02/ pdf/fs-037.pdf.

23. **Harden, T. M., et al.** 2011. *Flood-Frequency Analyses from Paleoflood Investigations for Spring, Rapid, Boxelder, and Elk Creeks, Black Hills, Western South Dakota*. USGS Scientific Investigations Report, Scientific Investigations Report. U.S. Department of the Interior, U.S. Geological Survey. http://pubs.usgs.gov/sir/2011/5131/pdf/sir2011- 5131.pdf.

24. **Carlisle, D. M., Wolock, D. M., and Meader, M. R.** 2010. "Alteration of Streamflow Magnitudes and Potential Ecological Consequences: A Multiregional Assessment." *Frontiers in Ecology and the Environment*, U.S. Department of the Interior, U.S. Geological Survey. http://water.usgs.gov/ nawqa/pubs/Carlisleetal_FLowAlterationUS.pdf.

25. **Walter, R. C., and Merritts, D. J.** 2008. "Natural Streams and the Legacy of Water-Powered Mills." *Science* 319: 299–304. doi:10.1126/science.1151716.

26. **U. S. Army Corps of Engineers.** "National Inventory of Dams." Accessed August 20, 2012. https://nid.usace.army.mil.

27. **Lehner, B., et al.** 2011. "High-resolution Mapping of the World's Reservoirs and Dams for Sustainable River-Flow Management," *Frontiers in Ecology and the Environment* 9: 494–502. doi:10.1890/100125.

28. **Vörösmarty, C. J., et al.** 2003. "Anthropogenic Sediment Retention: Major Global Impact from Registered River Impoundments." *Global and Planetary Change* 39: 169–190. doi:10.1016/S0921-8181(03)00023-7.

29. **Webb, R. H.** 1999. *The Controlled Flood in Grand Canyon*. American Geophysical Union; Schmidt, J. C. 2001. "The 1996 Controlled Flood in Grand Canyon: Flow, Sediment Transport, and Geomorphic Change." *Ecological Applications* 11: 657. doi:10.1890/1051- 0761(2001)011[0657:TCFIGC]2.0.CO;2.

30. **Taylor, J. P., and McDaniel, K. C.** 1998. "Restoration of Saltcedar (Tamarix Sp.)-infested Floodplains on the Bosque Del Apache National Wildlife Refuge." *Weed Technology* 345–352.

31. **The Heinz Center.** 2002. *Dam Removal: Science and Decision Making*. Washington, D.C.: H. John Heinz III Center for Science Economics and the Environment.

32. **Graf, W. L.** 1999. "Dam Nation: A Geographic Census of American Dams and Their Large-Scale Hydrologic Impacts." *Water Resources Research* 35: 1305–1311. doi:10.1029/1999WR900016.

33. Doyle, M. W., Stanley, E. H., and Harbor, J. M. 2003. "Channel Adjustments Following Two Dam Removals in Wisconsin." *Water Resources Research* 39: 1011. doi:10.1029/2002WR001714.

Merritts, D., et al. 2001. "Anthropocene Streams and Base-level Controls from Historic Dams in the Unglaciated mid-Atlantic Region, USA." *Philosophical Transactions of the Royal Society A: Mathematical, Physical and Engineering Sciences* 369: 976–1009. doi:10.1098/rsta.2010.0335.

34. Duda, J.J., Warrick, J.A., and Magirl, C.S. 2011. "Elwha River Dam Removal—Rebirth of a River." U.S. Geological Survey. Accessed January 25, 2013. http://pubs.usgs.gov/fs/2011/3097/.

35. Hipple, J. D., Drazkowski, B., and Thorsell, P. M. 2005. "Development in the Upper Mississippi Basin: 10 Years After the Great Flood of 1993." *Landscape and Urban Planning* 72: 313–323. doi:10.1016/j.landurbplan.2004.03.012.

36. Pinter, N., Thomas, R., and Wlosinski, J. H. n.d. "Assessing Flood Hazard on Dynamic Rivers." *Eos* 82: 333. doi:200110.1029/01EO00199.

37. Dierauer, J., Pinter, N., and Remo, J. W. F. 2012. "Evaluation of Levee Setbacks for Flood-loss Reduction, Middle Mississippi River, USA." *Journal of Hydrology* 450–451: 1–8. doi:10.1016/j.jhydrol.2012.05.044.

38. Maryland High Court of Chancery and Bland, T. 1841. *Reports of Cases Decided in the High Court of Chancery of Maryland [1811-1832].* J. Neal.

39. U.S. Environmental Protection Agency. "Wetlands—Status and Trends." Accessed January 25, 2012. http://water.epa.gov/type/wetlands/vital_status.cfm.

40. Committee on Characterization of Wetlands, National Research Council. 1995. *Wetlands: Characteristics and Boundaries.* Washington, D.C.: National Academies Press.

41. U.S. Environmental Protection Agency. "Wetlands Protection." Accessed January 25, 2013. http://water.epa.gov/type/wetlands/protection.cfm.

42. The Ramsar Convention on Wetlands. Accessed January 25, 2013. http://www.ramsar.org.

43. Committee on Restoration of Aquatic Ecosystems: Science, Technology, and Public Policy, National Research Council. 1992. *Restoration of Aquatic Ecosystems: Science, Technology, and Public Policy.* Washington, D.C.: National Academies Press.

44. Collins, B. D., Montgomery, D. R., and Haas, A. D. 2002. "Historical Changes in the Distribution and Functions of Large Wood in Puget Lowland Rivers." *Canadian Journal of Fisheries and Aquatic Sciences* 59: 66–76. doi:10.1139/F01-199; Collins, B. D., and Montgomery, D. R. 2002. "Forest Development, Wood Jams, and Restoration of Floodplain Rivers in the Puget Lowland, Washington." *Restoration Ecology* 10: 237–247. doi:10.1046/j.1526-100X.2002.01023.x.

45. Montgomery, D. R. 2003. *Restoration of Puget Sound Rivers.* Seattle: University of Washington Press.

46. Walter, R. C., and Merritts, D. J. 2008. "Natural Streams and the Legacy of Water-Powered Mills." *Science* 18: 299–304. doi:10.1126/science.1151716; Merritts, D. J., et al. 2011. "Anthropocene Streams and Base-Level Controls from Historic Dams in the Unglaciated Mid-Atlantic Region, USA." Philosophical Transactions of the Royal Society. Series A. Mathematical, Engineering, and Physical Sciences. doi:10.1098/rsta.2010.0335.

47. See Legacy Sediment Removal at: U.S. Environmental Protection Agency. "Mid-Atlantic Wetlands, Stream and Watershed Restoration." Accessed January 25, 2013. http://www.epa.gov/reg3esd1/wetlands/restoration.htm.

48. South Florida Water Management District. Restoring the Kissimmee River. Accessed January 16, 2013. http://www.sfwmd.gov/portal/page/portal/xweb%20protecting%20and%20restoring/kissimmee%20river.

49. Gleick, P. H. 1996. "Basic Water Requirements for Human Activities: Meeting Basic Needs." *Water International* 21: 83–92. doi:10.1080/02508069608686494.

50. Kenny, J. F., Barber, N. L., Hutson, S. S., Linsey, K. S., Lovelace, J. K., and Maupin, M. A. 2009. *Estimated Use of Water in the United States in 2005.* US Geological Survey Reston, VA. Accessed June 23, 2013. http://pubs.usgs.gov/circ/1344/.

51. U.S. Food and Agriculture Organization's Information System on Water and Agriculture. "Aquastat." Food and Agriculture Organization. Accessed January 24, 2013. http://www.fao.org/nr/water/aquastat/main/index.stm.

52. Hoekstra, A. Y., and Mekonnen, M. M. 2012. "The Water Footprint of Humanity." *Proceedings of the National Academy of Sciences* 109: 3232–3237. doi:10.1073/pnas.1109936109.

53. Dalin, C., et al. 2012. "Evolution of the Global Virtual Water Trade Network." *Proceedings of the National Academy of Sciences* 109: 5989–5994. doi:10.1073/pnas.1203176109.

54. United Nations Department of Economic and Social Affairs. "Water for Life Decade: Water Scarcity." Accessed January 18, 2013. https://www.un.org/waterforlifedecade/scarcity.shtml.

55. Brown, A., and Matlock, M. D. 2011. "A Review of Water Scarcity Indices and Methodologies." *Sustainability Consortium White Paper* 106. http://www.sustainabilityconsortium.org/wp-content/themes/sustainability/assets/pdf/whitepapers/2011_Brown_Matlock_Water-Availability-Assessment-Indices-and-Methodologies-Lit-Review.pdf.

56. Vörösmarty, C. J., et al. 2010. "Global Threats to Human Water Security and River Biodiversity," *Nature* 467: 555–561. doi:10.1038/nature09440.

57. **United Nations Educational, Scientific and Cultural Organization.** "Water." Accessed January 20, 2013. http://www.unesco.org/new/en/natural-sciences/environment/water/.

58. **World Health Organization.** "Diarrhoeal Disease." Accessed January 20, 2013. http://www.who.int/mediacentre/factsheets/fs330/en/index.html.

59. **Makinia, J., Dunnette, D., and Kowalik, P.** 1996. "Water Pollution in Poland." *European Water Pollution Control* 6: 26–33.

60. **Absalon, D.** 2009. "Changes in Water Quality in the Upper Wisla (Vistula) River Basin." Pp 29 –32 in *Second International Conference on Environmental and Computer Science, 2009. ICECS '09.* doi:10.1109/ICECS.2009.80; Kathuria, V. 2006. "Controlling Water Pollution in Developing and Transition Countries—Lessons from Three Successful Cases." *Journal of Environmental Management* 78: 405–426. doi:10.1016/j.jenvman.2005.05.007.

61. **U.S. Environmental Protection Agency.** "Summary of the Clean Water Act." Accessed June 23, 2013. http://www2.epa.gov/laws-regulations/summary-clean-water-act.

62. **U.S. Environmental Protection Agency.** "Watershed Assessment, Tracking, and Environmental Results. National Summary of State Information." Accessed January 25, 2013. http://ofmpub.epa.gov/waters10/attains_nation_cy.control.

63. **Hartig, J. H.** 2010. *Burning Rivers: Revival of Four Urban Industrial Rivers That Caught Fire.* Essex, UK: Multi-Science Publishing Company.

64. **U.S. Environmental Protection Agency.** "Water Monitoring and Assessment: Turbidity." Accessed January 25, 2013. http://water.epa.gov/type/rsl/monitoring/vms55.cfm.

65. **U.S. Environmental Protection Agency.** "DDT—A Brief History and Status." Accessed January 15, 2013. http://www.epa.gov/pesticides/factsheets/chemicals/ddt-brief-history-status.htm.

66. **Gilliom, R. J.** 2007. "Pesticides in U.S. Streams and Groundwater." *Environmental Science & Technology* 41: 3408–3414. doi:10.1021/es072531u.

67. **Gilliom, R. J., et al.** 2006. The Quality of Our Nation's Waters: *Pesticides in the Nation's Streams and Groundwater, 1992–2001.* Circular 1291. United States Geological Survey. Accessed June 23, 2013. http://pubs.usgs.gov/circ/2005/1291/.

68. **Duhigg, C.** 2009. "Debating How Much Weed Killer Is Safe in Your Water Glass." *New York Times,* August 22. http://www.nytimes.com/2009/08/23/us/23water.html.

———. 2009. "Regulators Plan to Study Risks of Atrazine." *New York Times,* October 7. http://www.nytimes.com/2009/10/07/business/energy-environment/07water.html.

69. **Wu, M., et al.** 2010. "Still Poisoning the Well." Washington, D.C.: Natural Resources Defense Council. http://www.nrdc.org/health/atrazine/files/atrazine10.pdf.

70. **U.S. Environmental Protection Agency.** "Impaired Waters and Total Maximum Daily Loads." Accessed January 20, 2013. http://water.epa.gov/lawsregs/lawsguidance/cwa/tmdl/index.cfm.

71. **U.S. Environmental Protection Agency.** "Basic Information about Regulated Drinking Water Contaminants and Indicators."Accessed January 23, 2013. http://water.epa.gov/drink/contaminants/basicinformation/index.cfm.

72. **Diduch, M., Polkowska, Z., and Namieśnik, J.** 2012. "Factors Affecting the Quality of Bottled Water." *Journal of Exposure Science and Environmental Epidemiology.* doi:10.1038/jes.2012.101.

73. **National Wild and Scenic Rivers System.** "Explore Designated Rivers." Accessed June 18, 2013. http://www.rivers.gov/rivers/map.php.

74. **The American Presidency Project.** "Lyndon B. Johnson: Remarks Upon Signing Four Bills Relating to Conservation and Outdoor Recreation. October 2, 1968." Accessed July 22, 2012. http://www.presidency.ucsb.edu/ws/index.php?pid=29150#axzz20JoFDnT8.

75. **The Heinz Center.** 2002. *Dam Removal: Science and Decision Making.* Washington, D.C.: H. John Heinz III Center for Science Economics and the Environment; National Drought Mitigation Center. Accessed July 9, 2012. http://drought.unl.edu/.

76. **Barriopedro, D., et al.** 2011. "The Hot Summer of 2010: Redrawing the Temperature Record Map of Europe." *Science* 332: 220–224. doi:10.1126/science.1201224.

77. **U.S. Fish and Wildlife Service.** 2011. "Critical Reach of Spring Creek Set to Get Water Boost if Drought Hits Hard Again: Wells Drilled to Help Save Endangered Species." Accessed July 9, 2012. http://www.fws.gov/southeast/news/2011/11-078.html.

78. See background, overview, facts, timeline, and court rulings at: **Atlanta Regional Commission.** "Tri-State Water Wars." Accessed July 1, 2012. http://www.atlantaregional.com/environment/tri-state-water-wars.

79. An electronic copy of Georgia Senate Bill 370 can be found online at: http://www.legis.state.ga.us/legis/2009_10/sum/sb370.htm.

Chapter 10

1. **Cook-Anderson, G.** "NASA Satellites Unlock Secret to Northern India's Vanishing Water." NASA. Accessed October 30, 2012. http://www.nasa.gov/topics/earth/features/india_water.html.

2. **Rodell, M., Velicogna, I., and Famiglietti, J. S.** 2009. "Satellite-Based Estimates of Groundwater Depletion in India." *Nature* 460 (7258): 999–1002.

3. **Famiglietti, J.** June 28, 2012. "Rallying Around Our Known Unknowns: What We Don't Know Will Hurt Us." Accessed October 30, 2012. http://blog.ucchm.org

/2012/06/28/rallying-around-our-known-unknowns-what-we-dont-know-will-hurt-us/.

4. **Andrianov, N. I.** 1987. *The Superdeep Well of the Kola Peninsula,* p. 285. Berlin: Springer-Verlag.

5. **Fitts, C. R.** 2013. *Groundwater Science,* 2nd ed. New York: Academic Press.

6. **U.S. Environmental Protection Agency.** 2012. "Basic Information About Nitrate (Measured as Nitrogen) in Drinking Water." Accessed October 20, 2012. http://water .epa.gov/drink/contaminants/basicinformation/nitrite.cfm.

7. **World Health Organization.** 2001. "Arsenic in Drinking Water." *WHO Fact Sheet* 210. Accessed October 20, 2012. https://apps.who.int/inf-fs/en/fact210.html.

8. **Hussam, A., Ahamed, S., and Munir, A. K. M.** 2008. "Arsenic Filters for Groundwater in Bangladesh: Toward a Sustainable Solution." *The Bridge: Linking Engineering and Society,* pp. 14–23. National Academy of Engineering of the National Academies. http://www.nae.edu/ File.aspx?id=7423.

9. **Bethke, C. M., and Johnson, T. M.** 2008. "Groundwater Age and Groundwater Age Dating." *Annual Review of Earth and Planetary Sciences* 36 (1): 121–152. doi:10.1146/ annurev.earth.36.031207.124210.

10. **Cech, T. V.** 2005. *Principles of Water Resources: History, Development, Management, and Policy,* 2nd ed, p. 102. Hoboken, NJ: John Wiley and Sons.

11. **UN Food and Agriculture Organization.** AQUASTAT database. Accessed March 15, 2013. http://www.fao.org/nr/ water/aquastat/data/query/index.html.

12. **King, A., and Borchert, C.** 2010. "Annual Water Report, City of Santa Fe." Accessed October 21, 2012. http://www .santafenm.gov/DocumentCenter/Home/View/9370.

13. **Grant, P.** 2002. "Santa Fe River Watershed Restoration Action Strategy." Accessed October 21, 2012. http:// www.nmenv.state.nm.us/swqb/Santa_Fe_WRAS-2002.pdf.

14. "City of Santa Fe Water Conservation and Drought Management Plan." Accessed October 21, 2012. http:// www.santafenm.gov/DocumentCenter/Home/View/9834.

15. **Reilly, T. E., Dennehy, K. F., Alley, W. M., and Cunningham, W. L.** 2008. "Ground-Water Availability in the United States." *U.S. Geological Survey Circular* 1323:6.

16. **International Atomic Energy Agency Water Resources Programme.** 2012. "Nubian Aquifer Project." Accessed October 22, 2012. http://www-naweb.iaea.org/napc/ih/ IHS_projects_nubian_ancient.html.

17. **Sturchio, N. C., and Purtschert R.** 2011. "Kr-81 Case Study: The Nubian Aquifer (Egypt)." In *Dating Old Groundwater: A Guidebook,* edited by A. Suckow. Vienna: IAEA.

18. **Sturchio, N. C., Du, X., Purtschert, R., Lehmann, B. E., Sultan, M., Patterson, L. J., Lu, Z.-T., et al.** 2004. "One Million Year Old Groundwater in the Sahara Revealed by Krypton-81 and Chlorine-36." *Geophysical Research Letters* 31 (5): L05503. doi:10.1029/2003GL019234.

Wald, C. 2012. "Uncharted Waters: Probing Aquifers to Head Off War." *New Scientist* 2851. Accessed October 30, 2012. http://www.newscientist.com/article/ mg21328512.200-uncharted-waters-probing-aquifers-to-head-off-war.html?full=true.

19. **Reilly, T. E., Dennehy, K. F., Alley, W. M., and Cunningham, W. L.** 2008. "Ground-Water Availability in the United States." *U.S. Geological Survey Circular* 1323:57–60.

20. **Maupin, M. A., and Barber, N. L.** 2000. "Estimated Withdrawals from Principal Aquifers in the United States." *U.S. Geological Survey Circular* 1279:12.

21. **Fetter, C. W.** 2001. *Applied Hydrogeology,* 4th ed., p. 447. Upper Saddle River, NJ: Prentice Hall.

22. **Yechieli, Y.** 2009. *GLOCOPH Israel 2009 Field Trip Guidebook, Station 4: Collapse Sinkholes Along the Dead Sea Coast,* pp. 71–80. Jerusalem: Hebrew University of Jerusalem.

23. **Cohen, N.** 2008. "Israel's National Water Carrier." *Present Environment and Sustainable Development* 2: 15–27.

24. **Bookman, R., Bartov, Y., Enzel, Y. and Stein, M.** 2006. "Quaternary Lake Levels in the Dead Sea Basin: Two Centuries of Research." In *Geological Society of America Special Paper 401: New Frontiers in Dead Sea Paleoenvironmental Research,* edited by Y. Enzel, A. Agnon, M. and Stein, pp. 155–170.

25. **Abelson, M., Yechieli, Y., Crouvi, O., Baer, G., Wachs, D., Bein, A., and Shtivelman, V.** 2006. "Evolution of the Dead Sea Sinkholes." In *Geological Society of America Special Paper 401: New Frontiers in Dead Sea Paleoenvironmental Research,* edited by Y. Enzel, A. Agnon, and M. Stein, pp. 241–253.

26. **The Joint Academies Committee on the Mexico City Water Supply.** 1995. *Mexico City's Water Supply: Improving the Outlook for Sustainability,* pp. 11–16. Washington, DC: National Academy Press.

27. **Malkin, E.** 2010. "Fears that a Lush Land May Lose a Foul Fertilizer." *Mixquiahuala Journal* (blog), *New York Times,* May 5. Accessed October 24, 2012. http:// www.nytimes.com/2010/05/05/world/americas/05mexico .html?pagewanted=all&_r=0.

28. **Sonenshein, R. S.** 1996. "Delineation of Saltwater Intrusion in the Biscayne Aquifer, Eastern Dade County, Florida, 1995." *U.S. Geological Survey Water Resources Investigations Report* 96-4285. http://fl.water.usgs.gov/ Miami/online_reports/wri964285/.

29. **U.S. Environmental Protection Agency.** "Basic Information About Trichloroethylene in Drinking Water." Accessed October 30, 2012. http://water.epa.gov/drink/ contaminants/basicinformation/trichloroethylene.cfm.

30. **U.S. Department of Health and Human Services, Agency for Toxic Substances and Disease Registry.** 1997. "Toxicological Profile for Trichloroethylene." Accessed October 30, 2012. http://www.atsdr.cdc.gov/toxprofiles/tp19.pdf.

31. **U.S. Environmental Protection Agency.** "Underground Storage Tanks." Accessed October 30, 2012. http://www.epa.gov/oust/.

32. **U.S. Environmental Protection Agency.** "Underground Storage Tanks: Cleaning Up UST System Releases." Accessed October 30, 2012. http://www.epa.gov/oust/cat/releases.htm.

33. **U.S. Geological Survey Toxic Substances Hydrology Program.** 2010. "Cape Cod Toxic Substances Hydrology Research Site." Accessed October 20, 2012. http://ma.water.usgs.gov/MMRCape/toxics_overview.html.

———. 2006. "Cape Cod Toxic Substances Hydrology Research Site: Physical, Chemical, and Biological Processes that Control the Fate of Contaminants in Ground Water." *U.S. Geological Survey Fact Sheet* 2006-3096. Accessed October 20, 2012. http://ma.water.usgs.gov/MMRCape/toxics_fs_3096.pdf.

34. **Godwin, K. S., Hafner, S. D., and Buff, M. F.** 2003. "Long-Term Trends in Sodium and Chloride in the Mohawk River, New York: The Effect of Fifty Years of Road-Salt Application." *Environmental Pollution* 124 (2): 273–281.

Peters, N. E. & Turk, J. T. 1981. "Increases in Sodium and Chloride in the Mohawk River, New York from the 1950s to the 1970s Attributed to Road Salt." *Water Resources Bulletin* 17 (4): 586–597.

Chapter 11

1. **Ruddiman, W. F.** 2008. *Earth's Climate: Past and Future,* 2nd ed., p. 209–308. New York: W. H. Freeman.

Matthews, J. A., and Briffa, K. R. 2005. "The 'Little Ice Age': Re-evaluation of an Evolving Concept." *Geografiska Annaler* 87: 17–36.

2. **Kaufman, D. S., et al.** 2009. "Recent Warming Reverses Long-Term Arctic Cooling." *Science* 325: 1236–1239.

Wanner, H., et al. 2008. "Mid- to Late Holocene Climate Change: An Overview." *Quaternary Science Reviews* 27: 1791–1828.

3. **Farrington, B.** 2010. "Cold Weather Strengthens Grip in the South." *Atlanta Journal-Constitution,* January 10.

4. **CBSAtlanta.com.** 2010. "Water Mains Break, Home Pipes Could Be Next." January 5. http://www.cbsatlanta.com/story/14759468/water-mains-break-home-pipes-could-be-next-1-04-2010.

5. **Fletcher, P.** 2010. "UPDATE 4-Freeze Mauls Florida Citrus, Significant Damage Seen." *Reuters,* January 11. http://www.reuters.com/article/idUSN1114830920100111?type=marketsNews.

6. **NBC.** 2010. "Cold Threatens Farmers' Bottom Lines." *NBC Nightly News* video, January 10. http://www.nbcnews.com/video/nightly-news/34793741/#34793741.

7. **Koch, W.** 2010. "Cold Snap Horrifying for Homeless." *USA Today,* January 6. http://www.usatoday.com/news/nation/2010-01-05-weather-homeless_N.htm.

8. **Barria, C.** 2010. "Brr! Florida Manatees Warm Up at Power Plant Hot Tub." *Reuters,* January 7. http://www.reuters.com/article/idUSN0720661320100107?type=marketsNews.

9. **MSNBC.com.** 2010. "Fruits Freeze, Iguanas Drop from Trees in Florida." MSNBC.com, January 7. http://www.msnbc.msn.com/id/34746400/ns/weather/.

10. **Thompson, D. W. J.** 2007. "A Brief Introduction to the Annular Modes and Annular Mode Research." Annular Modes. Accessed July 5, 2010. http://www.atmos.colostate.edu/ao/introduction.html.

11. **Thompson, D. W. J., and Wallace, J. M.** 1998. "The Arctic Oscillation Signature in the Wintertime Geopotential Height and Temperature Fields." *Geophysical Research Letters* 25: 1297–1300.

12. **Revkin, A. C.** 2010. "Cold Arctic Pressure Pattern Nearly Off Chart." *Dot Earth* (blog). *New York Times,* January 4. http://dotearth.blogs.nytimes.com/2010/01/04/polar-pressure-pattern-driving-chill-nearly-off-chart/.

13. **National Snow & Ice Data Center.** 2010. "Extreme Negative Phase of the Arctic Oscillation Yields a Warm Arctic." Arctic Sea Ice News & Analysis, January 5. Accessed July 5, 2013. http://nsidc.org/arcticseaicenews/2010/010510.html.

14. **Stull, R. B.** 2000. *Meteorology for Scientists and Engineers,* 2nd ed., p. 8. Pacific Grove, CA: Brooks/Cole Thomson Learning.

15. **Ruddiman, W. F.** 2008. *Earth's Climate: Past and Future,* 2nd ed., p. 43–45. New York: W. H. Freeman.

16. **Sessions, A. L., Doughty, D. M., Welander, P. V., Summons, R. E., and Newman, D. K.** 2009. "The Continuing Puzzle of the Great Oxidation Event." *Current Biology* 19: R567–R574.

17. **Kump, L. R.** 2008. "The Rise of Atmospheric Oxygen." *Nature* 451: 277–278.

18. **Allwood, A. C., et al.** 2009. "Controls on Development and Diversity of Early Archean Stromatolites." *Proceedings of the National Academy of Sciences of the United States of America* 106: 9548–9555.

Buick, R. 2008. "When Did Oxygenic Photosynthesis Evolve?" *Philosophical Transactions of the Royal Society B* 363: 2731–2743.

19. **Konhauser, K. O., et al.** 2009. "Oceanic Nickel Depletion and a Methanogen Famine Before the Great Oxidation Event." *Nature* 458: 750–753.

20. **Campbell, I. H., and Squire, R. J.** 2010. "The Mountains That Triggered the Late Neoproterozoic Increase in Oxygen: The Second Great Oxidation Event." *Geochimica et Cosmochimica Acta* 74: 4187–4206.

21. **Metcalfe, S., and Derwent, D.** 2005. *Atmospheric Pollution and Environmental Change,* p. 1. New York: Oxford University Press.

22. **Stull, R. B.** 2000. *Meteorology for Scientists and Engineers,* 2nd ed., p. 109. Pacific Grove, CA: Brooks/Cole Thomson Learning.

23. **Metcalfe, S., and Derwent, D.** 2005. *Atmospheric Pollution and Environmental Change,* p. 35. New York: Oxford University Press.

24. **Stull, R. B.** 2000. *Meteorology for Scientists and Engineers,* 2nd ed., p. 13. Pacific Grove, CA: Brooks/Cole Thomson Learning.

25. **National Aeronautics and Space Administration.** 2003. "Strange Clouds." Accessed February 7, 2010. http://science.nasa.gov/headlines/y2003/19feb_nlc.htm.

O'Carroll, C. M. 2009. NASA's AIM Satellite and Models Are Unlocking the Secrets of Mysterious 'Night-Shining' Clouds." Accessed February 7, 2010. http://www.nasa.gov/mission_pages/aim/news/nlc-secrets.html.

26. **Thomas, G. E.** 1996. "Is the Polar Mesosphere the Miner's Canary of Global Change?" *Advances in Space Research* 18: 149–158.

27. **Ruddiman, W. F.** 2008. *Earth's Climate: Past and Future,* 2nd ed., p. 45. New York: W. H. Freeman.

28. **Ruddiman, W. F.** 2001. *Earth's Climate: Past and Future,* chap. 2. New York: W. H. Freeman.

29. **Lutgens, F. K., and Tarbuck, E. J.** 2012. The Atmosphere: An Introduction to Meteorology, 12th ed., p. 283. Upper Saddle River, NJ: Prentice Hall.

30. **National Aeronautics and Space Administration.** 2010. "A Lightning Primer: Characteristics of a Storm." Accessed February 11, 2010. http://thunder.nsstc.nasa.gov/primer/primer2.html.

31. **Galloway, J. N.** 1998. "The Global Nitrogen Cycle: Changes and Consequences." *Environmental Pollution* 102: 15–24.

32. **National Oceanic and Atmospheric Administration.** 1995. "Tornadoes...Nature's Most Violent Storms." Washington, D.C.: U.S. Department of Commerce.

33. **KAKE News.** 2007. "Greensburg Tornado—Fact Sheet." Accessed February 11, 2010. http://www.kake.com/news/headlines/7347256.html.

34. **National Climatic Data Center.** 2008. "U.S. Tornado Climatology." Accessed February 12, 2010. http://www.ncdc.noaa.gov/oa/climate/severeweather/tornadoes.html.

35. **Huffstutter, P. J.** 2007. "Tornado Left Little Behind in Kansas Village." *Pittsburgh Post-Gazette,* May 20, p. A13.

36. **Hoxit, L. R., and Chappell, C. F.** 1975. "Tornado Outbreak of April 3–4, 1974; Synoptic Analysis." NOAA Technical Report ERL 338-APCL 37.

37. **National Oceanic and Atmospheric Administration.** 2011. "2011 Tornado Information." Accessed June 20, 2011. http://www.noaanews.noaa.gov/2011_tornado_information.html.

38. **Raycom News Network.** 2011. "Southern Storms' Toll Surpasses 300 Deaths." WTVM.com, Columbus, GA. Accessed June 20, 2011. http://www.wtvm.com/Global/story.asp?S=14532637.

39. **National Weather Service Paducah, KY Forecast Office.** 2010. "NOAA/NWS 1925 Tri-State Tornado—Startling Statistics." Accessed February 12, 2010. http://www.crh.noaa.gov/pah/1925/ss_body.php.

40. **Bazar, E.** 2008. "The Greening of Greensburg, Kan.; A Year Later, Tornado-Devastated Town Finds New Life in Rebuilding Eco-Friendly." USA Today, May 2, p. 3A.

41. **U.S. Department of Energy.** 2009. "Greensburg, Kansas, Wind Farm to Break Ground in Summer 2009."

42. **Kansas, Office of the Governor, and U.S. Department of Homeland Security, FEMA Region VII.** 2007. "Long-Term Community Recovery Plan: Greensburg + Kiowa County, Kansas, August 2007." http://www.epa.gov/region07/cleanup/greensburg/pdf/GB_LTCR_PLAN_Final_HiRes070815.pdf.

43. **National Weather Service, National Hurricane Center.** 2013. "Saffir-Simpson Hurricane Wind Scale." Accessed July 9, 2013. http://www.nhc.noaa.gov/aboutsshs.shtml.

44. **Bramer, D., Wojtowicz, D., and Hall, S. E.** 1995. "WW2010: The Weather World 2010 Project at the University of Illinois, Urbana-Champaign: Hurricanes." Accessed February 13, 2010. http://ww2010.atmos.uiuc.edu/(Gh)/guides/mtr/hurr/grow/home.rxml.

45. **United Nations.** 2009. "Myanmar: UN Launches Post-Cyclone Recovery Plan." Scoop Media, February 10. Accessed July 9, 2013. http://www.scoop.co.nz/stories/WO0902/S00158.htm; AVISO. 2008. "Typhoon Nargis." Accessed February 13, 2010. http://www.aviso.oceanobs.com/en/data/data-access-services/live-access-server-las/lively-data/2008/jun-30-2008-typhoon-nargis/index.html.

National Aeronautics and Space Administration. 2008. "NASA Satellite Captures Image of cyclone Nargis Flooding in Burma." Accessed February 13, 2010. http://www.nasa.gov/topics/earth/features/nargis_floods.html.

46. **Lee, L.** 2009. "Wave After Wave of Disasters." *The Straits Times* (Singapore). December 9.

47. **U.N. News Centre.** 2009. "One Year After Myanmar Cyclone, International Support Still Critical, UN says." May 1. Accessed February 13, 2010. http://www.un.org/apps/news/story.asp?NewsID=30667&Cr=nargis&Cr1=.

48. **Kates, R. W., Colten, C. E., Laska, S., and Leatherman, S. P.** 2006. "Reconstruction of New Orleans after Hurricane

Katrina: A Research Perspective." *Proceedings of the National Academy of Sciences* 103: 14653–14660.

49. **Pielke, Jr., R. A., Gratz, J., Landsea, C. W., Collins, D., Saunders, M. A., and Musulin, R.** 2008. "Normalized Hurricane Damage in the United States: 1900–2005." *Natural Hazards Review* 9: 29–42.

50. **National Oceanic and Atmospheric Administration.** 2004. "NOAA History—Stories and Tales of the Weather Service/StormTales/Galveston Storm of 1900." Accessed July 13, 2012. http://www.history.noaa.gov/stories_tales/cline2.html.

51. **Singh, A., and Agrawal, M.** 2008. "Acid Rain and Its Ecological Consequences." *Journal of Environmental Biology* 29: 15–24.

52. **Borer, C. H., Schaberg, P. G., Dehayes, D. H., and Hawley, G. J.** 2004. "Accretion, Partitioning and Sequestration of Calcium and Aluminum in Red Spruce Foliage: Implications for Tree Health." *Tree Physiology* 24: 929–939.

53. **Stephens, F. J., and Ingram, M.** 2006. "Two Cases of Fish Mortality in Low pH, Aluminium Rich Water." *Journal of Fish Diseases* 29: 765–770.

54. **Metcalfe, S., and Derwent, D.** 2005. *Atmospheric Pollution and Environmental Change*, p. 65. New York: Oxford University Press.

55. **Environment Canada.** 2005. "Acid Rain FAQ." Accessed July 22, 2013. http://www.ec.gc.ca/air/default.asp?lang=En&n=7E5E9F00-1.

56. **Jenkins, J., Roy, K., Driscoll, C., and Buerkett, C.** 2005. *Acid Rain and the Adirondacks: A Research Summary*, p. 183. Ray Brook, NY: Adirondack Lakes Survey Corporation.

57. **Stenstrom, T.** 1989. "Cadmium and Lead in Well and Groundwater." *VATTEN* 45: 145–156.

Rosborg, I., Nihlgard, B., and Gerhardsson, L. 2003. "Inorganic Constituents of Well Water in One Acid and One Alkaline Area of South Sweden." *Water, Air, and Soil Pollution* 142: 261–277.

58. **Arbaugh, M. J., Miller, P. R., Carroll, J. J., Takemoto, B., and Procter, T.** 1998. "Relationships of Ozone Exposure to Pine Injury in the Sierra Nevada and San Bernardino Mountains of California, USA." *Environmental Pollution* 101: 291–301.

Takemoto, B. K., Bytnerowicz, A., and Fenn, M. E. 2001. "Current and Future Effects of Ozone and Atmospheric Nitrogen Deposition on California's Mixed Conifer Forests." *Forest Ecology and Management* 144: 159–173.

Siefermann-Harms, D., Boxler-Baldoma, C., von Wilpert, K., and Heumann, H-G. 2004. "The Rapid Yellowing of Spruce at a Mountain Site in the Central Black Forest (Germany). Combined Effects of Mg Deficiency and Ozone on Biochemical, Physiological, and Structural Properties

of the Chloroplasts." *Journal of Plant Physiology* 161: 423–437.

59. **Murphy, J. J., Delucchi, M. A., McCubbin, D. R., and Kim, H. J.** 1999. "The Cost of Crop Damage Caused by Ozone Air Pollution from Motor Vehicles." *Journal of Environmental Management* 55: 273–289.

60. **Molina, M. J., and Rowland, F. S.** 1974. "Stratospheric Sink for Chlorofluoromethanes: Chlorine Atom Catalysed Destruction of Ozone." *Nature* 249: 810–812.

61. **Crutzen, P. J., and Arnold, F.** 1986. "Nitric Acid Cloud Formation in the Cold Antarctic Stratosphere: A Major Cause for the Springtime 'Ozone Hole'." *Nature* 324: 651–655.

62. **Office of Air and Radiation, U.S. Environmental Protection Agency.** 2007. "Achievements in Stratospheric Ozone Protection—Progress Report." Washington, D.C., Report EPA-430-R-07-001.

63. **National Aeronautics and Space Administration.** 2010. "Ozone Hole Watch." Accessed February 15, 2010. http://ozonewatch.gsfc.nasa.gov/.

64. **Manney, G. L., et al.** 2011. "Unprecedented Arctic Ozone Loss in 2011." *Nature* 478: 469–475.

65. **U.S. Environmental Protection Agency.** 2008. "Ozone Science: The Facts Behind the Phaseout." Accessed February 15, 2010. http://www.epa.gov/Ozone/science/sc_fact.html.

66. **Ozone Secretariat, United Nations Environment Programme.** 2013. "Status of Ratification for the Montreal Protocol and the Vienna Convention." Accessed July 9, 2013. http://ozone.unep.org/new_site/en/treaty_ratification_status.php.

67. **U.S. Environmental Protection Agency.** 2007. "Montreal Protocol Backgrounder: The 20th Anniversary of the Montreal Protocol—A Landmark Environmental Treaty." http://www.epa.gov/Ozone/downloads/MP20_Backgrounder.pdf.

68. **De Angelo, L.** 2012. "London Smog Disaster, England." Encyclopedia of Earth. Accessed July 9, 2013. http://www.eoearth.org/view/article/154281/.

69. **California Environmental Protection Agency, Air Resources Board.** 2010. "Key Events in the History of Air Quality in California." Accessed February 15, 2010. http://www.arb.ca.gov/html/brochure/history.htm.

70. **Roman, A.** 2007. "Air Pollution Control Act of 1955, United States." Encyclopedia of Earth. Accessed July 9, 2013. http://www.eoearth.org/view/article/149927/.

71. **U.S. Environmental Protection Agency.** 2008. "History of the Clean Air Act." Accessed February 15, 2010. http://www.epa.gov/air/caa/caa_history.html.

72. **Johnson, A. H.** 1992. "The Role of Abiotic Stresses in the Decline of Red Spruce in High Elevation Forests of the Eastern United States." *Annual Reviews of Phytopathology* 30: 349–367.

73. **U.S. Environmental Protection Agency.** "Clean Air Act." Accessed July 9, 2013. http://www.epa.gov/air/caa/.

74. **U.S. Environmental Protection Agency.** 2010. National Enforcement Initiatives for Fiscal Years 2008–2010. Clean Air Act: New Source Review/Prevention of Significant Deterioration. Accessed July 22, 2013. http://www.epa.gov/compliance/data/planning/priorities/caansrpsd.html.

75. **U.S. Environmental Protection Agency.** 2012. "Healthier Americans: Mercury and Air Toxics Standards (MATS) for Power Plants" Accessed July 13, 2012. http://www.epa.gov/airquality/powerplanttoxics/health.html.

76. **Metcalfe, S., and Derwent, D.** 2005. *Atmospheric Pollution and Environmental Change,* p. 54, 99. New York: Oxford University Press.

77. **People's Daily Online.** 2007. China Has Over 20 Million Privately-Owned Cars." February 28. Accessed February 16, 2010. http://english.peopledaily.com.cn/200702/28/eng20070228_353091.html.

China Daily. 2011. "Number of Cars in China Hits 100m." September 17. Accessed July 13, 2012. http://www.chinadaily.com.cn/bizchina/2011-09/17/content_13725715.htm.

78. **Wines, M.** 2009. "Beijing's Air Is Cleaner, But Far from Clean." *New York Times,* October 16.

79. **United Nations.** 2003. "UNEP Environmental Observation and Assessment Strategy: Reference Paper Annexes." Accessed February 15, 2010. http://www.un.org/earthwatch/about/docs/unepstrx.htm#IV. CRITICAL DATA GAPS.

80. **Metcalfe, S., and Derwent, D.** 2005. *Atmospheric Pollution and Environmental Change,* p. 91. New York: Oxford University Press.

Chapter 12

1. **Myers, R. A., and Worm, B.** 2003. "Rapid Worldwide Depletion of Predatory Fish Communities." *Nature* 423: 280–283.

2. **Cheung, W. W. L., Lam, V. W. Y., Sarmiento, J. L., Kearney, K., Watson, R., and Pauly, D.** 2008. "Projecting Global Marine Biodiversity Impacts Under Climate Change Scenarios." *Fish and Fisheries* 10: 235–251.

3. **United Nations Millenium Project.** 2005. *Environment and Human Well-Being: A Practical Strategy,* p. 51. Report of the Task Force on Environmental Sustainability. London: Earthscan. Accessed February 19, 2010. http://www.unmillenniumproject.org/documents/Environment-complete-lowres.pdf.

4. **Hohn, D.** 2007. "Moby-Duck: Or, the Synthetic Wilderness of Childhood." *Harper's,* January: 39–62. Accessed February 20, 2010. http://harpers.org/archive/2007/01/0081345.

5. **Clerkin, B.** June 27, 2007. "Thousands of Rubber Ducks to Land on British Shores After 15 Year Journey." *Mail Online.* Accessed February 20, 2010. http://www.dailymail.co.uk/news/article-464768/Thousands-rubber-ducks-land-British-shores-15-year-journey.html.

6. **Ebbesmeyer, C. C., Ingraham, W. J., Royer, T. C., and Grosch, C. E.** 2007. "Tub Toys Orbit the Pacific Subarctic Gyre." *EOS* 88, no. 1: 1–12.

7. **Argo.** "About Argo." 2010. Accessed February 25, 2010. http://www.argo.ucsd.edu/index.html.

8. **Hohn, D.** 2007. "Moby-Duck: Or, the Synthetic Wilderness of Childhood." *Harper's,* January: 39–62. Accessed February 20, 2010. http://harpers.org/archive/2007/01/0081345.

9. **Ebbesmeyer, C.** 2001. "Leftovers/Evangelical Currents." *Cabinet* 4. Accessed February 20, 2010. http://www.cabinetmagazine.org/issues/4/evangelicalcurrents.php.

10. **Ebbesmeyer, C., and Scigliano, E.** 2009. *Flotsametrics and the Floating World.* New York: HarperCollins.

Gregory, M. R. 2009. "Environmental Implications of Plastic Debris in Marine Settings—Entanglement, Ingestion, Smothering, Hangers-On, Hitch-Hiking and Alien Invasions." *Philosophical Transactions of the Royal Society B* 364: 2013–2025.

11. **Barnes, D. K. A.** 2005. "Remote Islands Reveal Rapid Rise of Southern Hemisphere Sea Debris." *The Scientific World Journal* 5: 915–921.

12. **National Research Council.** 2009. *Tackling Marine Debris in the 21st Century,* p. 40. Washington, DC: National Academies Press.

13. **Gregory, M. R.** 2009. "Environmental Implications of Plastic Debris in Marine Settings—Entanglement, Ingestion, Smothering, Hangers-On, Hitch-Hiking and Alien Invasions *Philosophical Transactions of the Royal Society B* 364: 2013–2025.

14. **Sverdrup, K. A., and Armbrust, E. V.** 2008. *An Introduction to the World's Oceans,* 9th ed., p. 37. Boston, MA: McGraw Hill Higher Education.

15. **Beck, M. W., Brumbaugh, R. D., Airoldi, L., Carranza, A., Coen, L. D., Crawford, C., Defeo, O., et al.** 2011. "Oyster Reefs at Risk and Recommendations for Conservation, Restoration and Management." *BioScience* 61: 107–116.

16. **Sverdrup, K. A., and Armbrust, E. V.** 2008. *An Introduction to the World's Oceans,* 9th ed., p. 153. Boston, MA: McGraw Hill Higher Education.

17. **Garrison, T.** 2006. *Essentials of Oceanography,* 4th ed., p. 121. Belmont, CA: Thomson Higher Education.

18. **Sverdrup, K. A., and Armbrust, E. V.** 2008. *An Introduction to the World's Oceans,* 9th ed., p. 158. Boston, MA: McGraw Hill Higher Education.

19. **Poulton, A. J., Stinchcombe, M. C., and Quartly, G. D.** 2009. "High Numbers of *Trichodesmium* and Diazotrophic Diatoms in the Southwest Indian Ocean." *Geophysical Research Letters* 36: L15610. doi:10.1029/2009GL039719.

20. **Sverdrup, K. A., and Armbrust, E. V.** 2008. *An Introduction to the World's Oceans,* 9th ed., p. 163. Boston, MA: McGraw Hill Higher Education.

21. **Garrison, T.** 2006. *Essentials of Oceanography,* 4th ed., Chapter 8. Belmont, CA: Thomson Higher Education.

22. **Hebert, P. D. N., and Biodiversity Institute of Ontario.** 2008, December 12. "Pacific Ocean." *The Encyclopedia of Earth.* Accessed September 6, 2011. http://www.eoearth.org/article/Pacific_Ocean#gen14.

23. **Garrison, T.** 2007. *Oceanography: An Invitation to Marine Science,* 7th ed. Belmont, CA: Brooks-Cole.

24. **Sverdrup, K. A., and Armbrust, E. V.** 2008. *An Introduction to the World's Oceans,* 9th ed., p. 211. Boston, MA: McGraw Hill Higher Education.

25. **Stewart, R. H.** 2005. "Response of the Upper Ocean to Winds," in *Ocean World,* chap. 9. Accessed February 28, 2010. http://oceanworld.tamu.edu/resources/ocng_textbook/chapter09/chapter09_02.htm.

26. **National Oceanic and Atmospheric Administration.** "What Is an El Niño?" Accessed September 5, 2011. http://www.pmel.noaa.gov/tao/elnino/el-nino-story.html.

27. **Hudson, R. A., ed.** 1992. *Peru: A Country Study.* Washington: GPO for the Library of Congress. http://countrystudies.us/peru/.

28. **Garrison, T.** 2006. *Essentials of Oceanography,* 4th ed., p. 162. Belmont, CA: Thomson Higher Education.

29. **National Oceanic and Atmospheric Administration.** 2010. "The La Niña Story." Accessed March 1, 2010. http://www.pmel.noaa.gov/tao/elnino/la-nina-story.html.

30. **National Weather Service Climate Prediction Center.** 2005. "Frequently Asked Questions About El Niño and La Niña." Accessed March 2, 2010. http://www.cpc.noaa.gov/products/analysis_monitoring/ensostuff/ensofaq.shtml.

31. **Neibauer, M.** 2010. "Upside-Down Weather Patterns Mean More Snow Ahead for D.C." *Washington Examiner,* February 14. Accessed March 1, 2010. http://washingtonexaminer.com/upside-down-weather-patterns-mean-more-snow-ahead-for-d.c./article/17221.

32. **Casselman, A.** 2010. "Vancouver 2010 to Be Warmest Winter Olympics Yet." *National Geographic Daily News,* February 12. Accessed March 1, 2010. http://news.nationalgeographic.com/news/2010/02/100212-vancouver-2010-warmest-winter-olympics/.

33. **National Weather Service Climate Prediction Center.** 2010. "Cold & Warm Episodes by Season: Changes to the Oceanic Niño Index (ONI)." Accessed March 1, 2010. http://www.cpc.noaa.gov/products/analysis_monitoring/ensostuff/ensoyears.shtml.

34. **National Oceanic and Atmospheric Administration.** 2009. "El Niño Arrives; Expected to Persist Through Winter 2009-10," Accessed March 2, 2010. http://www.noaanews.noaa.gov/stories2009/20090709_elnino.html.

35. **Tropical Atmosphere Ocean Project.** 2010. "TAO Project Overview." Accessed March 2, 2010. http://www.pmel.noaa.gov/tao/proj_over/proj_over.html.

36. **Garrison, T.** 2006. *Essentials of Oceanography,* 4th ed., p. 166. Belmont, CA: Thomson Higher Education.

37. **Alley, R. B.** 2007. "Wally Was Right: Predictive Ability of the North Atlantic 'Conveyor Belt' Hypothesis for Abrupt Climate Change." *Annual Review of Earth and Planetary Sciences* 35: 241–272. doi:10.1146/annurev.earth.35.081006.131524.

38. **Broecker, W. S.** 2010. *The Great Ocean Conveyor: Discovering the Trigger for Abrupt Climate Change.* Princeton, NJ: Princeton University Press.

39. **Broecker, W. S.** 1991. "The Great Ocean Conveyor." *Oceanography* 4: 79–89.

40. **Vellinga, M., and Wood, R. A.** 2008. "Impacts of Thermohaline Circulation Shutdown in the Twenty-First Century." *Climatic Change* 91: 43–63.

41. **Kuhlbrodt, T., Rahmstorf, S., Zickfeld, K., Vikebo, F. B., Sundby, S., Hofmann, M., Link, P. M, et al.** 2009. "An Integrated Assessment of Changes in the Thermohaline Circulation." *Climatic Change* 96: 489–537.

42. **Hinrichsen, D.** 1998. *Coastal Waters of the World: Trends, Threats, and Strategies.* Washington DC: Island Press.

43. **National Oceanic and Atmospheric Administration.** 2008. "Tides and Water Levels: What Causes Tides?" Accessed July 9, 2012. http://oceanservice.noaa.gov/education/kits/tides/tides02_cause.html.

44. **Pilkey, O. H.** 1989. "The Engineering of Sand." *Journal of Geological Education* 37: 308–311.

45. **National Oceanic and Atmospheric Administration: Coastal Services Center.** 2010. "Beach Nourishment—Socioeconomic Factors." Accessed March 2, 2010. http://www.csc.noaa.gov/archived/beachnourishment/html/human/socio/change.htm.

46. **Pilkey, O. H., and Young, R. S.** 2005. "Will Hurricane Katrina Impact Shoreline Management? Here's Why It Should." *Journal of Coastal Research* 21: iii–ix.

47. **Greene, K.** 2002. "Beach Nourishment: A Review of the Biological and Physical Impacts." *ASMFC Habitat Management Series* 7. Washington, DC: Atlantic States Marine Fisheries Commission.

48. **Moore, L. J., and Griggs, G. B.** 2002. "Long-Term Cliff Retreat and Erosion Hotspots Along the Central Shores of the Monterey Bay National Marine Sanctuary." *Marine Geology* 181: 265–283.

49. **U.S. Geological Survey.** 2005. "Tectonic Summary: Magnitude 9.1 Off the West Coast of Northern Sumatra,

Sunday, December 26, 2004 at 00:58:53 UTC." Accessed March 3, 2010. http://neic.usgs.gov/neis/eq_depot/2004/eq_041226/neic_slav_ts.html.

50. **U.S. Geological Survey.** 2004. "Summary of Magnitude 9.0 Sumatra-Andaman Islands Earthquake & Tsunami, Sunday, December 26, 2004 at 00:58:53 UTC." Accessed March 3, 2010. http://earthquake.usgs.gov/earthquakes/eqinthenews/2004/usslav/eqsummary.html.

51. **AFP.** 2005. "Asian Tsunami Death Toll Passes 144,000." ABC News Online, January 3. Accessed March 3, 2010. http://www.abc.net.au/news/2005-01-03/asian-tsunami-death-toll-passes-144000/612154.

52. **National Oceanic and Atmospheric Administration.** 2010. "DART (Deep-ocean Assessment and Reporting of Tsunamis)." Accessed March 3, 2010. http://nctr.pmel.noaa.gov/Dart/.

53. **Pacific Disaster Center.** 2010. "Hawaii Tsunami Events." Accessed March 3, 2010. http://archive.is/VjPj.

54. **National Oceanic and Atmospheric Administration.** 2010. "US IOTWS (Indian Ocean Tsunami Warning System) Program." Accessed March 3, 2010. http://nctr.pmel.noaa.gov/education/IOTWS/.

55. **United Nations News Center.** 2008. "Pacific Rim Countries Stage UN-Initiated Tsunami Warning Drill." Accessed March 3, 2010. http://www.un.org/apps/news/story.asp?NewsID=28742&Cr=tsunami&Cr1#.Up9OIY0jhWs.

56. **Jaffe, B., Geist, E., and Gibbons, H.** 2004. "Indian Ocean Earthquake Triggers Deadly Tsunami." Accessed March 3, 2010. http://soundwaves.usgs.gov/2005/01/.

57. **Grotius. H.** 1609. *The Freedom of the Sea: Or, the Right Which Belongs to the Dutch to Take Part in the East Indian Trade.* Accessed February 23, 2010. http://upload.wikimedia.org/wikipedia/commons/7/7b/Grotius_Hugo_The_Freedom_of_the_Sea_(v1.0).pdf.

58. **Orbach, M. K.** 2002. "Beyond the Freedom of the Seas: Ocean Policy for the Third Millennium." Fourth Annual Roger Revelle Commemorative Lecture, National Academies of Science. Accessed February 23, 2010. http://community.middlebury.edu/~scs/docs/Orbach,%20Beyond%20Freedom%20of%20the%20Seas.pdf.

59. **Myers, R. A., and Worm, B.** 2003. "Rapid Worldwide Depletion of Predatory Fish Communities." *Nature* 423: 280–283.

60. **Stow, D.** 2006. *Oceans: An Illustrated Reference,* p. 222. Chicago: The University of Chicago Press.

61. **Van Bynkershoek, C.** 1737/1930. *Quaestionum juris publici libri duo,* vol. 2. Translated by T. Frank. London: Clarendon Press. Accessed February 24, 2010. http://www.constitution.org/bynk/bynk.htm.

62. **Jacobson, J. L., and Rieser, A.** 1998. "The Evolution of Ocean Law." *Scientific American Presents: The Oceans* 9, no. 3.

63. **United Nations.** 2008. "Table of Claims to Maritime Jurisdiction (as of 15 July 2011)." Accessed February 24, 2010. http://www.un.org/Depts/los/LEGISLATIONANDTREATIES/PDFFILES/table_summary_of_claims.pdf.

64. **DeSombre, E. R., and Barkin, J. S.** 2002. "Turbot and Tempers in the North Atlantic," in Matthew, R., Halle, M., and Switzer, J., eds, *Conserving the Peace: Resources, Livelihoods and Security,* pp. 325–349. Winnipeg, Manitoba: International Institute for Sustainable Development. http://www.iisd.org/pdf/2002/envsec_conserving_peace.pdf.

65. **Maddox, B.** 2009. "Russia Leads Arctic Race to Claim Northwest Passage." *The Times,* February 6. Accessed February 24, 2010. http://www.timesonline.co.uk/tol/comment/columnists/bronwen_maddox/article5671438.ece.

66. **Gillies, R.** 2007. "Canada Puts Muscle Behind Claim to Northwest Passage." *Seattle Times,* July 10. Accessed February 24, 2010. http://seattletimes.nwsource.com/html/nationworld/2003782182_canada10.html.

67. **Lee, K.** 2001. "The Global Dimensions of Cholera." *Global Change and Human Health* 2: 6–17.

68. **Estrada-Garcia, T., and Mintz, E. D.** 1996. "Cholera: Foodborne Transmission and Its Prevention." *European Journal of Epidemiology* 12: 461–469.

69. **U.S. Environmental Protection Agency.** 1988. "Ocean Dumping Ban Act of 1988." Available at http://www2.epa.gov/aboutepa/ocean-dumping-ban-act-1988.

70. **Mallin, M. A.** 2006. "Wading in Waste." *Scientific American* 294: 52–59.

71. **U.S. Environmental Protection Agency.** 2012. "EPA's BEACH Report: 2011 Swimming Season." Accessed July 9, 2012. http://water.epa.gov/type/oceb/beaches/upload/national_facsheet_2011.pdf.

72. **Bricker, S., Longstaff, B., Dennison, W., Jones, A., Boicourt, K., Wicks, C., and Woerner, J.** 2007. "Effects of Nutrient Enrichment In the Nation's Estuaries: A Decade of Change." *NOAA Coastal Ocean Program Decision Analysis Series,* no. 26. Silver Spring, MD: National Centers for Coastal Ocean Science.

73. **Rabalais, N.** 2002. "Beyond Science into Policy: Gulf of Mexico Hypoxia and the Mississippi River." *BioScience* 52: 129–142.

74. **U.S. Geological Survey.** 2009. "Graphics of Streamflow and Nutrient Flux for the Mississippi-Atchafalaya River Basin." Accessed March 6, 2010. http://toxics.usgs.gov/hypoxia/mississippi/flux_ests/delivery/graphics/index.html.

75. **Louisiana Universities Marine Consortium.** 2010. "What Is Hypoxia?" Hypoxia in the Northern Gulf of Mexico. Accessed March 6, 2010. http://www.gulfhypoxia.net/Overview/.

76. **National Ocean and Atmospheric Administration.** August 4, 2011. "NOAA-Supported Scientists Find

Large Dead Zone in Gulf of Mexico." Accessed September 5, 2011. http://www.noaanews.noaa.gov/stories2011/20110804_deadzone.html.

77. **Nixon, S. W.** 1998. "Enriching the Seas to Death." *Scientific American Presents: The Oceans*, pp. 48–53.

Schmiedeskamp, M. 1998. "Getting the Nutrients Out." *Scientific American Presents: The Oceans*, pp. 50–51.

78. **U.S. EPA Office of Wetlands, Oceans, and Watersheds.** 2010. "Mississippi River Gulf of Mexico Watershed Nutrient Task Force." Accessed March 6, 2010. http://www.epa.gov/msbasin/index.htm.

79. **U.S. Geological Survey.** 1998. "The Chesapeake Bay: Geologic Product of Rising Sea Level." *U.S. Geological Survey Fact Sheet* 102-98. http://pubs.usgs.gov/fs/fs102-98/.

80. **U.S. Environmental Protection Agency.** 1982. "Chesapeake Bay: Introduction to an Ecosystem." Washington, DC. Accessed March 4, 2010. Available at http://www.chesapeakebay.net/content/publications/cbp_13039.pdf.

81. **Chesapeake Bay Foundation.** 2010. "Bay 101: Facts and Figures." Accessed March 4, 2010. http://www.chesapeakebay.net/discover/bay101/facts.

82. **Chesapeake Bay Program.** 2010. *Bay Barometer: A Health and Restoration Assessment of the Chesapeake Bay and Watershed in 2010.* Accessed December 4, 2013. http://www.chesapeakebay.net/documents/cbp_59306.pdf.

83. **Chesapeake Bay Program.** 2010. *Bay Barometer: A Health and Restoration Assessment of the Chesapeake Bay and Watershed in 2010*, p. 9. Accessed December 4, 2013. http://www.chesapeakebay.net/documents/cbp_59306.pdf.

84. **Chesapeake Bay Program.** 2009. "Native Oyster Abundance." Accessed March 8, 2010. http://www.chesapeakebay.net/oysterharvest.aspx?menuitem=14701.

85. **Stamm, D.** 2009. "Medical Waste Washes Up on Jersey Beaches." NBC New York, July 2. Accessed March 6, 2010. http://www.nbcnewyork.com/news/local-beat/Medical-Waste-Washes-Ashore-New-Jersey.html.

Newmarker, C. 2008. "N.J. Towns Close Beaches After Medical Waste Washes Ashore." *USA Today*, August 29. Accessed March 6, 2010. http://www.usatoday.com/travel/news/2008-08-29-nj-beaches-medical-waste_N.htm.

86. **Convey, P., Barnes, D. K. A., and Morton, A.** 2002. "Debris Accumulation on Oceanic Island Shores of the Scotia Arc, Antarctica." *Polar Biology* 25: 612–617.

87. **Barnes, D. K. A., Galgani, F., Thompson, R. C., and Barlaz, M.** 2009. "Accumulation and Fragmentation of Plastic Debris in Global Environments." *Philosophical Transactions of the Royal Society B* 364: 1985–1998.

88. **National Oceanic and Atmospheric Administration.** "Frequently Asked Questions: What We Actually Know About Common Marine Debris Factoids." Accessed September 11, 2011. http://marinedebris.noaa.gov/info/faqs.html#1.

89. **Fowler, C.** 1983. "Status of Northern Fur Seals on the Pribilof Islands." Background paper submitted to the 26th Annual Meeting of the Standing Scientific Committee of the North Pacific Fur Seal Commission, March 28th–April 8th. Washington DC.

90. **Laist, D. W.** 1997. "Impacts of Marine Debris: Entanglement of Marine Life in Marine Debris Including a Comprehensive List of Species with Entanglement and Ingestion Records." In *Marine Debris, Sources, Impacts, and Solutions*, edited by J. M. Coe & D. B. Rogers, pp. 99–139. New York: Springer-Verlag.

91. **Mato, Y., Isobe, T., Takada, H., Kanehiro, H., Ohtake, C., and Kaminuma, T.** 2001. "Plastic Resin Pellets as a Transport Medium for Toxic Chemicals in the Marine Environment." *Environmental Science and Technology* 35, no. 2: 318–324.

92. **National Oceanic and Atmospheric Administration.** 2008. "State of the Science Fact Sheet: Ocean Acidification." Accessed January 24, 2014. http://oceanservice.noaa.gov/education/yos/resource/01state_of_science.pdf.

Woods Hole Oceanographic Institution. 2010. "FAQs About Ocean Acidification." Accessed September 5, 2011. http://www.whoi.edu/OCB-OA/FAQs/.

93. **Hoegh-Guldberg, O., Mumby, P. J., Hooten, A. J., Steneck, R. S., Greenfield, P., Gomez, E. Harvell, C. D., et al.** 2005. "Coral Reefs Under Rapid Climate Change and Ocean Acidification." *Science* 318: 1737–1742.

94. **Orr, J. C., Fabry, V. J., Aumont, O., Bopp, L., Doney, S. C., Feely, R. A., Gnanadesikan, A., et al.** 2005. "Anthropogenic Ocean Acidification over the Twenty-First Century and Its Impact on Calcifying Organisms." *Nature* 437: 681–686.

95. **Ries, J. B., Cohen, A. L., and McCorkle, D. C.** 2009. "Marine Calcifiers Exhibit Mixed Responses to CO_2-Induced Ocean Acidification." *Geology* 37: 1131–1134.

96. **Hardt, M. J., and Safina, C.** 2010. "Threatening Ocean Life from the Inside Out." *Scientific American* 303: 66–73.

97. **International Maritime Organization.** 2002. "Prevention of Pollution by Sewage from Ships." Accessed March 8, 2010. http://www.imo.org/Environment/mainframe.asp?topic_id=237.

98. **International Maritime Organization.** 2002. "Prevention of Pollution by Garbage from Ships" Accessed March 8, 2010. http://www.imo.org/Environment/mainframe.asp?topic_id=297.

99. **International Maritime Organization.** "London Convention and Protocol." Accessed September 8, 2011. http://www.imo.org/OurWork/Environment/SpecialProgrammesAndInitiatives/Pages/London-Convention-and-Protocol.aspx.

100. **U.S. Environmental Protection Agency.** 2009. "Ocean Dumping Ban Act of 1988." Accessed September 8, 2011. http://www.epa.gov/aboutepa/history/topics/mprsa/02.html.

101. **National Oceanic and Atmospheric Administration.** 2009. "Fishing for Energy Partners." Accessed March 8, 2010. http://marinedebris.noaa.gov/projects/fishing4energy.html.

102. **Salazar, J.** September 1, 2011. "Sylvia Earle: An Open Ocean in Arctic Summer." EarthSky. Accessed September 14, 2011. http://earthsky.org/earth/sylvia-earle-an-open-ocean-in-the-arctic-in-summer.

Chapter 13

1. **Roebroeks, W., and Villa, P.** 2011. "On the Earliest Evidence for Habitual Use of Fire in Europe." *Proceedings of the National Academy of Sciences* 108 (13) (March):5209–5214. doi:10.1073/pnas.1018116108. (Note: Some estimates of earliest fire use for African hominins are as early as 1.6 million years ago.)

2. **Lovins, A. B.** 2011. *Reinventing Fire: Bold Business Solutions for the New Energy Era.* White River Junction, VT: Chelsea Green Publishing.

3. **Flavin, C., and Lenssen, N.** 1992. "Designing a Sustainable Energy System." *Changing Environmental Ideologies* 1:51–88.

4. **Greene, D. L.** 2010. "Measuring Energy Security: Can the United States Achieve Oil Independence?" *Energy Policy* 38 (4) (April):1614–1621. doi:10.1016/j.enpol.2009.01.041.

5. **U.S. Energy Information Administration.** 2012. "Annual Energy Review 2011," p. 370. http://www.eia.gov/totalenergy/data/annual/pdf/aer.pdf.

6. **Höök, M., Bardi, U., Feng, L., and Pang, X.** 2010. "Development of Oil Formation Theories and Their Importance for Peak Oil." *Marine and Petroleum Geology* 27 (9) (October):1995–2004. doi:10.1016/j.marpetgeo.2010.06.005.

7. **Bjorlykke, K.** 2011. *Petroleum Geoscience: From Sedimentary Environments to Rock Physics.* 1st Edition. 2nd Printing. Springer.

8. **Hubbert, M. K.** 1981. "The World's Evolving Energy System." *American Journal of Physics* 49 (11):1007. doi:10.1119/1.12656. (Note: This and other works by Hubbert provide the analysis he used that supports this comment.)

9. **U.S. Energy Information Administration.** "International Energy Statistics." Accessed October 3, 2013. http://www.eia.gov/cfapps/ipdbproject/IEDIndex3.cfm?tid=5&pid=57&aid=6.

10. **Jones, J. C.** 2009. "Technical Note: Total Amounts of Oil Produced over the History of the Industry." *International Journal of Oil, Gas and Coal Technology* 2 (2):199–200.

11. **BP Statistical Review of World Energy.** 2012. http://www.bp.com/content/dam/bp/pdf/statistical-review/statistical_review_of_world_energy_2013.pdf.

12. **Gorelick, S. M.** 2009. *Oil Panic and the Global Crisis: Predictions and Myths.* 1st ed., p. 241. Oxford, UK: Wiley-Blackwell.

13. **Hubbert, M. K.** 1956. "Nuclear Energy and the Fossil Fuel." *Drilling and Production Practice.* American Petroleum Institute.

14. **Lenssen, N., and Flavin, C.** 1996. "Sustainable Energy for Tomorrow's World: The Case for an Optimistic View of the Future." *Energy Policy* 24(9) (September):769–781 (quote on 780). doi:10.1016/0301-4215(96)00060-2.

15. **Natural Resources Defense Council.** 2011. "Arctic Wildlife Refuge: Why Trash an American Treasure for a Tiny Percentage of Our Oil Needs?" Accessed October 3, 2013. http://www.nrdc.org/land/wilderness/arctic.asp.

16. **U.S. Fish and Wildlife Service.** Arctic National Wildlife Refuge. Accessed October 3, 2013. http://arctic.fws.gov/.

17. **Attanasi, E. D.** 2005. "Undiscovered Oil Resources in the Federal Portion of the 1002 Area of the Arctic National Wildlife Refuge: An Economic Update." Open-File Report 2005-1217. U.S. Department of the Interior, U.S. Geological Survey. http://royaldutchshellplc.com/wp-content/uploads/2010/08/2005-1217.pdf.

18. **National Research Council.** 2003. *Cumulative Environmental Effects of Oil and Gas Activities on Alaska's North Slope,* p. 288. Washington, DC: National Academies Press. Accessed April 19, 2013. http://www.nap.edu/openbook.php?record_id=10639&page=10.

19. **Attanasi, E. D., and Meyer, R. F.** 2010. "Natural Bitumen and Extra-Heavy Oil" In *2010 Survey of Energy Resources,* p. 123–140. World Energy Council. http://www.worldenergy.org/documents/ser_2010_report_1.pdf.

20. **U.S. Energy Information Administration.** 2012. "Canada—Analysis." Accessed October 3, 2013. http://www.eia.gov/countries/cab.cfm?fips=CA.

21. **U.S. Energy Information Administration.** 2013. "How Much Shale Gas Is Produced in the United States?—FAQ." Accessed October 3, 2013. http://www.eia.gov/tools/faqs/faq.cfm?id=907&t=8.

22. **Andrews, A., Folger, P., Humphries, M., Copeland, C., Tiemann, M., Meltz, R., and Brougher, C.** 2009. "Unconventional Gas Shales: Development, Technology, and Policy Issues," p.7, 23.? Congressional Research Service. http://www.fas.org/sgp/crs/misc/R40894.pdf.

23. **Haluszczak, L. O., Rose, A. W., and Kump, L. R.** 2012 "Geochemical Evaluation of Flowback Brine from Marcellus Gas Wells in Pennsylvania, USA." *Applied Geochemistry* 28:55–61. doi:10.1016/j.apgeochem.2012.10.002.

24. Malakoff, D. 2012. "BP Criminal Case Generates Record Payout for Science and Restoration." *Science* 338:1137–1137. doi:10.1126/science.338.6111.1137.

25. Horizon, Unified Command Deepwater. 2010. "US Scientific Teams Refine Estimates of Oil Flow from BP's Well Prior to Capping." *Gulf of Mexico Oil Spill Response* (2010).

26. Camilli, R., Reddy, C. M., Yoerger, D. R., Van Mooy, B. A. S., Jakuba, M. V., Kinsey, J. C., McIntyre, C. P., et al. 2010. "Tracking Hydrocarbon Plume Transport and Biodegradation at Deepwater Horizon." *Science* 330:201–204. *doi:10.1126/science.1195223.*

27. Sumaila, U. R., Cisneros-Montemayor, A. M., Dyck, A., Huang, L., Cheung, W., Jacquet, J., Kleisner, K., et al. 2012. "Impact of the Deepwater Horizon Well Blowout on the Economics of US Gulf Fisheries." *Canadian Journal of Fisheries and Aquatic Sciences* 69(3):499–510. doi:10.1139/f2011-171.

28. Mitra, S., Kimmel, D. G., Snyder, J., Scalise, K., McGlaughon, B. D., Roman, M. R., Jahn, G. L., et al. 2012. "Macondo-1 Well Oil-derived Polycyclic Aromatic Hydrocarbons in Mesozooplankton from the Northern Gulf of Mexico." *Geophysical Research Letters* 39(1). doi:10.1029/2011GL049505.

29. Elish, J. 2013. "'Dirty Blizzard' in Gulf of Mexico May Account for Missing *Deepwater Horizon* Oil." Florida State University.

30. U.S. Department of Justice. 2012. "BP Exploration and Production Inc. Agrees to Plead Guilty to Felony Manslaughter, Environmental Crimes and Obstruction of Congress Surrounding Deepwater Horizon Incident." Justice News, Office of Public Affairs, November 15. Accessed October 3, 2013. http://www.justice.gov/opa/pr/2012/November/12-ag-1369.html.

31. U.S. Department of Justice. 2013. "Transocean Agrees to Plead Guilty to Environmental Crime and Enter Civil Settlement to Resolve U.S. Clean Water Act Penalty Claims from Deepwater Horizon Incident." Justice News, Office of Public Affairs, January 3. Accessed October 3, 2013. http://www.justice.gov/opa/pr/2013/January/13-ag-004.html.

32. U.S. Environmental Protection Agency. 2011 "Emergency Management, Oil Spill Response Techniques." Accessed October 3, 2013. http://www.epa.gov/osweroe1/content/learning/oiltech.htm.

33. University of Delaware College of Earth, Ocean, and Environment. 2004. "How Do You Clean Up an Oil Spill?" Accessed October 3, 2013. http://www.ceoe.udel.edu/oilspill/cleanup.html.

34. Lattanzio, R. K. 2013. "Canadian Oil Sands: Life-Cycle Assessments of Greenhouse Gas Emissions." Congressional Research Service, 7-5700, R42537. http://www.fas.org/sgp/crs/misc/R42537.pdf.

35. Ohio Department of Natural Resources. 2012. "Preliminary Report on the Northstar 1 Class II Injection Well in the Seismic Events in the Youngston, Ohio, Area." Ohio Department of Natural Resources. http://media3.vindy.com/news/documents/2012/03/09/UICReport.pdf.

36. Rozell, D. J., and Reaven, S. J. 2011. "Water Pollution Risk Associated with Natural Gas Extraction from the Marcellus Shale." *Risk Analysis* 32(8):1382–93. doi:10.1111/j.1539-6924.2011.01757.x.

37. Natural Gas Subcommittee of the Secretary of Energy Advisory Board. 2011. "Improving the Safety and Environmental Performance of Hydraulic Fracturing." Natural Gas Subcommittee of the Secretary of Energy Advisory Board, U.S. Department of Energy. Accessed October 3, 2013. http://www.shalegas.energy.gov/.

38. A good resource for all aspects of coal is the International Energy Agency Clean Coal Centre (IEA CCC). The IEA CCC describes itself as the world's foremost provider of balanced, objective information (with no political or commercial bias) on the clean and efficient use of coal worldwide. http://www.iea-coal.org/site/2010/home.

39. BP Statistical Review of World Energy 2012. June 2012, Coal. http://www.bp.com/content/dam/bp/pdf/Statistical-Review-2012/statistical_review_of_world_energy_2012.pdf.

40. McQuaid, J. 2009. "The Razing of Appalachia: Mountaintop Removal Revisited." Yale Environment 360. Accessed April 12, 2011. http://e360.yale.edu/content/print.msp?id=2150.

41. During Senate hearing testimony on mountaintop removal and valley fill mining in 2009, EPA scientist John Randy Pomponio described the importance of headwater streams: "These little streams are like capillaries in your blood system.... They're what travel through the landscape and capture the pollutants, clean those pollutants. And we frankly don't know where the tipping point is in losing one stream, five streams, or 18 streams in a particular watershed."

42. U.S. Environmental Protection Agency. 2011. "The Effects of Mountaintop Mines and Valley Fills on Aquatic Ecosystems of the Central Appalachian Coalfields (2011 Final)." U.S. Environmental Protection Agency, EPA/600/R-09/138F. Accessed October 3, 2013. http://cfpub.epa.gov/ncea/cfm/recordisplay.cfm?deid=225743.

43. Pond, G. J., Passmore, M. E., Borsuk, F. A., Reynolds, L., and Rose, C. J. 2008. "Downstream Effects of Mountaintop Coal Mining: Comparing Biological Conditions Using Family- and Genus-Level Macroinvertebrate Bioassessment Tools." *Journal of the North American Benthological Society* 27:717–737.

44. Palmer, M. A., Bernhardt, E. S., Schlesinger, W. H., Eshleman, K. N., Foufoula-Georgiou, E., Hendryx, M. S., et al. 2010. "Mountaintop Mining Consequences." *Science* 327(5962):148–149.

45. U.S. Environmental Protection Agency. "Water: Discharge of Dredged or Fill Materials: Section (404) Permitting." Accessed October 3, 2013. http://water.epa.gov/lawsregs/guidance/cwa/dredgdis/.

46. **U.S. Environmental Protection Agency.** 2011. "Mid-Atlantic Mountaintop Mining, Spruce No. 1 Mine." Accessed October 3, 2013. http://www.epa.gov/region03/mtntop/spruce1.html.

47. **U.S. Environmental Protection Agency.** 2011. "Questions & Answers—Spruce Mine Final Determination." U.S. Environmental Protection Agency. http://water.epa.gov/lawsregs/guidance/cwa/dredgdis/upload/FINAL_Spruce_404c_QA_011311.pdf.

48. This topographic change example was prepared by Skytruth.org: "Visualizing Elevation Change: Mountaintop Removal Mining" (accessed September 24, 2013) http://blog.skytruth.org/2012/07/visualizing-elevation-change.html.

49. **Stracher, G. B.** 2010. "The Rising Global Interest in Coal Fires." Accessed May 12, 2013. http://www.earthmagazine.org/article/rising-global-interest-coal-fires.

50. **Institute for Energy Research.** 2012. "U.S. Energy-Related Carbon Dioxide Emissions Are Declining." Institute for Energy Research. Accessed June 20, 2013. http://www.instituteforenergyresearch.org/2012/07/20/u-s-energy-related-carbon-dioxide-emissions-are-declining/.

51. **Fioletov, V. E., McLinden, C. A., Krotkov, N., Moran, M. D., and Yang, K.** 2011. "Estimation of SO_2 Emissions Using OMI Retrievals." *Geophysical Research Letters* 38(21). doi:10.1029/2011GL049402.

52. **National Aeronautics and Space Administration.** 2011. "NASA Satellite Confirms Sharp Decline in Pollution from U.S. Coal Power Plants." Accessed July 1, 2013. http://www.nasa.gov/topics/earth/features/coal-pollution.html.

53. **U.S. Environmental Protection Agency.** 2012. "Clean Air Interstate Rule." Accessed October 3, 2013. http://www.epa.gov/cair/basic.html.

54. **Orr, F. M.** 2009. "Onshore Geologic Storage of CO_2." *Science* 325(5948):1656–1658.

55. **U.S. National Energy Technology Laboratory.** "Technologies: Carbon Storage." Accessed October 13, 2013. http://www.netl.doe.gov/technologies/carbon_seq/.

56. **Metz, B., Davidson, O., de Coninck, H., Loos, M., and Meyer, L., eds.** 2005. "IPCC Special Report on Carbon Dioxide Capture and Storage." Intergovernmental Panel on Climate Change and Cambridge University Press.

57. **U.S. Environmental Protection Agency.** 2012. "Geologic Sequestration of Carbon Dioxide." U.S. Environmental Protection Agency. Accessed October 3, 2013. http://water.epa.gov/type/groundwater/uic/wells_sequestration.cfm.

58. **International Geothermal Association.** Accessed October 3, 2013. http://www.geothermal-energy.org/index.html.

59. **U.S. Environmental Protection Agency.** 2011. "Uranium Mill Tailings." U.S. Environmental Protection Agency. Accessed June 20, 2013. http://www.epa.gov/radiation/docs/radwaste/402-k-94-001-umt.html.

60. **U.S. Nuclear Regulatory Commission.** 2012. "Fact Sheet on Uranium Mill Tailings." U.S. Nuclear Regulatory Commission. Accessed June 20, 2013. http://www.nrc.gov/reading-rm/doc-collections/fact-sheets/mill-tailings.html.

61. **Nuclear Energy Institute.** 2012. "Nuclear Waste Disposal." Nuclear Energy Institute. Accessed June 20, 2013. http://www.nei.org/resourcesandstats/documentlibrary/nuclearwastedisposal/factsheet/disposal-of-commercial-low-level-radioactive-waste/.

62. **World Nuclear Association.** 2010. "Storage and Disposal Options." World Nuclear Association. Accessed June 20, 2013. http://world-nuclear.org/info/Nuclear-Fuel-Cycle/Nuclear-Wastes/Appendices/Radioactive-Waste-Management-Appendix-2--Storage-and-Disposal-Options/.

63. **Werner, J. D.** 2012. "U.S. Spent Nuclear Fuel Storage." Congressional Research Service, 7-5700, R42513. Accessed June 20, 2013. http://www.fas.org/sgp/crs/misc/R42513.pdf.

64. **U.S. Energy Information Administration.** 2013. "Levelized Cost of New Generation Resources in the Annual Energy Outlook 2013." U.S. Energy Information Administration, U.S. Department of Energy.

65. **Jacobson, M. Z., and Delucchi, M. A.** 2009. "A Path to Sustainable Energy by 2030." *Scientific American* 301(5):58–65. doi:10.1038/scientificamerican1109-58.

66. **Jacobson, M. Z., and Delucchi, M. A.** 2011. "Providing All Global Energy with Wind, Water, and Solar Power, Part I: Technologies, Energy Resources, Quantities and Areas of Infrastructure, and Materials." *Energy Policy* 39(3):1154–1169. doi:10.1016/j.enpol.2010.11.040.

67. **Ivanpah Solar Electric Power Generating System.** 2013. "Ivanpah Project Facts." Accessed June 20, 2013. http://ivanpahsolar.com/about.

68. **Union of Concerned Scientists.** 2013. "Environmental Impacts of Solar Power." Union of Concerned Scientists. Accessed June 26, 2013. http://www.ucsusa.org/clean_energy/our-energy-choices/renewable-energy/environmental-impacts-solar-power.html.

69. **Subramanian, M.** 2012. "The Trouble with Turbines: An Ill Wind." *Nature* 486(7403):310–311. doi:10.1038/486310a.

70. **Jacobson, M. Z., and Delucchi, M. A.** 2011. "Providing All Global Energy with Wind, Water, and Solar Power, Part I: Technologies, Energy Resources, Quantities and Areas of Infrastructure, and Materials." *Energy Policy* 39(3):1154–1169. doi:10.1016/j.enpol.2010.11.040. See also more recent estimates at: **Jacobson, M. Z., and Archer, C. L.** 2012. "Saturation Wind Power Potential and Its Implications for Wind Energy." *Proceedings of the National Academy of Sciences* 109(39):15679–15684. doi:10.1073/pnas.1208993109. And more recent results at: **Adams, A. S., and Keith, D. W.** 2013. "Are Global Wind Power Resource Estimates Overstated?" *Environmental Research Letters* 8(1):015021. doi:10.1088/1748-9326/8/1/015021. And most recently: **Petersen, E. L., Troen, I., Jørgensen, H. E.,**

and Mann, J. 2013. "Are Local Wind Power Resources Well Estimated?" *Environmental Research Letters* 8(1):011005. doi:10.1088/1748-9326/8/1/011005.

71. U.S. Energy Information Administration. 2013. "China: Analysis." U.S. Energy Information Administration. Accessed June 15, 2013. http://www.eia.gov/countries/cab.cfm?fips=CH.

72. Lovins, A. B. 2011. *Reinventing Fire: Bold Business Solutions for the New Energy Era*. White River Junction, VT: Chelsea Green Publishing.

73. Lovins, A. B. 1976. "Energy Strategy: The Road Not Taken." *Foreign Affairs* 55:65.

74. LEED is the acronym for Leadership in Energy and Environmental Design, an internationally recognized rating system developed by the U. S. Green Building Council for high-performing green buildings, homes, and neighborhoods. See: http://new.usgbc.org/leed.

75. Lovins, A. B. 2012. "A Farewell to Fossil Fuels: Answering the Energy Challenge." *Foreign Affairs* 91:135.

Chapter 14

1. Anderson, R. Y., Nuhfer, E. B., and Dean, W. E. 1985. "Sedimentation in a Blast-Zone Lake at Mount St. Helens, Washington—Implications for Varve Formation." *Geology* 13: 348–352.

2. Jones, J. 2000. "Preparing for California's Next Drought: Changes since 1987–92." Sacramento, CA: State of California Department of Water Resources. http://www.water.ca.gov/waterconditions/drought/docs/Drought_Report_87-92.pdf.

3. Center for the Continuing Study of the California Economy. 2010. "California Poised to Move Up in World Economy Rankings in 2013." CCSCE. http://www.ccsce.com/PDF/Numbers-July-2013-CA-Economy-Rankings-2012.pdf.

4. Crossett, K. M., Culliton, T. J., Wiley, P. C., and Goodspeed, T. R. 2004. "Population Trends along the Coastal United States: 1980–2008." National Oceanic and Atmospheric Administration, NOAA's National Ocean Service, Management and Budget Office, Special Projects, Coastal Trends Report Series, p. 44. http://oceanservice.noaa.gov/programs/mb/supp_cstl_population.html.

5. Stine, S. 1994. "Extreme and Persistent Drought in California and Patagonia During Mediaeval Time." *Nature* 369:546–549.

6. Hodell, D. A., Curtis, J. H., and Brenner, M. 1995. "Possible Role of Climate in the Collapse of Classic Maya Civilization." *Nature* 375:391–394.

7. Benson, L. V., Berry, M. S., Jolie, E. A., Spangler, J. D., Stahle, D. W., Hattori, E. M. 2007. Possible Impacts of Early-11th-, Middle-12th-, and Late-13th-Century Droughts on Western Native Americans and the Mississippian Cahokians." *Quaternary Science Reviews* 26:336–350.

8. Stahle, D. W., Cleaveland, M. K., Blanton, D. B., Therrell, M. D., and Gay, D. A. 1998. "The Lost Colony and Jamestown Droughts." *Science* 280:564–567.

9. Barlow, L. K., Sadler, J. P., Ogilvie, A. E. J., Buckland, P. C., Amorosi, T., Ingimundarson, H. J., Skidmore, P., et al. 1997. "Interdisciplinary Investigations of the End of the Norse Western Settlement in Greenland." *The Holocene* 7:489–499.

10. City of Santa Barbara. 2008. "City of Santa Barbara—Government—City Departments—Public Works—Frequently Asked Questions." Accessed March 10, 2010. http://www.santabarbaraca.gov/Government/Departments/PW/FAQ.htm.

11. Kasting, J. F., Toon, O. B., and Pollack, J. B. 1988. "How Climate Evolved on the Terrestrial Planets." *Scientific American* xx:90–97.

12. FSD Webmaster. 2006. "The Planet Mercury." National Weather Service Weather Forecast Office Sioux Falls, South Dakota. Accessed August 27, 2013. http://www.crh.noaa.gov/fsd/?n=mercury.

13. Haqq-Misra, J. D., Domagal-Goldman, S. D., Kasting, P. J., and Kasting, J. F. 2008. "A Revised, Hazy Methane Greenhouse for the Archean Earth." *Astrobiology* 8:1127–1137.

14. Kasting, J. 2010. *How to Find a Habitable Planet*, p. 45. Princeton: Princeton University Press.

15. Knoll, A. 2003. *Life on a Young Planet: The First Three Billion Years of Evolution on Earth*. Princeton: Princeton University Press.

16. Kasting, J. 2010. *How to Find a Habitable Planet*, p. 71–77. Princeton: Princeton University Press.

17. Kasting, J. 2010. *How to Find a Habitable Planet*, p. 76. Princeton: Princeton University Press.

18. Kasting, J. 2010. *How to Find a Habitable Planet*, p. 72. Princeton: Princeton University Press.

19. Kasting, J. 2010. *How to Find a Habitable Planet*, pp. 148–49. Princeton: Princeton University Press.

20. Kasting, J. F., Toon, O. B., and Pollack, J. B. 1988. "How Climate Evolved on the Terrestrial Planets." *Scientific American* xx:90–97.

21. Kasting, J. 2010. *How to Find a Habitable Planet*, p. 144. Princeton: Princeton University Press.

22. Taylor, F. W. 1994. "The Atmospheres of the Inner Planets." *Current Science* 66:512–524.

23. Kasting, J. 2010. *How to Find a Habitable Planet*, p. 111. Princeton: Princeton University Press.

24. Kasting, J. 2010. *How to Find a Habitable Planet*, p. 108. Princeton: Princeton University Press.

25. Kasting, J. 2010. *How to Find a Habitable Planet*, pp. 65–71. Princeton: Princeton University Press.

26. Kasting, J. 2010. *How to Find a Habitable Planet*, pp. 77–79. Princeton: Princeton University Press.

27. Kasting, J. 2010. *How to Find a Habitable Planet,* p. 61. Princeton: Princeton University Press.

28. Green, O. R., Searle, M. P., Corfield, R. I., and Corfield, R. M. 2008. "Cretaceous-Tertiary Carbonate Platform Evolution and the Age of the India–Asia Collision along the Ladakh Himalaya (NW India)." *Journal of Geology* 116:331–353.

29. Zhisheng, A., Kutzbach, J. E., Prell, W. L., and Porter, S. C. 2001. "Evolution of Asian Monsoons and Phased Uplift of the Himalaya-Tibetan Plateau since Late Miocene Times." *Nature* 411:62–66.

30. Katz, M. E., Cramer B. S., Toggweiler, J. R., Esmay, G., Liu, C., Miller, K. G., Rosenthal, Y., et al. 2011. "Impact of Antarctic Circumpolar Current Development on Late Paleogene Ocean Structure." Science 332:1076–1079.

31. Oke, T. R. 1987. *Boundary Layer Climates,* 2nd ed., p. 12. Methuen and Company.

32. Hoffman, P. F., and Schrag, D. P. 2000. "Snowball Earth." *Scientific American* 282(1):68–75.

33. Lorentz, N. J., and Corsetti, F. A. 2007. "Another Test for Snowball Earth." *Geology* 35:383–384.

34. Klein, C. 2005. "Some Precambrian Banded Iron-Formations (BIFs) from around the World: Their Age, Geologic Setting, Mineralogy, Metamorphism, Geochemistry, and Origin." *American Mineralogist* 90:1473–1499.

35. Seton, M., Gaina, C., Müller, R. D., and Heine, C. 2009. "Mid-Cretaceous Seafloor Spreading Pulse: Fact or Fiction?" *Geology* 37:687–690.

36. Ruddiman, W. F. 2008. *Earth's Climate: Past and Future,* 2nd ed., p. 84. New York: W. H. Freeman.

37. Ruddiman, W. F. 2008. *Earth's Climate: Past and Future,* 2nd ed., p. 123. New York: W. H. Freeman.

38. Ruddiman, W. F. 2008., *Earth's Climate: Past and Future,* 2nd ed., p. 122. New York: W. H. Freeman.

39. Ruddiman, W. F. 2008. *Earth's Climate: Past and Future,* 2nd ed., p. 124–126. New York: W. H. Freeman.

40. Imbrie, J., and Imbrie, K. P. 1979. *Ice Ages: Solving the Mystery,* chap. 6–8. Cambridge, MA: Harvard University Press.

41. Imbrie, J., and Imbrie, K. P. 1979. *Ice Ages: Solving the Mystery,* chap. 8–9. Cambridge, MA: Harvard University Press.

42. Ruddiman, W. F. 2003. "The Anthropogenic Greenhouse Era Began Thousands of Years Ago." *Climatic Change* 61:261–293.

43. McGuffie, K., and Henderson-Sellers, A. 2005. *A Climate Modeling Primer,* 3rd ed., p. 36. Chichester, England: John Wiley and Sons.

44. Sunda, W., Kieber, D. J., Kiene, R. P., and Huntsman, S. 2002. "An Antioxidant Function for DMSP and DMS in Marine Algae." *Nature* 418:317–320.

45. Bradley, R. S. 1985. *Quaternary Paleoclimatology: Methods of Paleoclimatic Reconstruction,* p. 125. Boston: Unwin Hyman.

46. Alley, R. B. 2000. *The Two-Mile Time Machine,* p. 17. Princeton: Princeton University Press.

47. Alley, R. B. 2000. "Ice Core Evidence of Abrupt Climate Changes." *Proceedings of the National Academies of Science* 97:1331–1334.

48. EPICA Community Members. 2006. "One-to-One Coupling of Glacial Climate Variability in Greenland and Antarctica." *Nature* 444:195–198.

49. Knutti, R., Flückinger, J., Stocker, T. F., and Timmermann, A. 2004. "Strong Hemispheric Coupling of Glacial Climate through Freshwater Discharge and Ocean Circulation." *Nature* 430:851–856.

50. Braun, H., Christl, M., Rahmstorf, S., Ganopolski, A., Mangini, A., Kubatzki, C., Roth, K., et al. 2005. "Possible Solar Origin of the 1,470-Year Glacial Climate Cycle Demonstrated in a Coupled Model." *Nature* 438:208–211.

51. Benson, L., and Paillet, F. 2002. "HIBAL: A Hydrologic-Isotopic-Balance Model for Application to Paleolake Systems." *Quaternary Science Reviews* 21:1521–1539.

52. Asmerom, Y., Polyak, V. J., and Burns, S. J. 2010. "Variable Winter Moisture in the Southwestern United States Linked to Rapid Glacial Climate Shifts." *Nature Geoscience* 3:114–117.

53. Wilf, P. 2000. "Late Paleocene-Early Eocene Climate Changes in Southwestern Wyoming: Paleobotanical Analysis." *GSA Bulletin* 112:292–307; Markwick, P. J. 1998. "Fossil Crocodilians as Indicators of Late Cretaceous and Cenozoic Climates: Implications for Using Palaeontological Data in Reconstructing Palaeoclimate." *Palaeogeography, Palaeoclimatology, Palaeoecology* 137:205–271.

54. Cronin, T. M. 1999. *Principles of Paleoclimatology,* p. 323. New York: Columbia University Press.

55. Brachert, T. C., Reuter, M., Kroeger, K. F., and Lough, J. M. 2006. "Coral Growth Bands: A New and Easy to Use Paleothermometer in Paleoenvironment Analysis and Paleoceanography (Late Miocene, Greece)." *Paleoceanography* 21:PA4217.

56. Webb III, T., Anderson, K. H., Bartlein, P. J., and Webb, R. S. 1998. "Late Quaternary Climate Change in Eastern North America: A Comparison of Pollen-Derived Estimates with Climate Model Results." *Quaternary Science Reviews* 17:587–606.

57. Bradley, R. S. 1985. *Quaternary Paleoclimatology: Methods of Paleoclimatic Reconstruction,* chap. 10. Boston, MA: Unwin Hyman.

58. Zachos, J., Pagani, M., Sloan, L., Thomas, E., and Billups, K. 2001. "Trends, Rhythms, and Aberrations in Global Climate 65 Ma to Present." *Science* 292:686–693.

59. **Mikolajewicz, U., Maier-Reimer, E., Crowley, T. J., and Kim, K.** 1993. "Effects of Drake and Panamanian Gateways on the Circulation of an Ocean Model." *Paleoceanography* 8:409–426; Lyle, M., Barron, J., Bralower, T. J., Huber, M., Lyle, A. O., Ravelo, A. C., Rea, D. K., et al. 2008. "Pacific Ocean and Cenozoic Evolution of Climate." *Reviews of Geophysics* 46:RG2002. doi:10.1029/2005RG000190.

60. **McGuffie, K., and Henderson-Sellers, A.** 2005. *A Climate Modeling Primer*, 3rd ed., p. 47. Chichester, England: John Wiley and Sons.

61. **McGuffie, K., and Henderson-Sellers, A.** 2005. *A Climate Modeling Primer*, 3rd ed., p. 72–76. Chichester, England: John Wiley and Sons.

Chapter 15

1. **Erdman, J.** 2012. "Superstorm Sandy: By the Numbers." The Weather Channel, October 31. Accessed February 21, 2013. http://www.weather.com/news/weather-hurricanes/superstorm-sandy-by-the-numbers-20121030?pageno=4.

2. **Associated Press.** 2012. "Superstorm Sandy Deaths, Damage And Magnitude: What We Know One Month Later." The Huffington Post, November 29. Accessed February 21, 2013. http://www.huffingtonpost.com/2012/11/29/superstorm-hurricane-sandy-deaths-2012_n_2209217.html.

3. **United States Department of Agriculture.** 2012. "World Agricultural Supply and Demand Estimates." United States Department of Agriculture. Accessed February 21, 2013. http://usda01.library.cornell.edu/usda/waob/wasde//2010s/2012/wasde-05-10-2012.pdf.

4. **Mohadjerin, M.** 2012. "Drought (U.S. Drought of 2012)." *New York Times.* Accessed February 21, 2013. http://topics.nytimes.com/top/news/science/topics/drought/index.html.

5. **Fleck, J.** 2012. "Whitewater-Baldy Fire Now Biggest in N.M. History." ABQ Journal Online, May 30. Accessed February 21, 2013. http://www.abqjournal.com/main/2012/05/30/abqnewsseeker/whitewater-baldy-now-largest-fire-in-new-mexico-history.html.

6. **Fears, D.** 2012. "U.S. Runs Out of Funds to Battle Wildfires." *Washington Post,* October 7. Accessed February 21, 2013. http://www.washingtonpost.com/national/us-runs-out-of-funds-to-battle-wildfires/2012/10/07/d632df5c-0c0c-11e2-bd1a-b868e65d57eb_story.html.

7. **National Oceanic and Atmospheric Administration, National Climatic Data Center.** 2012. "State of the Climate: Wildfires for August 2012." NOAA. Accessed February 21, 2013. http://www.ncdc.noaa.gov/sotc/fire/2012/8.

8. **Esch, M.** 2012. "Drought, Dry Summer Means More Encounters with Hungry Bears." *USA Today,* August 9. Accessed February 21, 2013. http://www.usatoday.com/weather/drought/story/2012-08-09/drought-bear-encounters/56899122/1.

9. **Stolberg, S. G., and Santora, M.** 2012. "Storms Trailed by Power Loss and Stifling Heat." *New York Times,* July 1. Accessed February 21, 2013. http://www.nytimes.com/2012/07/01/us/storms-leave-2-million-without-power.html?pagewanted=all.

10. **Sutton, J.** 2012. "Coast Guard Halts Traffic on Low-water Stretch of Mississippi." *CNN,* August 20. Accessed February 21, 2013. http://www.cnn.com/2012/08/20/us/mississippi-river-traffic/index.html.

11. **Yingqi, C.** 2012. "Climate Change Will Bring More Heavy Rains." *China Daily,* August 10. Accessed February 21, 2013. http://usa.chinadaily.com.cn/china/2012-08/10/content_15657384.htm.

12. **Park, M.** 2012. "Fury After 37 Killed in Beijing Floods." CNN, July 23. Accessed February 21, 2013. http://www.cnn.com/2012/07/23/world/asia/china-flooding-reaction/index.html.

13. **Whiteman, H.** 2012. "China Doubles Beijing Flood Death Toll." CNN, July 26. Accessed February 21, 2013. http://www.cnn.com/2012/07/26/world/asia/china-beijing-flood/index.html.

14. **Fennessy, M. J., and Kinter III, J. L.** 2011. "Climatic Feedbacks during the 2003 European Heat Wave." *Journal of Climate* 24:5953–5967.

15. **Viñas, M.-J.** 2012. "Satellites See Unprecedented Greenland Ice Sheet Surface Melt." NASA, July. Accessed February 21, 2013. http://www.nasa.gov/topics/earth/features/greenland-melt.html.

16. **Bryner, J.** 2012. "Greenland Ice Sheet Melting Breaks 30-Year Record." *Huffington Post,* August 15. Accessed February 21, 2013. http://www.huffingtonpost.com/2012/08/15/greenland-ice-sheet-melting_n_1783063.html.

17. **Harrabin, R.** 2012. "Arctic Sea Ice Reaches Record Low, NASA Says." *BBC News,* August 27. Accessed February 21, 2013. http://www.bbc.co.uk/news/science-environment-19393075.

18. **Siegel, M.** 2013. "Record Heat Fuels Widespread Fires in Australia." *New York Times,* January 10. Accessed February 21, 2013. http://www.nytimes.com/2013/01/10/world/asia/record-heat-fuels-widespread-fires-in-australia.html?_r=0.

19. **BBC News.** 2012. "Australia Floods Leave Two Dead, Thousands Evacuated." *BBC News,* March 5. Accessed February 21, 2013. http://www.bbc.co.uk/news/world-asia-17254966.

20. **Allen, M. R., Frame, D. J., Huntingford, C., Jones, C. D., Lowe, J. A., Meinshausen, M., and Meinshausen, N.** 2009. "Warming Caused by Cumulative Carbon Emissions Towards the Trillionth Tonne." *Nature* 458:1163–1166.

21. **National Oceanic and Atmospheric Administration, National Climatic Data Center.** 2012. "State of the Climate: Global Analysis for Annual 2012." NOAA. Accessed February 21, 2013. http://www.ncdc.noaa.gov/sotc/global/.

22. Lynch, P. 2012. "NASA Finds 2011 Ninth Warmest Year on Record." National Aeronautics and Space Administration, January 19. Accessed February 21, 2013. http://www.giss.nasa.gov/research/news/20120119/.

23. Rockström, J., Steffen, W., Noone, K., Persson, Å., Chapin III, F. S., Lambin, E., Lenton, T. M., et al. 2009. "Planetary Boundaries: Exploring the Safe Operating Space for Humanity." *Ecology and Society* 14(2):32.

24. Lenton, T. M., Held, H., Kriegler, E., Hall, J. W., Lucht, W., Rahmstorf, S., and Schellnhuber, H. J. 2008. "Tipping Elements in the Earth's Climate System." Proceedings of the National Academy of Sciences of the United States of America 105:1786–1793.

25. Guterl, F. 2012. "Climate Armageddon: How the World's Weather Could Quickly Run Amok [Excerpt]." *Scientific American* (May). Accessed February 21, 2013. http://www.scientificamerican.com/article.cfm?id=how-worlds-weather-could-quickly-run-amok.

26. Park, S., Croteau, P., Boering, K. A., Etheridge, D. M., Ferretti, D., Fraser, P. J., Kim, K.-R., et al. 2012. "Trends and Seasonal Cycles in the Isotopic Composition of Nitrous Oxide since 1940," *Nature Geoscience* 5:261–265.

27. Driscoll, D. A., Felton, A., Gibbons, P., Felton, A. M., Munro, N. T., and Lindenmayer, D. B. 2012. "Priorities in Policy and Management When Existing Biodiversity Stressors Interact with Climate-Change." *Climatic Change* 111:533–557.

28. Brosnan, T., and Pelstring, L. 2001. "Fall 2001 Status Report on the Hudson River Natural Resource Damage Assessment: Assessing Fish Health." National Oceanic and Atmospheric Administration. Accessed February 21, 2013. http://www.darrp.noaa.gov/northeast/hudson/pdf/hrfish.pdf.

29. Clarke, R., and King, J. 2004. *The Water Atlas*, p. 24. New York, NY: The New Press.

30. Howard, B. C. 2013. "8 Mighty Rivers Run Dry from Overuse." National Geographic. Accessed February 21, 2013. http://environment.nationalgeographic.com/environment/photos/rivers-run-dry/#/freshwater-rivers-colorado-1_45140_600x450.jpg.

31. National Aeronautics and Space Administration Earth Observatory. 2001. "The Shrinking Aral Sea." Accessed February 21, 2013. http://earthobservatory.nasa.gov/IOTD/view.php?id=1396.

32. Small, E. E., Giorgi, F., Sloan, L. C., and Hostetler, S. 2001. "The Effects of Desiccation and Climatic Change on the Hydrology of the Aral Sea." *Journal of Climate* 14:300–322.

33. Oyama, M. D., and Nobre, C. A. 2003. "A New Climate-Vegetation Equilibrium State for Tropical South America." *Geophysical Research Letters* 30:2199. doi:10.1029/2003GL018600.

34. United Nations Environment Program. 2003. "Environmental Effects of Ozone Depletion and Its Interactions with Climate Change: 2002 Assessment." Secretariat for The Vienna Convention for the Protection of the Ozone Layer and The Montreal Protocol on Substances that Deplete the Ozone Layer United Nations Environment Programme. http://ozone.unep.org/Assessment_Panels/EEAP/eeap-report2002.pdf.

35. Union of Concerned Scientists. 2007. "Smoke, Mirrors & Hot Air: How ExxonMobil Uses Big Tobacco's Tactics to Manufacture Uncertainty on Climate Science." Union of Concerned Scientists. http://www.ucsusa.org/assets/documents/global_warming/exxon_report.pdf.

36. Committee to Review the Intergovernmental Panel on Climate Change. 2010. "Climate Change Assessments: Review of the Processes and Procedures of the IPCC." InterAcademy Council. http://www.ipcc.ch/pdf/IAC_report/IAC%20Report.pdf.

37. Singer, S. F., Idso, C., and Carter, R. M. 2011. "Climate Change Reconsidered: 2011 Interim Report of the Nongovernmental International Panel on Climate Change." The Heartland Institute. http://www.nipccreport.org/reports/2011/2011report.html.

38. Idso, C. D., and Idso, S. B. 2011. *The Many Benefits of Atmospheric CO$_2$ Enrichment*. Pueblo West, CO: Science and Public Policy Institute.

39. Doran, P. T., and Zimmerman, M. K. 2009. "Examining the Scientific Consensus on Climate Change." *EOS* 90:22.

40. United Nations Environment Program World Glacier Monitoring Service. 2008. "Global Glacier Changes: Facts and Figures." http://www.grid.unep.ch/glaciers/pdfs/glaciers.pdf.

41. Hall, M. H. P., and Fagre, D. B. 2003. "Modeled Climate Induced Glacier Change in Glacier National Park, 1850–2100." *Bioscience* 53:131–140.

42. Church, J. A., White, N. J., Konikow, L. F., Domingues, C. M., Cogley, J. G., Rignot, E., Gregory, J. M., et al. 2011. "Revisiting the Earth's Sea-Level and Energy Budgets from 1961 to 2008." *Geophysical Research Letters* 38, L18601. doi:10.1029/2011GL048794.

43. Trenberth, K. E., Jones, P. D., Ambenje, P., Bojariu, R., Easterling, D., Klein Tank, A., Parker, D., et al. 2007. "Observations: Surface and Atmospheric Climate Change." In *Climate Change 2007: The Physical Science Basis. Contribution of Working Group I to the Fourth Assessment Report of the Intergovernmental Panel on Climate Change*, edited by S. Solomon, D. Qin, M. Manning, Z. Chen, M. Marquis, K. B. Averyt, M. Tignor, and H. L. Miller. Cambridge, UK: Cambridge University Press.

44. Bernstein, L., Bosch, P., Canziani, O., Chen, Z., Christ, R., Davidson, O., Hare, W., et al. 2007. "Climate Change 2007: Synthesis Report." Intergovernmental Panel on Climate Change Fourth Assessment Report, pg. 30.

45. Ruddiman, W. F. 2001. *Earth's Climate: Past and Future*, p. 407. New York, NY: W. H. Freeman.

46. Zachos, J., Pagani, M., Sloan, L., Thomas, E., and Billups, K. 2001. "Trends, Rhythms, and Aberrations in Global Climate 65 Ma to Present." *Science* 292:686–693.

47. Li, Y.-X., Yu, Z., Kodama, K. P. 2007. "Sensitive Moisture Response to Holocene Millennial-Scale Climate Variations in the Mid-Atlantic Region, USA." *The Holocene* 17:3–8.

48. Le Treut, H., Somerville, R., Cubasch, U., Ding, Y., Mauritzen, C., Mokssit, A., Peterson, T., et al. 2007. "Historical Overview of Climate Change." In *Climate Change 2007: The Physical Science Basis. Contribution of Working Group I to the Fourth Assessment Report of the Intergovernmental Panel on Climate Change*, edited by S. Solomon, D. Qin, M. Manning, Z. Chen, M. Marquis, K. B. Averyt, M. Tignor, and H. L. Miller, p. 107. Cambridge, UK: Cambridge University Press.

49. Bernstein, L., Bosch, P., Canziani, O., Chen, Z., Christ, R., Davidson, O., Hare, W., et al. 2007. "Climate Change 2007: Synthesis Report." Intergovernmental Panel on Climate Change Fourth Assessment Report, p. 38–39.

50. Forster, P., Ramaswamy, V., Artaxo, P., Berntsen, T., Betts, R., Fahey, D. W., Haywood, J., et al. 2007. "Changes in Atmospheric Constituents and in Radiative Forcing." In *Climate Change 2007: The Physical Science Basis. Contribution of Working Group I to the Fourth Assessment Report of the Intergovernmental Panel on Climate Change*, edited by S. Solomon, D. Qin, M. Manning, Z. Chen, M. Marquis, K. B. Averyt, M. Tignor, and H. L. Miller, p. 140. Cambridge, UK: Cambridge University Press.

51. Forster, P., Ramaswamy, V., Artaxo, P., Berntsen, T., Betts, R., Fahey, D. W., Haywood, J., et al. 2007. "Changes in Atmospheric Constituents and in Radiative Forcing." In *Climate Change 2007: The Physical Science Basis. Contribution of Working Group I to the Fourth Assessment Report of the Intergovernmental Panel on Climate Change*, edited by S. Solomon, D. Qin, M. Manning, Z. Chen, M. Marquis, K. B. Averyt, M. Tignor, and H. L. Miller, p. 143. Cambridge, UK: Cambridge University Press.

52. Denman, K. L., Brasseur, G., Chidthaisong, A., Ciais, P., Cox, P. M., Dickinson, R. E., Hauglustaine, D., et al. 2007. "Couplings Between Changes in the Climate System and Biogeochemistry." In *Climate Change 2007: The Physical Science Basis. Contribution of Working Group I to the Fourth Assessment Report of the Intergovernmental Panel on Climate Change*, edited by S. Solomon, D. Qin, M. Manning, Z. Chen, M. Marquis, K. B. Averyt, M. Tignor, and H. L. Miller, p. 512. Cambridge, UK: Cambridge University Press.

53. Intergovernmental Panel on Climate Change. 2007. "Summary for Policymakers." In *Climate Change 2007: The Physical Science Basis. Contribution of Working Group I to the Fourth Assessment Report of the Intergovernmental Panel on Climate Change*, edited by S. Solomon, D. Qin, M. Manning, Z. Chen, M. Marquis, K. B. Averyt, M. Tignor, and H. L. Miller, p. 4. Cambridge, UK: Cambridge University Press.

54. Trenberth, K. E., Jones, P. D., Ambenje, P., Bojariu, R., Easterling, D., Klein Tank, A., Parker, D., et al. 2007. "Observations: Surface and Atmospheric Climate Change." In *Climate Change 2007: The Physical Science Basis. Contribution of Working Group I to the Fourth Assessment Report of the Intergovernmental Panel on Climate Change*, edited by S. Solomon, D. Qin, M. Manning, Z. Chen, M. Marquis, K. B. Averyt, M. Tignor, and H. L. Miller, p. 268. Cambridge, UK: Cambridge University Press.

55. Shakun, J. D., Clark, P. U., He, F., Marcott, S. A., Mix, A. C., Liu, Z., Otto-Bliesner, B., et al. 2012. "Global Warming Preceded by Increasing Carbon Dioxide Concentrations During the Last Deglaciation," *Nature* 484:49–54.

56. Ruddiman, W. F. 2007. *Earth's Climate: Past and Future,* 2nd ed., p. 180. New York, NY: W. H. Freeman.

57. Ruddiman, W. F. 2007. *Earth's Climate: Past and Future,* 2nd ed., p. 182. New York, NY: W. H. Freeman.

58. Martin, J. H. 1990. "Glacial-Interglacial CO_2 Change: The Iron Hypothesis." *Paleoceanography* 5:1–13.

59. Shakun, J. D., Clark, P. U., He, F., Marcott, S. A., Mix, A. C., Liu, Z., Otto-Bliesner, B., et al. 2012. "Global Warming Preceded by Increasing Carbon Dioxide Concentrations During the Last Deglaciation." *Nature* 484:49–54.

60. Randall, D. A., Wood, R. A., Bony, S., Colman, R., Fichefet, T., Fyfe, J., Kattsov, V., et al. 2007. "Climate Models and Their Evaluation." In *Climate Change 2007: The Physical Science Basis. Contribution of Working Group I to the Fourth Assessment Report of the Intergovernmental Panel on Climate Change*, edited by S. Solomon, D. Qin, M. Manning, Z. Chen, M. Marquis, K. B. Averyt, M. Tignor, and H. L. Miller, p. 597–599. Cambridge, UK: Cambridge University Press.

61. Nakicenovic, N., Davidson, O., Davis, B., Grübler, A., Kram, T., La Rovere, E. L., Metz, B., et al. 2000. "IPCC Special Report: Emissions Scenarios, Summary for Policymakers." Special Report of Intergovernmental Panel on Climate Change Working Group III. http://www.ipcc.ch/pdf/special-reports/spm/sres-en.pdf.

62. Solomon, S., Qin, D., Manning, M., Alley, R. B., Berntsen, T., Bindoff, N. L., Chen, Z., et al. 2007. "Technical Summary." In *Climate Change 2007: The Physical Science Basis. Contribution of Working Group I to the Fourth Assessment Report of the Intergovernmental Panel on Climate Change*, edited by S. Solomon, D. Qin, M. Manning, Z. Chen, M. Marquis, K. B. Averyt, M. Tignor, and H. L. Miller, p. 26. Cambridge, UK: Cambridge University Press.

63. Solomon, S., Qin, D., Manning, M., Alley, R. B., Berntsen, T., Bindoff, N. L., Chen, Z., et al. 2007. "Technical Summary." In *Climate Change 2007: The Physical Science Basis. Contribution of Working Group I to the Fourth Assessment Report of the Intergovernmental Panel on Climate Change*, edited by S. Solomon, D. Qin, M. Manning, Z. Chen, M. Marquis, K. B. Averyt, M. Tignor, and H. L. Miller, p. 53. Cambridge, UK: Cambridge University Press.

64. Solomon, S., Qin, D., Manning, M., Alley, R. B., Berntsen, T., Bindoff, N. L., Chen, Z., et al. 2007. "Technical Summary." In *Climate Change 2007: The Physical Science Basis. Contribution of Working Group I to the Fourth Assessment Report of the Intergovernmental Panel on Climate Change*, edited by S. Solomon, D. Qin, M. Manning, Z. Chen, M. Marquis, K. B. Averyt, M. Tignor, and H. L. Miller, p. 44. Cambridge UK: Cambridge University Press.

65. Bernstein, L., Bosch, P., Canziani, O., Chen, Z., Christ, R., Davidson, O., Hare, W., et al. 2007. "Climate Change 2007: Synthesis Report." Intergovernmental Panel on Climate Change Fourth Assessment Report, p. 45.

66. Pfeffer, W. T., Harper, J. T., and O'Neel, S. 2008. "Kinematic Constraints on Glacier Contributions to 21st-Century Sea-Level Rise." *Science* 321:1340–1343.

67. Bernstein, L., Bosch, P., Canziani, O., Chen, Z., Christ, R., Davidson, O., Hare, W., et al. 2007. "Climate Change 2007: Synthesis Report." Intergovernmental Panel on Climate Change Fourth Assessment Report, p. 47.

68. van de Berg, W. J., van den Broeke, M., Ettema, J., van Meijgaard, E., and Kaspar, F. 2011. "Significant Contribution of Insolation to Eemian Melting of the Greenland Ice Sheet." *Nature Geoscience* 4:679–683.

69. Solomon, S., Qin, D., Manning, M., Alley, R. B., Berntsen, T., Bindoff, N. L., Chen, Z., et al. 2007. "Technical Summary." In *Climate Change 2007: The Physical Science Basis. Contribution of Working Group I to the Fourth Assessment Report of the Intergovernmental Panel on Climate Change*, edited by S. Solomon, D. Qin, M. Manning, Z. Chen, M. Marquis, K. B. Averyt, M. Tignor, and H. L. Miller, p. 59. Cambridge, UK: Cambridge University Press.

70. Bernstein, L., Bosch, P., Canziani, O., Chen, Z., Christ, R., Davidson, O., Hare, W., et al. 2007. "Climate Change 2007: Synthesis Report." Intergovernmental Panel on Climate Change Fourth Assessment Report, p. 46.

71. Seager, R., Ting, M., Held, I., Kushnir, Y., Lu, J., Vecchi, G., Huang, H., et al. 2007. "Model Projections of an Imminent Transition to a More Arid Climate in Southwestern North America." *Science* 316:1181–1184.

72. Kehrwald, N. M., Thompson, L. G., Tandong, Y., Mosley-Thompson, E., Schotterer, U., Alfimov, V., Beer, J., et al. 2008. "Mass Loss on Himalayan Glacier Endangers Water Resources." *Geophysical Research Letters* 35, L22503. doi:10.1029/2008GL035556.

73. Fennessy, M. J., and Kinter III, J. L. 2011. "Climatic Feedbacks During the 2003 European Heat Wave." *Journal of Climate* 24:5953–5967.

74. Centers for Disease Control and Prevention. 2010. "Where Malaria Occurs." CDC. Accessed February 21, 2013. http://www.cdc.gov/malaria/about/distribution.html.

75. McMichael, A. J., and Lindgren, E. 2011. "Climate Change: Present and Future Risks to Health, and Necessary Responses." *Journal of Internal Medicine* 270:401–413.

76. Soverow, J. E., Wellenius, G. A., Fisman, D. N., and Mittleman, M. A. 2009. "Infectious Disease in a Warming World: How Weather Influenced West Nile Virus in the United States (2001–2005)." *Environmental Health Perspectives* 117:1049–1052.

77. Mohan, J. E., Ziska, L. H., Schlesinger, W. H., Thomas, R. B., Sicher, R. C., George, K., and Clark, J. S. 2006. "Biomass and Toxicity Responses of Poison Ivy (*Toxicodendron radicans*) to Elevated Atmospheric CO_2." *Proceedings of the National Academy of Sciences* 103:9086–9089.

78. Devi, N., Hagedorn, F., Moiseev, P., Bugmann, H., Shiyatov, S., Mazepa, V., and Rigling, A. 2008. "Expanding Forests and Changing Growth Forms of Siberian Larch at the Polar Urals Treeline During the 20th Century." *Global Change Biology* 14:1581–1591.

79. Isaak, D. J., Luce, C. H., Rieman, B. E., Nagel, D. E., Peterson, E. E., Horan, D. L., Parkes, S., et al. 2010. "Effects of Climate Change and Wildfire on Stream Temperatures and Salmonid Thermal Habitat in a Mountain River Network." *Ecological Applications* 20:1350–1371.

80. Visser, M. E., Both, C., and Lambrechts, M. M. 2004. "Global Climate Change Leads to Mistimed Avian Reproduction." *Advances in Ecological Research* 35:89–110.

81. Chakraborty, S., Tiedemann, A. V., and Teng, P. S. 2000. "Climate Change: Potential Impact on Plant Diseases." *Environmental Pollution* 108:317–326.

82. Pounds, J. A., Bustamante, M. R., Coloma, L. A., Consuegra, J. A., Fogden, M. P. L., Foster, P. N., La Marca, E., et al. 2006. "Widespread Amphibian Extinctions from Epidemic Disease Driven by Global Warming." *Nature* 439:161–167.

83. Australian Academy of Science. 2003. "Coral Bleaching—Will Global Warming Kill the Reefs?" Australian Academy of Science. Accessed February 21, 2013. http://science.org.au/nova/076/076key.html.

84. Parry, M. L., Canziani, O. F., Palutikof, J. P., and coauthors. 2007. "Technical Summary." In *Climate Change 2007: Impacts, Adaptation and Vulnerability. Contribution of Working Group II to the Fourth Assessment Report of the Intergovernmental Panel on Climate Change*, edited by M. L. Parry, O. F. Canziani, J. P. Palutikof, P. J. van der Linden, and C. E. Hanson, p. 23–78. Cambridge, UK: Cambridge University Press.

85. Parry, M. L., Canziani, O. F., Palutikof, J. P., and coauthors. 2007. "Technical Summary." In *Climate Change 2007: Impacts, Adaptation and Vulnerability. Contribution of Working Group II to the Fourth Assessment Report of the Intergovernmental Panel on Climate Change*, edited by M. L. Parry, O. F. Canziani, J. P. Palutikof, P. J. van

der Linden, and C. E. Hanson, p. 28. Cambridge, UK: Cambridge University Press.

86. **Committee on the Importance of Deep-Time Geologic Records for Understanding Climate Change Impacts.** 2011. *Understanding Earth's Deep Past: Lessons for Our Climate Future,* p. 16. Washington, DC: National Academies Press.

87. **Berner, R. A.** 1997. "The Rise of Plants and Their Effect on Weathering and Atmospheric CO_2." *Science* 276:544–546.

88. **Hansen, J., Sato, M., Kharecha, P., Beerling, D., Berner, R., Masson-Delmotte, V., Pagani, M., et al.** 2008. "Target Atmospheric CO_2: Where Should Humanity Aim?" *The Open Atmospheric Science Journal* 2:217–231.

89. **Bradsher, K.** 2009. "China Outpaces U.S. in Cleaner Coal-Fired Plants." *New York Times,* May 11. Accessed February 21, 2013. http://www.nytimes.com/2009/05/11/world/asia/11coal.html?_r=0.

90. **Zachos, J., Pagani, M., Sloan, L., Thomas, E., and Billups, K.** 2001. "Trends, Rhythms, and Aberrations in Global Climate 65 Ma to Present." *Science* 292: 686–693.

91. **Tarnocai, C., Canadell, J. G., Schuur, E. A. G., Kuhry, P., Mazhitova, G., and Zimov, S.** 2009. "Soil Organic Carbon Pools in the Northern Circumpolar Permafrost Region." *Global Biogeochemical Cycles* 23.

92. **Walter, K. M., Zimov, S. A., Chanton, J. P., Verbyla, D., and Chapin III, F. S.** 2006. "Methane Bubbling from Siberian Thaw Lakes as a Positive Feedback to Climate Warming." *Nature* 443:71–75.

93. **Anthony, K. W.** 2009. "Methane: A Menace Surfaces." *Scientific American* (December):68–75.

94. **Roach, J.** 2007. "Arctic Melt Opens Northwest Passage." *National Geographic News,* September 17. Accessed February 21, 2013. http://news.nationalgeographic.com/news/2007/09/070917-northwest-passage.html.

95. **National Highway Traffic Safety Administration.** 2011. "President Obama Announces Historic 54.5 MPG Fuel Efficiency Standard." NHTSA, July 29. Accessed February 21, 2013. http://www.

96. **One Hundred Tenth Congress of the United States of America.** 2007. "Energy Independence and Security Act of 2007." U. S. Congress. http://www.gpo.gov/fdsys/pkg/BILLS-110hr6enr/pdf/BILLS-110hr6enr.pdf.

97. **Raish, D.** 2013. "Solar Energy on the Rise in Germany." Deutsche Welle, January 1. Accessed February 21, 2013. http://www.dw.de/solar-energy-on-the-rise-in-germany/a-16490941.

98. **Ministry of Foreign Affairs of Denmark.** 2013. "Wind Energy." Accessed February 21, 2013. http://denmark.dk/en/green-living/wind-energy/.

99. **Perkowski, J.** 2012. "China Leads the World in Renewable Energy Investment." Forbes, July 27. Accessed February 21, 2013. http://www.forbes.com/sites/jackperkowski/2012/07/27/china-leads-the-world-in-renewable-energy-investment/.

100. **The Royal Society.** 2009. "Geoengineering the Climate: Science, Governance and Uncertainty." The Royal Society. http://royalsociety.org/uploadedFiles/Royal_Society_Content/policy/publications/2009/8693.pdf.

101. **Trenberth, K. E., and Dai, A.** 2007. "Effects of Mount Pinatubo Volcanic Eruption on the Hydrological Cycle as an Analog of Geoengineering." Geophysical Research Letters 34, L15702. doi:10.1029/2007GL030524, 2007.

102. **Morton, O.** 2007. "Is This What It Takes to Save the World?" Nature 447:132–136.

103. **World Resources Institute.** 2013. "Carbon Dioxide Capture and Storage (CCS)." Accessed February 21, 2013. http://www.wri.org/project/carbon-dioxide-capture-storage.

104. **Karl, T. R. Melillo, J. M., and Peterson, T. C., eds.** 2009. Global Climate Change Impacts in the United States, p. 109. Cambridge, UK: Cambridge University Press.

105. **Bates, T. B.** 2013. "N.J. Sandy Rebuilding Rules: Go Higher or Pay More." USA Today, January 25. Accessed February 21, 2013. http://www.usatoday.com/story/news/nation/2013/01/25/sandy-rebuilding-flood-maps/1863761/.

106. **Grunfeld, D.** 2011. "Hurricane Protection Lake Borgne Surge Barrier 1." The Times-Picayune, May 26. Accessed February 21, 2013. http://photos.nola.com/tpphotos/2011/05/hurricane_protectioin_lake_bor_3.html.

107. **Connor, S.** 2008. "Sea Levels Rising Too Fast for Thames Barrier." The Independent, March 22. Accessed February 21, 2013. http://www.independent.co.uk/environment/climate-change/sea-levels-rising-too-fast-for-thames-barrier-799303.html.

108. **Squires, N.** 2008. "'Moses Project' to Secure Future of Venice." The Telegraph, December 6. Accessed February 21, 2013. http://www.telegraph.co.uk/news/worldnews/europe/italy/3629387/Moses-project-to-secure-future-of-Venice.html.

109. **Woods, D. E.** 2012. "Hurricane Sandy Brings Relevance to Blue Acres Program." NJ.com, November 25. Accessed February 21, 2013. http://www.nj.com/cumberland/index.ssf/2012/11/hurricane_sandy_brings_relevan.html.

110. **Pennsylvania Department of Environmental Protection.** 2011. "Pennsylvania Climate Adaptation Planning Report: Risks and Practical Recommendations." Accessed February 21, 2013. http://www.elibrary.dep.state.pa.us/dsweb/Get/Document-92911/27000-RE-DEP4303%20%20Pennsylvania%20Climate%20Adaptation%20Planning%20Report.pdf.

111. **Official City Website for Chandler, Arizona.** 2013. "Rebate Programs." Accessed February 21, 2003. http://chandleraz.gov/default.aspx?pageid=746.

112. **City of Peoria, Arizona.** 2013. "Retrofit Kit Program." Accessed February 21, 2003. http://www.peoriaaz.gov/NewSecondary.aspx?id=18850.

113. **United Nations.** 1997. "UN Conference on Environment and Development (1992)." United Nations. Accessed February 21, 2013. http://www.un.org/geninfo/bp/enviro.html.

114. **Friedman, L.** 2012. "Climate Conference Renews Kyoto Protocol but Looks to Successor Treaty." Scientific American (December). Accessed February 21, 2013. http://www.scientificamerican.com/article.cfm?id=climate-conference-renews-kyoto-protocol-but-looks-to-successor.

115. **Hooke, R. L.** 1999. "Spatial Distribution of Human Geomorphic Activity in the United States: Comparison with Rivers." *Earth Surface Processes and Landforms* 24:687–692.

Page numbers in **boldface** indicate definitions; page numbers in *italics* indicate figures and tables.